Public Transit Planning and Operation

Modeling, Practice and Behavior

Second Edition

Public Transit Planning and Operation

Modeling, Practice and Behavior

Second Edition

Avishai Ceder

*Technion-Israel Institute of Technology
and University of Auckland,
New Zealand*

CRC Press
Taylor & Francis Group
Boca Raton London New York

CRC Press is an imprint of the
Taylor & Francis Group, an **informa** business

A SPON PRESS BOOK

CRC Press
Taylor & Francis Group
6000 Broken Sound Parkway NW, Suite 300
Boca Raton, FL 33487-2742

First issued in paperback 2019

© 2016 by Taylor & Francis Group, LLC
CRC Press is an imprint of Taylor & Francis Group, an Informa business

No claim to original U.S. Government works

ISBN-13: 978-0-4665-6391-9 (hbk)
ISBN-13: 978-0-367-867607 (pbk)

Visit the Taylor & Francis Web site at
http://www.taylorandfrancis.com

and the CRC Press Web site at
http://www.crcpress.com

To my late Dad, Samuel (who worked for a large bus agency
as a driver and treasurer for over 30 years), and to my
Mom, Anna, with wishes for many good years.

Contents

6 Vehicle scheduling I: Fixed schedules   157

7 Vehicle scheduling II: Variable schedules   201

Preface

PERSONAL MOTIVATION

The following story may serve to help understand the stimulus behind the first and this second edition of the book: A ship is sailing through a stormy ocean, and a little girl, who happens to be the captain's daughter, is playing on the deck when all of a sudden a large wave carries her overboard into the sea. The captain, who sees this from his post, right away orders his sailors to jump into the ocean and save the girl, but none dares for fear of risking his life. Desperately the captain turns to the passengers and asks them for help while promising that whoever saves his daughter will get anything he wants as a reward. Again, no one reacts. But then suddenly a man with a long beard who has stood by the railing lurches overboard into the sea. The sailors throw him a life preserver, and fortunately he manages to lift the little girl safely back on deck and into the arms of her father. The captain then hugs the man, who is thoroughly drenched, and says he will give him anything he wants, just name it. The hero's response: "I don't want anything. I just want to know who pushed me. ..."

What pushed me actually started between 1967 and 1971, when I was a bus driver for Egged, whose 4000 buses make it one of the largest bus companies in the world. Before gaining a bus driver's license, I had a theory about the way one should drive a bus; now I have a bus driver's license—and no theory. The second motivation for writing this book came from my consulting work at Egged from 1975 to 1985, when I was exposed daily to planning and operational problems in the public transit field. The third motivation came in 1981. I was at MIT in Boston, where, together with Professor Nigel Wilson, I was to give a new summer session course on transit operations planning. This course became an annual offering until 2003. From 2006 to date, I am delivering two short transit courses together with Professor Graham Currie of Monash University in Melbourne.

From all these foregoing activities, I internalized the following realization: experience is what you get when you are expecting something else. My teaching of transit operations planning has taken place at universities in Adelaide, Auckland, Boston, Brisbane, California, Hong Kong, Israel, Melbourne, Rome, Perth, Sydney, and Wellington. Indeed, it has been my more than 40 years of teaching, research, and hands-on experience that has pushed me to write the first and this second edition of the book and to construct it in such a way that it will help not only teachers, researchers, and students in the area, but also practitioners in the field.

PURPOSE AND INTENDED AUDIENCE

This second edition of the book is constructed from two-thirds of the first edition, published in 2007; one-third comprises updated and new material. This second edition uses the concise term *transit* to refer to public transit or public transportation or public transport, all three of which are in common use.

A major goal of this volume is to establish a sort of a bridge between the world of practitioners and the world of research and academia to narrow the gap between them. We hope that such a bridge will also lead to opportunities for collaboration and interaction in order to improve public transit service and, no less important, its image. Henry Ford once said, "Failure is only the opportunity to start all over again more intelligently." With this in mind, the book intends to introduce a few new ways of thinking about (a) already implemented and investigated transit themes while combining retrospective thoughts and cumulative experience and (b) new concepts and ideas.

One of the main features of the second edition of the book is its stand-alone (self-contained) capability, obviating the need to look back and forth at other publications for comprehending the text. At the same time, every chapter contains a "Literature Review and Further Reading" list. Practitioners may have some difficulty in comprehending the sections with mathematical notation, but hopefully they can grasp the substance of the material and its practical implications. Researchers and academic professionals may find some of the sections unnecessarily detailed, but they should be aware that the book is also aimed at practitioners and undergraduate students, thus requiring more explanation. This work, in sum, follows the notion that (a) it is better to ask twice than to lose your way once and (b) clarity is no less important than certainty.

ORGANIZATION

Each chapter opens with a section containing information and remarks for practitioners, called "Practitioner's Corner." In fact, one can never tell which way the train went by looking at the track: for a practical decision, one needs more. The purpose of these "corners" is to impart guidance about sections of the chapter that are appropriate and sections that are perhaps too difficult for practitioners so as to allow the less academically inclined to flow with the book and capture its substance.

The organization of the book is described in Chapter 1. Generally speaking, five themes are addressed:

- Overview of transit planning and data collection needs (Chapters 1 and 2)
- Design and optimization of transit timetables and of vehicle and crew scheduling (Chapters 3 through 9)
- Passenger demand and assignment analysis (Chapters 10 and 11)
- Transit service, network, and route design, encompassing scheduling and behavioral elements (Chapters 12 through 17)
- Transit reliability, operational strategies, and future operations planning developments (Chapters 18 through 20)

This second edition includes significantly revised chapters (Chapters 5, 8, 12, and 15) and completely new chapters (Chapters 13, 14, and 19).

All the quantitative chapters offer exercises for practicing the methods covered; of the book's 20 chapters, only Chapters 1, 2, and 20 are without exercises. The answers to these exercises appear at the end of the book.

The literature review of papers relevant to the topic(s) covered in a chapter appears as the last numbered section of each chapter, except for Chapters 1 and 7 (the review for which actually precedes it, in Chapter 6). The reason for this order, instead of the traditional pattern of starting a scientific article with a literature review, is to focus on each chapter's essence from the beginning and only at the end to give the reader who may wish to broaden his or her knowledge of the particular topic a kind of annotated reference list and an extended bibliography.

REMARKS

The success of a professional book can be evaluated by the extent to which it succeeds in introducing new and improved ideas and methods. It is not only a matter of learning the book's content; it has to do, as well, with how much the volume can inspire the reader's imagination to think further. That concept has served as the guideline for the development of the second edition of this book.

Finally, when lecturing this transit course, I tended to use humor at times because I believe in the insight captured by the English playwright George Bernard Shaw, who once said, "When a thing is funny, search it carefully for a hidden truth." More than once I have been asked to employ some of this humor (including the cartoons that I have also drawn) if I ever wrote a book. I have done this to some extent, especially in the Practitioner's Corners.

Avishai (Avi) Ceder
Auckland, New Zealand and Haifa, Israel

MATLAB® is a registered trademark of The MathWorks, Inc. For product information, please contact:

The MathWorks, Inc.
3 Apple Hill Drive
Natick, MA 01760-2098 USA
Tel: 508-647-7000
Fax: 508-647-7001
E-mail: info@mathworks.com
Web: www.mathworks.com

Acknowledgments

First, I would like to repeat my thanks to everyone who contributed to the first edition of this book from 2007. After all, this second edition is constructed from two-thirds of the first edition, published in 2007; one-third comprises updated and new material.

Second and foremost, many thanks are extended to both my former and my current PhD students at the University of Auckland. Over the last seven years, I have shared with these students the exploration and search for new research avenues with innovative and novel concepts and ideas. My appreciation and thanks to Dr. Subeh Chowdhury (now a staff member at the University of Auckland) for her thorough study of transit-user behavior at transfer points; Dr. Stephan Hassold (an OR analyst at Air New Zealand) for his outstanding research on timetabling and vehicle scheduling with different vehicle sizes; and Supun Perera (of Sydney, Australia), who contributed to the updated literature review and with whom I had many good and interesting conversations. My gratitude and appreciation are also extended to three of my current doctoral students for their contributions: Tao Liu, for his enthusiasm and good research on transit real-time control; Mahmood Mahmoodi Nesheli, for his hard work and advanced study on transit modeling of operational tactics; and Mahdi Amiripour, for his great interest, devotion, and solid research on transit network design problems.

Last, I offer my heartfelt thanks to my wife, Selina (Lan Shi), for her great encouragement, love, and understanding. Finally, a bouquet of affections goes to my three sons and daughters-in-law, Roy + Roni, Ohad + Liat, and Dror + Valerie, and to my seven grandchildren, Eran, Elad, Yael, Yahli, Naama, Gefen, and Jonathan, as well as to my Mom, Anna, and my brothers, Tuli and Hagai, and their families—all of whom helped me get through difficult periods when writing this second edition of the book.

I retain, of course, sole responsibility for any errors. I would be very pleased to receive feedback.

Author

Avishai (Avi) Ceder received his PhD in 1975 at the University of California at Berkeley, and he is a professor and chair in transportation at the University of Auckland, New Zealand, and at the Technion–Israel Institute of Technology. He frequently delivers courses on public transit operations planning in Asia, Europe, Australasia, and the United States. His research and implementation interests in the subject stem from being a bus driver for four years for a large bus company.

Chapter 1

Introduction to transit service planning

CHAPTER OUTLINE

PRACTITIONER'S CORNER

As stated in "Preface," each chapter opens with information and notes to guide practitioners. This introductory chapter will show that a successful transit service is not permanent and that failure in a new service need not be fatal. The chapter contains neither mathematical formulations nor a complicated literature review.

The motivation behind the writing of this book is to bridge the world of the researchers and the world of the practitioners. The chapter first presents a real-life description of the planning process for transit operations. It then provides an overview and a critique of currently used standards and guidelines for transit service, pointing out those standards that warrant research and those for which administrative decisions are sufficient. The main components affecting the viability of a transit service from both the passenger and agency perspectives are described, as is the link between some problematic elements and the possible solutions advanced in the book.

The chapter ends with an outline of the other chapters and the links between each chapter and the main core of the transit operations planning process. Our attempt to show how to deal with and to solve in a practical manner current transit service dilemmas and confusions follows Pablo Picasso, who said: "I do not seek, I find."

1.1 MOTIVATION

A basic premise of this volume is that using transit service should be like eating potato chips—once you start, it's hard to stop. The fundamental question, though, is how do we design and approach such a service while knowing, because of the competition offered by cars, that an adequate transit design must provide superlative service in order to be good enough. This second edition of the book, following its first edition (Ceder, 2007), will demonstrate methods and procedures not only for improving transit design through the provision of an attractive, viable service but also for reducing its cost and increasing efficiency from the transit agency perspective.

One of the main motivations for writing the first edition of this book (Ceder, 2007) is the persistent use by transit agencies (especially in the United States) of the Bus Scheduling Manual of August 1947 (!), which was reprinted in 1982 (Rainvile, 1947). The manual opens as follows: "This volume is viewed today [i.e., 1982—A.C.] as a classic reference on manual scheduling practices. It represents a comprehensive effort made over 30 years ago to pull together the state of the practice in this field, and many of those practices still remain valid today." The manual's cover and an inside picture of a device to measure distances appear in Figure 1.1. Those transit agencies that may still be using this manual might be asked a question: Is the treatment of passenger loads and running-time data, frequency determination, and vehicle and crew scheduling really *classic*? Or is it obsolete? The conclusion implied by the latter served as a trigger to investigate why the treatment of these elements in the manual is not classic and to submit operations planning options that exploit existing, computerized computation power and advanced modeling. On the positive side, it may be said, as it was in the 1947 manual, that some basic concepts behind the planning process for transit operations do indeed remain the same.

In 1998, the U.S. Transit Cooperative Research Program (TCRP) sponsored Report 30 (Pine et al., 1998) on transit scheduling with the idea presumably of advancing knowledge in the field for training purposes. However, Report 30 neither contains nor refers to any research on transit scheduling (despite the many studies that existed); furthermore, it does not mention a

Figure 1.1 Cover and a page of the well-known *1947 Bus Scheduling Manual,* wrongly viewed as a *classic* reference.

single reference. The report actually provides very basic training material for new schedulers; in concept, it is almost a replica of the 1947 manual, while emphasizing that scheduling is a craft.

All in all, the following story may sum up the motivation for writing this book: two marketing people from a European shoe factory are sent to investigate the potential sale of shoes in Africa. A few days later, faxes from each of them arrive at the manager's office. The first fax reads: "No chances, everyone is barefoot." The second fax urges: "Lots of chances, everyone is barefoot." This book will adopt the *lots of chances* view.

Another stimulus for the first edition of this book (Ceder, 2007) is a long list of standards and guidelines (see Section 1.3) for transit service planning, operations, and control that determine the quality of service seen in practice. In the event that current transit service quality has reached a satisfactory level, then these standards and guidelines are apparently effective. Nonetheless, many aspects of transit service can stand improvement, necessitating changes or replacement of some of these standards and guidelines. Like those in the 1947 manual. We can imagine that the majority of the standards and guidelines to be covered in the present volume are intuitive based or administrative based instead of being research based; hence, an opportunity is presented to develop new research tools and methods for improving some of these standards and guidelines (as discussed in Section 1.3).

This book uses the term *transit* as shorthand for public transit or public transportation or public transport, all in common use. We describe in this book transit operations planning problems, a summary of different approaches that have been proposed for their solution, and a description of new methods and approaches incorporating some of the positive aspects of prior work. In essence, the new methods and approaches are all applicable to the four major modes of transit operation: airline, railway, bus, and passenger ferry. Although the operations planning process for bus, railway, and passenger ferry is similar, that for airline has several special features, but these are not covered in this book. Moreover, railway and passenger ferry can be perceived, to some extent, as special cases of bus service because of the variety of activities and problems offered by the latter. Finally, the methods and approaches proposed are intended to be easier to implement and more sensitive to the risks of making changes than are those currently in use by practitioners and researchers.

1.2 OPERATIONAL PLANNING DECOMPOSITION PROCESS

The transit operations planning process commonly includes four basic activities, usually performed in sequence: (1) network route design, (2) timetable development, (3) vehicle scheduling, and (4) crew scheduling (Ceder and Wilson, 1986; Ceder, 2001, 2002, 2007). Figure 1.2 shows the systematic decision sequence of these four planning activities. The output of each activity positioned higher in the sequence becomes an important input for lower-level decisions. Clearly, the independence and orderliness of the separate activities exist only in the diagram; that is, decisions made further down the sequence will have some effect on higher-level decisions. It is desirable, therefore, that all four activities be planned simultaneously in order to exploit the system's capability to the greatest extent and maximize the system's productivity and efficiency. Occasionally, the sequence in Figure 1.2 is repeated; the required feedback is incorporated over time. However, since this planning process, especially for medium to large fleet sizes, is extremely cumbersome and complex, it requires separate treatment for each activity, with the outcome of one fed as an input to the next.

The quantitative treatment of the transit planning process is reflected in the welter of professional papers on these topics and in the development of numerous computer programs to automate (at least partially) these activities. In the last 30 years, a considerable amount of effort has been invested in the computerization of the four planning activities (routing, timetabling, vehicle scheduling, crew scheduling) in order to provide more efficient, controllable, and responsive schedules. The best summary of this effort, as well as of the knowledge accumulated, was presented at the 2nd through 11th international workshops on Computer-Aided Scheduling of Public Transport (CASPT); recently, this abbreviation was changed to Conference of Advanced Systems of Public Transport. The research presented at CASPT appears in the book by Ceder (2007) and the books edited by Wren (1981), Rousseau (1985), Daduna and Wren (1988), Desrochers and Rousseau (1992), Daduna et al. (1995), Wilson (1999), Voss and Daduna (2001), and Hickman et al. (2008); the research of the 10th (Leeds, United Kingdom, 2007), the 11th (Hong Kong, China, 2009), and the 12th CASPT (Santiago, Chile, 2012) appear now in proceeding volumes (not books), but the best of them appear/will appear in the new journal: *Public Transport: Planning and Operations* by Springer. To learn more of the CASPT2015 (2015), visit its website: http://www.caspt.org. In addition, there is a large number of software for transit scheduling. Some of the known software appears in Table 1.1.

Many planning software packages are based on static data and therefore only offer static schedules. Software needs to be developed in such a way that it includes the flexibility that is required when drawing up and continually modifying service and work schedules. Although many common aspects are involved in workforce planning, every organization has its own planning wishes and requirements. This is why the software needs to be fully customized to the specific business model and company logic. Collective labor agreement rules, legislation relating to working hours, agreed contract hours, and the individual preferences of employees can all be entered into the system. On the basis of advanced algorithms and built-in intelligence, the software is able to create an optimal schedule or support planners in the creation of a schedule. In addition, the software should offer a veritable collection of management information and analyses of deployment, availability, and planning trends. These form an optimal basis for better control of activities and greater efficiency during these activities. Moreover, as a result of its built-in intelligence, the software effortlessly handles multiresource planning, for planning people and resources with the help of a single integrated system.

These software packages concentrate primarily on the activities of vehicle and crew scheduling (activities 3, 4 in Figure 1.2) because, from the agencies' perspective, the largest single cost of providing service is generated by drivers' wages and fringe benefits. Focusing on activities 3 and 4 would seem to be the best way to reduce this cost. However, because some of

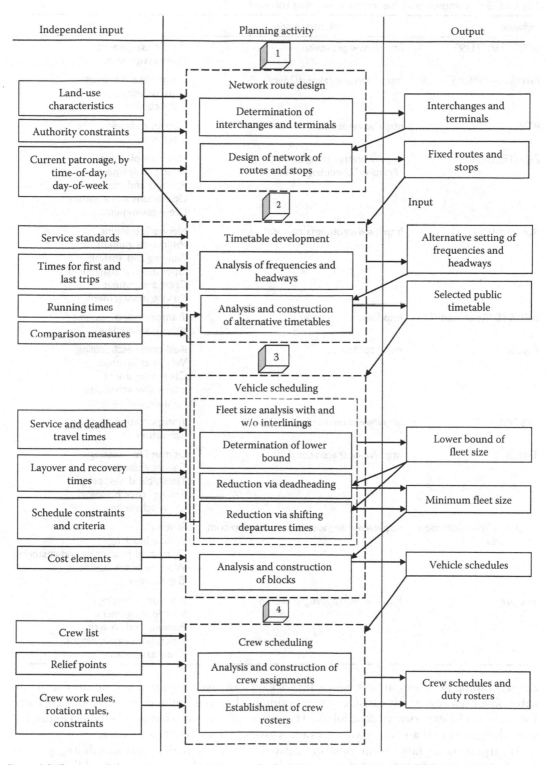

Figure 1.2 Functional diagram (system architecture) of a common transit operations planning process.

Table 1.1 Some known available transit-scheduling software

Software	Internet address	Features
GIRO "HASTUS"	http://www.giro.ca/en	Vehicle assignment Crew assignment
Merakas—"PIKAS"	http://www.merakas.lt/26en/	Timetable calculation Trips assignment Shifts cutting
PTV	http://www.ptv.de	Network planning Simulation
ROUTELOGIC	http://www.routelogic.com/ FeaturesScheduling.html	Network planning Vehicle assignment Planning and analysis Operations management Crew assignment
Routematch	http://www.routematch.com/	Network planning Vehicle assignment Planning and analysis Operations management Crew assignment Service management
NEMSYS "Routemate"	http://www.nemsys.it/	Transportation planning Vehicle scheduling
Optibus	www.optibus.co	Real-time rescheduling Vehicle assignment Crew assignment Interactive scheduling Planning and analysis
SYSTRA	http://www.systra.com	Transportation planning Simulation
Tracsis	http://www.tracsis.com/	Demand forecasting Capacity planning Timetable development Rolling stock planning Crew scheduling
Enghouse Transportation "TranSched"	http://www.enghousetransportation.com/	Scheduling Vehicle blocking Optimized runcutting and rostering Workforce management Trip planner
Trapeze	http://www.trapezegroup.com	Network planning Vehicle assignment Planning and analysis Operations management Crew assignment

the scheduling problems in these software packages are oversimplified and decomposed into subproblems, a completely satisfactory or optimal solution is not assured, thus making room for decisions by experienced schedulers. After all, experience is what we gain when expecting something else; said another way, the exam is given first and the lesson after.

An argument in favor of automating activities 3 and 4 is that this scheduling process is extremely cumbersome and time-consuming to undertake manually. In addition to the potential for more efficient schedules, the automated process enables services to be more controllable and more responsive. The cost and complexity of manual scheduling have served to discourage adjusting activities 1 and 2. Only with automated scheduling methods, which are becoming more widely accepted, it is feasible to focus on higher levels in the planning process. Nonetheless, a case can be made that these higher levels have received short shift by both researchers and practitioners.

The network route-design activity in Figure 1.2 is described in Chapters 12 and 15, which is also linked with Chapters 13 and 14. Planning practice in terms of transit route design focuses almost entirely on individual routes that, for one reason or another, have been identified as candidates for change. Occasionally, sets of interacting (e.g., overlapping or connecting) routes are subject to redesign, usually after a series of incremental changes to individual routes has resulted in a confusing, inefficient local system. Although it is difficult to predict the benefits that will result from redesigning any transit network without conducting a detailed assessment, it is reasonable to believe that they will be large compared with the benefits of additional efforts aimed just at problematic scheduling activities (2, 3, and 4 in Figure 1.2). The approach described mainly in Chapter 15 generates all feasible routes and transfers connecting each place (node) in the network to all others. From this vast pool of possible routes and transfers, smaller subsets are generated that maintain network connectivity. For each subset thus generated, transportation demand is met by calculating the appropriate frequency for each route. Next, prespecified optimization parameters are calculated for each subset. Based on the specific optimization parameter desired by the user, it is then possible to select the most suitable subset. This method has been designed as a tool for the planning of future transit networks as well as for the maintenance of existing networks. The method presented ensures flexibility by allowing the user either to input own data or to run the analysis automatically.

The timetable development activity in Figure 1.2 is described in Chapters 3 through 5 for establishing alternative frequencies and timetables. The aim of public timetables is to meet general public transportation demand. This demand varies during the hours of the day, the days of the week, from one season to another, and even from 1 year to another. It reflects the business, industrial, cultural, educational, social, and recreational transportation needs of the community. The purpose of this activity, then, is to set alternative timetables for each transit route in order to meet variations in public demand. Alternative timetables are determined on the basis of passenger counts, and they must comply with service frequency constraints. In Chapter 4, alternative timetables are constructed with either even headways, but not necessarily even loads on board individual vehicles at the peak-load section, or even average passenger loads on board individual vehicles, but not even headways. Average even loads on individual vehicles can be approached by relaxing the evenly spaced headways pattern (through a rearrangement of departure times). This dynamic behavior can be detected through passenger-load counts and information provided by road supervisors. The key word in the even-load cases is the ability to control the loading instead of being repeatedly exposed to an unreliable service resulting from an imbalance in loading situations.

The vehicle-scheduling activity in Figure 1.2 is described in Chapters 6 through 8 and is aimed at creating chains of trips; each is referred to as a vehicle schedule according to given timetables. This chaining process is often called vehicle blocking (a block is a sequence of revenue and nonrevenue activities for an individual vehicle). A transit trip can be planned either to transport passengers along its route or to make a deadheading trip in order to connect two service trips efficiently. The scheduler's task is to list all daily chains of trips (some deadheading) for each vehicle so as to ensure the fulfillment of both timetable and operator requirements (refueling, maintenance, etc.). The major objective of this activity is to minimize the number of vehicles required (Daduna and Paixao, 1995). Chapters 6 and 7 describe a highly informative graphical technique for the problem of finding the least number of vehicles. The technique used is a step function, which is introduced as far back as 30 years ago as an optimization tool for minimizing the number of vehicles in a fixed-trip schedule. The step function is termed deficit function, as it represents the deficit number of vehicles required at a particular terminal in a multiterminal transit system. In Chapter 6, the fixed-schedule case is extended to include variable trip schedules, in which given shifting tolerances allow for possible shifts in departure times. This opens up an opportunity to reduce fleet size further. The deficit function, because of its graphical characteristics, has been programmed and is available on a website. In this book, the deficit function is applied and linked to the following activities: vehicle scheduling with different vehicle types (Chapter 8), the design of operational transit parking spaces (Chapter 12), network route design (Chapters 12 and 15), and short-turn design of individual and groups of routes (Chapter 16). The value of embarking on such a technique is to achieve the greatest saving in number of vehicles while complying with passenger demand. This saving is attained through a procedure incorporating a man–computer interface allowing the inclusion of practical considerations that experienced transit schedulers may wish to introduce into the schedule.

The crew scheduling activity in Figure 1.2 is described in Chapter 9. Its goal is to assign drivers according to the outcome of vehicle scheduling. This activity is often called driver-run cutting (splitting and recombining vehicle blocks into legal driver shifts or runs). This crew-assignment process must comply with some constraints, which are usually dependent on a labor contract. A brief summary is given of the conceptual analytical tools used in the modeling and software of this complex, combinatorial problem. The crew-rostering component of this activity normally refers to priority and rotation rules, rest periods, and drivers' preferences. Any transit agency wishing to utilize its resources more efficiently has to deal with problems encountered by the presence of various pay scales (regular, overtime, weekends, etc.) and with human-oriented dissatisfaction. The crew-scheduling activity is very sensitive to both internal and external factors, a factor that could easily lead to an inefficient solution.

In Table 1.2, we list the essential inputs for the transit operations planning process illustrated in Figure 1.2. These mainly independent elements are arranged by activity number. It should be emphasized that their values differ by time of day and day of week.

1.3 SERVICE AND EVALUATION STANDARDS AND THEIR DERIVATIVES

Tremendous and monotonous in transit service can be perceived as synonymous. It is best to have a monotonous, automated transit system that will always be there for the passenger, for example, a modularized personal rapid transit system conducted like a horizontal elevator. This observation serves as an introduction to the need for transit service standards and guidelines. On the one hand, standards have to do with maintaining and improving existing service levels; on the other hand, they are often a source of fiscal pressure on transit agencies. Service standards are also linked to any evaluation effort aimed at improving the efficiency, effectiveness, and productivity of a transit service. The greater these measures, the higher is the level of service that can be offered.

1.3.1 Service standards

The need for dynamically updated standards in the transit industry is described in an article by Mora and Chernicoff (2005). This need deserves attention, especially because of the rapid introduction of advanced technologies in bus and rail transit and service. While standards,

Table 1.2 Typical input elements for the three schedule-planning activities

Planning activity	Input element
Timetable development	1. Route (line) number 2. Nodes (stops and timepoints on a route) 3. Pattern (sequence of nodes on a route) 4. Average passenger loads between adjacent stops on a route 5. Load factor (desired number of passengers on board the transit vehicle) 6. Policy headway (the inverse of the minimum frequency standard) 7. Vehicle type 8. Vehicle capacity 9. Average running time (travel time between stops/timepoints on a route)
Vehicle scheduling	1. Trip recovery-time tolerances (maximum and minimum time to be prepared for next trip) 2. Trip departure-time tolerances (maximum departure delay and maximum advance departure) 3. List of garages (names and locations) 4. List of trip start and end locations 5. Average deadhead times from garage locations to each trip start location (pull-outs) 6. Average deadhead times from trip end locations to garage locations (pull-ins) 7. Average deadhead time matrix between all trip end and start locations (by time-of-day)
Crew scheduling	1. Relief-point location (stops, trip start and end points, garages) 2. Average travel times between relief points 3. Trip-layover time (minimum and maximum rest times between two adjacent trips) 4. Type of duty (early, late, split, full, tripper, etc.) 5. Duty length (maximum spread time) 6. Number of vehicle changes on duty 7. Meal breaks 8. Duty composition 9. Other work rules 10. List of drivers by name and type (e.g., part-time, full-time, seniority) 11. Driver priority and equality rules 12. One-day-on, one-day-off work pattern

regulations, and best practices are in particular justified for supporting safety and security applications, they are also crucial for creating satisfactory transit service.

More than 50% of the transit agencies in the United States (METRO, 1984; TCRP Report, 1995) and in Europe (QUATTRO, 1998) employ formal standards in service planning. The three reports discussing standards and guidelines in the U.S. bus industry (METRO, 1984; TCRP Report, 1995) and in transit service in Europe (QUATTRO, 1998) are based solely on surveys conducted among transit agencies, for example, 109 of 345 agencies responded in the 1984 report and 11 of 297 agencies in the 1995 report. The objectives of each survey were predominately based on the *what* type of questions and only a few on the *how* type of question. (For example, what service standards are currently used? What data are collected and used to evaluate the various performance criteria utilized? How is it collected?) The surveys related only to the effectiveness of transit service evaluation efforts. They were conducted on the basis of how the agency perceived effectiveness (10 and 5 gradation scores in 1984 and 1995, respectively) in improving (1) service delivery, (2) equity, (3) ridership, and (4) operating costs. The highest ratings (i.e., standards perceived to be more effective) were given to (1) and (4), while (3) received the lowest score in 1984 and (2) the lowest in 1995.

The main standards currently utilized can be partitioned into two categories: (1) route design and (2) service design. These can be divided further into route level and network level standards and into planning level and monitoring level standards. The resulting 20 standards appear in Table 1.3 by category, group, number, standard name, typical criteria

Table 1.3 Available service standards and their typical criteria ranges

Category	Group	#	Standard item	Typical criterion range[a]	Remarks
Route design	Route level	1.	Route length	Max 40–100 min one-way	Longer length for larger agencies
		2.	Stop spacing	120–400 m (urban area)	Depends on pop. density, land use
		3.	Route directness[b]	Upper limit on deviation from car's shortest path, of 20%–50%	Seeking higher productivity on deviated segments
		4.	Short turn[b]	At peak period only	Aim at reducing operations cost
	Network level	5.	Route coverage	Min 800–1000 m route spacing (urban area); Max 400–800 m walk to stop	Min 50%–95% of residents to have below Max walk distance to stop
		6.	Route overlapping	Overlapping allowed only on approaches to CBD	Avoiding confusion and balancing route dispersion
		7.	Route structure[b]	Max of 2–3 branches per route/ loops around terminals	Reducing confusion by different route #
		8.	Route connectivity	Min of 1–3 routes that intersect a given route (transfer points)	Especially for a new route in an existing network
Service design	Planning level	9.	Span of service	Min operating hours by day-of-week, between 5–6 a.m. and 10 p.m. to 2 a.m.	Later ends for larger agencies
		10.	Load (crowding) level	Max load factor 125%–150% of seat capacity at peak hours/ segments; 100% at off-peak	Higher load factors (150%–175%) for short-hand service (Shuttle, Feeder)
		11.	Standees[b]	Max 50% of # of seats	Depends on the interior vehicles configuration
		12.	Headway, upper limit[b]	Max (policy) headway, 15–30 min at peak period; otherwise, 20-60 min	Varies by type of service and day-of-week
		13.	Headway, lower limit	Min headway, 2–3 min	More use in smaller agencies
		14.	Transfers	Max of 1–3 transfers for any O-D	Larger agencies permit more transfers
		15.	Passenger shelters[b]	Min of 65–100 daily boarding pass.	Attention to locations of elderly, and centers, such as hospital
	Monitoring level	16.	Schedule adherence[b]	Min 80% "on time" (0–5 min behind schedule) at peak period, 90% otherwise	This guideline is usually relaxed for short headway
		17.	Timed transfer	Max of 3–8 min vehicle wait at transfer point	More use in smaller agencies
		18.	Missed trips[b]	Min 90%–95% of scheduled trips are OK	Missed trips can also not comply with trip reliability criterion

(Continued)

Table 1.3 (Continued) Available service standards and their typical criteria ranges

Category	Group	#	Standard item	Typical criterion range[a]	Remarks
		19.	Passenger safety[b]	Max 6–10 pass. accidents per 10^6 pass.; Max. 4–8 accidents per $1.6 \cdot 10^5$ vehicle-km	Depends on updated safety data
		20.	Public complaints	Limits on # of complaints per driver/pass./time period	Public comments and complaints always received

[a] Reflects data mainly from the United States.
[b] Standards commonly in use.

range, and remarks. The figure, like elsewhere in the book, uses abbreviations for compactness of presentation: passengers (pass.), population (pop.), minimum (min), maximum (max), meter (m), origin–destination (O-D), and central business district (CBD). Table 1.3 shows typical bounds for each standard, which can be used informally as a guideline. In "Remarks," we provide extra information for each standard that will enhance its implementation characteristics.

Although the standards in Table 1.3 are survey based, using informative questions, we think that two, more elaborated and important questions are missing: (1) What are the specific purposes of a given standard? (2) Are the purposes in (1) fully and optimally attained? In answering these two questions, the transit agency may find out that the purposes are not clear enough to warrant a standard. Furthermore, even for well-defined purposes, the transit agency may realize that answering questions requires some search/research. The latter case indeed deserves well-founded research to justify the use of standards in terms of their range. We attempt in Table 1.4 to outline, for each standard, whether it is necessary to undergo research or whether it is sufficient to determine its criteria administratively. We found that 6 standards necessitated research, 3 standards could be administrative based, and research findings could be input for administrative decisions for the remaining 11 standards.

Under the research-based column in Table 1.4, we listed the recommended objective for 17 of the 20 standards. Under the administrative-based column, we inserted the recommended criteria for 15 of the 20 standards. Finally, under the remarks column, we introduced some helpful notes for each analysis recommended.

This book illustrates some of the required research findings and methodologies for certain important service standards. Chapter 3 demonstrates standards 10 (load-crowding level), 11 (standees), and 12 (headway, upper limit). Chapters 5 and 13 show standard 17 (timed transfer). Chapters 12 and 15 show standards 1 (route length), 3 (route directness), and 15 (transfers). Chapter 16 illustrates standard 4 (short turn), and Chapter 18 illustrates standard 16 (schedule adherence).

1.3.2 Evaluation standards

An assessment of ridership productivity and financial performance of any transit agency largely relies on five variables, determined on a route basis: (a) vehicle-hours, (b) vehicle-kilometers, (c) passenger measures, (d) revenue, and (e) operating cost. These five variables form the base for seven economic and productivity standards that are in use in the United States and Europe (METRO, 1984; TCRP Report, 1995; QUATTRO, 1998): (1) passengers per vehicle-hour, (2) passengers per vehicle-kilometer, (3) passengers per trip, (4) cost per passenger, (5) cost–recovery ratio, (6) subsidy per passenger, and (7) relative performance.

Table 1.4 Suitability of standards to be determined by research results, administrative decision, or both

#	Standard item	Research-based suitability	Administrative-based suitability	Remarks
1.	Route length	Max utilization	—	User and operator variables
2.	Stop spacing	Max utilization	—	User and operator variables
3.	Route direction[a]	Demand sensitivity to route deviation	Max % deviation from car's shortest path	Demand contains potential users
4.	Short turn[a]	Max vehicle saving with Min short turns	—	Operator and user variables
5.	Route coverage	—	Max walking distance to stop for majority of residents	Adequate route spacing will result
6.	Route overlapping	Max utilization	—	To comply with O-D demand
7.	Route structure[a]	Max utilization	—	To comply with O-D demand
8.	Route connectivity[a]	Max utilization	—	To allow for all O-D movements
9.	Span of service	—	Min operation hours, by day-of-week	Can be extended through lowering the frequency
10.	Load (crowding) Level	Determine acceptable criteria	Max load at the max load-point, by time-of-day	Flexible range for certain loading figures
11.	Standees[a]	Determine acceptable criteria	Max standees as % of seats for each vehicle type	Flexible range for certain loading figures
12.	Headway upper limit[a]	Determine criteria	Max (policy) headway, by time-of-day	Depends on previous and next headways
13.	Headway lower limit	Determine criteria	Min headway, by time-of-day	Depends on size of vehicles
14.	Transfers	Determine acceptable criteria	Max # of transfers for a given O-D	Depends on transfer smoothness and waiting time
15.	Passenger shelters[a]	Determine criteria	Min daily boarding to warrant shelter	Site specific (% of elderly, handicapped, etc.)
16.	Schedule adherence[a]	Determine criteria through modeling	Min *on-time* performance	Depends on real-time information provided
17.	Time transfer	Determine acceptable criteria	Max waiting time at transfer points	Depends on real-time information provided
18.	Missed trips[a]	Determine criteria	Max allowed % of missed trips	Vehicle breakdown separate from reliability problems
19.	Passenger safety[a]	Determine acceptable criteria	Max pass.-accidents per pass. and km driven	Depends on safety data
20.	Public complaints	—	Max # of complaints per driver/pass./time period	Unimportant separate from serious complaints

[a] Standards commonly in use.

The main evaluation standards currently utilized can be divided into two categories: (1) passenger based and (2) cost based. The first relates to ridership productivity criteria and the second to financial criteria. We inserted these seven standards in Table 1.5 by category, number, standard name, typical criteria range, and remarks. We added two more abbreviations, for vehicles (veh) and for kilometers (km). Whereas the service standards are

Table 1.5 List of evaluation standards and their typical criteria range

Category	#	Standard item	Typical criteria ranges[a]	Remarks
Passenger based	1.	Pass. per[b] Veh-hour (PVH)	Min of 8–40 PVH; Min 50%–100% systemwide PVH average	Depends on type of service, time of day, and day of week
	2.	Pass. per[b] Veh-km (PVK)	Min of 0.6–1.5 PVK; Min 60%–80% systemwide PVK average	Depends on type of service, time of day, and day of week
	3.	Pass. per trip	Min 5–15 riders per trip; Min 15 passengers average load for all routes	Min average load on express trip, where this standard is most useful, is 20–30 passengers
Cost based	4.	Cost per passenger	Max 1.4 of system average	This standard is often included in a composite score with other criteria
	5.	Cost–recovery ratio[b]	Min 0.15–0.30 ratio; Min 1.0 ratio for express-type services	Larger agencies differentiate between types of services and use a composite score
	6.	Subsidy per passenger[b]	Based on revenue per pass. with min 25%–33% of system average	Used as a difference between cost and revenue, it is simpler than cost to explain to the public
	7.	Relative performance	Min 10%–20% across all routes of a composite productivity score	Route performance is measured against that of other routes

[a] Reflects data mainly from the United States.
[b] Standards commonly in use.

self-explanatory, the evaluation standards need interpretation. Standard 1 in Table 1.5, pass./veh-hour, is the most widely used productivity criterion, mainly because of the fact that the operating budget is paid out on an hourly basis. It is based, to the greatest extent possible, on unlinked passenger trips (i.e., each boarding adds one to the amount of passengers) and service (revenue) hours. Standard 2, pass./veh-km, reflects the number of riders boarding a vehicle along a unit of distance rather than a unit of time; it, too, is based on unlinked passenger trips and on service kilometers. Standard 3, pass./trip, is the number of boarding passengers per single (one-way) trip; its advantage lies in its simplicity. Standard 4, cost/pass., is a financial criterion attempting to ascertain the productivity of a route. Standard 5, cost–recovery ratio, is the ratio between direct operating costs (wages, benefits, and maintenance costs) and the share recovered by the fares paid by the route's riders. Standard 6, subsidy/pass., is usually the difference between cost/pass. and revenue/pass. The revenue/pass. criterion reflects different fares and is used as a measure for one of the comparisons of routes. The last standard in Table 1.5 is the relative performance of a route compared to other routes usually having the same characteristics, such as its percentile rank in either an overall ranking of system routes or in a group of routes associated with the same type of service. The exact measure of this ranking varies across transit systems. Table 1.5 shows typical criteria ranges for each evaluation standard that can be used informally or in a more formal manner. "Remarks" provides extra information on each standard that will enhance implementation simplicity.

1.4 VIABILITY PERSPECTIVES

Following our discussion of the framework and derivatives of service and evaluation standards that focus on tools to set adequate and improved transit service, we will now present a more general view of how the basic transit goals should be approached. Current practice shows that transit agencies are experiencing increasing fiscal pressures caused by a decline in transit patronage, increased operational costs, and decreased government support. In response to this trend, transit agencies have more often been reexamining the manner in which their limited resources are allocated. We shall show here that this concentration on savings should not overlook some amenities (Ceder, 2002, 2007) that are concerned with the viability of service.

The decline in transit patronage is the result of two main factors: (a) poor level of service and (b) better competitors. New roads, bridges, and tunnels serve the automobile and to some extent the railroad, whereas the investment in transit enhancements has been at a relatively much lower level. On one hand, there is no need to promote transit service in a free market environment; on the other hand, transit has the best land-transportation safety record, and it can relieve some traffic congestion as well as help preserve the environment.

Transit viability can be looked at through the perspectives of the passenger and the agency. Orderliness is the key for success. In Figure 1.3, we show three cases when passenger will use transit service: (1) when there is no alternative, (2) when the service offered is acceptable, and

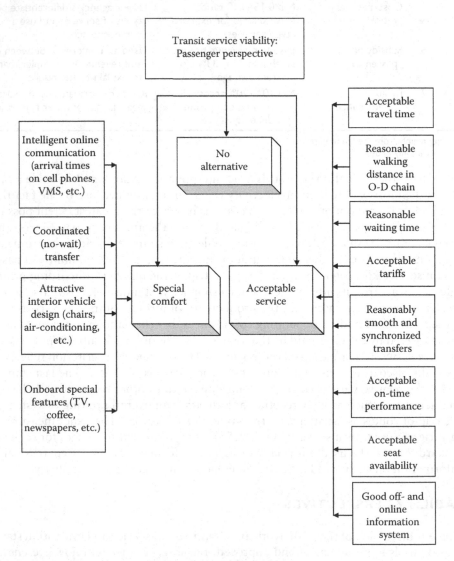

Figure 1.3 Viability of transit service from passenger perspectives.

(3) when the service offers more comfort than does the automobile. Acceptable services, as shown in Figure 1.3, rely basically on reasonable and acceptable travel time, walking distance to and from the transit stop, waiting time, tariff, transfers, and timetable adherence, as well as seat availability and reliability and availability of information. Overall walking distance in the O-D or door-to-door chain and comfortable transfers are central points in understanding passenger travel behavior, which has to do with the integration of transit service end points and the passenger's origin and destination points. Comfortable transfers (vehicle to vehicle, using either the same transit mode or different modes) ought to be smooth (escalator, same platform, etc.) and synchronized (a match between arrival and departure times as stated by online information). Case (3) in Figure 1.3 refers to passengers who regularly use their automobile, but would switch to transit if given special comfort features. Such features can include intelligent online communication, coordinated transfers without waiting, attractive transit vehicle interior design (a la advanced aircraft), and onboard services (again a la aircraft), none of which can be obtained in an ordinary automobile.

Figure 1.4 presents the agency perspective for viability. The provision of subsidy and commercial concessions is self-explanatory; however, the approaches to attain increased patronage and revenue and a reduction in operating costs deserve some elaboration. Increased patronage and revenue can be realized through remedying the current transit service illness, that is, obtaining a better match between service and demand, reliable service, coordinated network (à la a network of routes in a good metro system), and improved comfort and information. This book presents some remedies for a better match between service and demand in Chapter 4, for solving reliability problems in Chapter 18, and for coordination

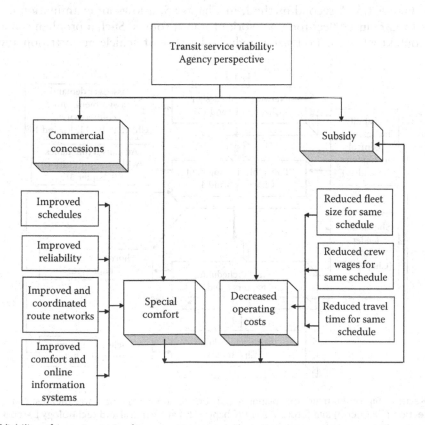

Figure 1.4 Viability of transit service from agency perspectives.

among transit routes in Chapters 5, 13, and 14. Decreased operating costs can be attained by reducing fleet size (less capital cost), total drivers' wages, and the average travel time in the transit network system without changing the service (timetables) offered. We will introduce techniques and methods to reduce fleet size in Chapters 6 through 8, and 12, to reduce total drivers' wages in Chapter 9, and to reduce travel time through a better match between service and demand in Chapter 4.

1.5 OUTLINE OF OTHER CHAPTERS

This introductory chapter has emphasized four major activities, as seen schematically in Figure 1.2, as the core of transit planning. Figure 1.5 describes the remaining themes of the book around these four activities; pertaining to the first edition of this book (Ceder, 2007), it includes significantly revised Chapters 5, 8, 12, and 15 and completely new Chapters 13, 14, and 19. Each block in Figure 1.5 lists the chapter(s) dealing with the content specified. As we see it, the appetite to change transit service across the board ought to start with a small board.

Chapter 2 covers transit data collection systems including manual and automated techniques, automatic vehicle monitoring sampling considerations, and some notes on passenger surveys. Chapter 3, which is part of the second activity in Figure 1.5, analyzes passenger-load and running-time data and four methods of frequency, and Chapters 4 and 5 presents advanced methods of constructing timetables. One of these (in Chapter 5) creates timetables with even passenger loads at different max load points for individual vehicles as opposed to a single max load point (across all vehicles in 1 h). Chapter 5 ties in with the second planning activity in Figure 1.5. A second method, in Chapter 5, allows an examination of optimal timetables for a reduced fleet for a network of transit routes. Such a problem can arise in a practical context when the fleet size is reduced because of vehicle-age attrition and budget

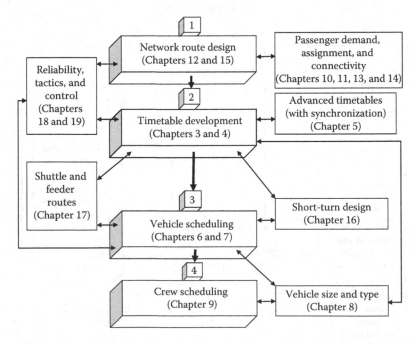

Figure 1.5 Relationship of the main four planning activities to the remaining book themes. Note: Data collection (Chapter 2) and future transit (Chapter 20) are general and technology based and, therefore, related to all planning activities.

policy decisions. A third method, in Chapter 5, describes optimal procedures for designing timetables with maximum simultaneous arrivals at given stops. This synchronization solution is based on a given range of the headways determined.

Chapter 8 introduces different vehicle types with scheduling requirements, examines various scheduling scenarios, and presents new methods to create the best schedules with vehicle-type constraints. In addition, Chapter 8 describes different models to determine the required size of a vehicle. This chapter ties in with both the second and third activity in Figure 1.5. Chapter 9, which is the fourth activity in Figure 1.5, demonstrates some of the optimization concepts behind the problem of assigning crew (mainly drivers) to vehicle schedules with the minimum cost involved. Following a review of some existing models, we present a method to create a good framework for optimal duties. This chapter interacts to some extent with procedures described in the previous chapter for different vehicle types.

Chapters 10 and 11 provide overview on passenger demand, route choice, and assignment. In Chapter 10, we give an overview of the factors affecting passenger demand and their sensitivity (elasticity) and methods designed to predict passenger demand; we also show estimation methods for predicting O-D matrices and how best to forecast ridership. Chapter 11 presents passengers' dilemma in route choice among alternative routes; the route choice is then incorporated into transit assignment modeling at the network level. The modeling in Chapters 10 and 11 interacts with the first activity in Figure 1.5.

Chapters 12 through 15 constitute the first planning activity outlined in Figure 1.5. Chapter 12 presents basic service design elements for different purposes and introduces a new method for dealing optimally with stop placement on streets with the consideration of topography. In addition, this chapter includes a theoretical approach of optimal stop location on a network of roads. Chapter 12 introduces also a new method to determine optimally operational parking limitations. This on-street bound on maximum parking spaces calls for solutions to avoid traffic congestion at some route end points. Chapters 13 and 14 are new chapters, in this second edition, brought up to enhance the importance of the need for coordination and connectivity. Chapter 13 presents new measures and analysis for transit connectivity, transfer based, including a method to detect bottlenecks of coordination in networks of transit routes. Chapter 14 delivers the behavioral aspects of transit coordination and connectivity; it is based on surveys and behavioral modeling showing how people appraise routes with transfers, with some advantages, compared with routes without transfers. Chapter 15 establishes objective functions for designing a network of transit routes while complying with passenger, agency, and government perspectives. This chapter delivers a complete solution for the creation of transit routes while taking into account the activities of vehicle scheduling and timetabling. Chapter 15 describes also some metaheuristic methods to attain a reliable transit network including a scenario in which the demand is varied across seasons. Overall these four chapters, Chapters 12 through 15, construct routes and transfers, assigning O-D demand and initial frequencies, and provide optimal criteria and best route network solutions for decision makers.

Chapter 16 defines a cost-effectiveness approach for operating single routes with short turns (not all trips start and end at the end points of the route). Employing some of the developments resulting from the second and third activities in Figure 1.5, this chapter provides an optimal method to maintain the required level of service while minimizing the fleet size through creating timetables with minimum short-turn trips and minimum required vehicles. Chapter 17 presents the need for and the concept behind an integrated transit system, with an emphasis on shorthand transit service (shuttle and feeder routes). This integrated service is presented as an attractive, reliable, rapid, smooth, and synchronized system. Ten different routing strategies are developed for bus shuttle and feeder routes while using a simulation tool for analysis along with real-time communication possibilities. Chapter 17 interacts with the second and third activities in Figure 1.5.

Chapters 18 and 19, especially the latter, focus on new technology principles to assist in improving transit operations planning. Chapter 18 discusses the essential part of transit reliability and its impact on operations planning. It covers the subjects of variability of concern among passengers and in the agency, the bus-bunching phenomenon, methods to improve reliability, the calculating of passenger waiting time and vehicle running time, and features and benefits of automatic vehicle location (AVL) systems. Naturally, Chapter 18 interacts with elements of the first three activities in Figure 1.5. In Chapter 19, we present a bridge between new technologies and transit operations using in particular operational tactics, the concept of flexible routing based on distributed computing, and electronic user–operator communication. The book ends with an outline of some future research needs.

REFERENCES

CASPT2015 (2015). Erasmus School of Economics and Rotterdam School of Management, Erasmus University, http://www.caspt.org. (Accessed March 12, 2015).

Ceder, A. (2001). Public transport scheduling. In *Handbooks in Transport—Handbook 3: Transport Systems and Traffic Control* (D. Hensher and K. Button, eds.), pp. 539–558, Pergamon Imprint, Elsevier Science, New York.

Ceder, A. (2002). Urban transit scheduling: Framework, review, and examples. *ASCE Journal of Urban Planning and Development*, **128**(4), 225–244.

Ceder, A. (2007). *Public Transport Planning and Operation—Theory, Modelling and Practice*, Elsevier, Butterworth-Heinemann, Oxford, U.K.

Ceder, A. and Wilson, N.H.M. (1986). Bus network design. *Transportation Research*, **20B**(4), 331–344.

Daduna, J.R., Branco, I., and Paixao, J.M.P. (eds.) (1995). *Computer-Aided Transit Scheduling*. Lecture Notes in Economics and Mathematical Systems, Vol. 430, Springer-Verlag, Berlin, Germany.

Daduna, J.R. and Paixao, J.M.P. (1995). Vehicle scheduling for public mass transit—An overview. In *Computer-Aided Transit Scheduling* (J.R. Daduna, I. Branco, and J.M.P. Paixao, eds.), pp. 76–90, Springer-Verlag, Berlin, Germany.

Daduna, J.R. and Wren, A. (eds.) (1988). *Computer-Aided Transit Scheduling*. Lecture Notes in Economics and Mathematical Systems, Vol. 308, Springer-Verlag, Berlin, Germany.

Desrochers, M. and Rousseau, J.M. (eds.) (1992). *Computer-Aided Transit Scheduling*. Lecture Notes in Economics and Mathematical Systems, Vol. 386, Springer-Verlag, Berlin, Germany.

Hickman, M., Mirchandani, P., and Voss, S. (eds.) (2008). *Computer-Aided Scheduling of Public Transport*. Lecture Notes in Economics and Mathematical Systems, Vol. 600, Springer-Verlag, Berlin, Germany.

Metropolitan Transit Authority of Harris County (METRO) (1984). *Bus Service Evaluation Methods: A Review*, Urban Mass Transportation Administration, Washington, DC, DOT-1-84-49.

Mora, J.G. and Chernicoff, W.P. (2005). Transit's complex route to improved standards and codes. *TR News*, Transportation Research Board, **236**, 3–7.

Pine, R., Niemeyer, J., and Chisholm, R. (1998). Transit scheduling: Basic and advanced manuals. TCRP Report 30, Transportation Research Board, Washington, DC.

Quality Approach in Tendering Urban Public Transport Operation in Europe (QUATTRO, 1998). *Transport Research Fourth Framework Programme*, Urban Transport, VII-51, Office for Official Publications of the European Communities, Luxembourg, Europe.

Rainvile, W.S. (1947). *Bus Scheduling Manual: Traffic Checking and Schedule Preparation*. Reprinted 1982, American Public Transit Association, U.S. Department of Transportation, Washington, DC, DOT-1-82-23.

Rousseau, J.M. (ed.) (1985). *Computer Scheduling of Public Transport 2*, North-Holland, Amsterdam, the Netherlands.

TCRP Report (1995). Bus Route Evaluation Standards, Synthesis of Transit Practice 10. Transit Cooperative Research Program, Transportation Research Board, Washington, DC.

Voss, S. and Daduna, R. (eds.) (2001). *Computer-Aided Scheduling of Public Transport*. Lecture Notes in Economics and Mathematical Systems, Vol. 505, Springer-Verlag, Berlin, Germany.

Wilson, N.H.M. (ed.) (1999). *Computer-Aided Scheduling of Public Transport*. Lecture Notes in Economics and Mathematical Systems, Vol. 471, Springer-Verlag, Heidelberg, Germany.

Wren, A. (ed.) (1981). *Computer Scheduling of Public Transport: Urban Passenger Vehicle and Crew Scheduling*, North Holland, Amsterdam, the Netherlands.

Chapter 2

Data requirements and collection

CHAPTER OUTLINE

PRACTITIONER'S CORNER

This chapter offers a number of potential benefits to transit agencies in the areas of data acquisition, data merits, and data utilization. The major objective is to help develop a comprehensive, statistically based data collection plan that will enable transit agencies to collect proper data and the right amount in a cost-effective manner.

The chapter starts with the three core keys representing the objectives of this book: (1) to know, (2) to plan and decide wisely, and (3) to operate in a smart manner. Achieving the first key, through the gathering of comprehensive data, is the objective of this chapter. The first section describes the various data collection techniques and methods, including definitions and interpretations. The second section continues with the linkage of the data collection method, the results expected from the data analysis, and enhanced service elements gained from the results. These two sections of the chapter are most suitable for practitioners, who usually understand the difficulties and sometimes even conflicts involved in real-life data collection as is well represented in the following real anecdote.

While trying to maximize their comfort, drivers are more sophisticated than one can imagine. The following anecdote concerns gaining extra planned travel time by arriving early and having extended layover time. The checkers in a bus agency who boarded the vehicle used to announce to the driver that they were measuring average travel time with a rate factor. If the bus is too slow (i.e., encountering more overtaking vehicles than it overtook), the rate is below 10, otherwise more than 10. One driver found *out* when this check was going to take place and asked members of his family to drive their cars at a slow pace just in front of his bus in order for him to earn a rate of below 10 and to extend the planned travel time.

The third and longest section of this chapter introduces basic statistical tools that are employed in almost every phase of the data collection undertaking. There is no prerequisite for understanding the statistical tools, though practitioners may want to skip the mathematical formulations and concentrate on the examples.

The chapter ends with a literature review and recommendations for further reading. Overall, the data collection effort should be handled professionally, with a precise understanding of the purpose of each data element. Finally, since we want not only high-quality data, but not in excessive amounts, it is good to recall what Galileo said: "Science proceeds more by what it has learned to ignore than what it takes into account."

2.1 INTRODUCTION

Current practice in transit agencies shows that sufficient data seldom exist for service operations planning. Manual data collection efforts are costly and, consequently, must be used sparingly. Automated data collection systems, though growing rapidly, are not yet perfectly linked to the requirements of planning data. Extraction of data from an automated system, which is ostensibly a simple task, may turn out in actuality to be rather complex. Having too much data is often as bad as having too little. The only concept we can trust is that in any collection of data, the figure most obviously correct beyond all need for checking is the mistake. At the same time, the data are essential for responding to basic passenger needs, namely, the route: where is the closest stop, what time should I be at the stop, and more. The data are certainly crucial, too, for responding to the operations planning needs of each transit agency: How can the network of routes, stops, and terminals be improved? How can each

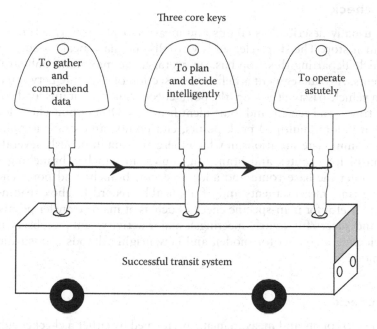

Figure 2.1 The three important keys for a successful transit system.

route be improved? What is the best timetable to deliver? How can fleet size be minimized while maintaining the same level of service? How can crew cost be minimized without service changes? No doubt, a common element for all transit agencies is their pursuit of data to aid in improving the efficiency, productivity, and effectiveness of their systems.

This chapter demonstrates data requirements and applications for enhancing transit-service productivity and efficiency. Fundamentally, there are three keys for achieving a successful transit service: (a) gathering and comprehending adequate data, (b) using the data collected for intelligent planning and decisions, and (c) employing the plans and decisions for astutely conducted operations and control. These three keys appear schematically in Figure 2.1. This chapter, which introduces the framework for attaining the first key, consists of four parts: first, different data collection techniques; second, the data requirements, with attention paid to their coordination; third, basic tools in statistics needed to comprehend the mechanism of data analysis; and fourth, a literature review and list for further reading.

2.2 DATA COLLECTION TECHNIQUES

We can divide transit data collection techniques required for operations planning into three categories: (1) manual-based methods, (2) automated-based methods, and (3) automatic vehicle location (AVL)-based methods. Fundamentally, there are only two types of methods, manual and automated. However, AVL or automatic vehicle monitoring (AVM) systems furnish more accurate information, especially in time and space, than other item-specific automated methods and, therefore, can be looked at as a separate category. The AVL-based technique will be covered in Chapter 17 in connection with linking the technique to remedies for reliability problems.

Five primary techniques for collecting transit data may be identified: point check, ride check, deadhead check, passenger survey, and population surveys.

2.2.1 Point check

Point check is usually described as counts and measurements performed by a checker stationed at a transit stop. The stop selected is virtually the max (peak) load point, at which the transit vehicle departing this stop has, on average, the maximum onboard load across all route segments. A route segment is defined as a section of the route between two adjacent stops. For each vehicle passing the stop, the point check usually contains load counts, arrival and departure times, and vehicle and route identifications. Other point-check locations than the peak stop are more (multiple) peak points, end points, and strategic points. Multiple point checks accommodate situations in which there are simultaneously several peak points and a situation of long routes and branching routes, in which a branching route is one that stretches along the base route and adds a certain branch. End point checks accommodate running-time measurements and, if applicable, record farebox readings. Strategic point checks are useful for item-specific checks, such as at major transfer points (measuring transfer time and successful vehicle meetings), major activity centers (observing passenger behavior in selecting a competitor mode), and new neighborhoods (measuring changes in passenger demand).

2.2.2 Ride check

Ride check refers to counts and measurements performed by either a checker riding the transit vehicle along the entire route or an automated instrument (hence, replacing the human checker). The ride check contains mainly on-and-off passenger counts, from which one can derive the onboard passenger load for each route segment, arrival and departure times for each stop, and sometimes item-specific surveys or measurements (vehicle running speed, boarding by fare category, gender of passengers, and baggage size) and record farebox readings. The common automated instrument for ride checks, called automated passenger counter (APC), can perform the main ride-check tasks. It cannot, however, replace the checker in counting boarding by fare category and in surveying passengers. A special ride check is one performed by the operator (driver). It usually involves the interaction of the driver and a farebox, with the driver inserting into the machine information related to fare category and origin–destination (O-D) per passenger. The action, however, increases the dwell time at the stop.

Figure 2.2 depicts passenger demand as interpreted by hourly passenger load, using both point-check and ride-check methods. This is a vital data element affecting operations planning. We can obtain from point check only the load for the peak segment; with ride check, we can illustrate the entire load profile by either time or space. Figure 2.2 will be used as space-based (passenger-km) illustrations in the next chapter, which is concerned with a determination of vehicle frequencies and headways. The time-based (passenger-hour) load profile will be used in Chapters 13 and 14, which deal with route design service and network. Figure 2.2 shows the location of the max load point (stop), from which 320 passengers are transported during given hour j along the max (peak) load segment. In addition, Figure 2.2 demonstrates the different visual appearances of the space- and time-load profiles. Whereas the x-axis of the space-based profile is fixed for a given route direction, the x-axis of a time-based profile is set according to average values.

2.2.3 Deadhead check

Deadhead check refers to the average vehicle running time between an arrival point on one route and a departure point on another route. This deadheading time is required in a transit

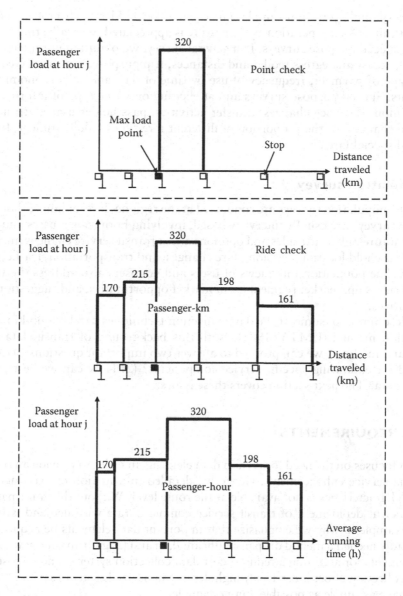

Figure 2.2 Main information obtained for operations planning through point check and ride check.

system with interlining routes. It is measured mainly by agency cars traveling along the shortest path (in time) between the two route end points. This shortest path varies by time of day, day of week, and type of day.

2.2.4 Passenger survey

Transit passenger surveys are conducted in essence while directly confronting the passengers. The known survey methods of this type are on board, at stop, at terminal, and mailback (postage-free forms). The most common is the onboard survey. All the surveys are carried out by agency checkers or drivers or by especially trained staff who either distribute forms to fill out or ask the questions in person. One way to increase the response incentive is to hand out a symbolic gift, for example, a nice pen or key holder. Such a token may open up more opportunities for cooperation as the gesture is appreciated. Basically, there are general surveys and special purpose surveys. In a general survey, we obtain multitype information, such as O-D, access and egress modes and distances, trip purpose, routes selected on a trip, fare paid, type of payment, frequency of use by time of day, and socioeconomic and attitude elements. Special purpose surveys aim at eliciting only *one type* of information, such as O-D, opinion of service changes, transfer activities, pass-holder usage, attitude toward possible fare changes, or the proportion of different fare types (adult, student, free passes, transfer, and special fare).

2.2.5 Population survey

Generally, the population surveys are conducted at the regional level at home, shop, or workplace. These surveys are usually interview based, involving both transit users and nonusers, in order to capture public attitudes and opinions about transit service changes (including the impact on household location decision), fare changes, and transportation, traffic, and land-use projects. The population interviews of users and nonusers also address the vital issues of potential ridership, market segmentation, market opportunities, and suggestions for new transit initiatives.

Further description of some of the data collection techniques can be found in the transit data collection manual (UMTA, 1985). With this background of transit data collection techniques and methods, we can proceed to answer two important questions: (1) What data are required for enhancing specific service components? (2) How can we best exploit and analyze the data? The next section covers these issues.

2.3 DATA REQUIREMENTS

This section focuses on the need for certain data elements in order to provide an opportunity for achieving service enhancements. Generally, detailed information on passenger demand and service characteristics is not available at the route level. Without this information, however, the efficient deployment of transit service commensurate with demand is impossible. This is an example of why we emphasize that important data elements be acquired so as to (1) obtain badly needed data and drop insignificant data and (2) gain an overview of the entire data requirements for attaining a reduced-cost data collection system. Since a cost appraisal of the data collection effort is ostensibly high, let us emphasize Einstein's saying: "Everything should be made as simple as possible, but not simpler."

There is a common thread of data needs across all transit agencies. This thread exists as long as agencies share these objectives: (a) improving service and operations, (b) improving productivity and efficiency by better matching supply and demand, (c) improving levels of service through increased reliability as a result of better control and response, and (d) reducing data gathering, processing, and reporting costs. Broadly speaking, the data items gathered by the techniques described earlier are useful for one or more aspects of route and service design, scheduling, information system, marketing, deficit allocation, monitoring management, and external reporting. This section will show, however, only the data elements pertaining to operations planning.

Figure 2.3 portrays in flowchart form three operations planning categories: an adequate data collection method, the results required from a data analysis, and relevant, enhanced service elements. Clarification of data collection methods appears in the previous section. Figure 2.3 refers to three planning levels: regional, group of routes, and route. At the

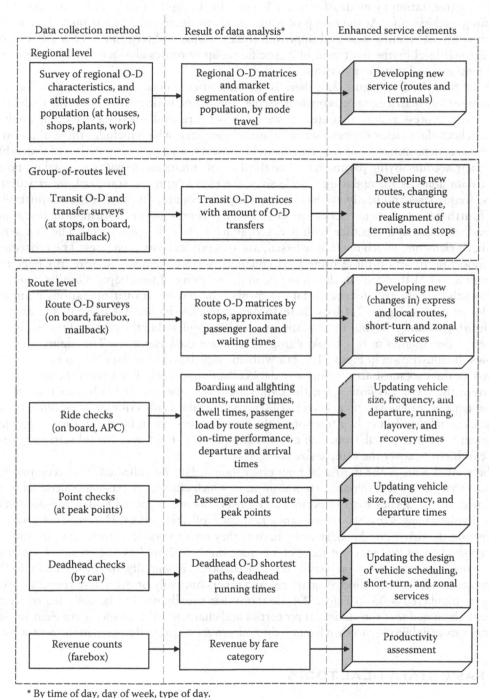

Data collection method | Result of data analysis* | Enhanced service elements

Regional level

Survey of regional O-D characteristics, and attitudes of entire population (at houses, shops, plants, work) → Regional O-D matrices and market segmentation of entire population, by mode travel → Developing new service (routes and terminals)

Group-of-routes level

Transit O-D and transfer surveys (at stops, on board, mailback) → Transit O-D matrices with amount of O-D transfers → Developing new routes, changing route structure, realignment of terminals and stops

Route level

Route O-D surveys (on board, farebox, mailback) → Route O-D matrices by stops, approximate passenger load and waiting times → Developing new (changes in) express and local routes, short-turn and zonal services

Ride checks (on board, APC) → Boarding and alighting counts, running times, dwell times, passenger load by route segment, on-time performance, departure and arrival times → Updating vehicle size, frequency, and departure, running, layover, and recovery times

Point checks (at peak points) → Passenger load at route peak points → Updating vehicle size, frequency, and departure times

Deadhead checks (by car) → Deadhead O-D shortest paths, deadhead running times → Updating the design of vehicle scheduling, short-turn, and zonal services

Revenue counts (farebox) → Revenue by fare category → Productivity assessment

* By time of day, day of week, type of day.

Figure 2.3 Data collection methods and resultant analysis and service elements.

regional level, we focus on developing new or improved transit service by means of routes and terminals. To accomplish this, we need the entire regional population O-D matrices and market segmentation by mode of travel. Chapters 10 through 15 cover some aspects of this planning undertaking. At the group-of-routes level, we focus solely on transit service with the objective of creating new and improved routes, stops, and terminals. This differs from the regional level by the treatment of a specific group of routes characterized by their own O-D matrices and transfer activities.

At the route level in Figure 2.3, there is the bulk of enhanced service elements. We start with route O-D matrices, approximate passenger loads, and waiting times in order to develop or improve express and local routes, as well as short-turn and zonal services. The analysis of ride-check data yields the next set of enhanced elements, which includes the evaluation of vehicle size, determination of frequencies and headways, and the construction of timetables with their accompanying parameters. The third set of enhanced elements is based on point-check data analysis for updating vehicle sizes, for example, mini, standard, or articulated buses, frequencies and headways based on peak point counts, and their derived timetables. The fourth set of elements relies on O-D average running-time information pertaining to deadheading trips, together with a description of O-D shortest paths. The last set of enhanced elements, which is revenue based, assesses productivity measures. The procedures and modeling to attain these route-level enhanced service elements appear in Chapters 3 through 8 and 11 through 18, that is to say, in the majority of the chapters in this book.

Altogether, there are 16 crucial data elements required for transit operations planning. The lower part of Figure 2.4 lists these 16 data elements. Figure 2.4 presents the abstract relationships among the four main planning activities depicted in the center of the figure and the remaining themes of this book, citing the relevant data elements. This figure virtually follows the illustration in Figure 1.5, but with an emphasis on the data elements.

Two vital issues must accompany any data collection system. The first is the *sample size* required; the second is *how often* the data should be collected. It is obvious that not all transit trips or transit passengers can be observed in the data collection process. Since transit agencies usually collect only a fraction of the data, it is uncertain how well the sample data represent the entire population. The next section provides the fundamental statistical tools that can broadly cover the sample size issue.

The second issue is the matter of how often should data be collected. We recognize that transit service exists in a dynamic environment, in which changes in passenger demand occur regularly. Transit agency road inspectors and supervisors are supposed to deliver information on noticeable changes that would warrant a new data collection effort. Unfortunately, we cannot rely on the drivers to do so, since by nature, they try to maximize their comfort and will not report changes that may put more of a burden on their shoulders (e.g., reduced passenger load—lots of empty seats—that justifies trip cancellations). Essentially, there are data items that are recommended to be collected quarterly, and others annually or every few years. The data elements numbered 8–13 in Figure 2.4 are those that usually need to be collected quarterly. It should be noted that the statistical properties and characteristics of each data element—for example, its variability—can shed light on the *how often* issue as will be seen in the next section.

2.4 BASIC STATISTICAL TOOLS

Statistical techniques play an important role in achieving the objectives of proper transit data acquisition, handling, and interpretation. The *Merriam-Webster Collegiate Dictionary* (2000) defines *statistics* as follows: "(1) a branch of mathematics dealing with a collection,

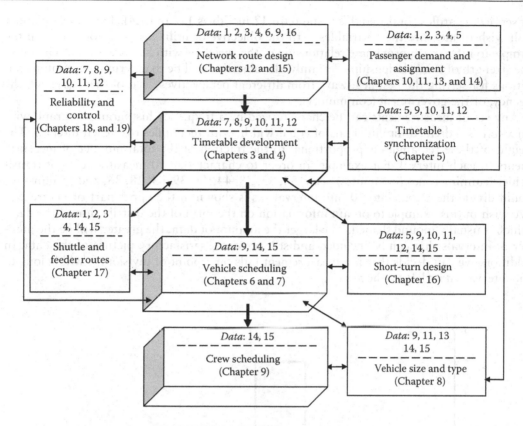

Figure 2.4 Data requirements by book chapter themes.

analysis, interpretation of masses of numerical data; (2) a collection of quantitative data."
Basically, the two definitions possess common elements implying a collection of data with
inference or *inference making* as the objective. A large body of data is called *population*,
and the subset selected from it is a *sample*. In any transit data collection system, the objective of the statistics is to infer (estimate or decide) something about the population, based on
the information contained in a sample.

2.4.1 Histogram and distribution

Consider a study whose aim is to determine important variables affecting headways between
transit vehicles moving on the same route and in the same direction. Among these variables
are existing passenger demand, potential or hidden passenger demand, and a minimum level

of service regardless of demand (see standard 12 in Tables 1.3 and 1.4). The transit planner will wish to measure these variables with the objective of utilizing the information in the sample to infer the approximate relationship of the variables with headways and to measure the strength of this relationship. Certainly, the objective will be to determine optimum conditions for setting the best headways from different perspectives, that of the passenger, the agency, or the government (community).

Any set of measurements can be characterized by a frequency histogram—a graph with an x-axis subdivided into intervals of equal width, with rectangles over each interval. The height of these rectangles is proportional to the fraction of the total number of measurements in each interval. For example, in order to characterize 10 measurements of transit vehicle running times (in minutes)—say, 36, 52, 38, 44, 48, 39, 63, 28, 36, and 55 min—we could divide the x-axis into 10 min intervals as is shown in the upper part of Figure 2.5. We wish in this example to obtain information on the form of the distribution of the data, which is usually mound shaped. The larger the amount of data, the greater will be the number of intervals that can be included and still present a satisfactory picture of the data. In addition, we can decide that if a measurement falls on a point of division, it will belong to the interval on its left on the x-axis.

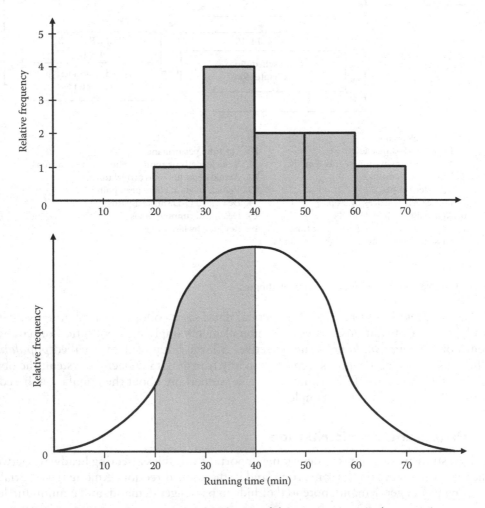

Figure 2.5 Relative frequency histogram and distribution of the running times in the example.

If we can assume that the measurements were selected at random from the original set, we can move one step further to a probabilistic interpretation. The probability that a measurement would fall in a given interval is proportional to the area under the histogram that lies over the interval. For example, the probability of selecting a running-time measurement over the 20–40 min interval in Figure 2.5 is 0.5. This interpretation applies to the distribution of any set of measurements, including populations. If the lower part of Figure 2.5 gives the frequency distribution of running times (of a given transit route, direction of travel, and time period), the probability that the running time will lie in the 20–40 interval is proportional to the shaded area under the distribution curve. In statistical terms, the distribution function for a random variable running time is denoted $F(t')$ and given by $F(t') = P(T' \leq t')$ for $-\infty < t' < \infty$, where $P(T' \leq t')$ is the probability of T' being less than or equal to the value of t' and T' is continuous. The function $F(t')$ is called the distribution function (or cumulative distribution function), and its derivative (if existing) is $f(t') = dF(t')/dt'$. The function $f(t')$ is called a probability density function (pdf) for the random variable T'. The distribution that appears in the lower part of Figure 2.5 is $f(t')$.

Since many similar histograms could be formed from the same set of measurements, we need to rigorously define quantities with which to measure the sample data, n. The most common measure of a central tendency used in statistics is the *arithmetic mean* or *average*, the latter being the term we use in this book. The most common measure of variability used in statistics is the *variance*, which is the function of deviations of the sample measurements from their average. The square root of the variance is called *standard deviation*. These three measures—variance, deviation, and standard deviation—appear in expressions (2.1) and (2.2) for a set of n measured responses $x_1, x_2, ..., x_n$, where $\bar{x}$, s^2, and s are the average, variance, and standard deviation, respectively. The corresponding three measures of population are denoted μ, σ^2, and σ:

$$\bar{x} = \frac{1}{n}\sum_{i=1}^{n} x_i \tag{2.1}$$

$$s^2 = \frac{1}{n}\sum_{i=1}^{n} (x_i - \bar{x})^2 \tag{2.2}$$

For the example of Figure 2.5, we obtain $\bar{x} = 43.9$ min, $s^2 = 98.578$ min², and $s = 9.929$ min. It may be noted that many distributions of transit data (as is virtually all other data in real life) are mound shaped, which can be approximated by a bell-shaped frequency distribution. This is known as a *normal* curve, with a normal density function, and may be expressed as follows:

$$f(x) = \frac{e^{-(x-\mu)^2/2\sigma^2}}{\sigma\sqrt{2\pi}}, \quad \sigma > 0, -\infty < \mu < \infty, -\infty < x < \infty \tag{2.3}$$

where
 e is the base of a natural logarithm (= 2.7182818 ...)
 π is the ratio of the length of the circle to its diameter (= 3.14159 ...)
 μ and σ are the average and standard deviation, respectively, and hence locate the center of the distribution and measure its spread

Data possessing the normal type of distribution have the following empirical rule: $\mu \pm i\sigma$ contains approximately 68%, 95%, and 99.7% of the measurements for $i = 1, 2, 3$, respectively.

In practical terms, we find that most random variables observed in nature lie within 1, 2, or 3 standard deviations of their average, with probabilities in the vicinity of 0.68, 0.95, and 0.997, respectively.

2.4.2 Estimation

We will now return to the objective of statistics—making an inference about the population, based on information contained in a sample. Usually, this inference is about one or more population parameters. For example, we might wish to estimate the fraction of trips that would have running times of less than or equal than 20 min. Let T' equal the running time; the parameter of interest is $P (T' \leq 20)$, which is the area under the pdf (lower part of Figure 2.5) over the interval $t' \leq 20$. Suppose we wish to estimate the average of T' and to use the sample in Figure 2.5, with its average estimate of 43.9 min. One way to measure the goodness of this estimation is in terms of the distance between the estimate and its real (unknown) population value. This distance, which varies in a random manner in repeated sampling, is called the *error of estimation*, and we certainly would like it to be as small as possible. If this error is given in percentages, it is often called precision (± given %). If it is in actual values (minutes for T'), it is often called tolerance (± given minutes).

Assuming that we want this average running time of 43.9 min to have a precision of ±10%, we then ask, "What is the probability that the average of T' will fall in the interval between 39.51 and 48.29 min?" This probability is called *confidence level* or *confidence coefficient*, and the interval is called *confidence interval*. The objective here is to find the tolerance that generates a narrow interval that encloses the average of T' with a high confidence level (probability) of usually 0.95%, or 95%. A standard normal random variable with a normal distribution as expressed in Equation 2.3 is usually denoted Z. If we want to find a confidence interval that possesses a confidence coefficient equal to $(1 - \alpha)$ for Z, we can use the following probability statement:

$$P(-z_{\alpha/2} < Z < z_{\alpha/2}) = 1 - \alpha \qquad (2.4)$$

where
   P is the probability
   $z_{\alpha/2}$ and $-z_{\alpha/2}$ are two tail end values of the standard normal distribution (the probability in between the tails is $1 - \alpha$)

For instance, Z has a 0.95 probability ($\alpha = 0.5$) of being between –1.96 and +1.96 (based on normal distribution tables). In this case, Equation 2.4 becomes $P (-1.96 < Z < 1.96) = 0.95$. That is, $Z_{1-\alpha/2} = Z_{1-0.5/2} = Z_{0.975} = 1.96$. More interpretations and description of this subject can be found in numerous books on statistics, such as Washington et al. (2003).

The transit data collection manual (UMTA, 1985) contains a reference to the U.S. national database specification. In this specification, the precision is ±10%, and the confidence level is at 95% for reporting annual passenger-boarding counts and passenger miles. This is a system-wide requirement by the Urban Mass Transit Administration (UMTA) for an average weekday and for Saturday and Sunday.

Other useful measures of the error of estimation described in the transit data collection manual (UMTA, 1985) are absolute tolerance (AT) and absolute equivalent tolerance (AET). AT and AET represent the error of estimation of a *proportion of observations* rather than the average of the data. The UMTA manual states as an example that schedule adherence is the proportion of trips lying in the categories early, on time, and late.

For instance, an estimated proportion of on-time performance may be 0.8 with AT = ±0.1. However, when the estimated proportion approaches 0 or 1.0, the AT resulting from a given sample size is small compared to the AT value for an estimated proportion in the vicinity of 0.5. Thus, the second measure, AET, is established as the absolute (equivalent) tolerance that can be attained with the following formula (UMTA, 1985) if the estimated proportion is 0.5:

$$\text{AET} = \frac{0.5\,\text{AT}}{\sqrt{p(1-p)}} \qquad (2.5)$$

where p is the expected proportion.

2.4.3 Selecting sample size

The sampling procedure affects the quantity of information per measurement. This procedure, together with the sample size, n, controls the total amount of relevant information in the sample. Start with the simplest sampling situation, in which random sampling is conducted among a relatively large population. For an independent, continuous variable X, we define the average or the arithmetic mean as the expected value, E(X), with a variance of V(X). It is known (see, e.g., Washington et al., 2003) that for x_1, x_2, ..., x_n independent random variables (or trials of an experiment), with $E(x_i) = \mu$ and $V(x_i) = \sigma^2$, we will obtain $E(\bar{X}) = \mu$ and $V(\bar{X}) = \sigma^2/n$, where $\bar{X}$ is distributed with an average of μ and a standard deviation of $\sigma_{\bar{x}} = \sigma/\sqrt{n}$.

Suppose we wish to estimate the average vehicle load at peak hour μ', and we wish the tolerance (error of estimation) to be fewer than eight passengers, with a confidence level (probability) of 95%. This is for a given route and travel direction during weekdays. Since approximately 95% of the sample averages, in terms of the number of trips to be sampled, will fall within $2\sigma_{\bar{x}}$ of μ' in repeated sampling, we can ask that $2\sigma_{\bar{x}}$ equal eight passengers, or $2\sigma_{\bar{x}} = 2\sigma/\sqrt{n} = 8$, from which we obtain the sample size n = n = $\sigma^2/16$. Certainly, we do not know the population standard deviation, σ; hence, we must use an estimate from a previous sample or knowledge of the range in which the measurement will fall. An empirical rule is that this range is approximately 4σ, and in this way, we will find σ. For our example, suppose the range is 50 passengers (average load during peak hours lies between 20 and 70 passengers). Thus, n = $(50/4)^2/16$ = 9.77. Accordingly, we will sample 10 trips during the peak hours of a given route and direction of travel spread out over several weekdays. We would then be reasonably certain (with a probability of 95%) that our estimate will lie within $2\sigma_{\bar{x}} = 8$ passengers of the true average of the vehicle load. The general sample size formula for this case is as follows:

$$n = Z_{1-\alpha/2}^2 \frac{\sigma^2}{(TO/2)^2} \qquad (2.6)$$

where $Z_{1-\alpha/2}$ is the $(1-\alpha/2)$ percentile normal deviate and TO is the tolerance. Based on the empirical rule cited earlier, we can use the formula in a practical situation

$$n = \frac{4s^2}{(TO)^2} \qquad (2.7)$$

where s is the estimate of σ.

In effect, we would expect in the example used that the tolerance would be fewer than eight passengers. According to the empirical rule, there is a probability of 68% that the tolerance is less than $\sigma_{\bar{x}} = 4$ passengers. Note that these probabilities of 95% and 68% are inexact, because σ was approximated.

2.4.4 Practical sample size for origin–destination survey

The importance of O-D data in improving transit service is apparent from Figure 2.3 for all operations planning levels. O-D matrices ought to be constructed for developing new transit routes, terminals, and stops, changing route structure and strategies, and amending time-tables and vehicle and crew schedules. Therefore, this section will be devoted to a practical method of finding an adequate sample size for O-D surveys.

While sampling *from where* people want to go and *to where* (by time of day, purpose of trip, etc.), we enter a process of interaction between events from a statistical perspective. We want to survey N people, with the outcome of each falling into one of the O-D cells (each cell is a specific O-D). The experiment of such N independent trials is called a *multinomial experiment* (generalization of the binomial experiment) and has a useful probability (multinomial) distribution for discrete random variables. The proper O-D matrix is, in fact, a contingency table based on a multinomial distribution, in which contingency means the relationship between O-D classifications. That is, there are B_1, B_2, ..., B_k possible O-D relationships with the corresponding θ_1, θ_2, ..., θ_k probabilities, such that $\sum_i \theta_i = 1$. We are interested in knowing the probability that, in N trials, B_1 will be observed x_1 times, B_2 will be observed x_2 times, ..., and B_k will be observed x_k times where $\sum_i x_i = N$. Because of independency, the following is obtained: $P(B_1) = \theta_1^{x_1}, P(B_2) = \theta_2^{x_2}, ..., P(B_k) = \theta_k^{x_k}$, and

$$P(x_1, x_2, ..., x_k) = \frac{N! \theta_1^{x_1} \cdot \theta_2^{x_4} ... \theta_k^{x_k}}{x_1! x_2! ... x_k!} \quad \text{(multinomial distribution)} \tag{2.8}$$

where P represents probability.

Consider now x_i as a binomial variable with average $N\theta_i$ and variance $N\theta_i(1 - \theta_i)(1 - n/N)$, where n is the sample size. For $n \ll N$, the fraction n/N approaches zero. Further, for large n, say over 50, x_i can be considered normally distributed as in Equation 2.3: $x_i \sim$ Normal$[\mu = n\theta_i, \sigma^2 = n\theta_i(1 - \theta_i)]$.

Let $p_i = x_i/n$ represent a group i proportion within the population. Thus, $p_i \sim$ Normal $[\theta_i, \theta_i(1 - \theta_i)/n]$. The sample size in Equations 2.6 and 2.7 yields $n = Z_{1-\alpha/2}^2 s^2 / TO_i^2$, where s, the estimate of σ, can be represented by $\sqrt{p_i(1 - p_i)}$. Since p_i is the group proportion, $p_i(1 - p_i)$ can never go beyond 0.25 (the case in which $p_i = 0.5$). Hence, it can be used as an upper bound for the sample size. The required sample size, therefore, is

$$n = Z_{1-\alpha/2}^2 \frac{p_i(1 - p_i)}{TO_i^2}, \quad i = 1, 2, ..., k \tag{2.9}$$

In cases in which we cannot consider $n \ll N$, we can as a practical step use the following expression for n_o sample size: $n_o = n/(1 - n/N)$

We can now establish a stepwise procedure for choosing the sample size:

1. Ratio estimation in each cell of the O-D matrix. That is, what is the ratio, R, between the number of passengers traveling from a given origin to a given destination and the population (which ratio can be estimated by performing a small pretest)?
2. Determination of the required precision in each cell of the O-D matrix.
3. Determination of the confidence interval.
4. Determination of the base ratio for the sample size selection procedure. It should be noted that the selection of the smallest ratio in the O-D matrix will result in the largest sample size, n_m. However, an average ratio will result in a smaller sample size than n_m, but with reduced confidence intervals for cells having lower ratios than the average.
5. Selection of the sample size based on given precision tables (see example in the following text).

The end of this chapter presents the precision table for commonly used sample sizes for an O-D survey: 500, 1,000, 2,000, 5,000, and 10,000. For example, there are in Table 2.1 three origins and three destinations, **a**, **b**, and **c**, and the estimated ratios, R, between the number of passengers for each O-D cell and the true population. We then assume for all the cells a precision of $TO_i/p_i = \pm 10\%$ and a confidence interval of 95%. The base ratio $\bar{R}$ is selected to be the average, or close to the average, of all the ratios in Table 2.1, $\bar{R} = 0.12 \ (\approx 0.7/6)$.

Table 2.2 extracts the relevant portion of the entire precision table at the end of the chapter and emphasizes the precisions for $\bar{R} = 0.12$ for all commonly used n. That is,

For n = 500, the precision is $\pm 23.74\%$.
For n = 1,000, the precision is $\pm 16.78\%$.
For n = 2,000, the precision is $\pm 11.87\%$.
For n = 5,000, the precision is $\pm 7.51\%$.
For n = 10,000, the precision is $\pm 5.31\%$.

It may be seen that none of the precision levels is $\pm 10\%$. In order to attain greater precision, we can decide on $\pm 7.51\%$ (the closest to $\pm 10\%$ from below) and a sample of 5000 people. Consequently, for n = 5000, and based on Table 2.2, the precision is ± 12.08 for the cell with R = 0.05%.

For 0.10, the precision is $\pm 8.32\%$.
For 0.15, the precision is $\pm 6.60\%$.
For 0.20, the precision is $\pm 5.54\%$.

These cell-based precisions are also emphasized in Table 2.2. Finally, more transit-related sample size formulae appear in the transit data collection manual UMTA (1985).

Table 2.1 Example of ratios between each origin–destination cell and the true population

| | From | | |
To	a	b	c
a	—	0.05	0.20
b	0.15	—	0.10
c	0.15	0.05	—

Table 2.2 Precision table for the example

| | 95% Confidence interval | | | | |
| | n = Sample size | | | | |
R = Ratio	500	1,000	2,000	5,000	10,000
0.01	87.21	61.67	43.61	23.58	19.50
0.02	61.36	43.39	30.68	19.40	13.72
0.03	49.84	35.24	24.92	15.76	11.14
0.04	42.94	30.36	21.4	13.58	9.60
0.05	38.21	27.02	19.10	12.08	8.54
0.06	34.69	24.53	17.35	10.97	7.76
0.07	31.95	22.59	15.97	10.10	7.14
0.08	29.72	21.02	14.86	9.40	6.65
0.09	27.87	19.71	13.94	8.81	6.23
0.10	26.30	18.59	13.15	8.32	5.88
0.11	24.93	17.63	12.47	7.88	5.58
0.12	23.74	16.78	11.87	7.51	5.31
0.13	22.68	16.03	11.34	7.17	5.07
0.14	21.72	15.36	10.86	6.87	4.86
0.15	20.87	14.75	10.43	6.60	4.67
0.16	20.08	14.20	10.04	6.35	4.49
0.17	19.37	13.70	9.68	6.12	4.33
0.18	18.71	13.23	9.35	5.92	4.18
0.19	18.10	12.80	9.05	5.72	4.05
0.20	17.53	12.40	8.77	5.54	3.92

2.5 LITERATURE REVIEW AND FURTHER READING

This section reviews papers that deal with the collection of transit data, focusing on data on service performance and ridership. These data are necessary for operations planning in the short run and for network design in the long run. The collection of general travel behavior data, passenger-trip diaries, or data concerning passenger preferences is not covered by this review.

According to the transit data collection design manual (UMTA, 1985), the route-specific data needed for effective basic operation planning include total boardings, loads at key points, running time, schedule adherence, revenues, and some passenger characteristics. At the design level, data include passenger origins and destinations, fare category distribution, boardings and alightings by stop, transfers between routes, passenger attitudes, and passengers' travel behavior. The manual divides the techniques for collecting transit data into four groups, according to the identity and location of the surveyor: ride checks, which are taken on board; point checks, taken by an observer standing at the road side; driver checks, with the data collected by the driver; and automated checks, in which the information is collected by machines. Eight different types of counts and readings are described: on/off passenger counts, counts of boardings only, passenger-load counts, farebox readings, revenue counts, transfer counts, O-D stop counts, and passenger surveys. The manual also presents sampling methods, techniques for sample size determination, calculation of conversion factors to enable survey results to represent a whole population, and data accuracy determination.

The review of transit data collection techniques by Booz-Allen and Hamilton (1985) documents the most common collection techniques. Nine leading, manual collection techniques are described: ride checks, point checks, boarding counts, farebox readings, revenue counts, speed and delay studies, running-time checks, transfer counts, and stop studies. The review describes each technique and procedures for its application, as well as the data items collected by the technique. Instructions are given on how to aggregate and manipulate the data.

Bamford et al. (1984) assert that onboard passenger information can be gathered either by using self-completed questionnaires or by an interviewer. They examine the benefits of a data collection method based on a very short questionnaire that can be self-completed by means of *rub-off*, within no more than 1 min and without the use of a pen. They claim that this method is cheaper than most other onboard surveying methods and yields higher response rates.

A modular approach to onboard automatic data collection systems (NCTRD report, 1986) defines standard requirements for such systems. The authors argue that these systems offer advantages over manual systems in many transit applications: improved data-processing time, lower collection costs, and better quality data. They describe a standard microprocessor-based system that provides an electronic means of gathering passenger, fare, and schedule data. Step-by-step guidelines for selecting and implementing the data collection system are presented, as is a discussion of the cost of such a system. The paper also enumerates the requirements for equipping a bus with an automatic data collection system and details a technique for determining the necessary number of equipped buses, which works out to almost 10% of the fleet. Perhaps the most important conclusion derived from equipping a bus fleet with automatic data collection devices is the need for coordination with the existing dispatching procedures. The cooperation of dispatchers and drivers, particularly in regard to modifications of dispatching practices, is crucial.

Furth and McCollom (1987) focus on the calculation of conversion factors, which are used for converting survey results (number of boardings, alightings, peak load, revenue, etc.) into representative values. The authors discuss statistical aspects of the conversion factor approach, including sample size estimation and a determination of accuracy measures. They outline several scenarios concerning available input data, since preliminary information about mean values is not always available. An optimal sampling plan to minimize costs is offered.

Macbirar (1989) discusses data retrieval from electronic ticket machines. These machines constitute a potential source of information not only on fares and revenues but also on passenger characteristics that change with fare, origins and destinations, and riding times. The paper discusses in detail the importance of keeping a simple, well-arranged enumeration system for trips, stops, etc., which is necessary for documenting the data. The paper also discusses data needs for reimbursing concession-fare schemes, that is, the way operators receive compensation for being forced to provide reduced prices for specific sectors.

Weinstein and Albom (1998) discuss techniques for collecting data on qualitative features of transit performance that are difficult to measure, such as station/vehicle cleanliness. They present a methodology, including rating criteria, for estimating these features.

Richardson (1998) describes a survey technique that has been used to measure ticket-usage rates. This information should assist in the allocation of farebox revenue to operators of a private transit system. Because of the unrepresentative sampling method, a principal component of the methodology presented is a weighting procedure, based on data partially collected beforehand. A combination of questionnaires and observational surveys is used in order to obtain accurate bias estimates.

Navarrete (1999) describes a methodology for using digital assistants instead of paper records when inspectors collect field data. Criteria for the efficiency of the digital data collection device are developed; among them are ease of data entry and retrieval, utilization of existing technologies, and effective handwriting recognition. A suitable device is chosen and tested. According to field tests, the most difficult problem encountered is the learning curve for the handwriting-recognition program.

Barua et al. (2001) discuss the use of geographical information systems (GIS) for transit O-D data analysis following an onboard transit survey. GIS is used not just for geocoding origin and destination locations but also as a tool for checking survey data quality and for validating the data analysis. A suggested methodology for weighing survey results and converting them into a representative trip matrix is based on spatial criteria. The GIS-based approach for validating the quality of the results includes a comparison of transit link volumes with data from passenger counts and a parallel validation of subarea transit demand characteristics.

Some of the literature in the field of data collection presents the features of specific software or hardware systems, without detailing methodologies. For example, Barnes and Urbanik (1990) describe a software for a computerized data collection process. Rossetti (1996) reports on a transit-monitoring system based on radio frequency identification that integrates AVL with automatic passenger counting. Other papers on AVL and AVM are discussed in Chapters 18 and 19.

Morency et al. (2007) have analyzed the potential of smart card data for measuring the variability of urban public transit network use. A discussion on the organization of data using an object-oriented approach is presented, and then the measures of spatial and temporal variability of transit use for various types of card are defined and estimated using the datasets presented. Data-mining techniques have also been occupied to identify transit use cycles and homogenous days and weeks of travel among card segments and at various times of the year. This study shows that smart card data have the potential to give a continuous profile of transit use by various types of cards. The aforementioned text has been concluded using well-formatted data that were obtained with the help of an object-oriented approach to identify all the objects of the system. From the experiments, the authors were able to conclude that it is possible to observe regularity indicators by using raw information on boardings only, even though little individual information was available.

Wilson et al. (2009) describe two applications on automated data collection system usage and passenger behavior, which have been developed jointly between Massachusetts Institute of Technology (MIT) and the Chicago Transit Authority (CTA), using CTA's automatic fare collection (AFC) and AVL systems. The two applications put forth by the authors illustrate the ability of engaging inference methods to estimate the passenger O-D travel patterns and rail path choice behavior without using any survey data. The authors conclude by suggesting possible extensions to the proposed approaches, which will ultimately enable comprehensive coverage of multimodal public transport systems. Such work will eventually provide valuable insights on the multidimensional performance aspects of the public transit services and its impact on passenger travel behaviors.

Mandelzys and Hellinga (2010) present a methodology for identifying bus stops that are unable to meet schedule adherence performance standards and the factors causing inadequate performance. The methodology is designed to be automated and therefore can be applied efficiently to AVL/APC data. The analysis presented in this paper can be ultimately conducted for an entire transit network to identify those bus stops that are not meeting the expected performance standards. The authors suggest that the use of this proposed method will enable transit agencies to more efficiently identify service quality issues and their respective root causes. Furthermore, this information, in conjunction with other visualization techniques, can provide useful insights for transit network planners to make effective decisions at a low cost.

Pelletier et al. (2011) present a review of the smart card data use in public transit. The authors emphasize the importance of smart card AFC systems on collecting large quantities of very detailed data on onboard transactions. These data can be very useful to transit planners, from the day-to-day operation of the transit system to the strategic long-term planning of the network. This review covers several aspects of smart card data use in the public transit context by first presenting the available technologies and legal issues related to the dissemination of smart card data, data storage, and encryption. Then, the various uses of the data at three levels of management are described: strategic (long-term planning), tactical (service adjustments and network development), and operational (ridership statistics and performance indicators). A number of issues such as data/journey validation, economic feasibility, new modeling and analysis methods, and technological improvements are identified as potential future challenges for the smart card transit operators. Notwithstanding privacy concerns, the authors state that smart card data can provide a rich continuous source of data, which can be used to understand the behavior of transit users, compare planned and implemented schedules, undertake adjustments to the systematic schedules, and apply the survival models to ridership.

Lee and Hickman (2014) use the extensive farecard transaction data for deriving useful information about transit passenger behavior, namely, trip purpose or activity. They illustrate how the farecard data can be used to infer trip purpose and to reveal travel patterns in an urban area. A case study is used to demonstrate the process of trip purpose inference based on farecard data from Metro Transit in the Minneapolis–St. Paul metropolitan area. The proposed approach only concerns a small set of transit users, although in conjunction with O-D pairs, it is possible to combine an unbiased source of transit demand data with an inferred trip purpose—which is typically a difficult task to undertake through traditional surveys. The authors state that future research efforts should focus on validation of the data with onboard surveys and subsequent interpretations to gain further insights to the transit users' trip purposes.

Munizaga and Palma (2012) present a methodology for estimating a public transport O-D matrix from smartcard and GPS data for Santiago, Chile. The proposed method is applied to two 1-week datasets obtained for different time periods. From the data available, detailed information about the time and position of boarding public transportation are obtained, and an estimation of time and position of alighting for over 80% of the boarding transactions is generated. The results are available at any desired time–space disaggregation. After some postprocessing and after incorporating expansion factors to account for unobserved trips, public transport O-D matrices are built. The main contributions of this paper are moving from a trip segment perspective to a trip O-D estimation and using generalized time rather than physical distance to estimate alighting point for transfers and destinations. Even though some information are missing, such that evasion and trip segments in nonintegrated transport modes and validation with exogenous data needs to be conducted, the magnitude of the achieved success rates suggests that this new source of information has a great potential and can actually replace an important part of the large O-D surveys.

Table 2.3 summarizes the main features and characteristics of the literature reviewed by categories. Having presented the overview of the first key in Figure 2.1, we will proceed in the next 14 chapters to deal with the second key: how to plan and decide intelligently. The third key will be discussed in Chapters 17 and 18.

Table 2.3 Summary of the main features of the literature reviewed, by category

Source	Type of data collection	Data collection method	Contribution of the paper	Includes methodology for weighting survey results?	Includes discussion of data accuracy?
Transit Data Collection Design Manual (1985)	Various	Various	Review	Yes	Yes
Review of Transit Data Collection Techniques (1985)	Various	Various	Review	Yes	Yes
Bamford et al. (1984)	Various	On board questionnaires	Technical features of the questionnaire	No	No
Modular Approach to On Board Automatic Data Collection Systems (1986)	Various	On board automatic	Definition of system requirements	No	No
Furth and McCollom (1987)	Various	Any method	Analytical method	Yes	Yes
Macbirar (1989)	Various	On board automatic (through ticket machine)	Review/ discussion	No	No
Weinstein and Albom (1998)	Qualitative data, such as cleanliness	Observation (on board or at station)	Analytical method	No	Yes
Richardson (1998)	Ticket-usage rates	On board, combination of questionnaires and observations	Analytical method	Yes	Yes
Navarrete (1999)	Various	Digital assistants	New technology	No	No
Barua et al. (2001)	Passenger origin– destination	Any (focus is on analysis method)	Analytical method	Yes	Yes
Barnes and Urbanik (1990)	Various	On board automatic	New technology	No	No
Rossetti (1996)	Various	Combination of AVL and onboard automatic counters	New technology	No	No
Morency et al. (2007)	Variability of urban public transit network use	Smart card	Review/ discussion	No	Yes

(Continued)

Table 2.3 (Continued) Summary of the main features of the literature reviewed, by category

Source	Type of data collection	Data collection method	Contribution of the paper	Includes methodology for weighting survey results?	Includes discussion of data accuracy?
Wilson et al. (2009)	Various	On board automatic	Analytical method	No	No
Mandelzys and Hellinga (2010)	Various	Automatic vehicle location and automatic passenger counting (APC)	Analytical method	No	No
Pelletier et al. (2011)	Various	Smart card	Review	No	No
Lee and Hickman (2014)	Passenger-trip purpose or activity	Farecard transactions	Analytical method	No	Yes
Munizaga and Palma (2012)	Passenger origin–destination	Smart card and GPS	Analytical method	No	No

Precision table for five survey sample sizes

	90% Confidence interval					95% Confidence interval					99% Confidence interval				
								n							
R	500	1,000	2,000	5,000	10,000	500	1,000	2,000	5,000	10,000	500	1,000	2,000	5,000	10,000
0.01	73.19	51.75	36.60	23.15	16.37	87.21	61.67	43.61	27.58	19.50	114.62	81.05	57.31	36.25	25.63
0.02	51.49	36.41	25.75	16.28	11.51	61.36	43.39	30.68	19.40	13.72	80.64	57.02	40.32	25.50	18.03
0.03	41.83	29.58	20.91	13.23	9.35	49.84	35.24	24.92	15.76	11.14	65.50	46.32	32.75	20.71	14.65
0.04	36.04	25.48	18.02	11.40	8.06	42.94	30.36	21.47	13.58	9.60	56.43	39.90	28.22	17.85	12.62
0.05	32.06	22.67	16.03	10.14	7.17	38.21	27.02	19.10	12.08	8.54	50.21	35.51	25.11	15.88	11.23
0.06	29.12	20.59	14.56	9.21	6.51	34.69	24.53	17.35	10.97	7.76	45.60	32.24	22.80	14.42	10.20
0.07	26.81	18.96	13.41	8.48	6.00	31.95	22.59	15.97	10.10	7.14	41.99	29.69	20.99	13.28	9.39
0.08	24.95	17.64	12.47	7.89	5.58	29.72	21.02	14.86	9.40	6.65	39.06	27.62	19.53	12.35	8.74
0.09	23.39	16.54	11.70	7.40	5.23	27.87	19.71	13.94	8.81	6.23	36.63	25.90	18.31	11.58	8.19
0.10	22.07	15.60	11.03	6.98	4.93	26.30	18.59	13.15	8.32	5.88	34.56	24.44	17.28	10.93	7.73
0.11	20.92	14.80	10.46	6.62	4.68	24.93	17.63	12.47	7.88	5.58	32.77	23.17	16.38	10.36	7.33
0.12	19.92	14.09	9.96	6.30	4.45	23.74	16.78	11.87	7.51	5.31	31.19	22.06	15.60	9.86	6.98
0.13	19.03	13.46	9.51	6.02	4.26	22.68	16.03	11.34	7.17	5.07	29.80	21.07	14.90	9.42	6.66
0.14	18.23	12.89	9.12	5.77	4.08	21.72	15.36	10.86	6.87	4.86	28.55	20.19	14.28	9.03	6.38
0.15	17.51	12.38	8.76	5.54	3.92	20.87	14.75	10.43	6.60	4.67	27.42	19.39	13.71	8.67	6.13
0.16	16.85	11.92	8.43	5.33	3.77	20.08	14.20	10.04	6.35	4.49	26.39	18.66	13.20	8.35	5.90
0.17	16.25	11.49	8.13	5.14	3.63	19.37	13.70	9.68	6.12	4.33	25.45	18.00	12.73	8.05	5.69
0.18	15.70	11.10	7.85	4.96	3.51	18.71	13.23	9.35	5.92	4.18	24.59	17.39	12.29	7.78	5.50
0.19	15.19	10.74	7.59	4.80	3.40	18.10	12.80	9.05	5.72	4.05	23.78	16.82	11.89	7.52	5.32
0.20	14.71	10.40	7.36	4.65	3.29	17.53	12.40	8.77	5.54	3.92	23.04	16.29	11.52	7.29	5.15
0.21	14.27	10.09	7.13	4.51	3.19	17.00	12.02	8.50	5.38	3.80	22.34	15.80	11.17	7.07	5.00
0.22	13.85	9.79	6.93	4.38	3.10	16.50	11.67	8.25	5.22	3.69	21.69	15.34	10.85	6.86	4.85
0.23	13.46	9.52	6.73	4.26	3.01	16.04	11.34	8.02	5.07	3.59	21.08	14.90	10.54	6.67	4.71
0.24	13.09	9.26	6.55	4.14	2.93	15.60	11.03	7.80	4.93	3.49	20.50	14.50	10.25	6.48	4.58

(Continued)

Precision table for five survey sample sizes

	90% Confidence interval					95% Confidence interval					99% Confidence interval				
								n							
R	500	1,000	2,000	5,000	10,000	500	1,000	2,000	5,000	10,000	500	1,000	2,000	5,000	10,000
0.25	12.74	9.01	6.37	4.03	2.85	15.18	10.74	7.59	4.80	3.39	19.95	14.11	9.98	6.31	4.46
0.26	12.41	8.78	6.20	3.92	2.77	14.79	10.46	7.39	4.68	3.31	19.43	13.74	9.72	6.15	4.35
0.27	12.10	8.55	6.05	3.82	2.70	14.41	10.19	7.21	4.56	3.22	18.94	13.39	9.47	5.99	4.24
0.28	11.80	8.34	5.90	3.73	2.64	14.06	9.94	7.03	4.44	3.14	18.47	13.06	9.24	5.84	4.13
0.29	11.51	8.14	5.75	3.64	2.57	13.71	9.70	6.86	4.34	3.07	18.02	12.75	9.01	5.70	4.03
0.30	11.24	7.95	5.62	3.55	2.51	13.39	9.47	6.69	4.23	2.99	17.60	12.44	8.80	5.56	3.93
0.31	10.97	7.76	5.49	3.47	2.45	13.08	9.25	6.54	4.14	2.92	17.19	12.15	8.59	5.43	3.84
0.32	10.72	7.58	5.36	3.39	2.40	12.78	9.03	6.39	4.04	2.86	16.79	11.87	8.40	5.31	3.75
0.33	10.48	7.41	5.24	3.31	2.34	12.49	8.83	6.24	3.95	2.79	16.41	11.61	8.21	5.19	3.67
0.34	10.25	7.25	5.12	3.24	2.29	12.21	8.64	6.11	3.86	2.73	16.05	11.35	8.02	5.08	3.59
0.35	10.02	7.09	5.01	3.17	2.24	11.94	8.45	5.97	3.78	2.67	15.70	11.10	7.85	4.96	3.51
0.36	9.81	6.94	4.90	3.10	2.19	11.69	8.26	5.84	3.70	2.61	15.36	10.86	7.68	4.86	3.43
0.37	9.60	6.79	4.80	3.04	2.15	11.44	8.09	5.72	3.62	2.56	15.03	10.63	7.52	4.75	3.36
0.38	9.40	6.64	4.70	2.97	2.10	11.20	7.92	5.60	3.54	2.50	14.71	10.40	7.36	4.65	3.29
0.39	9.20	6.51	4.60	2.91	2.06	10.96	7.75	5.48	3.47	2.45	14.41	10.19	7.20	4.56	3.22
0.40	9.01	6.37	4.50	2.85	2.01	10.74	7.59	5.37	3.39	2.40	14.11	9.98	7.05	4.46	3.15
0.41	8.82	6.24	4.41	2.79	1.97	10.51	7.44	5.26	3.33	2.35	13.82	9.77	6.91	4.37	3.09
0.42	8.64	6.11	4.32	2.73	1.93	10.30	7.28	5.15	3.26	2.30	13.54	9.57	6.77	4.28	3.03
0.43	8.47	5.99	4.23	2.68	1.89	10.09	7.14	5.05	3.19	2.26	13.26	9.38	6.63	4.19	2.97
0.44	8.30	5.87	4.15	2.62	1.86	9.89	6.99	4.94	3.13	2.21	13.00	9.19	6.50	4.11	2.91
0.45	8.13	5.75	4.07	2.57	1.82	9.69	6.85	4.85	3.06	2.17	12.74	9.01	6.37	4.03	2.85
0.46	7.97	5.64	3.99	2.52	1.78	9.50	6.72	4.75	3.00	2.12	12.48	8.83	6.24	3.95	2.79
0.47	7.81	5.52	3.91	2.47	1.75	9.31	6.58	4.65	2.94	2.08	12.23	8.65	6.12	3.87	2.74
0.48	7.66	5.41	3.83	2.42	1.71	9.12	6.45	4.56	2.88	2.04	11.99	8.48	5.99	3.79	2.68
0.49	7.50	5.31	3.75	2.37	1.68	8.94	6.32	4.47	2.83	2.00	11.75	8.31	5.88	3.72	2.63
0.50	7.36	5.20	3.68	2.33	1.64	8.77	6.20	4.38	2.77	1.96	11.52	8.15	5.76	3.64	2.58
0.51	7.21	5.10	3.61	2.28	1.61	8.59	6.08	4.30	2.72	1.92	11.29	7.98	5.65	3.57	2.52
0.52	7.07	5.00	3.53	2.23	1.58	8.42	5.95	4.21	2.66	1.88	11.07	7.83	5.53	3.50	2.47
0.53	6.93	4.90	3.46	2.19	1.55	8.25	5.84	4.13	2.61	1.85	10.85	7.67	5.42	3.43	2.43
0.54	6.79	4.80	3.39	2.15	1.52	8.09	5.72	4.04	2.56	1.81	10.63	7.52	5.32	3.36	2.38
0.55	6.65	4.70	3.33	2.10	1.49	7.93	5.61	3.96	2.51	1.77	10.42	7.37	5.21	3.30	2.33
0.56	6.52	4.61	3.26	2.06	1.46	7.77	5.49	3.88	2.46	1.74	10.21	7.22	5.11	3.23	2.28
0.57	6.39	4.52	3.19	2.02	1.43	7.61	5.38	3.81	2.41	1.70	10.01	7.07	5.00	3.16	2.24
0.58	6.26	4.43	3.13	1.98	1.40	7.46	5.27	3.73	2.36	1.67	9.80	6.93	4.90	3.10	2.19
0.59	6.13	4.34	3.07	1.94	1.37	7.31	5.17	3.65	2.31	1.63	9.60	6.79	4.80	3.04	2.15
0.60	6.01	4.25	3.00	1.90	1.34	7.16	5.06	3.58	2.26	1.60	9.41	6.65	4.70	2.97	2.10
0.61	5.88	4.16	2.94	1.86	1.32	7.01	4.96	3.50	2.22	1.57	9.21	6.51	4.61	2.91	2.06
0.62	5.76	4.07	2.88	1.82	1.29	6.86	4.85	3.43	2.17	1.53	9.02	6.38	4.51	2.85	2.02
0.63	5.64	3.99	2.82	1.78	1.26	6.72	4.75	3.36	2.12	1.50	8.83	6.24	4.41	2.79	1.97
0.64	5.52	3.90	2.76	1.74	1.23	6.57	4.65	3.29	2.08	1.47	8.64	6.11	4.32	2.73	1.93
0.65	5.40	3.82	2.70	1.71	1.21	6.43	4.55	3.22	2.03	1.44	8.45	5.98	4.23	2.67	1.89
0.66	5.28	3.73	2.64	1.67	1.18	6.29	4.45	3.15	1.99	1.41	8.27	5.85	4.13	2.61	1.85
0.67	5.16	3.65	2.58	1.63	1.15	6.15	4.35	3.08	1.95	1.38	8.08	5.72	4.04	2.56	1.81
0.68	5.05	3.57	2.52	1.60	1.13	6.01	4.25	3.01	1.90	1.34	7.90	5.59	3.95	2.50	1.77
0.69	4.93	3.49	2.47	1.56	1.10	5.88	4.15	2.94	1.86	1.31	7.72	5.46	3.86	2.44	1.73
0.70	4.82	3.41	2.41	1.52	1.08	5.74	4.06	2.87	1.81	1.28	7.54	5.33	3.77	2.38	1.69

(Continued)

Precision table for five survey sample sizes

	Precision table														
	90% Confidence interval					95% Confidence interval					99% Confidence interval				
								n								
R	500	1,000	2,000	5,000	10,000	500	1,000	2,000	5,000	10,000	500	1,000	2,000	5,000	10,000
0.71	4.70	3.32	2.35	1.49	1.05	5.60	3.96	2.80	1.77	1.25	7.36	5.21	3.68	2.33	1.65
0.72	4.59	3.24	2.29	1.45	1.03	5.47	3.87	2.73	1.73	1.22	7.18	5.08	3.59	2.27	1.61
0.73	4.47	3.16	2.24	1.41	1.00	5.33	3.77	2.67	1.69	1.19	7.01	4.95	3.50	2.22	1.57
0.74	4.36	3.08	2.18	1.38	0.97	5.20	3.67	2.60	1.64	1.16	6.83	4.83	3.41	2.16	1.53
0.75	4.25	3.00	2.12	1.34	0.95	5.06	3.58	2.53	1.60	1.13	6.65	4.70	3.33	2.10	1.49
0.76	4.13	2.92	2.07	1.31	0.92	4.93	3.48	2.46	1.56	1.10	6.47	4.58	3.24	2.05	1.45
0.77	4.02	2.84	2.01	1.27	0.90	4.79	3.39	2.40	1.51	1.07	6.30	4.45	3.15	1.99	1.41
0.78	3.91	2.76	1.95	1.24	0.87	4.66	3.29	2.33	1.47	1.04	6.12	4.33	3.06	1.93	1.37
0.79	3.79	2.68	1.90	1.20	0.85	4.52	3.20	2.26	1.43	1.01	5.94	4.20	2.97	1.88	1.33
0.80	3.68	2.60	1.84	1.16	0.82	4.38	3.10	2.19	1.39	0.98	5.76	4.07	2.88	1.82	1.29
0.81	3.56	2.52	1.78	1.13	0.80	4.25	3.00	2.12	1.34	0.95	5.58	3.95	2.79	1.76	1.25
0.82	3.45	2.44	1.72	1.09	0.77	4.11	2.90	2.05	1.30	0.92	5.40	3.82	2.70	1.71	1.21
0.83	3.33	2.35	1.66	1.05	0.74	3.97	2.81	1.98	1.25	0.89	5.21	3.69	2.61	1.65	1.17
0.84	3.21	2.27	1.61	1.02	0.72	3.83	2.71	1.91	1.21	0.86	5.03	3.55	2.51	1.59	1.12
0.85	3.09	2.19	1.55	0.98	0.69	3.68	2.60	1.84	1.16	0.82	4.84	3.42	2.42	1.53	1.08
0.86	2.97	2.10	1.48	0.94	0.66	3.54	2.50	1.77	1.12	0.79	4.65	3.29	2.32	1.47	1.04
0.87	2.84	2.01	1.42	0.90	0.64	3.39	2.40	1.69	1.07	0.76	4.45	3.15	2.23	1.41	1.00
0.88	2.72	1.92	1.36	0.86	0.61	3.24	2.29	1.62	1.02	0.72	4.25	3.01	2.13	1.35	0.95
0.89	2.59	1.83	1.29	0.82	0.58	3.08	2.18	1.54	0.97	0.69	4.05	2.86	2.02	1.28	0.91
0.90	2.45	1.73	1.23	0.78	0.55	2.92	2.07	1.46	0.92	0.65	3.84	2.72	1.92	1.21	0.86
0.91	2.31	1.64	1.16	0.73	0.52	2.76	1.95	1.38	0.87	0.62	3.62	2.56	1.81	1.15	0.81
0.92	2.17	1.53	1.08	0.69	0.49	2.58	1.83	1.29	0.82	0.58	3.40	2.40	1.70	1.07	0.76
0.93	2.02	1.43	1.01	0.64	0.45	2.40	1.70	1.20	0.76	0.54	3.16	2.23	1.58	1.00	0.71
0.94	1.86	1.31	0.93	0.59	0.42	2.21	1.57	1.11	0.70	0.50	2.91	2.06	1.46	0.92	0.65
0.95	1.69	1.19	0.84	0.53	0.38	2.01	1.42	1.01	0.64	0.45	2.64	1.87	1.32	0.84	0.59
0.96	1.50	1.06	0.75	0.47	0.34	1.79	1.27	0.89	0.57	0.40	2.35	1.66	1.18	0.74	0.53
0.97	1.29	0.91	0.65	0.41	0.29	1.54	1.09	0.77	0.49	0.34	2.03	1.43	1.01	0.64	0.45
0.98	1.05	0.74	0.53	0.33	0.23	1.25	0.89	0.63	0.40	0.28	1.65	1.16	0.82	0.52	0.37
0.99	0.74	0.52	0.37	0.23	0.17	0.88	0.62	0.44	0.28	0.20	1.16	0.82	0.58	0.37	0.26

REFERENCES

Bamford, C.G., Carrick, R.J., and MacDonald, R. (1984). Public transport surveys: A new effective technique of data collection. *Traffic Engineering and Control*, 25(6), 318–319.

Barnes, K.E. and Urbanik, T. (1990). Automated transit ridership data collection software development and user's manual. Texas Transportation Institute, Report UMTA-TX-08-1087-91-1, Texas, USA.

Barua, B., Boberg, J., Hsia, J.S., and Zhang, X. (2001). Integrating geographic information systems with transit survey methodology. *Transportation Research Record*, 1753, 29–34.

Booz-Allen and Hamilton (1985). *Review of Transit Data Collection Techniques*. Report of Booz-Allen and Hamilton. Philadelphia, PA.

Furth, P.G. and McCollom, B. (1987). Using conversion factors to lower transit data collection costs. *Transportation Research Record*, 1144, 1–6.

Lee, S.G. and Hickman, M. (2014). Trip purpose inference using automated fare collection data. *Public Transport: Planning and Operation*, 6(1–2), 1–20.

Macbirar, I.D. (1989). Current issues in public transport data collection. In *Proceedings of Seminar C held at the 17th PTRC Transport and Planning Summer Annual Meeting* P318, vol. 99, pp. 219–229, London, U.K.

Mandelzys, M. and Hellinga, B. (2010). Identifying causes of performance issues in bus schedule adherence with automatic vehicle location and passenger count data. *Transportation Research Record*, **2143**, 9–15.

Merriam-Webster. (2000). *Merriam-Webster's Collegiate Dictionary*, 10th ed., Merriam-Webster, Springfield, MA.

Modular Approach to On-Board Automatic Data Collection Systems. (1986). National Cooperative Transit Research & Development Program, Report No. 9, *Transportation Research Board*, Washington, DC.

Morency, C., Trepanier, M., and Agard, B. (2007). Measuring transit use variability with smart-card data. *Transport Policy*, **14**(3), 193–203.

Munizaga, M.A. and Palma, C. (2012). Estimation of a disaggregate multimodal public transport Origin–Destination matrix from passive smartcard data from Santiago, Chile. *Transportation Research Part C: Emerging Technologies*, **24**, 9–18.

Navarrete, G. (1999). In the palm of your hand: Digital assistant's aid in data collection. *Journal of Management in Engineering*, **15**(4), 43–45.

Pelletier, M.P., Trépanier, M., and Morency, C. (2011). Smart card data use in public transit: A literature review. *Transportation Research*, **19C**, 557–568.

Richardson, T. (1998). Public transport ticket usage surveys: A methodological design. *Australian Transport Research Forum*, **22**, 697–712.

Rossetti, M.D. (1996). Automatic data collection on transit users via radio frequency identification. Report of investigation, Transit-IDEA Program, No. 10, Washington, DC.

Urban Mass Transportation Administration. (1985). *Transit Data Collection Design Manual*, U.S. Department of Transportation, Washington, DC.

Washington, S.P., Karlaftis, M.G., and Mannering, F.L. (2003). *Statistical and Econometric Methods for Transportation Data Analysis*, CRC Press LLC, Boca Raton, FL.

Weinstein, A. and Albom, R. (1998). Securing objective data on the quality of the passenger environment for transit riders: Redesign of the passenger environment measurement system for the Bay Area Rapid Transit District. *Transportation Research Record*, **1618**, 213–219.

Wilson, N.H., Zhao, J., and Rahbee, A. (2009). The potential impact of automated data collection systems on urban public transport planning. In *Schedule-Based Modeling of Transportation Networks* (N.H.M. Wilson and A. Nuzzolo, eds.), pp. 1–25, Kluwer Academic, Dordrecht, the Netherlands.

Chapter 3

Frequency and headway determination

High service frequency for low passenger demand is like trying to hit an apple on the system's head using a cannon

CHAPTER OUTLINE

3.1 Introduction
3.2 Max load (point-check) methods
3.3 Load-profile (ride-check) methods
3.4 Criterion for selecting point check or ride check
3.5 Conclusion (two examples)
3.6 Literature review and further reading
Exercises
References

PRACTITIONER'S CORNER

This chapter addresses the topic of determining frequencies on transit routes, a problem that must receive attention, either explicitly or implicitly, several times a year. The chapter provides insights and a solution for intelligently integrating resource saving and an effective level of service. The basic premise here is that such an integration is achievable.

The importance of ridership information has led transit agencies to introduce automated surveillance techniques or, alternatively, to increase the amount of manually collected data. Naturally, the transit agencies are expected to gain useful information for operations planning by obtaining more accurate passenger counts. This chapter describes and analyzes several appropriate data collection approaches in order to set frequencies and headways efficiently. Four different methods are presented for deriving frequency, two based on point-check (max load) data and two on ride-check (load-profile) data. A ride check provides more complete information than does a point check, but at greater cost; there is a question whether the additional information gained justifies the expense. The four methods, all of which are based on available old profiles, provide the planner with adequate guidance in selecting the type of data collection procedure. In addition, the planner or scheduler can evaluate the minimum expected vehicle runs when the load standard is eased and avoid overcrowding (in an average sense) at the same time.

The chapter starts by describing the significance of proper frequency calculation, followed by a presentation and interpretation of the four methods. The fourth section established a criterion, the profile-based load, for selecting either the point-check or the ride-check data collection technique. In Section 3.5, two examples demonstrate the four methods, one of the examples employing real-life, heavy bus-route data. The chapter ends with a literature review and exercises for practicing the methods. The majority of the chapter is suitable for practitioners; however, the end of Sections 3.2 and 3.4 may be skipped to some extent. They can be replaced by the procedure depicted, in flowchart form, in Figure 3.10 in Section 3.5.

Dealing with data collection to affect the number of vehicle runs brings to mind the following real-life anecdote. A chief planner in a bus agency once observed that there was approved extra, but unnecessary, runs on one route. He then found out that this run was performed at the end of a duty day in order to bring the driver close to his home. The chief planner immediately ordered the local planner to eliminate this run. Six months later, the chief planner learned to his surprise that this run still existed. Angrily, he then called the local planner and asked for an explanation, to which the latter replied: "Oh, I told this driver that on the day the checker is there, not to make this run."

We can wrap up this corner with some pithy advice and admonition: the chains of traditional methods are too weak to be felt until they are too strong to be broken. In transit operations planning, therefore, there should always be a window for flexibility, new initiatives, and changes.

3.1 INTRODUCTION

One of the major foci in determining transit service is the selection of the most suitable frequency (vehicles/hour) for each route in the system, by time of day, day of week, and day type. This chapter provides an initial focus on the subject of the second key in Figure 2.1, intelligent decision making when planning transit service. This second key will accompany us through Chapter 16.

Figure 1.2 showed that frequency determination was in essence of paramount importance for creating transit timetables. Furth and Wilson (1981) wrote that transit agencies typically used service standards as the basis for setting frequencies, while combining this action with experience, judgment, and passenger counts. The service standards in normal use appear in Table 1.2: crowding level, allowed maximum standees, and upper (policy) and lower limits on headways. Table 1.3 emphasizes that these standards cannot be rules of thumb; they must be based on determined criteria. This is especially true because of the high cost involved in providing higher frequencies, an act that may not always be necessary. Prudent transit management requires a balance between increasing frequency and the cost of its implementation. Often, we observe that some transit agencies, following the belief that last week's frequency is good for this week, do not adjust their service to fluctuating demand. To those with this philosophy, we can only say: it is better to wear out by changing than to rust out by believing that what you have is good enough.

This chapter presents different methods that allow not only for efficient frequency setting but also for a sensitivity analysis of possible changes. Basically, the objectives of this chapter are twofold: one, to set vehicle frequencies in order to maintain adequate service quality and minimize the number of required vehicle runs; two, to construct an evaluation tool that will efficiently allocate the cost of gathering appropriate passenger-load data at the route level.

The main input data for this chapter will be point-check and ride-check data. These types of passenger counts are explicitly described in Section 2.2 and Figure 2.2; they will be combined in the analysis with crowding level and minimum frequency (inverse of policy headway) standards. It is common for load-profile data to be gathered annually or every few years along the entire length of the transit route (ride check). Usually, the most recent passenger-load information will come from one or more selected stops along that part of the route where the vehicle carries its heaviest loads (point check). Point-check information is routinely surveyed several times a year for the purpose of possible schedule revisions, which can range from completely new timetables for new or revised routes to daily adjustments to accommodate changes in Central Business District (CBD) and industrial area working hours and school-dismissal times. A ride check provides more complete information than a point check but is more expensive because of the need for either additional checkers to provide the required data or an automated surveillance system (e.g., automatic passenger counter [APC], AVL). The question is whether the additional information justifies the expense of gathering it. This chapter explores ways in which a transit agency can use the old profile to determine which method, ride check or point check, is more appropriate and less costly for collecting the new data.

Following Furth and Wilson (1981), transit agencies use methods to set headways that are commonly based on existing service standards of crowding level and policy headway. These standards are based on two requirements: (1) adequate space will be provided to meet passenger demand and (2) an upper bound value is placed on the headways to assure a minimum frequency of service. The first requirement is appropriate for heavily traveled route hours (e.g., peak period) and the second for lightly traveled hours. The first requirement is usually met by a widely used *peak-load factor* method (point check), which is similar to the max load procedure—both will be explained in Sections 3.2 and 3.3, respectively. The second requirement is met by *policy headway*, which usually does not exceed 60 min and is restricted in some cases to 30 min. Occasionally, a lower-bound value is set on the headway, based on productivity or revenue/cost measures. There are also mathematical programming techniques to approach the problems of route design and service frequency simultaneously. Such a technique was adopted by Furth and Wilson (1981) to find the appropriate headway that would maximize social benefit, subject to the constraints of total subsidy, fleet size, and bus-occupancy levels. Nevertheless, mathematical programming models are hardly employed, since they cannot incorporate practical operational considerations in the optimization analysis. A further review of the literature appears at the end of the chapter.

This chapter consists of three major parts. First, two methods are presented to derive vehicle frequencies based on point-check (max load) data. Second, two further methods are proposed, using ride-check (load-profile) data. Third, a criterion is established for determining the appropriateness of each data collection method. All four methods are applied—and analyzed—on a route basis, following Ceder (1984). These three parts are succeeded by two examples (one real life): a literature review and exercises.

3.2 MAX LOAD (POINT-CHECK) METHODS

One of the basic objectives in the provision of transit service is to ensure adequate space to accommodate the maximum number of onboard passengers along the entire route over a given time period. Let us denote the time period (usually an hour) as j. Based on the peak-load factor concept, the number of vehicles required for period j is

$$F_j = \frac{\overline{P}_{mj}}{\gamma_j \cdot c} \tag{3.1}$$

where
$\overline{P}_{mj}$ is the average maximum number of passengers (max load) observed on board in period j
c represents the capacity of a vehicle (number of seats plus the maximum allowable standees)
γ_j is the load factor during period j, $0 < \gamma_j \leq 1.0$

For convenience, let us refer to the product $\gamma_j \cdot c$ as d_{oj}, the desired occupancy on the vehicle at period j. The standard γ_j can be set so that d_{oj} is equal to a desired fraction of the capacity (e.g., d_{oj} = number of seats). It should be noted here that if $\overline{P}_{mj}$ is based on a series of measurements, one can take its variability into account. This can be done by replacing the average value in Equation 3.1 with $\overline{P}_{mj} + b \cdot S_{pj}$, where b is a predetermined constant and S_{pj} is the standard deviation associated with $\overline{P}_{mj}$.

The max load data are usually collected by a trained checker, who stands and counts at the transit stop located at the beginning of the max load section(s). This stop is usually determined from the old ride-check data or from the information given by a mobile supervisor. Often, the checkers are told to count at only one stop for the entire day instead of switching among different max load points, depending on period j. Certainly, it is less costly to position a checker at one stop than to have several checkers switching among stops. Given that a checker is assigned to one stop, that which apparently is the heaviest *daily* load point along the route, we can establish the so-called Method 1 for determining the frequency associated with this single stop

$$F_{1j} = \max\left(\frac{P_{mdj}}{d_{oj}}, F_{mj}\right), \quad j = 1, 2, \ldots, q \tag{3.2}$$

$$P_{md} = \max_{i \in S} \sum_{j=1}^{q} P_{ij} = \sum_{j=1}^{q} P_{i*j}$$

$$P_{mdj} = P_{i*j}$$

where

F_{mj} is the minimum required frequency (reciprocal of policy headway) for period j; there are q time periods

S represents the set of all route stops i excluding the last stop

i* is the *daily max load point*

P_{ij} is a defined statistical measure (simple average or average plus standard deviation) of the total number of passengers on board all the vehicles departing stop i during period j

The terms P_{mdj} and P_{md} are used for the (average) observed load at the daily max load point at time j and the total load observed at this point, respectively.

Figure 3.1 exhibits an example of passenger counts (Example 1) along a 10 km route with six stops between 6:00 and 11:00 a.m. The second column in the table in this figure presents the distances, in km, between each stop. The desired occupancy and minimum frequency are same for all hours, and hence their time period subscript is dropped: $d_o = 50$ passengers and $F_m = 3$ vehicles, respectively. The set of stops S includes five is, j = 1, 2, ..., 5, each period of 1 h being associated with a given column. The last column in the table represents $\sum_{j=1}^{5} P_{ij}$ in which each entry in the table is P_{ij} (an average value across several checks). Thus, i* is the third stop with $P_{md} = 1740$, and P_{mdj} in Equation 3.2 refers only to those entries in the third row.

The second point-check method, or *Method 2*, is based on the max load observed in each time period. That is,

$$F_{2j} = \max\left(\frac{P_{mj}}{d_{oj}}, F_{mj}\right), \quad j = 1, 2, \ldots, q \tag{3.3}$$

where $P_{mj} = \max_{i \in S} P_{ij}$, which stands for the maximum observed load (across all stops) in each period j.

In the table in Figure 3.1, the values of P_{mj} are circled, and a rectangle is placed around P_{md}. Figure 3.1 also illustrates passenger counts in three dimensions (load, distance, and period), from which the hourly max load is observed for the first 3 h. The results of Equations 3.1 and 3.2 applied to Example 1 appear in Table 3.1 for both frequency (F_{kj}) and headways (H_{kj}) rounded to the nearest integer, where k = 1, 2. The only nonrounded headway is $H_{kj} = 7.5$ min, since it fits the so-called clock headways: these have the feature of creating timetables that repeat themselves every hour, starting on the hour. Practically speaking, $H_{kj} = 7.5$ can be implemented in an even-headway timetable by alternating between $H_{kj} = 7$ and $H_{kj} = 8$.

We will also retain nonrounded F_{kj}s and show in the next chapter how to use these determined values for constructing timetables with and without even headways.

Stop #	Distance (km) to next stop	Average observed load (passengers), by hour					Total load (passengers)
		6:00–7:00	7:00–8:00	8:00–9:00	9:00–10:00	10:00–11:00	
1	2	50	136	245	250	95	776
2	1	100	⟨510⟩	310	208	122	1250
3	1.5	⟨400⟩	420	⟨400⟩	⟨320⟩	200	⟨1740⟩
4	3	135	335	350	166	⟨220⟩	1206
5	2.5	32	210	300	78	105	725

Notes: (1) Route length is 10 km, and stop #6 is the last stop
(2) For all hours, $d_o = 50$, $c = 90$ passengers, $F_m = 3$ veh/h

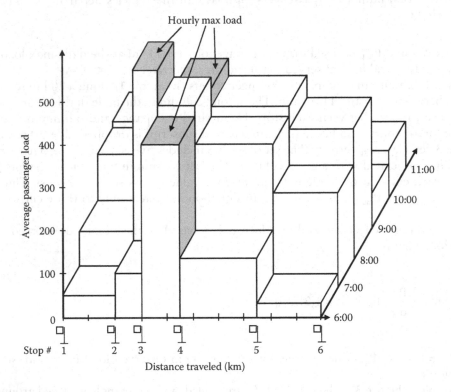

Figure 3.1 Example 1: Five-hour load profiles, with indications of hourly and daily max load points.

Table 3.1 Frequency and headway results for Example 1 in Figure 3.1, according to Methods 1 and 2

Period j	Method 1 (daily max load point)		Method 2 (hourly max load point)	
	F_{1j} (veh/h)	H_{1j} (min)	F_{2j} (veh/h)	H_{2j} (min)
6:00–7:00	8.0	7.5	8.0	7.5
7:00–8:00	8.4	7	10.2	6
8:00–9:00	8.0	7.5	8.0	7.5
9:00–10:00	6.4	9	6.4	9
10:00–11:00	4.0	15	4.4	14

Although Method 1 has an advantage over Method 2 through cost saving in data gathering, it cannot be traded off with accuracy of the results. That is, it is less costly and more convenient to retain a checker at one transit stop throughout the entire working day than to assign the same checker or others to a different stop at every period j. Consequently, we will now concentrate on a comparison of the two methods: if the difference is statistically not significant, then the routine use of Method 1 will be preferred. This comparison may be carried out by the chi-square statistic, χ^2, for testing the hypothesis concerning the population variances of the data supplied for each method. We assume that the random sample of passenger counts comes from a normal distribution (see Section 2.4 in this book for basic statistics), with averages μ_1, μ_2 and variances σ_1^2, σ_2^2, for Methods 1 and 2, respectively. The hypothesis that $\sigma_1^2 = \sigma_2^2$ will be examined, using the χ^2 test while knowing that σ_2^2 is the theoretical (expected) variance with which it is to be compared. The statistic χ^2, $\chi^2 = (n - 1) s^2/\sigma_2^2$ has a chi-square probability distribution with $(n - 1)$ degrees of freedom, where s^2 is computed from the sample. If $\sigma_1^2 > \sigma_2^2$, both s^2 and χ^2 will be larger than expected. The hypothesis will be rejected if $\chi^2 > \chi_\alpha^2$, where χ_α^2 is chosen so that $P(\chi^2 > \chi_\alpha^2) = \alpha$ (see Section 2.4); the χ_α^2 tables can be found, for example, in the website http://www.stat.lsu.edu/exstweb/statlab/Tables/TABLES98-Chi.html, (LSU Experimental Statistics, 2015).

It should be noted that there is often a misjudgment of statistical data when chi-square tests are incorrectly utilized (Ceder and Dressler, 1980). For example, it is incorrect to compare the frequency or headway results given by the two methods. Instead, the actual observations should be compared, because any transformation of the actual units distorts the χ^2 test. If the observed (O) and the expected (E) data are multiplied by k, then the calculated χ^2 is equal to k times the corrected χ^2 value: $\sum_i[(kO_i - kE_i)^2/kE_i] = k\sum_i[(O_i - E_i)^2/E_i]$. In our case, the observed items are the passenger counts of Method 1 and the expected items are the counts of Method 2, using the following calculation:

$$\chi^2 = \sum_{j=1}^{q}\left[\frac{(P_{i*j} - P_{mj})^2}{P_{mj}}\right] \tag{3.4}$$

where $i*$ is the daily max load point.

A comparison of the two methods can now be applied to Example 1 using Equation 3.4. This yields $\chi^2 = (420 - 510)^2/510 + (200 - 220)^2/220 = 17.7$; for $\alpha = 0.05$, we obtain $\chi_\alpha^2 = 9.45$ from the chi-square table, with a degree of freedom of four. Hence, the possibility of assigning a checker only to stop 3 of Example 1 must be rejected, and we can skip over the Method 1 results.

3.3 LOAD-PROFILE (RIDE-CHECK) METHODS

The data collected by ride check enable the planner to observe the load variability among the transit stops, or what is termed the *load profile*. Usually, a recurrent, unsatisfactory distribution of loads will suggest the need for possible improvements in route design. The most common operational strategy resulting from observing the various loads is short turning (shortlining). A start-ahead and/or turn-back point(s) after the start and/or before the end of the route may be chosen, creating a new route that overlaps the existing route. This short-turn design problem is covered in Chapter 15. Other route design–related actions using load data are route splitting and route shortening, both of which are dealt with in Chapter 13. This section will use the ride-check data for creating more alternatives to derive adequate frequencies, while assuming that the route remains same. Nevertheless, we know that in practice, the redesign of an existing route is not an activity often undertaken by transit agencies.

Two examples of load profiles are illustrated in Figure 3.2. These profiles are extracted from Example 1 in Figure 3.1 for the first and third hours. It may be noted that in most available transit-scheduling software (see, e.g., Section 1.2), these load profiles are plotted with respect to each stop without relating the x-axis to any scale. A more appropriate way to plot the loads is to establish a passenger-load profile with respect to the distance traveled from the departure stop to the end of the route. It is also possible to replace the (deterministic) distance by the average running time; in the latter case, however, it is desirable for the running time to be characterized by low and persistent variations. These plots furnish the important evaluation measures of passenger-km and passenger-hour as is also shown in Figure 3.2.

Let us observe the area marked by dashed lines in Figure 3.2. If a straight line is drawn across the load profile where the number of passengers is equal to the observed average hourly max load, then the area below this line but above the load profile is a measure of superfluous productivity. When Method 2 is used to derive the headways, this area represents

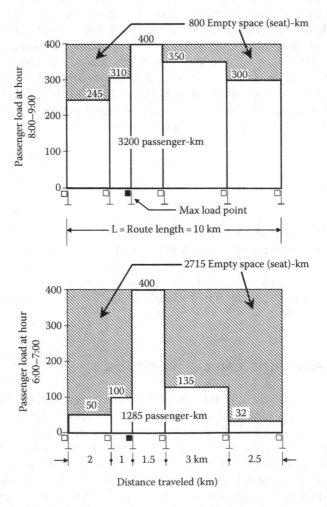

Figure 3.2 Two load profiles from Example 1 with the same 8-vehicle frequency, but with different passenger- and empty space-km.

empty space-kilometers. Furthermore, if d_{oj} in Method 2 is equal to the number of seats—often, this is the desired occupancy or load factor used—then this measure is empty-seat-kilometers. In light of this measure of unproductive service, we can see in Figure 3.2 that the 8:00–9:00 load profile is more than twice as productive as the 6:00–7:00 profile, though both have the same (max load point–based) frequency. We can now use the additional information supplied by the load profile to overcome the problem exhibited in Figure 3.2 when using Method 2. This can be done by introducing frequency determination methods based on passenger-km rather than on a max load measure. The first load-profile method considers a lower-bound level on the frequency or an upper bound on the headway, given the same vehicle capacity constraint. We call this *Method 3*, and it is expressed as follows:

$$F_{3j} = \max\left[\frac{A_j}{d_{oj} \cdot L}, \frac{P_{mj}}{c}, F_{mj}\right] \tag{3.5}$$

$$A_j = \sum_{i \in S} P_{ij} \cdot \ell_i, \quad L = \sum_{i \in S} l_i$$

where
  l_i is the distance between stop i and the next stop (i + 1)
  A_j is the area in passenger-km under the load profile during time period j
  L is the route length

The other notations were previously defined in Equations 3.1 through 3.3.

One way to look at Method 3 is to view the ratio A_j/L as an average representative of the load P_{ij} (regardless of its statistical definition), as opposed to the max load (P_{mj}) in Method 2. Method 3 guarantees, on the average basis of P_{ij}, that the onboard passengers at the max load route segment will not experience crowding above the given vehicle capacity constraint. This method is appropriate for cases in which the planner wishes to know the number of vehicle runs (frequency) expected, while relaxing the desired occupancy standard constraint and, at the same time, avoiding situations in which passengers are unable to board the vehicle in an average sense. Using the results of Method 3 allows planners to handle (1) demand changes without increasing the available number of vehicles, (2) situations in which some vehicles are needed elsewhere (e.g., breakdown and maintenance problems or emergencies), and (3) occasions when there are fewer drivers than usual (e.g., owing to budget cuts or problems with the drivers' union). On the other hand, Method 3 can result in unpleasant travel for an extended distance in which the load (occupancy) is above d_{oj}.

To eliminate or control the possibility of such an undesirable phenomenon, we introduce another method, called *Method 4*. This Method 4 establishes a level-of-service consideration by restricting the total portion of the route length having loads greater than the desired occupancy. Method 4 takes the explicit form

$$F_{4j} = \max\left[\frac{A_j}{d_{oj} \cdot L}, \frac{P_{mj}}{c}, F_{mj}\right] \tag{3.6}$$

$$\text{subject to (s.t.)} \sum_{i \in I_j} l_i \leq \beta_j \cdot L,$$

where mathematically $I_j = \{i:(P_{ij}/F_j) > d_{oj}\}$; in other words, I_j is the set of all stops i in time period j, such that the load P_{ij} exceeds the product of d_{oj} times the frequency F_{4j}, and β_j is the allowable portion of the route length at period j in which P_{ij} can exceed the product $F_{4j} \cdot d_{oj}$. The other notations in Equation 3.6 were previously defined. By controlling parameter β_j, it is possible to establish a level-of-service criterion. We should note that for $\beta_j = 0$ and $\beta_j = 1.0$, Method 4 converges to Methods 2 and 3, respectively.

The load profile of Example 1 (see Figure 3.1) is presented in Figure 3.3 pertaining to the hour 9:00–10:00. The *considered load level* associated with Method 3 is simply the area under the load profile, divided by L = 10 km, or 188.1 passengers in this case. This is an average load profile. However, all loads between stops 1 and 4 (4.5 km), which stretch out for 45% of the route length, exceed this average. To avoid the load exceeding the desired profile for more than a predetermined percentage of the route length, Method 4 can be introduced. If this percentage is set at 40% of the route length, the considered load will be 208 passengers, thus allowing only the stretch between stops 1 and 2 (2 km) and that between stops 3 and 4 (1.5 km) to have this excess load. We term this situation Method 4 (40%). Setting the percentage to 20% results in an average of 250 passengers; in the case of 10% (1 km), the considered load level converges to the Method 2 average of 320 passengers.

Figure 3.4 illustrates the fundamental trade-off between the load profile and max load concepts. We will show it for the case of Method 4 (20%) and the 9:00–10:00 h. Based on Figure 3.1 data and Equation 3.6, with $\beta_j = 0.2$, we attain $F_4 = \max(250/50, 320/90, 3) = 5$ veh/h from raising the considered load level from 188.1 to 250 passengers.

The five assigned departures will, in an average sense, carry 320/5 = 64 passengers between stops 3 and 4 (14 more than the desired load of 50). Nonetheless, this excess load for 1.5 is traded off with $(320-250) \cdot 8.5 = 595$ empty space-km as is shown in Figure 3.4. This trade-off can be interpreted economically and perhaps affect the ticket tariff.

In case that the calculated frequency in Equations 3.5 and 3.6 is the result of P_{mj}/c, then the considered load will be determined by the product $(P_{mj}/c) \cdot d_{oj}$. Table 3.2 shows the results for Methods 3 and 4, in which the percentage of route length allowed to have an excess load in Method 4 is set at 10%, 20%, and 30%.

We may observe that in the Method 3 results in Table 3.2, the first hour relies on P_{mj}/c or, specifically, 400/90 = 4.44 veh/h. When we turn to Method 4 for this first hour, the results

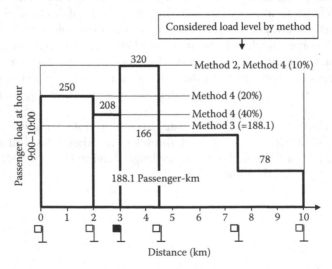

Figure 3.3 Load profile from Example 1 between 9:00 and 10:00 with considered load levels for three methods and Method 4 standing for 10%, 20%, and 40% of the route length.

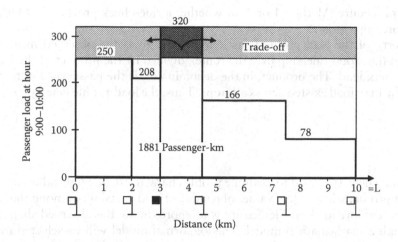

Figure 3.4 Load profile from Example I (9:00–10:00) using Method 4 (20%), with an indication of trade-off between this method and Method 2 (more crowding in return for less empty space-km).

Table 3.2 Frequency and headway results for Example I in Figure 3.1 for Methods 3 and 4

| | | | Method 4 | | | | | |
| | Method 3 | | 10% | | 20% | | 30% | |
Period j	F_{3j} (veh/h)	H_{3j} (min)	F_{4j} (veh/h)	H_{4j} (min)	F_{4j} (veh/h)	H_{4j} (min)	F_{4j} (veh/h)	H_{4j} (min)
6:00–7:00	4.44	14	8.00	7.5	4.44	14	4.44	14
7:00–8:00	5.88	10	8.40	7	8.40	7	6.70	9
8:00–9:00	6.40	9	8.00	7.5	7.00	9	7.00	9
9:00–10:00	3.72	16	6.40	9	5.00	12	5.00	12
10:00–11:00	3.07	20	4.40	14	4.40	14	4.00	15

of Method 2 for 10% are attained, since the max load stretches along more than 10% of the route length. For 20% and 30%, the vehicle capacity constraint still governs. In the second hour, 7:00–8:00, the following is obtained for Method 3: $F_3 = \max(2942/50 \cdot 10, 510/90, 3) = 5.88$ veh/h. Continuing in the second hour for Method 4 (10%), the average max load of 294.2 passengers rises to 420, resulting in $F_4 = 8.40$ veh/h. Table 3.2 continues to be fulfilled in the same manner. It should be noted that, as in Table 3.1, all the headways are rounded to their nearest integer.

Although we aim at a resource saving using Methods 3 and 4, there is a question as to whether this saving justifies the additional expense involved in using ride check as opposed to point check. The next section attempts to answer this question by constructing a criterion suggesting when to use the point check or, otherwise, the ride-check data collection technique.

3.4 CRITERION FOR SELECTING POINT CHECK OR RIDE CHECK

This section will test an assumption that particular load-profile characteristics suggest the data collection technique to be used. The basic idea is to provide the transit agency with adequate preliminary guidance in selecting the type of technique based on old load profiles. The assumption will be investigated whether a relatively flat profile suggests the use of a point-check procedure (Method 1 or 2) or whether a ride-check procedure (Method 3 or 4) would be more appropriate.

One property of the load profile is its density, ρ. This is the observed measure of total passenger-km (i.e., total ridership over the route), divided by the product of the length of the route and its max load. The product, in the denominator, is the passenger-km that would be observed if the max load existed across all stops. Thus, the load-profile density for hour j, ρ_j, is

$$\rho_j = \frac{A_j}{P_{mj} \cdot L} \tag{3.7}$$

The load-profile density is used to examine profile characteristics. High values of ρ indicate a relatively flat profile, whereas low values of ρ indicate load variability among the route stops.

One way to explore load-profile density is to approximate the observed shapes of profile curves through a mathematical model. The lognormal model will be selected for this purpose, since it provides a family of curves that can be controlled by varying the two parameters, μ and σ. The lognormal model takes the form

$$f(x) = \frac{1}{x \cdot \sigma \sqrt{2\pi}} \, e^{(-(\ln x - \mu)2/2\sigma)^2}; \quad x > 0 \tag{3.8}$$

The equation satisfying the optimum conditions, $df(x)/dx = 0$, is $x_o = e^{\mu - \sigma^2}$, in which Equation 3.8 reaches maximum. In our case, the maximum coincides with the max load. This continuous model can only approximate some of the observed load profiles, since it has only one peak and represents monotonically increasing and decreasing functions before and after this peak, respectively. Nonetheless, this model is useful for observing some general differences between the ride-check and point-check methods.

In order to compare the methods, $f(x)$ will be used as a normalized load (the load divided by the max load) and x as a normalized distance (the distance from the initial route stop divided by the length of the route). At a given time interval of 1 h, j, the considered max load is $P_{mj} = 650$ passengers. Given that $d_{oj} = 65$, c = 100, and $F_m = 2$, the determined frequency and headway for both Methods 1 and 2 are $F_j = 10$ and $H_j = 6$. By applying this information to Methods 3 and 4, and using a variety of lognormal curves, we will obtain the frequencies and headways shown in Table 3.3. The results in this table are arranged in order of increasing density. For Method 3, the capacity constraint determines the values of F and H up to and including $\rho = 0.64$ and up to different ρ values (if any) for Method 4. Examples of the lognormal normalized curves are shown in Figures 3.5 and 3.6, the latter showing the determination of the frequency for one curve ($\mu = -1.5$, $\sigma = 1.0$). Given that the normalized route length is 1.0, then the different percentages (for this length) in Figure 3.6 are set for the Method 4 criterion, with an explicit calculation example for the 10% case.

From Table 3.3, it appears that, for Method 3, the ride-check data result in the same rounded headway as the point-check data for $\rho > 0.84$. For Method 4, the ride-check and point-check information tend to yield the same headways for $\rho > 0.34$ (10% case), $\rho > 0.50$ (20% case), and $\rho > 0.68$ (30% case).

In practice, the transit agency wishes to save vehicle runs and, eventually, to perform the matching between demand and supply with fewer vehicles. As we will show in Chapters 4 and 6, different headway values do not necessarily save vehicle runs or reduce the required fleet size. However, the analysis of the profile-density measure can be used by the transit agency as a preliminary check before embarking on a more comprehensive analysis. The following are practical observations: (1) for densities below 0.5, savings are likely to result

Table 3.3 Frequencies (F), rounded headways (H), and density (ρ) for different load-profile configurations (lognormal model based), using Methods 3 and 4

| Lognormal model parameters | | Profile density | Method 3 | | Method 4 | | | | | |
| | | | | | 10% | | 20% | | 30% | |
μ	σ	ρ	F	H	F	H	F	H	F	H
−1.0	0.2	0.18	6.50[a]	9	7.60	7	6.50[a]	9	6.50[a]	9
−1.5	0.5	0.25	6.50[a]	9	8.46	7	6.50[a]	9	6.50[a]	9
−1.5	1.0	0.32	6.50[a]	9	8.36	7	6.50[a]	9	6.50[a]	9
−1.0	1.5	0.34	6.50[a]	9	7.55	7	6.50[a]	9	6.50[a]	9
−1.0	1.2	0.43	6.50[a]	9	9.00	6[b]	7.05	8	6.50[a]	9
−1.0	0.6	0.44	6.50[a]	9	9.50	6[b]	8.10	7	6.50[a]	9
−0.8	0.5	0.47	6.50[a]	9	9.55	6[b]	8.40	7	6.50[a]	9
−1.0	0.9	0.48	6.50[a]	9	9.45	6[b]	8.05	7	7.05	8
−0.4	1.5	0.50	6.50[a]	9	9.05	6[b]	7.35	8	6.50	9
0.0	0.5	0.56	6.50[a]	9	9.92	6[b]	9.67	6[b]	9.27	9
−0.5	0.5	0.57	6.50[a]	9	9.76	6[b]	9.11	6[b]	8.16	6
−0.4	0.5	0.59	6.50[a]	9	9.81	6[b]	9.31	6[b]	8.46	7
−0.4	1.2	0.62	6.50[a]	9	9.65	6[b]	8.85	6[b]	7.80	7
−0.5	1.0	0.64	6.50[a]	9	9.79	6[b]	9.04	6[b]	8.19	7
−0.4	0.8	0.68	6.77	8	9.87	6[b]	9.42	6[b]	8.72	7
0.5	1.5	0.75	7.46	8	9.86	6[b]	9.36	6[b]	8.76	6[b]
0.0	1.0	0.76	7.63	7	9.93	6[b]	9.68	6[b]	9.23	6[b]
0.5	1.0	0.78	7.77	7	9.97	6[b]	9.87	6[b]	9.72	6[b]
1.0	1.5	0.84	8.41	7	9.96	6[b]	9.76	6[b]	9.46	6[b]
1.5	1.5	0.87	8.72	6	9.97	6[b]	9.92	6[b]	9.82	6[b]

Note: For Methods 1 and 2, F = 10, H = 6, where d_o = 65, c = 100.

[a] Whenever F = 6.5, H = 9, the capacity constraint is met.
[b] Whenever H = 6, ride check and point check yield the same headway.

by gathering the load-profile information and using ride-check methods (alternatively, the profile can be examined with such low ρ values for short-turn strategies, to be discussed in Chapter 14); (2) for densities between 0.5 and 0.85, actual comparisons can be made between the point-check and ride-check methods, along with Table 3.3 frequencies (F), rounded headways (H), and density (ρ) for different load-profile configurations (lognormal model based), using Methods 3 and 4 further saving considerations (for constructing time-tables and vehicle schedules); and (3) for densities above 0.85, it is likely that the majority of the required information for the headway calculation can be obtained from a point-check procedure (either Method 1 or 2). A further simplified and explicit practical criterion is to use Methods 3 and 4 for ρ ≤ 0.5, and Method 1 and 2 otherwise. It may be argued that for the range 0.5 < ρ ≤ 0.85, the use of load-profile methods cannot produce significant gains over the max load methods, although theoretically it may be justified. This argument is supported by the relatively small amount of passenger-km in the trade-off situation exhibited in Figure 3.4.

3.5 CONCLUSION (TWO EXAMPLES)

Another simple example, which we call Example 2, will now ascertain fully the procedures employed. This example will be followed by a real-life example that includes the results of

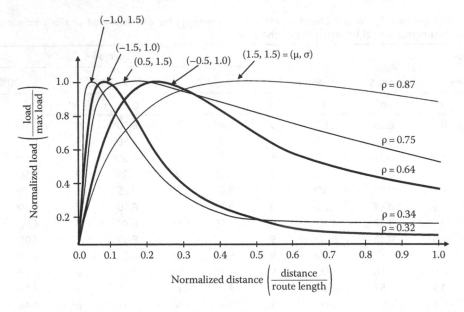

Figure 3.5 Five approximated load profiles based on the lognormal model with normalized scales.

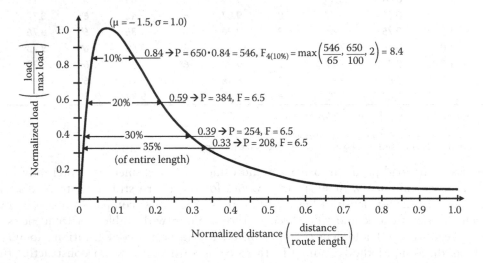

Figure 3.6 An approximated load profile based on the lognormal model, with indications of different (route) length percentages associated with Method 4.

the four methods. This section will end by portraying the methodology proposed in flow-chart form, with an orientation toward practice.

Example 2, which is presented and illustrated in Figure 3.7, will also be used throughout Chapters 4 and 5. The basic required input consists of (1) distance between stops, (2) desired occupancy per hour (for every planned vehicle type), (3) minimum frequency per hour (could be the inverse of policy headway), (4) number of observed (scheduled usually) vehicles each hour, (5) observed load (an average value or a consideration of its variability) between each two adjacent stops for each hour, and (6) vehicle capacity (for every planned vehicle type). A 2 h operation period (06:00–08:00) is chosen for simplicity

Stop #	Distance (km) to next stop	Average observed load (passenger), by scheduled departure time					Average load, by hour*		Total hourly load (passengers)
		6:15	6:50	7:15	7:35	7:50	6:00–7:00	7:00–8:00	
1	2	22	25	50	60	45	67	135	202
2	1.5	52	40	90	87	75	128	(216)	344
3	4.5	35	65	85	44	83	(134)	178	312
Number of observed scheduled vehicles							2	2	
Desired occupancy							50	60	
Minimum frequency							1	2	Calculated
Vehicle capacity (seats + allowed standees)							90	90	
Area under the load profile (passenger-km)							929	1395	

* See calculation in Figure 3.8 under the 7:15 departure.

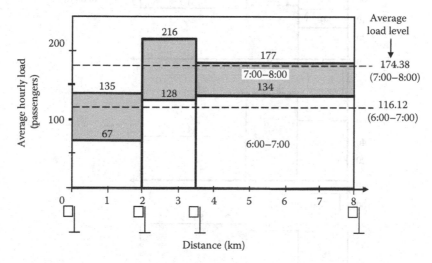

Figure 3.7 Example 2: three-hour data for frequency and headway determination by each of the four methods.

in this example. Furthermore, as a matter of decision, all vehicles depart on the hour pertaining to this next hour.

Let us recall the assumption that the observed loads in each hour are based on a uniform passenger-arrival rate (demand). That is, the number of passengers carried by the first vehicle in each hour is divided proportionally to reflect the demand in two time intervals: (1) the start of the hour and the departure time of the first vehicle that hour and the (2) last departure time in the previous hour and the start of the considered hour.

Hence, vehicles departing on the hour carry a demand from the previous hour. Figure 3.8 interprets this proportionality for the 7:15 departure; in this figure, five load profiles are shown for each observed vehicle. We assume that the load on the 7:15 vehicle contains 2/5 of the 6:00–7:00 demand (10 min from 6:50 to 7:00 and 15 min from 7:00 to 7:15). This proportion is marked in Figure 3.8 and inserted in the table of Figure 3.7 under average load by hour.

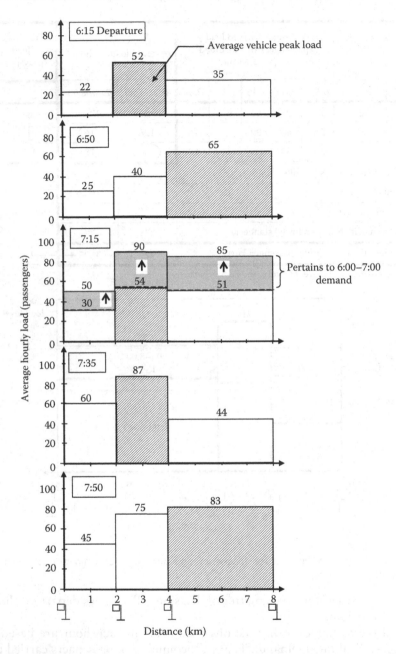

Figure 3.8 Example 2 data by vehicle load, showing the demand (passenger-km) that pertains to the 6:00–7:00 period, assuming uniform passenger-arrival rates.

We can see from the table in Figure 3.7 that the daily max load point is stop 2, with stops 3 and 2 being the hourly max points for the 6:00–7:00 and the 7:00–8:00 h, respectively. The load profiles are depicted on the same scales in Figure 3.7. The last column, last row, and "average load by hour" column in Figure 3.7 are calculated items. The difference between Method 1 and 2 is simply between the set of counts (128, 216) and (134, 216), respectively. Using Equation 3.4, we obtain $\chi^2 = 0.269$, where $\chi^2 = 3.84$ for $\alpha = 0.05$ with a single degree of freedom. Hence, Method 1 can substitute for Method 2 for future

Table 3.4 Frequency (F) and rounded headway (H) results for Example 2, by each of the four methods

Hour	Method 1		Method 2		Method 3		Method 4 (20%)	
	F	H	F	H	F	H	F	H
6:00–7:00	2.56	23	2.68	22	2.32	26	2.68	22
7:00–8:00	3.60	17	3.60	17	2.91	21	2.95	20
Calculation								
6:00–7:00	$\max\left(\dfrac{128}{50},1\right)=2.56$		$\max\left(\dfrac{134}{50},1\right)=2.68$		$\max\left(\dfrac{929}{50\cdot 8},\dfrac{134}{90},1\right)=2.32$		$\max\left(\dfrac{134}{50},\dfrac{134}{90},1\right)=2.68$	
7:00–8:00	$\max\left(\dfrac{216}{60},2\right)=3.60$		$\max\left(\dfrac{216}{60},2\right)=3.60$		$\max\left(\dfrac{1395}{60\cdot 8},\dfrac{216}{90},2\right)=2.91$		$\max\left(\dfrac{177}{60},\dfrac{216}{90},2\right)=2.95$	

passenger counts. Moreover, by using Equation 3.7 and Figure 3.7, we obtain $\rho_{6-7}=0.87$ and $\rho_{7-8}=0.81$, meaning that the max load-point data collection procedure is acceptable.

The results of Example 2 for all four methods are shown in Table 3.4 along with their calculations. Equations 3.2 and 3.3 may be applied for Methods 1 and 2, respectively, and Equations 3.5 and 3.6 for Methods 3 and 4, respectively. Table 3.4 considers, for Method 4, the constraint that no more than 20% of the route length is allowed to have an excess load, provided that the considered load level for this excess load is equal to or below the hourly max load divided by vehicle capacity. The headways in Table 3.4 are rounded off to their nearest integer. The frequencies in this table will be used in the next chapter (using the same example) as noninteger numbers while attempting to construct alternative (and automated) timetables.

For a real-life situation, we will demonstrate the procedures developed when using ride-check data from an old Los Angeles Metro (previously called SCRTD) bus route. This route, 217, was considered a heavy route and was characterized by 60 stops. All the trips on this route cross the daily max load point. The complete northbound ride-check input for route 217 appears in Table 3.5. The content of this table is comprehensive: it includes the distances (in km) to the next stop and the stop name in the first and second columns, respectively; the last column represents the total load for the whole day. The entries in Table 3.5 are based on rounded average values following several checks. Vehicle capacity, required for Methods 3 and 4, is 80 passengers.

An automated program is constructed for the four methods to be commensurate with Equations 3.2, 3.3, 3.5, and 3.6. The intermediate result of this program, which concerns max load information, is illustrated in Table 3.6. The daily max load point is the Fairfax/Rosewood stop, with a total of 4413 observed passengers over the entire day. The hourly max load points for each hour appear in Table 3.6. Occasionally, there are multiple-stop results; for example, at 10:00–11:00 and 22:00–23:00. The results of the load-profile density measure reveal that all $\rho < 0.5$; hence, the ride-check count is the data collection technique appropriate for this route. The frequency and headway results for route 217 are summarized in Table 3.7 for all four methods. The chi-square comparison between Method 1 and 2 for every hour over all hours, using $\alpha = 0.05$, shows that the hypothesis concerning equal methods is rejected; therefore, the hourly max load data collection technique should be applied for this data set. Nonetheless, as we mentioned before, the ride check is the appropriate technique. The full results, in Table 3.7, are presented solely for comparison purposes; they show that from 6:00 to 7:00 and after 19:00, route 217 relies on its minimum frequency. Methods 2 and 3 serve as upper and lower bounds on the frequency, respectively; and Method 4 ranges between the two, depending on the percentage of route length allowed to have excess load.

A graphical comparison of the frequency results of Methods 2, 3, and 4 (20%) for both directions of route 217 is presented in Figure 3.9. This figure also contains the actual observed

Table 3.5 Ride-check data for LA bus route 217 northbound

Time period	6:00–7:00	7:00–8:00	8:00–9:00	9:00–10:00	10:00–11:00	11:00–12:00	12:00–13:00	13:00–14:00	14:00–15:00	15:00–16:00	16:00–17:00	17:00–18:00	18:00–19:00	19:00–20:00	20:00–21:00	21:00–22:00	22:00–23:00	23:00–24:00	24:00–25:00	Total
# of buses observed	6	10	8	8	8	7	8	8	8	10	8	9	6	5	3	3	2	2	2	
Minimum frequency (veh/h)	2	2	2	2	2	2	2	2	2	2	2	2	2	2	2	2	2	2	2	
Desired occupancy (passengers)	60	70	70	60	50	50	50	50	50	60	70	70	60	60	50	50	50	50	50	

Dist. (km)	Stop name	6:00–7:00	7:00–8:00	8:00–9:00	9:00–10:00	10:00–11:00	11:00–12:00	12:00–13:00	13:00–14:00	14:00–15:00	15:00–16:00	16:00–17:00	17:00–18:00	18:00–19:00	19:00–20:00	20:00–21:00	21:00–22:00	22:00–23:00	23:00–24:00	24:00–25:00	Total
0.16	Adams / Washingt	13	27	115	24	11	11	6	4	10	9	16	23	18	8	2	3	2	5	1	308
0.26	Fairfax / Adams	21	68	148	39	16	27	26	8	16	19	24	38	23	12	3	7	2	5	1	503
0.27	Fairfax / Washingt	22	89	160	48	22	32	29	15	23	20	35	40	23	15	3	8	6	5	1	596
0.26	Fairfax / Apple	25	101	163	51	26	34	29	16	23	23	40	45	24	14	3	8	6	5	1	637
0.27	Fairfax / Venice	29	112	183	53	37	43	37	22	26	34	60	51	32	19	3	8	9	5	2	765
0.27	Fairfax / Venice F	37	124	217	84	52	44	51	38	36	48	78	60	37	21	7	8	10	6	3	961
0.24	Fairfax / 18TH	40	119	188	83	51	45	52	39	37	57	78	61	37	20	6	8	9	6	3	939
0.24	Fairfax / Airdrome	40	131	192	89	52	49	54	42	38	60	77	65	35	19	8	11	9	7	3	981
0.24	Fairfax / Pickford	45	151	195	94	48	50	54	42	40	60	78	67	35	19	9	12	9	7	3	1018
0.24	Fairfax / Saturn	48	167	211	97	58	50	55	45	39	60	73	63	31	18	9	13	9	7	3	1056
0.24	Fairfax / Pico	56	217	246	116	83	78	84	95	73	94	105	97	49	30	11	12	14	8	4	1472
0.26	Fairfax / Packard	59	228	252	120	91	80	87	98	76	96	105	98	47	30	11	12	15	8	4	1517
0.26	Fairfax / Whitwort	63	250	257	125	98	87	90	100	78	96	104	98	53	31	14	12	15	9	4	1584
0.00	Fairfax / Olympic	59	244	275	130	108	98	103	120	99	86	122	99	54	38	16	12	18	10	5	1696
0.24	Olympic / Ogden	59	332	278	152	123	99	116	142	103	104	129	115	64	39	16	12	18	10	5	1916
0.22	Fairfax / San Vice	70	355	325	171	144	124	141	181	142	149	151	134	73	40	17	15	19	12	6	2269
0.24	Fairfax / 8th St	66	357	330	175	150	128	145	188	164	156	170	158	77	42	18	16	19	13	6	2378
0.29	Fairfax / Wilshire	54	363	349	236	288	250	354	356	342	397	382	343	190	91	40	24	36	18	8	4121
0.29	Fairfax / 6th St	55	369	351	238	291	257	363	365	347	398	386	343	189	94	41	27	34	18	8	4174
0.29	Fairfax / Drexel	54	376	355	234	291	258	381	378	350	398	391	339	191	93	42	27	34	18	9	4219
0.29	Fairfax / 3rd St	48	401	370	234	261	258	338	381	367	412	454	367	205	101	56	37	38	19	9	4356
0.29	Fairfax / 1st St	48	400	366	232	265	256	339	385	373	422	440	368	206	104	54	37	39	19	9	4362
0.34	Fairfax / Beverly	44	392	354	232	249	252	314	362	353	409	459	377	218	98	59	44	44	20	9	4289
0.35	Fairfax / Oakwood	52	391	351	237	271	282	315	366	352	416	467	370	220	106	57	48	51	22	9	4383
0.34	Fairfax / Rosewood	49	370	326	246	288	292	327	376	367	418	481	371	214	102	57	50	51	20	8	4413
0.35	Fairfax / Melrose	46	113	175	165	245	265	308	378	365	427	439	370	201	101	51	47	50	20	8	3774
0.34	Fairfax / Willowgh	44	112	173	155	238	246	291	358	344	406	411	353	191	91	51	45	48	19	8	3584
0.35	Fairfax / Santa Mo	33	90	140	148	154	174	216	292	273	322	353	299	171	97	45	50	51	19	4	2931
0.35	Fairfax / Fountain	35	92	134	153	149	172	208	281	244	300	321	283	157	89	42	49	49	20	4	2782

(Continued)

Table 3.5 (Continued) Ride-check data for LA bus route 217 northbound

Dist (km)	Time period / Stop name	6:00–7:00	7:00–8:00	8:00–9:00	9:00–10:00	10:00–11:00	11:00–12:00	12:00–13:00	13:00–14:00	14:00–15:00	15:00–16:00	16:00–17:00	17:00–18:00	18:00–19:00	19:00–20:00	20:00–21:00	21:00–22:00	22:00–23:00	23:00–24:00	24:00–25:00	Total
	# of buses observed	6	10	8	8	8	7	8	8	8	10	8	9	6	5	3	3	2	2	2	
	Minimum frequency (veh/h)	2	2	2	2	2	2	2	2	2	2	2	2	2	2	2	2	2	2	2	
	Desired occupancy (passengers)	60	70	70	60	50	50	50	50	50	60	70	70	60	60	50	50	50	50	50	
0.29	Fairfax / Sunset	32	70	95	132	127	144	172	249	210	270	280	264	144	80	44	48	48	20	4	2433
0.29	Sunset / Genesee	31	69	90	133	125	143	170	248	206	269	287	259	146	80	42	48	47	20	4	2417
0.29	Sunset / Stanley	30	71	92	135	127	135	171	248	206	260	288	253	145	77	43	46	48	18	4	2397
0.29	Sunset / Gardner	31	73	85	132	117	124	165	240	210	249	274	241	138	83	43	46	45	17	4	2317
0.29	Sunset / Martel	32	75	82	131	110	116	164	228	208	236	263	228	128	82	40	47	45	17	4	2236
0.29	Sunset / Poinsett	33	73	77	127	118	115	171	220	193	232	260	222	129	81	38	55	46	18	4	2212
0.30	La Brea / Sunset	36	65	70	122	113	106	163	211	177	222	252	218	132	86	39	54	45	17	4	2130
0.18	La Brea / Hollywood	36	62	68	118	116	98	156	199	172	213	242	206	127	81	36	56	42	16	2	2045
0.18	Hollywood / Sycamore	36	65	67	118	116	104	155	192	165	208	225	194	122	78	38	56	45	17	1	2002
0.18	Hollywood / Orange	32	62	66	112	112	107	152	176	166	197	204	179	113	75	36	54	43	17	1	1904
0.18	Hollywood / Highland	19	33	49	84	90	94	130	147	154	185	146	146	88	65	33	55	35	16	2	1571
0.18	Hollywood / Las Palm	16	26	45	77	83	89	120	126	140	168	141	139	85	58	31	55	34	15	2	1450
0.18	Hollywood / Whiley	15	22	42	62	67	76	111	92	116	144	129	116	68	43	27	53	31	13	2	1229
0.18	Hollywood / Wilcox	14	21	40	54	62	66	105	84	108	131	119	98	58	38	18	44	27	13	2	1102
0.19	Hollywood / Cahuenga	11	20	36	44	61	66	90	80	96	123	107	85	50	40	16	41	25	13	1	1005
0.19	Hollywood / Ivar	9	16	33	33	50	57	73	61	84	111	98	84	47	35	14	29	24	10	1	869
0.19	Hollywood / Vine	3	14	19	17	31	39	35	33	68	87	73	67	47	25	14	24	17	9	1	623
0.19	Argyle / Hollywood	4	20	17	20	31	35	32	44	69	91	83	66	44	20	16	24	15	10	0	644
0.19	Argyle / Yucca	3	20	17	19	30	28	32	44	68	81	79	64	25	15	13	21	11	9	0	598
0.27	Franklin Argyle	2	16	17	16	21	22	29	34	47	64	59	47	10	11	10	17	8	6	0	451
0.11	Gower / Franklin	2	8	13	11	10	15	21	2	27	53	38	30	7	8	8	6	0	0	0	282
0.16	Beachwood / Franklin	2	11	12	8	8	6	9	12	25	49	21	24	7	4	6	3	0	0	0	219
0.18	Beachwood / Midway	2	11	11	11	8	6	14	12	25	48	19	19	5	2	5	3	0	0	0	201
0.18	Beachwood / Scenic	2	10	11	10	7	5	13	11	16	44	17	18	4	2	5	3	0	0	0	178
0.16	Beachwood / Temple H	2	9	10	9	7	4	10	5	14	44	14	13	3	3	3	3	0	0	0	156
0.18	Beachwood / Winans	1	7	10	9	7	4	5	6	13	39	14	12	1	1	3	3	0	0	0	134
0.18	Beachwood / Cheremoy	1	7	9	6	7	3	5	4	13	34	13	7	1	1	3	3	0	0	0	112
0.18	Beachwood / Glen Ald	1	7	7	6	6	3	4	4	8	30	11	5	1	1	2	1	0	0	0	96
0.19	Beachwood / Glen Dak	1	7	7	6	6	3	2	4	8	26	10	4	1	1	2	0	0	0	0	87
0.21	Beachwood / Westshir	0	7	6	2	2	3	2	3	0	5	8	2	0	0	0	0	0	0	0	42
0.00	Beachwood / Westsh f	—	—	—	—	—	—	—	—	—	—	—	—	—	—	—	—	—	—	—	—

Table 3.6 Max load information for LA route 217 northbound, using Methods 1 and 2

Time period	Max load point	Max load (passengers)
06:00–07:00	Fairfax / San Vice	70
07:00–08:00	Fairfax / 3rd St	401
08:00–09:00	Fairfax / 3rd St	370
09:00–10:00	Fairfax / Rosewood	246
10:00–11:00	Fairfax / 6th St	291
	Fairfax / Drexel	
11:00–12:00	Fairfax / Rosewood	292
12:00–13:00	Fairfax / Drexel	381
13:00–14:00	Fairfax / 1st St	385
14:00–15:00	Fairfax / 1st St	373
15:00–16:00	Fairfax / Melrose	427
16:00–17:00	Fairfax / Rosewood	481
17:00–18:00	Fairfax / Beverly	377
18:00–19:00	Fairfax / Oakwood	220
19:00–20:00	Fairfax / Oakwood	106
20:00–21:00	Fairfax / Beverly	59
21:00–22:00	La Brea / Hollywood	56
	Hollywood / Sycamore	
22:00–23:00	Fairfax / Oakwood	51
	Fairfax / Rosewood	
	Fairfax / Santa Monica	
23:00–24:00	Fairfax / Oakwood	22
24:00–25:00	Fairfax / Drexel	9
	Fairfax / 3rd St	
	Fairfax / 1st St	
	Fairfax / Beverly	
	Fairfax / Oakwood	
All day	Fairfax / Rosewood	4413

(scheduled) frequency. We can easily see that the observed frequencies in both directions represent an excessive amount of bus runs. The transit agency provided the input, including the desired occupancy, set forth by the planners. The method used by the agency's planning department overlapped with Method 2. The observed frequency, therefore, poses an apparent enigma: why does the scheduled frequency not overlap, or at least come close to, the results of Method 2? More in-depth investigation can provide an answer, which has to do with the objective function of the drivers who attempt to maximize their comfort. When a bus run is overloaded repeatedly, the driver requests reinforcement; that is, an added run. When a bus run is almost empty, silence becomes golden. With time, the number of bus runs only increases, and this is the main reason behind the phenomenon exhibited in Figure 3.9. Incidentally, the agency's schedulers fixed their schedule following the presentation of the Figure 3.9 results. Nonetheless, the actual data support the use of Method 4 (20%) from a resource-saving perspective. It is interesting to note that Method 4 (20%) results in a much lower frequency than does Method 2, particularly in the southbound direction. The lower bound on the frequency needed to accommodate the passenger load while neglecting the load factor (desired occupancy) is exhibited by the results of Method 3. In most hours, it is as much as half the observed frequency.

Table 3.7 Frequency and headway results for LA route 217 northbound, by each of the four methods

Time period	Method 1 F (veh/h)	Method 1 H (min)	Method 2 F	Method 2 H	Method 3 F	Method 3 H	Method 4 10% F	Method 4 10% H	Method 4 20% F	Method 4 20% H	Method 4 30% F	Method 4 30% H
06:00–07:00	2.00	30	2.00	30	2.00	30	2.00	30	2.00	30	2.00	30
07:00–08:00	5.28	11	5.72	10	5.01	12	5.41	11	5.11	12	5.01	12
08:00–09:00	4.65	13	5.28	11	4.62	13	5.02	12	4.72	13	4.62	13
09:00–10:00	4.09	15	4.09	15	3.07	20	3.97	15	3.07	20	3.07	20
10:00–11:00	5.75	10	5.82	10	3.63	17	5.43	11	4.93	12	3.63	17
11:00–12:00	5.83	10	5.83	10	3.65	16	5.25	11	5.05	12	3.65	16
12:00–13:00	6.53	9	7.61	8	4.76	13	6.76	9	6.16	10	4.76	13
13:00–14:00	7.51	8	7.69	8	4.81	12	7.61	8	7.21	8	5.01	12
14:00–15:00	7.33	8	7.46	8	4.66	13	7.06	8	6.96	9	4.66	13
15:00–16:00	6.96	9	7.11	8	5.33	11	6.93	9	6.73	9	5.33	11
16:00–17:00	6.87	9	6.87	9	6.01	10	6.31	10	6.01	10	6.01	10
17:00–18:00	5.30	11	5.38	11	4.71	13	5.31	11	4.91	12	4.71	13
18:00–19:00	3.56	17	3.66	16	2.75	22	3.45	17	3.25	18	2.75	22
19:00–20:00	2.00	30	2.00	30	2.00	30	2.00	30	2.00	30	2.00	30
20:00–21:00	2.00	30	2.00	30	2.00	30	2.00	30	2.00	30	2.00	30
21:00–22:00	2.00	30	2.00	30	2.00	30	2.00	30	2.00	30	2.00	30
22:00–23:00	2.00	30	2.00	30	2.00	30	2.00	30	2.00	30	2.00	30
23:00–24:00	2.00	30	2.00	30	2.00	30	2.00	30	2.00	30	2.00	30
24:00–25:00	2.00	30	2.00	30	2.00	30	2.00	30	2.00	30	2.00	30

Finally, the procedures described and interpreted in this chapter will be demonstrated, using a flowchart (Figure 3.10), which provides a pragmatic overview leading to the use of either max load-point method (Method 1 or 2) or load-profile Method 4. The method chosen has a direct impact on the technique required for data acquisition. From a practitioner's perspective, we may note that the noticeable importance of ridership information has led transit agencies to introduce automated surveillance techniques mixed with manually collected data.

We are also aware of the current considerations by many transit agencies worldwide that want to implement APCs. Such systems are based on (1) infrared beam interruption, (2) pressure-sensitive stairwell mats, (3) counts by weighing the load on the vehicle, and (4) ultrasonic beam interruption. Naturally, the transit agencies expected to gain useful information for operations planning by obtaining more accurate passenger counts. However, as we mentioned earlier, there is always a point at which the additional accuracy is not worth the accompanying costs of data collection and analysis. Finally, in light of the methods proposed, as opposed to existing methods in use, the following rule may apply: forgetting is as important as remembering in the practical use of what is right to do.

3.6 LITERATURE REVIEW AND FURTHER READING

Only papers that focus on the determination of frequencies or headways with a linkage to the creation of timetables are reviewed in this section. Models in which headways or frequencies are set as part of a comprehensive network-design process are reviewed in Chapter 14. Methods for setting headways and frequencies for feeder routes are discussed in Chapter 16.

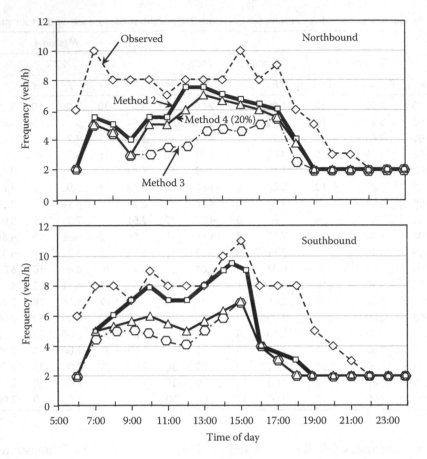

Figure 3.9 Comparison of frequency results of Methods 2, 3, and 4 (20%) and the observed frequency of route 217 in Los Angeles.

Furth and Wilson (1981) mention four common approaches for determining the headways of transit routes: policy headways, which are not directly derived from passenger demand; headways determined according to peak passenger load and vehicle capacity; headways designed such that the revenue/cost ratio will not exceed a preset value; and headways designed to achieve a desired ratio of passenger miles or vehicle miles per hour. An optimization (nonlinear programming) model and a solution algorithm are developed for setting frequencies in a given route network. The model assumes that the demand on each line is elastic; that is, sensitive to frequency changes. However, there are no demand relationships between different routes. The model allocates available buses to routes, subject to subsidy, fleet size, and vehicle-loading-level constraints. Multiple day periods are taken into account. The frequencies determined depend on fares, subsidies, fleet size, and the value of waiting time.

Koutsopoulos et al. (1985) develop a programming problem for determining frequencies in a transit network with a demand that varies along daily periods. Operating costs and travel times are assumed to be time varying, as well. The optimization is subject to subsidy, fleet size, and vehicle capacity constraints. The model is not based on the common assumption of a half-headway average waiting time, but on a more sophisticated waiting-time submodel; for example, the situation in which some passengers cannot board the first arriving bus because of crowding is considered. Another submodel measures passenger inconvenience, which is included in the total cost that the model aims to minimize. Since the

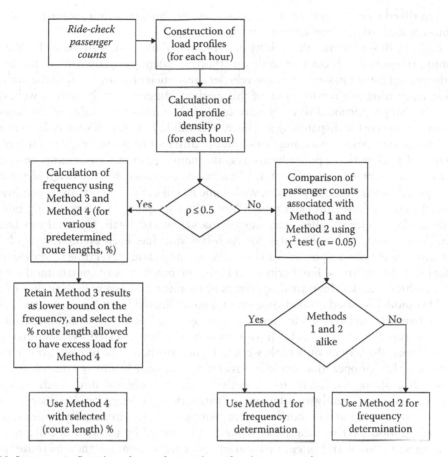

Figure 3.10 Summary in flowchart form of procedures for determining frequency.

solution procedure of the nonlinear problem presented may be complicated, the authors propose a simplified version of the initial formulation, in which the daily periods are divided into subperiods, during which headways are constant. The simplified problem can be solved by, for example, using linear programming.

LeBlanc (1988) introduced a model for determining frequencies, using a modal-split assignment programming model with distinct transit routes. The author shows how to refine conventional modal-split assignment models to include features of the proposed methodology. Multimodal considerations, such as the proportion of transit riders and the influence of the determined frequencies on road congestion, are taken into account.

Banks (1990) presents a model for setting headways in a transit route system. A case in which route demands vary with frequency is compared to a fixed-demand case; in the frequency-dependent demand case, there is no interline dependency, and demand functions are assumed to be unknown. An optimal solution is discussed in an unconstrained case and in several cases with subsidy, fleet size, and capacity constraints. In the unconstrained case, a complete analytical solution is presented, showing that the optimal headway is proportional to the square root of operating cost and of travel time. It is also inversely proportional to the square root of waiting time and cost and of the number of passengers. No analytical expressions for optimal headway are developed for the constrained cases.

Although Khasnabis and Rudraraju (1997) do not propose a methodology for frequency determination, they use simulation experiments to show that preemption strategies that are used in signalized intersections along the bus route are an important factor that should be taken into consideration when setting service headway.

Wirasinghe (2003) examines the validity of a traditional method, formulated by Newell, for determining frequencies. According to this method, frequency is proportional to the square root of the arrival rate of passengers if the vehicles are sufficiently large and to the arrival rate, otherwise. According to a modification of this method, suggested by Newell as well, optimal frequency is also proportional to the square root of the ratio of the value of the passengers' waiting time to the cost of dispatching a vehicle. The validity of Newell's formulae is examined in several situations, based on assumptions that are different from the original: uniform headway during off-peak periods, policy headways, alternative waiting-time assumptions, stochastic demand, and many-to-many demand. Wirasinghe concludes that the original *square root policy* proposed by Newell is applicable, with some modifications, under most conditions.

Kim et al. (2009) determine the optimal bus service frequency by considering bus operation costs to the service supplier, passenger demand, and so forth. Optimal bus frequency has been determined, and a timetable for each bus stop has been created by applying the demand- and travel time–responsive model or the demand- and travel time–responsive model for critical scheduling areas. Furthermore, a real-time bus headway control model has been developed, which enables the transit operators to monitor and assess punctuality of a given service. This model is based on real-time event data, collected from buses in service, such as bus stop departures and arrivals for buses operating on a line-based timetable, constructed for each bus stop. Subsequently, this information has been converted to traffic cost per bus stop, from which the service punctuality could be determined. The model's ability to evaluate the punctuality of operation made it possible to transmit headway control instructions (when to decelerate or accelerate) to a bus driver via an onboard unit. Such an approach allows transmission of headway adjustment instructions, which could reduce traffic cost per bus stop by aiming at key bottleneck bus stops—where punctuality issues have been experienced. A model verification process was implemented by using data collected from a bus management system and integrated transit fare card system for the bus route in Seoul. To evaluate the reliability and uncertainty of the optimal solution, a sensitivity analysis was implemented for the various parameters and assumptions used in the models.

Ceder et al. (2013a) describe a methodology to approach even-headway and even-load timetables by utilizing different available bus sizes. The composition and size of the fleet of buses can be either given or with a flexibility of introducing new sizes. Two objectives were set. First, minimize the deviation of the determined headways from a desired even headway. Second, minimize the deviation of the observed passenger loads from a desired even-load level of the buses at the maximum-load point. Two heuristics have been developed to attain these objectives: (1) determination of the best headway with respect to minimizing the load discrepancy from an even-load level on all buses at the max load point and (2) construction of the best timetable to approach planned even headway and even load at the max load point, with the selected headway of (1) and based on a range of alternative timetables represented in a Pareto frontier. The methodology has been applied to a case study in Auckland, New Zealand. Overall, the algorithms developed provide significantly better results than the commonly used timetable. The case study, with even headway, shows that the approach of this work can reduce the observed 38% load discrepancy, from a desired load at the max load point to a discrepancy between only 0%–15% when utilizing properly available sizes of buses. This result is attained together with only between 0% and 7% relative time deviation of departure times from those of even-headway timetable.

Ceder et al. (2013b) also present the creation of bus timetables using different vehicle sizes, but with the use of simulation. This work attempts to bridge between the timetables that are designed with even headways and timetables with even passenger loads on the vehicles at the maximum-load points, but with uneven headways. The analysis contains both of these strategies through the use of the incorporation of a mixed fleet size, running in conjunction. The timetables have been constructed using two key concepts: assigning capacity and shifting departure times. The methodology for the creation of timetables was applied to a real-life example from Auckland, New Zealand. Several instances of timetables have been created and tested using simulation. The simulation showed encouraging results for all cases. Comparing the proposed timetable to a currently operated one, the waiting time of passengers could be reduced by 15%. At the same time, the number of empty-seat hours can be significantly improved by 47%, while the number of standee hours did not increase dramatically.

Table 3.8 summarizes by different categories the main features and characteristics of the literature reviewed. Although we are finished with the four methods described in

Table 3.8 Characteristics of the literature covered in Section 3.6

Source	Demand features	Constraints	Other features	Variables in the frequency formula derived
Furth and Wilson (1981)	Elastic frequency, but without interline influence; multiperiod	Subsidy, fleet size, vehicle-loading level		Fares, subsidy, fleet size, value of waiting time.
Koutsopoulos et al. (1985)	Multiperiod	Subsidy, fleet size, vehicle capacity	Waiting-time submodel, passenger inconvenience submodel	
LeBlanc (1988)			Multimodal considerations	
Banks (1990)	Elastic/fixed cases examined	Subsidy, fleet size, vehicle capacity (several cases examined, including unconstrained cases)		Operating cost, travel time, value of waiting time, demand level (in the unconstrained case).
Khasnabis and Rudraraju (1997)			Emphasis on the need-to-consider signal preemption strategies	
Wirasinghe (2003)	Among the cases discussed: stochastic demand, one-to-many demand, many-to-many demand		Alternative waiting-time assumptions	Operating cost, value of waiting time, passenger-arrival rate.
Kim et al. (2009)	Microscopic demand	Subsidy, fleet size, bus-occupancy levels		Bus operation costs to the service. Supplier and passenger demand.
Ceder et al. (2013a)	Based on ride-check data	Number of buses of each size	Heuristic algorithm; sensitivity analysis	Multisize vehicles.
Ceder et al. (2013b)	Boarding and alighting data based on smart cards	Number of buses of each size	Use of simulation	Timetable incorporated the use of a mixed fleet size.

this chapter, we should not count the chickens before they are hatched. That is, we still need to check whether the results fit the other parts of the operational planning process. Consequently, following the calculation of frequencies and headways, we will proceed to the next chapter, which deals with the construction of alternative public timetables.

EXERCISES

3.1 Given a bus route with 6 stops and all buses on the route having 50 seats and the same total capacity of 80 passengers. The following table provides average passenger-load counts in each of the 5 h between 06:00 and 11:00, the number of buses observed, desired occupancy (passengers), and minimum frequency (bus/h):

Stop #	Distance to next stop (km)	Time period				
		06:00–07:00	07:00–08:00	08:00–09:00	09:00–10:00	10:00–11:00
1	2.2	72	161	65	182	138
2	0.8	90	328	199	318	193
3	1.4	85	468	365	300	222
4	0.6	68	397	388	212	166
5	1	54	286	140	147	84
Number of observed buses		2	6	5	4	3
Desired occupancy (passengers)		50	60	60	50	50
Minimum frequency		2	2	2	2	2

Construct a load profile for each hour and calculate its area; determine the future data collection technique proposed for each hour. Explain:

(a) Calculate the frequency and headway for each hour with each of the four methods, Method 4 being associated with a maximum of 20% of the route length having an excess load.

(b) Compare the results of the four methods, and draw conclusions.

(c) What is the trade-off between excess load (in passenger-km) using Methods 3 and 4 (20%) and the reduced empty space-km for each hour?

(d) If the cost of a single empty space-km is 3¢/km (operations cost), what is the range of a reduced single tariff per km for route segments suffering excess load that yield a total saving (cost associated with reduced empty space-km) larger than the total reduced (farebox) income?

3.2 Given a transit route and 2 h data with a bus capacity of 80 passengers. The basic ride-check data are shown in the following:

Stop #	Distance to next stop (km)	Period and departure times			
		06:00–07:00		07:00–08:00	
		06:15	06:45	07:10	07:40
1	2	22	25	50	60
2	3	52	40	60	75
3	3	35	65	80	45
Desired occupancy (passengers)		40		50	
Minimum frequency (veh/h)		2		3	

(a) Determine the hourly max loads and max load points.
(b) Determine the frequency for each hour, using Method 2.
(c) Determine the frequency using Method 4 (40%), in which a maximum of 40% of the route length has excess load.

REFERENCES

Banks, J.H. (1990). Optimal headways for multi-route transit systems. *Journal of Advanced Transportation*, **24**, 127–154.

Ceder, A. (1984). Bus frequency determination using passenger count data. *Transportation Research*, **18A**(5/6), 439–453.

Ceder, A. and Dressler, O. (1980). A note on the χ^2 test with applications and results of road accidents in construction zones. *Accident Analysis and Prevention*, **12**, 7–10.

Ceder, A., Hassold, S., and Dano, B. (2013a). Approaching even-load and even-headway transit time-tables using different bus sizes. *Public Transport: Planning and Operation*, **5**(3), 193–217.

Ceder, A., Hassold, S., Dunlop, C., and Chen, I. (2013b). Improving urban public-transport service using new timetabling strategies with different vehicle types. *International Journal of Urban Sciences*, **17**(2), 239–258.

Furth, P.G. and Wilson, N.H.M. (1981). Setting frequencies on bus routes: Theory and practice. *Transportation Research Record*, **818**, 1–7.

Khasnabis, S. and Rudraraju, R.K. (1997). Optimum bus headway for preemption: A simulation approach. *Transportation Research Record*, **1603**, 128–136.

Kim, W., Son, B., Chung, J.H., and Kim, E. (2009). Development of real-time optimal bus scheduling and headway control models. *Transportation Research Record*, **2111**, 33–41.

Koutsopoulos, H.N., Amedeo, R.O., and Wilson, N.H.M. (1985). Determination of headways as a function of time varying characteristics on a transit network. In *Computer Scheduling of Public Transport 2* (J.M. Rousseau, ed.), pp. 391–413, Elsevier Science, Amsterdam, the Netherlands.

LeBlanc, L.J. (1988). Transit system network design. *Transportation Research*, **22B**, 383–390.

LSU Experimental Statistics (2015). Department of Experimental Statistics, Louisiana State University, USA. http://www.stat.lsu.edu/exstweb/statlab/Tables/TABLES98-Chi.html, March 22, 2015.

Wirasinghe, S.C. (2003). Initial planning for urban transit systems. In *Advanced Modeling for Transit Operations and Service Planning* (W.H.K. Lam and M.G.H. Bell, eds.), pp. 1–29, Elsevier Science, Oxford, U.K.

Chapter 4

Timetable development

Science is organized receipts: wisdom is organized time

Even headway

Even load

A.C.

CHAPTER OUTLINE

PRACTITIONER'S CORNER

The aim of this chapter is to create and present approaches and procedures for deriving prudent public timetables. These timetables and their compliance mirror the quality of the transit service provided. Hence, vehicles departing too early or ahead of schedule need to be restrained, just as those leaving late must be scheduled or rescheduled to be on time. Because of existing problems of transit reliability, there is a need to improve the correspondence of vehicle departure times with passenger demand instead of assuming that passengers will adjust themselves to given timetables (excluding situations characterized by short headways). The following riddle may shed some light on the substance of adequate, and accurate, timetables.

Two identical shopping malls, M and K, are joined by a road connecting them. A bus departs every 10 min (headway) from each of the malls to the other mall (M to K and K to M); the travel time in each direction is the same. A bus stop is located at exactly the midpoint between the two malls. Passengers arrive randomly at this bus stop and take the first bus to arrive, regardless of the mall to which it is headed. Observations at this bus stop reveal that 9 out of every 10 passengers board the bus that goes from M to K. Is this possible? If so, how? (The answer appears at the end of this Practitioner's corner.)

The chapter starts with an explication of the advantage of having optional or alternative public timetables. This feature gives rise to improved service quality and resource saving, using derived comparison measures that constitute a basis for comparing the alternative timetables. A procedure for calculating departure times is provided, based on even headways and a smooth transition between time periods (usually hours), which allows for even average loads on the last and first vehicles in each period. Thereafter, another procedure is presented for calculating departure times based on even average passenger loads (as required by the standard), rather than on even headways, at the hourly max load point. The chapter considers four frequency-determination methods and gives examples (one of them is real life) used in Chapter 3, along with possible practical decisions on headways and number of departures. The chapter ends with a literature review and practical exercises.

Practitioners are encouraged to read the entire chapter, but to concentrate on the examples and their figures while studying the two procedures, rather than attempting to fully comprehend the methodological arguments. The overall analysis of the optional timetables also contains a direct planner/scheduler intervention in setting frequencies. Such an intervention is required in situations known only to the planner and does not rely on passenger loads.

We can now return to the riddle. Indeed, the situation described can be realized if the timetables from M to K and K to M are set to differ by 1 min difference. Specifically, the bus departs from M to K at 7:00, 7:10, 7:20, and so on, whereas the bus leaves from K to M at 7:01, 7:11, 7:21, etc. Ostensibly, the average waiting time for the M to K bus is nine times that for the opposite direction; proportionally, therefore, more passengers will be on hand for the M to K bus. This example shows how important it is to set timetables adequately. Situations similar to that of the riddle, but for a network of routes, undoubtedly imply an unreliable transit service. Improvement in creating timetables can offer a remedy.

4.1 INTRODUCTION

Public transit timetables constitute the most profound bridge between the transit agency (and/or the community) and the passenger seeking a reliable service. Inadequate and/or inaccurate timetables not only confuse the passengers but also reinforce the bad image of public transit as a whole. There is a saying: "A stitch in time would have confused Einstein." Along these lines, one can say that many stitches in a transit timetable would confuse those (the passengers) who want to use the service.

The assumption that passengers will adjust themselves to a given timetable (with headways of, say, longer than 10 min), instead of planners' adjusting the timetable to passenger demand, constitutes a major source of unreliable service. When passenger demand is not met, transit vehicles slow down (i.e., increase dwell time), traveling behind schedule and entering the inevitable process of further slowing down. This situation will eventually lead to the known bunching phenomenon with the vehicles that follow. In contrast, a situation of overestimating demand may result in transit vehicles running ahead of time. Neither situation is observed when the service is frequent and characterized by a low variance in headway distribution.

The products of the derived frequencies and headways given in Chapter 3 yield the timetables for the public, for the drivers, and for the supervisors and inspectors. Once the timetables are constructed, as we see in Figure 1.2, it is possible to initiate the task of scheduling vehicles and crews for the previously determined trips. Naturally, the transit agency wishes to utilize its resources more efficiently by minimizing the number of required vehicles and crew costs. To accomplish this, the planner (or an automated procedure) examines optional timetables during the vehicle and crew assignment processes. Optional timetables are derived by shifting departure times and/or reducing the number of departures without prudent consideration of the load profile. Often, because of the uncertainty involved, some planners/schedulers prefer to shift departure times in small increments instead of adjusting them to the data-based demand–supply point. This can follow the rule that if you put aside a small amount for savings every day, you will be surprised to learn how little was accumulated in a year. All in all, it is desirable to extend the analysis of deriving appropriate headways to an evaluation of optional timetables in conjunction with the required resources.

This chapter, and the next one as well, will continue with an explication of the second planning activity in Figure 1.2, timetable development. The basic challenge is how to improve the correspondence of vehicle departure times with fluctuating passenger demand. It is known that passenger demand varies even within the space of 1 h. This dynamic behavior can be detected through passenger-load counts and information provided by road inspectors. A more balanced load timetable is achievable by adjusting departure times. The two procedures presented in this chapter, and one in the next, can be applied to both single and interlining transit routes, and they can be carried out in an automated manner. While the first procedure yields departure times (a timetable) for vehicles with even scheduled headways, the second procedure produces departure times for vehicles having even average loads at the hourly max load point. A third procedure, discussed in Chapter 5, refers to balanced (even) loads on individual vehicles at their (individual) peak-load segments. The key point here is to be able to control the loading, instead of repeatedly exposing passengers to unreliable service resulting from an imbalance in loading situations.

This chapter consists of three major parts. First, a spectrum of optional timetables is exhibited, along with utilization measures, thus enabling a comparison of various timetables.

Second, a procedure is presented for constructing departure times in which the headways are evenly spaced. In this context, smoothing techniques are developed in the transition segments between adjacent time periods, usually hourly segments. Third, another procedure is offered, this one for a case in which the headways are allowed to be unevenly spaced; here, the departure times are shifted so as to obtain uniform average loads at the hourly max load point, instead of even headways. All the procedures are applied on a route basis; they are similar, corrected, and improved over those presented earlier by Ceder (1986, 2003). A literature review and exercises follow these three parts.

4.2 OBJECTIVES, OPTIONAL TIMETABLES, AND COMPARISON MEASURES

A cost-effective and efficient transit timetable embodies a compromise between passenger comfort and service cost. A good match between vehicle supply and passenger demand occurs when transit schedules are constructed so that the observed passenger demand is accommodated while the number of vehicles used is minimized. This approach helps in minimizing the agency cost in terms of drivers' wages and capital costs required to purchase vehicles. This cost-effective concept led us to establish five objectives in creating public transit timetables:

1. Evaluate optional timetables in terms of required resources.
2. Improve the correspondence of vehicle departure times with passenger demand while minimizing resources.
3. Permit, in the timetable construction procedure, direct vehicle-frequency changes for possible exceptions (known to the planner/scheduler) that do not rely on passenger-demand data.
4. Allow the construction of timetables with headway-smoothing techniques (similar to those performed manually) in the transition segments between adjacent time periods.
5. Integrate different frequency-setting methods and different timetable construction procedures.

This chapter attempts to show how to fulfill these five objectives. The next chapter continues this attempt.

4.2.1 Elements in practice

Different transit agencies use different scheduling strategies based on their own experience. Consequently, it is unlikely that two independent transit agencies will use exactly the same scheduling procedures, at least at the level of detail. In addition, even within the same transit agency, planners may use different scheduling procedures for different groups of routes. Therefore, we believe that when developing computerized procedures, there is need to supply the schedulers with schedule options, along with an interpretation and explanation of each option. Undoubtedly, it is desirable that one of the options should coincide with the scheduler's manual/semimanual procedure. In this way, the scheduler will be in a position not only to expedite manual/semimanual tasks but also to compare procedures and methods in regard to the trade-off between the passengers' comfort and the operating cost.

The number of vehicle runs determined by the timetable and eventually the number of vehicles required are sensitive to the procedure that the scheduler uses to construct the departure times. Some transit agencies routinely round off the frequency F_j to the next highest integer and then calculate the appropriate headways for the time period. By doing so, they increase the number of daily departures beyond what is needed to appropriately match demand with supply. Such a procedure may result in nonproductive runs (many empty seat-km). For example, Table 3.7, in Chapter 3, lists the number of required daily departures $\sum_{j=1}^{19} F_j$ for Los Angeles County Metropolitan Transportation Authority (LACMTA) route 217: 86.52 and 78.11 for Methods 2 and 4 (20% case), respectively. When the quantity F_j is *rounded up*, we obtain, respectively, 92 and 85 daily departures for these two methods. Obviously, by rounding off F_j to the next highest integer, the level of passenger comfort is increased but at the expense of unnecessary operating costs. However, in some cases, the *rounding-off* procedure may be justified if P_{ij} (see Equation 3.2) is used as an average load when the load variance is high. In this case (provided that additional runs are made by rounding off F_j upward), possible overcrowding situations may be reduced as opposed to increasing average empty seat-km. Nonetheless, to overcome the problem of highly variable loads, we can use a statistical load measure that considers its variance as input to a frequency method; this was explained under Equation 3.2.

4.2.2 Optional timetables

The five objectives enumerated in the previous text and current timetable construction procedures provide the basis to establish a spectrum of optional timetables. Three categories of options may be identified: (1) type of timetable, (2) method or combination of methods for setting frequencies, and (3) special requests. These three groups of options are illustrated in Figure 4.1. A selected path in this figure provides a single timetable. Hence, there are a variety of timetable options.

The first category in Figure 4.1 concerns alternative types of timetables. The even-headway type simply means constant time intervals between adjacent departures in each time period, or the case of evenly spaced headways. Even average load refers to unevenly spaced headways in each time period, but the observed passenger loads at the hourly max load point are similar on all vehicles. A second type of timetables entails situations in which even headways will result in significantly uneven loads. Such uneven-load circumstances occur, for example, around work and school dismissal times, but they may in fact occur on many other occasions. Figure 4.1 shows that the average even load can be managed either at the hourly max load point (even loads on all vehicles at that point) or at each individual vehicle's max load point. The average even load at the hourly max load point type is discussed in this chapter, and the other type in the next chapter.

In the second category of options, it is possible to select different frequency or headway-setting methods. This category allows for the selection of one method or for combinations of methods for different time periods. The methods considered, and indicated in Figure 4.1, are the two-point check and the two-ride check, both described in Chapter 3. In addition, there might be procedures used by the planner/scheduler that are not based on data but on observations made by road supervisors and inspectors or other sources of information.

In the third category of selections, we allow for special scheduling requests. One characteristic of existing timetables is the repetition of departure times, usually every hour. These easy-to-memorize departure times are based on so-called clock headways: 6, 7.5, 10, 12, 15, 20, 30, 40, 45, and 60 min. Ostensibly, headways less than or equal to 5 min are not thought to influence the timing of passenger arrivals at a transit stop. The clock headway is obtained by rounding the derived headway down to the nearest of these *clock* values. Consequently,

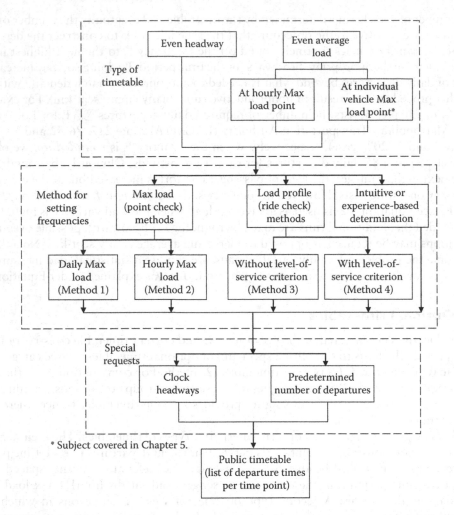

Figure 4.1 Optional public timetables.

similar to the *rounding off upward* of frequencies, clock headways require a higher number of departures than what is actually necessary to meet the demand.

A second possible special request is to allow the scheduler to predetermine the total number of vehicle departures during any time period. This request is most useful in crises, when the agency needs to supply a working timetable for an operation based on tightly limited resources (vehicles and/or crews). By controlling the total number of departures while complying with other requests, the scheduler achieves better results than by simply dropping departures without any systematic procedure. Furthermore, there might be cases in which the agency would like to increase the level of service by allowing more departures in the belief that passenger demand can be increased by providing improved (more frequent) service. Certainly, this special request can also be approached through varying the desired occupancy (load factor) standard; however, this option can be a compulsory standard.

Finally, it should be emphasized that not all the paths concerning clock headways in Figure 4.1 are meaningful. The selection of the even average load type of timetable cannot be performed if there is a clock-headway constraint. Moreover, the number of departures cannot be predetermined for clock headways because of the specific time restrictions on those headways.

4.2.3 Comparison measures

With computerized timetable construction, the transit agency can assess a variety of optional timetables rather than being limited to examining one or a few. Two interrelated measures may be useful for the agency to compare optional timetables: (1) number of required runs (departures) and (2) required single-route fleet size.

The first comparison measure, total number of departures, can serve as an indicator of the number of vehicles required and also whether or not it is possible to save vehicle runs.

The second comparison measure refers to each route separately and provides an estimate of the required fleet size at the route level. In a large transit agency, an efficient arrangement of vehicle blocks includes interlining (switching a vehicle from one route to another) and dead-heading trips. Hence, fleet size is not determined at the route level but at the network level (see Chapters 7 and 8). The second comparison measure, however, is based on a simple formula derived by Salzborn (1972) for a continuous time function and explicitly shown by Ceder (1984) for discrete time points. This formula states that if T is the round-trip time, including layover and turnaround time, then the minimum fleet size is the largest number of vehicles departing at any time interval during T. This value is adequate for a single route with a coinciding departure and arrival location. Consequently, the second comparison measure can be used for each direction separately as well as for both directions when selecting the maximum of two derived values. This single-route fleet size formula will be elaborated in Chapter 7.

4.2.4 Anchoring the timetable to a single time point

A public timetable usually consists of lists of vehicle departure times at all route stops or at selected stops called time points. Occasionally, this public timetable is given at just a single point—the route departure point or a major boarding stop. The running time between adjacent time points may vary from one time period to another, based on ride-check information. In essence, the timetable can initially be constructed at only one point, and be referred to as such, and then extended forward and backward using the average running-time information. That is, in order to ensure an appropriate transit service to meet the variations in passenger demand, it suffices to construct the timetable at one point. This observation is stated in the following proposition:

Proposition:

For a route-based transit timetable consisting of more than one time point, the association of max observed load in each time period with only a single time point ensures that the average vehicle load on each route segment is less than or equal to the desired occupancy.

Proof and interpretation:

For simplicity sake, we refer to the single time point as the route departure point. The derived frequency is greater than or equal to the maximum required frequency (across all route segments) in each time period. The proof for frequencies determined by Method 2 was provided in Chapter 3; however, this can easily be converted to, and shown by, any of the four methods. As a note, when only the daily max load point is considered, this does not necessarily imply that the observed max load in each period occurs at that point. We can then treat the problem similarly to the 3D time–load–distance representation in Figure 4.2, based on Example 2 (which appears in Figure 3.7 in Chapter 3). That is, the shaded two-part area in Figure 4.2 describes the max loads of Example 2 in each hour: 134 (between 06:00

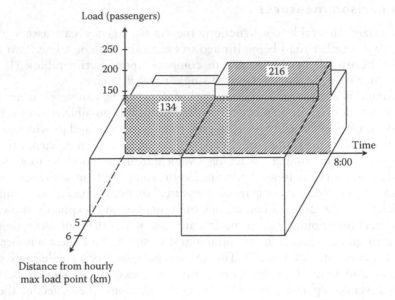

Figure 4.2 Three-dimensional illustration of Example 2 load profiles positioned relative to the distance from the hourly max load point.

and 07:00) and 216 passengers (07:00–08:00). Calculation of the frequency at distance = 0 in Figure 4.2 ensures compliance with Method 2.

Since Example 2 is used throughout this chapter, Table 4.1 presents its entire input data required for constructing optional timetables. Table 4.1 also contains information on (1) T, the time required for the same vehicle, at the same departure point, to pick up another departure and (2) the calculated data on an individual vehicle basis. For the sake of clarity, Figure 4.3 shows five time–space vehicle trajectories of Example 2. Because of possible different average running times in each period, the headways at each stop do not necessarily coincide as seen in Figure 4.3. However, since the hourly max load is used for calculating the frequency, the timetable, set only at one time point, reflects the maximum number of required vehicles at the observed hourly max load point. There is, then, a question of whether the determined frequency at the observed hourly max load point complies with the desired occupancy constraint.

Table 4.1 Part of the Example 2 (Figure 3.7) data and calculated complementary data required for illustrating the optional timetable procedure

Time period	Departure time	Average load (passengers) at the max load point, by method number		Area under the load profile (passenger-km)	Frequency (veh/h) and headway (min, in parenthesis), by method number		
		I	*2*		*I*	*2*	*4 (20%)*
6:00–7:00	06:00[a]	—	—	929	2.56 (23)	2.68 (22)	2.68 (22)
	06:15	52	35				
	06:50	40	65				
7:00–8:00	07:15	90	90	1395	3.60 (17)	3.60 (17)	2.95 (20)
	07:35	87	87				
	07:50	75	75				

Added data: T (round-trip time + layover and turnaround time) = 60 min.

[a] First predetermined trip.

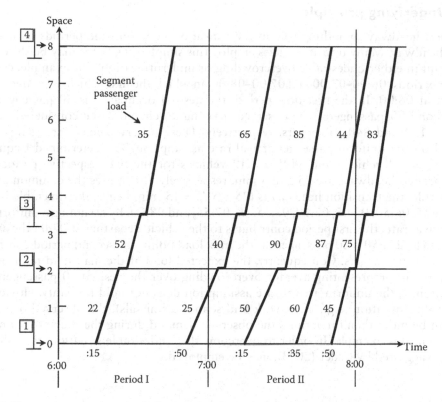

Figure 4.3 Time–space trajectories of all vehicles in Example 2, including segment loads.

It is important to note that the running-time information must rely on the fact that in an average sense, vehicles do not overtake each other. Average running times should be determined not only from ride-check data but also from the requirement that the first vehicle to depart cannot be the second to arrive at any time point. Thus, the time–space trajectories in Figure 4.3 cannot cross each other in an average (deterministic) context.

Based on possible differences in actual average headways at each stop, the associated time span at the observed hourly max load point, covering all trips across the time period at the route's departure point, can be shorter or longer than the time span of the route's departure point. For example, we see in Figure 4.3 that the time span of the last three departures, 35 min (7:15–7:50), is larger than the time span of these trips at the hourly max load point of period II, stop 3. Fortunately in either case, what governs the frequency calculation is the observed max load. Hence, by construction, the result is that as long as the time–space trajectories do not cross one another, the requested frequency at each stop must be less than or equal to the frequency determined at the route's departure point.

Finally, it may be noted that the proposition in the previous text can also be applied to Methods 1, 3, and 4—each with its own loading standards and constraints.

4.3 EVEN HEADWAYS WITH SMOOTH TRANSITIONS

One characteristic of existing timetables is the repetition of the same headway in each time period. However, a problem facing the scheduler in creating these timetables is how to set departure times in the transition segments between adjacent time periods. This section addresses the issue.

4.3.1 Underlying principle

A common headway-smoothing rule in the transition between time periods is to use an average headway. Many transit agencies employ this simple rule, but it may be shown that it can result in either undesirable overcrowding or underutilization. For example, consider two time periods, 06:00–07:00 and 07:00–08:00, in which the first vehicle is predetermined to depart at 06:00. In the first time period, the desired occupancy is 50 passengers and in the second 70 passengers. The observed maximum demand to be considered in these periods is 120 and 840 passengers, respectively. These observed loads at a single point are based on the uniform passenger-arrival-rate assumption. The determined frequencies are 120/50 = 2.4 vehicles and 840/70 = 12 vehicles for the two respective periods, and their associated headways are 25 and 5 min, respectively. If one uses the common average headway rule, the transition headway is (25 + 5)/2 = 15 min; hence, the timetable is set to 06:00, 06:25, 06:50, 07:05, 07:10, 07:15, ..., 07:55, and 08:00. By assuming a uniform passenger arrival rate, the first period contributes to the vehicle departing at 07:05 the average amount of $(10/25) \cdot 50 = 20$ passengers at the max load point; the second period contributes $(5/5) \cdot 70 = 70$ passengers. Consequently, the expected load at the max load point is 20 + 70 = 90, a figure representing average overcrowding over the desired 70 passengers after 7:00. Certainly, the uniform arrival-rate assumption does not hold in reality. However, in some real-life situation (e.g., after work and school dismissals), the observed demand in 5 min can be more than three times the observed demand during the previous 10 min, as is the case in this example. In order to overcome this undesirable situation, the following principle, suggested by Ceder (2003), may be employed.

Principle 4.1:

Establish a curve representing the cumulative (noninteger) frequency determined versus time. Move horizontally for each departure until intersecting the cumulative curve, and then vertically; this will result in the required departure time.

Proposition 4.1:

Principle 4.1 provides the required evenly spaced headways with a transition load approaching the *average desired occupancies* of d_{oj} and $d_{o(j+1)}$ for two consecutive time periods, j and j + 1.

Proof: Figure 4.4 illustrates Principle 4.1 using the frequency results from Example 2 (see Figure 3.7) appearing in Table 3.4 (Chapter 3). Since the slopes of the lines are 2.68 and 3.60 for j = 1 and j = 2, respectively, the resultant headways are those required. The transition load is the load associated with the 7:05 departure, which consists of arriving passengers during 16 min for j = 1 and of arriving passengers during 5 min for j = 2. Therefore, $(16/22) \cdot 50 + (5/17) \cdot 60 = 54$ approximately. This transition load is not the exact average between $d_{o1} = 50$ and $d_{o2} = 60$, since departures are made in integer minutes. That is, the exact determined departure after 7:00 is $(3 - 2.68) \cdot 60/3.60 = 5.33$ min. Inserting this value, instead of the 5 min mentioned earlier, yields a value that is closer to the exact average. Basically, the proportions considered satisfy the proof by construction of Proposition 4.1.

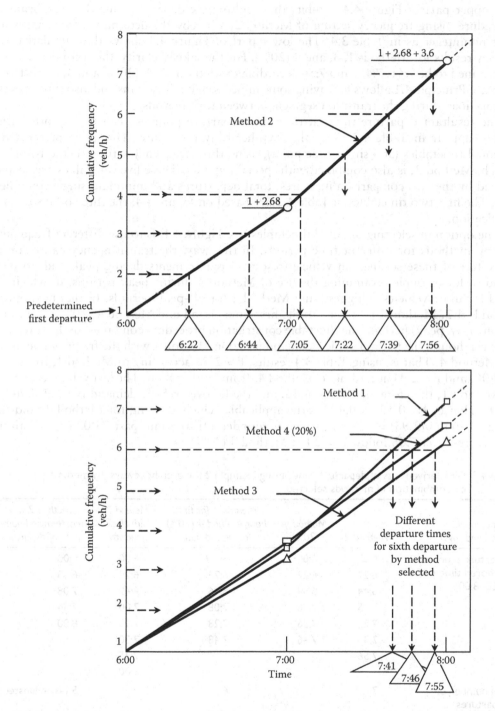

Figure 4.4 Determination of departure times for evenly spaced headways and for all methods in Example 2, including a smoothing process between periods.

4.3.2 Interpretation using Example 2

The upper part of Figure 4.4 exhibits the resulting six departures for the even-headway procedure, using frequency setting of Method 2, whereby the determined frequencies are kept noninteger as in Table 3.4. The lower part of Figure 4.4 shows the cumulative frequency results for Methods 1, 3, and 4 (20%). For the sake of clarity, the sixth departure is determined to be 7:41, 7:46, and 7:56 according to Methods 1, 4, (20%) and 3, respectively. Overall, Principle 4.1 allows for saving some unnecessary vehicle runs and also stabilizes the average load during the transition segment between time periods.

The resultant departure times at the route departure point—it can be any other time point—appear in Table 4.2 using the even-headway procedure. This table presents five optional timetables (at a single time point), using three frequency-setting methods, one of which, Method 2, is also combined with special requests. These five timetables are accompanied by the two comparison measures: total departures and minimum single-route fleet size. The first two timetables in Table 4.2 are based on Figure 4.4; the three others deserve explication.

One option in selecting optional timetables (see Figure 4.1) is to use different frequency-setting methods for different time periods. In this way, the transit agency can examine the effect of these settings on vehicle (resource) requirements during peak and off-peak periods, for example, examining the use of Method 4 during peak periods, in which the need for more vehicles is highest, and Method 2 for off-peak periods. In the third option in Table 4.2, Method 1 is applied to the first time period and Method 4 (20% case) to the second (say, peak) period. The timetable construction procedure combines, in the transition between hours, the leftover vehicle demand utilizing Method 1 with the frequency required by Method 4. That is, using Table 3.4 results, F = 2.56 according to Method 1, from 6:00 to 7:00, and F = 2.95 according to Method 4, from 7:00 to 8:00. The last vehicle according to Method 1, therefore, departs at 6:46, and the leftover vehicle demand is 14 (2.56/60) = 0.597. Hence, 1 − 0.597 = 0.403 is the applicable vehicle demand for Method 4, and this yields 0.403(60/2.95) = 8.2 min, which is rounded off to 8 min past 7:00. The remaining departures after 7:08 follow H = 20 of Method 4 (20%).

Table 4.2 Five derived sets of departure times using Example 2 for even headways, by method or combination of methods selected

Timetable characteristics (even headways)	Method 2	Method 4 (20%)	Method 1 for first hour, Method 4 (20%) for second hour	Method 2 with clock headway	Method 2 with predetermined number (=5) of departures
Departure times at anchored time point (6:00–8:00)	6:00[a]	6:00[a]	6:00[a]	6:00[a]	6:00[a]
	6:22	6:22	6:23	6:20	6:35
	6:44	6:44	6:46	6:40	7:08
	7:05	7:06	7:08	7:00	7:34
	7:22	7:26	7:28	7:15	8:00
	7:39	7:46	7:48	7:30	
	7:56			7:45	
				8:00	
Total number of departures	7	6	6	8	5 (as requested)
Minimum single-route fleet size (T = 60 min)	4	3	3	4	3

[a] First predetermined trip.

The fourth timetable option in Table 4.2 is based on Method 2 with clock headways. In this case, we simply round down the headway determined by Method 2 to its nearest clock headway. Thus, from 6:00 to 7:00, H = 22 (see Table 3.4), which is rounded to 20, and H = 17, from 7:00 to 8:00, which is rounded down to 15.

The last timetable option is based on five predetermined departures (including the first 6:00 departure) while using Method 2. The total number of required vehicles using Method 2 (excluding 6:00) is 2.68 + 3.60 = 6.28. Since only four departures have to be constructed, the frequencies are modified proportionally by the ratio 4/6.28 = 0.637. The procedure continues the same way as that without the special request. That is, F = 1.71 (= 2.68·0.637) with H = 35 for 6:00–7:00, and F = 2.29 with H = 26 for 7:00–8:00. Certainly, if the demand remains same, the scheduler should recognize the potential risk of overcrowding when restricting the total number of departures. Nevertheless, the purpose of this special request is to have a systematic computerized procedure to manage both crisis situations (limited resources) and situations in which additional passenger demand can possibly be attracted (i.e., requesting more departures than calculated).

This explication of Table 4.2 ends with the derivation of the single-route fleet size using the round-trip time information of T = 60 min. The derivation of this comparison measure will be illustrated for the first option in Table 4.2 (Method 2). Adding 60 min to the first 6:00 departure reveals that the vehicle associated with this departure can perform the 7:05 departure. Hence, three vehicles are required for this step (to perform 6:00, 6:22, and 6:44 departures). By continuing to add 60 to 6:22, this will enable the vehicle to pick up the 7:22 departure (again, three vehicles are required). Adding 60 to 6:44 shows that the next pickup is executed by the 7:56 departure, resulting in four independent departures (6:44, 7:05, 7:22, and 7:39). Continuing to count the number of departures in each 60 min window leads to finding the maximum number, which is the minimum single-route fleet size required. For the first option in Table 4.2, this number ostensibly is four.

4.4 HEADWAYS WITH EVEN AVERAGE LOADS

This section opens with the following premise: transit managers/planners/schedulers who believe that problems related to attracting more transit users and reliability problems are drowned in the *ocean* of even-headway timetables should be told that these problems know how to swim. In other words, even-headway timetables do not necessarily deliver the merchandise (satisfactory transit service) to the customer (passengers).

We have already noted that passenger demand varies even within a single time period, hence resulting for even headways in an imbalanced load on individual vehicles at the hourly max load point. On heavy-load routes and short headways, the even-headway timetable suffices. However, in the course of reducing reliability problems, we may occasionally prefer to use the even-load instead of the even-headway procedure. Moreover, the availability of automatic passenger counts provides a framework in which to investigate systematically the variation in passenger demand. With the anticipated vast amount of passenger-load data, we can then better match vehicle departure times with variable demand. Two procedures carry out this endeavor. The first, addressed in this section, deals with even average load on individual vehicles at the hourly (or daily) max load point. The second procedure, addressed in the next chapter, ensures an even average load at each individual vehicle max load point. In this section, the procedures described by Ceder (1986, 2003) will be corrected and improved.

4.4.1 Underlying principle

A simple example is presented here to illustrate the underlying load-balancing problem. Consider an evenly spaced headway timetable in which vehicles depart every 20 min between 07:00 and 08:00, that is, at 07:20, 07:40, and 08:00. The observed load data consistently show that the second vehicle, which departs at 07:40, has significantly more passengers than the third vehicle. The observed (average) max load during this 60 min period is 150 passengers, and the desired occupancy is 50 passengers. Hence, using Method 2, three vehicles are required to serve the demand as in the case of the evenly spaced headways timetable. The average observed loads at the hourly max load point on the three vehicles are 50, 70, and 30 passengers, respectively. Given that these average loads are consistent, then the transit agency can adjust the departure times so that each vehicle has a balanced load of 50 passengers on the average at the hourly max load point. The assumption of a uniform passenger-arrival rate results in $70/20 = 3.5$ passengers/min between 07:20 and 07:40 and $30/20 = 1.5$ passengers/min between 07:40 and 08:00. If the departure time of the second vehicle is shifted by X min backward (i.e., an early departure), then the equation $3.5X = 70-50$ yields the balanced schedule, with $X = 5.7 \approx 6$ min, or departures at 07:20, 07:34, and 08:00. The third departure will add this difference of 20 passengers at the hourly max load point. The even-headway setting assures enough vehicles to accommodate the hourly demand, but it cannot guarantee balanced loads for each vehicle at the peak point. In order to avoid this imbalanced situation, the following principle should be exploited.

Principle 4.2:

Construct a curve representing the cumulative loads observed on individual vehicles at the hourly max load points. Move horizontally per each d_{oj} for all j, until intersecting the cumulative-load curve, and then vertically; this results in the required departure times.

Proposition 4.2:

Principle 4.2 results in departure times such that the average max load on individual vehicles at the hourly jth max load point approaches the desired occupancy d_{oj}.

Proof: Figure 4.5 illustrates Principle 4.2 for the Example 2 problem appearing in Figure 3.7 and Table 4.1. Method 2 will be used in the upper part of Figure 4.5 in which the derived departure times are unevenly spaced to obtain even loads at stop 3 for j = 1 and at stop 2 for j = 2. These even loads are constructed on the cumulative curve to approach $d_{o1} = 50$ and $d_{o2} = 60$. If we assume a uniform passenger-arrival rate between each two observed departures, it can be shown that the load (at stop 3) of the first derived departure (6:23) consists of the arrival rate between 6:00 and 6:15 ($35/15 = 2.33$) and the rate between 6:15 and 6:50 ($65/35 = 1.86$). Thus, $2.33 \cdot 15 + 1.86 \cdot 8 \approx 50$. In the transition between j = 1 and j = 2 (in the upper part of Figure 4.5), the value of $d_2 = 60$ is considered, since the resultant departure comes after 7:00. The load of the vehicle departing at 7:07 at its hourly max load point, stop 2, is simply $17 \cdot (90/25) = 61.2$ from rounding off the departure time to the nearest integer. That is, $(10 + y) \cdot (90/25) = 60$ results in y = 6.67 min. This completes the proof by construction of Proposition 4.2.

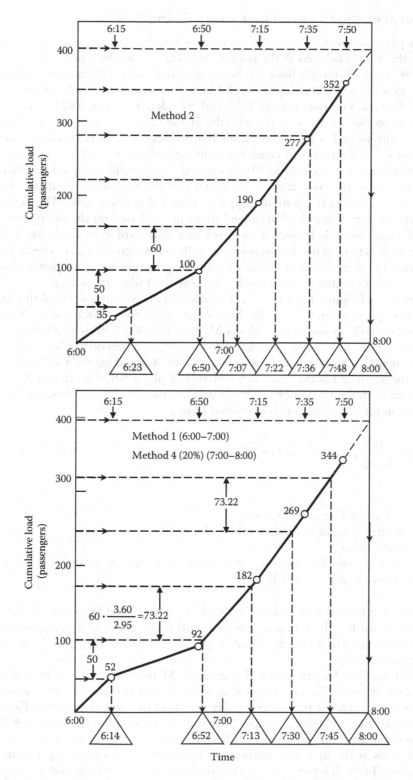

Figure 4.5 Determination of Example 2 departure times with even load at the hourly max load point using Method 2 and at the daily max load point using combination of Methods 1 and 4 (20%).

4.4.2 Interpretation of even load using Example 2

In its upper part, Figure 4.5 shows the resulting seven departures for the even-load procedure, using the loads observed at the hourly (Method 2) max load points (as in Table 4.1). A cumulative curve of straight lines can be drawn from observed departure times and max loads $(35 + 65 + 90 + \cdots)$. The slope of each line will represent the uniform arrival rate. For example, the arrival rate between the 6:50 and 7:15 departures is $90/25 = 3.6$ passengers/min. The time on the curve associated with the first desired occupancy, $d_{o1} = 50$ passengers, is 06:23, and the second is 6:50; this means that the second departure is unchanged, since $35 + 65 = 50 + 50$. We then can check the cumulative-load level of 150 and learn that its associated time on the curve is after 7:00. Hence, $d_{o2} = 60$ will be applied for the rest of the example. The cumulative-load curve also ensures that the first vehicle to depart in any time period will accommodate the desired occupancy assigned to that period. The only exception in this process occurs when the determined frequency is based on the *minimum frequency standard* of Equation 3.3. In such a case, we may disregard even loads for a given time period j and switch to even headways, since virtually no overcrowding is expected. However, if we still want to maintain the even-load situation for the minimum frequency case, then d_{oj} must be replaced according to the formula shown in the following section.

The lower part of Figure 4.5 shows the derivation of an even average load timetable at the daily max load point as opposed to the hourly max load point described for Method 2. In this example, we will combine the results of Method 1 (6:00–7:00) with Method 4 using the 20% criterion (7:00–8:00). The cumulative-load curve is based on the loads observed only at stop 2 (daily max load point) and appearing in Table 4.1 and Figure 3.7. While the desired occupancy for Method 1 is the same as for Method 2 ($d_{o1} = 50$), the value of d_{o2} needs to be commensurate with the frequency result given by Method 4 (20%). Thus, we can calculate the substitution for d_{o2} simply in proportional terms:

$$d_{ij} = d_{oj} \cdot \frac{F_{ij}}{P_{mj}/d_{oj}}, \quad i = 3,4 \text{ and } j = 1,2,\ldots,q \tag{4.1}$$

where
  d_{ij} is the adjusted *desired occupancy*
  i is the frequency determination method number
  j is the time period
  F_{ij}, P_{mj}, and d_{oj} are the frequency, maximum observed load, and desired occupancy, respectively as defined in Chapter 3

Equation 4.1 also holds for situations in which the determined frequency is the minimum frequency standard for all methods. This standard is the result of Equations 3.3 and 3.5, that is, in the case of Method 2, $F_{2j} = \max(P_{mj}/d_{oj}, F_{mj}) = F_{mj}$, where F_{mj} is the minimum frequency in time period j.

Based on Equation 4.1, and using Example 2, Method 4 (20%) yields the following: $d_{4,2} = 60(3.60/2.95) = 73.22$. This adjusted quantity is used in Figure 4.5 for constructing the timetable from the route's departure point. The results of the two timetables of Figure 4.5 and of other optional timetables for Example 2 are shown in four columns in Table 4.3.

When applying the load-profile-based (frequency) methods, we can expect higher than the desired loads at the max load points as a trade-off for less empty space-km. The second column of Table 4.3 presents the resulting timetable at the hourly max load points according to Method 4 (20%). This timetable is derived according to Principle 4.2, with the same cumulative-load curve as in the upper part of Figure 4.5, but an adjusted d_{ij} based on

Table 4.3 Four derived sets of departure times using Example 2 for an even average load
(at a single time point), by method or combination of methods selected

Timetable characteristics (even average load)	Method 2	Method 4 (20%)	Method 1 (6:00–7:00), Method 4 (20%) (7:00–8:00)	Method 2 with predetermined number (=5) of departures
Point where average even load is expected	Hourly max load	Hourly max load	Daily max load	Hourly max load
Departure times at anchored time point (6:00–8:00)	6:00[a]	6:00[a]	6:00[a]	6:00[a]
	6:23	6:23	6:14	6:38
	6:50	6:50	6:52	7:10
	7:07	7:10	7:13	7:33
	7:22	7:28	7:30	7:52
	7:36	7:44	7:45	
	7:48	7:58	8:00	
	8:00			
Total number of departures	8	7	7	5 (as requested)
Minimum single-route fleet size (T = 60 min)	5	4	4	3

[a] First predetermined trip.

Equation 4.1. Since the frequencies given by Methods 2 and 4 (20%) for the first 6:00–7:00 period coincide, $d_{i1} = 50$ for both. For the 7:00–8:00 period, $d_{i2} = 73.22$, with both d_{i1} and d_{i2} used for the hourly max loads. That is, the first departure after 7:00 will be y min from 6:50 with a load of 73.22; hence, y = 73.22·25/90 = 20.33 ≈ 20, thereby determining the 6:50 + 20 = 7:10 departure.

The last timetable option in Table 4.3 is based on five predetermined departures (including the first 6:00 departure) while using Method 2. The desired occupancy is modified proportionally in the same manner as was done in Section 4.3.2, by the ratio 4/6.28 = 0.637. Consequently, the two periods in Example 2 will use the load quantities 50/0.637 = 78.5 and 60/0.637 = 94.2, respectively, instead of $d_{o1} = 50$ and $d_{o2} = 60$. With these quantities, Principle 4.2 will be applied on the upper cumulative curve of Figure 4.5 to arrive at the results. It should be understood that the special request of a clock headway cannot be applied to even-load procedures. Ostensibly, timetables with clock headways cannot incorporate uneven headways.

The comparison measures in Table 4.3 are derived in the same manner as for the even-headway procedure as explained in Section 4.3.2. The comparison measures of number of departures and single-route fleet size should be virtually similar for the even-headway and even-load procedures. Nonetheless, because of different usages of the end of the schedule (8:00), this expected similarity almost disappears. The next section, using LACMTA real-life data, will show that this similarity indeed exists, as expected, since both types of procedures rely on the same frequencies.

4.5 AUTOMATION, TEST RUNS, AND CONCLUSION

The outcome of Chapters 3 and 4 can be jointly automated. Practically speaking, a set of computer programs was created that perform (1) conversion from transit agency files to adequate input files, (2) setting of frequencies and headways by four methods, and

(3) construction of a public timetable at all route time points. We are aware of the need, in practice, to actually test optional timetables prior to drawing a definite conclusion. Often, the following saying is true: logic is a systematic method of coming to the wrong conclusion with confidence. Hence, only real testing, using a before and after study, can reveal the magnitude—if at all—of the improvement.

Figure 4.6 exhibits a summary of the procedures that can take place in any possible automation. This flowchart is fed by the data required before embarking on frequency-setting and timetable construction procedures. Compliance with the third objective of this chapter (see Section 4.2) is assured, and the possibility is furnished of direct frequency insertion and the manual construction of timetables.

For a real-life example and test runs, the ride-check data from the old Los Angeles (LA) Metro route 217 (in Los Angeles), which has already been described and employed in Chapter 3, were used. Route 217 was characterized by 60 stops and 8 time points in each direction. In our set of programs, the user can request various optional timetables

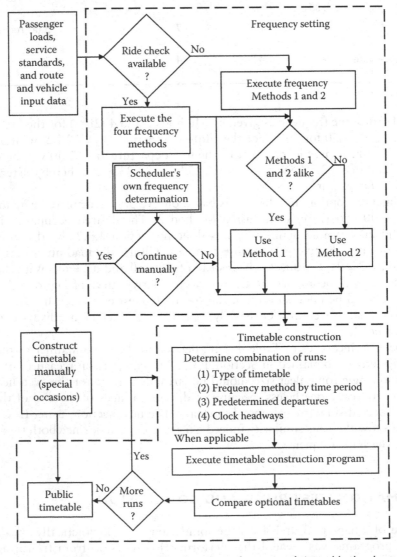

Figure 4.6 Overall automated procedure, in flowchart form, for optional timetable development.

according to the combinations indicated in Figure 4.6. For each computer run, the user simply keypunches requests as follows:

1. Type of timetable—"1" for evenly spaced headways and "2" for even average max load (at hourly max load point)
2. *Number* of method to be used (among the inserted frequency-setting methods)
3. For each used method, the user specifies
   a. Method *number*
   b. The time period *number* in which to start using the method
   c. The last time-period *number* in which to use the method in the combination being considered (i.e., the same method can be used several times for different time periods, but each combination must be specified)
4. Clock headway—"0" for not required and "1" for required
5. Predetermined number of departures — "0" for no need and *Given number* of departures for using the constraint

Eighteen different combinations of runs for each direction of travel are shown in Table 4.4 as the results of the comparison measures for each run. Tables 4.5 through 4.7 display the results of a variety of runs at the northbound departure point of route 217. Table 4.5 shows a comparison of Method 2 and 4 (20%) results for both even-headway and even-load procedures. In Table 4.6, there are two types of comparisons: between Method 2 and 4 (20%) for even clock headways and between even-headway and even-load procedures using a combination of Method 2 and 4 (20%). Table 4.7, as well, presents two types of comparisons: between predetermined 70 and 140 departures using the even-load procedure and between Method 1 and Method 3 results using the even-headway procedure.

As was noted earlier, the comparison measures across all the real-life results in Tables 4.5 through 4.7 are the same for the two types of timetables (even headway, even load). Where they differ is in the frequency method selected and the special requests (clock headway, predetermined number of departures). See, for example, the comparison measures for the lower and upper bounds of route 217 northbound. Specifically, Method 3 delivers a lower bound of 67 and 11 for the number of departures and the single-route fleet size, respectively. This contrasts with the upper bound, given by Method 2 and a clock headway, of 89 departures and 16 vehicles required. Finally, for the sake of clarity, Table 4.8 presents a complete timetable (across all eight time points), this time for route 217 southbound, using Method 4 (20%) and an even-headway procedure.

Admittedly, the large number and variety of timetables may complicate the decision process at the transit agency. However, it offers an opportunity to examine different timetable and frequency scenarios rapidly. We anticipate that the skilled scheduler/planner will select only a few options for comparison while recognizing the full potential of the procedures.

In conclusion, the consequence of this chapter can be described generally in the light of the five study objectives set forth in Section 4.2. Using passenger-load data, the procedures developed provide optional timetables in terms of vehicle departure times at all specified time points. Each timetable is accompanied by two comparison measures. These evaluation measures fulfill the first objective. One set of options when selecting the procedure to construct the timetables is referred to as balancing passenger loads on individual vehicles while allowing unevenly spaced headways. The underlying approach in that set of options is to shift the departure times of individual vehicles so as to obtain even average loads while relaxing the even-headway requirement. The latter accomplishes the second objective.

Table 4.4 Combination of requests for test runs using Los Angeles route 217 ride-check data

Run number	Direction of travel: South (S) and North (N)	Type of timetable: 1 = Even Headway; 2 = Even Load	Method #	Clock headway: 1 = With; 0 = Without	Predetermined number of departures	Results of comparison measures	
						Number of departures	Minimum fleet size
1	N	1	1	0	0	84	15
2	N	1	2	0	0	87	15
3	N	1	3	0	0	67	11
4	N	1	4	0	0	78	14
5	N	1	2, 4[a]	0	0	85	15
6	N	1	2	1	0	80	16
7	N	1	4	1	0	78	14
8	N	1	2	0	70	70	12
9	N	1	2	0	140	140	24
10	N	2	1	0	0	84	15
11	N	2	2	0	0	87	15
12	N	2	3	0	0	67	11
13	N	2	4	0	0	78	14
14	N	2	2, 4[a]	0	0	85	15
15	N	2	2	0	70	70	12
16	N	2	2	0	140	140	24
17	S	1	1	0	0	88	19
18	S	1	2	0	0	91	19
19	S	1	3	0	0	69	13
20	S	1	4	0	0	73	13
21	S	1	2, 4[a]	0	0	89	19
22	S	1	2	1	0	91	20
23	S	1	4	1	0	74	14
24	S	1	2	0	70	70	15
25	S	1	2	0	140	140	30
26	S	2	1	0	0	88	19
27	S	2	2	0	0	91	20
28	S	2	3	0	0	69	13
29	S	2	4	0	0	73	13
30	S	2	2, 4[a]	0	0	89	20
31	S	2	2	0	70	70	15
32	S	2	2	0	140	140	30

[a] Method 2 is used for off-peak periods # (1), (4, 5, 6, 7, 8, 9, 10, 11) and (14, 15, 16, 17, 18, 19); Method 4 is used for peak periods # (2, 3) and (12, 13)—a total of five combinations.

The third objective of the study is to allow for direct vehicle-frequency changes. This is achieved in the process described by Figure 4.6, whereby the schedulers can either interject their own set of intuitive or experience-based frequencies or they can substitute some of the derived frequencies. In this way, possible scheduling exceptions (e.g., special passenger demand because of a sports event) could be inserted. The common manual process in creating timetables often encounters a problem in smoothing the headways in the transition segments between adjacent time periods. In both types of timetables that have been described (even headway and even load), the procedures developed ensure, in an average sense, the fulfillment of the desired occupancy standard.

Table 4.5 Comparison of Methods 2 and 4 (20% criterion) using the two types of timetables for Los Angeles route 217 northbound

Characteristics	Even headways, Method 2						Even headways, Method 4 (20%)					
Departure times at route departure point	06:00	09:31	12:11	14:10	16:15	19:28	06:00	10:04	12:54	15:04	17:36	23:59
	06:30	09:45	12:19	14:18	16:24	19:58	06:30	10:16	13:03	15:13	17:48	00:30
	07:00	10:00	12:27	14:26	16:33	20:28	07:00	10:28	13:12	15:22	18:01	01:00
	07:11	10:10	12:35	14:34	16:42	20:58	07:12	10:41	13:20	15:31	18:19	
	07:21	10:21	12:43	14:42	16:50	21:29	07:24	10:53	13:29	15:40	18:38	
	07:32	10:31	12:51	14:50	16:59	21:59	07:36	11:05	13:37	15:49	18:57	
	07:42	10:42	12:59	14:58	17:10	22:29	07:48	11:17	13:46	15:58	19:25	
	07:53	10:52	13:07	15:07	17:21	22:59	08:00	11:30	13:54	16:08	19:56	
	08:04	11:02	13:15	15:15	17:33	23:29	08:13	11:42	14:03	16:18	20:26	
	08:15	11:13	13:22	15:24	17:44	24:00	08:26	11:54	14:11	16:28	20:57	
	08:27	11:23	13:30	15:32	17:55	00:30	08:39	12:05	14:20	16:38	21:27	
	08:38	11:33	13:38	15:41	18:09	01:00	08:51	12:15	14:29	16:49	21:57	
	08:49	11:44	13:46	15:49	18:26		09:07	12:25	14:38	16:59	22:28	
	09:01	11:54	13:54	15:58	18:42		09:26	12:34	14:46	17:11	22:58	
	09:16	12:03	14:02	16:06	18:59		09:46	12:44	14:55	17:23	23:29	
Total number of departures	87						78					
Minimum single-route fleet size	15						14					

	Even average load, Method 2						Even average load, Method 4 (20%)					
Departure times at route departure point	06:00	09:25	12:07	14:08	16:14	19:28	06:00	10:03	12:51	15:04	17:32	23:59
	06:38	09:39	12:14	14:15	16:26	19:58	06:36	10:17	13:03	15:15	17:51	00:12
	07:03	10:00	12:25	14:29	16:33	20:23	07:03	10:31	13:07	15:20	18:02	01:00
	07:23	10:12	12:33	14:35	16:43	20:59	07:21	10:41	13:15	15:31	18:17	
	07:35	10:24	12:40	14:44	16:49	21:34	07:35	10:54	13:24	15:37	18:36	
	07:42	10:35	12:49	14:53	16:59	21:58	07:44	11:06	13:33	15:46	18:55	
	07:50	10:44	12:58	14:59	17:09	22:26	07:52	11:16	13:46	15:58	19:26	
	07:56	10:54	13:05	15:08	17:19	22:59	08:00	11:30	13:54	16:07	19:56	
	08:02	11:04	13:10	15:16	17:27	23:17	08:07	11:39	14:02	16:20	20:19	
	08:10	11:13	13:17	15:23	17:46	23:59	08:17	11:53	14:08	16:30	20:57	
	08:21	11:20	13:25	15:34	17:56	00:13	08:33	12:04	14:16	16:42	21:26	
	08:36	11:32	13:34	15:40	18:12	01:00	08:44	12:12	14:30	16:49	21:57	
	08:44	11:43	13:44	15:48	18:25		09:03	12:24	14:37	16:59	22:26	
	09:01	11:54	13:53	15:58	18:41		09:20	12:34	14:48	17:10	22:58	
	09:11	12:03	14:02	16:05	18:57		09:37	12:41	14:55	17:20	23:16	
Total number of departures	87						78					
Minimum single-route fleet size	15						14					

Table 4.6 Comparison of Methods 2 and 4 (20% criterion) using the two types of timetables and of the two types of timetables for a combination of these methods, for Los Angeles route 217 northbound

Characteristics	Even clock headways, Method 2						Even clock headways, Method 4 (20%)					
Departure times at route departure point	06:00	09:30	12:07	14:00	15:52	18:45	06:00	10:00	12:50	15:00	17:36	00:00
	06:30	09:45	12:15	14:07	16:00	19:00	06:30	10:12	13:00	15:10	17:48	00:30
	07:00	10:00	12:22	14:15	16:10	19:30	07:00	10:24	13:07	15:20	18:00	01:00
	07:10	10:10	12:30	14:22	16:20	20:00	07:12	10:36	13:15	15:30	18:20	
	07:20	10:20	12:37	14:30	16:30	20:30	07:24	10:48	13:22	15:40	18:40	
	07:30	10:30	12:45	14:37	16:40	21:00	07:36	11:00	13:30	15:50	19:00	
	07:40	10:40	12:52	14:45	16:50	21:30	07:48	11:12	13:37	16:00	19:30	
	07:50	10:50	13:00	14:52	17:00	22:00	08:00	11:24	13:45	16:10	20:00	
	08:00	11:00	13:07	15:00	17:12	22:30	08:12	11:36	13:52	16:20	20:30	
	08:12	11:10	13:15	15:07	17:24	23:00	08:24	11:48	14:00	16:30	21:00	
	08:24	11:20	13:22	15:15	17:36	23:30	08:36	12:00	14:10	16:40	21:30	
	08:36	11:30	13:30	15:22	17:48	00:00	08:48	12:10	14:20	16:50	22:00	
	08:48	11:40	13:37	15:30	18:00	00:30	09:00	12:20	14:30	17:00	22:30	
	09:00	11:50	13:45	15:37	18:15	01:00	09:20	12:30	14:40	17:12	23:00	
	09:15	12:00	13:52	15:45	18:30		09:40	12:40	14:50	17:24	23:30	
Total number of departures	89						78					
Minimum single-route fleet size	16						14					

Characteristics	Even headways, combination of Methods 2 and 4 (20%)						Even average load, combination of Methods 2 and 4 (20%)					
Departure times at route departure point	06:00	09:48	12:20	14:19	16:25	20:28	06:00	09:43	12:16	14:17	16:27	20:23
	06:30	10:02	12:28	14:27	16:34	20:59	06:38	10:02	12:26	14:30	16:33	20:59
	07:00	10:12	12:36	14:35	16:43	21:29	07:01	10:15	12:34	14:36	16:44	21:35
	07:12	10:22	12:44	14:43	16:51	21:59	07:21	10:27	12:41	14:46	16:50	21:58
	07:24	10:33	12:52	14:51	17:00	22:29	07:35	10:36	12:50	14:54	17:00	22:27
	07:36	10:43	13:00	15:00	17:13	22:59	07:43	10:46	13:00	15:00	17:11	22:59
	07:47	10:54	13:08	15:08	17:25	23:29	07:52	10:55	13:05	15:11	17:22	23:17
	07:59	11:04	13:16	15:16	17:37	00:00	08:00	11:07	13:11	15:16	17:35	00:00
	08:12	11:14	13:24	15:25	17:50	00:30	08:07	11:14	13:18	15:24	17:53	00:13
	08:25	11:25	13:31	15:33	18:03	01:00	08:16	11:21	13:26	15:35	18:07	01:00
	08:38	11:35	13:39	15:42	18:22		08:33	11:34	13:35	15:41	18:20	
	08:50	11:45	13:47	15:50	18:40		08:43	11:45	13:46	15:50	18:38	
	09:04	11:56	13:55	15:59	18:59		09:02	11:56	13:54	15:59	18:58	
	09:18	12:05	14:03	16:08	19:28		09:14	12:03	14:02	16:06	19:29	
	09:33	12:12	14:11	16:16	19:58		09:27	12:08	14:09	16:16	19:58	
Total number of departures	85						85					
Minimum single-route fleet size	15						15					

Table 4.7 Comparison of timetables associated with a predetermined number of departures and of timetables using Methods 1 and 3 for Los Angeles route 217 northbound

Characteristics	Even average load, 70 departures, Method 2				Even average load, 140 departures, Method 2							
Departure times at route departure point	06:00	10:57	14:07	17:21	06:00	08:37	10:56	12:35	14:05	15:38	17:18	20:57
	06:49	11:11	14:16	17:36	06:32	08:42	11:02	12:40	14:09	15:42	17:22	21:20
	07:13	11:19	14:31	17:54	06:49	08:49	11:10	12:46	14:14	15:47	17:28	21:39
	07:32	11:34	14:39	18:12	06:55	09:01	11:14	12:51	14:23	15:54	17:39	21:52
	07:42	11:48	14:52	18:30	07:12	09:06	11:18	12:57	14:29	15:59	17:48	22:11
	07:51	12:01	15:00	18:49	07:24	09:15	11:22	13:02	14:33	16:04	17:55	22:28
	07:59	12:07	15:15	19:21	07:31	09:23	11:32	13:05	14:36	16:08	18:03	22:45
	08:08	12:16	15:19	19:59	07:38	09:32	11:39	13:08	14:41	16:14	18:13	23:04
	08:21	12:28	15:33	20:30	07:42	09:41	11:46	13:12	14:49	16:23	18:20	23:14
	08:38	12:37	15:41	21:18	07:47	09:55	11:53	13:16	14:54	16:28	18:31	23:39
	08:50	12:48	15:53	21:51	07:51	10:03	11:59	13:21	14:57	16:32	18:41	00:02
	09:06	13:00	16:02	22:27	07:55	10:11	12:03	13:27	15:01	16:37	18:51	00:09
	09:24	13:06	16:12	23:04	07:58	10:19	12:06	13:32	15:08	16:44	19:06	00:32
	09:44	13:14	16:27	23:38	08:03	10:27	12:09	13:38	15:15	16:47	19:24	01:00
	10:04	13:24	16:35	00:09	08:07	10:33	12:14	13:45	15:16	16:53	19:44	
	10:20	13:35	16:46	01:00	08:12	10:38	12:21	13:50	15:21	16:59	20:01	
	10:34	13:48	16:58		08:19	10:44	12:27	13:55	15:29	17:04	20:15	
	10:46	13:58	17:10		08:31	10:51	12:31	14:01	15:34	17:12	20:32	
Total number of departures	70 (as requested)				140 (as requested)							
Minimum single-route fleet size	12				24							

Characteristics	Even headways, Method 1						Even headways, Method 3				
Departure times at route departure point	06:00	09:32	12:06	14:06	16:04	18:41	06:00	09:50	13:21	16:11	19:55
	06:30	09:47	12:15	14:14	16:13	18:58	06:30	10:09	13:34	16:21	20:26
	07:00	10:01	12:24	14:22	16:22	19:27	07:00	10:25	13:46	16:31	20:56
	07:12	10:12	12:33	14:30	16:31	19:57	07:13	10:42	13:59	16:41	21:27
	07:23	10:22	12:43	14:39	16:40	20:28	07:25	10:59	14:12	16:51	21:57
	07:35	10:33	12:52	14:47	16:48	20:58	07:37	11:16	14:25	17:02	22:28
	07:46	10:43	13:01	14:55	16:57	21:28	07:49	11:32	14:38	17:15	22:58
	07:57	10:54	13:09	15:03	17:08	21:58	08:01	11:49	14:51	17:28	23:29
	08:10	11:04	13:17	15:12	17:19	22:29	08:14	12:04	15:04	17:41	23:59
	08:23	11:14	13:25	15:21	17:31	22:59	08:28	12:17	15:15	17:54	00:30
	08:36	11:25	13:33	15:29	17:42	23:29	08:41	12:30	15:27	18:12	01:00
	08:49	11:35	13:41	15:38	17:53	00:00	08:54	12:43	15:38	18:34	
	09:02	11:45	13:49	15:47	18:07	00:30	09:11	12:56	15:49	18:56	
	09:17	11:56	13:57	15:56	18:24	01:00	09:31	13:08	16:01	19:25	
Total number of departures	84						67				
Minimum single-route fleet size	15						11				

Table 4.8 Complete timetable with even headway, Method 4 (20%), at all time points of route 217 southbound in Los Angeles

Departure #	Beachwood, Westshire	Gower, Franklin	Hollywood, Vine	La Brea, Sunset	Fairfax, Santa Monica	Fairfax, Beverly	Fairfax, Olympic	Adams, Washington
1	05:37	05:41	05:44	05:50	05:56	06:00	06:05	06:11
2	06:07	06:11	06:14	06:20	06:26	06:30	06:36	06:42
3	06:33	06:38	06:41	06:47	06:55	07:00	07:06	07:13
4	06:46	06:51	06:54	07:01	07:08	07:13	07:19	07:26
5	06:59	07:04	07:07	07:14	07:21	07:26	07:32	07:39
6	07:12	07:17	07:20	07:27	07:35	07:40	07:46	07:52
7	07:25	07:30	07:33	07:40	07:48	07:53	07:59	08:05
8	07:35	07:40	07:43	07:51	08:00	08:05	08:11	08:19
9	07:47	07:52	07:55	08:03	08:12	08:17	08:23	08:31
10	07:59	08:04	08:07	08:15	08:24	08:29	08:35	08:43
11	08:11	08:16	08:19	08:27	08:36	08:41	08:47	08:55
12	08:23	08:28	08:31	08:39	08:48	08:53	08:59	09:07
13	08:34	08:39	08:42	08:50	08:59	09:04	09:10	09:18
14	08:45	08:50	08:53	09:01	09:10	09:15	09:21	09:29
15	08:57	09:02	09:05	09:13	09:22	09:27	09:33	09:41
16	09:08	09:13	09:16	09:24	09:33	09:38	09:44	09:52
17	09:19	09:24	09:27	09:35	09:44	09:49	09:55	10:03
18	09:30	09:35	09:38	09:46	09:55	10:00	10:07	10:15
19	09:40	09:45	09:48	09:56	10:05	10:10	10:17	10:25
20	09:50	09:55	09:58	10:06	10:15	10:21	10:27	10:36
21	10:00	10:05	10:08	10:16	10:25	10:31	10:37	10:46
22	10:11	10:16	10:19	10:27	10:36	10:41	10:48	10:56
23	10:21	10:26	10:29	10:37	10:46	10:51	10:58	11:06
24	10:31	10:36	10:39	10:47	10:56	11:02	11:09	11:18
25	10:42	10:47	10:50	10:58	11:07	11:13	11:20	11:29
26	10:53	10:58	11:01	11:09	11:18	11:24	11:31	11:40
27	11:04	11:09	11:12	11:20	11:29	11:35	11:42	11:51
28	11:15	11:20	11:23	11:31	11:40	11:46	11:53	12:02
29	11:26	11:31	11:34	11:42	11:51	11:57	12:04	12:13
30	11:38	11:43	11:46	11:54	12:03	12:09	12:16	12:27
31	11:50	11:55	11:58	12:06	12:15	12:21	12:28	12:39
32	12:02	12:07	12:10	12:18	12:27	12:33	12:40	12:52
33	12:15	12:20	12:23	12:31	12:40	12:46	12:53	13:04
34	12:27	12:32	12:35	12:43	12:52	12:58	13:05	13:16
35	12:37	12:43	12:46	12:54	13:03	13:09	13:16	13:25
36	12:48	12:54	12:56	13:05	13:14	13:20	13:27	13:36
37	12:59	13:04	13:07	13:16	13:25	13:31	13:38	13:47
38	13:10	13:15	13:18	13:26	13:35	13:41	13:48	13:57
39	13:21	13:26	13:29	13:37	13:46	13:52	13:59	14:08
40	13:29	13:34	13:37	13:47	13:57	14:03	14:11	14:20
41	13:38	13:43	13:46	13:56	14:06	14:12	14:20	14:29
42	13:47	13:52	13:55	14:05	14:15	14:21	14:29	14:38

(Continued)

Table 4.8 (Continued) Complete timetable with even headway, Method 4 (20%), at all time points of route 217 southbound in Los Angeles

Departure #	Beachwood, Westshire	Gower, Franklin	Hollywood, Vine	La Brea, Sunset	Fairfax, Santa Monica	Fairfax, Beverly	Fairfax, Olympic	Adams, Washington
43	13:57	14:02	14:05	14:15	14:25	14:31	14:39	14:48
44	14:06	14:11	14:14	14:24	14:34	14:40	14:48	14:57
45	14:16	14:21	14:24	14:34	14:44	14:50	14:58	15:07
46	14:25	14:30	14:33	14:43	14:53	14:59	15:07	15:16
47	14:34	14:39	14:42	14:52	15:02	15:08	15:16	15:25
48	14:42	14:47	14:50	15:00	15:10	15:16	15:24	15:33
49	14:51	14:56	14:59	15:09	15:19	15:25	15:33	15:42
50	14:59	15:04	15:07	15:17	15:27	15:33	15:41	15:50
51	15:08	15:13	15:16	15:26	15:36	15:42	15:50	15:59
52	15:16	15:21	15:24	15:34	15:44	15:50	15:58	16:07
53	15:25	15:30	15:33	15:43	15:53	15:59	16:07	16:16
54	15:41	15:46	15:49	15:59	16:09	16:15	16:23	16:32
55	15:58	16:03	16:06	16:16	16:26	16:32	16:40	16:49
56	16:15	16:20	16:23	16:33	16:43	16:49	16:57	17:06
57	16:35	16:40	16:43	16:53	17:03	17:09	17:17	17:26
58	17:00	17:04	17:07	17:17	17:27	17:33	17:41	17:50
59	17:24	17:29	17:32	17:42	17:52	17:58	18:06	18:15
60	17:57	18:01	18:04	18:13	18:22	18:27	18:34	18:43
61	18:27	18:31	18:34	18:43	18:52	18:57	19:04	19:13
62	18:59	19:04	19:07	19:15	19:23	19:28	19:34	19:42
63	19:29	19:34	19:37	19:45	19:53	19:58	20:04	20:12
64	20:00	20:05	20:08	20:16	20:23	20:28	20:34	20:42
65	20:30	20:35	20:38	20:46	20:53	20:58	21:04	21:12
66	21:01	21:06	21:09	21:17	21:24	21:29	21:35	21:43
67	21:31	21:36	21:39	21:47	21:54	21:59	22:05	22:13
68	22:01	22:06	22:09	22:17	22:24	22:29	22:35	02:43
69	22:31	22:36	22:39	22:47	22:54	22:59	23:05	23:13
70	23:01	23:06	23:09	23:17	23:24	23:29	23:35	23:43
71	23:32	23:37	23:40	23:48	23:55	00:00	00:06	00:14
72	00:02	00:07	00:10	00:18	00:25	00:30	00:36	00:44
73	00:32	00:37	00:40	00:48	00:55	01:00	01:06	01:14

This basically fulfills the smoothing necessity expressed in the fourth objective. The fifth and last objective is attained by allowing the transit agency to request a selection of different frequency-setting methods for different time periods. The scheduler/planner can then select for peak periods those methods that are more sensitive to resource saving (e.g., Method 4) and for off-peak periods methods that are more sensitive to passenger comfort (e.g., Method 2).

4.6 LITERATURE REVIEW AND FURTHER READING

The problem of finding the best dispatching policy for transit vehicles on fixed routes has a direct impact on constructing timetables. This dispatching-policy problem, which has been dealt with quite extensively in the literature, can be categorized into four groups: (1) models for an idealized transit system, (2) simulation models, (3) mathematical programming models, and (4) data-based models.

The first group, idealized transit systems, was investigated by, for example, Newell (1971), Osana and Newell (1972), Hurdle (1973), Wirasinghe (1990, 2003), and De Palma and Lindsey (2001). Newell (1971) assumed a given passenger-arrival rate as a smooth function of time, with the objective of minimizing total passenger waiting time. He showed analytically that the frequency of transit vehicles with large capacities (in order to serve all waiting passengers) and the number of passengers served per vehicle each varied with time approximately as the square root of the arrival rate of passengers. Osana and Newell (1972) developed control strategies for either holding back a transit vehicle or dispatching it immediately, based on a given number of vehicles, random round-trip travel times with known distribution functions, and uniform passenger-arrival rates with a minimum waiting-time objective. Using dynamic programming, they found that the optimal strategy for two vehicles and a small coefficient of variation of trip time retained nearly equally spaced dispatch times. Hurdle (1973), investigating a similar problem, used a continuum fluid-flow model to derive an optimal dispatching policy while attempting to minimize the total cost of passenger waiting time and vehicle operation.

Wirasinghe (1990, 2003) examined and extended Newell's dispatching policy while considering the cost components initially used by Newell (1973). Wirasinghe considered the average value of a unit waiting time per passenger (C_1) and the cost of dispatching a vehicle (C_2) to show that the passenger-arrival rate in Newell's square root formula is multiplied by ($C_1/2C_2$). Wirasinghe also showed how to derive the equations of total mean cost per unit of time by using both uniform headway policy and Newell's variable-dispatching policy.

De Palma and Lindsey (2001) develop a method for designing an optimal timetable for a single line with only two stations. The method is suitable for a situation in which each rider has a precise time in which the person wants to travel; traveling earlier or later than desired increases the total cost. The objective is to minimize riders' total schedule-delay costs. Two cases are analyzed with respect to passenger preferences. In the first case, all passengers treat a unit of delay equally. The second case assumes several rider groups, with different levels of delay costs ascribed to riders from the different groups. In addition, the researchers compared two models: a *line* model, in which preferred travel times are uniformly distributed over part of the day and trips cannot be rescheduled between days, and a *circle* model, in which preferred travel times are uniformly distributed over the full 24 h day and trips can be rescheduled between days. Optimal timetables are derived for each of the models.

In the second group, simulation models were studied by, for example, Marlin et al. (1988), Adamski (1998), and Dessouky et al. (1999). Marlin et al. (1988) developed a simulation model for dispatching transit vehicles every day. They checked the feasibility of the results and used mathematical programming for vehicle assignments in an interactive computer-support system. Adamski (1998) employed a simulation model for real-time dispatching control of transit vehicles while attempting to increase the reliability of service in terms of on-time performance. His simulation implemented optimal stochastic control with linear feedback. The use of intelligent transportation systems was applied by Dessouky et al. (1999) in a study of bus dispatching at timed transfer points. The researchers used a simulation analysis to show that the benefit of knowing the location of the bus was most significant when the bus was experiencing a significant delay, especially when there was a small number of connecting buses at transfers point.

Mathematical programming methods, the third group for determining frequencies and timetables, have been proposed by Furth and Wilson (1981), Koutsopoulos et al. (1985), Ceder and Stern (1984), Eberlein et al. (1998), Gallo and Di-Miele (2001), and Peeters and Kroon (2001). Furth and Wilson sought to maximize the net social benefit, consisting of ridership benefit and waiting-time saving, subject to constraints on total subsidy, fleet size, and passenger-load levels. Koutsopoulos et al. extended this formulation by incorporating crowding-discomfort costs into the objective function and treating the time-dependent character of transit demand and performance. Their initial problem consisted of a nonlinear optimization program relaxed by linear approximations. Ceder and Stern addressed the problem with an integer programming formulation and a heuristic person–computer interactive procedure. Their approach focuses on reconstructing timetables when the available vehicle fleet size is restricted. Eberlein et al. (1998) studied a special dispatching problem for the purpose of introducing deadheading trips in high-frequency transit operations. They solved their dispatching strategy optimally; they also determined the number of stops that could be skipped in order to minimize total passenger cost in the system.

Gallo and Di-Miele (2001) produced a model for the special case of dispatching buses from parking depots. Their model is based on the decomposition of generalized assignments and design, noncrossing, and matching subproblems. It can be extended to a case in which there is an overlap between arrival and departure vehicle flows. Peeters and Kroon (2001) present a procedure for planning an optimal cyclic railway timetable, that is, a timetable in which trains leave at the same minute every hour. The problem is represented through a constraint graph, in which each node is an event that needs to be scheduled; cycles are examined according to a calculation of tensions and potentials. The model is formulated as a mixed-integer nonlinear program with the objectives of minimizing passenger time, maximizing timetable robustness, and minimizing the number of required trains. A solution procedure is suggested, by which the nonlinear part of the formulation is transformed into a mixed-integer linear problem that is an approximation of the original problem; further actions are taken in order to reduce the number of constraints.

Louwerse and Huisman (2012) consider the problem of adjusting a railway timetable in case of large disruptions. They present integer programming formulations based on event activity networks for the situations of a partial and a complete blockade. The disruption measures included, to convert the normal timetable into a disposition timetable, delaying trains, cancelling trains, and reversing trains at stations adjacent to the blockade. The models determine which trains need to be cancelled and defined the disposition timetable of the trains that are not cancelled. The authors also present how the models can be modified in case a regular disposition timetable is preferred. Two real-world cases are used to test the models presented. The results show a balance between the number of trains that can be operated in the disposition timetable and the maximum delay that is allowed. Also, the authors show the effect of limited start inventories at the number of trains that can be operated. By presenting also the regular disposition timetable, a trade-off is offered between a disposition timetable that is more regular or one that deviates less from the normal timetable. For all instances, the time needed to compute the disposition timetable is at most a few seconds, which is acceptable in real-time disruption management.

Hassold and Ceder (2012) demonstrate how to make public-transit services more attractive by using two simultaneous objectives: minimizing the average estimated passenger waiting time and minimizing the discrepancy from a desired occupancy level on the vehicles. The first objective aims at improving the service and attract more users, and the second objective ensures economical operation. A network-based procedure is used to create timetables with multiple vehicle types to solve this biobjective problem. In addition, the authors suggest that this methodology allows the synchronization of the arrival times with the departure times of other modes and services to enable easier and more convenient transfers. The methodology developed is applied to a case study in Auckland, New Zealand, where more than 46% of passenger waiting time was achieved while, at the same time, attaining an acceptable passenger load on all vehicles. Moreover, for each of the desired arrival times, which can be either given as a list or determined by a fixed headway, the timetables show departures that arrive within a certain tolerance at the interchange point. The suggested methodology demonstrates that a significant improvement is attainable in terms of efficiency and use of resources where, at the same time, this improvement can be combined with reducing the waiting time for passengers.

Hassold and Ceder (2013) present an extension to a previously introduced multiobjective methodology to create timetable using multiple vehicle types. The extension has been made to base the creation of timetables on the data at different maximum load points. The methodology used allows evaluating the timetables according to multiple criteria. It has been applied to a case study from a major bus line in Auckland, New Zealand. The Pareto frontiers presented in this study indicate that the timetables created using the hourly and daily maximum load points are very limited in their range to achieve different objective function combinations. Moreover, they violate capacity constraints and show a number of standees on the vehicles. The authors state that comparison of the results for the hourly and daily max load points restricts conclusions to be derived for a universal recommendation. However, in this study, the hourly max load point timetables consistently indicate a significantly higher number of standee minutes, so that the daily max load point method should be preferred in most of the cases in the case study presented. Although none of the solutions from the individual max load point timetables dominate the other solutions, the authors say that it is the only method that ensures that the constraints are held at all times. Moreover, it is the only method that allows more flexibility to reach certain objective value combinations if required.

Ceder et al. (2013a) continue the study of different available bus sizes and describe a new heuristic-based methodology to approach even-headway and even-load timetables. Ceder et al. (2013b) also present the creation of bus timetables using different vehicle sizes but with the use of simulation. Both of these articles are reviewed in the previous chapter and are elaborated in the next chapter.

In the fourth and last group, the data-based models described in this chapter are based on Ceder (1986, 2003). Advanced models pertaining to this group will be presented in the next chapter.

EXERCISES

4.1 During a morning time period of 90 min (07:00–08:30), ride-check data were collected on a given bus route (average across several days). The route is 7 km long, and there is only one stop (B) between the first (departure) stop (A) and last (arrival) stop (C), the distance from A to B being 3 km. The average of the data gathered and information on three trips are as follows:

	Average load (passengers) when departing the stop		Desired occupancy (passengers)	Minimum frequency (veh/period) (07:00–08:30)	Bus capacity (passengers)
Departure time	A	B			
07:20	50	25	50	2	80
07:55	70	49			
08:30	40	30			

Given also a predetermined departure time of 07:00,

(a) Construct an even-load timetable at point A using (i) Method 3 and (ii) Method 4 (30%) for frequency determination, with Method 4 having an excess load along a maximum of 30% of its route length

(b) Repeat (a) the aforementioned step, but with a bus capacity of 60 (instead of 80) for all trips

4.2 Based on the data and results of Exercise 3.1 (Chapter 3),
   (a) Construct evenly spaced headway timetables using Method 2 and 4 (20%) at the route departure point
   (b) Construct clock-headway timetables using Method 2 and 4 (20%) at the route departure point
   (c) Find the comparison measures for each of the four timetables in (a) and (b) given that the complete round-trip time is 45 min, that is, total number of departures and minimum required single-route fleet size

4.3 Based on the data and results of Exercise 3.2 (Chapter 3),
   (a) Construct an even average load timetable according to Method 2
   (b) Use the results of (a) to answer the following: (i) Are the loads at each stop, on each vehicle, in each hour, the same? (ii) If the individual loads are not same, at which stop and for what bus will the average load exceed the desired occupancy?

4.4 Using the following data and a requirement that the average max load on the last vehicle between 07:00 and 08:00 be the same as on the first vehicle between 08:00 and 09:00, or 44 = (64 + 24)/2 passengers,

Time	Average max load (passengers)	Desired occupancy (passengers)	Passengers' arrival-rate pattern	Minimum required frequency (veh/h)
07:00–08:00	240	64	Uniform	2
08:00–09:00	180	24	Uniform	4

   (a) Construct even-headway timetable at the max load point (the same for both hours), assuming that passenger demand starts at 07:00; include the first departure after 09:00, assuming that demand after 09:00 is same as that from 08:00–09:00
   (b) What departure times and average loads are derived by Method 2 using an even-headway procedure in the transition between the 2 h? Explain differences from (a) and draw conclusions

REFERENCES

Adamski, A. (1998). Simulation support tool for real-time dispatching control in public transport. *Transportation Research*, **32A**(2), 73–87.

Ceder, A. (1984). Bus frequency determination using passenger count data. *Transportation Research*, **18A**(5/6), 439–453.

Ceder, A. (1986). Methods for creating bus timetables. *Transportation Research*, **21A**(1), 59–83.

Ceder, A. (2003). Public transport timetabling and vehicle scheduling. In *Advanced Modeling for Transit Operations and Service Planning* (W. Lam and M. Bell, eds.), pp. 31–57, Pergamon Imprint, Elsevier Science, New York, NY.

Ceder, A., Hassold, S., and Dano, B. (2013a). Approaching even-load and even-headway transit timetables using different bus sizes. *Public Transport—Planning and Operation*, **5**(3), 193–217.

Ceder, A., Hassold, S., Dunlop, C., and Chen, I. (2013b). Improving urban public-transport service using new timetabling strategies with different vehicle types. *International Journal of Urban Sciences*, **17**(2), 239–258.

Ceder, A. and Stern, H.I. (1984). Optimal transit timetables for a fixed vehicle fleet. In *Proceedings of the 10th International Symposium on Transportation and Traffic Theory* (J. Volmuller and R. Hammerslag, eds.), pp. 331–355, UNU Science Press, Den Haag, the Netherlands.

De Palma, A. and Lindsey, R. (2001). Optimal timetables for public transportation. *Transportation Research*, **35B**, 789–813.

Dessouky, M., Hall, R., Nowroozi, A., and Mourikas, K. (1999). Bus dispatching at timed transfer transit stations using bus tracking technology. *Transportation Research*, 7C(4), 187–208.

Eberlein, X.J., Wilson, N.H.M., and Barnhart, C. (1998). The real-time deadheading problem in transit operations control. *Transportation Research*, 32B(2), 77–100.

Furth, P.G. and Wilson, N.H.M. (1981). Setting frequencies on bus routes: Theory and practice. *Transportation Research Board*, 818, 1–7.

Gallo, G. and Di-Miele, F. (2001). Dispatching buses in parking depots. *Transportation Science*, 35(3), 322–330.

Hassold, S. and Ceder, A. (2012). Multiobjective approach to creating bus timetables with multiple vehicle types. *Transportation Research Record*, 2276, 56–62.

Hassold, S. and Ceder, A. (2013). Public-transport timetabling based on different max-load points using multi-size vehicles. *Transportation Research Record*, 2352, 104–113.

Hurdle, V.F. (1973). Minimum cost schedules for a public transportation route. *Transportation Science*, 7(2), 109–157.

Koutsopoulos, H.N., Odoni, A., and Wilson, N.H.M. (1985). Determination of headways as function of time varying characteristics on a transit network. In *Computer Scheduling of Public Transport 2* (J.M. Rousseau, ed.), pp. 391–414, Elsevier Science, North Holland, Amsterdam, the Netherlands.

Louwerse, I. and Huisman, D. (2012). Adjusting a railway timetable in case of partial or complete blockades. Report EI 2012-23, Erasmus School of Economics (ESE), Econometric Institute, Erasmus University Rotterdam, pp. 1–26, Rotterdam, the Netherlands.

Marlin, P.G., Nauess, R.M, Smith, L.D., and Rhoades, M. (1988). Computer support for operator assignment and dispatching in an urban transit system. *Transportation Research*, 22A(1), 13–26.

Newell, G.F. (1971). Dispatching policies for a transportation route. *Transportation Science*, 5, 91–105.

Newell, G.F. (1973). Scheduling, location, transportation and continuous mechanics: Some simple approximations to optimization problems. *SIAM Journal of Applied Mathematics*, 25(3), 346–360.

Osana, E.E. and Newell, G.F. (1972). Control strategies for an idealized public transportation system. *Transportation Science*, 6(1), 52–72.

Peeters, L. and Kroon, L. (2001). A cycle based optimization model for the cyclic railway timetabling problem. In *Computer-Aided Scheduling of Public Transport*. Lecture Notes in Economics and Mathematical Systems, vol. 505 (S. Voss and J. R. Daduna, eds.), pp. 275–296, Springer-Verlag, Heidelberg, Germany.

Salzborn. F.J.M. (1972). Optimum bus scheduling. *Transportation Science*, 6, 137–148.

Wirasinghe, S.C. (1990). Re-examination of Newell's dispatching policy and extension to a public transportation route with many to many time varying demand. In *Transportation and Traffic Theory* (M. Koshi, ed.), pp. 363–378, Elsevier Ltd, Tokyo, Japan.

Wirasinghe, S.C. (2003). Initial planning for an urban transit system. In *Advanced Modeling for Transit Operations and Service Planning* (W. Lam and M. Bell, eds.), pp. 1–29, Pergamon Imprint, Elsevier Science Ltd, New York, NY.

Chapter 5

Advanced timetables

Optimization and synchronization

Synchronization comes before work only in the dictionary

A.C.

CHAPTER OUTLINE

5.1 Introduction
5.2 Even max load on individual vehicles
5.3 Both—Even headways with even average loads
5.4 Timetables based on a multiobjective approach
5.5 Optimization, operations research, and complexity
5.6 Minimum passenger-crowding timetables for a fixed vehicle fleet
5.7 Maximum synchronization formulation and OR model
5.8 Literature review on synchronization and further reading
Exercises
References

PRACTITIONER'S CORNER

The term *advanced* in the title of this chapter has already surrendered its level of computation. Indeed not all the sections may interest practitioners. Nonetheless, this chapter and the next present a practical framework and tools for improving and optimizing the construction and setting of transit timetables. Practitioners can, almost effortlessly, study the procedures with the aid of graphical schemes, though the process of constructing timetables themselves requires some mathematical treatment.

The present chapter contains four main, independent parts. First, it picks up where Chapter 4 left off, with the construction of timetables, while assuring an even average max load at each vehicle's max load point. This contrasts with Chapter 4's even average max load at a specific time point, usually the hourly max load point. Second, the possibility of mixed fleet sizes is introduced, in timetabling, to attain (1) even headways and even average load and (2) minimum empty-seat km and minimum passenger wait time. Then, Section 5.5 presents the principles of optimization to aid the remaining parts of this chapter. The third part covers optimal timetables, in which passenger-crowding situations are minimized and are derived for a reduced fleet size. In other words, an answer is given to the question: What are the best timetable frequencies that can be assigned for a given fleet size when trying to minimize expected overcrowding situations? The fourth main part of this chapter addresses and formulates the problem of generating headway-based timetables, in which the number of simultaneous vehicle arrivals at connection (transfer) points is maximized. This is a maximum synchronization problem in time and space that is meant to enable the transfer of passengers from one route to another with minimum waiting time at the transfer points. Heuristic procedures (i.e., procedures or algorithms that do not result for sure in an optimum solution) are proffered, in which the setting of departure times follows efficient rules. These results are then improved (i.e., increasing the number of meetings between vehicles on different routes) by allowing a shift in departure time within given boundaries. This improvement uses defined efficient rules imbedded in the second heuristic procedure. We provide detailed examples to illustrate the two proposed procedures explicitly. These examples follow the two procedures step-by-step and can help practitioners gain a better grasp of the analysis and results. Overall, Sections 5.7 asserts that, basically, a noncoordinated transit network is simply the unfolding of a miscalculation.

After reading Section 5.1, practitioners are encouraged to follow carefully the implementation of Example 2, used initially in Chapters 3 and 4 in the case of even loads at different vehicle-max load points. This implementation takes place in Section 5.2.3, which includes an interpretation and explanation of the procedure developed. Section 5.3 attempts the creation of timetables featuring both even headways and even loads; this is, in particular, relevant for a fleet consisting of different vehicle sizes or if a new fleet is to be purchased. In addition, practitioners may be interested in Sections 5.6.1 and 5.6.3, and possibly 5.6.4, as they discuss this special problem of optimal timetables for a fixed fleet size. Practitioners are encouraged to read Section 5.7 to get a better understanding of what a synchronized service means. It is also useful to solve the exercises at the end of the chapter by means of the procedures described.

The following is a short story about people who do not know how to keep up with a timetable. Three persons arrive at a train station and ask for the next train to some destination. The conductor in the booth replies that the train has just left and the next one will leave in 1 h. The three look at one another and say: "Let's go and have a beer, we have time." When they return, the conductor tells them that they missed the train by 2 min and they will have to wait 1 h again. The three go off again to have another beer. For the third time, they miss the train, by 1 min. They are then told that the next train will be the last one of the day and will be leaving in 90 min, so they better be on time. Ninety minutes, the three nod, and go to have more beer. Arriving back at the station, they see the train starting to pull out. All three run after it, two succeeding to hang on while the third person remains standing and laughing on the platform. The conductor asks him: "Why are you laughing?" He replies: "You see the two people hanging onto the train; they just wanted to escort me to the train." We hope in this chapter to offer some planning tools to help passengers to be on time and also to put to rest the following saying: "The only way to catch a bus/train is to miss the one before it."

5.1 INTRODUCTION

Although the even load procedure described in Chapter 4 ensures even average loads of d_{oj} at the jth hourly max load point, it does not guarantee that the average load for individual vehicles at other stops will not exceed d_{oj}; therefore, the result may be overcrowding. In other words, the hourly max load point represents an average peak point at j, whereas the max load point for an individual vehicle can come at another stop and exceed d_{oj}. For a given time period, then, each vehicle may have a different max load point and a different observed average load across the entire vehicle route. The purpose of the first part of this chapter is to derive a timetable such that, on average, all vehicles will have even loads (equal to the desired occupancy) at the max load stop for each vehicle. The adjustments in the timetable are not intended for highly frequent urban services, in which the headway may be less than, say, 10 min or an hourly frequency of about six vehicles or more. The objective of the first part of this chapter (based on Ceder, 2001 and Ceder and Philibert, 2014) is to construct a timetable so as to avoid passenger-overcrowding situations (i.e., loads greater than d_{oj}).

The second part of this chapter introduces another variable, namely, the size of vehicles, to attain timetables that result with both even headways and average even load. Certainly, the implication of different vehicle sizes refers also to the component of vehicle scheduling, and this will be dealt with in Chapter 8. This new approach has the advantage of having even headway timetables with balanced passenger loads resulting in an improved reliability of service.

The third part of the chapter will examine optimal timetables for a network of transit routes with reduced fleets. In a practical context, this problem can arise when the fleet size is reduced because of vehicle-age attrition and budget policy decisions. Although the next two chapters cover vehicle scheduling and minimization of fleet size, they assume that the frequencies or departure times are given. The purpose of the second part of this chapter is to construct timetables while assuring that (1) the difference between the observed max load at hour j and d_{oj} is minimized for all routes and (2) the given (available) fleet size is sufficient. This is a problem of optimal multiterminal timetable construction for different fixed sizes of a vehicle fleet. The formulation used in this part follows Ceder and Stern (1984).

The fourth and last main part of this chapter focuses on the problem of how to construct a timetable with a maximum number of simultaneous vehicle arrivals at connection (transfer) points. From the user perspective, a global approach to establishing these timetables usually considers the minimization of travel and waiting (and possibly walking) times. This, then, becomes a transit-network scheduling problem, which utilizes O-D (Origin–Destination) data. We assume, as is done in practice, that there is an existing transit network of routes with a certain passenger demand according to the time of the day. Maximum synchronization is a rather important objective from both the agency's and the user's perspectives. The process described in this chapter follows Ceder et al. (2001) and Ceder and Tal (2001).

5.2 EVEN MAX LOAD ON INDIVIDUAL VEHICLES

This section continues from Sections 4.3 and 4.4, but employs a procedure to assure average even load at each vehicle's max load point. This is in contrast to Principle 4.2 and Proposition 4.2 (Section 4.4.1), in which even loads are practicable only at the hourly max load point. To have an even desired load at the (individual) vehicle's max load is an additional notion in the attempt to further reduce overcrowding; the procedure can be implemented in certain transit operation scenarios by route and time period. We will continue to use Example 2 as it appeared in Figure 3.7 and Table 4.1. Recently, this procedure was published by Ceder and Philibert (2014).

5.2.1 Underlying principle

The results of applying Proposition 4.2 on Example 2 are shown in Figure 4.5 and Table 4.3. The upper part of Figure 4.5 and the first column of Table 4.3 show the derived departure times, based on even load at Stop 3 (6:00–7:00) and Stop 2 (7:00–8:00), using Method 2 for frequency determination. Take, for example, the resultant departure at 7:48 (Figure 4.5 and Table 4.3, using Method 2), with 60 average passengers on board at Stop 2 (hourly max load point). This (desired) load of 60 is derived by construction. However, if we check the average load of this 7:48 departure at Stop 3, we will find a larger load than 60. That is, from Figure 3.7, the average passenger-arrival rate between 7:35 and 7:50 is 83/15 = 5.53 passenger per minute. Hence, between 7:36 and 7:48 (the results based on Proposition 4.2), this rate holds and will yield $12 \cdot 5.53 \approx 66$ passengers on the 7:48 bus at Stop 3. In order to overcome such undesirable overcrowding (above the desired load of 60), the following principle is employed:

Principle 5.3

Construct a cumulative passenger-load curve at each stop (except for the arrival point) moving horizontally per each d_{oj} (desired occupancy at time period j) for all j on each curve until each of the cumulative-load curves is intersected, and then vertically to establish a departure time for each curve. The required departure time is the *minimum* time across all curves. Using the last determined departure time, set the corresponding loads across all curves; add to these loads d_{oj} or the next d_{oj} (in the transition between time periods) and move horizontally and vertically, as in the first step, to derive the next departure time. Repeat until the end of the time span.

Proposition 5.3

Principle 5.3 results in departure times such that the observed average max load on individual vehicles approaches the desired occupancy d_{oj}.

Proof: Proposition 5.3 is proved, as in Chapter 4, by construction. Fortunately Principle 5.3, similar to Principles 4.1 and 4.2, can be graphically interpreted to ease the proof. Figure 5.1 illustrates Principle 5.3 for Example 2 in Figure 3.7. Figure 5.2 is an enlarged part of Figure 5.1 for the beginning times at Stops 2 and 3. Figure 5.1 shows the load profiles of the five departures and three cumulative-load curves at three stops. The curves at Stops 2 and 3 are shifted by 8 and 14 min, respectively, to allow for an equal time basis at the route's departure point. At the initialization, the value of $d_{o1} = 50$ is coordinated with the

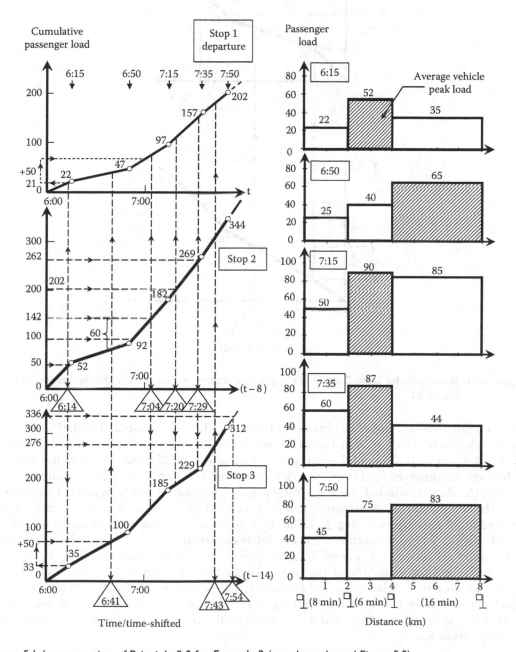

Figure 5.1 Interpretation of Principle 5.3 for Example 2 (see also enlarged Figure 5.2).

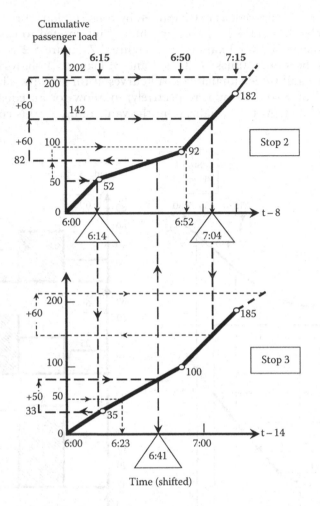

Figure 5.2 Enlarged part for Stops 2 and 3 of the graphical interpretation of the algorithm balance shown in Figure 5.1.

three cumulative-load curves to obtain: 6:51.5 at Stop 1, 6:14 at Stop 2, and 6:41 at Stop 3. Although this is clearly seen in Figure 5.3 for Stops 2 and 3, the determination at Stop 1 is based on an arrival rate of 2 passengers per minute (=50/25) and an additional load of 3 beyond the observed load of 47 at 6:50. According to Principle 5.3, the minimum time among the three is selected: the departure at 6:14 (emphasized in the figures). This means that the scheduled 6:15 departure is shifted backward by 1 min to have an average of 50 instead of 52 passengers at Stop 2. Then, $d_{o1} = 50$ is added to the value of the cumulative-load curves at 6:14 to attain three more candidate departure times: 7:02 (see following note), 6:52, and 6:41 for Stops 1, 2, and 3, respectively. Hence, 6:41 is selected. If the candidate time is beyond the boundary of the time period (7:00), another d_{oj}, $d_{o2} = 60$, should be applied; therefore, 7:02 becomes 7:07 for Stop 1. The procedure continues with $d_{o2} = 60$ and results in 7:04 as the third departure having an average even 60-passenger load at Stop 2 and four more departures as illustrated in Figure 5.1. This completes the proof-by-construction of Proposition 5.3.

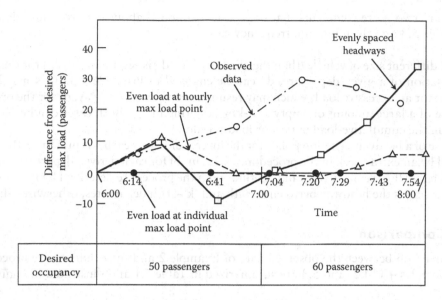

Figure 5.3 Differences between individual vehicle's max load and the desired occupancy according to the procedure selected in Example 2.

The basic assumptions of Proposition 5.3 are as follows: (1) the change of departure times will not affect the arrival pattern of passengers and (2) the change of departure times (with the same frequency) will not affect passenger demand. These assumptions are not critical to the procedure described, since changes in passenger-arrival patterns and demand will be captured frequently using updated data.

5.2.2 Minimum frequency standard

The procedure described by Principle 5.3, similar to Principle 4.2, does not guarantee that the minimum frequency standard (the inverse of the headway policy), F_{mj}, for each time interval, j, will always be met. Usually, F_{mj} is necessary at the beginning and at the end of the day. Therefore, during the process of Principle 5.3, F_{mj} needs to be checked in the transition between time intervals—that is, whenever a new derived departure is in advance of the time boundary. Using N_j, which is the number of departures derived by Principle 5.3 during time interval j – 1, F_{mj} is checked in the following manner:

1. Check for each j, $N_j > F_{mj}$?; if yes—END, otherwise continue j = 1, 2, ..., q.
2. Calculate the alternative desired occupancy, d_{om}, for the minimum frequency situation: $d_{om} = \max_{i=1,2,...,n} \lfloor L_i(t_j) - L_i(t_{j-1}) \rfloor / (F_{mj} + 1)$, and return to the procedure based on Principle 5.3 at the (j – 1)th transition time, where there are n stops, t_j is the jth transition time (between time periods), and $L_i(t_j)$ is the cumulative-load curve at Stop i at t_j; change d_{oj} to d_{om} and go to (1).

The procedure described by steps (1) and (2) ensures that the minimum frequency criterion will always be met. If the procedure described by Principle 5.3 results in time interval j with a frequency of less than F_{mj}, then the maximum difference in passenger loads on the cumulative-load curve between t_j and t_{j-1} determines a new desired occupancy, d_{om}. This d_{om} then replaces d_{oj} until the derivation of the first departure time at interval j + 1.

There are two more comments that should be mentioned about the procedure described by Principle 5.3 and the minimum frequency case:

1. If a different size of vehicle (than the one for which d_{oj} is set) is considered for the determination of a given departure, d_{oj} can be changed in Principle 5.3. This may be the case for an excessive load, which may result in too short a headway, or for the opposite case of a large amount of empty seat-km, resulting in F_{mj}. Both cases can be observed from the cumulative-load curves or load profiles.
2. If use of a headway criterion policy (or the inverse of F_{mj} for each j) is preferred, then there will be an extra check in the procedure for F_{mj} in (b) for each derived departure time k at j: if $h_{kj} > 1/F_{mj}$, set $h_{kj} = 1/F_{mj}$ and continue with the procedure described by Principle 5.3, where h_{kj} is the headway between the kth and $(k-1)$th departures, otherwise—END.

5.2.3 Comparison

The comparison between the observed data of Example 2 and the results of the procedures using Principles 4.1, 4.2, and 5.3 are summarized in Table 5.1 and illustrated in Figure 5.3.

Table 5.1 Departure times for Example 2 route-departure points (three even headways and even load procedures), accompanied by their associated max load and its starting stop, using frequencies in Method 2

Trip characteristics associated with route-departure point		Observed data	Procedures		
			Evenly spaced headways	Even average load at hourly max load point	Even average load at individual max load point
1st[a] departure	Depart. time	6:15	6:22	6:23	6:14
	max load	52	60	61	50
	max load point	Stop 2	Stop 2	Stop 2	Stop 2
2nd		6:50	6:44	6:50	6:41
		65	41	50	50
		Stop 3	Stop 3	Stop 3	Stop 3
3rd		7:15	7:05	7:07	7:04
		90	61	61	60
		Stop 2	Stop 2	Stop 2	Stop 2
4th		7:35	7:22	7:22	7:20
		87	66	59	60
		Stop 2	Stop 2	Stop 2	Stop 2
5th		7:50	7:39	7:36	7:29
		83	77	62	60
		Stop 3	Stop 2	Stop 2	Stop 2
6th		—	7:56	7:48	7:43
			94	66	60
			Stop 3	Stop 3	Stop 3
7th		—	—	8:00	7:54
				66	60
				Stop 3	Stop 3

[a] Subsequent to a predetermined 6:00 departure.

Table 5.1 shows the associated *individual* average max load under each departure; its corresponding stop appears under max load.

Figure 5.3 shows the diversity of the individual max loads across all the procedures and the observed ones. Certainly, this comparison applies only to the specific Example 2, as it will vary from one situation to another. In situations in which the hourly max load point usually coincides with the individual max load, the results of Principle 4.2 will be close to those of Principle 5.3 as is the case of Example 2. In the examples that were presented by Ceder (2001), the differences are more significant.

5.2.4 Case study of even headway and the two even load procedures

This section is based on Ceder and Philibert (2014) in which real data from Auckland, New Zealand, are used for the comparison between even headway and even load at hourly max load point and even load at individual max load timetables. Two types of analyses are considered: (1) use of calculated stop-based demand pattern exhibited by the data and (2) use of a demand derived by simulation. In other words, Analysis (1) examines passenger loads on the vehicles for each of the three timetables derived and Analysis (2) examines the same, but introduces stochastic consideration of travel time with different passenger-boarding rates for the three timetables.

Consider the even headway procedure as EHP, the even load procedure based on hourly max loads as ELHP, and the even load procedure based on individual-vehicle max loads as ELIP. The data were collected for 3 weeks on 2012 by both intelligent transportation systems (ITS) (GPS-based) information and checkers. The three timetables by the three procedures were calculated utilizing MATLAB®. The values of the desired occupancy and minimum frequency standards follow those of the Auckland company, that is, $d_{ou} = 37$ (all day), $F_{mj} = 4$ (6 a.m.–10 p.m.), and $F_{mj} = 2$ (10 p.m.–midnight).

The results of Analysis (1), without simulation, and Analysis (2) with simulation, are shown in Table 5.2 and Figures 5.4 and 5.5, respectively. These table and figures refer to the description of overcrowding at the max load point of each vehicle (each departure) in comparison with the desired occupancy. Table 5.2 shows the percentage of departures facing overcrowding by each procedure and Figure 5.4 of the load differences of Analysis (1). We note that Analysis (2) employs simulation (Vissim Release 5.40) for the examination of overcrowding situations with the use of variations of the averages of passenger-boarding rates and travel times, as is shown in Figure 5.5. The results of both analyses clearly demonstrate that utilizing EHP leads to overcrowded vehicles. However, the difference between the results of ELHP and ELIP is much less than with EHP. The stochastic impact of variations of passenger-boarding rates and travel times certainly cause increase in the overcrowding percentages.

Analysis (2) enables the comparison between the scheduled and simulated departure times at each stop. This comparison of schedule adherence has a direct implication of the reliability

Table 5.2 Results of overcrowding by Analysis (1) compared with Analysis (2)

Characteristic	EHP(%)	ELHP(%)	ELIP(%)
Percent of overcrowded vehicles at the max load point using calculation (Analysis (1))	16	1.2	0
Percent of overcrowded vehicles at the max load point using simulation (Analysis (2))	19.1	9.4	7

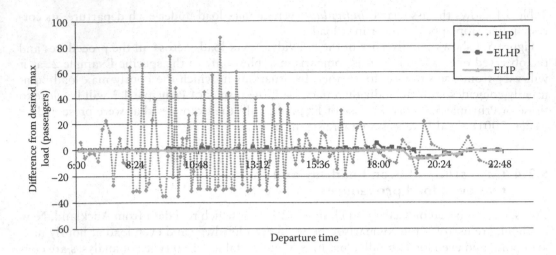

Figure 5.4 Differences between the Max and desired loads for individual vehicles using Analysis (1) without simulation.

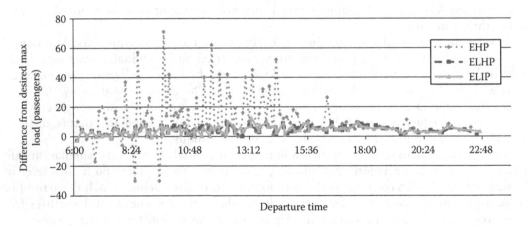

Figure 5.5 Differences between the Max and desired loads for individual vehicles using Analysis (2) with variations of boarding rates and travel times.

of service, or in other words, on the image of the transit service. Figure 5.6 illustrates these differences. Careful observation of the results of Figure 5.7, and an analysis of the standard deviations of these differences, shows that the ELJP results in the minimum discrepancy in most of the stops, thus with the best reliability expected.

5.3 BOTH—EVEN HEADWAYS WITH EVEN AVERAGE LOADS

This section continues from Sections 4.3 and 4.4 in which timetables with even headways and even average (passenger) loads, respectively, were introduced. The even headway feature reduces the flexibility of the scheduler to accommodate fluctuations of demand within a given time period. This lack of flexibility may result in undesirable operational scenarios

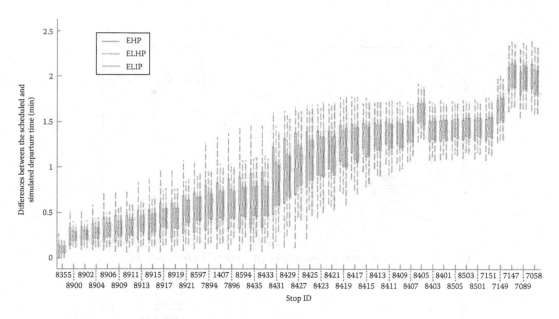

Figure 5.6 Results of scheduled vs. simulated headways using Analysis (2).

such as overcrowding or vehicles running almost empty. Uneven loads lead to either passenger discomfort, in the case of overcrowding, or uneconomical and energy inefficient operation of the vehicles in the latter case (Potter, 2003). However, the even load timetables can lead to long and exceedingly irregular headways and thus increase waiting times for passengers who arrive randomly. The time passengers spend at the bus stop waiting for a bus to arrive is largely recognized as a major factor of satisfaction of the transit user. The waiting time interpretation and formulation is described in Section 18.4.

This section introduces an approach to overcome the disadvantages of both even headway and even load by using different types and sizes of vehicles. Figure 5.7 illustrates how multiple vehicle sizes can be used to achieve such timetables. Nonetheless, in practice, this is not always possible. Hence, the quality of the timetables is based on two criteria: (1) load discrepancy on the vehicles from a desirable load and (2) time discrepancy from a desirable headway. The process described follows Ceder et al. (2013b).

5.3.1 Graphical approach

The focus of the following analysis is to determine the appropriate headway and departure times so as to come near to even headway and even load timetables. This is done by subsequent heuristics.

A range of feasible headways for multiple vehicle sizes are calculated using the inverse of the frequency based on Method 2 or on Equation 3.2; this is to find the most suitable headway. It is done by calculating a lower bound, Hmin, and an upper bound, Hmax, for the headway by using the smallest and largest vehicle sizes, respectively. Any headway outside this range would result in either underutilized service or overcrowding situation. Given a length of a time period T_j, $j = 1, 2, ..., q$, a desired occupancy d_{okj} for vehicle size S_k,

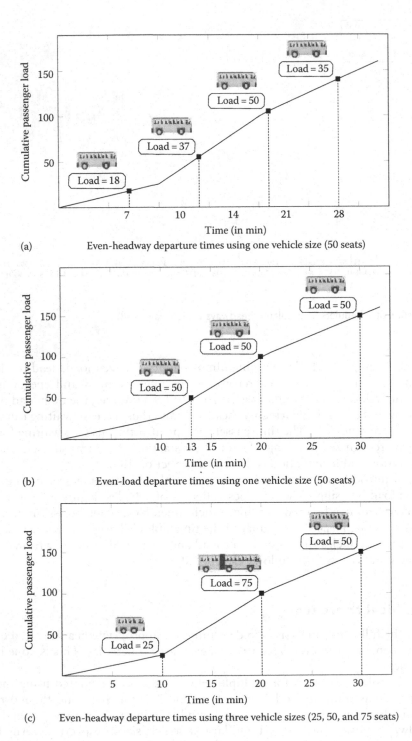

Figure 5.7 Illustration of departure times with their corresponding passenger loads in three cases with (a) even headways (b) even load, and (c) both, by three vehicle sizes.

$k = 1, 2, \ldots, K$, and period j, and hmax the inverse of Fmin, the range for the best headway, Bhw, is formulated in the following equation:

$$\text{Hmin} = \max\left(\frac{d_{o1j} \cdot T_j}{P_{mj}}, \text{hmin}\right) \leq \text{Bhw} \leq \text{Hmax} = \min\left(\frac{d_{okj} \cdot T_j}{P_{mj}}, \text{hmax}\right) \tag{5.1}$$

where

d_{o1j} and d_{okj} are, respectively, the desired occupancy of the smallest and the largest vehicles for period j

hmin is the minimum headway, from a productivity perspective

5.3.1.1 Heuristic procedure for finding the headway

Let the observed max load of passengers, traversing the max load segment, during period T_j be P_{mj}. The following procedure (Ceder et al., 2013b) describes how to determine the preferred (best) headway Bhw. Two criteria are used: LD and %LD, the total of the load discrepancy and the total of the relative load discrepancy, respectively. The load discrepancy, designated Disc(t) for a departure at t, is the difference between the expected load on a given vehicle size and its d_{okj} at the max load point. The relative load discrepancy, designated RLD(t) for a departure at t, is the percentages of discrepancy from d_{okj} of a given vehicle size.

The best headway can be either based on the load-discrepancy criterion, designated BhwLD, or on the relative load-discrepancy criterion, designated Bhw%LD. The best headway is determined from the minimum between its related total load discrepancy LD(BhwLD) and its related total relative load discrepancy %LD(BhwLD). In practice, timetables are provided in integer minutes. Thus, all time values used such as the headway, increments, and shifts of departure times, are integers. Figure 5.8 summarizes the procedure to find the best headway; this procedure is called Algorithm 5.1.

5.3.1.2 Heuristic procedure for creating timetables

Based on Algorithm 5.1, the expected passenger load for a vehicle departing at time t is designated L(t), and it is based on the cumulative-load curve at the max load point. Following is a heuristic procedure to create timetables based on the previously determined preferred headway. The service hours are assumed to be split into q adjacent periods T_j, $j = 1, 2, \ldots, q$, where T_j, starts at time t_{js} and ends at t_{je}. In this procedure, vehicles are assigned departure times based on even headways while complying with the requirement to handle the observed max load of passengers.

The procedure to create the departure times is called Algorithm 5.2. To enable a comparison base, the last departure of any timetable, *dep*, is modified to coincide with the length of T_j. The transition between subsequent periods is done by modifying their end and start times. In case that the last departure does not coincide with the end of the period and would fall into the next period, the departure is cancelled and the end of the current period is set to the time of the last departure. Following a change of the end of period T_j and the start of T_{j+1}, Algorithm 5.1 can be executed again using the modified period T_{j+1} to determine the best headway.

A flowchart of Algorithm 5.2 is given in Figure 5.9; this figure also highlights some macrosteps such as the *Next t*-function to determine the next departure time. Those macrosteps represent a range of different actions and outcomes, determined by a certain strategy chosen

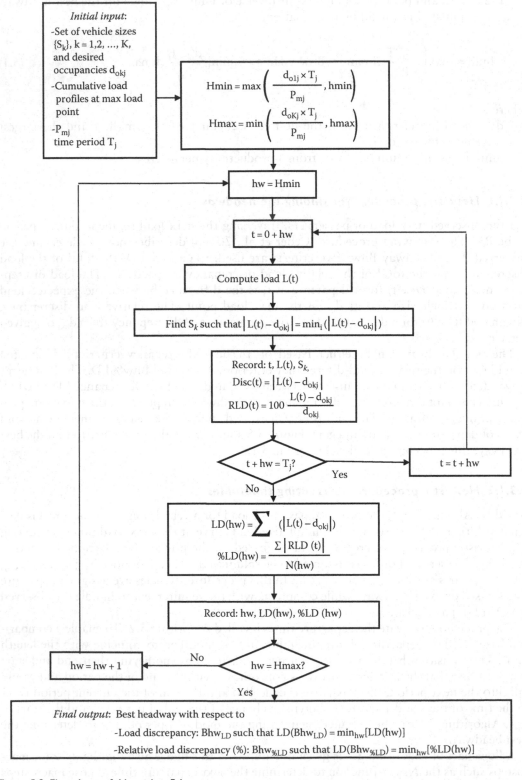

Figure 5.8 Flowchart of Algorithm 5.1.

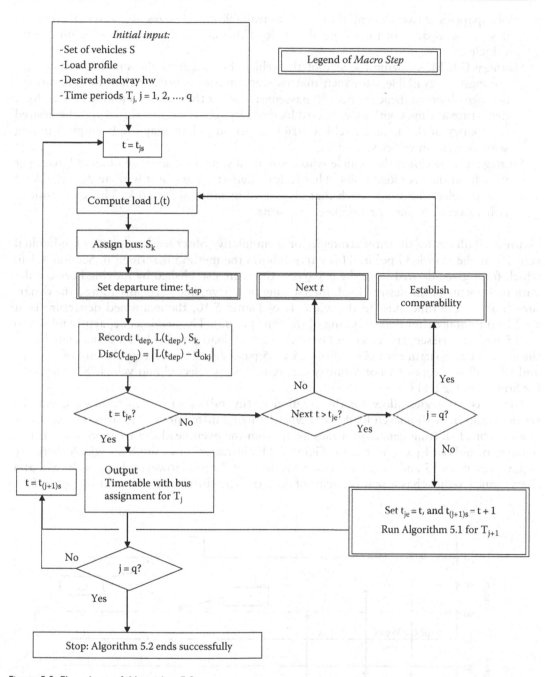

Figure 5.9 Flowchart of Algorithm 5.2.

to deal with the choices of vehicle size. The following three main strategies are considered in the first macrostep to assign vehicle size.

Strategy C1: Minimizing the size of the bus by assigning the largest bus size amongst all available sizes such that its seat capacity is less than or equal to the average observed (hence expected) passenger load, at the max load point. That is, for a departure at t with an expected load of $L(t)$, $d_{o(k-1)j} \leq L(t) < d_{okj}$, where $d_{o(k-1)j}$ and d_{okj} are the desired occupancies of two sequential (w.r.t. size) available vehicle sizes; $d_{o(k-1)j}$ will then be the desired occupancy of the assigned vehicle. This may imply overcrowding on certain vehicles.

Strategy C2: Maximizing the size of the vehicle by assigning the smallest vehicle size amongst all available sizes such that its seat capacity is greater than or equal to the average observed (hence expected) passenger load, at the max load point. That is, for a departure at time t with an expected load of $L(t)$, $d_{o(k-1)j} < L(t) \leq d_{okj}$; d_{okj} is the desired occupancy of the assigned vehicle size k in period j. This may imply empty running seats on certain vehicles.

Strategy C3: Selecting the vehicle whose size is closest to the average observed load, per vehicle, at the max load point. That is, for a departure at time t with an expected load of $L(t)$, select the size S_k such that $|d_{okj} - L(t)|$ is minimal for all k. This can result in either overcrowding or running empty seats.

Figure 5.10 illustrates the three strategies on a cumulative observed-load curve of individual vehicles at the max load point. This curve follows the method described in Section 4.4 in which for a given desired load, the departure times are established by moving horizontally with increments of the desired load, intersecting the curve, and vertically to find the departure time on the time axis. In the example of Figure 5.10, the examined departure is at t = 24 (in minutes after the beginning of the time period). The previous departure relates to t = 15 with 25 passengers (pax) on board at the max load point. The load associated with the examined departure is $L(24) = 60 - 25 = 35$ pax. Vehicle sizes available are of S1 = 25 and S2 = 40 seat capacity. For Strategy C1, vehicle S1 is selected and vehicle S2 is selected for Strategies C2 and C3.

The three strategies allow for the creation of timetables with even headways, but the results could be with uneven loads even when utilizing different vehicle sizes. Hence, there is a benefit of shifting departure times away from the even headway to approach a better balance of onboard passenger load. Figure 5.11 illustrates an example in which there are departures at t = 15 and t = 22 with even headway of 7 min. However, $L(22) = 30$ pax and the assigned vehicle has a seat capacity of 35 pax. The shifting to the right in Figure 5.11

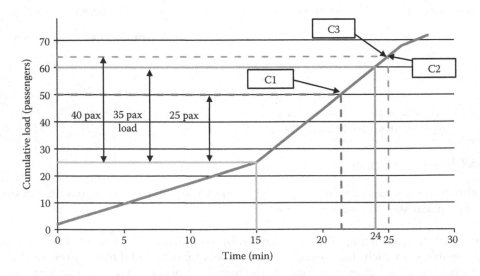

Figure 5.10 Strategies for selecting the size of vehicle.

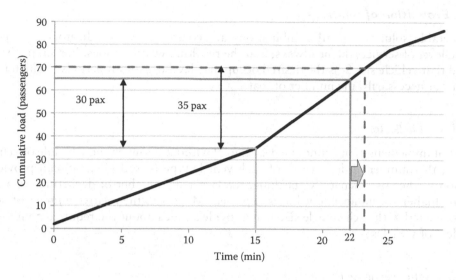

Figure 5.11 Shifting procedure.

(see arrow) will make the headway 8 min, but will provide a more efficient service without harming, in an average sense, the quality of service.

Shifting tolerance can be introduced to maintain the headways as even as possible. Following the calculation of the departure time by the allowed shifting, or without shifting, the next departure examined will be determined by applying the *Next t* function of Algorithm 5.2 in Figure 5.9. Assuming that an initial departure at t_0 is being shifted by a small time increment Δt to $t_0 + \Delta t$, the *Next t* function can have two possible outcomes. First, the *Next t* follows the original timetable before shifting and sets the next departure to $t_0 + hw$. We note that in certain cases Δt might be larger than the desired headway hw to have the next departure as $t_0 + m*hw$ with m being the first integer to comply with $\Delta t < m*hw$. Second, the *Next t* strictly adheres to the desired headway with the next departure time examined be $t_0 + \Delta t + hw$.

Combining the two options for *Next t* with different shifting policies allows for the construction of methods tailored for different preferences. However, one might not be able to predetermine which strategy is the best for the given scenario. In this case, different scenarios should be tested and examined. Five basic methods can be defined and termed as A, B, C, D, and E based on the setting shown in Table 5.3. Each of these methods can use vehicle selection strategies C1, C2, or C3, thus 15 combinations for constructing a set of feasible timetables can be examined. Note that setting up the limited shift to different values can increase the number of possibilities; in Ceder et al. (2013) seven, not five, methods are shown including these possibilities. Table 5.3 shows all combinations.

Table 5.3 All combinations used in the selection of best timetable

	Method				
Strategy	A	B	C	D	E
Bus-selection strategy	C1, C2, or C3	C1, C2, or C3	C1, C2, or C3	C1, C2, or C3	C1, C2, or C3
Setting departure time	No shift	Unlimited shift	Unlimited shift	Limited shift	Limited shift
Next t function	NA	$t + \Delta t + hw$	$t + hw$	$t + hw$	$t + \Delta t + hw$

5.3.1.3 Evaluation of solutions

The timetable solutions of all combinations are compared with indicators of expected cost, the level of service, the headways, and the resulting waiting times. For simplicity, it is assumed that vehicle size S_k is the desired occupancy level d_{okj} for all periods j, which is usually, but not necessarily, the number of seats.

5.3.1.3.1 Cost indicator

The cost of implementation cannot be determined without knowing the schedule of vehicles. However, the number of departures of each vehicle type (size-dependent) can provide an indication of the cost. Generally speaking, the larger the number of departures in a timetable, the higher the number of vehicles expected. Moreover, the relative number of departures associated with each vehicle size may provide an idea about the required composition of the fleet of vehicles.

5.3.1.3.2 Level-of-service indicators

Level-of-service measures are associated with the headways and the deviation of the max load from the standard desired load level. Because the headways are based on the max load observed, the measures used are load-discrepancy dependent. Four measures are used in the analysis: (1) total load discrepancy, (2) average relative (%) load discrepancy, (3) average percentage of standees, and (4) level of service.

5.3.1.3.2.1 TOTAL LOAD DISCREPANCY

The load discrepancy refers to the difference between the average number of onboard passengers at the max load point and the desired occupancy for a given vehicle size. It is Disc(t) = $L(t) - d_{okj}$ for a bus of size S_k in period j and L(t) passengers at the max load point; if positive, it represents overcrowding or average number of standees, otherwise underutilization or average number of empty seats at the heaviest load point. The total load discrepancy for a timetable Tt with N departures at times t_i, i = 1, 2, ..., N, is

$$LD(Tt) = \sum_{i=1}^{N} |Disc(t_i)| \tag{5.2}$$

5.3.1.3.2.2 AVERAGE RELATIVE LOAD DISCREPANCY

In contrast to Disc(t), the relative load discrepancy is the ratio between Disc(t) and the desired occupancy of vehicle S_k in percentage. That is, given that the desired occupancy is the number of seats, a load discrepancy of five passengers in a vehicle with 75 seats has a relative load discrepancy of 6.7% compared with 25% for a vehicle size of 20 seats. Thus, it is in percentage RLD(t) = $100((L(t) - d_{okj})/d_{okj})$. The average relative load discrepancy for a timetable Tt with N departures at times t_i is

$$\%LD(Tt) = \frac{\sum_{i=1}^{N} |RLD(t_i)|}{N} \tag{5.3}$$

5.3.1.3.2.3 AVERAGE PERCENTAGE OF STANDEES

Equation 5.3 provides the average relative load discrepancy in absolute terms, but does not give any insight of the nature of the discrepancy. The average percentage of standees can add further information to enable drawing more solid conclusions. The average percentage of standees for a timetable Tt with N departures at times t_i is

$$\%\text{Stand}(Tt) = \frac{\sum_{i=1}^{N} \max\left[\text{RLD}(t_i), 0\right]}{N} \tag{5.4}$$

5.3.1.3.2.4 AVERAGE LEVEL OF SERVICE

The average level of service (LOS) refers to the loading situation of all departures at the daily max load point. Given a timetable Tt with N departures at times t_i and vehicles of sizes S_{ki} (with a desired occupancy of $d_{o(ki)j}$) carrying $L(t_i)$ passengers at the max load segment, for $i = 1, 2, ..., N$ (assuming theoretically that for each departure another size of a bus can be assigned), the average level of service is defined as

$$\text{LOS} = \frac{100}{N} \sum_{i=1}^{N} \frac{L(t_i)}{d_{o(ki)j}} \tag{5.5}$$

That is, the desired occupancy level at every departure provides an LOS of 100%. Passengers experience overcrowding for LOS greater than 100% and vehicles are underutilized for LOS less than 100%. This indicator represents only an average figure, thus cannot be used on its own.

5.3.1.3.3 Time indicators

The quality of the timetable with respect to the headway is determined by applying the following indicators.

5.3.1.3.4 Average headway

Computation of the average headway of the final timetable.

5.3.1.3.5 Expected waiting time

The expected waiting time, of randomly arriving passengers, is dependent on the variance of the headway as is shown in Section 18.4. Because of the coupling of regularity and the average headway criteria, this is a better indicator than the average headway. However, to make different timetables comparable the consideration of both indicators needs to take place.

5.3.1.3.6 Time-discrepancy ratio

The time-discrepancy ratio (TDR) is defined as the ratio between the expected waiting time and the average headway. This normalizes the waiting time and makes different timetables comparable. It provides information about the evenness of the headways. For simplicity this indicator is set between 0% and 50%, where 0% presents the perfect even headway timetable without any time discrepancy and 50% the case in which the headway is exponentially distributed. This TDR indicator is termed %TDR and shown in the following equation:

$$\%TDR = \left(\frac{Wt}{hw} - \frac{1}{2}\right)100 = 50\left(\frac{Var(hw)}{hw^2}\right) \tag{5.6}$$

Further details and a case study of this even headway and even load graphical-based approach appear in Ceder et al. (2013). The following different approach is based on the optimization of the timetables to attain at the same time (called multiobjective approach) minimum empty-seat km and minimum passenger waiting time. The interpretation of *optimization* and its principles appear in Section 5.4; those who are unfamiliar with optimization are advised to read Section 5.5 before Section 5.4.

5.4 TIMETABLES BASED ON A MULTIOBJECTIVE APPROACH

The methodology described in this section is based on Hassold and Ceder (2012). It is a network-based approach using a multiobjective label correcting algorithm to find the set of dominant solutions considering minimizing a penalty for running empty seats (ES) on the vehicles and minimizing the estimated passenger waiting time (WT).

Figure 5.12 illustrates a sample network of the problem for a given bus line. The network consists of nodes representing feasible departure times of buses and directed arcs connecting certain nodes, which represent feasible sequences of departures. The source S and sink T are the beginning and end of the operating period of the bus service. The service is using multiple vehicle types/sizes, represented by different arcs in the network. A dotted line between departures D2 and D5 implies that D5 is performed by a standard bus, if a path is selected from S to T via this arc. For another path, for example, a path containing D3 the departure D5 is handled by a different vehicle type, a minibus. Moreover, each arc has two cost elements associated with the objectives of the problem. These costs represent waiting times for passengers arriving between the two departures, and a penalty for empty seats. The WT is calculated as passenger waiting minutes, namely, the sum of passenger waiting times; in the example it is assumed that passengers are arriving randomly at a uniform arrival rate

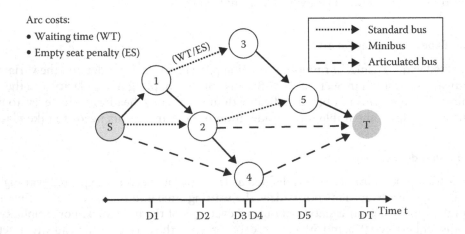

Figure 5.12 Sample network.

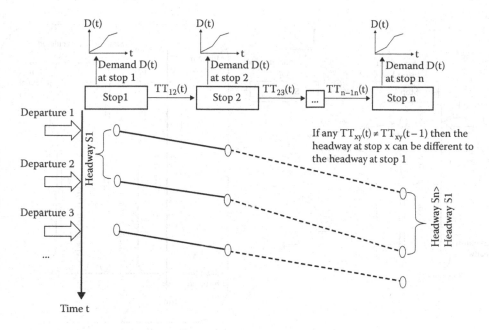

Figure 5.13 Base information to create the network.

though any demand function can be considered. ES is the empty seats on the bus multiplied by the distance traveled. In Figure 5.12, the arc D2-T and D4-T share the same end node and same demand pattern. Because the headway D2-T is greater than for D4-T, the total cost WT will be greater for D2-T. However, both arcs represent the same vehicle type such that the higher occupancy for D2-T results in a smaller ES than for D4-T.

The procedure of creating the network for a transit line is illustrated in Figure 5.13. The cumulated demand function $D(t)$ is given for each stop of the line. Naturally, the demand functions shown vary along the line. In addition, the procedure considers the dependent travel times between stops $TT(t)$ visualized by different slopes in the connecting arcs between two nodes. It is to note that these travel times can vary significantly by the time of the day. In the example of Figure 5.13, a greater travel time is causing the headway between departure 1 and 2 to be shorter at Stop 2 than at Stop n. This also impacts the waiting times for arriving passengers at Stop n. The nodes on the left side of Figure 5.13 represent departure times at the first stop and each sequence of connected nodes has to satisfy the capacity constraint of the vehicle at all individual stops; otherwise the selected departure time is invalid. In addition, there are minimum and maximum headway constraints providing a certain time window for each departure time to take place. The choice between different vehicles types (minibus, standard bus, and articulated bus) results in different feasible departure-time ranges for each vehicle type as is illustrated in Figure 5.14. The solution algorithm starts at the source node S and finds the feasible departure times for each given vehicle type that are then added to the network as individual nodes. These nodes (see Figure 5.12) are connected to S by a directed arc with the information about the type of vehicle used and the associated costs. Separate arcs are inserted for nodes reached by different vehicle types. The algorithm continues from the next (earliest) node in the network, and ends once all nodes within the operating hours have been processed. Hassold and Ceder (2012) provide further elaboration of this process to be explicated in the following.

Let $G = (N, A)$ be a directed network with $N = \{1, ..., n\}$ representing the set of nodes and S and T (of Figure 5.12) and $A = \{(i_1, j_1), ..., (i_m, j_m)\} \subseteq N \times N$ is the set of arcs. The set of

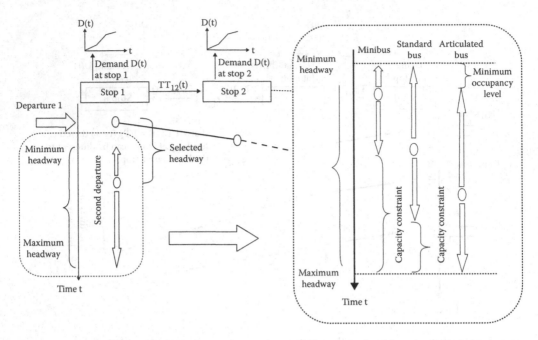

Figure 5.14 Headway ranges for different vehicle types.

arcs consists of all feasible arcs added during the solution algorithm. Moreover, two positive arc costs $c_{ij} = (c_{ij}^1, c_{ij}^2) \in N \times N$ are defined for the network for waiting time and the empty seat penalty.

Finding a feasible timetable can now be interpreted as finding a path from the source $S \in N$ to the sink $T \in N$ in the completed network G. Optimal timetables refer to the two costs and can be formulated as a shortest path problem; see Appendix 9.A for the solution of this problem. For instance, the formulation of timetable with the minimal waiting time (defined as the arc cost c_{ij}^1) is

$$\min z(x) = \sum_{(i,j)\in A} c_{ij}^1 x_{ij} \tag{5.7}$$

$$\text{s.t.} \sum_{(i,j)\in A} x_{ij} - \sum_{(j,i)\in A} x_{ji} = \begin{cases} 1, & \text{if } i = S \\ 0, & \text{if } i \neq S, T \\ -1, & \text{if } i = T \end{cases} \tag{5.8}$$

$$x_{ij} \in \{0,1\}, \quad \forall (i,j) \in A \tag{5.9}$$

This is in fact a network flow formulation (see Appendix 6.A) of the shortest path problem with the objective function in Equation 5.7 aiming at minimizing the sum of the costs for all arcs used. Equations 5.8 and 5.9 make sure that the inbound flows equal the outbound flows at every node and that every arc can be used either exactly once or not at all. Nodes S generates and Node T absorbs one unit of flow, thus these two nodes are exceptions. Standard shortest path algorithms, such as Dijkstra's (1959) can be used to find the timetable with the least waiting time or the least empty-seat km. See Appendix 9.A for these algorithms.

However, we are dealing with two objectives at the same time and, thus, we face a biobjective problem. Equation 5.10 shows the biobjective formulation, and to this we add the constraints of Equations 5.8 and 5.9:

$$\min z(x) = \begin{cases} z_1 = \sum_{(i,j) \in A} c_{ij}^1 x_{ij} \\ z_2 = \sum_{(i,j) \in A} c_{ij}^2 x_{ij} \end{cases} \tag{5.10}$$

The objective function vector for a very feasible solution in X consists of two elements, one for each objective. In contrast to the single objective case this makes it impossible to compare solutions with a simple *greater/equal/smaller than* relation. To do so the following, Definition BI is introduced:

Definition BI: A solution $\hat{x} \in X$ with objection function values $z_1(\hat{x})$ and $z_2(\hat{x})$ is called efficient or dominant if there is no other solution x' with $z_1(x') \leq z_1(\hat{x})$ while at the same time $z_2(x') \leq z_2(\hat{x})$. Hence every x' that satisfies either of the two equation pairs dominates solution $\hat{x}$.

Based on Definition BI, the different solutions can now be compared with. If a solution x' dominates solution $\hat{x}$, then solution x' is better and solution $\hat{x}$ can be discarded. It implies that a multitude of dominant solutions can exist; this set of dominant solutions is referred in the following as Pareto frontier.

The problem of creating timetables that consider both waiting time and empty seat penalty can be solved using a biobjective label-correcting algorithm that works like its single-objective counterparts; the main difference being that one node can carry a set of dominant labels instead of only one value. Based on the comparison of solution strategies for biobjective shortest path problems by Raith and Ehrgott (2009), we use the approach of Brumbaugh-Smith and Shier (1989) as follows.

The algorithm starts by labeling the source S with its set of labels consisting only of one element Labels(S) = {(0, 0)}. The sets of labels are extended to the succeeding nodes, which means a particular node j carries all labels of its preceding nodes i, increased by the cost of the incoming arcs (i, j). However, only dominant labels are added to the set of labels; dominated labels are discarded. This operation is called merging; it is the most time-consuming step of the algorithm. Using ordered label sets allows a computational effort of $O(|L| + |M|)$ when merging label sets L and M. See Section 5.4 for understanding the $O(\cdot)$ notation. Every time the algorithm changes a label set the set is marked for follow-up. Marked sets are processed in the order the changes occurred until the set of marked nodes is empty and the algorithm ends as shown in the following description.

Algorithm: Bi-Objective Label correcting

---

followUp={} //set of nodes to check again, in order of entering
Labels(i)=Ø //starting with an empty set of labels at each node
Label(S)={(0,0)} //adding label to node S
append *S* to *followUp*//making S the first node to consider
while *followUp* is nonempty **do**
    remove first node *i* from *follow Up*
    for all outgoing arcs *(i,j)* **do** //extend the set of labels to all nodes
        merge *(Labels(i) + c_{ij}, Labels(j))*// extend the label set from *i* to *j*
        if (changes in *Labels(j)* **AND** *j* ∉ *followUp*) **then**
            append *j* to *follow Up*
        endif
    end for
end while

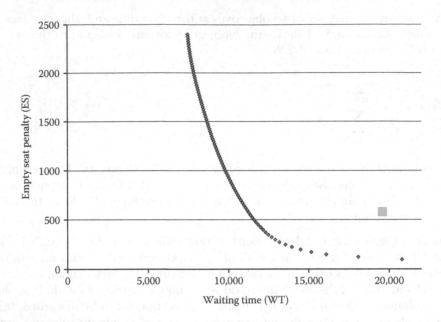

Figure 5.15 Pareto frontier of optimal solutions compared with the current timetable (gray square).

After the algorithm terminates successfully, the label set at T contains the lengths of all the shortest paths between S and T. The corresponding paths can then be found by backtracking through the network. The algorithm then results with a set of Pareto-optimal timetables exhibited, for example, in Figure 5.15; the data used for Figure 5.15 are explicated in Hassold and Ceder (2012). That is, there are no better timetables in terms of [ES, WT] pairs than those belonging to the Pareto-optimal set shown by the curve in Figure 5.15; any better, but impossible, timetable will have a solution to the left and/or below the Pareto frontier curve. The alternative solutions with different [ES, WT] pairs, along the curve, enable the transit planner to decide on the desirable timetable for implementation. This multiobjective approach also provides benefits for the vehicle-scheduling phase when using different vehicle sizes; see Chapter 8 for details. Finally, the next section introduces the principles of *optimization* to improve the comprehension of this section and of Sections 5.6 through 5.8 as well as of other chapters.

5.5 OPTIMIZATION, OPERATIONS RESEARCH, AND COMPLEXITY

This section introduces the notion of optimization, using operations research (OR) principles. Optimization usually means finding the best solution to some problem from a set of alternatives. Following Ceder (1999), OR means a scientific approach to decision-making or formulating problems so they can be solved quantitatively. Hence, optimization involves dealing with mathematical models of situations or phenomena that exist in the real world. The development of powerful computers has had a very strong impact on the motivation to develop complex, sophisticated OR models.

The basic elements in OR models include the following: (1) *Variables* (decision, policy, independent, or dependent variables); these are quantities that can be manipulated to achieve some desired objective or outcome. (2) *Objective function* (profit, revenue, or cost function); this is a measure of effectiveness (efficiency, productivity, or utility value) associated with some particular combination of the variables and is a function that is to be optimized (minimized or maximized). (3) *Constraints* (feasibility conditions); these are equations or inequalities that the variables must satisfy in addition to providing a minimum or maximum value of the objective function.

Mathematically speaking, the variables are often represented as x_1, x_2, ..., x_{1n} or the vector of the variables $\mathbf{x} = (x_1, x_2, ..., x_{1n})$. The objective function, as a single-value function, is represented as $z = f(x_1, x_2, ..., x_{1n})$, or $z = f(\mathbf{x})$. The constraints are also represented as a function of $\mathbf{x}$, $g_v(\mathbf{x})$, with v as a constraint number. The general variable optimization problem, with m constraints, is to

$$\text{minimize/maximize} \quad z = f(\mathbf{x})$$
$$\text{subject to (s.t.)} \quad g_v(\mathbf{x})\{\geq, =, \leq\}0, \quad v = 1, 2, ..., m \tag{5.11}$$

It is apparent that $\min[f(\mathbf{x})] = -\max[-f(\mathbf{x})]$, and hence, each minimization problem can be treated as a maximization problem and vice versa. Notations that correspond to OR and are used in the formulations in this book contain the symbols $\forall$ (for all), $\in$ (belong to), $\notin$ (not belong to), and Φ (empty set).

Occasionally, we cannot prove that optimality is attained, nor can we prove that optimality cannot be attained in all those cases. In these situations, the OR model is termed a *heuristic procedure*. In certain cases, a mathematical formulation may possibly be too complex to be handled even by advanced computers, thereby leading to the intentional construction of a heuristic procedure. This is actually the case with timetable synchronization, to be dealt with in due course.

A framework for developing a mathematical OR model of a system's problem is shown in Figure 5.16 for a deterministic (average-based) study. It relies on an eight-step procedure. OR analysts should first define the system's (organization's) problem while searching for its actual objectives, then look for the entire system's limitations and constraints. The third step, in Figure 5.16, translates the first two steps into a formulation similar to Equation 5.11. Once the formulation is correctly established, a check can be made of the existence of feasible solutions or solutions that comply fully with all the constraints. If feasibility does not exist, return to the first two steps for possible changes and/or relaxation of some constraints. Otherwise, select a software package or develop an algorithm to solve the OR formulation automatically. The sixth step will involve examining the important issue of the complexity (described in the following) of the algorithm to verify whether indeed the system's problem can be solved by a computer within a reasonable amount of time. If the algorithm is too complex, thus requiring too long a running time, a simplified heuristic procedure may be constructed to solve the problem. Otherwise, a check will have to be conducted of the sensitivity of the problem's objective value to different input data. The final step appropriately presents the results along with the sensitivity analysis and the problem's limitations. In the course of a heuristic procedure, it becomes apparent that optimality cannot be guaranteed.

One important characteristic of an OR model/procedure/algorithm is the efficiency with which the formulation may be run by all the various computing resources required. The solution time is usually checked for the worst-case scenario or worst-case complexity, which is well described by Ahuja et al. (1993). It becomes commonplace to use the notation $O(f(k, q))$, an operation of $f(k, q)$, for expressing the running-time complexity function, which is dependent on the input parameter size k and q, where $f(k, q)$ is the function of those two parameters. For example, if the running time is $f(k, q) = (10k + 0.01k^3) \cdot q^5$, then for all $k \geq 100$ (as we always look for the worst case), the second term in the parenthesis

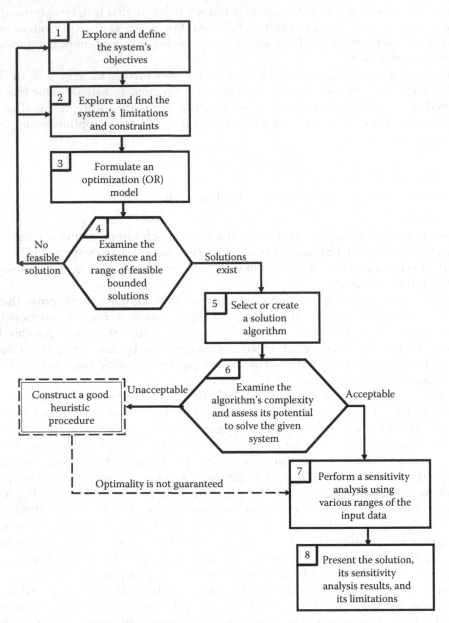

Figure 5.16 Schematic framework of an OR process to solve a deterministic optimization problem analytically.

dominates, and the complexity of this function is $O(k^3 \cdot q^5)$. The $O(\cdot)$ notation provides, for sufficiently large input values, an indication of the number of elementary mathematical operations required. Following Ahuja et al. (1993), the complexity of OR optimization algorithms is often interpreted as *efficient* if its worst case is bounded by a polynomial function of the input parameters; for example, algorithms with $O(k^3 + q^5)$, $O(kq + q^3\log(k^2))$. Nonetheless, the algorithm found in many combinatorial optimization problems is of the type called *exponential-time algorithm* with, for instance, $O(k!)$, $O(3^q)$, $O(k^{\log(q)})$. Some of these hard-to-solve problems belong to a class known as *NP-Complete* (nonpolynomial complete). Certainly, it is a challenge to develop a polynomial-time algorithm for a problem in this class. In these NP-complete problems, we may depart from a completely proven optimization solution and begin developing a heuristic procedure.

The previous Section 5.4, the next two sections, as well as several subsequent chapters, will use the OR optimization concept as a framework for achieving the best results.

5.6 MINIMUM PASSENGER-CROWDING TIMETABLES FOR A FIXED VEHICLE FLEET

The problem addressed in this section is that of optimal multiroute timetable construction for a transit agency. The motivation for this task lies in the challenge facing schedulers/planners in reconstructing timetables when the available vehicle fleet is reduced. The purpose of this section is to present the formulation to solve the problem optimally by using OR tools (i.e., integer programming). Because of the complexity involved in solving integer programming formulation, a heuristic approach designed for a person–computer interactive procedure will also be proffered in the next chapter.

5.6.1 Problem description

More concretely, the problem is as follows: Given a fixed fleet size (number of vehicles) and passenger load at the max load segment between a set of terminals (predefined routes), what is the frequency of vehicle departures that will maximize some measure of passenger satisfaction? A fixed set of terminals and a fixed planning horizon are given.

The time horizon for each terminal is partitioned into a fixed number of q time periods (usually hours, but not necessarily equal). For each time period j, j = 1, 2, ..., q and route, the max load is given. The number of vehicle departures, F_{jk}, for a given terminal k and for each time period j must lie within a bounded range, the two standards of the lower and upper bounds of the frequency (see Figure 1.4). The lower bound is based on a predetermined policy headway and the upper bound on a minimum passenger-loading factor. The objective is to minimize the sum of all service measures that might be interpreted as a minimum passenger-disservice cost, or crowding, at the max load segments.

5.6.2 Formulation

Let d(k, t) represent the total number of departures less the total number of trip arrivals at terminal k up to and including time t. The maximum value of d(k, t) over the schedule horizon $[T_1, T_2]$ is designated D(k). If we denote the set of all terminals as T, the sum of D(k) $\forall k \in T$ is equal to the minimum number of vehicles required to service the set T. This is known as the fleet size formula. It was independently derived by Bartlett (1957), Gertsbach and Gurevich (1977), and Salzborn (1972). Mathematically, for a given fixed schedule:

$$N = \sum_{k \in T} D(k) = \sum_{k \in T} \max_{t \in [T_1, T_2]} d(k, t) \qquad (5.12)$$

where N is the minimum number of vehicles to service the set T. Further interpretation and analysis using Equation 5.12 appears in Chapter 6.

Let T_k be the set of all departure times from terminal k. Let (j, k_1, k_2) $\equiv$ represent a possible trip bundle departing during period j from terminal k_1 to terminal k_2.

Define a 0–1 variable as

$$x^F(\cdot) = \begin{bmatrix} 1, & \text{if F departures are selected during period j from terminal } k_1 \text{ to } k_2 \\ 0, & \text{otherwise} \end{bmatrix}$$

where F is an index running from $L(\cdot)$ to $U(\cdot)$; that is, $F = L(\cdot), L(\cdot) + 1, L(\cdot) + 2, ..., U(\cdot) - 1, U(\cdot)$. Using the definition of Equation 3.2, let $P_m(\cdot)$ be the max load and $d_o(\cdot)$ the desired occupancy—both definitions are associated with period j and the route from terminal k_1 to k_2. The crowding measure (i.e., cost) associated with $x^F(\cdot)$ is defined as

$$c^F(\cdot) = \max[P_m(\cdot) - Fd_o(\cdot), 0] \tag{5.13}$$

and the total crowding defining the objective function, Z:

$$Z = \sum_{\forall(\cdot)} \sum_{F=L(\cdot)}^{U(\cdot)} c^F(\cdot) \cdot x^F(\cdot)$$

The mathematical programming formulation, which is inspired from Salzborn's note (1972) on fleet routing models for transportation systems is stated as follows:

$$\text{minimize} \sum_{\forall(j,k_1,k_2)} \sum_{F=L(j,k_1,k_2)}^{U(j,k_1,k_2)} c^F(j, k_1, k_2) \cdot x^F(j, k_1, k_2) \tag{5.14}$$

s.t.

Bundle departure constraints:

$$\sum_{F=L(\cdot)}^{U(\cdot)} x^F(\cdot) = 1, \quad \forall(\cdot) \tag{5.15}$$

Assigned vehicle bounds:

$$\begin{Bmatrix} \text{the net number of departures less arrivals} \\ \text{that occur before or at t at terminal k as} \\ \text{determined by the value of } x^F(\cdot) \end{Bmatrix} \leq D(k), \quad t \in T_k, k \in T \tag{5.16}$$

where $D(k)$ is the maximum number of vehicles assigned to terminal k.

Resource constraint (total fleet size):

$$\sum_{k \in T} D(k) \leq N_o \tag{5.17}$$

where N_o is the total (fixed and given) fleet size.

Variable constraints:

$$x^F(\cdot) = 0,1 \ \forall F$$
$$D(k) \geq 0 \ \forall k \in T \tag{5.18}$$

In the overall formulation, $D(k)$ will be an integer in any feasible solution.

Constraint (5.15) ensures that only one bundle of departures is selected for a given terminal pair (route) and time period. Constraint (5.16) ensures that the number of vehicles using a given terminal k up to time t does not exceed the number of vehicles, $D(k)$, assigned to terminal k. The left-hand side of constraint (5.16) can be represented as a linear function of the $x^F(\cdot)$ variable (as will be demonstrated in the example in the next section). Not all departure times need be considered, as some lead to redundant equations; this can occur when there is a sequence of departure times unbroken by intervening arrivals. Constraint (5.17) indicates that the sum of vehicles assigned to all terminals is no greater than the given fleet size N_0. Finally, constraint (5.8) defines $x^F(\cdot)$ as a binary and the boundary of $D(k)$.

The optimal solution $\left\{x_o^F(\cdot), D_o(k)\right\}$ yields both the assignment of vehicles to terminals, $D_o(k)$ and the optimal number of departures within each time period. Thus, an optimal timetable may be constructed from $x_o^F(\cdot)$.

5.6.3 Example

In order to gain further understanding of the underlying structure of the mathematical formulation, an example consisting of two terminals and two time period will now be presented. It follows Ceder and Stern (1984). The basic input data of the example problem, along with the 0–1 variables and their associated cost, $c^F(\cdot)$, are indicated in Table 5.4. For each trip bundle $(\cdot)$, the upper and lower bounds on the number of departures are shown in Table 5.3. For example: $L(1, b, a) = 2$ and $U(1, b, a) = 3$. The indicated cost for each $(\cdot)$ is calculated in accordance with Equation 5.13; that is, max $(P_m - F \cdot d_o, 0)$. There are three 0–1 variables for the first trip bundle $(1, a, b)$ and two variables for every other trip bundle. Altogether there are 11 variables—9 x's, $D(a)$, and $D(b)$.

The construction of the bounds for the constraint (5.16) is based on a determination of the arrival and departure times for each of the 9 x's. Table 5.5 contains the information, with the headways equally spaced for each set of departures. This assumes an arrival rate like that shown in Table 5.4.

The example problem can now be constructed in terms of integer programming:

$$\text{minimize } \{Z = 80x_1 + 15x_2 + 28x_4 + 30x_6 + 23x_8\}$$

Table 5.4 Basic input data for the example

Terminals (k_1, k_2)	Travel time	Period j	Time span	No. of passengers P_m	Desired occupancy d_o	$F = 1$ c^F	$F = 1$ Variable	$F = 2$ c^F	$F = 2$ Variable	$F = 3$ c^F	$F = 3$ Variable
(a, b)	4	1	0–4	145	65	80	x_1	15	x_2	0	x_3
		2	4–8	75	47	28	x_4	0	x_5	—	—
(b, a)	3	1	0–4	160	65	—	—	30	x_6	0	x_7
		2	4–8	70	47	23	x_8	0	x_9	—	—

Table 5.5 Departure and arrival times for each set of departures

Departure terminal				*a*							*b*				
Arrival terminal				*b*							*a*				
Variable	x_1	x_2		x_3	x_4	x_5		x_6		x_7		x_8		x_9	
Departure time	4	2	4	$1\frac{1}{3}$	$2\frac{2}{3}$	4	8	6	8	2	4	$1\frac{1}{3}$	$2\frac{2}{3}$	4	8	6	8
Arrival time	8	6	8	$5\frac{1}{3}$	$6\frac{2}{3}$	8	12	10	12	5	7	$4\frac{1}{3}$	$5\frac{2}{3}$	7	11	9	11

s.t.

$$\begin{aligned} x_1 + x_2 + x_3 &= 1 \\ x_4 + x_5 &= 1 \\ x_6 + x_7 &= 1 \\ x_8 + x_9 &= 1 \end{aligned}$$ (i)

$$\begin{aligned} x_1 + 2x_2 + 3x_3 &\le D(a) \\ x_1 + 2x_2 + 3x_3 + x_5 - x_6 - 2x_7 &\le D(a) \\ x_1 + 2x_2 + 3x_3 + x_4 + 2x_5 - 2x_6 - 3x_7 &\le D(a) \\ 2x_6 + 3x_7 &\le D(b) \\ -x_2 - x_3 + 2x_6 + 3x_7 + x_9 &\le D(b) \\ -x_1 - 2x_2 - 3x_3 + 2x_6 + 3x_7 + x_8 + 2x_9 &\le D(b) \end{aligned}$$ (ii)

$$D(a) + D(b) \le N_o$$ (iii)

$$\begin{aligned} x_i &= 0, 1; \quad i = 1, 2, \dots, 9 \\ D(a), D(b) &\ge 0 \end{aligned}$$ (iv)

The constraints in (i) to (iv) are based on Equations 5.15 to 5.16, respectively, and on the information given in Tables 5.4 and 5.5. In constraints (ii), each possible combination of the net number of departures for a given terminal is restricted so as not to exceed the number of vehicles assigned to that terminal. For example, the first constraint in (ii) refers to $0 \le t \le 4$ and the second-to $0 \le t \le 6$, in regard to the net number of departures in terminal *a*. For $0 \le t \le 6$ in terminal *a*, we take three possible departures of x_3, two departures of x_2, and one departure of x_1 and x_5, as opposed to two arrivals of x_7 and one arrival of x_6.

The known OR package MPSX has been used to solve this simple example for $N_o = 7, 6, 5, 4$, 3, 2. It is possible to obtain the solutions by relaxing the integrality constraint on the x's and, if necessary, round off any fractions to the nearest integer. The results are presented in Table 5.6. Note that the right-hand side on N_o may be used to directly obtain a curve of fleet size versus minimum cost in a single computer run. This trade-off is shown as the solid line in Figure 5.17. In the next section, the problem is resolved by allowing small departure times between vehicles.

5.6.4 Variable scheduling

Practical vehicle scheduling often involves shifting departure times in order to better match vehicle assignment with a given set of trips. This practice is mentioned in Chapter 1 and described in Chapter 8. In this case, the departure times are allowed to vary over prespecified limits.

Table 5.6 Optimal results for different fleet sizes

Fleet size no.	Sets of departures in solutions $x_i = 1$	D(a)	D(b)	Minimum cost (Z)
7	x_3, x_5, x_7, x_9	3	3	0
6	x_3, x_5, x_7, x_9	3	3	0
5	x_2, x_5, x_7, x_9	2	3	15
4	x_2, x_5, x_6, x_9	2	2	45
3	x_1, x_5, x_6, x_8	1	2	133
2	Infeasible solution			—

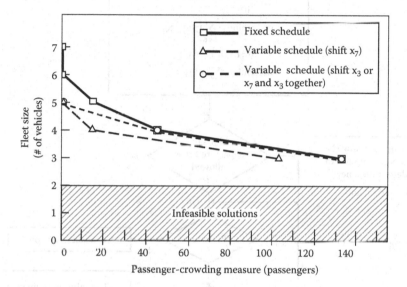

Figure 5.17 Optimal results of the example problem for a minimum passenger-crowding timetable.

A precomputation analysis can be carried out of the bounds on the number of departures. The steps of this analysis are shown in flowchart form in Figure 5.18. The functions d(k, t) are first constructed for the minimum and maximum number of departures, L(·) and U(·), for all the trip bundles (·). The lower and upper bounds on the fleet size are then determined as N_L and N_U, respectively. Before executing the integer program, two tests should be performed: (1) feasibility and (2) appropriateness of the U(·) solution. These two tests are shown in Figure 5.15. Applying this precomputation stage (without shifting departure times) to the example problem obviates the need to execute the integer programming for three fleet size values $N_o = 7$, 6, and 2, since $N_L = 3$ and $N_U = 6$.

Assume that the departure-time tolerance for the example problem, described in Tables 5.4 and 5.5, is $\Delta(\cdot) = \pm(1/3)$ time units for all trip bundles. The analysis of d(k, t), carried out in Chapter 7, reveals that N_U can be reduced by one by varying individual departure times, whereas N_L remains unchanged. The value of N_U can be reduced from 6 to 5 through three alternative shift procedures:

(1) Decrease the first departure time of x_7, from 1⅓ to 1.0.
(2) Increase the third departure time of x_3 from 4.0 to 4⅓.
(3) Shift both departure times in (1) and (2) in opposite directions i.e., from 1⅓ to 1⅙ and from 4.0 to 4⅙.

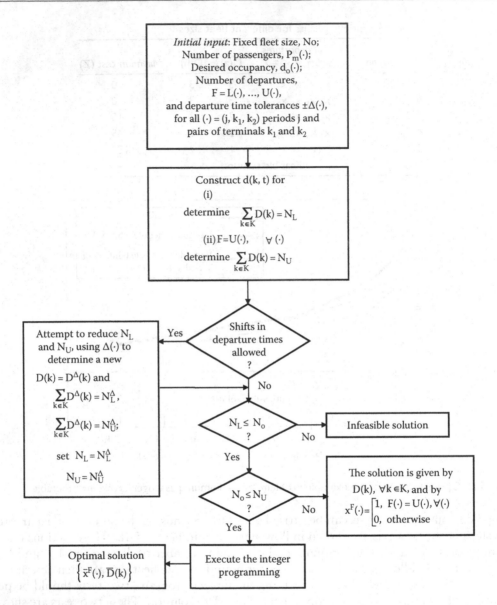

Figure 5.18 A precomputation analysis.

Note that in all three cases, it is possible to maintain even headways by shifting all departures in each bundle by the same amount, but this is not true in general.

The integer programming for these possibilities is similar to that explained in Section 5.6.3, except for the following:

For case (1), the first constraint in (2) is changed into

$$3x_3 + 2x_2 + x_1 - x_7 \le D(a)$$

For cases (2) and (3), the first and last constraints in (2) are changed into

$$2x_3 + 2x_2 + x_1 \le D(a)$$
$$3x_7 + 2x_6 - 2x_3 - 2x_2 - x_1 + x_8 + 2x_9 \le D(b)$$

Table 5.7 Optimal results for the example variable scheduling problem

Case	Fleet size no.	Sets of departures in the solution, $x_i = 1$	D(a)	D(b)	Minimum cost
(I)	≥5	x_3, x_5, x_7, x_9	2	3	0
	4	x_2, x_5, x_7, x_9	1	3	15
	3	x_1, x_5, x_7, x_8	0	3	103
(II) and (III)	≥5	x_3, x_5, x_7, x_9	2	3	0
	4	x_2, x_5, x_6, x_9	2	2	45
	3	x_1, x_5, x_6, x_8	1	2	133

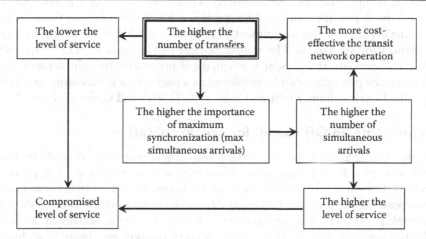

Figure 5.19 Trade-offs dependent on transit-passenger transfers.

and an additional constraint is added to (2):

$$3x_3 + 2x_2 + x_1 - x_7 \leq D(a)$$

The results of the modified problem are presented in Table 5.7.

The precomputation analysis shown in Figure 5.19 has been applied to this modified problem. Consequently, the MPSX package was used (while relaxing the integrality constraints) only for $N_0 = 4, 3$ in all shifting cases. The optimal results exhibited in Tables 5.6 (without shifting) and 5.7 (after shifting) are compared in Figure 5.17. This figure demonstrates the trade-off between passenger-crowding (disservice) cost and fleet size. Such a graphical representation can be used as an evaluation tool by the transit planner/scheduler.

5.7 MAXIMUM SYNCHRONIZATION FORMULATION AND OR MODEL

Ceder and Wilson (1986), in a study of transit route design at the network level, emphasized the importance of eliminating a large number of transfer points because of their adverse effect on the user. From the operating-cost perspective, a trade-off no doubt exists between this elimination and the efficiency of the transit route network. This trade-off is depicted schematically in Figure 5.19.

In order to allow for an adequate or compromised level of service, the planners/schedulers face a synchronization task to ensure maximum smooth transfers involving the switching of passengers from one route to another without waiting time. This task is extensive, but minimizes the waiting time for passengers who require connections. The result is that the planner/scheduler creates a more attractive transit system, one that generates the opportunity for increasing the number of riders. Figure 5.16 illustrates how the concept of synchronized transfers mitigates any adverse impact on the level of service while achieving a more cost-effective transit network.

Actually, synchronization is one of the most difficult, but vital, scheduling tasks. A poor transfer can cause the user to stop using the transit service. It is analogous to the Chinese proverb: "Two leaps [in our case, two trips with a transfer] per chasm are fatal." Nevertheless, synchronization is addressed almost intuitively. Generally speaking, the planner/scheduler attempts to fix the departure times in the timetable while conforming to three elements: (1) required frequency, (2) efficient assignment of trips to a single vehicle chain, and (3) maximum synchronization of arrivals. The second element will be dealt with in Chapters 6 and 7. This chapter will present an efficient mathematical procedure to help achieve maximum synchronization; the procedure can be employed as a useful tool for creating timetables. The process described in this chapter follows Ceder et al. (2001) and Ceder and Tal (2001).

5.7.1 Formulating an OR model for synchronization

Chapters 4 and 5 established optional timetables in conjunction with adequate frequency and satisfactory loads for a single transit route. When two or more routes interact, particularly at connection (transfer) points, the issue of coordination becomes critical. In order to gain a better approach to coordination at connection points, there is a need to relax some of the rigorously defined timetable parameters. The parameter chosen that can offer a measure of flexibility is the headway. Fortunately, only certain routes, at certain time periods, are able to be included in this optimal search for reducing waiting time at transfer points. This section follows the meaning of optimization as a framework for mathematical treatment, as described earlier in this chapter, and formulates the problem of maximizing simultaneous vehicle arrivals at transfer points.

The following OR model, based on Ceder et al. (2001), is presented in five parts: (1) notation of known (given) data, (2) decision variables, (3) objective function, (4) constraints, and (5) assumptions and notes.

5.7.1.1 Notation of known (given) data

The given transit network is presented by

$$G = \{A, \bar{N}\}$$

where
A is a set of directed arcs representing the traveling path of transit routes
$\bar{N}$ is a set of transfer nodes in network G

The problem data are the following:
T = planning horizon (departure times can be set in the interval [0, T], which is a discrete interval)
M = number of transit routes in the network
N = number of transfer nodes in the network
$Hmin_k$ = minimum headway (agency's requirements) between two adjacent departures on route k ($1 \leq k \leq M$)
$Hmax_k$ = maximum headway (policy headway) permitted between two adjacent bus departures on route k ($1 \leq k \leq M$)

F_k = number of departures to be scheduled for route k during the interval $[0, T]$ ($1 \le k \le M$)

T_{kj} = travel time from the starting point of route k to node j ($1 \le k \le M$, $1 \le j \le N$)
   Travel times are considered to be deterministic and can be referred to as mean travel times

5.7.1.2 Decision variables

1. X_{ik} represents the *i*th departure time on route k ($1 \le i \le F_k$).
2. Z_{ikjqn} is a binary variable that yields the value 1 if the vehicle of the ith departure on route k meets the vehicle of the jth departure of route q at node n; otherwise, it yields the value 0.

Let $A_{kq} = \{n: 1 \le n \le N, T_{kn} \ge 0, T_{qn} \ge 0\}$.

5.7.1.3 Objective function

$$\text{Max} \sum_{k=1}^{M-1} \sum_{i=1}^{F_k} \sum_{q=k+1}^{M} \sum_{j=1}^{F_q} \sum Z_{ikjqn} \tag{5.19}$$

5.7.1.4 Constraints

$$X_{1k} \le H\max_k, \quad 1 \le k \le M \tag{5.20}$$

$$X_{F_k k} \le T, \quad 1 \le k \le M \tag{5.21}$$

$$H\min_k \le X_{(i+1)k} - X_{ik} \le H\max_k, \quad 1 \le k \le M, 1 \le i \le F_k - 1 \tag{5.22}$$

$$Z_{ikjqn} = \text{Max}\left[1 - \left|(X_{ik} + T_{kn}) - (X_{jq} + T_{qn})\right|, 0\right] \tag{5.23}$$

Constraint (5.20) ensures that the first departure time will not begin beyond the maximum headway from the start of the time horizon, while constraint (5.21) ensures that the last departure is executed within the planning horizon. Constraint (5.22) indicates the headway limits, and constraint (5.23) defines the binary variable of the objective function.

5.7.1.5 Assumptions and notes

1. It is assumed that the first departure on each route k must take place in the interval $[0, H\max_k]$.
2. The problem is impractical unless the following constraints hold for each k:

(a) $H\max_k \ge H\min_k$ $\hspace{2cm}$ (5.24)

(b) $T \ge (F_k - 1) \cdot H\min_k$ $\hspace{2cm}$ (5.25)

(c) $T \le F_k \cdot H\max_k$ $\hspace{2cm}$ (5.26)

3. The case of a route k that does not pass through a node j is represented by $T_{kj} = -1$.

The OR model given earlier can be simplified by defining a variable, Y_{kq}, representing the overall number of simultaneous arrivals of vehicles on route k with vehicles on route q. The model is changed to

$$\text{Max} \sum_{k-1}^{M-1} \sum_{q=k+1}^{M} Y_{kq} \tag{5.27}$$

$$\text{s.t.} \quad Y_{kq} = \sum_{n \in Akq} \sum_{i=1}^{F_k} \sum_{j=1}^{F_q} \text{Max}\left[1 - \left|(X_{ik} + T_{kn}) - (X_{jq} + T_{qn})\right|, 0\right] \tag{5.28}$$

Constraints (5.20) through (5.22) remain unchanged.

The last formulation represents a nonlinear programming problem. It can be reformulated as a mixed-integer linear programming (MIP) problem, which can be solved (up to certain sizes) by several software packages. The nonlinear constraint is (5.28). Let D_{nijkq} denote a binary variable (defined over the same domain as Z_{ikjqn}) and B a large number ($B = T + \text{Max}_{i,j} T_{ij}$). The constraint in (5.28) is exchanged with the following constraints:

$$B \cdot D_{nijkq} \geq X_{ik} + T_{kn} - (X_{jq} + T_{qn}) \tag{5.29}$$

$$B \cdot D_{nijkq} \geq X_{jq} + T_{qn} - (X_{ik} + T_{kn}) \tag{5.30}$$

$$Y_{kq} < \sum_{n \in Akq} \sum_{i=1}^{F_k} \sum_{j=1}^{F_q} (1 - D_{nijkq}) \tag{5.31}$$

If $X_{ik} + T_{kn} = X_{jq} + T_{qn}$, there is a simultaneous arrival of the vehicle of the ith departure on route k with the vehicle of the jth departure of route q at node n. The variable D_{nijkq} can yield the value 0 and Y_{kq} is increased by one, according to (5.31).

If $X_{ik} + T_{kn} \neq X_{jq} + T_{qn}$, the arrivals do not coincide, and D_{nijkq} must yield the value in order to satisfy constraints (5.29) and (5.30). The number of simultaneous arrivals between vehicles of routes k and q (Y_{kq}) is not increased in (5.31).

The following is an upper bound on the number of possible simultaneous arrivals in a given transit network:

$$Z^* = \sum_{k=1}^{M-1} \sum_{q=k+1}^{M} \sum_{n \in Akq} \text{minimum}(F_k, F_q) \tag{5.32}$$

The number of integer variables in an MIP problem is generally a good index of its complexity (represented by the computerized processing time required). The variable D_{nijkq} represents the simultaneous arrival of the vehicle of the ith departure of route k and the vehicle of the jth departure of route q at node n. This means an integer variable for every combination of two trips on different routes that intersect at node n. Let F be $\text{Max}(F_k)$; the number of integer variables in the worst case is $O(NM^2F^2)$, which is a very large number. However, in a more realistic setting, the number is $O(M^2F^2)$, where N can be replaced by the average number of nodes commonly shared by any two routes.

The problem formulated by (5.27) through (5.31) is a large, mixed-integer, linear-programming problem. Running small network examples (five routes, five nodes), using general algebraic modeling system (GAMS) software on a PC requires hours, even days.

This time-consuming process provided the motivation to develop heuristic procedures that would solve such problems within a reasonable time. Two heuristic procedures were implemented in Turbo Pascal, and many examples were checked and compared to the optimal solutions obtained using the GAMS. The first procedure is based on the selection of nodes within the network. In each step, the next node is selected, provided that not all the departure times have been determined for that node. After the departure time is resolved, all its corresponding arrival times are set. The second procedure attempts to improve the solution through possible shifts in departure times. The following two sections present these two heuristic procedures.

5.7.2 Synchro-1 procedure

Definition 5.1:

A node is defined as *possible* if

1. At least one transit route that passes through the node, and not all the departure times for that route are set
2. More simultaneous arrivals can be created at the node

Definition 5.2:

A node is defined as *new* if no arrival times have been set for it.

The Synchro-1 procedure uses several components, as shown in the flowchart in Figure 5.20. Its steps are as follows.

> *Step 1*: Initialization; check whether the problem is feasible and set the data structure. Mark all nodes as possible.
> *Step 2*: Select the next node, NO, from the possible nodes.
> *Step 3*: If NO is new, perform component FIRST, otherwise perform component MIDDLE.
> *Step 4*: If there is any possible node, go to *Step 2*; if there are any more routes, perform component CHOOSE and go to *Step 2*, otherwise stop.

Step 1 contains a check of whether the problem is feasible; if so, toward that end, two data structures are built:

1. A structure called *route* for each transit route i, which includes $Hmin_i$, $Hmax_i$, F_i, the number of nodes the route passes through, and the departure times that have already been set.
2. A structure called *node* for each node n, which includes two elements: (1) the number of routes passing through the node and (2) the route with the maximum traveling time to the node; it also includes the number of simultaneous arrivals at the node at each time point in interval $[0, T + MaxT_{ij}]$, where the max is over all i, j.

All the nodes are marked possible.

In Step 2, the next node, NO, is selected from among the possible nodes. Node NO must satisfy the following conditions:

1. The number of different vehicle-arrival times at the node is at its maximum; in such a node, there is a greater probability that another vehicle departure can be set so that it will arrive at NO by any one of the (already set) arrival times.

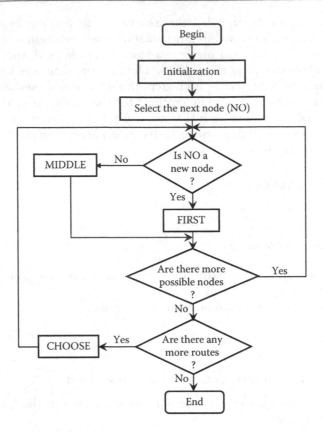

Figure 5.20 Synchro-1 procedure flowchart.

   2. Among the nodes satisfying the preceding condition, NO is that through which a maximum number of routes pass; in such a node, there is a greater potential for simultaneous arrivals.

   3. Among the nodes satisfying the preceding condition, NO minimizes the maximum travel time of all routes from their origin to the node (after the departure times of vehicles are set in order to meet at NO, the potential for simultaneous arrivals at distant nodes still exists).

A distinction is made in Step 3 between a new node and another node. For a new node, the component FIRST attempts to set the departure times of vehicles that pass through it, such that the vehicles will arrive at the node at the earliest time possible and simultaneous arrivals will continue to be created at the node according to the Hmin and Hmax of the routes. For example, transit vehicles on Routes 2 and 3 arrive simultaneously at a certain node at time t_o; if the next departure time for each of these routes can be set at a fixed difference, d minutes, from the last departure time of the route [for all i = 1, 2, and 3, Max ($Hmin_i$) ≤ d ≤ minimum ($Hmax_i$)], there will be additional simultaneous arrivals at the node at time t_o + d. Component FIRST finds the minimum d possible. If parameter d cannot be set ($Hmin_i$ > $Hmax_j$), the next departure times of vehicles on these routes will not be resolved in this step.

   For a node that is not *new*, an attempt is made to set the vehicle-departure times on routes passing through it, such that the vehicles will arrive at the node at the earliest of all arrival times already set in that node. If no more simultaneous arrivals are available at the node, the node is marked as *not possible*. This component is called MIDDLE.

Step 4 tests whether any more nodes are possible. If not, there may be routes on which not all the vehicle departure times were set. In such cases, the route that passes through the maximum number of nodes is chosen, and its next departure time is set by using the difference $Hmin_i$ from the last departure. In this way, the procedure sets additional vehicle arrivals for the maximum number of nodes possible. All the nodes through which route i passes are marked possible, and the procedure returns to Step 2. This component is called CHOOSE.

The complexity of the Synchro-1 procedure is, in the worst case, $O(NTFM^2)$, where N is the number of transfer nodes, M is the number of routes, T is the planning horizon, and F is the maximum number of scheduled departure times across all routes.

5.7.3 Synchro-2 procedure

The results of Synchro-1 could be improved by allowing a shifting of departure times in the timetable obtained by the procedure. That is, for each node and for each two time points, t_1 and t_2 ($t_1 < t_2$) at which vehicles arrive at the node, an attempt can be made to introduce a shift in all the vehicle departure times for vehicles arriving at the node at t_1 so that they will arrive at time t_2. If this succeeds, the timetable is changed accordingly, and the number of simultaneous arrivals increases. It should be noted that in order to shift a single departure time, the following must be checked:

1. After the shifting is done, the constraints on Hmin and Hmax must still hold. Otherwise, additional departure times on the route should be shifted.
2. As a result of shifting the departure time of vehicle i, its arrival time for all nodes through which it passes is changed. Therefore, the departure time of each vehicle that arrives simultaneously with vehicle i at any node must also be shifted. For each shift, the constraints on Hmin and Hmax must be checked, and so on. This process is recursive, and an attempt to shift a single departure time may cause the shifting of all network departure times.
3. A shift of Δt minutes in the departure time of vehicle i may result in changing other departure times, which ultimately will lead to changing the departure time of vehicle i by more than Δt. A check must eliminate this situation.

The shifting procedure can be added to the Synchro-1 procedure with two components, MERGE and MOVE, and a process TRY-MOVE as shown in Figure 5.21.

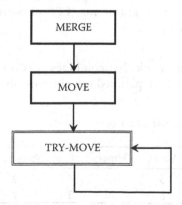

Figure 5.21 Synchro-2 procedure flowchart.

5.7.3.1 Component MERGE

Component MERGE identifies two departure times, t_1 and t_2 ($t_1 < t_2$), for a given node NO and delivers three elements to component MOVE: the transit routes arriving at t_1 and t_2 and the required shift $t_2 - t_1$.

5.7.3.2 Component MOVE

The MOVE component attempts to shift t_1 by $\Delta t = t_2 - t_1$, where t_2 is changed by route number. This component also contains the number of vehicles that need to be shifted, and a vehicle array with all possible vehicles that need to be checked for shifting. Each vehicle is identified by route number and vehicle code. Component MOVE uses the process TRY-MOVE, which allows for individual shifts to be checked.

TRY-MOVE is a recursive process that attempts to shift the departure time of a given vehicle and indicates whether MOVE is *successful* or *false*. This TRY-MOVE process checks whether the required new departure time is within [O, T] and whether or not the resultant headway exceeds Hmax.

5.7.3.3 Synchro-1 and Synchro-2 combined

Synchro-1 and Synchro-2 can be combined to shift Synchro-1 results. Further, the first selected node NO can change N times: NO = 1, 2, ..., N; that is, in order to run Synchro-1 N times. Such N time runs maybe required when the criterion of Synchro-1 for selecting NO cannot be justified; for example, in a case in which the node with the maximum routes passing through it is far from the origin. These N runs over different NOs can also be combined with Synchro-2. Example 3 in the following examines four variations: (1) Synchro-1, (2) N runs of Synchro-1 (first node varies), (3) Synchro-2, and (4) Shifts in the results of N runs of Synchro-1.

The complexity (running time as a function of network size parameters in the worst case) of each variation is certainly not the same. Synchro-1 has the lowest complexity, $O(NTFM^2)$, followed by N runs of Synchro-1, $O(N^2TFM^2)$, which is less than the complexity of Synchro-2, $O(N^2T^3F^2M^2)$. The most complex run is the one with N changes in the first node while running Synchro-2, $O(N^3T^3F^2M^2)$.

5.7.4 Examples

Three examples of transit networks will now be described, following Ceder et al. (2001) and Ceder and Tal (2001). Two of the examples are presented in detail for the sake of clarity. An optimal procedure with the linear programming (LP) software GAMS was applied to each example for comparison purposes.

Detailed Example 1

Figure 5.22 presents a simple network that combines two transfer points with two routes. The numbers on the arcs are average planned travel times (in minutes). A demonstration of Synchro-1 follows.

Step 1: Building the *route* data structure

I	Hmin$_i$	Hmax$_i$	F$_i$	No. of nodes
I	5	15	4	2
II	8	20	3	2

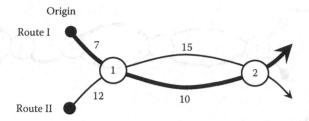

Figure 5.22 Example I basic network.

and the *node* data structure:

Node n	No. of routes	Route with Max T_{in}
I	2	II
2	2	II

The data comply with the three constraints (5.24) through (5.26).

Step 2: Three criteria must be selected: NO with maximum arrival time, maximum routes crossing, and minimum of all routes' maximum travel time (min–max criterion) from origin to NO. Thus, the number of departure (arrival) times equals 0 for both nodes. The number of crossing routes equals 2 for both nodes. The maximum travel time for node 2 is Max(17, 27) = 27; for node 1, it is Max(7, 12) = 12; the minimum of the two is 12. The selected NO is, therefore, node 1.

Step 3: The earliest time possible is set for node 1. Continuing in this node, the synchronization is based on Max(5, 8) $\leq$ d $\leq$ minimum(15, 20) $\rightarrow$ d_{min} = 8. This is the FIRST component, which provides the following results:

Departure time		
Route I	Route II	Meeting time
5	0	12
13	8	20
21	16	28

where meeting refers to a simultaneous arrival. The number of departures for route II complies with F_2 = 3.

Step 4: Because node 2 has yet to be examined, Step 2 is selected.

Step 3: Because node 2 is not new, component MIDDLE is applied. F_1 = 4 > 3 (currently created). Thus, one more departure time is set for route I based on the already created departure times for route II (at 0, 8, 16) so that the synchronization is made at node 2. The result is the additional departure time for route I (at 26, which meets the route II vehicle's departure at 16).

Step 4: No more routes are left to examine, and the algorithm ends with the final results as follows:

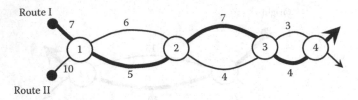

Figure 5.23 Example 2 basic network.

Departure time		Meeting time		
Route I	Route II	At node I	At node 2	Total meetings
5	0	12		
13	8	20		4
21	16	28		
	26		43	

The optimal and heuristic procedures coincide in this simple example.

Detailed Example 2

Figure 5.23 presents the second example with a case of $Hmin_i > Hmax_j$ for a two-route, four-node network. The numbers on the arcs are average planned travel times (in minutes). Both Synchro-1 and Synchro-2 may be used for this example.

The data for example 2 are as follows:

i	$Hmin_i$	$Hmax_i$	F_i	T
Route I	6	10	4	30
Route II	3	5	6	

in which $Hmin_1 > Hmax_2$.

Using Synchro-1

Step 1: In the *route* structure data, the number of nodes through which each route passes is four.

Node data structure:

Node n	No. of routes	Route with Max T_{in}
I	2	II
2	2	II
3	2	II
4	2	I, II

Step 2: NO is selected such that (1) the number of departure (arrival) times equals 0 for all nodes, (2) the number of crossing routes equals 2 for all nodes, and (3) the route's maximum travel times for the nodes are 10, 16, 20, and 23, respectively.

The minimum is 10, so NO = 1.

Step 3: Component FIRST is applied. The first meeting time possible at node 1 is 10. The component cannot set the parameter d. Therefore, procedure FIRST results in only the first departure time for each route. That is, departure times 3 and 0 for routes I and II, respectively, to meet at 10.

Step 4: With more possible nodes, Synchro-1 returns to step 2.

Step 2: Number of arrival times is 1, 2, 2, and 2, respectively, for nodes 1, 2, 3, and 4. NO = 2 (min–max travel time is 16).

Step 3: Component MIDDLE sets a new meeting at node 2, at time 27, by setting the third departure time for route I to 15 and the fourth departure time for route II to 11.

Steps 2, 3: NO = 3. Component MIDDLE sets new meetings at node 3, at times 28 and 34, by setting the second departure time for route I to 9, the third departure time for route II to 8, and the fifth departure for route II to 14.

Steps 2, 3: NO = 4. No meetings are possible.

Synchro-1 continues to perform component MIDDLE until it ends with the following results:

Departure time		Meeting time at node 1	Meeting time at node 2	Meeting time at node 3	Total meetings
Route I	Route II				
3	0	10			5
9	3			28	
15	8		27	34	
21	11	28			
	14				
	18				

The meeting time is shown in the same row as its associated departure for route I.

Using Synchro-2

Component MERGE and MOVE check with the possible shifting of TRY-MOVE. The only successful shifting is the second departure for route II from 3 to 5. The new results are as follows:

Departure time		Meeting time at node 1	Meeting time at node 2	Meeting time at node 3	Total meetings
Route I	Route II				
3	0	10			
9	5		21	28	6
15	8		27	34	
21	11	28			
	14				
	18				

The optimal solution is six meetings, coinciding with the solution given by Synchro-2.

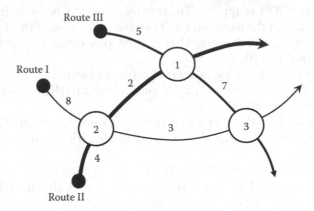

Figure 5.24 Example 3 basic network.

Example 3

Figure 5.24 exhibits the third example, in which the numbers on the arcs are average planned travel times (in minutes).

The following data apply for Figure 5.24:

i	$Hmin_i$	$Hmax_i$	F_i	T
Route I	3	8	20 combinations	
Route II	4	7	(see Table 6.1)	20
Route III	5	12		

This third example was examined with four variations of procedures: Synchro-1, Synchro-1 with N changes of first node, Synchro-2, and Synchro-2 with N changes of first node. The results of these four variations appear in Table 5.8, including a comparison of the optimal results. Table 5.7 shows 20 combinations of frequencies F_1, F_2, and F_3 for routes I, II, and III, respectively.

The results of Table 5.8 are as follows: (1) for the most complex variation, with Synchro-2 and N changes at first node (of Synchro-1), there are 18 (out of 20) optimal results; (2) Synchro-1 and Synchro-1 with N runs result in 13 identical results (in other words, the decision about the first node in Synchro-1 was the best selection in 13 of the 20 cases); and (3) based on example 3, it is not obvious that Synchro-2 provides better results than Synchro-1 + N runs.

5.7.5 Real-life example

The study by Ceder et al. (2001) selected a real-life synchronization problem in Israel for testing Synchro-1 and Synchro-2. The real-life network, shown in Figure 5.25, consists of a main road that has three major transfer nodes. There are seven bus routes traveling in each direction on this network and a total of 14 routes that meet at three nodes. (Route numbers appear in Figure 5.25.) The time span for the synchronization problem was 3 h during morning rush hours. The best results (using all four variations) were 240 meetings using Synchro-1 with N changes for the first selected node. The real-life example could not be checked against the optimal solution because of its large size.

Table 5.8 Maximum number of meetings produced by 4 variations of Synchro-1 (S1) and Synchro-2 (S2) with 20 combinations of vehicle frequencies

Combination no.	F_1	F_2	F_3	Procedure variation				
				S1	S1 + N runs	S2	S2 + N runs (with shifts)	Optimal
1	1	1	1	2	2	2	2	2
2	1	1	2	2	2	2	2	3
3	1	2	1	3	3	3	3	3
4	1	2	2	3	3	4	4	4
5	2	1	1	2	3	2	3	3
6	2	1	2	2	3	3	4	4
7	2	2	1	3	3	3	4	4
8	2	2	2	4	4	5	5	5
9	3	1	1	2	3	2	3	3
10	3	1	2	3	3	3	4	4
11	3	2	1	3	3	3	4	4
12	3	2	2	4	5	5	6	6
13	3	3	1	4	4	4	5	5
14	3	3	2	5	5	6	6	7
15	3	1	3	3	4	3	4	4
16	3	2	3	4	5	6	7	7
17	3	4	4	7	7	10	10	10
18	4	4	4	8	8	11	11	11
19	5	5	5	8	8	8	8	8
20	6	6	5	4	7	4	7	7

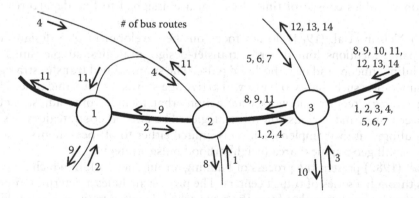

Figure 5.25 A real-life example of bus routes that need synchronization at three points.

The approach presented in this section can serve as a useful tool to assist the transit scheduler in one of the most difficult scheduling tasks. The problem of a successful connection is reflected in the following definition of a missed transfer: the same thing happened today that happened yesterday, but usually to different passengers. Thus, the more successful connections that can be created, the better and more reliable will be the service.

This planning tool needs to be complemented with the on-time performance of transit vehicles. The latter is expected to be realized when using new automatic vehicle location (AVL) and communications technologies. All together, the procedures described in this section may offer a step forward in changing the image of transit services and, ultimately, in view of increasing traffic congestion, lead to increased ridership. The next section provides a literature review of the research conducted in this area.

5.8 LITERATURE REVIEW ON SYNCHRONIZATION AND FURTHER READING

The advanced timetabling related literature is already covered in the previous chapter. This section is about the literature of transit synchronization. The need to transfer between routes generates a major cause of discomfort for transit users. Designing routes and schedules with a minimum amount of waiting time during a transfer may decrease the level of inconvenience. This section reviews methodologies for the design of such synchronized transit services.

Rapp and Gehner (1976) describe a graphic, interactive tool for minimizing transfer delays in a transit network. Optimal system-wide timetable coordination, under operating-cost constraints, is achieved through determining deviations from the original departure times. The diversity of headways along different routes and the interdependence of layover times in terminals are taken into account. Seeking an optimal solution involves an iterative assignment of passenger demand on the route network.

Salzborn (1980) presents a methodology for scheduling a system that includes several urban feeder lines and one interurban line. The latter passes by a few terminals and its timetable is coordinated with the arrival times of passengers in all terminals. Two separate scheduling processes are described, one for the feeder lines and one for the interurban line. The objective is to minimize both passenger delays and the number of required vehicles. The decision variables consist of time slots that are assigned to line departures from the terminals.

Although Nelson et al. (1981) do not focus on the development of a design methodology, their paper mentions four different transfer-design strategies: simple timed transfer, pulse scheduling, lineup, and neighborhood pulse. The simple timed transfer strategy means adjusting timetables such that two buses will arrive at a station at the same time. The pulse-scheduling strategy requires network-wide actions that include route adjustments, headways, layover times, and various control strategies. The two other strategies are similar to pulse scheduling, but their implementation is limited either to off-peak hours (lineup strategy) or to a small geographical area (neighborhood pulse strategy).

Kyte et al. (1982) present the process of building a route network in which a main trunk line passes through a series of transit centers. The process includes a determination of clock headways; that is, headways that form the same schedule every hour, with the objective of coordinating transfer times at the transit centers.

Schneider et al. (1984) provide a detailed list of criteria for choosing a proper site for the location of a timed-transfer transit center.

Hall (1985) develops a model for schedule coordination at a single transit terminal between a set of feeder routes and the line that they feed. The travel time on each of the routes is assumed to be subject to random delay. The optimized variable is the slack time between feeder arrivals and the main-line departure.

Abkowitz et al. (1987) relate to the simple case of two bus routes that meet at one spot. Their analysis aims at determining the conditions under which time transfer provides an improved service level compared with unscheduled transferring. A computer program is developed that simulates four transfer strategies: unscheduled transfers, scheduled transfers without vehicle waiting, scheduled transfers in which the lower-frequency bus is held until the higher-frequency vehicle arrives, and scheduled transfers in which both buses are held until a transfer event occurs. Mathematical expressions for waiting times are developed for each of the four cases. Simulation results of the four cases are analyzed to arrive at a generalization of situations for which each of the transfer strategies is suitable.

Klemt and Stemme (1988) formulate an integer-programming problem that attempts to minimize transfer times in a network, represented as a graph. The authors describe rules for the construction of a route graph on which the solution procedure is based.

Voss (1990) discusses two cases of a schedule-synchronization problem formulated as programming problem. The first case is similar to the one presented by Klemt and Stemme; namely, a relaxed version of a quadratic assignment problem. The transit routes in this case meet at a given set of transfer points. The second case, a modification of the first one, represents a transit system in which different routes partly use the same streets or tracks. Solution approaches are discussed and initial feasible solutions generated; they are then improved by using a Tabu search algorithm.

Desilets and Rousseau (1990) propose a model that is basically equivalent to Klemt and Stemme, but they also include an extended discussion of the definition of the cost that the model aims at minimizing. Several alternative cost definitions are suggested, such as a definition that focuses on service reliability and a definition based on choosing a group of connections for which synchronization is especially desirable. The paper also describes a heuristic solution based on an initial random solution and on a local search technique for solution improvement.

Lee and Schonfeld (1991) introduce two models for optimizing the slack time between a train line and a bus line that interchange passengers at adjoining terminals. The models are analytical, but are solved numerically or by using simulation. The first model assumes that train-arrival times are always the same, whereas bus arrivals are according to a given probability function. In the second model, both train and bus arrivals are probabilistic. Transfer-cost functions are developed, showing relationships between slack times and headways, transfer volumes, passenger-time values, bus operating costs, and the variability of bus and train arrivals. Schedule coordination between the two routes is found to be not worth attempting when the standard deviation of arrivals exceeds certain levels.

Adamski (1993) develops several transfer-optimization models. The first is a static model for a single transfer point, under the assumption of random vehicle travel time. Optimal timetable offsets are calculated with the objective of minimizing passenger disutility, which is a function of waiting time. Four alternative forms of the disutility function are examined. Cases with and without a holding control are analyzed. In the second model, optimal offset times are determined in the case of different lines that use a common road segment. The third model that Adamski develops is a dynamic model for real-time transfer optimization. A measure of off-schedule deviations is used to determine the value of control variables, such as holding times at control points.

Daduna and Voss (1993) discuss the optimization of transfers, using a quadratic semiassignment problem formulated as a programming problem. The authors focus on heuristic solution procedures rather than on a new synchronization methodology. A regret heuristic is proposed for finding an initial solution. Tabu search and simulated annealing strategies for improving the solution are described.

Adamski and Chmiel (1997) formulate two schedule-synchronization problems, which they solve by using genetic algorithms; the first for the synchronization problem of transfers at transfer points and the second for the synchronization of different transit lines that travel a common road segment. Deterministic and stochastic cases are compared. Decision variables are offsets entered into timetables in order to minimize waiting times.

Clever (1997) discusses the differences between systems with a single-mode transfer center and systems with multimodal, multicenter transfer timing. Characteristics of both types of networks are presented and compared. Among the aspects discussed are service symmetry, relations between service and passenger demand, minimal headway, and economic aspects.

Shih et al. (1998) present a heuristic model for the design of a coordinated network with transfer centers. The routes are not predetermined, but an output of the model. In addition, this model determines the appropriate vehicle size for each route. The proposed design process is composed of four major procedures. The first procedure generates a set of routes. In the second procedure, the network is analyzed and frequencies are set; this procedure incorporates a trip-assignment model. In the third procedure, transfer-center locations are determined. The last procedure involves network improvements. Algorithms are described qualitatively, using flowcharts for each of the four procedures.

Becker and Spielberg (1999) present principles for designing and implementing a multi-center-based timed-transfer network. To construct such a network, a prime headway is chosen, and all headways in the network are set to be multiples of this headway. Transfer centers are located in such a manner that the distances between them are also multiples of the prime headway. Other adjustments are made in order to improve network coordination, such as connecting routes at common ends and connecting feeder routes to the transfer centers.

Maxwell (1999) discusses four transfer strategies: (1) close headways with short waiting times, so that synchronization is not needed; (2) main-line trunk with timed-transfer branch lines; (3) a single hub with spoke lines, so that all transfers occur at the same spot; and (4) several transfer hubs with fixed-interval timed transfers. Advantages and disadvantages of each strategy are discussed and compared. A methodology for using the fourth strategy is presented in detail. Transfer points on a symmetrical route are shown to be located in spots that divide the route into an integer amount of equal segments. It is also shown that other locations can be chosen if speeds are adjusted accordingly. The use of a schedule map and symmetrical train graphs as analysis tools is demonstrated.

Ibarra-Rojas and Rios-Solis (2012) studied the bus network of Monterrey, Mexico, which is a large bus network where passenger transfers must be favored, almost evenly spaced departures are sought, and bus bunching of different lines must be avoided. The authors formulate the timetabling problem of this network with the objective of maximizing the number of synchronizations to facilitate passenger transfers and avoid bus bunching along the network. These synchronizations are defined as the arrivals of two trips with a separation time within a time window to make a flexible formulation. This flexibility is identified as a critical aspect for the bus network, since travel times vary because of reasons such as driver speed, traffic congestion, and accidents. By proving that the subject problem is NP-hard, the authors answer a 10-year-old open question about the NP-hardness of similar problems present in literature. Next, the study presents an analysis of the structural properties of the feasible solution space of the model. This analysis leads to a preprocessing stage that eliminates numerous decision variables and constraints. Moreover, this preprocessing defines feasible synchronization and arrival time windows that are used in a new metaheuristic algorithm. Finally, this study states that empirical experimentation shows that the proposed algorithm obtains high-quality solutions for real-size instances in less than 1 min.

Ibarra-Rojas et al. (2012) focus on the synchronization bus timetabling problem (SBT) that allows to maximize passenger transfers, avoids bus bunching, and includes exibility that is crucial for many transit networks in Latin America. The authors state that a timetabling solution that is not close to the optimum has strong repercussions in the vehicle and crew scheduling problems since sequential resolution approaches are often required for solving the entire planning process. As such, the authors develop five families of valid inequalities that drastically strengthen the integer linear programming model of SBT. Three of them manage to bound the number of synchronizations while the other two are lifting ones. Experimental results presented in this study indicate that the enhanced mixed-integer linear programming yields high quality solutions (less than 1% of relative deviation from the optimal solution) in minutes for large instances based on real transit networks (200 bus lines and 40 synchronization nodes).

Ceder et al. (2013) analyzed how to use selected operational tactics in public-transport networks for increasing the actual occurrence of scheduled transfers. A model is developed to determine the impact of instructing vehicles to either hold at or skip certain stops, on the total passenger travel time and the number of simultaneous transfers. The model is comprised of two components. First, a simulation of public-transport network examines the two tactics for maximizing the number of transfers. Second, an ILOG optimization model is used for optimal determination of the combination of the two tactics to achieve the maximum number of simultaneous transfers. A bus network has been created, as a case study, in Auckland, New Zealand, to verify the impact of the model's application. The results show that applying online operational tactics dramatically improves the frequency of simultaneous transfers by more than 100%. The authors conclude that the proposed concept has large potential for increasing the efficiency and attractiveness of public-transport networks that involve scheduled transfers. Further studies on this type of synchronization using tactics appear in Chapter 19.

It should be noted that several guidelines for the design or the implementation of synchronized transit systems are mentioned repeatedly in most of the papers reviewed. Some of these guidelines are as follows:

- For an efficient synchronized system, route design and schedule design should be effected simultaneously.
- The success of timed-transfer services relies heavily on the use of real-time control. Schedule adherence has a crucial role.
- The effort in introducing synchronized services may not be worthwhile when frequencies are relatively high. Most papers mention a 30 min headway as an example of a headway that justifies transfer timing.

The main characteristics of the models reviewed are described in Table 5.9.

Table 5.9 Characteristics of works related to transit synchronization covered in Section 5.7

Source	Network structure/ transferring strategies	Location of synchronized transfer	Special features
Abkowitz et al. (1987)	Two bus routes that meet at one spot. Four transfer strategies: unscheduled, scheduled without waiting, scheduled when the less-frequent bus waits, scheduled when both buses wait	A single meeting point	
Adamski (1993)	1st model: several routes that meet at a terminal. 2nd model: several routes that traverse a common road segment. 3rd model: real-time control strategies	1st model: terminals. 2nd model: a road segment along which several routes pass. 3rd model: control points	Random vehicle travel time; several objective-function alternatives in the 1st model
Adamski and Chmiel (1997)	1st model: several routes that meet at a terminal. 2nd model: several routes that traverse a common road segment	1st model: terminals. 2nd model: a road segment along which several routes pass	Deterministic and stochastic cases
Becker and Spielberg (1999)	Any route structure	Multiple centers	
Ceder et al. (2001) and Ceder and Tal (2001)	Any route structure	All meeting points	Several headway-design options and several objective-function alternatives
Ceder et al. (2013)	Any route structure		Potential to apply online real-time tactics in the real world
Clever (1997)	Any route structure	1st option: single transfer center. 2nd option: multiple centers	
Daduna and Voss (1993)	Any route structure	All meeting points	
Desilets and Rousseau (1990)	Any route structure	All meeting points	Several objective function alternatives
Hall (1985)	Several feeder lines and one main line	A single terminal	Random delay
Ibarra-Rojas and Rios-Solis (2012)	Lines with *sub-lines* that share a common route segment but have variations at the beginning and at the end of the routes	At bunching nodes and transfer nodes	
Ibarra-Rojas et al. (2012)	Any route structure	Case (a): At bunching nodes. Case (b): At transfer nodes	
Klemt and Stemme (1988)	Any route structure	All meeting points	

(Continued)

Table 5.9 (Continued) Characteristics of works related to transit synchronization covered in Section 5.7

Source	Network structure/ transferring strategies	Location of synchronized transfer	Special features
Kyte et al. (1982)	A main trunk line with complementary services	Several centers through which the main line passes	
Lee and Schonfeld (1991)	A train line and a bus line that meet at one spot	A single meeting point	Two models with different assumptions regarding delay variability
Maxwell (1999)	Four transfer strategies: close headways without synchronization, main-line trunk with timed-transfer branch lines, a single hub and spoke, several transfer hubs with fixed-interval timed transfers		
Nelson et al. (1981)	Four transfer strategies: simple timed transfer, pulse scheduling, line-up, neighborhood pulse		
Rapp and Gehner (1976)	Any route structure	All meeting points	
Salzborn (1980)	Several feeder lines and one main line	Several centers through which the main line passes	Objective includes minimizing the number of required vehicles
Schneider et al. (1984)		Detailed discussion of transfer-center location	
Shih et al. (1998)	Any route structure (routes are generated by the model)	Multiple centers	Transfer-center locations and optimal vehicle sizes are determined
Voss (1990)	Any route structure	1st case: meeting points. 2nd case: road segments along which several routes pass	

EXERCISES

5.1 Ride-check data, shown in the following table, were collected on a given bus route (average across several days). There are two stops (B and C) between the first (route departure) stop (A) and the last (route arrival) stop (D). The data cover a 2 h morning period (6:00–8:00). Average travel times between stops are as follows: 8 min (A to B), 14 min (B to C), and 22 min (C to D). Minimum frequency is 2 veh/h for both hours.

Using Method 2 for frequency determination, *find* and *compare* the departure times obtained by (a) the existing (given) data, (b) the even headway procedure, (c) the even load procedure at the hourly max load point, and (d) the even load procedure at the individual vehicle max load point. In this comparison, *provide* the average max load and its associated max load point for each departure and procedure used. Note: only one departure having a desired occupancy of 50 passengers will be derived after 8:00.

Departure time (at A)	Average load (passengers) when departing stop			Desired occupancy (passengers)
	A	B	C	
6:00	25	18	12	
6:35	65	72	48	40
7:10	67	82	35	
7:45	84	33	47	60
8:00	75	41	57	50

5.2 The simple network given combines two transfer points with three routes. The numbers on the arcs are travel times (average in minutes).

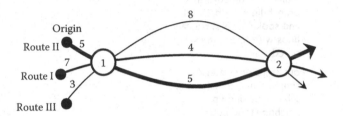

The data for the network:

i	Hmin$_i$	Hmax$_i$	F$_i$	T
Route I	7	13	2	20
Route II	7	15	2	
Route III	—	20	1	

Note that there is no Hmin for Route III, since only one vehicle is assigned.

Find the number of meetings, their upper-bound Z*, and their associated departure and meeting times using Synchro-1 and Synchro-2. *Suggestion*: Run an optimization software for optimal (maximum number of meetings) results.

5.3 The network given combines five transfer points with four routes. The numbers on the arcs are travel times (average in minutes).

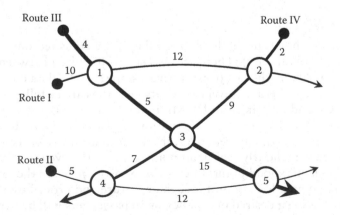

The data for the network:

i	Hmin$_i$	Hmax$_i$	F$_i$	T
Route I	5	10	3	30
Route II	6	10	3	
Route III	7	8	3	
Route IV	10	15	3	

Find the number of meetings, their upper-bound Z^*, and their associated departure and meeting times using Synchro-1 and Synchro-2. *Suggestion*: Run an optimization software for optimal (maximum number of meetings) results.

5.4 Synchro-1 and Synchro-2 are intended to maximize the number of simultaneous arrivals (meetings). Given the average number of passengers that needs to be transferred between routes, construct a heuristic procedure that, instead of maximizing the number of meetings, will maximize the number of passenger meetings (each passenger transfer, without waiting time, counts as 1). The new objective function weighs each passenger transfer the same.

REFERENCES

Abkowitz, M., Josef, R., Tozzi, J., and Driscoll, M.K. (1987). Operational feasibility of timed transfer in transit systems. *Journal of Transportation Engineering*, 113(2), 168–177.

Adamski, A. (1993). Transfer optimization in public transport. In *Computer-Aided Transit Scheduling*. Lecture Notes in Economics and Mathematical Systems, vol. 430 (J.R. Daduna, I. Branco, and J.M.P. Paixao, eds.), pp. 23–38, Springer-Verlag, Berlin, Germany.

Adamski, A. and Chmiel, W. (1997). Optimal service synchronization in public transport. In *Transportation Systems*, vol. 3 (M. Papageorgiou and A. Pouliezos, eds.), pp. 1283–1287, Elsevier Science, Oxford, U.K.

Ahuja, R.K., Magnanti, T.L., and Orlin, J.B. (1993). *Network Flows*. Prentice Hall, Englewood Cliffs, New Jersey.

Bartlett, T.E. (1957). An algorithm for the minimum number of transport units to maintain a fixed schedule. *Navel Research Logistics Quarterly*, 4, 139–149.

Becker, A.J. and Spielberg, F. (1999). Implementation of a timed transfer network at Norfolk, Virginia. *Transportation Research Record*, 1666, 3–13.

Brumbaugh-Smith, J. and Shier, D. (1989). An empirical investigation of some bicriterion shortest path algorithms. *European Journal of Operational Research*, 43(2), 216–224.

Ceder, A. (1999). *Systems Analysis as an Introduction to Operations Research*, Michlol Publication Technion—Israel Institute of Technology, Haifa, Israel.

Ceder, A. (2001). Bus timetables with even passenger loads as opposed to even headways. *Transportation Research Record*, 1760, 28–33.

Ceder, A., Golany, B., and Tal, O. (2001). Creating bus timetables with maximal synchronization. *Transportation Research*, 35A(10), 913–928.

Ceder, A., Hadas, Y., McIvor, M., and Ang, A. (2013a). Transfer synchronization of public-transport networks. *Transportation Research Record*, 2350, 9–16.

Ceder, A., Hassold, S., and Dano, B. (2013b). Approaching even-load and even-headway transit time-tables using different bus sizes. *Public Transport—Planning and Operation*, 5(3), 193–217.

Ceder, A. and Philibert, L. (2014). Transit timetables resulting in even Max-load on individual vehicles. *IEEE Transactions on Intelligent Transportation Systems*; 15(6), 2605–2614.

Ceder, A. and Stern, H.I. (1984). Optimal transit timetables for a fixed vehicle fleet. In *Transportation and Traffic Theory* (J. Volmuller and R. Hammerslag, eds.), pp. 331–355, UNU Science Press, Den Haag, the Netherlands.

Ceder, A. and Tal, O. (2001). Designing synchronization into bus timetables. *Transportation Research Record*, **1760**, 3–9.

Ceder, A. and Wilson, N.H.M. (1986). Bus network design. *Transportation Research*, **20B**(4), 331–344.

Clever, R. (1997). Integrated timed transfer: A European perspective. *Transportation Research Record*, **1571**, 109–115.

Daduna, J.R. and Voss, S. (1993). Practical experiences in schedule synchronization. In *Computer-Aided Transit Scheduling*. Lecture Notes in Economics and Mathematical Systems, vol. 430 (J.R. Daduna, I. Branco, and J.M.P. Paixao, eds.), pp. 39–55, Springer-Verlag, Berlin, Germany.

Desilets, A. and Rousseau, J.M. (1990). SYNCHRO: A computer-assisted tool for the synchronization of transfers in public transit networks. In *Computer-Aided Transit Scheduling*. Lecture Notes in Economics and Mathematical Systems, vol. 386 (M. Desrochers and J.-M. Rousseau, eds.), pp. 153–166, Springer-Verlag.

Dijkstra, E.W. (1959). A note on two problems in connexion with graphs. *Numerische mathematik*, **1**(1), 269–271.

Gertsbach, I. and Gurevich, Y. (1977). Constructing an optimal fleet for a transportation schedule. *Transportation Science*, **11**, 20–36.

Hall, R.W. (1985). Vehicle scheduling at a transportation terminal with random delay en route. *Transportation Science*, **19**, 308–320.

Hassold, S. and Ceder, A. (2012). Multiobjective approach to creating bus timetables with multiple vehicle types. *Transportation Research Record*, **2276**, 56–62.

Ibarra-Rojas, O.J., Fouilhoux, P., Kedad-Sidhoum, S., and Rios-Solis, Y.A. (2012). Valid inequalities for the synchronization bus timetabling problem. Technical report PISIS-2012-02 78, Graduate Program in System Engineering, UANL, San Nicolas delos Garze, Mexico.

Ibarra-Rojas, O.J. and Rios-Solis, Y.A. (2012). Synchronization of bus timetabling. *Transportation Research Part B*, **46**(5), 599–614.

Klemt, W.D. and Stemme, W. (1988). Schedule synchronization for public transit networks. In *Computer-Aided Transit Scheduling*. Lecture Notes in Economics and Mathematical Systems, vol. 308 (J.R. Daduna and A. Wren, eds.), pp. 327–335, Springer-Verlag, Berlin, Germany.

Kyte, M., Stanley, K., and Gleason, E. (1982). Planning, implementing, and evaluating a timed-transfer system in Portland, Oregon. *Transportation Research Record*, **877**, 23–29.

Lee, K.K.T. and Schonfeld, P. (1991). Optimal slack time for timed transfers at a transit terminal. *Journal of Advanced Transportation*, **25**(3), 281–308.

Maxwell, R.R. (1999). Intercity rail fixed-interval, timed-transfer, multihub system: Applicability of the "Integraler Taktfahrplan" strategy to North America. *Transportation Research Record*, **1691**, 1–11.

Nelson, M., Brand, D., and Mandel, M. (1981). Use and consequences of timed-transfers on U.S. transit properties. *Transportation Research Record*, **197**, 50–55.

Potter, S. (2003). Transport energy and emissions: Urban public transport. In *Handbook of Transport and the Environment*, vol. 4 (D.A. Hensher and K.J. Button, eds.), Elsevier, Oxford, U.K.

Raith, A. and Ehrgott, M. (2009). A comparison of solution strategies for biobjective shortest path problems. *Computers and Operations Research*, **36**(4), 1299–1331.

Rapp, M.H. and Gehner, C.D. (1976). Transfer optimization in an interactive graphic system for transit planning. *Transportation Research Record*, **619**, 27–33.

Salzborn, F.J.M. (1972). Optimum bus scheduling. *Transportation Science*, **6**, 137–148.

Salzborn, F.J.M. (1980). Scheduling bus systems with interchanges. *Transportation Science*, **14**, 211–220.

Schneider, J.B., Deffebach, C., and Cushman, K. (1984). The timed-transfer/transit center concept as applied in Tacoma/Pierce County, Washington. *Transportation Quarterly*, **38**(3), 393–402.

Shih, M.C., Mahmassani, H.S., and Baaj, M.H. (1998). Planning and design model for transit route networks with coordinated operations. *Transportation Research Record*, **1623**, 16–23.

Voss, S. (1990). Network design formulations in schedule synchronization. In *Computer-Aided Transit Scheduling*. Lecture Notes in Economics and Mathematical Systems, vol. 386 (M. Desrochers and J.-M. Rousseau, eds.), pp. 137–152, Springer-Verlag, Berlin, Germany.

Chapter 6

Vehicle scheduling I

Fixed schedules

CHAPTER OUTLINE

6.1 Introduction
6.2 Fleet size required for a single route
6.3 Example of an exact solution for multiroute vehicle scheduling
6.4 Max-flow technique for fixed vehicle scheduling
6.5 Deficit-function model with deadheading trip insertion
6.6 Depot-constrained vehicle scheduling
6.7 Literature review and further reading
Exercises
Appendix 6.A: Maximum-flow (Max-flow) problem
References

PRACTITIONER'S CORNER

Given demand and coordinated timetables, the next phase is to establish chains of daily trips or vehicle blocs. Each chain or bloc constitutes a vehicle's schedule for a day. The problem of minimizing the number of chains and, at the same time, fulfilling agency requirements (refueling, maintenance, etc.) is complex and cumbersome for medium and large-scale transit agencies. This chapter provides an overview of the problem, outlines specific solution procedures and tools, and describes some of the experience with a single transit agency.

The chapter contains five main parts following an introductory section. Section 6.2 exhibits a simple method for ascertaining the minimum number of vehicles required for a given single route without interlining. Section 6.3 presents exact mathematical-programming solutions for the case of multiroute vehicle scheduling. Practitioners can skip the math formulation here but should concentrate on the experience of a 4000-bus agency that led to the adoption of different approaches (described in the fourth part). In Section 6.4, the problem is converted to a known network-flow problem for enriching the quantitative interpretation of the underlying required analysis. Practitioners can skip this third part. The fourth part, Section 6.5, is the core of the chapter, proffering a graphical person–computer interactive approach based on a step function called deficit function. This approach provides a highly informative graphical technique that is simple to interact with and use. Practical suggestions can be interjected by the scheduler followed immediately by describing the effects of the suggestions on the vehicle's schedule. Section 6.6, the fifth part, suggests a treatment for an often imperative postulate—maintaining a balance in the number of vehicles starting and ending at a depot. This part shows the formulation of both balancing depots and complying with limitations on the number of overnight spaces at depots or parking facilities. The chapter ends with literature review and exercises.

Practitioners may skip all of Section 6.4 and Appendix 6.A, as well as the math treatment (formulation) in Sections 6.3, 6.5, and 6.6. They should, though, pay special attention to the examples and procedures depicted in Figures 6.6 through 6.12.

The subjects of vehicle and crew scheduling, the latter to be dealt with in Chapter 10, do receive attention in available software. This is not to say that those agencies using the software are fully happy with the results. Nonetheless, the existing competition among the various software companies is healthy and brings to mind the following story. A large branch of Chase Manhattan Bank in New York displayed a sign with the company's then slogan: "You have a friend at Chase." Bank Leumi Le-Israel (BLL) decided to open one of its large branches adjacent to this branch of Chase. To compete with it, BLL also put up a huge sign: "With BLL you have a FAMILY." Three weeks later, Chase responded with another sign: "With such a family you really need a FRIEND."

6.1 INTRODUCTION

Vehicles, especially buses, are often shifted from one route to another (interlining) within a large-scale transit system, which frequently has them perform deadheading (DH) trips in order to operate a given timetable with the minimum vehicles required. As was noted in explaining Figure 1.2, it is desirable to analyze simultaneously the procedures for constructing timetables and for scheduling vehicles. However, because of the complexity of this analysis, the two procedures are treated separately. This chapter has two major foci: (1) to describe the task of vehicle scheduling and possible operations research (OR) solutions for both a single transit route and a network of routes and (2) to proffer a graphical technique that is easy to interact with and that responds to practical concerns.

The motivation for the second foci arises from a problematic component of Egged Israel's national bus carrier cooperative, which is presently engaged in scheduling about 4000 buses over some 2000 routes. Egged operates over an extensive geographical network composed of urban, suburban, regional, and intercity routes and performs an average of 50,000 daily trips—one of the world's largest schedules. Egged's crucial component of scheduling buses to trips was performed manually by about 60 schedulers using Gantt charts. A scheduler's duty was to list all daily chains (some DH) for a bus, ensuring the fulfillment of the timetable and the operator's requirements (refueling, maintenance, etc.). However, because of frequent changes in the schedule and frequent additional imposed trips, the skilled schedulers were not capable of handling the bus-scheduling tasks efficiently. Consequently, the Egged management decided to test a fully computerized system. Their experience with this system is shown in Section 6.3. The experience led to the development of an informative graphical method, to be explicated in Sections 6.5 and 6.6.

In Adelaide, Australia, the newspaper headline on April 14, 2000, shown in Figure 6.1, captures our eyes. We might think that a reverse headline, "Fewer Buses, No Waiting," would be the prudent one deserving a scoop. This chapter seeks ways of minimizing the number of vehicles and, at the same time, maintaining the best level of service set forth by the decided timetable.

Approaches to vehicle scheduling worldwide range from primitive decision-making to computerized mathematical-programming techniques. The latter is reviewed in Section 6.3. As for primitive approaches, an example was observed in Santo Domingo. The drivers for a 40-bus company would gather very early each morning and create vehicle schedules like it was a lottery. That is, all trips in the timetable were represented by small pieces of papers (each trip with a different number), and the drivers simply picked up their piece of paper containing the chain of trips to be performed that day as though they were drawing a lucky number from a hat. Some drivers, of course, went back to sleep.

A common practice in vehicle scheduling is to use time–space diagrams similar to the one from Europe appearing in Figure 6.2. Each line in the diagram represents a trip moving over time (x-axis) at the same average commercial speed represented by the slope of the line. Although many schedulers became accustomed to this description, it is cumbersome, if not impossible, to use these diagrams to make changes and improvements in the scheduling. It is impossible to use them for interlining because the space (y-axis) refers only to a single route, as can be seen in the figure. It is also difficult to use different average speeds for different route segments, in which the lines in the time–space diagram can cross one another; this is not to mention the inconvenience of using these diagrams manually for inserting DH trips and/or shifting departure times. These limitations of the time–space diagram caused us to look more closely into a more appealing approach—the one presented in Sections 6.5 and 6.6.

6.2 FLEET SIZE REQUIRED FOR A SINGLE ROUTE

This section considers the case in which interlinings and DH trips are not allowed and each route operates separately. Given the average round-trip time and chosen layover time, the minimum fleet size for a radial route can be found according to the formula derived by

Figure 6.1 Main headline of an Adelaide newspaper on April 14, 2000. Wouldn't a real scoop be the reverse: "Fewer Buses, No Waiting"?

Salzborn (1972). Specifically, let T_r be the average round-trip time, including layover and turn-around times, of a radial route r (departure and arrival points are same). The minimum fleet size is equal to the largest number of vehicles that departs within T_r.

Although Salzborn's modeling provides the base for fleet size calculation, it relies on three assumptions that do not hold up in practice: (1) vehicle departure rate is a continuous function of time, (2) T_r is the same throughout the period under consideration, and (3) route r is a radial route starting at a major point (e.g., CBD). In practice, departure times are discrete (see Chapters 4 and 5), average trip time is usually dependent on time-of-day, and a single transit route usually has different timetables for each direction of travel. For that reason, we will now broaden Salzborn's model to account for practical operations planning.

Let route r have two end points: a and b. Let T_{ria} and T_{rjb} be the average trip time on route r for vehicles departing at t_{ia} and t_{jb} from a and b, respectively, including layover time at their respective arrival points. Let n_{ia} be the number of departures from a between t_{ia}, in which departure ia is included, and $t_{i'a}$ in which departure i'a is excluded. Thus, ia arrives to terminal b, then continues with trip jb, the latter being the *first* feasible departure from

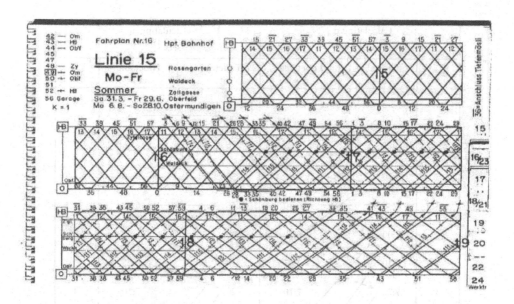

Figure 6.2 Typical time–space diagram of transit trips used to describe and change vehicle scheduling in many transit agencies.

b to a at a time greater than or equal to the time $t_{ia} + T_{ria}$; $t_{i'a}$ is the *first* feasible departure from a to b at a time greater than or equal to $t_{jb} + T_{rjb}$. Similarly n_{jb} may be defined for a trip j from b.

Lemma 6.1

In the case of no DH trips, n_{ia} departures must be performed by different vehicles at a, and n_{jb} must be performed by different vehicles at b, for all ia and jb in the timetables of r.

Proof: The proof is actually based on a contradiction. Let us assume that the same vehicle can perform two departures included in n_{ia} at a. However, in order to complete a full round trip, including layover times, this vehicle can only pick up the i'a departure at a, which is not included (by definition) in n_{ia}; hence, it is impossible for the same vehicle to execute two departures within n_{ia}.

Theorem 6.1

In the case with no interlining (between routes) and no DH trips, the minimum fleet size required for route r is

$$N^r_{min} = max\{max_i n_{ia},\ max_j n_{jb}\} \tag{6.1}$$

Proof: Based on Lemma 6.1, $max_i n_{ia}$ and $max_j n_{jb}$ represent the minimum number of vehicles required to execute the timetables at a and b, respectively. Consequently, the minimum fleet size for r is the greater of the two.

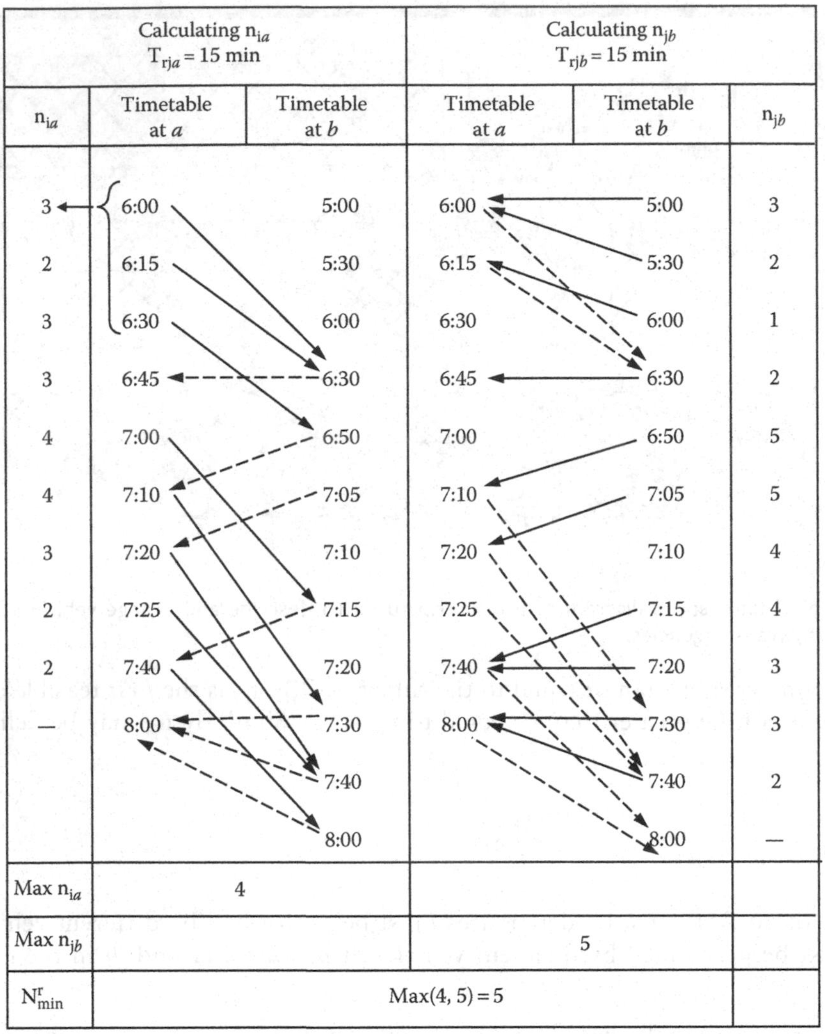

	Calculating n_{ia} $T_{rja} = 15$ min		Calculating n_{jb} $T_{rjb} = 15$ min		
n_{ia}	Timetable at a	Timetable at b	Timetable at a	Timetable at b	n_{jb}
3	6:00	5:00	6:00	5:00	3
2	6:15	5:30	6:15	5:30	2
3	6:30	6:00	6:30	6:00	1
3	6:45	6:30	6:45	6:30	2
4	7:00	6:50	7:00	6:50	5
4	7:10	7:05	7:10	7:05	5
3	7:20	7:10	7:20	7:10	4
2	7:25	7:15	7:25	7:15	4
2	7:40	7:20	7:40	7:20	3
—	8:00	7:30	8:00	7:30	3
		7:40		7:40	2
		8:00		8:00	—
Max n_{ia}	4				
Max n_{jb}			5		
N^r_{min}		Max(4, 5) = 5			

Figure 6.3 Example of derivation of single-route fleet size with no deadheading trips.

An example of deriving the required fleet size for a single transit route r is shown in Figure 6.3. In this figure, a single average travel time $T_{ria} = T_{rjb} = 15$ min is used throughout the timetable for both directions of r. The timetables contain 12 departures at b and 10 at a. The calculations for n_{ia} and n_{jb} are shown by arrows—starting with the departures at a for n_{ia} (using $T_{ria} = 15$), and starting at b for n_{jb}. The solid line in Figure 6.3 represents the direction from the starting time to the first feasible connection (after 15 min), and the dashed line in the opposite direction links to the first feasible connection (also after 15 min) from the starting point. This leads to a determination of both n_{ia} and n_{jb}, and eventually the minimum fleet size, $N^r_{min} = 5$, according to Equation 6.1. It should be mentioned that the same T_{ria} and T_{rjb} are used throughout the example only for the sake of simplicity. Varied T_{ria} and T_{rjb} can be utilized in the same manner for each departure.

Vehicle chains (blocks) can be constructed by using the FIFO (first-in, first-out) rule. That is, a block will start at a depot for the first assigned scheduled trip, and then will make the first feasible connection with a departure (based on the route's timetable) at the other end point of the route, and so on. The block usually ends with a trip back to the depot.

The trips to and from the depot are often DH trips. In the example illustrated in Figure 6.3, the five blocks can be constructed starting with the first departure (5:00) at b and using the FIFO (first-feasible connection) rule, then deleting the departures selected and continuing with another block until all departures are used. At the start of each step (at b), a check is made to see whether the next (in time) departure can be connected to an earlier unused departure at a and, if so, whether this connection can be allowed. The five blocks, therefore, are as follows: $[5:00(\text{at } b)–6:00(a)–6:30(b)–6:45(a)–7:05(b)–7:20(a)–7:40(b)–8:00(a)]$; $[5:30(b)–6:15(a)–6:50(b)–7:10(a)–7:30(b)]$; $[6:00(b)–6:30(a)–7:10(b)–7:25(a)–8:00(b)]$; $[7:00(a)–7:15(b)–7:40(a)]$; $[7:20(b)]$. An earlier connection, linking the 7:15 departure at b to the 7:00 departure at a is possible only in the fourth block. The preceding FIFO process can certainly start at a as well. Note that the last block has only one trip; the five blocks can undergo changes, including swapping trips, between blocks. Each block can start and end at a depot or can be used as a segment in a larger block.

Finally, when DH trips between the ends of two routes and/or slightly shifting departure times are allowed, it is more complex to use the formulation developed earlier. Instead, the solution can be found using the graphical method presented in Section 6.5 and in Chapter 7.

6.3 EXAMPLE OF AN EXACT SOLUTION FOR MULTIROUTE VEHICLE SCHEDULING

Practical vehicle scheduling usually involves a network of transit routes with interlining between routes. The switch from one route to another can be executed either by a service trip when the two routes share the same end point (terminal) or by a DH (empty) trip. The most problematic part of the vehicle-scheduling task for a network of routes is to minimize the number of vehicles required to carry out the timetables. It is basically a cost-minimization problem. This section provides an example of the implementation of one solution.

The problem of scheduling vehicles in a multiterminal (multiroute) scenario is known as the Multi-Depot Vehicle Scheduling Problem (MDVSP). This problem is complex (NP-complete, which is explained in Section 5.5), and considerable effort is devoted to solving it in an exact way. A review and description of some exact solutions can be found in Desrosiers et al. (1995), Daduna and Paixao (1995), and in the literature review at the end of this chapter.

An example of the formulation of the MDVSP is as follows:

Objective function: $\displaystyle \underset{y}{\text{Min}} \sum_{i=1}^{n+1} \sum_{j=1}^{n+1} c_{ij} y_{ij}$ (6.2)

Subject to

$$\sum_{i=1}^{n+1} y_{ij} = 1, j = 1,2,\ldots,n \quad \text{and} \quad \sum_{j=1}^{n+1} y_{ij} = 1, i = 1,2,\ldots,n$$

$$\sum_{j=1}^{n+1} y_{n+1,j} = 1 \quad \text{and} \quad \sum_{i=1}^{n+1} y_{i,n+1} = 1$$

Finally, there are also constraints ensuring that trip end i can link to trip start j, and that y_{ij} is integer and greater than or equal to zero, where
   i is the end of a trip at time t_i,
   j is the start of a trip at time t_j,

$$y_{ij} = \begin{cases} 1, & \text{ending is a connection to start} \\ 0, & \text{otherwise} \end{cases}$$

For i = n + 1, then $y_{n+1,j} = 1$ if a depot supplies a vehicle for the jth trip. For j = n + 1, then $y_{i,n+1} = 1$ if, after end of the ith trip, the vehicle returns to a depot and $y_{n+1,n+1} =$ number of vehicles remaining unused at a depot. The cost function c_{ij} takes the form:

$$c_{ij} = \begin{cases} K; & i = n+1; j = 1, 2, ..., n \\ 0; & i = 1, 2, ..., n; j = n+1 \\ L_{ij} + E_{ij}; & i, j = 1, 2, ..., n \end{cases} \tag{6.3}$$

where
   K is the saving incurred by reducing the fleet size by one vehicle
   L_{ij} is the DH cost from event i to j
   E_{ij} is the cost of a driver's idle time between i and j

This formulation, appearing in a similar form in Gavish et al. (1978), covers the chaining of vehicles in a sequential order from the depot to the transit routes, alternating with idle time and DH trips, and back to the depot. In OR terms (see Section 5.5), this is a zero-one integer-programming problem in that it can be converted to a large-scale assignment problem. The objectives are to minimize fleet requirements (minimize linkages of routes) and minimize DH and idle time (maximize vehicle and driver utilization). In addition, the assignment of vehicles from the depots to the vehicle schedule generated in the foregoing chaining process can be formulated as a "transportation problem," as this is known in the OR literature. The objectives of the assignment stage are to minimize the driving cost from the depot to the first trip on the schedule, and from the last trip to a depot.

The formulation presented partially in Equations 6.2 and 6.3 was applied by Egged (the Israeli national bus carrier). As mentioned earlier, Egged has about 4000 buses and tried to gain more efficient schedule handling based on Equations 6.2 and 6.3. Egged examined the algorithm based on several sets of bus-scheduling data. It is worth noting that the program is capable of solving bus-scheduling problems involving up to 2500 bus trips. The full test results and discussion of this examination were reported by Ceder and Gonen (1980). For example, the results of five sets are shown in Figure 6.4. Only the DH and idle times of the first data set are not available (indicated by N/A). These examples were selected for peak-hour bus schedules, and at a first look the results seemed quite attractive. However, when there was an attempt to implement the results, it became evident that several significant constraints remained unfilled. Applying these constraints results in a solution that is considerably different from that obtained by the optimal method. In Egged, as in most transit agencies, each bus is used by the same drivers. During the manual scheduling of buses to trips, it was necessary for the scheduler to consider some of the drivers' constraints, as well, although that is of secondary importance at this stage.

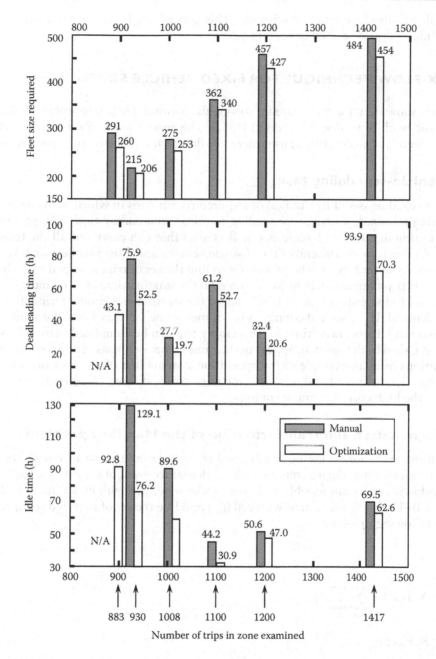

Figure 6.4 Comparison of manual and optimization (programmable) solutions for vehicle-scheduling planning at a large bus agency.

In summary, the major limitations of the programmable algorithm were that it could not consider the following: (1) integration of more than 2500 trips, (2) availability of adequate bus type for each trip, (3) need for bus refueling, (4) need for driver's meals, and (5) location and availability of drivers. The Egged schedulers agreed that using software as a black box was blind as opposed to understanding the software and interacting with it, which becomes a real eye-opener. In order to satisfy these limitations, it was decided to search for a procedure that would allow the inclusion of practical considerations that experienced schedulers might wish to introduce into the schedule. This procedure, based on the deficit-function graphical method, is described in Section 6.5.

6.4 MAX-FLOW TECHNIQUE FOR FIXED VEHICLE SCHEDULING

This section shows that the procedure for solving the minimum fleet size problem has its roots in the classic work of Ford and Fulkerson (1962). The approach described is based on Ceder (1978) and Stern and Ceder (1983a) using network-flow techniques well known in the OR field.

6.4.1 Vehicle-scheduling task

Consider a schedule (set of timetables) of required transit trips in which each trip is defined by a starting terminal, starting time, ending terminal, and ending time. The problem is to find the minimum number of vehicles (the fleet size) that can carry out all the trips in the schedule. As this is most efficiently done if single vehicles are allowed to service several trips in succession, a second part of the problem is to find the set of trips assigned to each vehicle in the fleet. If trip j immediately follows i, then (a) the starting time of trip j must be greater than or equal to the ending time of trip i and (b) the starting and ending terminal of j and i must be identical. If the second condition is not met, some systems allow the vehicle to run empty from i to j. If the travel time for this empty trip can be completed before the starting time of trip j, then both trips may appear in the same chain and be assigned the same vehicle. Bus operations offer an example of transportation systems that often carry out such empty or DH trips. On the other hand, airline companies rarely, if ever, "deadhead" their aircraft because of the high cost of running empty.

6.4.2 Formulation and transformation of the Max-flow problem

A trip-joining array for S may be constructed by associating the ith row with the arrival event of the ith trip, and the jth column with the departure event of the jth trip. Cell (i, j) will be admissible if i and j can feasibly be joined. Otherwise, (i, j) will be an inadmissible cell. Let x_{ij} be a 0–1 variable associated with cell (i, j) and I be the set of required trips; consider, then, the following problem:

Problem P1

$$\text{Max} \, Z1 = \sum_{i \in I} \sum_{j \in I} x_{ij} \tag{6.4}$$

Subject to:

$$\sum_{j \in I} x_{ij} \leq 1, \quad i \in I \tag{6.5}$$

$$\sum_{i \in I} x_{ij} \leq 1, \quad j \in I \tag{6.6}$$

$$\left. \begin{array}{l} x_{ij} \in \{0,1\}, \quad \text{all}(i, j) \text{ admissible} \\ x_{ij} = 0, \quad \text{all}(i, j) \text{ inadmissible} \end{array} \right\} \tag{6.7}$$

A solution with $x_{ij} = 1$ indicates that trips i and j are joined. The objective function maximizes the number of such joinings. Constraint (6.5) insures that each trip may be joined with, at most, one successor trip. Similarly, constraint (6.6) indicates that each trip may be joined with, at most, one predecessor trip. The following theorem states that maximizing (6.4) in Problem P1 is tantamount to minimizing the number of chains for a trip schedule of size n.

Theorem 6.2

Let N and n denote the number of chains and trips, respectively. Then, Min N = n – Max Z1.

Proof of Theorem 6.2 can be found in Ford and Fulkerson (1962).

The Problem P1 is equivalent to a special arrangement of the maximum-flow (Max-flow) problem. The Max-flow algorithm that solves the vehicle-scheduling problem with DH trips is called an augmenting-path algorithm. It appears, in an extensive treatment, in the classic book by Ford and Fulkerson (1962). A complete description of the augmenting-path algorithm, which will be used for further applications in this volume, is shown in Appendix 6.A at the end of this chapter.

The vehicle-scheduling problem can be transformed to a unit-capacity bipartite network, in which the solution time has the complexity of $O(n^{1/2}m)$ with n nodes (departure times) and m arcs. A bipartite network is a network in which the set of nodes is partitioned into two subsets of nodes such that directed arcs exist only between (not within) the two subsets. This solution time, first shown by Even and Tarjan (1975), was explicated by Ahuja et al. (1993). The complexity function is explained in Section 5.5.

We will use a simple numerical example as an explanatory tool to describe and interpret the construction and solution of the vehicle-scheduling problem transformed into a Max-flow network problem. Consider the three-terminal problem defined by the data in Tables 6.1 and 6.2. The data in Table 6.1 are transformed into the upper part of the network-flow representation in Figure 6.5, which has two dummy nodes: a *source* node s and a *sink* node t. The nodes, being connected from s, are the arrival times of the example, with an indication, in parenthesis, of the arrival terminal. The nodes connected to t are the departure times, with an indication of the departure terminal. Feasible connections between the arrival and departure times, utilizing the DH data in Table 6.2, establish the arcs between the left-side and right-side nodes, based on Equation 6.4. Each arc capacity represents the number of connections that can "flow" through the arc. In our case, there is only a unit capacity assigned to each arc, because only one connection (if any) between a given arrival time and terminal and a given departure time and terminal is possible. The more flow

Table 6.1 Trip schedule S for the example problem

Trip number (i)	Departure terminal (p^i)	Departure time (t_e^i)	Arrival terminal (q^i)	Arrival time (t_s^i)
1	b	6:00	b	6:30
2	a	7:05	c	8:05
3	c	7:10	a	8:00
4	b	8:30	a	9:20
5	a	9:00	b	9:45

Table 6.2 Average DH travel-time (minutes) matrix for the example in Table 6.1

Departure terminal	Arrival terminal		
	a	b	c
a	0	30	50
b	35	0	45
c	45	40	0

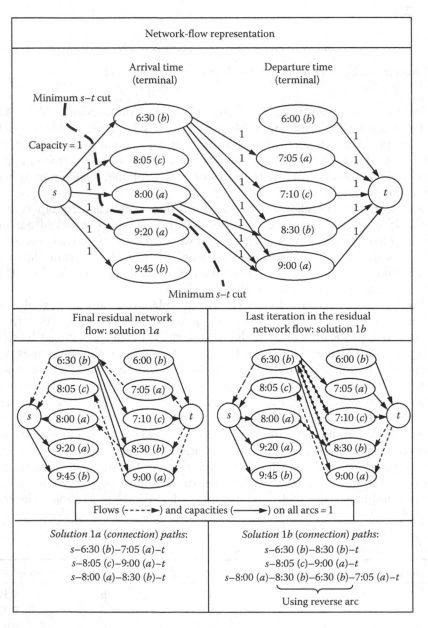

Figure 6.5 Solutions 1a and 1b of the vehicle-scheduling example using the augmenting-path algorithm appearing in Appendix 6.A.

created, the fewer chains will be required as stated by Theorem 6.2. The objective function Z1 in Problem P1 equals the flow to be created at *s* and absorbed at *t*. The maximum flow of four in Figure 6.5 means that all trips can be handled by a single chain, with four linkages between arrivals and departures.

The augmenting-path algorithm described in Appendix 6.A is applied for the example problem. We look for the minimum-arc path between *s* and *t*, assign a flow = 1, and create a residual network in which all arcs utilized have a reverse arc (dashed line in the two residual networks in Figure 6.5). All capacities and flows in the residual networks have a unit value. We then search for the next minimum-arc path between *s* and *t*, assigning to it flow = 1 on the residual network, and construct an updated residual network. The process continues until no *s*–*t* path is found. Two solutions, 1a and 1b, leading to the same optimal result (Max-flow = 3) are shown in Figure 6.5. In solution 1b, we show a case in which a reverse arc is used in the augmenting-path algorithm. Based on Theorem 6.A.4 (in the appendix), Max-flow = minimum *s*–*t* cut in the original network-flow. This minimum *s*–*t* cut, explicated in Appendix 6.A, is shown in the upper part of Figure 6.5.

The result of the example is Max-flow = Max Z = 3, and, following Theorem 6.2, Min N = 5 – 3 = 2 chains. That is, the timetable in Table 6.1 can be carried out by a minimum of two vehicles having the connections shown in Figure 6.5. Explicitly the two blocs, by their trip number in Table 6.1, are [1–2–5] and [3–4]. These two blocs have three DH trips for connecting arrival and departure terminals: *b*–*a* and *c*–*a* (in the first bloc) and *a*–*b* (in the second bloc), with the total of, respectively, 35 + 45 + 30 = 110 min DH time.

6.5 DEFICIT-FUNCTION MODEL WITH DEADHEADING TRIP INSERTION

The minimum fleet size problem may be referred to with or without DH trips. When DH is allowed, we can reach the counterintuitive result of decreasing the required resources (fleet size) by introducing more work into the system (adding DH trips). This approach assumes that the capital cost of saving a vehicle far outweighs the cost of any increased operational cost (driver and vehicle travel cost) imposed by the introduction of DH trips. This section offers a highly informative, graphical person–computer interactive technique based on a step function, called deficit function (DF), which is simple to use and interact with. The lesson that is learned is that it is easier to deadhead vehicles or shift their departure time (see next chapter) when the problem is seen graphically, even if it is an almost mission impossible. Perhaps the inspiration for this approach is the saying: "It is easier for an elephant to pass through the eye of a needle if it is lightly greased." Following is explanatory background on a step-function approach, described by Ceder and Stern (1981) and Ceder (2003), for allocating the minimum number of vehicles to a given timetable.

6.5.1 Definitions and notations

Let I = {i: i = 1, ..., n} denote a set of required trips. The trips are conducted between a set of terminals K = {*k*: *k* = 1, ..., *q*}, each trip to be serviced by a single vehicle, and each vehicle able to service any trip. Each trip i can be represented as a 4-tuple (p^i, t_s^i, q^i, t_e^i), in which the ordered elements denote departure terminal, departure (start) time, arrival terminal, and arrival (end) time. It is assumed that each trip i lies within a schedule horizon $[T_1, T_2]$, that is, $T_1 \le t_s^i \le t_e^i \le T_2$. The set of all trips $S = \{(p^i, t_s^i, q^i, t_e^i) : p^i, q^i \in K, i \in I\}$ constitutes the timetable. Two trips, i and j, may be serviced sequentially (feasibly joined) by the same vehicle if and only if (a) $t_e^i \le t_s^j$ and (b) $q^i = p^j$.

If i is feasibly joined to j, then i is said to be the predecessor of j, and j the successor of i. A sequence of trips $i_1, i_2, \ldots, i_w$ ordered in such a way that each adjacent pair of trips satisfies (a) and (b) is called a chain or block. It follows that a chain is a set of trips that can be serviced by a single vehicle. A set of chains in which each trip i is included in I exactly once is said to constitute a vehicle schedule. The problem of finding the minimum number of chains for a fixed schedule S is defined as the minimum fleet size problem.

Let us define a DH trip as an empty trip from some terminal p to some terminal q in time $\tau(p, q)$. If it is permissible to introduce DH trips into the schedule, then conditions (a) and (b) for the feasible joining of two trips, i, j, may be replaced by the following:

$$t_e^i + \tau(q^i, p^i) \le t_s^j \tag{6.8}$$

Now let us introduce a deficit-function-based model, which will be utilized and employed throughout most of the remaining chapters of the book.

A *deficit function* (DF) is a step function defined across the schedule horizon that increases by one at the time of each trip departure and decreases by one at the time of each trip arrival. This step function is called a DF because it represents the deficit number of vehicles required at a particular terminal in a multiterminal transit system. To construct a set of DFs, the only information needed is a timetable of required trips. The main advantage of the DF is its visual nature. Let $d(k, t, S)$ denote the DF for terminal k at time t for schedule S. The value of $d(k, t, S)$ represents the total number of departures minus the total number of trip arrivals at terminal k, up to and including time t. The maximum value of $d(k, t, S)$ over the schedule horizon $[T_1, T_2]$, designated $D(k, S)$, depicts the deficit number of vehicles required at k.

The DF notations are presented in Figure 6.6, in which $[T_1, T_2] = [5:00, 8:30]$. It is possible to partition the schedule horizon of $d(k, t)$ into a sequence of alternating hollow and maximum intervals $(H_0^k, M_1^k, H_1^k, \ldots, H_j^k, M_{j+1}^k, \ldots, M_{n(k)}^k, H_{n(k)}^k)$. Note that S will be deleted when it is clear which underlying schedule is being considered. Maximum intervals $M_j^k = [s_j^k, e_j^k]$, $j = 1, 2, \ldots, n(k)$ define the intervals of time over which $d(k, t)$ takes on its maximum value. Index j represents the jth maximum intervals from the left; $n(k)$ represents the total number of maximal intervals in $d(k, t)$, where s_j^k is the departure time for a trip leaving terminal k and e_j^k is the time of arrival at terminal k for this trip. The one exception occurs when the DF reaches its maximum value at $M_{n(k)}^k$ and is not followed by an arrival, in which case $e_{n(k)}^k = T_2$.

A hollow interval H_j^k, $j = 0, 1, 2, \ldots, n(k)$ is defined as the interval between two maximum intervals: this includes the first hollow, from T_1 to the first maximum interval, $H_0^k = [T_1, s_1^k]$; and the last hollow, which is from the last interval to T_2, $H_{n(k)}^k = [e_{n(k)}^k, T_2]$. Hollows may contain only one point; if this case is not on the schedule horizon boundaries (T_1 or T_2), the graphical representation of $d(k, t)$ is emphasized by a clear dot.

The sum of all DFs over k is defined as the overall DF, $g(t) = \sum_{k \in K} d(k, t)$. This function $g(t)$ represents the number of trips that are simultaneously in operation, that is, a count, from a bird's-eye view at time t, of the number of transit vehicles in actual service over the entire transit network of routes. The maximum value of $g(t)$, $G(S)$, is exploited for a determination of the lower bound on the fleet size (see Chapter 7). An example of a

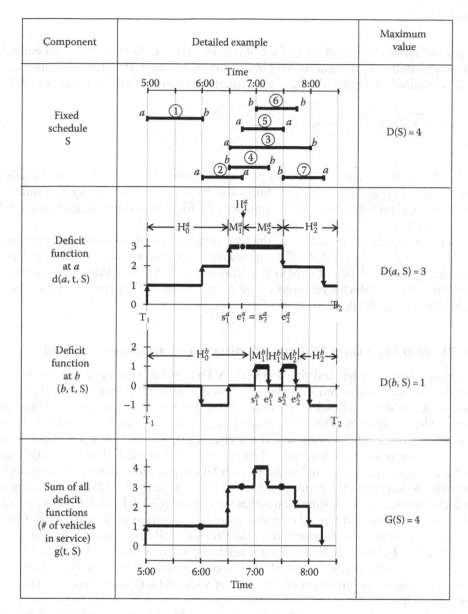

Figure 6.6 Illustration of two-terminal fixed schedule with associated deficit functions and their sum, including notations and definitions.

two-terminal operation, a fixed schedule of trips, and the corresponding set of DFs and notations is illustrated in Figure 6.6.

6.5.2 Fleet size formula

Determining the minimum fleet size, D(S), from the set of DFs is simple enough—one merely adds up the deficits of all the terminals. In the example in Figure 6.6 without DH trips, D(S) = D(a) + D(b) = 4. This result was apparently derived independently by Bartlett (1957), Salzborn (1972, 1974), and Gertsbach and Gurevich (1977). It is formally stated as Theorem 6.3.

Theorem 6.3

If, for a set of terminals K and a fixed set of required trips I, all trips start and end within the schedule horizon $[T_1, T_2]$ and no DH insertions are allowed, then the minimum number of vehicles required to service all trips in I is equal to the sum of all the maximum deficits:

$$\text{Min N} = \sum_{k \in K} D(k) = \sum_{k \in K} \max_{t \in [T_1, T_2]} d(k, t) \tag{6.9}$$

Proof: Let F_k be the number of vehicles present in terminal k at the start of the schedule horizon T_1; let $s(k, t)$ and $e(k, t)$ be the cumulative number of trips starting and ending at k from T_1 up to and including time t. The number of vehicles remaining at k at time $t \geq T_1$ is $F_k - s(k, t) + e(k, t)$.

In order to service all trips leaving k, the preceding expression must be nonnegative, that is, $F_k \geq s(k, t) - e(k, t)$, $T_1 \leq t \leq T_2$. The minimum number of vehicles required at k is then equal to the maximum deficit at k: $\text{Min } F_k = \text{Max}_t[s(k, t) - e(k, t)] = \text{Max}_t d(k, t)$. Hence, the minimum number of vehicles required for all terminals in the system is equal to the total deficit $\text{Min N} = \sum_{k \in K} \text{Min } F_k = \sum_{k \in K} D(k) = D(S)$.

6.5.3 DH trip insertion: Effect and initial fleet size lower bound

This section follows Ceder and Stern (1981). A DH trip is an empty trip between two termini that is usually inserted into the schedule in order to (1) ensure that the schedule is balanced at the start and end of the day, (2) transfer a vehicle from one terminal where it is not needed to another where it is needed to service a required trip, and (3) refuel or undergo maintenance. This section will consider case (2), and Section 6.6 will comment on case (3). We start by asking where and when are such trips needed. Usually, a trip schedule received from operating personnel includes such DH trips, and it is easy to apply the fleet size formula to determine the minimum fleet size, followed by the FIFO rule to construct each vehicle's schedule. The assumption is that the trip schedule S has been purged of all DH trips, leaving only required trips. From this point, the question of how to insert DH trips into the schedule in order to further reduce the fleet size will be examined. At first, it seems counter-intuitive that this can be achieved, since it implies that increased work (adding trips to the schedule) can be carried out with decreased resources (fewer vehicles). This section will show through an examination of the effect of such DH trip insertions on DFs that this is indeed possible.

Consider the example in Figure 6.7. In its present configuration, according to the fleet size formula, four vehicles are required at terminal a, 0 at terminal b, and 1 at terminal c for a fleet size of five. The dashed arrows in the figure represent the insertion of DH_1 trip from b to a and DH_2 from c to b. After the introduction of these DH trips into the schedule, the DFs at all three terminals are shown updated by the dotted lines. The net effect is a reduction in fleet size by one unit at terminal a. It is interesting to examine the particular circumstances under which this reduction was achieved. After adding an arrival point in the first hollow of terminal a before s_1^a, the maximal interval when using DH_1 is reduced by one unit, causing a unit decrease in the deficit at a. This arrival point becomes, therefore, e_1^a. Since the DH_1 departure point is added in the middle hollow of terminal b, at e_1^b, it is necessary to introduce a second DH trip, which will arrive at the start of the second maximum interval of b. Fortunately, this DH_2 trip departs from the last hollow of c, where it could no longer

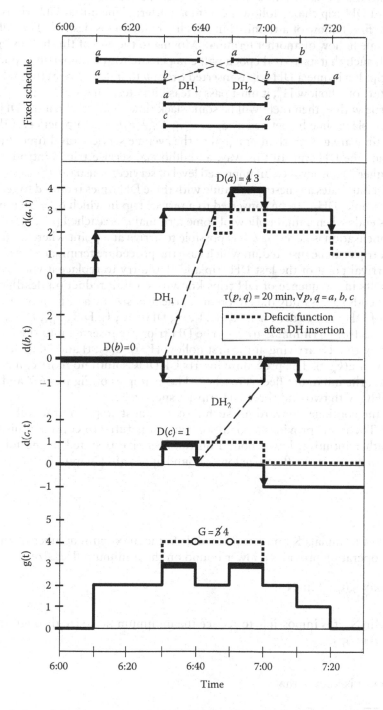

Figure 6.7 Description of six-trip, two-terminal example in which the fleet size is reduced by one using a chain of two DH trips (URDHC) and in which g(t) is changed.

affect the deficit at c. In general, it is possible to have a string of DH trips to reduce the fleet size by one unit: one *initiator trip* and the others *compensating trips*.

All successful DH trip chains follow a common pattern. The initial DH trip is introduced to arrive in the first hollow of a terminal in which a reduction is desired. This DH trip must depart from some hollow of another terminal. Moving to the end of this hollow, another DH trip is inserted, such that its arrival epoch will compensate for the departure epoch added by the first DH trip. Each time a DH trip is inserted (from p to q) to arrive at the end of a hollow H_i^q from the start of a hollow H_j^p, it must pass a feasibility test, that is, $e_j^p + \tau(p, q) \leq s_{i+1}^q$. If the inequality is true with $<$, then there will be some slack time, during which the DH trip can be shifted. Let this slack time be defined as $\delta_{pq} = s_{i+k}^q - [e_j^p + \tau(p, q)]$. In practice, if the DH time plus the slack time are greater than or equal to the average service travel time, then a service trip may replace the DH trip. In this way, an additional service trip is introduced, thereby resulting in higher frequency (i.e., an improved level of service) at usually the same operational cost. Figure 6.8 illustrates a nine-trip schedule with three DH times required to reduce $D(b)$ by one. In this example, DH_2 can be converted to a service trip in which $\delta_{ac} = 30$ min.

The process ends when a final hollow of some terminal q is reached (i.e., $H_i^q = H_{n(q)}^q$), after which no compensation is necessary. It is possible to arrive at a point where no feasible compensating DH trips can be inserted, in which case the procedure terminates or one may back track to the arrival point of the last DH trip added and try to replace it with another. This procedure results in a sequence of DH trips known as a unit reduction deadheading chain (URDHC) if it ends successfully (i.e., if it reduces the fleet size by a unit amount). A URDHC is a sequence of DH trips of the form $(k_1, DH_1, k_2, DH_2, \ldots, k_j, DH_j, k_{j+1}, DH_w, k_{w+1})$, where DH_j is a DH trip from terminal k_j to k_{j+1}. The DH trips are inserted into the DFs from hollow to hollow, with DH_1 arriving at the first hollow $H_0^{k_2}$ of $d(k_2, t)$ and DH_w departing from the last hollow of $d(k_w, t)$. The procedure inserts URDHCs until no more can be found or a lower bound on the minimum fleet is reached. The examples of Figures 6.7 and 6.8 include a single URDHC with two and three DH trips, respectively.

Obviously, the continued reward for such a search must stop, and the following "Initial Lower Bound Theorem" provides a condition when it is futile to continue this search. The initial, and rather intuitive, lower bound on the fleet size is stated in the following text. Further improvements in the fleet size lower bound appear in Chapter 7.

Theorem 6.4

(a) For a set of terminals K and a trip schedule S, the maximum number of trips in simultaneous operation provides a lower bound on the minimum fleet size

$$G = \max_{t \in [T_1, T_2]} g(t) \leq \text{Min} \, N \qquad (6.10)$$

(b) If $G = \text{Min} \, N$, it is impossible to reduce the minimum fleet size through the introduction of DH trips.

Proof

(i) In general it is known that

$$\max_{t \in [T_1, T_2]} \sum_{k \in K} d(k, t) \leq \sum_{k \in K} \max_{t \in [T_1, T_2]} d(k, t)$$

From the definition of g(t) and (6.9), $\max_{t \in [T_1, T_2]} g(t) \leq \text{Min} \, N$

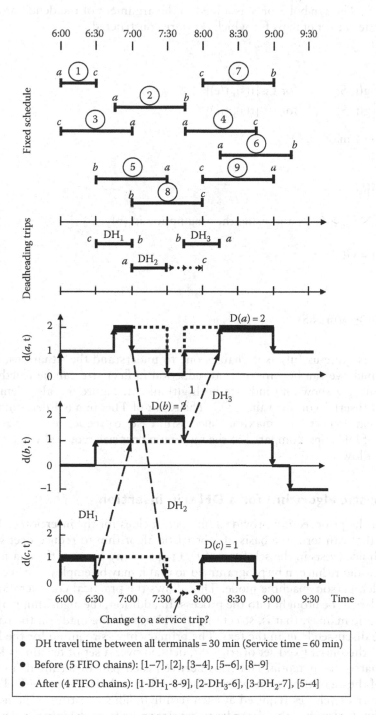

Figure 6.8 Example of a URDHC with three DH trips in which DH_2 can be converted to a service trip.

(ii) Let S represent the original schedule of given trips. Add some trip i starting at t(s) and ending at time t(e) to the schedule (this can be a DH trip). Call the new schedule (with i) S'. The symbol S or S' is added to the argument of the deficit and overall DFs to indicate the set of trips for which they are constructed.

Since

$$g(t, S') = \begin{cases} g(t, S), & \text{for } t \notin [t(s), t(e)] \\ g(t, S) + 1, & \text{for } t \in [t(s), t(e)] \end{cases}$$

$$\max_t g(t, S') \geq \max_t g(t, S)$$

$$G(S') \geq G(S)$$

From (i) Min N(S') ≥ G(S') and using the assumption in (b)

$$\text{Min } N(S) = G(S)$$

Therefore,

$$\text{Min } N(S') \geq \text{Min } N(S)$$

This theorem is quite useful, as it enables one to understand the situations, based on the deficits, in which we can be sure that no reduction in fleet size can be (further) achieved. Figures 6.6 and 6.7 show and indicate both g(t) and G. Figure 6.7 also demonstrates the impact of DH insertion on the value of G. The result of Theorem 6.4 determines the extent to which we can expect (in a maximal successful case) to reduce a fleet size through the introduction of DH trips, compared to the minimum fleet size required when DH trip insertions are not allowed.

6.5.4 Heuristic algorithm for a DH trip insertion

The results of the prior section provide a number of clues for the insertion of DH trips into the schedule that can form the basis of a heuristic algorithm to reduce fleet size. As much freedom of choice exists in the selection of DH trips at any point in the chain construction procedure, the algorithm can be programmed so that it may be employed in a conversational (manual mode), person–machine mode. This allows the practical considerations of experienced schedulers to be brought into the process. In addition, the algorithm may operate in a purely automatic manner, that is, selections of DH trips being made on the basis of several criteria, to be discussed later in the text. The primary inputs required for the heuristic procedure are (1) the initial fixed set of trips, S°, defined over a set of terminals K, and (2) the travel time matrix for potential DH trips between each pair of terminals in the schedule. The output of the algorithm is a new trip schedule that includes the set of DH trips inserted and the number of vehicles required at each terminal. This constitutes the fleet size. Final DF information is also available. In order to construct the trip schedule of each vehicle in the fleet, a second phase is required. This can be done by applying the chain construction rules described in the next section.

A general framework of the heuristic algorithm is shown in a flow diagram form, following Ceder and Stern (1981), in Figure 6.9a and b. The parentheses {} denote a set of elements. The result of the lower-bound fleet-reduction theorem suggests that $G(S^0)$ should be examined before trying to insert DH trips into the schedule. This is called a lower-bound termination test in Figure 6.9a. Note that after DH trip insertions are made, $G(S^0)$ may be increased as shown in the example of Figure 6.6. It is thus important to evoke the lower-bound termination test after each DH chain insertion. Nonetheless in Chapter 7, $G(S)$ is improved (can be increased) to $G''(S'')$; hence, the lower-bound termination test will have $G''(S'')$, a non-updated value, instead of $G(S)$ to compare with $D(S)$. In the latter case, the updated $G(S)$ is kept in order to know the number of vehicles in service at any time.

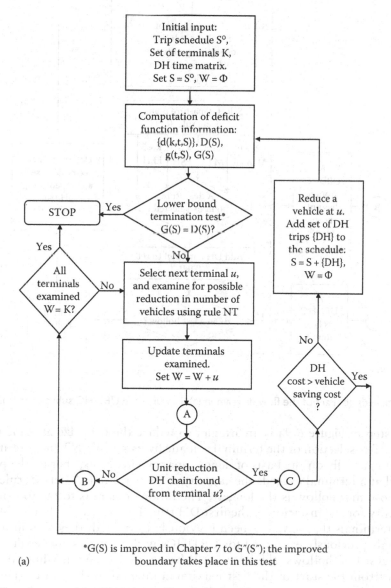

(a)

*$G(S)$ is improved in Chapter 7 to $G''(S'')$; the improved boundary takes place in this test

Figure 6.9 (a) Flow diagram of the DH insertion heuristic algorithm.   *(Continued)*

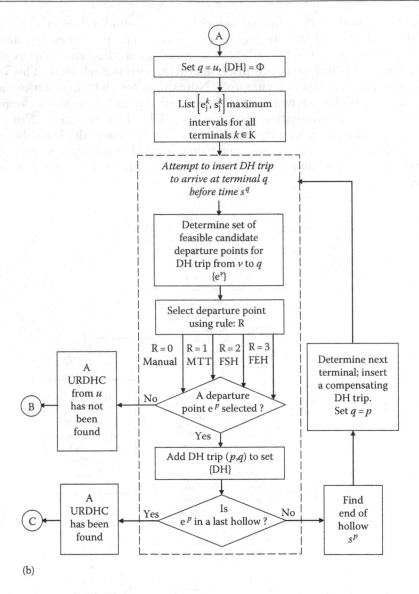

(b)

Figure 6.9 (Continued) (b) Part of the flow diagram in (a), in which the URDHC subroutine is illustrated.

The next step in Figure 6.9a is an attempt to reduce the fleet size at some terminal in the schedule. The selection of the terminal is made by using rule NT (next terminal). This may be done manually on the basis of depot capacity violations, that is, the present fleet size required at a terminal exceeds the capacity of the terminal. A default rule selects the terminal whose first hollow is the longest. The rationale here is to try to offer the largest opportunity for the insertion of the first DH trip. This selection should, of course, be made from terminals that have not been previously examined; they should not be made from terminals previously examined for URDHCs with unsuccessful results. In cases of multiple longest-first hollows, the selection is based on a terminal whose overall maximum region (from the start of the first maximum interval to the end of the last one) is the shortest.

Once a terminal, u, is selected, the algorithm searches for a URDHC from that terminal. This is done in the sub-routine URDHC, which appears with a dashed line in Figure 6.9b. If a URDHC is not found from u, this terminal is added to the set W of terminals examined. If all terminals have been examined (W = K), then the algorithm stops. Otherwise, a new terminal is selected and the search for a URDHC continues. If a URDHC has been found from u, in practice another test takes place. This concerns the DH cost involved as compared with the cost saving of a single vehicle; the comparison is usually computed by accountants. If the DH cost is higher than the saving cost, the URDHC is cancelled. Otherwise, the set of DH trips {DH} is added to the previous schedule; all DF information is updated for this new schedule, and a new iteration initiated. The algorithm must terminate after a finite number of steps because the fleet size in each iteration is reduced by one unit until either it reaches its lower bound or a URDHC cannot be found. Since the search for a URDHC is conducted many times in the algorithm, it has been placed in a sub-routine, called URDHC, which will now be described.

The sub-routine URDHC in Figure 6.9b starts by setting a dummy computer variable q to u. The number taken on by q represents the terminal for which the next DH trip is to be inserted. This trip is to arrive before time s_{j+1}^q, that is, before the end of the hollow H_j^q (in the case of $q = u$, $H_j^q = H_0^u$). The computer then determines a set of feasible candidate departure points $\{e^v\}$ from the set of terminals $V = \{u\}$, $q \notin V$. These points are identified by the start of hollows that lie to the right of the last hollow examined. That is, the selection of one of the candidate trips is made using rule R. When R = 0, the computer allows this selection to be inserted manually into a conversational mode. Otherwise, an automatic selection is made by the computer using one of the following criteria:

R = 1: The minimum DH trip time (MTT). Insert the candidate DH trip with the least travel time.

R = 2: The furthest start of a hollow (FSH). Insert a candidate DH trip whose hollow starts furthest to the right.

R = 3: The farthest end of a hollow (FEH). Insert a candidate DH trip whose hollow ends furthest to the right.

Figure 6.10 shows the effect of each criterion on three examples. For each example, the value of the total travel time of the DH trips in the URDHC is found, and the minimum value across all criteria circled. These examples show that the minimum DH travel time associated with the URDHC can be found by any of the criterion; no one criterion dominates the others for this performance measure.

If a DH trip cannot be selected and the completion of a URDHC is blocked, the algorithm backs up to the last candidate list with the last selection deleted; it then proceeds to select another DH trip from the candidate list. In the manual mode, the scheduler may choose a new terminal and initiate a new search for a URDHC instead of backing up. The option is also available in the automatic mode to forego the backup feature when blocks occur in order to reduce excessive computation time. If the backup feature is allowed to continue, one may return to the initial hollow of terminal u, after which a new terminal may be selected to reinitiate the search for a URDHC. When the full backup feature is used, the procedure will find the optimal solution, which is equivalent to Ford and Fulkerson's (1962) Max-flow procedure, which is explained in Section 6.4. If a final hollow is reached, a URDHC has been found, and it is returned to the main program for introduction into the schedule. Otherwise, the search for a compensating trip is repeated until one trip is found that departs from a last hollow.

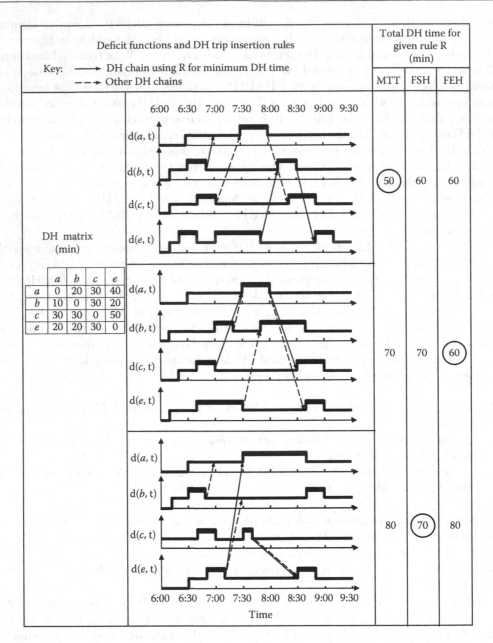

Figure 6.10 The effect on total deadheading time of three alternative heuristic rules for selecting candidate DH trip-departure terminals.

6.5.5 Constructing vehicle schedules (chains/blocks)

At the end of the heuristic algorithm, all trips, including the DH trips, are chained for constructing the vehicle schedule (blocks). Two rules can be applied for creating the chains: FIFO and a chain extraction procedure described by Gertsbach and Gurevich (1977). The FIFO rule simply links the arrival time of a trip to the nearest departure time of another trip (at the same terminal); it continues to create a schedule until no connection can be made. The trips considered are deleted, and the process continues.

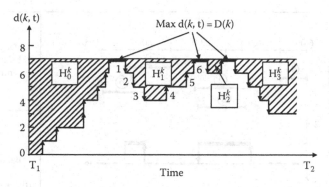

- FIFO set of connections: [(1–4), (2–5), (3–6)]
- Other sets of connections: [(1–4), (2–6), (3–5)],
  [(1–5), (2–6), (3–4)], [(1–6), (2–4), (3–5)],
  [(1–5), (2–4), (3–6)], [(1–6), (2–5), (3–4)]

Figure 6.11 Example of creating trips connections within one hollow, H_i^k, using the FIFO rule and all other possibilities while maintaining the minimum fleet size attained.

The chain extraction procedure allows an arrival–departure connection for any pair within a given hollow (on each DF). The pairs considered are deleted, and the procedure continues. Figure 6.11 illustrates one DF at k. This $d(k, t)$ has four hollows, H_j^k, j = 0, 1, 2, 3, with H_1^k having arrivals of Trips 1, 2, and 3 and departures of Trips 4, 5, and 6. Figure 6.11 are the FIFO connections (within this hollow) as well as other alternatives; in all, the minimum fleet size at k, $D(k)$, is maintained. In addition, two full FIFO chains are shown in Figure 6.8 before and after DH insertion procedure.

6.6 DEPOT-CONSTRAINED VEHICLE SCHEDULING

This section combines the description of previous sections with the methodology advanced by Stern and Ceder (1983a). A vehicle schedule (chain/block) is said to be balanced if for every terminal k, the number of vehicles starting their schedules from k equals the number of vehicles ending their schedules at k (not necessarily the same vehicles). Otherwise, the vehicle schedule is unbalanced. In practice, transit agencies seek this balanced condition at terminals with overnight parking (usually called depots).

6.6.1 Deficit function formulation

An example of a balanced schedule appears in Figure 6.12 for a solution without DH trips, using the data of Table 6.1. The solution indicates that exactly one vehicle chain starts and ends at each of the three terminals (albeit not the same chain for each terminal). However, a solution with DH trips, which saved a vehicle at terminal a, resulted in an unbalanced schedule (see lower part of Figure 6.12). A DH trip from a to c after 9:20 will balance the schedule. To determine whether a schedule is balanced, it is not necessary to carry out the actual chain constructions. The assessment may be made through an examination of the DFs of the schedules. If vs(k) and ve(k) are defined as the number of vehicles required at the start and remaining at the end of the schedule horizon, respectively, for terminal k, then the schedule is balanced if vs(k) = ve(k) for all k.

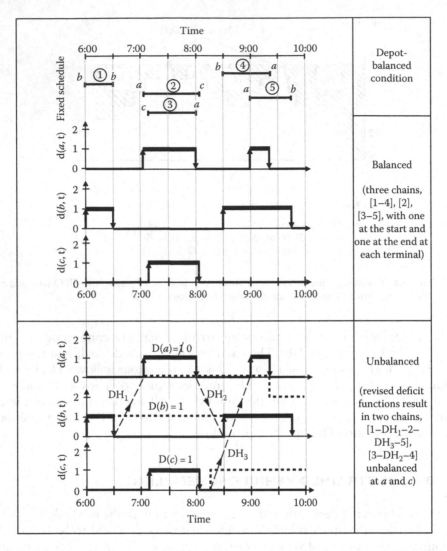

Figure 6.12 Deficit-function solution, with depot-balanced evidence, for the example in Tables 6.1 and 6.2.

Theorem 6.5

A vehicle schedule is balanced if and only if the value of each DF at the end of the schedule horizon is zero, that is,

$$d(k, T_2) = 0, k \in K$$

Proof: $D(k)-d(k, T_2)$ is the net number of arrivals (arrivals minus departures) from the end of the last maximum interval. As these arrivals are not followed by departures in any chain construction, they represent the number of chain ends at k, that is, $ve(k) = D(k)-d(k, T_2)$. Furthermore, it follows from Theorem 6.5 that $vs(k) = D(k)$. The requirement from a balanced schedule, $vs(k) = ve(k)$, implies $d(k, T_2) = 0$.

The conclusions drawn from Theorem 6.5 are as follows: (1) if $d(k, T_2) < 0$, then a surplus of $d(k, T_2)$ vehicles is present at k at the end of the day; (2) if $d(k, T_2) > 0$, then a shortage of $d(k, T_2)$ vehicles occurs at k at the end of the day; and (3) the following identity is true under a system that conserves vehicles:

$$\sum_{k \in K_1} d(k, T_2) = -\sum_{k \in K_2} d(k, T_2) \tag{6.11}$$

in which K_1 and K_2 represent the sets of shortage and surplus terminals, respectively.

One may notice, however, that in order to balance the schedule, it may not be enough to satisfy practical situations in which limitations exist on the number of overnight depots or parking facilities. Let $Q(k)$ represent such a limitation at terminal k. Then a feasible schedule must also satisfy the depot constraints:

$$vs(k) \leq Q(k), \quad ve(k) \leq Q(k) \text{ for all } k$$

If the schedule is balanced, this reduces to $D(k) = Q(k)$, and the depot-constrained balance-schedule problem may be stated in DF form as follows:

Problem P2

Insert as many DH trips as necessary into the schedule of required trips in order to

$$\text{Min} Z2 = \sum_{k \in K} D(k) \tag{6.12}$$

Subject to:

$$d(k, T_2) = 0, \quad k \in K \tag{6.13}$$

$$D(k) \leq Q(k), \quad k \in K \tag{6.14}$$

Alternatively, using the identities of the proof of Theorem 6.5, the problem may be restated as

$$\text{Min} Z2 = \sum_{k \in K} vs(k) \tag{6.15}$$

Subject to:

$$vs(k) - ve(k) = 0, \quad k \in K \tag{6.16}$$

$$0 \leq vs(k) \leq Q(k), \quad k \in K \tag{6.17}$$

$$0 \leq ve(k) \leq Q(k), \quad k \in K \tag{6.18}$$

In this case, either (6.17) or (6.18) is redundant, but is still included in order to facilitate the development of the next section.

6.6.2 Mathematical programming formulation

In order to extend Problem P1 (Section 6.4.2) to include depot and vehicle-balancing constraints, a number of dummy overnight trips may be defined that travel backward in time. For each terminal k, let $n + k$ represent an undisclosed number of trips of the form $(n + k, t_s^{n+k}, n + k, t_e^{n+k})$, such that $t_e^{n+k} \le T_1 < T_2 \le t_s^{n+k}$. The times t_e^{n+k} and t_s^{n+k} represent the latest time at which a vehicle can arrive at depot k after servicing its last required trip and the earliest time a vehicle can leave depot k to service its first required trip, respectively. Such times may be stipulated in a driver-union contract. The following additional variables are now introduced:

$Y_{n+k,j} = 1$ if a vehicle departs for depot k (after completing dummy trip $n + k$) to service j, its first required trip; otherwise, equals 0.

$Y_{i,n+k} = 1$ if a vehicle services its last required trip as trip i before arriving at depot k to park for the night (start dummy trip $n + k$); otherwise equals 0.

z_{kk} is an integer variable whose value represents the number of unused spaces at depot k $(0 \le z_{kk} \le Q(k), k \in K)$.

Let the travel time of the morning and evening DH trips from and to depot k be defined as $\tau(n + k, p^j)$ and $\tau(q^i, n + k)$, respectively. The joining $(n + k, j)$ will be considered admissible if $t_e^{n+k} + \tau(n+k, p^j) \le t_s^j$ for $j \in I$, $k \in K$. Otherwise, $(n + k, j)$ will be inadmissible, and $Y_{n+k,j} = 0$. The joining $(i, n + k)$ is admissible if $t_e^i + \tau(q^i, n + k) \le t_s^{n+k}$ for $i \in I$, $k \in K$. Otherwise, $(i, n + k)$ is inadmissible, and $Y_{i,n+k} = 0$. Note that $(n + k, j)$ and $(i, n + k)$ are always admissible if $p^j = k$ and $q^i = k$, respectively. Let A represent the set of admissible joinings of the preceding type in addition to those between required trips (i, j) as described earlier. The mathematical programming version of this problem with DH trips may now be stated.

Problem P3

$$\text{Max} \, Z3 = \sum_{i \in I} \sum_{j \in I} x_{ij} \tag{6.19}$$

Subject to:

$$\sum_{j \in I} x_{ij} + \sum_{k \in K} y_{i,n+k} = 1, \quad i \in I \tag{6.20}$$

$$\sum_{i \in I} x_{ij} + \sum_{k \in K} y_{n+k,j} = 1, \quad j \in I \tag{6.21}$$

$$\sum_{j \in I} y_{n+k,j} + z_{kk} = Q(k), \quad k \in K \tag{6.22}$$

$$\sum_{i \in I} y_{i,n+k} + z_{kk} = Q(k), \quad k \in K \tag{6.23}$$

$$z_{kk} \in \{0, 1, ..., Q(k)\}, \quad k \in K \tag{6.24}$$

$$\left.\begin{array}{ll} x_{ij} \in \{0, 1\} & (i, j) \in A \\ \\ x_{ij} = 0 & (i, j) \notin A \end{array}\right\} \tag{6.25}$$

$$\left.\begin{array}{ll} y_{i,n+k} \in \{0, 1\} & (i, n+k) \in A \\ \\ y_{i,n+k} = 0 & (i, n+k) \notin A \\ \\ y_{n+k,j} \in \{0, 1\} & (n+k, j) \in A \\ \\ y_{n+k,j} = 0 & (n+k, j) \notin A \end{array}\right\} \tag{6.26}$$

The set A represents the set of admissible trip joinings. The variables x_{ij} are the same as defined in Problem Pl, in which a solution with $x_{ij} = 1$ indicates that trips i and j are joined. Constraint (6.20) insures that each required trip i is joined to exactly one successor trip. The successor trip may be either another required trip j or a dummy trip n + k. The latter implies that trip i is the last trip serviced by a vehicle before returning to depot k. Constraint (6.21) insures that each required trip j must be joined to exactly one predecessor trip. The predecessor trip may be either a required trip i or a dummy trip n + k. If joined to the latter trip, j is the first trip serviced by a vehicle departing from depot k. The objective function (6.19) is identical to that of Problem Pl. From the definitions of $y_{n+k,j}$ and $y_{i,n+k}$, the first terms of (6.22) and (6.23) represent the following:

$\sum_{j \in I} y_{n+k,j}$ is the number of vehicles that start their daily schedule from depot k

$\sum_{i \in I} y_{i,n+k}$ is the number of vehicles that end their daily schedule at depot k

Both terms are also equal to the number of occupied overnight parking spaces (used capacity) at depot k.

Theorem 6.6

Problems P2 and P3 are equivalent. The solution to Problem P3 minimizes fleet size, subject to depot-capacity and depot-balance constraints.

Proof: Make the following substitutions in Problem P3:

$$\sum_{j \in I} y_{n+k,j} = vs(k), \quad k \in K$$

$$\sum_{i \in I} y_{i,n+k} = ve(k), \quad k \in K$$

Any solution satisfying (6.22) and (6.23) satisfies the balance constraint, since subtracting (6.23) from (6.22) for each k yields Equation 6.16:

$$vs(k) - ve(k) = 0, \quad k \in K$$

Constraints (6.22) and (6.23) are also equivalent to the depot constraints (6.17) and (6.18) when the nonnegative slack variables z_{kk} are dropped. Summing up Equation 6.21 for all j shows that the objective functions for both problems are equivalent, that is,

$$\max \sum_{j \in I} \sum_{i \in I} x_{ij} = n + \max \left(- \sum_{j \in I} \sum_{k \in K} y_{n+k,j} \right) = n - \min \sum_{k \in K} vs(k)$$

It should be noted that Problem P3 is a capacitated transportation problem, known in the OR field. As such, it has a bipartite graph representation similar to the example in Figure 6.5, but without s,t. The set of the first-column (supply) nodes are in one-to-one correspondence with the arrival epochs of trips i = 1, 2, ..., n and n + 1, n + 2, ..., n + k, ..., n + q. Similarly, the set of second-column (demand) nodes are in one-to-one correspondence with the departure epochs of each of the required and dummy trips. Since the supplies and demands are integer, the solution will be integer upon relaxing the integer requirements on the variables. Any of the standard transportation algorithms (see, e.g., Ahuja et al., 1993) may be used to solve Problem P3. However, it is possible that Problem P3 is not feasible, since not every possible arc is admissible from the supply node set to the demand node set. Feasibility conditions are stated in the multiterminal, supply–demand theorem found in Ford and Fulkerson (1962).

6.7 LITERATURE REVIEW AND FURTHER READING

Vehicle scheduling refers to the problem of determining the optimal allocation of vehicles to carry out all the trips in a given transit timetable. A chain of trips is assigned to each vehicle although some of them may be DH or empty trips in order to reach optimality. The number of feasible solutions to this problem is extremely high, especially in the case in which the vehicles are based in multiple depots. Much of the focus of the literature on scheduling procedures is, therefore, on computational issues.

Dell Amico et al. (1993) developed several heuristic formulations, based on a shortest-path problem, that seek to minimize the number of required vehicles in a multiple-depot schedule. The algorithm presented is performed in stages, in each of which the duty of a new vehicle is determined. In each such stage, a set of forbidden arcs is defined, and then a feasible circuit through the network is sought that does not use any of the forbidden arcs. Computational efficiency is obtained by searching for the shortest path across a subset of all arcs in the network, rather than searching the entire network. Several modifications to the basic algorithms are offered that save computer time by substituting parts of the full problem with problems of a reduced size. These modifications include, for instance, solving the reassignment of trips as a single-depot problem, an attempt to swap parts of duty segments, and an internal reassignment of trips within each pair of vehicles associated with different depots.

Löbel (1998, 1999) discussed the MDVSP and its relaxation into a linear programming formulation that can be tackled using the branch-and-cut method. A special multicommodity flow formulation is presented, which, unlike most other such formulations, is not arc-oriented. A column-generation solution technique is developed, called Lagrangian pricing; it is based on two different Lagrangian relaxations. Heuristics are used within the procedure to determine the upper and lower bounds of the solution, but the final solution is proved to be the real optimum.

Kwan and Rahin (1999) described an object-oriented approach for bus scheduling, based on the VAMPIRES algorithm for iterative improvement of the solution presented by Smith and Wren in 1981. A key feature of VAMPIRES is the attempt to swap links at each stage of the solution in order to improve the current schedule; Kwan and Rahin improved this feature by refining the swapping criteria. In addition, the methodology presented introduces a hierarchical classification of auxiliary activities: trip, layover, relocation, invalid layover, invalid relocation, depot return, depot start, depot end. This classification scheme for vehicle activities enables planners to improve the current solution more efficiently.

Mesquita and Paixao (1999) used a tree-search procedure, based on a multicommodity network flow formulation, to obtain an exact solution for the multidepot vehicle-scheduling problem. The methodology employs two different types of decision variables. The first type describes connections between trips in order to obtain the vehicle blocks, and the other relates to the assignment of trips to depots. The procedure includes creating a more compact, multicommodity network flow formulation that contains just one type of variable and a smaller amount of constraints, which are then solved using a branch-and-bound algorithm.

Banihashemi and Haghani (2000) and Haghani and Banihashemi (2002) focused on the solvability of real-world, large-scale, MDVSPs. The case presented includes additional constraints on route time in order to account for realistic operational restrictions such as fuel consumption. The authors proposed a formulation of the problem and the constraints, as well as an exact solution algorithm. In addition, they described several heuristic solution procedures. Among the differences between the exact approach and the heuristics is the replacement of each incorrect block of trips with a legal block in each iteration of the heuristics. Applications of the procedures in large cities are shown to require a reduction in the number of variables and constraints. Techniques for reducing the size of the problem are introduced, using such modifications as converting the problem into a series of single-depot problems.

Freling et al. (2001b) discussed the case of single-depot with identical vehicles, concentrating on quasi-assignment formulations and auction algorithms. A quasi-assignment is a reduced-size, linear problem in which some of the nodes and their corresponding arcs are not considered. An auction algorithm is an iterative procedure in which neither the primal nor the dual costs are obliged to show an improvement after each iteration. The authors proposed four different algorithms and compared their performance: an existing auction algorithm for the asymmetric assignment problem; a new auction algorithm for the quasi-assignment problem; an alternative, two-phase, asymmetric assignment formulation (valid in a special case), in which vehicle blocks are determined first and combined afterwards; and a core-oriented approach for reducing the problem size.

Freling et al. (2001a) and Huisman et al. (2005) presented an integrated approach for vehicle and crew scheduling for a single bus route. The two problems are first defined separately; the vehicle-scheduling problem is formulated as a network-flow problem, in which each path represents a feasible vehicle schedule, and each node a trip. In the combined version, the network problem is incorporated into the same program with a set partitioning formulation of the crew-scheduling problem (see Chapter 10).

Haase et al. (2001) formulated another problem that incorporated both crew and vehicle scheduling. For vehicle scheduling, the case of a single depot with a homogenous fleet is considered. The crew-scheduling problem (see Chapter 10) is a set partitioning formulation that includes side constraints for the bus itineraries; these constraints guarantee that an optimal vehicle assignment can be derived afterward.

Haghani et al. (2003) compared three vehicle-scheduling models: one multiple-depot (presented by Banihashemi and Haghani, 2000) and two single-depot formulations, which are special cases of the multiple-depot problem. The analysis showed that a single-depot vehicle-scheduling model performed better under certain conditions. A sensitivity analysis with respect to some important parameters is also performed; the results indicated that the travel speed in the DH trip was a very influential parameter.

Huisman et al. (2004) proposed a dynamic formulation of the multidepot vehicle-scheduling problem. The traditional, static vehicle-scheduling problem assumes that travel times are a fixed input that enters the solution procedure only once; the dynamic formulation relaxes this assumption by solving a sequence of optimization problems for shorter periods. The dynamic approach enables an analysis based on other objectives except for the traditional minimization of the number of vehicles, that is, by minimizing the number of trips starting late and minimizing the overall cost of delays. The authors showed that a solution that required only a slight increase in the number of vehicles could also satisfy the minimum late starts and minimum delay-cost objectives. To solve the dynamic problem, a "cluster re-schedule" heuristic was used; it started with a static problem in which trips were assigned to depots, and then it solved many dynamic single-depot problems. The optimization itself was formulated through standard mathematical programming in a way that could use standard software.

Pepin et al. (2009) compare the performance of five different heuristic approaches for solving the MDVSP. The approaches considered include a truncated branch-and-cut method, a Lagrangian heuristic, a truncated column generation method, a large neighborhood search (LNS) heuristic using truncated column generation for neighborhood evaluation, and a tabu search heuristic. The first three methods are adaptations of existing methods, while the last two are new in the context of this problem. The presented comparisons reveal that the column generation heuristic produces highest quality solution given sufficient computational time and stability. In contrast, the LNS method, which utilizes truncated column generation for neighborhood evaluation, provides a faster solution without adversely compromising the solution quality. The comparisons and results illustrate that using mathematical programming tools in conjunction with a metaheuristic framework can offer a well-balanced solution quality and computational efficiency for the MDVSP.

Steinzen et al. (2010) discuss the integrated vehicle- and crew-scheduling problem in public transit with multiple depots. The authors present a new modeling approach that is based on a time–space network representation of the underlying vehicle- and crew-scheduling problem. The suggested approach is based on a Lagrangian heuristic in conjunction with column generation, where the pricing problem is formulated as a resource-constrained shortest-path problem. It has been illustrated that this proposed network representation for the pricing problem produces favorable results compared to the other approaches available in literature. The results presented in this study indicate that medium-sized instances, with around 640 trips and 4 depots, can be solved efficiently. The presented method outperforms other methods from the literature, in terms of the solution quality and computational time.

Naumann et al. (2011) present a new stochastic programming approach for robust vehicle scheduling in bus transit. The presented approach uses typical disruption scenarios during the optimization to minimize the expected sum of planned costs and costs caused by disruptions. The schedule is represented as a time–space network with all connecting arcs to enable independent penalization of every connection between two consecutive service trips. The results indicate that stochastic programming for the vehicle-scheduling problem with disruptions leads to solutions that are of higher quality in terms of total expected costs compared to a simple approach with fixed buffer times. Notwithstanding the consideration of complete delay-propagation, the presented stochastic programming approach finds the optimal solution for the given data, which outperforms the simple approach. The authors have also created a set of Pareto-optimal solutions in terms of maximum robustness and minimum planned costs. Considering the entire delay propagation, the stochastic programming solutions still provide solutions that are of higher quality in terms of total expected costs compared to the simple approach adding fixed buffer times. It is therefore concluded that despite the increased computational complexity of the approach presented, small and medium-sized real-world instances can be solved.

Lin and Kwan (2012) propose a two-phase approach for the train unit scheduling problem. The first phase assigns and sequences train trips to train units temporarily ignoring some infrastructure details at train stations. The solution of the first phase is near-operable. The second phase focuses on satisfying the remaining constraints with respect to detailed station layouts. Real-world scenarios such as compatibility among traction types and time allowances for coupling/decoupling activities are also considered, so that the solutions would be fully operable. The first phase is modeled as a large-scale integer multicommodity network flow (MCNF) problem. A branch-and-price integer linear programming approach is proposed. The authors state that the preliminary experiments have shown that the approach would not yield solutions in reasonable time for problem instances exceeding about 1,000 train trips. It is stated in this study that the train company collaborating in this research operates over 2,400 train trips on a typical weekday with 10 different types of train units in use. Hence, a two-phase approach is proposed where the problem is first addressed from a network wide perspective, subsequently focusing on finer details at the individual train station level. A heuristic has been designed by the authors for hybridization of the first phase solver by compacting the problem instance to a much smaller size, before the integer linear programming (ILP) solver is applied. The process is iterative with evolving compaction based on the results from the previous iteration thereby converging to near optimal results. The second phase is modeled as a multidimensional matching problem with a mixed integer linear programming (MILP) formulation. Finally, the authors acknowledge that further testing and refinement of the models with feedback from the railway company are ongoing.

Borndörfer et al. (2012) provide a generic formulation for rolling stock planning problems in the context of intercity passenger traffic. The main contributions are a graph theoretical model and a mixed-integer-programming formulation that integrate all main requirements of the considered vehicle-rotation-planning problem (VRPP). The authors show that it is possible to solve this model for real-world instances using modern algorithms and computers. The results from this study show the expected behavior of algorithms that try to solve the VRPP. The authors conclude that the train composition and regularity aspects are not too hard to tackle while the maintenance constraints and moreover capacity constraints increase the complexity of the VRPP significantly. The computational study presented in this paper demonstrates that the solution approach developed can be used to produce high-quality solutions for large-scale vehicle-rotation-planning problems.

Main features of the works reviewed are illustrated in Table 6.3.

Table 6.3 Characteristics of work covered in Section 6.7 concerning literature review and further reading

Source	Depots	Exact or heuristic	Special features
Dell Amico et al. (1993)	Multiple	Heuristic	
Löbel (1998, 1999)	Multiple	Both	
Kwan and Rahin (1999)	Single	Heuristic	Object-oriented program
Mesquita and Paixao (1999)	Multiple	Exact	
Banihashemi and Haghani (2000), Haghani and Banihashemi (2002)	Multiple	Both	
Freling et al. (2001b)	Single	Heuristic	
Freling et al. (2001a), Huisman et al. (2005)	Only one route	Exact	Vehicle and crew scheduling
Haase et al. (2001)	Single	Exact	Vehicle and crew scheduling
Haghani et al. (2003)	Multiple and single	Heuristic	Comparison of three formulations
Huisman et al. (2004)	Multiple	Heuristic	Dynamic scheduling
Pepin et al. (2009)	Multiple and single	Heuristic	Comparison of five different heuristic approaches
Steinzen et al. (2010)	Multiple	Heuristic	
Naumann et al. (2011)	Multiple	Exact	Stochastic programming
Lin and Kwan (2012)	Multiple	Heuristic	
Borndörfer et al. (2012)	Multiple	Heuristic	

EXERCISES

6.1 Given a bus system consisting of only two cyclical routes (performing round trips) departing from and arriving at the same terminal, the mean trip time for *both* routes is given as *3 h*. The headways for each route are determined based on passenger demand. These headways are given by hour of day in the following table:

Route 1		Route 2	
Hour of day	Headway (min)	Hour of day	Headway (min)
6 a.m.–9 a.m.	7.5	6 a.m.–8 a.m.	6
9 a.m.–12 noon	12	8 a.m.–12 noon	7.5
12 noon–2 p.m.	10	12 noon–3 p.m.	10
2 p.m.–4 p.m.	15	3 p.m.–6 p.m.	7.5
4 p.m.–6 p.m.	6	6 p.m.–11 p.m.	10
6 p.m.–11 p.m.	10		

What is the *minimum fleet size* required for this bus system? *Do not* use DF. Comment on the generalization of your approach.

Note: Buses may be switched from one route to another if necessary.

6.2 Given the following eight-trip, four-terminal timetable and DH matrix, find the minimum number of vehicles required to carry out the timetable, including their chains (blocks), by using the Max-flow augmenting-path algorithm.

Trip number (i)	Departure terminal	Departure time	Arrival terminal	Arrival time
1	a	7:00	b	7:25
2	d	7:00	a	7:35
3	b	7:15	c	8:15
4	c	7:30	d	8:35
5	a	7:35	d	8:00
6	d	8:00	a	8:30
7	a	8:35	d	9:05
8	c	9:05	d	10:00

Average DH travel time (minutes) matrix

	Arrival terminal			
Departure terminal	a	b	c	d
a	0	15	30	25
b	20	0	40	30
c	25	35	0	45
d	20	25	35	0

6.3 The following problem uses the DF solution approach. A schedule S is given between 5:00 (T_1) and 9:30 (T_2), with two terminals (a, b) and one depot. An eight-trip schedule is shown in the following table. Note that trips 6 and 7 are conducted at the same time and in the same direction because of specific passenger demand. The DH average travel time for all trips between a and b (both directions of travel) is 30 min, and *there are no* DH trips between either a or b and the depot (both directions). The DH trips are to be assigned to the latest time possible.

(i) Construct the DFs for terminals a and b only; find the minimum number of vehicles required at terminals a and b using DH trip-insertion procedures between the two terminals.

(ii) Construct the sum function g(t) for the entire schedule S; is G(S) the minimum fleet size required for S? Explain.

(iii) Find the minimum fleet size required for S and construct vehicle schedules (chains, blocks) using the FIFO rule.

Trip number	Departure terminal	Departure time	Arrival terminal	Arrival time
1	Depot	5:00	b	6:00
2	Depot	5:30	a	6:30
3	a	6:00	b	6:30
4	Depot	6:40	a	8:00
5	a	7:00	Depot	9:00
6	a	7:30	Depot	9:30
7	a	7:30	Depot	9:30
8	b	8:30	a	9:30

Note: Further exercises involving DFs, shifting departure times, and DH-insertion procedures appear in the Exercises sections of Chapters 7, 8, and 16.

APPENDIX 6.A: MAXIMUM-FLOW (MAX-FLOW) PROBLEM

The following description is based on Ceder (1978). Further reading can be found in the fine book by Ahuja et al. (1993). The general augmenting-path algorithm described can solve the vehicle-scheduling problem. This appendix supplements material for Section 6.4, and Chapter 13.

6.A.1 Definitions

In the Max-flow problem, one considers a directed graph (network) $G = \{N, A\}$, with a single node s as *source* and another node t as a *sink*. Each arc (i, j) has a capacity $c(i, j)$ = the maximum flow that can be traversed from i to j. This can be associated with a flow of people, vehicles, trains, bits of information, dollars, water, etc. An *s–t* flow is the amount of the flow that leaves s and arrives at t, provided that:

(a) $0 \le f(i, j) \le c(i, j)$ ∀ (for all) $(i, j) \in$ (belong to) A,
(b) And to maintain flow conservation, $\sum_{j|(i,j)\in A} f(i, j) = \sum_{k|(k,i)\in A} f(k, i) \, \forall i \in N, i \ne s, t$, where $f(i, j)$ is the arc (i, j)'s flow.

Lemma 6.A.1

Let f be an s–t flow on $G = \{N, A\}$. Then: $f(s, N) - f(N, s) = f(N, t) - f(t, N)$, where $f(U, V)$ is the flow between two subsets, U and V, such that $(U, V) = \{(i, j) \in A | i \in U, j \in V\}$ and $f(U, V) = \sum_{(i,j)\in(U,V)} f(i, j)$

Proof: $f(N, N) = f(s, N) + f(t, N) + \sum_{\substack{i\in N \\ i\ne s,t}} f(i, N) = f(N, s) + f(N, t) + \sum_{\substack{i\in N \\ i\ne s,t}} f(N, i)$.

Since the flow is conserved at each node, $f(i, N) = f(N, i) \, \forall i \in N, i \ne s, t$. Therefore, we obtain $V(f) = f(s, N) - f(N, s) = f(N, t) - f(t, N)$, where $V(f)$ is called the *value* of the flow.

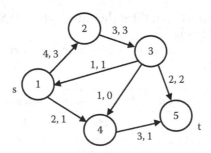

Example:
The first number on each arc (see network on the right) represents its capacity and the second the flow. In this example, the s–t flow = 3 and can be measured at node 1 or 5.

s–t cut: If in a directed graph $G = \{N, A\}$ the set of nodes N is partitioned into two sets, X and $\overline{X}$ (i.e., $X \cap \overline{X} = \phi$ and $X \cup \overline{X} = N$), such that $s \in X$ and $t \in \overline{X}$, then $(X, \overline{X})$ = the set of all arcs in A connecting nodes in X to nodes in $\overline{X}$ is called the *s–t cut*. The capacity of the cut is defined as $c(X, \overline{X})$

Example: In the preceding example network:

$$X = \{1, 2, 3\} \quad (X, \overline{X}) = \{(1, 4), (3, 4), (3, 5)\}, \quad c(X, \overline{X}) = 5$$
$$\overline{X} = \{4, 5\} \Rightarrow (\overline{X}, X) = \phi \quad\quad\quad\quad\quad c(\overline{X}, X) = 0$$

Lemma 6.A.2

$$V(f) = f(X, \overline{X}) - f(\overline{X}, X)$$

Proof:

$$V(f) = f(s, N) - (N, s) + \sum_{\substack{i \in X \\ i \neq s}} [f(i, N) - f(N, i)] = f(X, N) - f(N, X)$$

$$= [f(X, \overline{X}) - f(X, X)] - [f(\overline{X}, X) + f(X, X)] = f(X, \overline{X}) - f(\overline{X}, X)$$

Theorem 6.A.1

For any s–t flow f and any s–t cut $(X, \overline{X})$:

$$V(f) \leq c(X, \overline{X})$$

Proof: For every f and $(x, \overline{x})$: $f(X, \overline{X}) \leq c(X, \overline{X})$ and $f(X, \overline{X}) = 0$.
    Therefore, by Lemma 6.A.2 → $V(f) = f(X, \overline{X}) - f(\overline{X}, X) \leq c(X, X)$.

Theorem 6.A.1 leads to the fundamental result in network-flow theory: a flow exists that is equal to the minimum capacity cut; hence, it is the maximum flow. The latter claim will be shown following the description of the maximum-flow (Max-flow) algorithm in the next section.

6.A.2 Augmenting-path algorithm

For a given $G = \{N, A\}$ with $c\{i, j\}$ $\forall$ $(i, j) \in A$, and some s–t flow f (it is possible to start with $f(i, j) = 0$ $\forall$ $(i, j) \in A$):

Step 1: Construct a *residual network* $G(f) = \{N, A(f)\}$ in which each arc is labeled with number $\alpha(i, j)$ as follows:

    a For $f(i, j) < c(i, j)$, then $(i, j) \in A(f)$ and $\alpha(i, j) = c(i, j) - f(i, j)$; (i, j) is called a *forward arc*.
    b For $f(i, j) > 0$, then $(j, i) \in A(f)$ and $\alpha(j, i) = f(i, j)$; (j, i) is called a *reverse arc*.

Step 2: If there is no path from s to t in $G(f)$, then terminate; f is the maximum s–t flow (see Theorem 6.A.3).

Step 3: Let P be the shortest path (least number of arcs), and let $\alpha = \min_{(i,j) \in P} \alpha(i, j)$. Define a new flow f' on G (see Theorem 6.A.2):

  a If (i, j) is a forward arc on P, let f'(i, j) = f(i, j) + α
  b If (i, j) is a reverse arc on P, let f'(i, j) = f(i, j) – α
  c If (i, j) does not belong to P, let f'(i, j) = f(i, j)

Step 4: Let f:= f'; go to Step 1.

6.A.3 Labeling procedure to find minimum-arc paths

In *Step* 3 of the augmentation algorithm, we seek to find the shortest path from s to t. (An arbitrary path can also fit *Step* 3, but usually with more computations than the shortest path. The latter has a finite number, and normally fewer computations; hence, it is an improved version). The labeling procedure is simple: at each stage, every node i $\in$ N is in one of three states: (i) unlabeled (indicated by a blank), (ii) labeled but not scanned (indicated by a single label, l(i)), and (iii) labeled and scanned (indicated by l(i) and marked by ✓).

The procedure:

Step L_1: Label the node s as l(s) = s.
Step L_2: If node t is labeled, stop; [l(t), t] is the last arc on the shortest path from s to t. [l(l,(t)),l(t)] is the next-to-last arc, and so on until reaching node s.
Step L_3: If all labeled nodes are scanned, stop; no path exists from s to t.
Step L_4: If all labeled, but non-scanned nodes i, use the "first labeled, first scanned" policy. Label each unlabeled node j, provided (j, i) $\in$ A, with l(j) = i. Mark i as a scanned node with a ✓.
Step L_5: Go to *Step L_2*.

6.A.4 Example of the augmenting-path algorithm

A network G = {N, A} is given in which the first number on each arc represents the capacity, and the second the flow.

First iteration

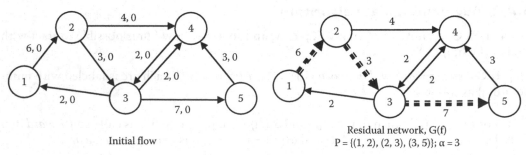

Initial flow

Residual network, G(f)
P = {(1, 2), (2, 3), (3, 5)}; α = 3

In order to demonstrate the labeling procedure, we will show how one can find P in the first iteration (it is used similarly in all other iterations).

		Iteration		
Node	First	Second	Third	Fourth
s = 1	1 = l(1)	1 ✓	1 ✓	1 ✓
2		1 = l(2)	1 ✓	1 ✓
3			2 = l(3)	2 ✓
4			2 = l(4)	2
t = 5				3 = l(5)

The shortest path is then reconstructed backward: $5 \to 3 \to 2 \to 1$ (see P earlier).

2nd iteration

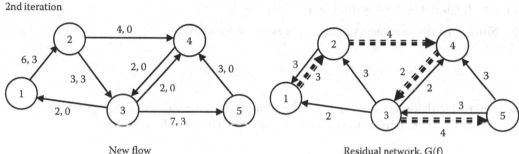

New flow

Residual network, G(f)
P = {(1, 2), (2, 4), (4, 3), (3, 5)}; α = 2

3rd iteration

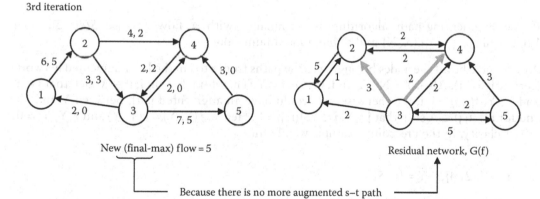

New (final-max) flow = 5

Residual network, G(f)

Because there is no more augmented s–t path

The final and optimum result is Max-flow = 5. The bottleneck (minimum cut; see Theorems 6.A.3 and 6.A.4) consists of this Max-flow and is emphasized by a heavy gray line in the last residual network.

Note: When an augmenting path that includes a reverse arc (j*, i*) is detected, we can use it for increasing the s–t flow. However, according to *Step* 3(b) of the algorithm, we have to subtract the value of α from the previous flow on (i*, j*) in constructing the new flow.

6.A.5 Additional theorems

The following theorems assure that the augmenting-path algorithm indeed solves the Max-flow problem.

Theorem 6.A.2

(a) The function f' (see *Step 3*) is an s–t (i.e., $0 \leq f'(i, j) \leq c(i, j)$ $\forall$ (i, j) $\in$ A and $f'(i, N) =$ $f'(N, i)$ $\forall$ i $\in$ N, i $\neq$ s, t)

(b) $V(f') > V(f)$

Proof: (a) The values of $\alpha(i, j)$ and $\alpha(j, i)$ are the maximum amount of flow with which we can increase or decrease $f(i, j)$, $(i, j) \in A$, respectively. The function f' is different from f only for $(i, j) \in P =$ augmenting path. Since $\alpha = \min_{(i,j) \in P} \alpha(i, j) = \alpha(q, k)$, then $f'(q, k) \leq c(q, k)$ and the capacity constraints hold for all other arcs in the augmenting path. Hence, $0 \leq f'(i, j) \leq c(i, j)$. The conservation-flow required retains because for every node i along P (except s, t), the number of arcs entering i = number of arcs leaving i; for (i, j) as a forward arc $f''(i, j) = f(i, j) + \alpha$; for (i, j) as a reverse arc $f''(i, j) = f(i, j) - \alpha$.

(b) Similar to the arguments in (a), it is easy to show that:

$$f''(s, N) - f'(N, s) = f(s, N) - f(N, s) + \alpha \quad \text{and since } \alpha > 0 \rightarrow V(f') > V(f)$$

Theorem 6.A.2 demonstrates the feasibility criterion in part (a) and the improvement criterion in part (b). The optimality criterion is shown in Theorem 6.A.3.

Theorem 6.A.3

If the augmenting-path algorithm is terminated with a flow f* (see *Step 2*), then $V(f^*) = \min_{s-t \text{ cuts}(X,\bar{X})} c(X, \bar{X})$ and f* has a maximum value.

Proof: Let Y be all the nodes belonging to the paths from s to i in the last augmented network (certainly i $\neq$ t). Also, $\bar{Y} = N - Y$ and, hence, t $\in$ $\bar{Y}$. Therefore, for all (i, j) $\in$ A such that i $\in$ Y and j $\in$ $\bar{Y}$, f(i, j) = c(i, j) (otherwise, we would have an augmented path between Y and $\bar{Y}$); and for all (i, j) $\in$ A such that j $\in$ $\bar{Y}$, j $\in$ Y, f(i, j) = 0. Hence, $f(Y, \bar{Y}) = c(Y, \bar{Y})$ and $f(\bar{Y}, Y) = 0$.

For clarity, in the preceding example we obtain:

$$Y = \{1, 2, 4\}, \quad \bar{Y} = \{3, 5\}$$

$$f(Y, \bar{Y}) = c(Y, \bar{Y}) = f(2,3) + f(4,3) = 5$$

$$f(\bar{Y}, Y) = f(3,1) + f(3,4) + f(5,4) = 0$$

Because $(Y, \bar{Y})$ is an s–t cut, then according to Lemma 6.A.2:

$$V(f^*) = f(Y, \bar{Y}) - f(\bar{Y}, Y) = c(Y, \bar{Y}) \geq \min_{s-t \text{ cuts}(X,\bar{X})} c(X, \bar{X}).$$

Thus, using Theorem 6.A.1 ($V(f^*) \leq c(X, \bar{X})$ $\forall(X, \bar{X})$), we can show that $(Y, \bar{Y})$ is a minimum s–t cut and f has a maximum value.

Theorem 6.A.4

$\max_{s-t\text{flows},f} V(f) = \min_{s-t\text{cuts}(X,\bar{X})} c(X, \bar{X})$, which is the known maximum-flow minimum-cut theorem.

Proof: By applying Theorems 6.A.1 and 6.A.3, the proof is straightforward.

6.A.6 Linear programming (LP) formulation

For the sake of simplicity, let us add to $G = \{N, A\}$ an arc (t, s) with $c(t, s) = \infty$. This forms $G' = \{N, A'\}$ in which $A' = A \cup (t, s)$. If an arc (t, s) already exits, it can be removed without changing the problem. The LP formulation is as follows:

$$\max\{Z = f(t, s)\}$$

s.t.

$$f(i, N) - f(N, i) = 0, \quad \text{for all } i \in N$$

$$f(i, j) - c(i, j), \quad \text{for all } (i, j) \in A$$

$$f(i, j) \geq 0, \quad \text{for all } (i, j) \in A'$$

This LP has $|N|$ conservation constraints (for all $i \in N$), $|A|$ capacity constraints (one for each arc in G), and $|A'| = |A| + 1$ nonnegativity constraints.

The LP formulation enables us to solve the Max-flow problem by the simplex method (or one of its variants—see OR literature). However, we note that the coefficient matrix of the conservation constraints has a very special structure (this matrix is called the incidence matrix of G'). The rows and columns in this matrix correspond to the nodes and arcs in G', respectively. Every column of this matrix contains exactly two nonzero elements: 1 and –1. Because of this special structure of the matrix, it is possible to reduce significantly the computational effort required by the simplex method.

6.A.7 Useful extensions

Undirected arcs
An undirected arc (i, j) with capacity $c(i, j)$ can be replaced by a pair of directed arcs (i, j) and (j, i), each with a capacity of $c(i, j)$.

Unlimited capacities
Some arcs may have unlimited capacities, and the augmenting-path algorithm might end with an infinite flow. Therefore, in this case we calculate primarily $\max_{s-t\text{ paths}}\{\min \alpha\}$.

Nodes with capacities
It is possible to establish an upper bound on the amount of flow traversing given nodes. In this case, the node i with capacity c(i) is replaced by two new nodes i' and i" with an arc having the capacity c(i) = c(i', i") in between.

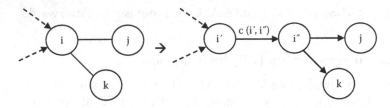

Lower bounds on the flow
For cases with a lower bound on the flow (e.g., assuring minimum flow to justify the permission to use the arc), the flow constraints are changed to $l(i, j) \le f(i, j) \le c(i, j)$, where $l(i, j)$ is the lower bound on the flow on arc (i, j). This problem can be solved by converting the original network into an artificial network (satisfying both upper and lower bounds) called a transshipment scheme. Its solution appears in OR literature, such as in Ahuja et al. (1993).

REFERENCES

Ahuja, R.K., Magnanti, T.L., and Orlin, J.B. (1993). *Network Flows*. Prentice Hall, Englewood Cliffs, New Jersey.
Banihashemi, M. and Haghani, A. (2000). Optimization model for large-scale bus transit scheduling problems. *Transportation Research Record*, 1733, 23–30.
Bartlett, T.E. (1957). An algorithm for the minimum number of transport units to maintain a fixed schedule. *Naval Research Logistics Quarterly*, 4, 139–149.
Borndörfer, R., Reuther, M., Schlechte, T., and Weider, S. (2012). Vehicle rotation planning for intercity railways. In *CD-ROM of the Conference on Advanced Systems for Public Transport (CASPT12)*, Santiago, Chile, July 23–27, 2012.
Ceder, A. (1978). *Network Theory and Selected Topics in Dynamic Programming*. Dekel Academic Press (in Hebrew), Tel-Aviv, Israel.
Ceder, A. (2003). Public transport timetabling and vehicle scheduling. In *Advanced Modeling for Transit Operations and Service Planning* (W. Lam and M. Bell, eds.), pp 31–57, Pergamon Imprint, Elsevier Science, New York, NY.
Ceder, A. and Gonen, D. (1980). The operational planning process of a bus company. *UITP Review*, 29, 199–218.
Ceder, A. and Stern, H.I. (1981). Deficit function bus scheduling with deadheading trip insertion for fleet size reduction. *Transportation Science*, 15(4), 338–363.
Ceder, A. and Stern, H.I. (1985). The variable trip procedure used in the AUTOBUS vehicle scheduler. In *Computer Scheduling of Public Transport 2*, (J.M. Rousseau, ed.), pp. 371–390. North Holland, Amsterdam, the Netherlands.
Daduna, J.R. and Paixao, J.M.P. (1995). Vehicle scheduling for public mass transit-an overview. In *Computer-Aided Transit Scheduling*. Lecture Notes in Economics and Mathematical Systems, vol. 430 (J.R. Daduna, I. Branco, and J.M.P. Paixao, eds.), pp.76–90, Springer-Verlag, Berlin, Germany.
Dell Amico, M., Fischett, M., and Toth, P. (1993). Heuristic algorithms for the multiple depot vehicle scheduling problem. *Management Science*, 39(1), 115–123.
Desrosiers, J., Dumas, Y., Solomon, M.M., and Soumis, F. (1995). Time constrained routing and scheduling. In *Network Routing*. Handbooks in Operations Research and Management Science, vol. 8 (M.O. Ball, T.L. Magnanati, C.L. Monma, and G.L. Nemhauser, eds.), pp. 35–39, Elsevier Science B.V., Amsterdam, the Netherlands.

Even, S. and Tarjan, R.E. (1975). Network flow and testing graph connectivity. *SIAM Journal on Computing*, 4, 507–518.

Ford, L.R. Jr. and Fulkerson, D.R. (1962). *Flows in Networks*. Princeton University Press, New Jersey.

Freling, R., Huisman, D., and Wagelmans, A.P.M. (2001a). Applying an integrated approach to vehicle and crew scheduling in practice. In *Computer-Aided Scheduling of Public Transport*. Lecture Notes in Economics and Mathematical Systems, vol. 505 (S. Voss and J.R. Daduna, eds.), pp. 73–90, Springer-Verlag, Berlin, Germany.

Freling, R., Wagelmans, A.P.M., and Paixao, J.M.P. (2001b). Models and algorithms for single-depot vehicle scheduling. *Transportation Science*, 35(2), 165–180.

Gavish, B., Schweitzer, P., and Shlifer, E. (1978). Assigning buses to schedules in a metropolitan area. *Computers and Operations Research*, 5, 129–138.

Gertsbach, I. and Gurevich, Y. (1977). Constructing an optimal fleet for transportation schedule. *Transportation Science*, 11, 20–36.

Gertsbach, I. and Stern, H.I. (1978). Minimal resources for fixed and variable job schedules. *Operations Research*, 26, 68–85.

Haase, K., Desauliniers, G., and Desrosiers, J. (2001). Simultaneous vehicle and crew scheduling in an urban mass transit system. *Transportation Science*, 35(3), 286–303.

Haghani, A. and Banihashemi, M. (2002). Heuristic approaches for solving large-scale bus transit vehicle scheduling problem with route time constraints. *Transportation Research*, 36A, 309–333.

Haghani, A., Banihashemi, M., and Chiang, K.H. (2003). A comparative analysis of bus transit vehicle scheduling models. *Transportation Research*, 37B, 301–322.

Huisman, D., Freling, R., and Wagelmans, A.P.M. (2004). A robust solution approach to the dynamic vehicle scheduling problem. *Transportation Science*, 38(4), 447–458.

Huisman, D., Freling, R., and Wagelmans, A.P.M. (2005). Models and algorithms for integration of vehicle and crew scheduling. *Transportation Science*, 39, 491–502.

Kwan, R.S.K. and Rahin, M.A. (1999). Object oriented bus vehicle scheduling—The BOOST system. In *Computer-Aided Transit Scheduling*. Lecture Notes in Economics and Mathematical Systems, vol. 471 (N.H.M. Wilson, ed.), pp. 177–191, Springer-Verlag, Berlin, Germany.

Lin, Z. and Kwan, R.S. (2012). A two-phase approach for real-world train unit scheduling. In *CD-ROM of the Conference on Advanced Systems for Public Transport* (CASPT12), Santiago, Chile, July 23–27, 2012.

Löbel, A. (1998). Vehicle scheduling in public transit and Lagrangean pricing. *Management Science*, 44(12), 1637–1649.

Löbel, A. (1999). Solving large-scale multiple-depot vehicle scheduling problems. In *Computer-Aided Transit Scheduling*. Lecture Notes in Economics and Mathematical Systems, vol. 471 (N.H.M. Wilson, ed.), pp. 193–220, Springer-Verlag, Berlin, Germany.

Mesquita, M. and Paixao, J.M.P. (1999). Exact algorithms for the multi-depot vehicle scheduling problem based on multicommodity network flow type formulations. In *Computer-Aided Transit Scheduling*. Lecture Notes in Economics and Mathematical Systems, vol. 471 (N.H.M. Wilson, ed.), pp. 221–243, Springer-Verlag, Berlin, Germany.

Naumann, M., Suhl, L., and Kramkowski, S. (2011). A stochastic programming approach for robust vehicle scheduling in public bus transport. *Procedia-Social and Behavioral Sciences*, 20, 826–835.

Pepin, A.S., Desaulniers, G., Hertz, A., and Huisman, D. (2009). A comparison of five heuristics for the multiple depot vehicle scheduling problem. *Journal of Scheduling*, 12(1), 17–30.

Salzborn, F.J.M. (1972). Optimum bus scheduling. *Transportation Science*, 6, 137–148.

Salzborn, F.J.M. (1974). Minimum fleet size models for transportation systems. In *Transportation and Traffic Theory* (D.J. Buckley, ed.), pp. 607–624, Reed, Sydney, Australia.

Steinzen, I., Gintner, V., Suhl, L., and Kliewer, N. (2010). A time-space network approach for the integrated vehicle-and crew-scheduling problem with multiple depots. *Transportation Science*, 44(3), 367–382.

Stern, H.I. and Ceder, A. (1983a). The garage constrained-balance vehicle schedule minimum fleet size problem. In *the Proceeding of the Ninth ISTTT—International Symposium on Transportation & Traffic Theory* (V.F. Hurdle, E. Hauer, and G.N. Stewart, eds.), pp. 527–556, University of Toronto Press, Toronto, Ontario, Canada.

Stern, H.I. and Ceder, A. (1983b). An improved lower bound to the minimum fleet size problem. *Transportation Science*, 17(4), 471–477.

Chapter 7

Vehicle scheduling II

Variable schedules

A.C.

Knowing it is not understanding; understanding is trying to improve

CHAPTER OUTLINE

7.1 Introduction
7.2 Fleet size lower bound for fixed schedules
7.3 Variable trip-departure times
7.4 Fleet size lower bound for variable schedules
7.5 Fleet-reduction procedures
7.6 Experiences with bus schedules
7.7 Examination and consideration of even-load timetables
Exercises
Appendix 7.A: Example of vehicle-scheduling software
References

PRACTITIONER'S CORNER

This chapter extends the deficit-function (DF) approach to include possible modifications in the creation and editing of trip timetables and vehicle schedules (blocks). In addition to dead-heading (DH) trip insertions, schedulers may consider a variable, instead of a fixed schedule, that is, shifting departure times (SDTs) based on some acceptable tolerances in the process of minimizing fleet size. Admittedly, this shifting effort is a very time-consuming task and usually not done in a systematic manner. This chapter will show how to use the graphical person–machine interactive approach to assist vehicle schedulers in selecting the most efficient shifts (in trip-departure times), with and without the inclusion of DH trip insertions. To that end, we can say that this will not be hard work or involve complex software that will wear out the scheduler. It is the knowledge that a small change can do the job.

The chapter contains six main parts, following an introductory section. Section 7.2 presents DF procedures for fixed schedules in order to find rapidly the lower bound on the fleet size without creating vehicle schedules. This lower-bound calculation can help decision makers in evaluating the number of vehicles required for any given change. Section 7.3 introduces the concept of maximum advance (early) and delay (late) from the scheduled departure time, given shifting tolerances (early, late) for each trip. Left (early) and right (late) shift limits are established, including the possibility of simultaneous left and right shifts in opposite directions. Section 7.4 provides an analysis and method of deriving the fleet size lower bound when SDTs are allowed within their tolerances. Section 7.5 exhibits procedures in which a single or chain (multiple) of shifting is required to reduce the fleet size, with and without its integration with DH trip insertions. Section 7.6 illustrates the process of implementation through an example in which variable schedules are used to construct vehicle blocks in a large bus agency. Section 7.7 looks into the effect of shifting scheduled departure times on even-load timetable. A criterion for maximum allowable change in the even load is introduced. The chapter ends with exercises.

Although this chapter contains math notations and proofs of derived conclusions, all sections are appropriate for practitioners. We especially recommend looking at the examples, illustrated in figures, to capture the interactive person–machine process. We end this corner, as we do almost customarily, with a story. A scheduler who couldn't insert DH trips because of only a few minutes difference and who made an effort not to increase the number of buses went to his supervisor. "Guess what solution I found (shifting)?" he asked. The supervisor told him that he reminded him that a new father once told his friend: "I have a new born baby, guess what?" "A boy," replied the friend. "Guess again" said the father. "A girl." The new father exclaimed: "Who told you?"

7.1 INTRODUCTION

Chapter 6 dealt with fixed schedules, in which departure times could not be changed. In practical transit scheduling, however, schedulers should attempt to allocate vehicles in the most efficient manner possible, including the employment of small shifts in departure times. For example, in Egged, Israel's national bus carrier, the schedulers consider a variable instead of a fixed schedule in addition to DH trip insertions. That is, they mull shifting trip-departure times (SDTs), based on some acceptable tolerances. Although confident of their work efficiency, the

Egged schedulers agree that variable scheduling with DH trip insertions is a very complex process, not systematic, and a very time-consuming task; it can turn a good schedule into a soured one. This situation is analogue to the following truth. If you put a spoonful of wine (an efficient chain of trips) into a barrel of sewage (problematic schedule), you get sewage. If you put a spoonful of sewage (a chain of trips impossible to execute) in a barrel full of wine (implemental schedules), you still get sewage (problematic schedule).

This chapter presents an extension of the DF concept in an effort to provide an interactive computer technique to assist the scheduler in the planning task. Usually, only the timetable construction and vehicle-scheduling (generation of chains or blocks) portions of the transit planning process are demonstrable. Given a multiterminal trip schedule, the object is to minimize the required fleet size, subject to a number of constraints, including trip-departure time tolerances and insertable DH trips. Once the fleet size is determined, a number of alternative block constructions may be generated.

Another important feature, and one that will open this chapter, is the derivation of a fleet size lower bound for both fixed and variable trip schedules. The DF lower-bound approach introduced can be used across almost all operational planning components. The importance of knowing how many vehicles are actually required applies to many transit-design and planning problems. Five such problems are as follows: (1) vehicle scheduling with different vehicle types (see Chapter 8), (2) crew scheduling (see Chapter 9), (3) design of operational transit-parking spaces (see Chapter 12), (4) design of a new transit network or redesign of an existing one (see Chapter 15), and (5) design of efficient short-turn trips (see Chapter 16).

Finally, Appendix 7.A, of this chapter, contains an example of vehicle-scheduling software including its key features and expected saving. The software is one of those presented in Table 1.1.

7.2 FLEET SIZE LOWER BOUND FOR FIXED SCHEDULES

7.2.1 Overview and example

The initial lower bound on the fleet size with DH trip insertions was found by Ceder and Stern (1981) to be the sum of all DFs, $G(S)$, as shown in Figures 6.6 and 6.7. An improved lower bound was established later by Stern and Ceder (1983), based on extending each trip's arrival time to the time of the first feasible departure of a trip to which it may be linked or to the end of the finite time horizon. Further, lower-bound improvements for fixed and variable schedules appear in Ceder (2002, 2005). The direct calculation of the fleet size lower bound enables schedulers and transit decision makers to ascertain more promptly how much the fleet size can be reduced by DH trip insertions and allowing shifts in departure times.

Figure 7.1 presents a nine-trip example with four terminals (a, b, c, and d). Table 7.1 shows the data required for the simple example used in this section and Sections 7.3 and 7.4 for demonstrating the lower-bound methods. Four DFs are constructed along with the overall DF. According to the next terminal (NT) procedure (see Chapter 6), terminal d (whose first hollow is the longest) is selected for a possible reduction in $D(d)$. The DH insertion process depicted in Figure 6.9a and b continues using the criterion $R = 2$. The first unit reduction DH chain (URDHC) is $DH_1 + DH_2$, and the second DH_3. The result is that $D(c)$ and $D(d)$ are reduced from 1 to 0 and from 2 to 1, respectively; hence, $N = D(S) = 5$ and G is increased from 3 to 4 using three inserted DH trips. The five FIFO-based blocks are as follows: [1–5], [2-DH_1-7], [3-DH_2-9], [4-DH_3-6], and [8].

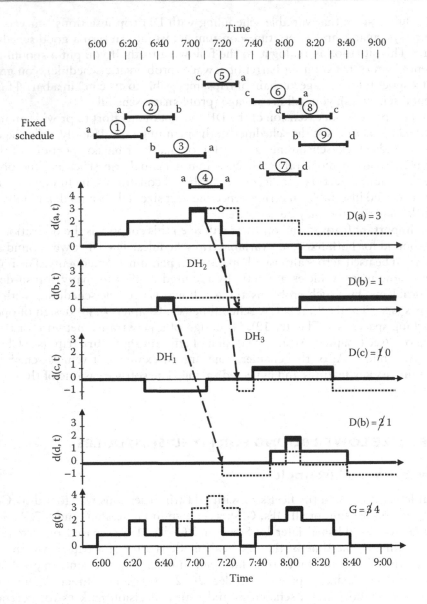

Figure 7.1 Nine-trip example with DH trip insertions for reducing fleet size.

7.2.2 Stronger fleet size lower bound

While Stern and Ceder (1983) extended each unlinked trip's departure time (i.e., one that cannot be linked to any trip's arrival time) to both T_1 and T_2, Ceder (2002, 2005) proved that an extension only to T_2 was sufficient. The extension to the time of the first feasible departure time of a trip with which it may be linked, or to T_2, results in an S' schedule and an overall DF, $g'(t, S')$, with its maximum value $G'(S')$.

While S' is being created, it is possible that several trip-arrival points are extended forward to the same departure point that is their first feasible connection. However, in the final solution of the minimum fleet size problem, only one of these extensions will be linked to the single departure point. This observation provides an opportunity to look into further

Table 7.1 Input data for the problem illustrated in Figure 7.1

Trip no.	Departure terminal	Departure time	Arrival terminal	Arrival time	Deadheading (DH) trips	
					Between terminals	DH time (same for both directions) (min)
1	a	6:00	c	6:30	a and b	20
2	a	6:20	b	6:50		
3	b	6:40	a	7:10	a and c	10
4	a	7:00	a	7:20		
5	c	7:10	a	7:30	a and d	60
6	c	7:40	a	8:10		
7	d	7:50	d	8:10	b and c	30
8	d	8:00	c	8:30		
9	b	8:30	d	9:00	b and d	30
					c and d	20

artificial extensions of certain trip-arrival points without violating the generalization of requiring all possible combinations for maintaining the fleet size at its lower bound.

Figure 7.2 illustrates three cases of multiple extensions to the same departure point. Case (i) shows two extensions, Trips 1 and 2, both with the same arrival point b, which is their first feasible connection at point a of Trip 3. Because only one of the two trips will be connected to Trip 3, the question is, "Which one can be extended further?" It is clear that Trip 1 has better DH chances to be connected to Trip 4 than to Trip 2 because of its longer DH time. Hence, Trip 1 can be further extended (second extension) to the start of Trip 4 if it is feasible. Case (ii), Figure 7.2, shows that Trips 1 and 2 do not end at the same point and that Trip 4 has different points than in Case (i). The argument of Case (i) cannot hold here, since the DH time differs between each two different points. In this case, the second feasible connection for Trip 1 is T_2. By using the Case (i) argument, one can then create three possible chains, [1], [2–3], and [4], instead of two chains, [1–3] and [2–4]. Case (iii) shows an opposite situation to that of Case (ii), with multiple extensions from different arrival points. If we link, in Case (iii), Trip 1 (longest DH time to the common departure point) and Trip 3 and extend Trip 2 to Trip 4, we have another multiple extension case like Case (i), this one concerning the start of Trip 4 (linked to Trips 2 and 3). Following the Case (ii) argument, Trip 3 will be linked to Trip 4, and Trip 2 will have its third extension. This results in three possible chains, [1-3-4], [2], and [5], instead of two, [1–5] and [2-3-4]. Cases (ii) and (iii) show why it is impossible to apply any general rule to a multiple extension of different arrival epochs. Consequently, further improvement of $G'(S')$ can be made only for Case (i) situations.

The following is the procedure for finding a stronger fleet size lower bound:

1. Establish S'.
2. Select a case in which more than one extension is linked to the same departure time t_{sk}^j of trip j at terminal k. If no more such cases—*stop*. Otherwise, select a group (two or more) of extensions with the same scheduled arrival terminal, u, and apply the following steps:
   a. Find a trip that fulfills $\min_{i \forall i \in E_u} \left(t_{sk}^j - t_{eu}^i \right)$, where E_u = set of all trips arriving at u and extended to t_{sk}^j and t_{eu}^i is the arrival epoch of trip i at terminal u.
   b. Perform the second feasible extension for all trips $i \in E_u$, except the one selected in Step 2a. Go to Step 2.

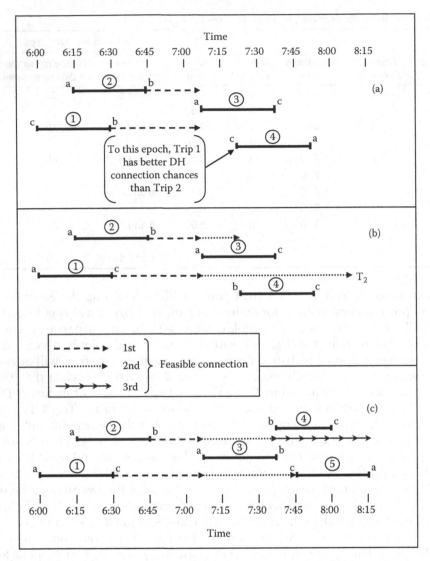

Figure 7.2 Part (a) shows why one should select the Trip 2 extension, Part (b) shows that the argument in (a) cannot be used in case of multiple connections from different terminals, and Part (c) shows another case in which multiple connections cannot be applied for constructing the lower bound.

Using this procedure, define the overall DF of the extended S' schedule by g''(t, S'') with the maximum value G''(S''). The following theorem and its proof establish that G''(S'') is a stronger lower bound than G'(S').

Theorem 7.1

Let $N_o(S)$ be the minimum fleet size for S with DH insertions. Let G'(S') and G''(S'') be the maximum value of the overall DF for S' and S'', respectively. Then,

(i) $G''(S'') \geq G'(S')$.

(ii) $G''(S'') \leq N_o(S)$.

Proof

(i) The new overall DF, $g''(t, S'')$, has more extensions than $g'(t, S')$; that is, $g''(t, S'') \geq g'(t, S')$. Therefore, $G''(S'') \geq G'(S')$.

(ii) According to the definition of S'', at any time t in which $g''(t, S'') = G''(S'')$, there exist $G''(S'') - g'(t, S')$ trip extensions over S'. The additional extensions in S'' represent multiple extensions (second, third, etc.), given that each extended trip is associated with another trip having the same arrival epoch and terminal and has only one extension. In the optimal chain solution, a departure time t_s^* may or may not be linked to its nearest feasible arrival epoch (t_e^*) across all other points representing the same arrival terminal. Linkage to t_e^* complies with the procedure to construct S''. Otherwise, t_e^* in S'' is further extended either to another trip or to T_2, while t_s^* is linked to $t_e^{**} < t_e^*$. We should note that t_e^{**} is linked to t_s^{**} when using the procedure described. Because t_e^* to t_s^* is the shortest link, the additional extension of t_e^* cannot be linked to a trip that starts before t_s^{**} (otherwise, t_e^{**} too will be linked to it and not to t_s^*). Therefore, the additional extension of t_e^* in the optimal chain solution, $N_o(S)$, results in a greater overlap between trips (when constructing $g''(t, S'')$). Hence, $G''(S'') \leq N_o(S)$.

Figure 7.3 presents the schedule of Figure 7.1, with S' in its upper part, S'' in its middle part, and three overall DFs—$g(t, S)$, $g'(t, S')$, and $g''(t, S'')$—in the lower part. For S', it may be observed that Trips 3, 4, and 5 are extended to the same departure point as Trip 6 from the same arrival terminal a. According to the procedure for constructing S'', the extension of Trip 5 is selected, and Trips 3 and 4 are further extended to the departure time of Trip 9. These additional extensions create another multiple connection associated with Trips 3 and 4, in which Trip 4 is the selected extension and Trip 3 is further (third time) extended. The initial lower bound is $G = 3$, the first improved lower bound is $G' = 4$, and the proposed improved lower bound is $G'' = 5$, which happens to be the optimal solution (see Figure 7.1).

7.3 VARIABLE TRIP-DEPARTURE TIMES

A small amount of shifting in scheduled departure times becomes almost common in practice when attempting to minimize fleet size or the number of vehicles required. This section presents methods, mostly according to the DF, to realize a variable trip schedule in an efficient manner. It may be recalled that Chapters 3 and 4 provided methods to improve the correspondence of vehicle departure times with passenger demand. The construction of timetables in Chapters 4 and 5 was based on either even headways or even average loads, entailing situations in which even headways result in uneven passenger loads. SDTs, therefore, may unbalance these desirable features in the timetable while favoring resource (vehicle) saving. Nonetheless, the last section of this chapter introduces a method to eliminate, at least to some extent, the possibility of too drastic changes in the even-load timetable requirement. Similar to the subject order of Chapter 6, we will start this section with reference to a single transit route and then continue with a minimum fleet size analysis for a network of routes.

7.3.1 Single-route minimum fleet size

Section 6.2 describes a method for ascertaining the minimum fleet size required for a given single route without interlining. We will now extend this method to account for possible shifting in departure times for given backward and forward shifting tolerances (in minutes) for each trip. This additional flexibility, which is employed in practice, can reduce the fleet

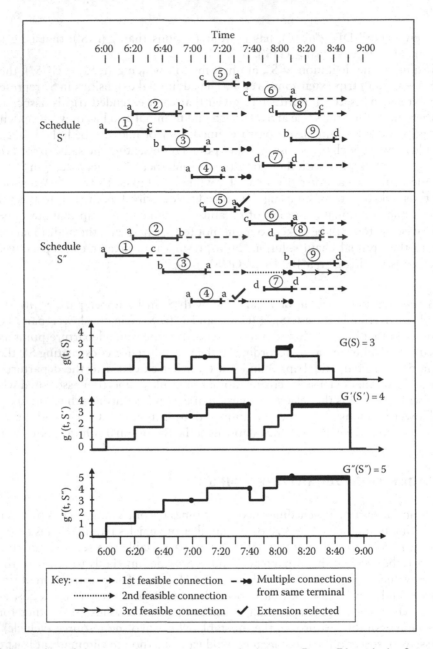

Figure 7.3 Lower-bound determination using the example shown in Figure 7.1, with the first and second improvement procedures.

size required as the primary objective. A secondary objective is to minimize the length of the shifting within their given tolerances.

In practice, departure times are shifted without any systematic method. Shifting tolerances are usually determined by rule of thumb although it makes sense to correlate them with the headways between departures. A proposed method appears in Table 7.2, in which the length of the shifting tolerance is headway dependent. That is to say, the longer the headway, the shorter is the tolerance. If the shifting is backward, the preceding headway is considered as H; if it is forward, the next headway is considered.

Table 7.2 Shifting tolerances as headway dependent

Headway (H, in min)	Percentage of H for tolerance determination (%)	Tolerance length as H dependent (min)
<10	50	0.5H
10–20	40	0.4H
21–40	30	0.3H
>40	20	0.2H

A new process needs to be designed for applying the shifting capability of departure times on single routes. This process simply attempts, through possible shifting of the relevant departure times, to reduce the minimum fleet size required. We will use the same notation for route r as in Section 6.2, but the symbol r is deleted, because it is clear which underlying route is being referred to. Thus, a and b are the end points; T_{ia} and T_{jb} are the average trip times on the route for vehicles departing at t_{ia} and t_{jb} from a and b, respectively, including layover time at their respective arrival points; and n_{ia} is the number of departures from a between t_{ia}, in which departure ia is included, and $t_{i'a}$, in which departure i'a is excluded. Trip ia arrives at terminal b, then continues with trip jb, the latter being the *first* feasible departure from b to a at a time greater than or equal to the time $t_{ia} + T_{ia}$; and $t_{i'a}$ is the *first* feasible departure from a to b at a time greater than or equal to $t_{jb} + T_{jb}$. Similar notations are defined for a trip starting from b.

Let $[t_{ia} - \Delta^{ia(-)}, t_{ia} + \Delta^{ia(+)}]$ be the tolerance time interval of the departure time of trip ia, in which $\Delta^{ia(-)}$ = maximum advance of the trip's scheduled departure time (the case of an early departure) and $\Delta^{ia(+)}$ = maximum delay from the scheduled departure time (the case of a late departure). Note that $t_{ik} + \Delta^{ik(+)} < t_{(i+1)k}$ and $t_{ik} - \Delta^{ik(-)} > t_{(i-1)k}$, for all $k \in K$. The minimum fleet size, N_{min}, is then attained by construction, using the procedure illustrated in a flow diagram in Figure 7.4a and b. The procedure described fits the case of Equation 6.1, in which $N_{min} = \max_i n_{ia}$. For the case in which $N_{min} = \max_j n_{jb}$ (determined by a trip starting from b), the same procedure is applied, but with b replacing a and j replacing i. The procedure first identifies the departure ia (or one of a few) referring to $N_{min} = n_{ia}$; then it attempts through shifting t_{ia} to arrive at b before or at $t_{(j-1)b}$, and most importantly to arrive before or at $t_{(i'1)a}$. If the process manages to reduce n_{ia} by one or more units, it looks for the next $n_{ia} = N_{min}$ or $n_{jb} = N_{min}$ to continue. A successful process is that in which N_{min} is reduced. In addition, the procedure depicted in Figure 7.4a and b minimizes the length of SDTs, except for the shifting of the first departure, t_{ia}.

The interpretation of the shifting procedure may be assisted by the example in Figure 6.3. Here, $N_{min} = 5$, resulting from the fifth and sixth departures from b. When b replaces a and j replaces i in Figure 7.4a and b, we can then use the procedure described and start with the 6:50b departure. Given $\Delta^{6:50b(-)} = 5$ min, then $\Delta^{7:00a(+)} = \Delta^{7:20b(+)} = \Delta^{7:15b(+)} = \Delta^{7:10b(+)} = 3$ min. The first check in the Figure 6.3 example results in shifting 6:50–6:45 from b having $\Delta^{6:45b(-)} = 0$ min. Then from a, the first feasible connection is at 7:03 (including a forward tolerance). The second check, $7:00 \geq 6:45 + 15$, results in setting the departure time from a at 7:00 with $\Delta^{7:00a(-)} = 0$. The third check, $7:00 + 15 \leq 7:20 + 3$, leads to finding the first feasible connection to 7:00 to be from a. That is, min $[7:20 + 3, 7:15 + 3, 7:10 + 3] \geq 7:15$ is 7:18. In the fourth check, $t_{(j-1)b} = 7:15$. Hence, $n_{6:45b} = 3$, instead of the previously $n_{6:50b} = 5$.

We now move to the 7:05 departure from b, in which $n_{7:05b} = 5$. Given are $\Delta^{7:05b(-)} = 0$ min and $\Delta^{7:10a(+)} = \Delta^{7:20b(-)} = \Delta^{7:30b(+)} = 5$ min. In the first check, an early departure from *a* is impossible. The second check, $7:20 - 0 + 15 \leq 7:30 + 5$, results in setting $t_{(j'-1)b} = 7:35$ and $n_{7:05b} = 4$. The result is a multicase of $N_{min} = n_{ia} = n_{jb} = 4$, but without any further possibility of improvement, because $n_{7:05b}$ cannot be further reduced; this case is based on the procedure constructed in Figure 7.4a and b.

7.3.2 Variable scheduling using deficit functions

The transit scheduler who employs shifting in trip-departure times is not always aware of the consequences that could arise from these shifts. This section develops a formal algorithm to handle the complexities of SDTs. The algorithm is intended for both automatic and man–computer conversational modes.

According to the definitions in Section 6.5.1, s_j^k and e_j^k are the start and end of the jth maximal interval, M_j^k, $j = 1, 2, ..., n(k)$, at terminal k, $k \in K$, and N is the number of vehicles (chains, blocks) determined. If the length of M_j^k is denoted as $\bar{M}_j^k = e_j^k - s_j^k$, then s_j^k and e_j^k are associated with t_s^i and $t_e^{i'}$, respectively. That is, s_j^k refers to the departure time of a trip designated by I and e_j^k to the arrival time of a trip designated by i' (where i, i' can be selected from several trips that depart at time s_j^k and arrive at e_j^k, respectively). The shifting terms defined in the foregoing section will also be used.

The nine-trip example illustrated in Figure 7.1 is used for possible SDTs in Figure 7.5, which employs the DF display. The tolerances of this example are $\Delta^{i(+)} = \Delta^{i(-)} = 5$ min for all

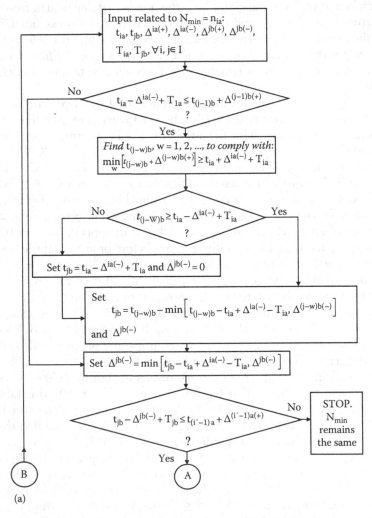

(a)

Figure 7.4 (a) Flow diagram of the shifting departure-time process for reducing the minimum fleet size N_{min} determined at terminal A (for terminal B, the same process is used with a change of symbols).

(Continued)

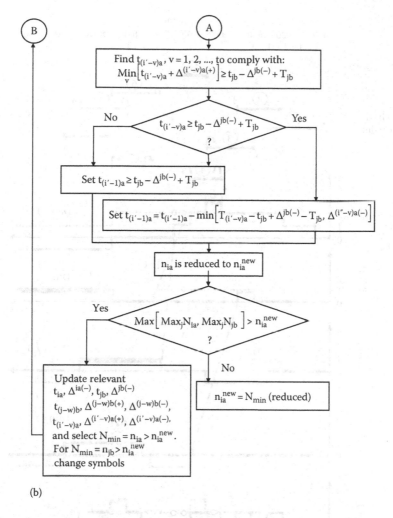

(b)

Figure 7.4 (Continued) (b) flow diagram continued from (a).

trips in the schedule. Starting with shifting Trip 3 backward and Trip 4 forward by 5 min results in reducing D(a) from 3 to 2. This may be continued with shifting Trips 7 and 8 to reduce D(d) from 2 to 1.

Because no further shifting in departure times is feasible for the given tolerances, the process becomes one of searching for URDHC using DH trip insertion. This yields three DH trips resulting in Min $N = G(S^{new}) = 3$, in which S^{new} is the new schedule. The three blocks are determined by FIFO: [1-5-DH$_2$-9], [2-DH$_1$-7-8], and [3-4-DH$_3$-6]. In case a DH trip insertion is not allowed, the shifting process will end with Min $N = 5$ and the FIFO-based blocks: [1–5], [2–9], [3–4], [7–8], [6].

Another seven-trip example is depicted in Figure 7.6. It demonstrates the possible chaining effect of SDTs. In this example, $\Delta^{i(+)} = \Delta^{i(-)} = 10$ min. Hence, it is possible at the outset to reduce D(b) by one unit through the shifts of t_s^3 to the right and t_s^4 to the left. However, these shifts increase D(a), and the net saving is zero. Consequently, another iteration is needed in which t_s^7 is shifted to the right. Only then can a total saving be obtained of one vehicle. Given the desire to reduce a maximal interval M_j^k by shifting a maximum of two trips, let us

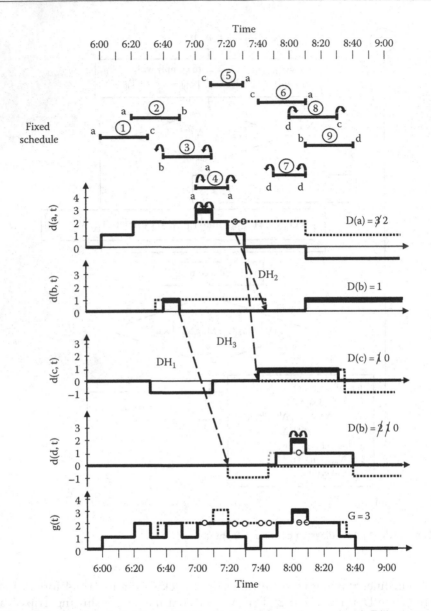

Figure 7.5 The nine-trip example (of Figure 7.1), first with shifting and second with DH trip insertion, for reducing fleet size.

consider the following three cases: (A) shift only trip i to the right, (B) shift only trip i′ to the left, and (C) shift both trips i and i′ in opposite directions (see Figure 6.6 for definitions used).

7.3.2.1 Case A: Right-shift limit

Let trip i (between terminals k and m), which starts at $t_s^i = s_j^k$, arrive at hollow H_{q-1}^m (preceding M_q^m) at time t_e^i. Shift trip i to the right as close to e_j^k as possible without increasing the maximal interval M_q^m or without exceeding $\Delta^{i(+)}$. Let this right-shift limit be defined as

$$\delta(+) = \min\left\{s_q^m - t_e^i, \Delta^{i(+)}, \overline{M}_j^k\right\} \tag{7.1}$$

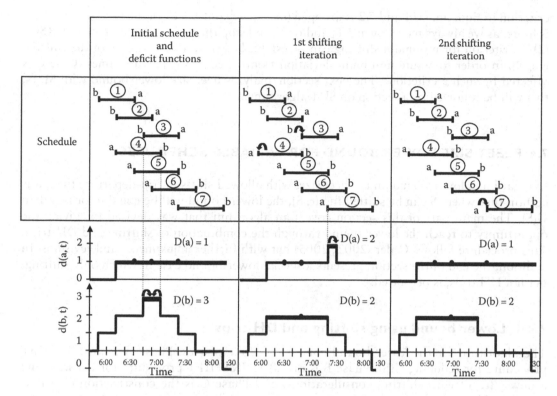

Figure 7.6 Example of two shifting iterations to reduce the required fleet size from 4 to 3.

If $\delta = s_q^m - t_e^i$, then the shift has reached M_q^m and any further right shift will increase $D(m)$. If $\delta(+) = \Delta^{i(\cdot)}$, then the shift is stopped by the tolerance limit of trip i. If $\delta(+) = \overline{M}_j^k$, then a successful shift has occurred, and $D(k)$ is reduced by one.

7.3.2.2 Case B: Left-shift limit

Let trip i′ (between terminals m and k) depart from hollow H_{q-1}^m at time $t_s^{i'}$. Shift trip i′ to the left as close to s_j^k as possible, without increasing the maximal interval M_{q-1}^m or without exceeding $\Delta^{i'(-)}$. Let this left-shift limit be defined as

$$\delta(-) = \min\left\{ t_s^{i'} - e_{q-1}^m, \Delta^{i'(-)}, \overline{M}_j^k \right\} \tag{7.2}$$

If $\delta(-) = t_s^{i'} - e_{q-1}^m$, then the shift has reached M_{q-1}^m, and any further shift left will increase $D(m)$. If $\delta(-) = \Delta^{i'(-)}$, then the shift is stopped by the tolerance limit of Trip i′. If $\delta(-) = \overline{M}_j^k$, then $D(k)$ is reduced by one.

7.3.2.3 Case C: Shift both trips

Without loss of generality, $D(k)$ can be reduced by shifting both trips i and i′ in opposite directions. Assume that the procedure starts with an attempt to shift trip i to the right and is unsuccessful, $\delta(+) < \overline{M}_j^k$. Now perform Case B, with the length of $\overline{M}_j^{\prime k} = \overline{M}_j^k - \delta(+)$ reduced from $\overline{M}_j^k$. Similarly, the procedure can start with Case B and continue with Case A.

These three cases can be incorporated into a formal SDT algorithm. Two points should be noted: (1) If a unit reduction shifting chain (URSC) is allowed, as in Figure 7.6, the first criterion in Equations 7.1 and 7.2 is dropped (i.e., $s_q^m - t_e^i$ and $t_s^{i'} - e_{q-1}^m$); the process for URSC is finite, as we always move toward T_1 and/or T_2 and stop there (for an unsuccessful URSC). (2) A criterion for minimum shifting can be established (i.e., a least square of the shifting length) in order to assure minimum deviation from scheduled departure times (Case C is affected by such a criterion). The next section analyzes fleet size lower bound with SDTs; this will be followed by a section on SDT algorithms.

7.4 FLEET SIZE LOWER BOUND FOR VARIABLE SCHEDULES

In regard to a lower bound on the fleet size, with allowed shifts in trip-departure time, it is certain that where S can be shifted in g(t, S), the lower bound will be equal to or less than G(S). The main part of this section covers an algorithm that was devised for giving the opportunity to reach the lower bound through the combination of shifting and DH trips. This procedure follows Ceder (2002, 2005) but with further adjustments and revisions. In addition, the end of the section presents a special lower-bound case in which only shifting, but not DH trips, is permitted.

7.4.1 Lower bound using shifting and DH trips

The algorithm for the lower bound consists of three phases. Phase A is an LB-SHIFT procedure for reducing G(S). Phase B is an LB-DH&SHIFT procedure for constructing a new g'(t, S') with shifting considerations, and Phase C is the construction of a new g"(t, S"). Phase A involves the construction of a new temporary schedule, with shifts to be termed S_{sf}, which will end with $G(S_{sf}) \leq G(S)$. Phase B constructs a new temporary schedule with shifts combined with DH trips, to be termed S'_{sf} that will end with $G'(S'_{sf}) \leq G'(S_{sf}) \leq G'(S')$. Note that $G'(S_{sf})$ is an extended schedule that follows the construction of g'(t, S') in Section 8.2 but with an S_{sf}, in which no shifts are made in constructing g'(t, S_{sf}). Finally, Phase C applies the procedure for constructing g"(t, S") in Section 7.2 to S'_{sf} without any additional shifts. All notations used here are applicable to the procedures that follow.

7.4.1.1 Phase A: LB-SHIFT procedure

Step 0 (initialization): Define $M_j = [s_j, e_j]$, j = 1, 2, ..., n(g), and $\overline{M}_j = e_j - s_j$ as the *j*th maximum interval and its length, respectively, of g(t, S_{sf}). Let $s_j = t_s^i$, $e_j = t_e^{i'}$, and S = S_{sf}, in which s_j refers to the departure time of a trip designated by i and e_j to the arrival time of a trip designated by i'. Set j = 0.

Step 1 (selecting the next maximum interval): Let j = j + 1; if j > n(g), stop; otherwise, continue.

Step 2 (feasibility check): If $\overline{M}_j \leq \Delta^{i(+)} + \Delta^{i'(-)}$, continue; otherwise, go to *Step 1*.

Step 3 (right shift): Compute $\delta(+) = \min\{\Delta^{i(+)}, \overline{M}_j\}$ for g(t, S_{sf}); if $\delta(+) = \overline{M}_j$, go to *Step 5*; otherwise, continue.

Step 4 (left shift): Compute $\delta(-) = \min\{\Delta^{i'(-)}, \overline{M}_j\}$ for g(t, S_{sf}); if $\delta(-) = \overline{M}_j$, go to *Step 5*; otherwise, if $\Delta^{i'(-)} \geq \overline{M}_j - \Delta^{i(+)}$ (Case C for g(t, S_{sf}) in Section 8.3), set $\delta(-) = \overline{M}_j - \Delta^{i(+)}$, $\delta(+) = \Delta^{i(+)}$, and go to Step 5. Otherwise, go to *Step 1*.

Step 5 (reduce the lower bound): Update S_{sf} and g(t, S_{sf}) with the confirmed shifts; go to *Step 1*.

7.4.1.2 Phase B: LB-DH&SHIFT procedure

Step 0 (initialization): Construct $g'(t, S_{sf})$ from $g(t, S_{sf})$. Define $M'_j = [s'_j, e'_j]$, $j = 1, 2, ..., n(g')$ and $\overline{M}'_j = e'_j - s'_j$ as the jth maximum interval and its length, respectively, of $g'(t, S_{sf})$. Let $s'_j = t^i_s$, $e'_j = t^{i'}_e$, and $S_{sf} = S'_{sf}$, in which s'_j refers to the departure time of a trip designated by i and e'_j to the arrival time of a trip designated by i'. Set $j = 0$.

Step 1 (selecting the next maximum interval): Let $j = j + 1$; if $j > n(g')$, stop; otherwise, continue.

Step 2 (right-shift check): If trip i of $s'_j = t^i_s$ can be shifted by $\Delta^{i(+)}$ (was not shifted in Phase A) and results in an increase in $G'(S_{sf})$, go to *Step 4*; otherwise, continue.

Step 3 (artificial shift): Let t^i_s be shifted to the right by $\Delta^{i(+)}$ and included in S'_{sf}; set $t^i_s = t^i_s + \Delta^{i(+)}$.

Step 4 (trip extensions with shifts): Let E be the set of all trips in which their artificial extension overlaps with M'_j, $i \notin E$; examine each trip u, $u \in E$, to determine whether its first feasible linkage point (departure time of a trip with which it may be linked) is t^i_s by performing a shift to the left ($\leq \Delta^{u(-)}$); if no linkage exists, stop; no improvement can be made. Otherwise, continue.

Step 5 (new shifts and extensions): Perform for all $u \in E$ that can be shifted to the left (and linked to t^i_s) the new shifts and extensions to be included in S'_{sf}; go to *Step 1*.

If Phase B indeed improves the lower bound, $G'(S'_{sf}) < G'(S_{sf})$, the foregoing S'_{sf} is then subject to Phase C. The latter has further extensions of arrival epochs if more than a single extension exists from the same terminal; this results in construction S''_{sf} from S'_{sf} according to the analysis of the construction of S'' in Section 7.2.

To better comprehend Phases A and B, the example problem of Figures 7.1 and 7.3 is presented. This problem undergoes these phases in Figure 7.7, with $\Delta^{i(+)} = \Delta^{i(-)} = 5$ min for all $i \in S$. Figure 7.7 shows the schedule and shifting required to construct $g(t, S_{sf})$ and $g'(t, S'_{sf})$. The process of constructing $g(t, S_{sf})$ exhibits Phase A, in which two trips are shifted in opposite directions, Trips 7 and 8, to allow for $G(S_{sf}) = 2 < G(S) = 3$; this change is represented by a dotted line. The function $g'(t, S'_{sf})$ exhibits Phase B; the dotted line in $g'(t, S'_{sf})$ shows that the departure time of Trip 5 is the start of the only maximum interval that exists in $g'(t, S_{sf})$ following the two shifts in Phase A. Shifting Trip 5 to the right by 5 min will open up the possibility for Trips 2 and 3 to be extended to Trip 5 while being shifted to the left by 5 min, instead of extended as it was previously to Trip 6. This is not the case with Trip 4, whose original extension overlaps, too, with the maximum interval. Finally, it should be observed that by shifting Trip 5 to the right, it can no longer be extended to Trip 6, and its new extension is to Trip 9. However, this new extension does not create a new maximum interval (see Step 2 in Section 7.4.1.2).

The final Phase C applies the procedure to attain $G''(S''_{sf})$ of the function $g''(t, S''_{sf})$. That is, to start with schedule S'_{sf} and allow further extensions of arrival points if more than a single extension exists from the same terminal. In the example shown in Figure 7.7, Phase C cannot be implemented. Assuming, however, that Trip 3 starts and ends at same terminal *b*, then Trip 2 could be further extended to Trip 6 (the shortest extension of Trip 3 is the one selected to remain). Nonetheless, the latter extension of Trip 2, based on the earlier assumption, will not increase $G'(S'_{sf}) = 3$. This final lower bound of three is the optimal number of chains (blocks) that can be derived in the example problem. This compares with five chains with only a DH trip insertion. The shifting of departure times in the LB-SHIFT and LB-DH&SHIFT procedures can be used in the example to create the optimal three trip chains: [1–4–6], [3(*shifted*)–5(*shifted*)–9], and [2–7(*shifted*)–8(*shifted*)], in which the shift of Trip 2 in Figure 7.7 is not required, because it links to Trip 7 not Trip 5.

A special scheduling case occurs when only shifting, but not a DH trip insertion, is permitted. In practice, this situation arises in transit systems without interlining and in which

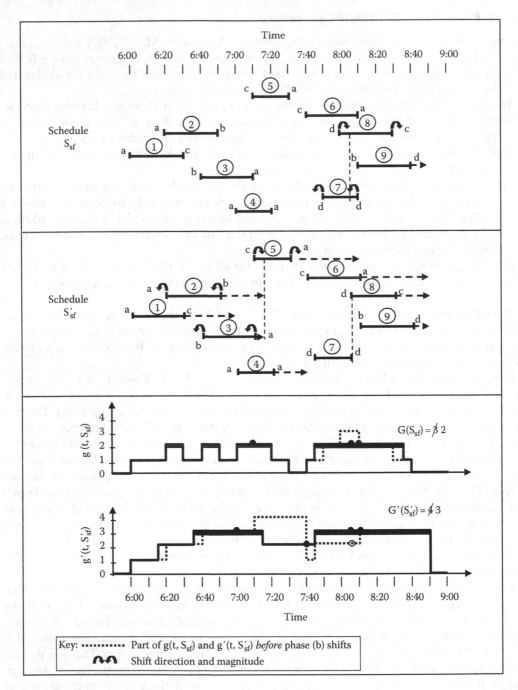

Figure 7.7 Lower-bound determination using the example in Figure 7.1 with LB-SHIFT and LB-DH&SHIFT procedures.

passenger demand does not differ greatly in either direction of the route (e.g., a rail operation). The lower-bound algorithm for this case follows the LB-DH&SHIFT procedure, but without a DH extension consideration. That is, each trip's arrival time is extended to the time of the first feasible departure time of a trip to which it may be linked, by having the same arrival and departure terminal as the extension, or to the end of the finite time horizon.

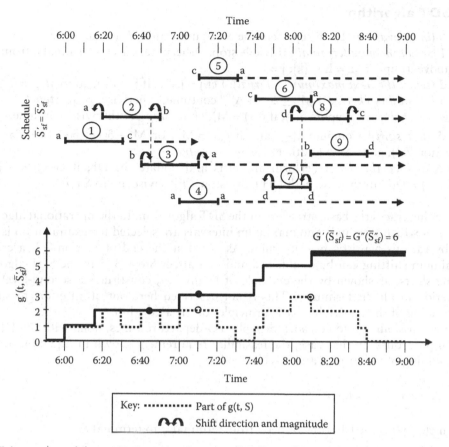

Figure 7.8 Lower-bound determination using the example in Figure 7.1 with the LB-EXT SHIFT procedure.

Let $\overline{S}'_{sf}$ and $\overline{S}''_{sf}$ be the extended schedules that are similar to S'_{sf} and S''_{sf}, but with a possible extension to link only the same terminals or to the end of the time horizon. We then obtain $G'(\overline{S}'_{sf}) \geq G'(S'_{sf})$ and $G''(\overline{S}''_{sf}) \geq G''(S''_{sf})$ and of course $G''(\overline{S}''_{sf}) \geq G'(\overline{S}'_{sf})$. Figure 7.8 uses the example in Figure 7.1 for constructing $\overline{S}'_{sf}$, which has the same results as $\overline{S}''_{sf}$. In this example, therefore, $G''(\overline{S}''_{sf}) = G'(\overline{S}'_{sf}) = 6$.

7.5 FLEET-REDUCTION PROCEDURES

This section will describe some of the considerations in incorporating the SDT algorithm into the fleet-reduction procedures. The primary inputs required for the heuristic procedures are the following:

1. An initial set of fixed trips, S^o, defined over a set of terminals K
2. A tolerance matrix of trip-departure times
3. A travel time matrix of potential DH trips (if applicable)

The output of the algorithm will be a new vehicle schedule, including the shifts of trip-departure times, the set of DH trips inserted (if allowed), and the required number of vehicles at each terminal. The trips assigned to each vehicle can be constructed in a second phase using the FIFO rule or the chain extraction procedure explained in Section 6.5.5.

7.5.1 SDT algorithm

Step 0 (initialization): Let E represent the set of unexamined terminals; set $E = K$.

Step 1 (select the next terminal): If $E = \Phi$, stop. Otherwise, select a terminal k from E and remove it; update $E = E - \{k\}$; $j = 1$.

Step 2 (select the next maximum interval): Let $j = j + 1$; if $j > n(k)$, go to *Step 6*.

Step 3 (feasibility check): If $\overline{M}_j^k \leq \Delta^{i(+)} + \Delta^{i'(-)}$, continue; otherwise, go to *Step 1*.

Step 4 (right shift): Compute $\delta(+)$; if $\delta(+) = \overline{M}_j^k$, go to *Step 6*; otherwise, continue.

Step 5 (left shift): Compute $\delta(-)$; and if $\delta(-) = \overline{M}_j^k$, or $\overline{M}_j^k - \delta(+) \leq \delta(-)$ (see Case C in Section 7.3), go to *Step 6*; otherwise, go to *Step 1*.

Step 6 (reduce the fleet): Perform all shifts and update the DF; if $G(S_{sf}) = D(S)$, or $G''(S_{sf}'') = D(S)$ in the case with DH trips, stop. Otherwise, go to *Step 1*.

The foregoing traces the basic structure of the SDT algorithm. In the operational algorithm, the order in which terminals and maximum intervals are selected for examination is determined by various heuristic priority rules. As noted at the end of Section 7.3, a criterion for minimum shifting can be established and will affect Steps 3–5 in the SDT algorithm. Chaining shifts, as shown by the example of Figure 7.6, constitutes a second-level phase examination of all the terminals. This is not described here out of an effort to simplify understanding of the underlying basic principles of the algorithm.

Another possibility is to consider variable trip-departure times along with the DH trip-insertion procedure. In that case, the feasibility requirement shown in Equation 6.8 for a feasible joining of two trips, i and j, is changed to

$$t_e^i - \Delta^{i(-)} + \tau(q^i, p^i) = t_s^j + \Delta^{j(+)} \tag{7.3}$$

in which $\tau(q^i, p^i)$ is the DH time of trip i from terminal *q* to terminal *p*.

7.5.2 Order of subroutines

In the variable-scheduling problem, two major subroutines are identified in order to reduce the fleet size: (a) SDT within their tolerances and (b) searching for a URDHC. In some practical cases, DH trips are not allowed and only an SDT algorithm is utilized. The URDHC procedure could be made both with and without consideration of the SDT subroutine. The following is a discussion of these subroutines.

The selection of the SDT subroutine needs to be examined primarily in light of the maximum possible saving of vehicles. Figure 7.9 illustrates an example of three terminals using the DF representation. The first column, *URDHC Start*, begins with the URDHC procedure and ends with a modified URDHC (mixed with the SDT) procedure. The second column, *SDT Start*, uses first the SDT procedure and second the modified URDHC procedure. In this example, $\Delta^{i(+)} = \Delta^{i(-)} = 5$ min for all trips i in the schedule; the DH trip time for all possible DH trips is 10 min. It is also assumed that the departure times for trips shifted to the left and the arrival times for trips shifted to the right are outside the timescale for the SDT procedure.

The first-column solution succeeds in reducing D(a) from 4 to 2 in Step 1 of Figure 7.9. In Step 2, D(b) is reduced by one through four shifts of trip-departure times and two DH trips. The final schedule can be constructed from the DFs in the last step, in which D(S), the fleet size, has been reduced from 11 to 8 vehicles. In the second-column solution (SDT Start), D(b) is reduced from 4 to 3 by four trip-departure time shifts. Then in Step 2, D(a) is reduced from 4 required vehicles to 2 through two shifts of the trip-departure time and two DH trips. The final schedule has D(S) = 8.

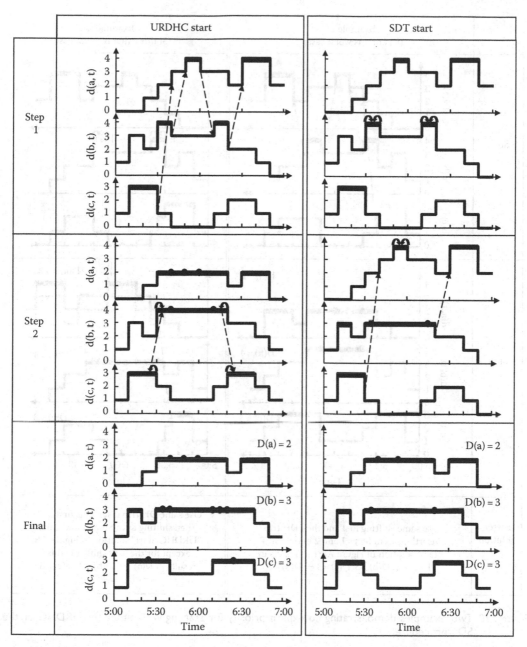

Figure 7.9 Two solutions for reducing fleet size; the first starts with the URDHC subroutine and the second with the SDT subroutine; both continue with the modified URDHC procedure.

Note that in the final DF configuration in both columns of Figure 7.9, no further reductions in fleet size can be achieved based on the formal SDT algorithm and on the feasibility requirement for DH trip insertion indicated in Equation 7.3. In consideration of the secondary objectives, it appears that the second column has priority over the first column (four shifts and two DH trips as against four shifts and six DH trips, respectively). Intuitively, it seems worthwhile to start with the SDT procedure. However, the next example shows that this is not always the case.

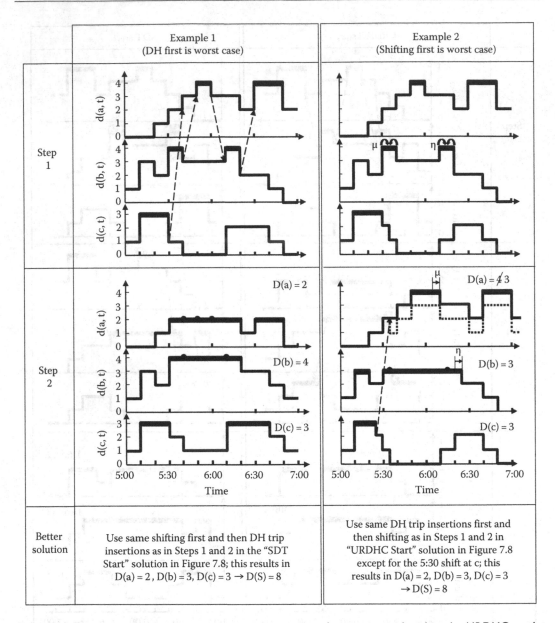

Figure 7.10 Two examples demonstrating no explicit priority for starting with either the URDHC or the SDT procedure.

Figure 7.10 presents two examples with shift tolerances and a DH trip time as in the example in Figure 7.9. In Example 1, $d(a, t)$ and $d(b, t)$ are the same as in Figure 7.9, but a small change is introduced in $d(c, t)$. The trip, which departs at 6:20, moves to depart at 6:10 from terminal c. In Example 1, only two vehicles are saved (see Figure 7.10) by starting with the URDHC procedure. This is in comparison with a saving of three vehicles if we begin with the SDT procedure (see the better solution indicated under Example 1 in Figure 7.10). In Example 2, $d(a, t)$ and $d(b, t)$ are the same as in Figure 7.9; and in $d(c, t)$, one trip arrives at 5:35 instead of 5:30. In this example, we relax the assumption that the SDT procedure influences departure and arrival times outside the time scale (though this assumption holds in Example 1).

In Step 1 of Example 2, in Figure 7.10, four shifts are illustrated. The shifts that are denoted by μ and η (forward shifts) affect the DFs at terminals a and b, respectively. Shift μ refers to a trip between terminals b and a with a trip time of 30 min, and shift η refers to a round trip from terminal b with a length of 10 min. Thus, a DH trip between terminals c and a reduces D(a) by one. Consequently, we can see in Step 2 (Example 2) in Figure 7.10 that by starting with the SDT procedure, it is possible to save only two vehicles at terminals b and a. This compares to a saving of three vehicles when applying the better solution indicated in Example 2 in Figure 7.10.

A DH departure has the effect of increasing the departure terminals' DF and also of reducing the deficit of some maximum intervals at their destination terminals (either by reducing the fleet requirement or by compensating for an earlier DH trip departure). In this way, maximum intervals are leveled and expanded at both the departure and the arrival terminals. This reduces the opportunities for a further reduction of the fleet size through the SDT procedure. On the other hand, similar consequences are observed when starting with the shifting of trip-departure times. That is, the SDT procedure generally reduces the opportunities for a further reduction of the fleet size through DH trips (though some shifts might open up new opportunities because of a larger time interval for a DH trip insertion).

7.5.3 Description of the procedure

For the fleet-reduction procedure, when DH trip insertion is allowed, we use a heuristic rule and start with the SDT subroutine, followed by the modified URDHC (mixed with SDT) subroutine. This rule can represent to some extent the viewpoint of the transit agency that wishes to minimize operational costs.

The basic fleet-reduction procedure that contains only the SDT algorithm is presented in the flow diagram in Figure 7.11 and is designed for a person–computer interactive system. It does not interact with the DH trip-insertion process. The selection of a terminal u can be made by the scheduler by inspecting the DFs on a graphical display. The search for a reduction in the fleet requirement at terminal u (see Figure 7.10) can be performed manually by the scheduler or by procedures based on the formal SDT algorithm. If a URSC is found, all DF information is updated for this new schedule, and a new iteration initiated. The SDT subroutine continues until either the lower-bound test is successful or all terminals are examined.

The modified URDHC (mixed with SDT) subroutine, shown schematically in a flow diagram in Figure 7.12, is based on the URDHC procedure described in Section 6.5.4. The upper part of the flow diagram is the SDT subroutine described in Figure 7.11. In the lower part of Figure 7.12, the lower-bound determination suggests that $G''(S''_{sf})$ should be examined before trying to insert DH trips into the schedule. In this modified procedure, the feasibility requirement for DH trip insertion is based on Equation 7.3. Another point that should be mentioned is that DH trips added to schedule S should include a shifting tolerance for the next iterations of the SDT and modified URDHC procedures. Finally, if a URDHC has been found, the DH chain cost involved is compared with the saving cost of a single vehicle. If the DH cost is higher than the saving cost, the URDHC is cancelled. Otherwise, the set of DH trips, designated {DH}, is added to the previous schedule, and all DF information, including shifting, is updated for a new schedule. The modified URDHC subroutine continues with updated S until $G''(S''_{sf}) = D(S)$ or until all terminals are examined.

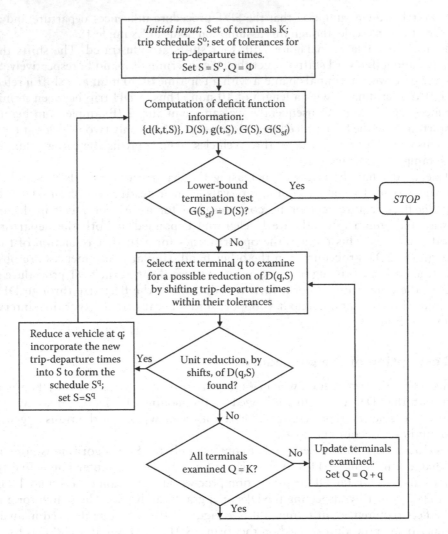

Figure 7.11 Flow diagram of a fleet-reduction procedure involving only shifting of departure times (SDT algorithm).

7.6 EXPERIENCES WITH BUS SCHEDULES

Before sharing some experiences, it is advised to look at Appendix 7.A. This appendix contains an example of vehicle-scheduling software including its key features and expected saving. The software is one of those presented in Table 1.1. Appendix 7.A especially demonstrating the saving one can expect using software of vehicle scheduling.

As mentioned in Section 7.1, fundamental problems in designing variable schedules arise from practice. An example of a transit agency that faced these problems is Egged, Israel's national bus carrier, which was discussed in Section 6.3. The Egged fleet of about 4,000 buses operates over a spread-out national network with some 2,000 routes; it performs an average of 50,000 daily trips, among them 12,000 DH trips. Bus schedules are produced by about 60 schedulers using Gantt charts through a trial-and-error approach. The need for a quicker response to timetable changes has led the Egged management to investigate the use of a fully computerized system. As is explained in Section 6.3, attempts to implement the

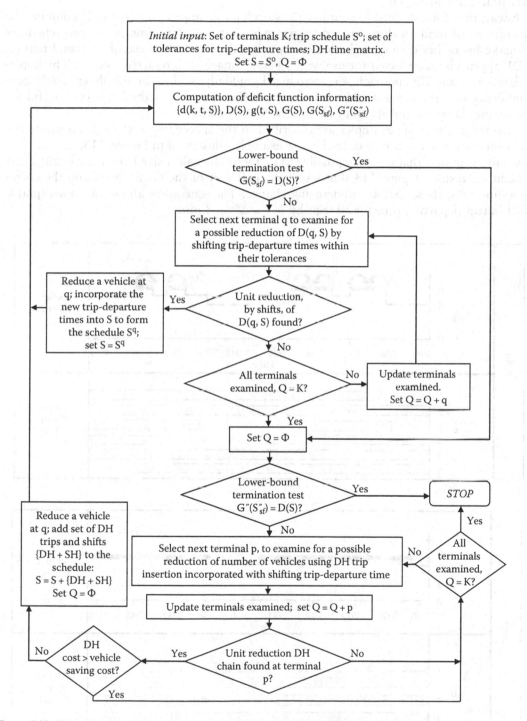

Figure 7.12 Flow diagram of fleet-reduction procedure involving shifting of departure times and DH insertions (modified URDHC subroutine).

computer-generated schedule have failed because of an inability to meet a number of necessary practical constraints.

It was, therefore, decided to continue the search for an approach that would combine the advantages of modern electronic computers while, at the same time, allow the scheduler to make his or her own contribution to the scheduling task. Because of its visual nature, a DF approach was selected for use with a person–machine interactive system. The implementation of the DF approach was introduced gradually so that the schedulers could gain confidence in this approach and reach the conclusion that this method was very useful for increasing the speed and accuracy of the scheduling tasks.

Two simple, real-life examples are described in the succeeding text to demonstrate the implementation stage at Egged. In the first example, illustrated in Figure 7.13, the schedulers claimed at first that it was impossible to further reduce fleet size from their Gantt chart scheduling results. Figure 7.13 shows only a small part of the Gantt chart and the corresponding DFs, those that are undergoing change. The schedulers allow for an acceptable shift in trip-departure time for all trips by $\Delta^{i(+)} = \Delta^{i(-)} = 3$ min.

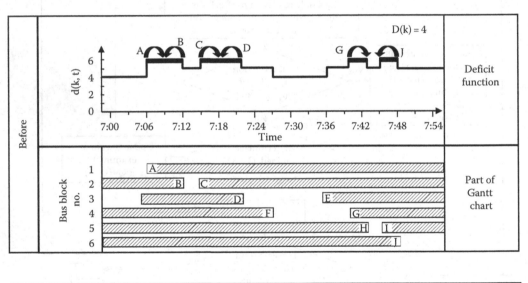

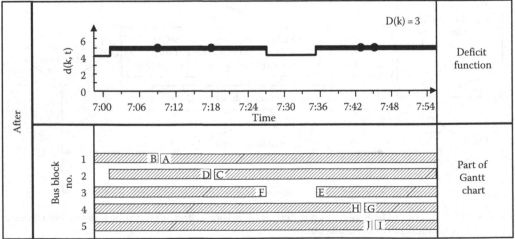

Figure 7.13 Demonstration of the superiority of deficit-function representation over the Gantt chart approach.

By illustrating d(k, t), however, we saved one bus by shifting six trips, each by 3 min. Before the changes in Figure 7.13, trips are identified in the Gantt chart by letters. This is for the sake of clarity when referring to shifts and for reconstructing the Gantt chart chains after the changes. From Figure 7.13, the problem appears easy to handle. However, before the changes in part of Figure 7.13, only 6 of 52 portions of the bus blocks were shown, and those 6 rows were spread among the 46 other rows in the Gantt chart.

Following this demonstration, the schedulers were still not wholly convinced. They argued that with a little more effort on their part, they too could have saved the bus as in Figure 7.13. Therefore, a more complex example was decided on, as shown in Figure 7.14. This second example refers to an afternoon schedule for two Egged branches: Ramle, terminal k, and Lod, terminal m. On the left-hand side of Figure 7.14, only trips that involved changes are exhibited in the before-and-after Gantt chart representation; trips are designated by letters. On the right-hand side, the DFs of the complete schedule are illustrated, including trips not shown in the left-hand side. The schedulers again claimed that no further reductions could be achieved from the D(k) + D(m) = 57 + 19 = 76 fleet size requirement. The given information was that $\Delta^{i(+)} = \Delta^{i(-)} = 2$ min and that the DH trip time between the k and m terminals (both directions) was 7 min.

As seen in Figure 7.14, 6 shifts in trip-departure times and a single DH trip were required in order to save 2 buses and to reduce the total fleet requirement to 74 buses. It was only after this second demonstration that the schedulers began to take a serious interest in the DF model. This was particularly due to its simplicity and visual nature. The schedulers expressed a positive feeling about the valuable aid of this gradual approach.

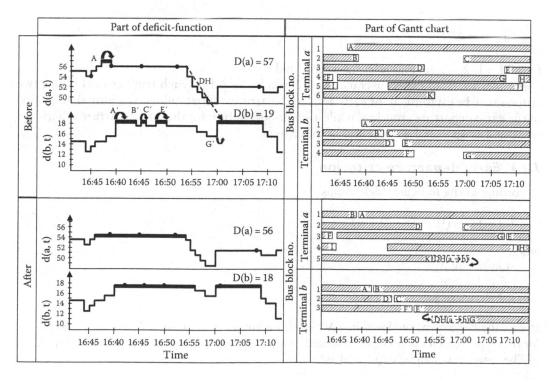

Figure 7.14 Two-terminal case, in which two buses are saved through shifting departure times and modifying URDHC procedures.

7.7 EXAMINATION AND CONSIDERATION OF EVEN-LOAD TIMETABLES

We noted in Section 7.3 that the construction of timetables in Chapters 4 and 5 was based on either even headways or even average loads, entailing situations in which even headways resulted in uneven passenger loads. Even average load can be formulated either at the hourly max load point or at the max load point of individual vehicles (Ceder and Philibert, 2014). SDTs, therefore, may unbalance these desirable features in the timetable while favoring cost (vehicle) saving. This section introduces the possibility of establishing a load-tolerance criterion for even-load timetables.

Following the notation in Section 5.2, let $SL_i(t)$ be the slope of $L_i(t)$ at t, in which $L_i(t)$ is the cumulative load curve at stop i, $i = 1, 2, ..., n$; Δd_τ is a given positive tolerance (in passengers) of the desired occupancy d_τ at time interval $\tau = 1, 2, ..., v$; $t_{1j}^* = t_{qj} - T_{q\tau}$ is a departure time determined by Principle 5.3 in Section 5.2 (even load on individual vehicles) at stop q, in which t_{qj} is the jth candidate departure time from stop q, $j = 1, 2, ..., m$; and $T_{q\tau}$ is the average service travel time between the departure terminal $q = 1$ and stop q when t_{qj} is at interval τ (note that by definition, $T_{1\tau} = 0$), and $\Delta^{j\tau(-)}$ and $\Delta^{j\tau(+)}$ are given positive tolerances (in minutes) for maximum shifting t_{1j}^* to the left (early departure) and to the right (late departure) for each interval τ, $j = 1, 2, ..., m$; $\tau = 1, 2, ..., v$.

When $d\tau$ is added along $L_i(t)$, the minimum time that intersects one of the cumulative load curves (see Principle 5.3 in Section 5.2) is determined at stop q. Hence,

$$d_\tau = L_q\left(t_{1j}^*\right) - L_q\left(t_{1,j-1}^*\right) \tag{7.4}$$

The shifts in departure times made by the DF model are defined as follows:

$$t_{1j}^- = t_{1j}^* - \Delta^{j\tau(-)} \tag{7.5}$$

$$t_{1j}^+ = t_{1j}^* + \Delta^{j\tau(+)}$$

for all $j = 1, 2, ..., m$, and relevant τ for t_{1j}^- and t_{1j}^+.

In order to avoid excess average loads beyond $d_\tau + \Delta d_\tau$ at each trip's critical point, two criteria may be established for early and late departures. What follows is a formal derivation of the early-departure criterion, while a similar derivation for the late-departure criterion is presented as an exercise at the end of the chapter.

7.7.1 Early departure criterion

According to the foregoing definitions,

$$L_i\left(t_{1j}^-\right) = L_i\left(t_{1j}^*\right) - \Delta^{j\tau(-)}SL\left(t_{1j}^*\right) \tag{7.6}$$

where
$SL_i(t_{1j}^*) = SL_i(t_{1j}^-)$
$i = 1, 2, ..., n$
t_{1j}^- belongs to τ

In case the slope is changed within a $\Delta^{j\tau(-)}$ shift for any stop i, Equation 7.6 should consider two (or more) decreased portions, each related to a different slope and its associated part of $\Delta^{j\tau(-)}$.

The loads at t_{1j}^- can be expressed as

$$L_i\left(t_{1j}^-\right) - L_i\left(t_{1,j-1}^*\right) < d_\tau, \quad i = 1, 2, ..., n, \text{ belongs to } \tau \tag{7.7}$$

These loads for the new t_{1j}^- departure across all stops are based on Principle 3 (Section 5.2) in which the desired occupancy d_τ is attained for t_{1j}^* only at q and the load is less than d_τ at all other stops. Using a $\Delta^{j\tau(-)}$ shift will further reduce these loads. However, the loads at each stop i for the departure times adjacent to t_{1j}^- at $t_{1,j+1}^*$ will increase the loads if the shifting takes place. This increase is $\Delta^{j\tau(-)}SL_i(t_{1j}^*)$ for all i and relevant τ, or it is the sum of portions of the slope that are changed within $\Delta^{j\tau(-)}$.

The increased new loads at $t_{1,j+1}^*$ need to be checked against $d_\tau + \Delta d_\tau$ across all stops. This check is applied for the maximum load increase, and hence the early-departure criterion for accepting $\Delta^{j\tau(-)}$ is as follows:

$$\max_{i=1,2,\ldots,n}\left[L_i\left(t_{1,j+1}^*\right)-L_i\left(t_{1j}^-\right)\right]\le d_\tau + \Delta d_\tau \tag{7.8}$$

where
  $t_{1,j+1}^*$ belongs to interval τ
  $L_i(t_{1j}^-)$ is obtained by Equation 7.6

Figure 7.15 illustrates a simple example of Principle 3 in Section 5.2. Given a transit line A → B → C, with average travel times of 15 min between A and B; three departures, at 6:15, 6:45, and 7:10; and a desired occupancy of 50 passengers. The average observed onboard loads on the 6:15 vehicle are 30 passengers at stop A and 65 at stop B; on the 6:45 vehicle, 80 and 35 passengers at A and B, respectively; and on the 7:10 vehicle, 25 and 80 passengers at A and B, respectively. Figure 7.15 shows $L_A(t)$ and $L_B(t - 15)$ as the cumulative load curves of the three vehicles, in which the curve at B is shifted by 15 min to allow for an equal time basis (at the route's departure point) in the analysis. The procedure explicated in Principle 5.3 in Section 5.2 results in individual even-load departures at 6:11, 6:31, and 6:56. This procedure is indicated by the lines with arrows in Figure 7.15.

Figure 7.16 presents the example of Figure 7.15 with a $\Delta^{j\tau(-)}$ shift in part (a) of the figure and a $\Delta^{j\tau(+)}$ shift in part (b), utilizing $\Delta d_\tau = 10$ passengers for both parts. In part (a), the shifting is $\Delta^{j\tau(-)} = 3$ min. The solid lines (with arrows) show graphically how to determine the new loads on the (new) 6:28 and 6:56 departures at both stops A and B. The slope at A between 6:28 and 6:31 is 8/3 and times 3 min; this results in an average load of eight more passengers on the 6:56 departure and eight fewer on the 6:28 departure. At stop B, the slope is 7/6, and the change in load is 3.5 passengers. The criterion in Equation 7.8 considers the load at A with 120 − 72 + 8 = 56 and at B with 134 − 84 + 3.5 = 53.5 passengers for the 6:56 departure. The maximum of the two is 56; because $\Delta d_\tau = 10$ (the average load can reach 60 passengers), the value of $\Delta^{j\tau(-)} = 3$ min is accepted.

Part (b) of Figure 7.16 presents an example of $\Delta^{j\tau(+)} = 3$ min for the 6:31 departure. The solid lines (with arrows) show graphically how to determine the new loads on the (new) 6:34 and 6:56 departures at both A and B. The slope at A is the same as for part (a) of Figure 7.16 and results in a difference of 8 passengers, while the outcome of the slope at B is 3.5 passengers. The increased load at A is 50 + 8 = 58, and at B is 84 + 3.5 − 50 = 37.5 passengers for the new 6:34 departure. Consequently, and similar to the early-departure case, the shift of $\Delta^{j\tau(-)} = 3$ min is accepted.

This section provided a procedure to integrate two operational planning components (timetabling and vehicle scheduling) based on the DF model and given tolerances. The outcome may be a set of efficient schedules from both the passenger and operator perspectives.

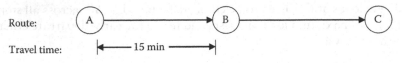

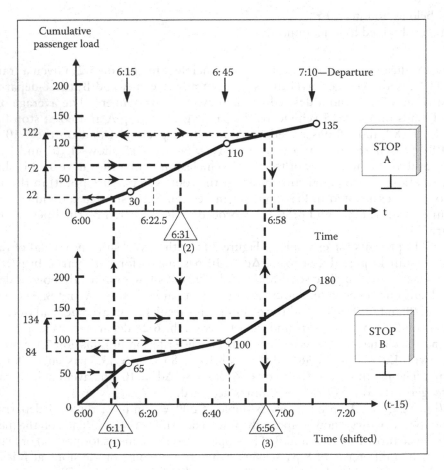

Figure 7.15 Example using the even max load procedure on individual vehicles (see Section 5.2).

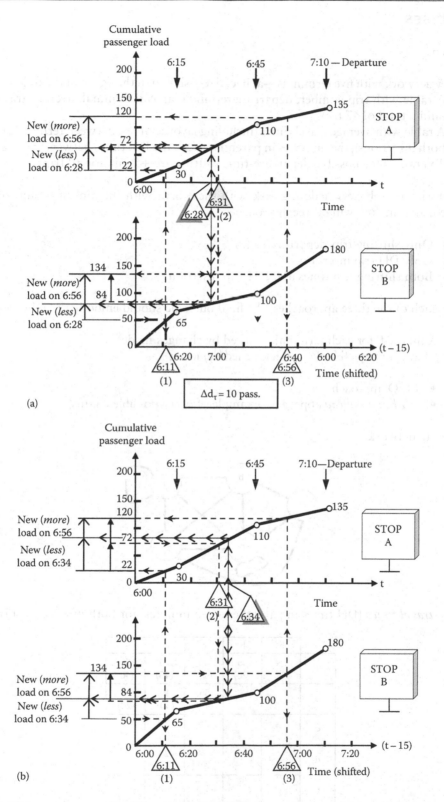

Figure 7.16 Effect on even max load when shifting the 6:31 departure backward (a) and forward (b) by 3 min.

EXERCISES

7.1 Given:

- A network with five terminals—a, b, c, d, e—shown in the figure of the following text.
- A table with trip number, departure terminal, arrival terminal, average travel times (and a total of 47 trips).
- A table with average travel times (including layover times) between terminals (same in both directions); the numbers in parentheses are the DH times between each terminal.
- Tolerances for possible departure-time shifting are ±5 min for all trips.

Perform the vehicle-scheduling task with these data with the aim of minimizing fleet size using the following three approaches:

(i)   Only shifting trip-departure times
(ii) Only DH trip insertion
(iii) Both shifting trip times and DH trip insertion

For each of the three approaches, the final outcome should embody the following:

(1) A final DF for each terminal, marked by changes made
(2) Chains of trips (blocks, vehicle schedule) using

- FIFO approach
- *Within a hollow* approach (example of other possible chains)

Basic route network

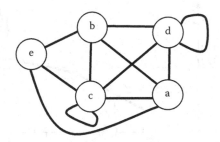

Average travel times (DH times in parentheses), in minutes, for both directions of travel

	a	b	c	d	e
a		35 (20)	60 (40)	70 (45)	50 (40)
b			55 (40)	40 (30)	25 (20)
c				25	35 (25)	30 (25)
d					20	
e					

Trip schedules

Trip #	Departure terminal	Arrival terminal	Departure time	Trip #	Departure terminal	Arrival terminal	Departure time
1	d	d	6:30	25	b	c	7:30
2	b	a	6:30	26	c	c	7:30
3	a	d	6:30	27	d	a	7:35
4	e	a	6:30	28	e	b	7:35
5	c	c	6:30	29	b	d	7:45
6	b	a	6:45	30	b	a	7:45
7	d	c	6:45	31	a	c	7:45
8	e	a	6:50	32	a	e	7:50
9	b	a	6:55	33	d	a	7:55
10	d	d	7:00	34	d	b	8:00
11	a	d	7:00	35	a	b	8:00
12	b	a	7:00	36	c	a	8:00
13	b	d	7:00	37	c	d	8:00
14	a	b	7:00	38	a	d	8:00
15	c	a	7:00	39	c	b	8:05
16	c	b	7:05	40	e	a	8:10
17	a	c	7:10	41	b	a	8:15
18	c	e	7:25	42	b	a	8:30
19	e	c	7:25	43	b	e	8:30
20	d	d	7:30	44	c	e	8:30
21	b	e	7:30	45	e	c	8:30
22	b	a	7:30	46	e	b	8:30
23	a	b	7:30	47	a	b	8:35
24	e	a	7:30				

7.2 Given the following is their trip schedule (where each route has the same schedule), coordinated with the train schedule:

	Shuttle route 1			Shuttle route 1	
Trip #	Departure time	Arrival time	Trip #	Departure time	Arrival time
1	6:30	7:05	6	7.55	8.15
2	6:45	7:25	7	6:45	7:35
3	7:00	7:28	8	7:00	7:50
4	7:20	7:50	9	7:20	8:00
5	7:40	8:05	10	7:50	8:15

(1) Find the minimum fleet size required to execute the entire trip schedule, using (1) the modeling of single routes and (2) the DF model.

(2) Repeat (1) with shifting possibilities in trip-departure times, in which for all trips, the forward shifting tolerance ($\Delta^{i(+)}$) is 3 min and the backward shifting tolerance ($\Delta^{i(-)}$) is 6 min.

(3) Construct the DF after the shifting; can the shifted schedule have an adverse effect on the Shuttle 2 route operation?

(4) What is the improved lower bound of the trip schedule? Is it needed?

(5) Apply the FIFO rule to create vehicle chains (blocks), using the shifted schedule determined in (2).

7.3 Find the three levels of lower bound—G(S), G'(S'), and G"(S")—for the 10-trip schedule, given as follows:

Trip #	Departure terminal	Departure time	Arrival terminal	Arrival time	Deadheading (DH) trips Between terminals	DH time (same for both directions) (min)
1	b	6:00	c	6:30	a and b	60
2	a	6:10	a	6:50		
3	d	6:10	b	7:10	a and c	30
4	b	7:00	b	7:30		
5	c	7:10	b	7:30	a and d	30
6	c	7:40	b	8:10		
7	d	7:50	a	8:30	b and c	10
8	a	8:00	a	8:30		
9	a	8:30	d	9:10	b and d	50
10	b	9:00	c	9:20	c and d	20

7.4 For the 10-trip schedule in Exercise 7.3, consider SDTs by 5 min for both forward and backward shifts ($\Delta^{i(+)} = \Delta^{i(-)} = 5$) for all trips.

(1) Find $G'(\bar{S}'_{sf})$ and $G''(S''_{sf})$ by constructing $g(t, \bar{S}'_{sf})$ and $g''(t, \bar{S}''_{sf})$.

(2) Find $G(S_{sf})$, $G'(S'_{sf})$, $G''(S''_{sf})$ using the LB-SHIFT and LB-DH&SHIFT procedures.

7.5 Derive a load-tolerance criterion for even-load timetables for the case of a late departure. End the derivation with a formulation similar to Equation 7.8.

APPENDIX 7.A: EXAMPLE OF VEHICLE-SCHEDULING SOFTWARE

There is no doubt more and more planning/scheduling software packages will have to include a real-time component for adjustments and changes. Table 1.1 lists some of the known software available with their Internet address. The software, called Optibus, specializes in *real-time rescheduling* and thus will be briefly described, in this appendix, as an example of such software.

Optibus is an algorithm-based Software as a Service (SaaS) platform optimizing scheduling of vehicles and crew in real time. Figure 7.A.1 of this system exposes a web interface in which the scheduler can generate optimized schedules while taking into account the constraints and preferences. Using this interactive interface, the scheduler can interact and incorporate his or her expertise in the generated schedule while receiving recommendations and feedback from the optimization engine in an interactive manner. This interactive system handles simultaneously both vehicle and crew scheduling.

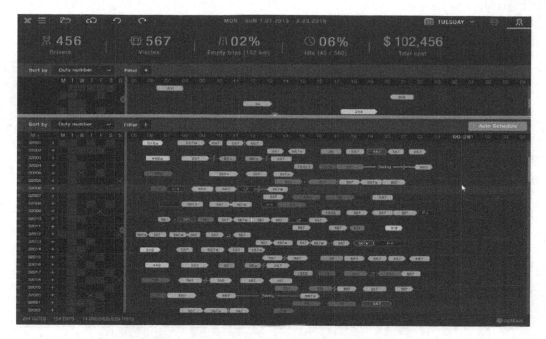

Figure 7.A.1 Interactive vehicle-scheduling optimization process.

7.A.1 Key features

The key features of the Optibus system are related to (1) fast generation of optimized vehicle and crew schedule while taking into consideration more than a few constraints and preferences, such as the following:

- Multiple vehicle types and their hierarchy
- Multiple depots, with different vehicle capacities, with the option to create schedules that return to the starting depot
- Crew-work regulation and labor constraints
- Similarity with existing schedules and between days of week

(2) Optimal allocation of vehicle parking and service depots to minimize costs; (3) an interactive Gantt-based interface (see Figure 7.A.1) visualizing the schedule and allows for scheduler's by adjustments; (4) real-time rescheduling integrated with automatic vehicle location (AVL) and automatic passenger counter (APC) systems to allow for anticipation of unexpected deviation from the original schedule and changes of passengers demand performed through suggestions given with the consideration of regulations, constraints, preferences, and minimizing costs; (5) generating different detailed reports for analysis, planning, and decisions to take place (see, for instance, the illustration in Figure 7.A.2 of the number of vehicles required by the optimal solution [of a given depot] every hour of the day with the upper and lower curves representing urban and interurban fleet requirements, respectively); and (6) use of computer-based components such as RESTful API and reusable HTTP for integrating/embedding the planning with other management needs such as a control center.

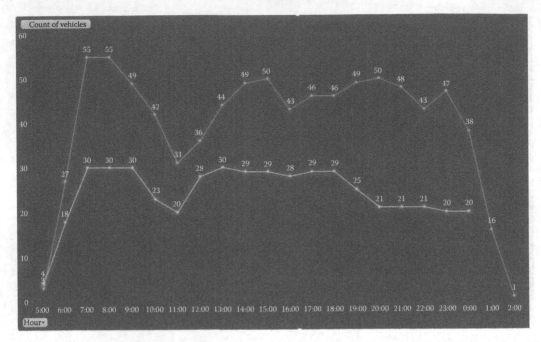

Figure 7.A.2 Example of a generated report of optimal urban (upper curve) and interurban (lower curve) vehicle requirements, by time of day.

7.A.2 Expected saving

Two experiments, done by the Optibus software, provide a rough idea of the saving expected. The comparison, shown in the succeeding text, in the first table is based on 975,713 yearly service trips with an accumulated distance of 19,865,410 km.

Source	Idle kilometers (thousands)	Idle trip hours (thousands)	Standby hours (thousands)
Available solution	3422	94	239
Optibus outcome	2,678 (i.e., −22%)	78 (i.e., −17%)	186 (i.e., −22%)

The second table shows an experiment done at a single operational branch with a different set of schedules comprised of 1,100 daily service trips.

Source	Single day idle kilometers	Single day nonservice paid hours (idle + standby)
Existing figures	3056	294
Optibus outcome	2545 (i.e., −17%)	240 (i.e., −18%)

REFERENCES

Ceder, A. (2002). A step function for improving transit operations planning using fixed and variable scheduling. In *Transportation and Traffic Theory* (M.A.P. Taylor, ed.), pp. 1–21, Elsevier Science, Oxford, U.K.
Ceder, A. (2005). Estimation of fleet size for variable bus schedules. *Transportation Research Record*, **1903**, 3–10.

Ceder, A. and Philibert, L. (2014). Transit timetables resulting in even max load on individual vehicles. *IEEE Transactions on Intelligent Transportation Systems*, 15(6), 2605–2614.

Ceder, A. and Stern, H.I. (1981). Deficit function bus scheduling with deadheading trip insertion for fleet size reduction. *Transportation Science*, 15, 338–363.

Stern, H.I. and Ceder, A. (1983). An improved lower bound to the minimum fleet size problem. *Transportation Science*, 17, 471–477.

Chapter 8

Vehicle type and size considerations in vehicle scheduling

The real competition is competing with yourself

A.C.

CHAPTER OUTLINE

8.1 Introduction
8.2 Overview and optimization framework
8.3 DF-based procedure for vehicle-type-scheduling problem
8.4 Examples of the DF-based VTSP
8.5 Min-cost flow procedure for multi-types vehicle-scheduling problem
8.6 Vehicle size determination
8.7 Optimal transit vehicle size: Literature review
Exercises
References

PRACTITIONER'S CORNER

This chapter contains three main, almost independent parts. First, it addresses the problem of how to allocate vehicles efficiently for carrying out all the trips in a given transit timetable, while taking into account the association between the characteristics of each trip (urban, peripheral, intercity, etc.), and its required vehicle type. Second, a methodology is introduced also for assigning vehicles of different types, but with the consideration of optimal timetables described in Section 5.4. Third, it provides an overview and tools for the determination of and a practical decision concerning transit vehicle size. Both parts are directly related to the prominent act of purchasing a vehicle or a fleet of vehicles.

Ostensibly when dealing with vehicle type there is need for more operational details. Paying attention to details is similar to answering the following riddle. (Do so without using a calculator.) You are driving a bus from terminal *a* to terminal *b*. At the first stop, 10 people get on the bus and 2 get off. At the second stop, 12 get on and 7 get off. At the third stop, 5 people get on and 10 get off. At the fourth stop, 11 people get off and 1 gets on. You then arrive at terminal *b*. What is the name of the driver? (See end of this corner.)

The chapter consists of six sections, following the introduction. Section 8.2 provides an overview of literature review and describes optimization framework for the vehicle-scheduling problem, in which categories of vehicle types are arranged in decreasing order of vehicle cost and comfort level. The assignment of vehicles to trips using DH (deadheading) and shifting procedures within the optimization framework is discussed in Section 8.3; the objective is to minimize the total purchasing cost. Section 8.4 presents two detailed examples (one of which is real life) of the optimization procedures developed. Section 8.5 introduces a method enabling, in assigning vehicles, to stipulate the use of a particular vehicle type for a trip or to allow for a substitution either by a larger vehicle or by a combination of smaller vehicles with the same or higher total capacity; it is also linked with the selection of optimal timetable. Section 8.6 provides basic tools for the determination of vehicle size, including a known square-root formula. Section 8.7 continues, more in depth, with the overview of the tools and formulation for vehicle-size determination. The chapter ends with exercises.

There is no doubt that the successful deployment of optimal or adequate vehicle types requires a considerable amount of analysis. After all, success comes before work only in the dictionary. Finally, as to the riddle, it said: "*You are* driving a bus ...."

8.1 INTRODUCTION

The vehicle-scheduling task described in Chapters 6 and 7 considers only one type of transit vehicle. In practice, however, more than one type is used; for example, a bus operation may employ minibuses, articulated and double-decker buses, and standard buses with varying degrees of comfort and different numbers of seats. Commonly, the consideration of vehicle type in transit operations planning involves two considerations: first, determining the suitable or optimal vehicle size; second, choosing vehicles with different comfort levels, depending on trip characteristics. Certainly, a multicriteria effort may treat both considerations simultaneously, but this is seldom done in practice. The issue of what vehicle type to consider arises when purchasing a vehicle or a fleet of vehicles, an undertaking that is not performed frequently. There is a saying that good judgment comes from experience, and experience comes from bad judgment. Another is that experience is not what happening to us, but what we do with what is happening to us. In our case, the experience accumulated focuses on the need for an analysis framework in order to decide how many and which type of vehicle to purchase. This chapter attempts to introduce (a) a cost-effective framework for choosing a vehicle's comfort level, (b) a method for optimal vehicle assignment integrated with the selection of optimal timetable, and (c) the basic tools for analyzing optimal vehicle size.

The purpose of the first part of the chapter is to address the vehicle-scheduling problem, while taking into account the association between the characteristics of each trip (urban, peripheral, inter-city, etc.) and the vehicle type required for the particular trip. This means complying with a certain level of service for that trip: degree of comfort, seat availability, and other operational features. The purpose of the second part is to introduce a methodology for optimal vehicle scheduling using multiple vehicle types. The methodology is based on a minimum-cost network flow model with an input of a set of Pareto-optimal timetables (see Section 5.4) for individual transit lines. The purpose of the third part of the chapter (Sections 8.6 and 8.7) is to review a formulation for trade-offs between vehicle size and operational variables that can be used as a tool for the determination of optimal vehicle size for a given frequency of service.

8.2 OVERVIEW AND OPTIMIZATION FRAMEWORK

Section 8.7 is the main literature review part of the chapter, but it focuses only on vehicle size and not on vehicle scheduling with different sizes. Thus, in what follows is a brief overview of optimization-related literature of what is known as multiple vehicle types vehicle scheduling problem (MVT-VSP).

Bunte and Kliewer (2009) provide a detailed literature review and discuss the modeling approaches for different kinds of vehicle-scheduling problems. In the MVT-VSP, vehicle types have different capacities and might have different fixed and variable cost. According to Lenstra and Kan (1981), the problem is NP-complete (see Section 5.5) for the single depot case. There have been numerous approaches published, for example, El-Azm (1985), all of which consider multiple vehicle types. A common approach by Bodin et al. (1983), used also by Costa et al. (1995), includes the use of a multigraph with a subnetwork for each vehicle type. However, if there are further restrictions such as certain trips requiring a certain vehicle type, the model has to be further extended. This was considered by Forbes et al. (1994), Löbel (1997), and Ceder (2011). It was also included in a time–space network approach by Kliewer et al. (2006). Finally, Kliewer et al. (2012) consider a multiple depot case for the simultaneous vehicle and crew scheduling with time windows. A time–space network formulation is used and shifting of trips is explicitly modeled using additional arcs in the network.

This and the two following sections use the notation and procedures of deficit function (DF) described in Chapters 6 and 7. The problem addressed concerns the assignment of vehicles to trips using DH and shifting procedures, in which categories of types are arranged in decreasing order of vehicle cost and comfort. The process described follows Ceder (1995, 2011). Section 8.5 provides another approach, not DF based, that uses the optimal timetable developed in Section 5.4.

The problem, entitled the vehicle-type-scheduling problem (VTSP), is based on given set S of trips (schedule) and set M of vehicle types. The set M is arranged in decreasing order of vehicle cost so that if $u \in M$ is listed above $v \in M$, it means that $c_u > c_v$, where c_u, c_v are the costs involved in employing vehicle types u and v, respectively. Each trip $i \in S$ can be carried out by vehicle type $u \in M$ or by other types listed prior to u in the aforementioned order of M.

The problem can be formulized as a cost-flow network problem, in which each trip is a node and an arc connects two trips if, and only if, it is possible to link them in a time sequence with and without DH connections. On each arc (i, j), there is a capacity of one unit and an assigned cost C_{ij}. If the cost of the lower-level vehicle type associated with trip i is higher than the cost of the vehicle type (even if of a lower level) required for trip j, then $C_{ij} = c_i$. That is, $C_{ij} = \max(c_i, c_j)$. The use of such a formulation was implemented by Costa et al. (1995), who employed three categories of solutions: (a) a multicommodity network flow; (b) a single-depot vehicle-scheduling problem; and (c) a set-partitioning problem with side constraints. The mixed-integer programming of these problems is known to be NP-complete (see Section 5.5) as may be seen, for example, in Bertossi et al. (1987). The math-formulation concepts for the third category are further explained in Chapter 9.

Because of the complexity involved in reaching an optimal solution for a large number of trips in S, a heuristic method is considered a more practical approach. The heuristic procedure developed is called the VTSP algorithm. It begins by establishing lower and upper bounds on fleet size. The upper bound is attained by creating different DFs, each associated with a certain vehicle type $u \in M$, which includes only the trips whose lower-level required vehicle type is u. Certainly, this scheduling solution reflects a high cost, caused by the large number of vehicles demanded. The lower bound on the fleet size is attained by using only one vehicle type: the most luxurious one with the highest cost that can clearly carry out any trip in the timetable. Between these bounds on fleet size, the procedure searches for the best solution, based on the properties and characteristics of DF theory.

This optimization framework is illustrated in Figure 8.1, with (C_1, N_1) and (C_2, N_2) representing the lower- and upper-boundary solutions, respectively. The following are added notations to DF theory (described in Chapters 6 and 7):

M is the set of all vehicle types u, u = 1, 2, ..., m, arranged in decreasing order of vehicle cost
S_u is the set of required trips (schedule) for vehicle types u, u = 1, 2, ..., m
$S = \cup_{u=1}^{m} S_u$, which is the combined schedule of all trips
t_{su}^i is the start time of trip $i \in S_u$ of type $u \in M$
t_{eu}^i is the end time of trip $i \in S_u$ of type $u \in M$
$\Delta_u^{i(+)}$ is the maximum delay tolerance from the scheduled departure time of trip $i \in S$ of type $u \in M$
$\Delta_u^{i(-)}$ is the maximum advance tolerance of the trip scheduled departure time of trip $i \in S$ of type $u \in M$

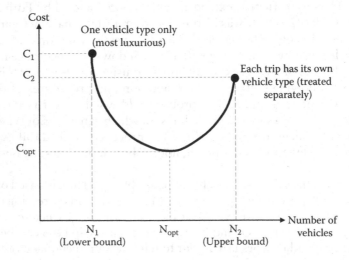

Figure 8.1 Trade-off between cost of purchasing vehicles and number of vehicles.

$d_u(k, t, S_u)$, $D(k, S_u)$, $D(S_u)$, $[s_{iu}^k, e_{iu}^k]$, $g(t, S_u)$, $G(S_u)$ = refer to the DF definitions (see Chapters 6 and 7), but only with respect to trips of type $u \in M$

c_u is the cost involved in employing a vehicle of type $u \in M$

N_1 is the minimum number of vehicles required to service all trips in S, using a single vehicle of type $u = 1$

n_{2u} is the minimum number of vehicles required to service all trips in S_u, $u \in M$

$N_2 = \sum_{u=1}^m n_{2u}$ = sum of all minimum numbers of vehicles required when treating each type separately

$C_1 = c_1 N_1$ is the total cost involved in employing N_1 vehicles of type 1 (most luxurious)

$C_2 = \sum_{u=1}^m c_u n_{2u}$ is the total cost involved in employing N_2 vehicles (for each type separately)

$C = \sum_{u=1}^m c_u n_u$ is the objective function (total cost) following algorithm VTSP, with n_u vehicles required of type u for all $u \in M$

$N = \sum_{u=1}^m n_u$ is the total number of vehicles incurring cost C

8.3 DF-BASED PROCEDURE FOR VEHICLE-TYPE-SCHEDULING PROBLEM

The algorithm VTSP developed is heuristic in nature while incorporating all DF components. Because of the graphical features associated with DF theory, the algorithm can be applied in an interactive manner or in an automatic mode, along with the possibility of examining its intermediate steps. The following is a general description of algorithm VTSP in a stepwise manner.

8.3.1 Algorithm VTSP

Step 0: Arrange the set of vehicle types M in decreasing order of vehicle cost (so that if $u \in M$ is listed above $v \in M$, it means that $c_u \geq c_v$).

Step 1: Solve the problem as a single-vehicle-type problem using DF theory, including the DH and shifting procedures (see Figures 6.9a and b and 7.12), to obtain N_1 vehicles, considered as type 1, with a total cost of C_1.

Step 2: Divide the trips by their associated type and apply the DF methodology with the DH and shifting procedures for each type separately. Add up the number of vehicles derived to obtain the total of N_2 vehicles with a total cost C_2.

Step 3: If $N_1 = N_2$, stop. Use the solution in *Step 2*.

Step 4: Consider $d_u(k, t)$ as in *Step 2* for all $k \in K$ and $u \in M$.

Step 5: Perform the URSC (shifting only) procedure for shifting departure times within their tolerances (see Figure 8.11).

Step 6: Find a URDHC (see Figures 6.9a and b and 7.12), such that each DH trip (with possible shifting) in this chain fulfills condition (a) and/or (b):

(a) The DH trip is from DF of vehicle type u to DF of type v, such that $u \leq v$, meaning that $c_u \geq c_v$ (see Proposition 8.2).

(b) The URDHC aims at saving a vehicle of type w and $-c_w + \sum_{\forall q, r \in E}(c_q - c_r) \leq 0$, in which the set E is composed of all DH trips included in the URDHC; each DH trip is from a DF of vehicle type r to a DF of vehicle type q, in which $q < r$ (see Proposition 8.3). If no URDHC can be found, stop.

Step 7: Examine whether the total cost of the URDHC (DH cost) is less than the cost of saving one vehicle (of the type considered). If it is not, delete this possibility and go to *Step 6*. Otherwise, update $d_u(k, t)$ for all $k \in K$ and $u \in M$.

Step 8: Apply the improved lower-bound check. If $D(S) = G''(S''_{sf})$, stop; otherwise, go to *Step 5*.

Among the eight steps of algorithm VTSP, the conditions specified in *Step 3* and particularly *Step 6* deserve further attention. The following four propositions clarify and interpret these conditions:

Proposition 8.1 (for *Step 3*): If $N_1 = N_2$, then $C_2 \leq C_1$.

Proof: Given $M \geq 2$ and $c_1 \geq c_2 \geq \ldots \geq c_m$, the proof is straightforward because

$$C_1 = N_1 c_1, \quad C_2 = \sum_{j=1}^{m} N_{2j} c_j, \quad N_2 = \sum_{j=1}^{m} N_{2j}, \quad \text{and} \quad N_1 = N_2$$

Proposition 8.2 (for *Step 6a*): Any DH trip connection from a DF of vehicle type u to a DF of type v, such that $u \leq v$ within any URDHC, does not increase C.

Proof: Any DH trip from u to v will link, in the final vehicle chain (block), an arrival epoch at u and a departure epoch at v, including possible idle times at u and v. Because $c_u \geq c_v$, this DH trip connection cannot lead to an upgrade of the vehicle type; therefore, it cannot increase the objective function C.

Proposition 8.3 (for *Step 6b*): Any URDHC that aims at saving a vehicle of type w does not increase C if $-c_w + \sum_{\text{for all } q,r \in E}(c_q - c_r) \leq 0$, where the set E is composed of all the DH trips included in the URDHC; each DH trip is from a DF of vehicle type r to a DF of type q in which $q < r$.

Proof: In the URDHC, there may be DH trips that comply with Propositions 8.1 and 8.2 and, therefore, do not increase C. Each of the other DH trips is from a DF of vehicle type r to a DF of type q, in which $q < r$, that is, $c_q \geq c_r$. This DH trip connection may upgrade the vehicle type (carrying out this DH trip) from r to q. That is, the result may be a saving of a vehicle of type w, along with several vehicle-type upgrades. In order to ensure that C does not increase, the condition set forth is that the cost saving of vehicle type w is greater than or equal to the additional cost required by the possible upgrades.

Algorithm VTSP is shown in Figure 8.2 in two parts. It starts in Figure 8.2a by calculating N_1 and C_1, and then N_2 and C_2. If $N_1 = N_2$, we stop with $N = N_2$ and $C = C_2$, indicating that each trip is carried out by its own designated type. Otherwise, the algorithm continues by using Figure 8.2b, in which *Step 6*, with its two conditions, is utilized. The algorithm ends once all terminals are examined.

8.3.2 Sensitivity analysis

The VTSP presents a vehicle-scheduling problem in which the objective is to minimize the cost involved; primarily, it aims at vehicle-purchase cost although indirectly it also affects operational cost. Often, we want to know the result of the solution when changes occur in vehicle cost. Alternatively, we may want to know the extent to which c_u for all $u \in M$ can change and still have the same scheduling solution. Certainly, changes in vehicle cost by type can shift the location of $u \in M$ in decreasing order of vehicle cost. The examination of the solution for these cost-change possibilities is usually called sensitivity analysis.

The first observation in algorithm VTSP is that as long as the arrangement in M of the decreasing order of vehicle cost is preserved, the scheduling solution will not change. That is, any change of c_u—it can be of more than one vehicle type—that will not change the order of vehicle cost will result (except in special cases) in a change of C, but not in the blocks attained. The proof of this observation is straightforward because of unchanged situations applied to conditions (a) and (b) in *Step 6* in algorithm VTSP.

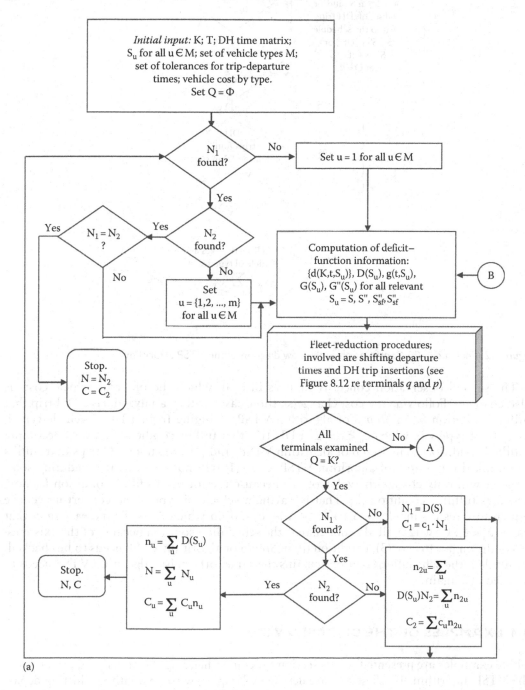

(a)

Figure 8.2 (a) Flow diagram of the VTSP algorithm.   (*Continued*)

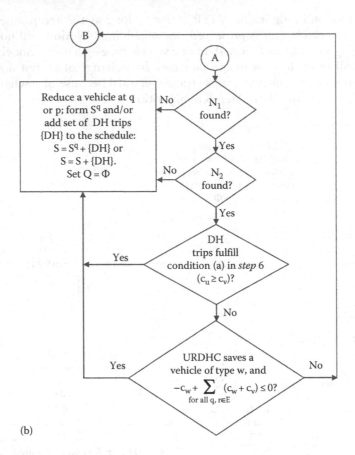

(b)

Figure 8.2 (Continued) (b) Continuation of the flow diagram of the VTSP algorithm.

The second observation concerns changes in c_u in which the order of vehicle costs is also changed. Following the cost change(s), three cases will be analyzed: (1) a DH trip that fulfills condition (a) in *Step 6* of algorithm VTSP, changing from a DF of vehicle type u to a DF of type v in which $c_u < c_v$; (2) a DH trip that fulfills condition (b), thus becoming fulfilled condition (a) in *Step 6* of algorithm VTSP; and (3) a chain of DH trips that fulfills condition (b) in *Step 6* of algorithm VTSP. Case (1) will not change the scheduling solution—it will only change the value of C—because the former fulfilled condition (a) now becomes fulfilled condition (b). That is, saving a vehicle of type v and converting vehicle type u to v results in—$(c_v + \Delta_v) + (c_v + \Delta_v) - c_u < 0$, in which Δ_v is the increase in c_v that yields $c_u < c_v$. Case (2) will also not change the scheduling solution because of the existence of condition (a). In case (3), condition (b) in *Step 6* of algorithm VTSP needs to be checked. If fulfilled, the scheduling solution remains the same; otherwise, algorithm VTSP needs to be executed again.

8.4 EXAMPLES OF THE DF-BASED VTSP

Three examples are presented in this section for comprehending the underlying principles of the VTSP algorithm. The first two are detailed examples, with and without shifting departure times; the third conveys the results of a real-life example.

8.4.1 Detailed Example 1

Example 1, which is illustrated in Figure 8.3, consists of 8 trips, 3 terminals (a, b, c), and 3 types of vehicles, with the cost of 12, 5, and 3 cost units, respectively. Figure 8.3a presents the simple network of the routes, in which the DH travel time between each two terminals is 20 min and shifts in departure times are not allowed. The timetable and trip travel times are shown in Figure 8.3b according to vehicle type. The DFs of *Step 1* of algorithm VTSP for Example 1 are depicted in Figure 8.3c; all trips, it should be recalled, are served by the same vehicle type (type 1). For inserting a DH trip, the NT rule (the first hollow is the longest) is applied; this results in the selection of terminal b (for the DH insertion procedure, see Section 6.5.4). The URDHC procedure with R = 2 (furthest start of a hollow) then results in three DH trips, in which DH_2 is used for maintaining the level of D(a). Figure 8.4 constructs g'(t, S') = g''(t, S''), which determines G'(S') = G''(S'') = 2. Thus, *Step 1* stops when D(S) = N_1 = 2. Two vehicle chains are then created, using the FIFO [1-3-DH_1-5-7] and [2-4-DH_2-6-DH_3-8], and the total cost is C_1 = 24.

Algorithm VTSP continues in *Step 2*, in which vehicle types are treated separately. Figure 8.5, which illustrates this step, is marked by (d) to show that this step follows Figure 8.3c. The maximum DF of types 1 and 2 are reduced by one, using DH_1 and DH_2, respectively;

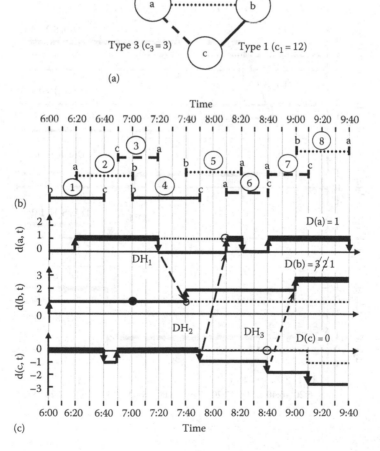

(a)

(b)

(c)

Figure 8.3 Example 1: (a) network of routes, vehicle types, and cost; (b) eight-trip schedule; (c) DFs with DH insertions for a single vehicle.

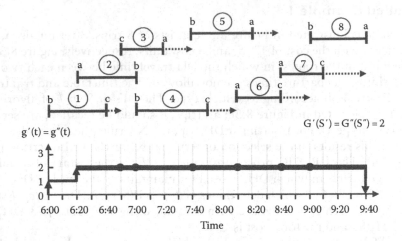

Figure 8.4 Determination of the lower bound corresponding to Step 1 of algorithm VTSP for Example 1.

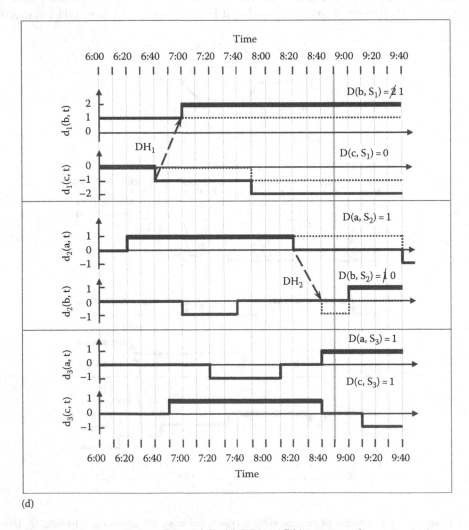

(d)

Figure 8.5 Example 1 (continued from Figure 8.3): (d) DFs and DH insertions for each vehicle type.

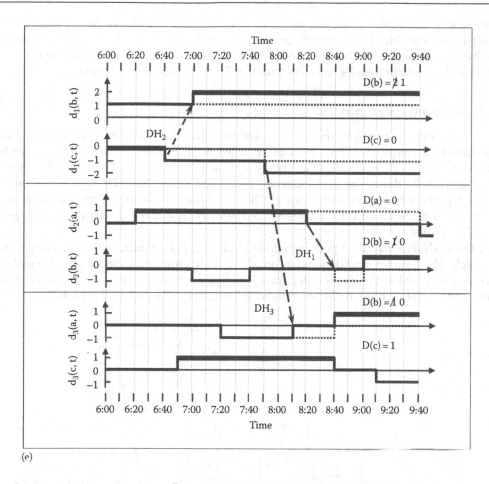

(e)

Figure 8.6 Example I (continued from Figures 8.3 and 8.5): (e) DFs and DH insertions for optimal solution.

the number of type 3 vehicles remains same. Thus, $N_2 = 1 + 1 + 2 = 4$, and the four following chains are derived by using the FIFO rule: [1-DH_1-4] (vehicle type 1), [2-5-DH_2-8] (vehicle type 2), [3–7], and [6] (vehicles of type 3); this results in a total cost of $C_2 = 23$.

The next step in algorithm VTSP compares $N_1 = 2$ with $N_2 = 4$, and then moves to the relevant *Step 6* because of not allowing shifting departure times. Figure 8.6 illustrates the process of *Step 6*, using its two conditions. Note that S_u will be deleted (as in Figure 8.6) when it is clear which underlying vehicle type is being considered. This step again applies the NT rule and the URDHC procedure with R = 2 (furthest start of a hollow), but this time (especially condition (a)) with the possibility of inserting any DH trip from a DF with a more expensive vehicle type to a DF with a less expensive type. Following the flow diagram in Figure 8.2, the first terminal selected is b, based on $d_2(b, t)$, from which DH_1 is determined from terminal a. The DFs are then updated and the next terminal is again b, but related to $d_1(b, t)$; DH_2 is inserted from terminal c. We continue with the next selected terminal c, based on $d_3(c, t)$; however, no DH trip can be inserted into its maximum-interval starting point, including the check for both conditions (a) and (b) of *Step 6*. Thus, a is selected next, based on $d_3(a, t)$, and DH_3 is inserted to arrive from c, based on the updated $d_1(c, t)$. This terminates *Step 6* and results in the three following (FIFO) chains: [1-DH_2-4-DH_3-6] (vehicle type 1), [2-5-DH_1-8] (vehicle type 2), and [3–7] (vehicle type 3), with a total cost of $12 + 5 + 3 = 20$.

For the sake of simplicity, we assume that a check of *Step* 7 of algorithm VTSP allows for the latter solution. Moving to *Step* 8, one may then see that D(S) > G''(S''), but the process stops by reiterating *Step* 6, because no URDHC can be found. At this stage, we may conclude that the 3-chain (blocks) solution with C = 20 is the best one attained.

8.4.2 Detailed Example 2

The example demonstrated in this section exploits shifts in departure times and condition (b) in *Step* 6 of algorithm VTSP. Example 2, illustrated in parts (a) and (b) in Figure 8.7, consists of 14 trips, 3 terminals, and 3 types of vehicles, with costs of 10, 3, and 2, respectively. Figure 8.7c shows the fleet-reduction procedure, using both DH trip insertion and shifting departure times. For simplicity, all DH travel times between the three terminals are 15 min, $\Delta_u^{i(+)} = \Delta_u^{i(-)} = 15$ min, for all $u \in M$, and priority in shifting departure times is given to late departures.

The DF analysis with only type 1, in Figure 8.7c, results in one shift, enabling a reduction of one vehicle; the lower-bound determination, shown in Figure 8.8, includes possible shifts of trip times combined with feasible extensions. The first analysis terminates with $N_1 = 6$ (lower bound) and $C_1 = 60$. The analysis proceeds with finding $N_2 = 7$ and $C_2 = 2 \cdot 10 + 4 \cdot 8 + 1 \cdot 6 = 58$,

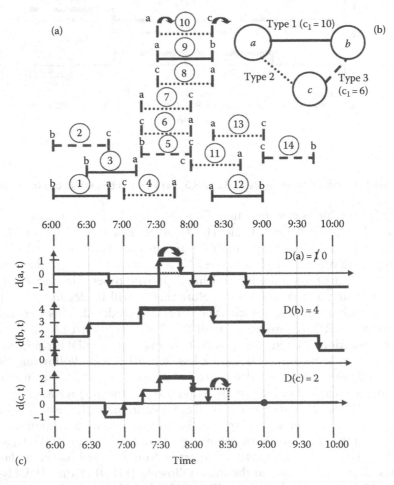

Figure 8.7 Example 2: (a) Basic input of a network of routes, vehicle types, and cost, (b) complete 14-trip schedule, and (c) DFs with Step 1 of algorithm VTSP.

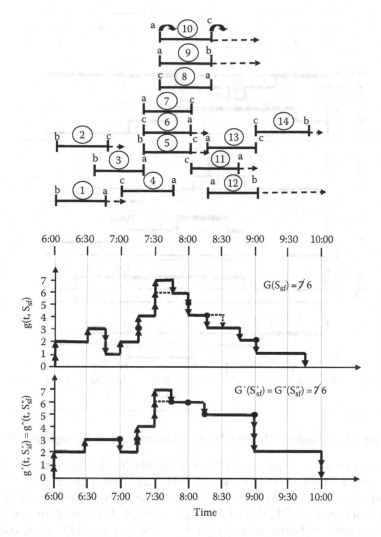

Figure 8.8 Determination of the lower bound corresponding to Step 1 of algorithm VTSP for Example 2.

in which $n_{21} = 2$, $n_{22} = 4$, and $n_{23} = 1$, using *Step 2* of algorithm VTSP. This is illustrated with one shift in departure time and one DH trip in Figure 8.9. Following the flow diagram in Figure 8.2a, we can now check whether $N_1 = N_2$ to obtain that $N_1 < N_2$; then continuing with *Steps 4* and *5* of the algorithm, resulting in a single shift forward (late departure) of trip 10 at 7:30 as is shown in Figure 8.10. Because $D(S_{sf}) = 8 \neq G''(S''_{sf}) = 6$, the algorithm advances to a search of URDHC, including possible shifts in departure times. The NT rule determines terminal c of type 2 as a candidate location for reducing a vehicle; this results in DH_1 following condition (a) of *Step 6* of the algorithm. However, in order to avoid the increase of $D(a)$ in $d_1(a, t)$, another DH trip is found, DH_2, but this time using condition (b) of *Step 6* of the algorithm. The search for another URDHC allows for DH_3 and DH_4, using conditions (b) and (a), respectively, of *Step 6* of the algorithm. The procedure then stops because the sum of all $D(k) = 2 + 2 + 2 = 6$, $k = a$, b, c, equals the lower bound attained.

Unlike Example 1, the types of vehicles corresponding to the resultant six blocks cannot be derived by the sum $D(k)$, $k = a$, b, c, associated with each type. First, the blocks need to be constructed using the inserted DH trips; second, the type of each block is determined by

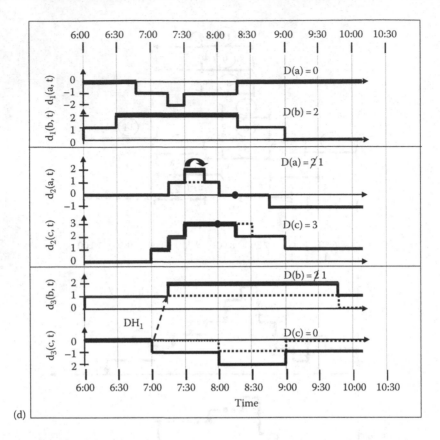

(d)

Figure 8.9 Example 2 (continued from Figure 8.7): (d) DFs describing *Step 2* of algorithm VTSP.

its highest (most expensive) type trip. Following the FIFO rule, for instance, the first three blocks are of type 1: [1–9], [3-DH_1-8], and [5-DH_2-12] because there is at least one trip of this highest type in each block (trips 8 and 5 are of types 2 and 3, respectively). The second three blocks are of type 2: [2-DH_3-7-11-DH_4-14], [6–13], [4–10] (trips 2 and 14 are of type 3). Thus, in the best solution found, $n_1 = 3$, $n_2 = 3$, $n_3 = 0$, N = 6, and C = 54, compared with $C_1 = 60$ and $C_2 = 58$.

8.4.3 A real-life example

Algorithm VTSP was used to examine a real-life scheduling problem. The problem selected pertains to the Egged bus company, which has three different bus routes departing from a main terminal in Haifa. The three routes are shown schematically in Figure 8.11a: intercity, peripheral, and urban. The daily timetable of the intercity route is characterized by 18 departures, 120 min being the average travel time and 105 min the average DH time between terminals A and B in each direction. The peripheral route has 22 daily departures, with 45 min the average travel time between terminals A and C, 24 min the average DH time between A and C, and 36 min between C and D. The urban route has 24 daily departures, with 30 min the average travel time and 15 min the average DH time between A and D. The relative costs of the intercity, peripheral, and urban vehicles are 1.6, 1.3, and 1.0, respectively. The allowed shifts are $\Delta_u^{i(+)} = \Delta_u^{i(-)} = 3$ min for all trips i and vehicle type u.

The VTSP algorithm results in the optimal solution shown in Figure 8.11b, with C = 19.4 and 14 vehicles required: 7 intercity, 4 peripheral, and 3 urban vehicles, respectively.

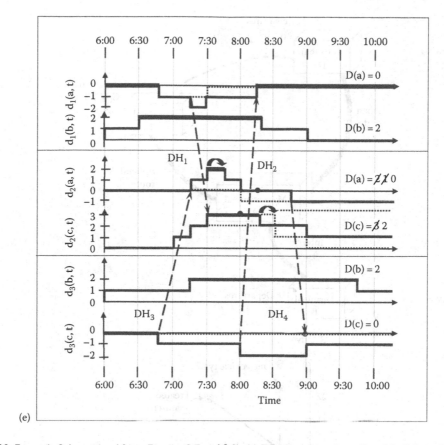

Figure 8.10 Example 2 (continued from Figures 8.7 and 8.9): (e) DFs describing *Steps 5 and 6* of algorithm VTSP.

Steps 1 and 2 of algorithm VTSP result in $C_1 = 22.4$ (14 intercity vehicles) and $C_2 = 22$ (7 intercity, 6 peripheral, and 3 urban vehicles), respectively. This outcome of the algorithm is circled in Figure 8.11b. The results of the heuristic method suggest that algorithm VTSP can be used, too, for large transit agencies, ensuring an efficient allocation of the different vehicles to trips while reducing the cost involved to a minimum level. The use of DF may be subject to the interference of the schedulers in the process whenever they think it justifiable.

8.5 MIN-COST FLOW PROCEDURE FOR MULTI-TYPES VEHICLE-SCHEDULING PROBLEM

This section is based on the study by Hassold and Ceder (2014a) with a methodology using minimum-cost network flow model. The input is set of Pareto-optimal timetables explicated in Section 5.4 for individual transit lines.

8.5.1 Description of concepts

In the optimization-related literature review of MTV-VSP, in Section 8.2, the problems are solved by dividing the vehicle-scheduling problem into sub-problems—one for each vehicle type—and optimize it independently. However, it does not necessarily lead to the best global solution. This is shown by an example depicted (with notations) in Figure 8.12, in which trips between Stations A and B are to be covered.

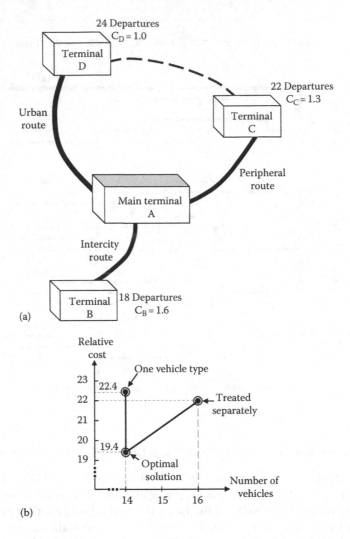

Figure 8.11 (a) Schematic description of a real-life, four-terminal, three-vehicles type example and (b) trade-off situation between total number of vehicles and relative cost associated with five disparate solutions.

Interaction between the sub-problems in Figure 8.12 can certainly reduce the number of vehicles required while complying with the capacity constraints. This is somewhat similar to what was done in Sections 8.3 and 8.4 using the DF principles. Moreover, optimality can be attained using a set of Pareto efficient timetables following the analysis of Section 5.4. This possibility of selecting a slightly different, yet optimal, timetable to improve the vehicle schedule is an added value to the whole process. This approach is illustrated in Figure 8.13 with trips for two routes shown in Figure 8.13a. There is only one timetable for Route 1 while there are two slightly different timetables for Route 2. Figure 8.13b and c shows possible vehicle schedules based on the selection of Timetable 1 and Timetable 2, respectively. Three vehicles are required for Timetable 2, one standard-bus and two minibuses. Two vehicles are required for Timetable 1, one standard-bus and one minibus. Thus, the fleet size can be reduced by selecting Timetable 1 for Route 2. In this solution, the capacity of

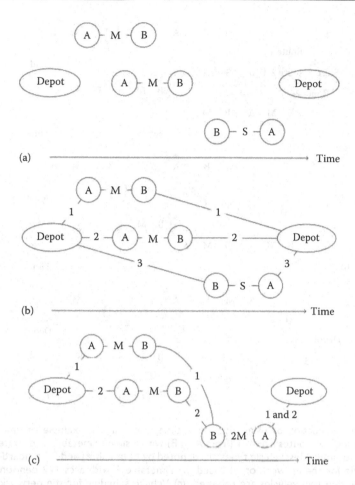

Figure 8.12 Vehicle requirements for a simple transit network. (a) An example of time–space network for a timetable to serve Stations A and B consisting of two minibus trips (M) and one standard-bus trip (S). (b) Optimal solution for the network of (a) using separate problems for each vehicle type with arcs 1–3 denoting vehicle IDs. (c) Optimal solution for the network of (a) allowing for a replacement of a standard-bus by two minibuses (2M) given that the capacity of two minibuses is greater than or equal to that of one standard-bus.

the individual trips remains roughly the same enabling same level of passenger comfort. The following methodology is founded on these two observations with the objective to minimize the total fleet size and operating cost.

8.5.2 Layers of network used

Following the concepts of Figures 8.12 and 8.13, we use network G that consists of set of nodes N and set of arcs A, as a network with flows of vehicles (see a similar approach used in Section 6.4 for vehicle scheduling). For the sake of clarity, G is separated into two compatible layers, the *logical* layer G_L and the *transportation* layer G_T. Figure 8.14 illustrates these two layers. In addition, Table 8.1 introduces notations and definitions used in describing the solution process; Table 8.1 follows the notations of Hassold and Ceder (2014a), and not necessarily those used in the sections given earlier.

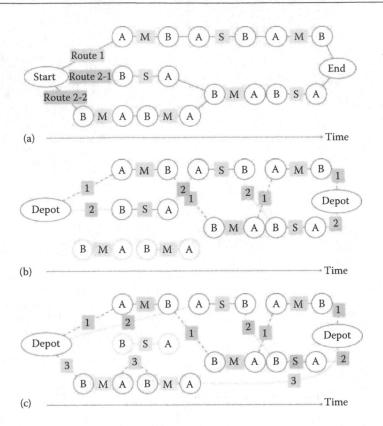

Figure 8.13 Fleet size reduction by using alternative timetables. (a) An example of time–space network illustrating two routes: route 1 (from A to B) with a single timetable, and route 2 (from B to A) with two timetables; vehicle types are denoted by M (minibus) and S (standard-bus). (b) Vehicle schedule for the network of (a) based on Timetable 1 with arcs 1–2 denoting vehicle IDs; in this solution two vehicles are required. (c) Vehicle schedule for the network of (a) based on Timetable 2 with arcs 1–3 denoting vehicle IDs; in this solution three vehicles are required.

Figure 8.14 shows that the *logical* layer contains the routes and the corresponding timetables, but does not represent the actual vehicle flows. The *transportation* layer represents the actual vehicle flows on the network. A flow of one unit represents the operation of a single vehicle on the associated arc. The *transportation* and *logical* layers consist of the following elements:

Circulation-arcs: Arcs added in the *transportation* layer between the sink node and the source node of the same vehicle type. The number of vehicles required can be determined by the flow on the circulation-arcs.

Departure and arrival nodes: Each such node stores the information of departure or arrival times of trips. The set of arrival nodes is called N_A the set of departure nodes is N_D; N_A, $N_D \subset N$. These nodes are contained in both layers.

Logical-arcs: Arcs of the *logical* layer that establish a logical connection between trips belonging to the same timetable. For a chosen timetable, a flow of one unit is transported along the logical-arcs and along trip-arcs as shown in Figure 8.14.

Logical-circulation-arcs: Arcs that create a feasible network structure; these arcs are a certain type of logical-arcs connecting the source node S_l and sink node T_l. A lower bound (LB) and upper bound (UB) of one unit limits the number of selected timetables for a single line to be exactly one.

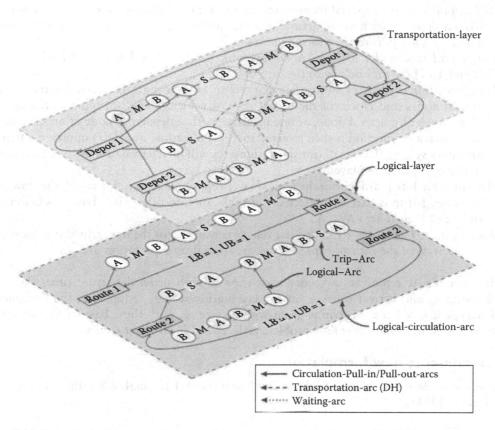

Figure 8.14 Illustration of the separate network layers: transportation layer and logical layer.

Table 8.1 Notations and definitions

General			Network			Layers
Tt	Timetable	G	Network		G_L	Logical layer
m	Trip ∈ Tt	N	Set of nodes		G_T	Transportation layer
d_m	Departure time of trip m	A	Set of arcs		A_L	Set of logical arcs
a_m	Arrival time of trip m	N_A	Set of arrival nodes		A_T	Set of transportation arcs
s_m	First stop of trip m	N_D	Set of departure nodes		A_{L_m}	Logical arcs for trip m
e_m	Last stop of trip m	S_0	Source node		A_{T_m}	Transportation arcs for trip m
L	Set of bus-lines l	S_l	Source node for line l		A_T^k	Transportation layer for vehicle type k
VT	Set of bus-types k	T_0	Sink node		$A_{T_m}^k$	Transportation arcs for trip m vehicle type k
v_k	Capacity of bus-types k	T_l	Sink node for line l		*Others:*	
TT_{uv}	Travel-time between station u and v	LB l_{ij}	Lower bound Lower bound arc ij		MS	Minibus Standard-bus
m·α·n	Compatibility relation: $a_m + TT_{uv}(t) \leq s_n$	UB l_{ij}	Upper bound Upper bound arc ij		c_k c_{ij}	Capital cost for bus-type k Cost for arc ij

Pull-in/pull-out-arcs: Special transportation-arcs representing the movement of a vehicle from a depot to its first departure location (pull-out), or back to the depot (pull-in). These arcs are part of the *transportation* layer.

Source and sink nodes of lines: A sub-network for each line $l \in L$ is formed between S_l and T_l. These arcs belong to the *logical* layer.

Transportation-arcs: Arcs running in parallel to each trip-arc between departure and arrival nodes and represent the actual flow of vehicles. The transportation-arcs can also connect an arrival node, at one location, with a departure node at another location; such a connection is then referred to as deadheading-arc (*DH*) implying a trip of an empty vehicle. The set of transportation-arcs will be referred to as $A_T \subset A$ and is part of the *transportation* layer.

Trip-arcs: Each trip-arc within the *logical* layer represents trip m, for which the flow can be either 1 if m is part of a selected timetable, or zero otherwise. Trip-arcs belong to the set of logical-arcs $A_L \subset A$.

Waiting-arcs: These arcs are added if a vehicle is waiting for the next trip at one location. These arcs, of the *transportation* layer, are special transportation-arcs.

Each timetable Tt represents a sequence of trips *m* with given departure times d_m, first station/stop s_m and arrival times a_m at the last station/stop e_m. Moreover, the designated vehicle type $k \in VT$ for this trip is given with its size v_k. With these basic definitions, the network-based minimum cost flow problem is described in the next section.

8.5.3 Min-cost flow formulation

The min-cost flow formulation is shown by Equations 8.1 through 8.8 following Hassold and Ceder (2014a).

$$\min \sum_{(i,j) \in A} c_{ij} x_{ij} + e \tag{8.1}$$

s.t.

$$\sum_{i:(i,j) \in A} x_{ij} - \sum_{i:(j,i) \in A} x_{ji} = 0 \quad \forall j \in N \tag{8.2}$$

$$\sum_{i:(i,j) \in A_L} x_{ij} - \sum_{i:(j,i) \in A_L} x_{ji} = 0 \quad \forall j \in N \tag{8.3}$$

$$\sum_{i:(i,j) \in A_T^k} x_{ij} - \sum_{i:(j,i) \in A_T^k} x_{ji} = 0 \quad \forall k \in VT, \forall j \in N \tag{8.4}$$

$$\sum_{k \in K} \sum_{i:(i,j) \in A_{T_m}^k} x_{ij} v_k \geq \sum_{i:(i,j) \in A_{L_m}} x_{ij} cap_m \quad \forall m \in Tt \tag{8.5}$$

$$\sum_{m \in Tt} \left(\sum_{k \in VT} \sum_{(i,j) \in A_{T_m}^k} x_{ij} v_k - \sum_{(i,j) \in A_{L_m}} x_{ij} cap_m \right) = e \tag{8.6}$$

$$l_{ij} \leq x_{ij} \leq u_{ij} \quad \forall (i,j) \in A \tag{8.7}$$

$$x_{ij} \geq 0 \text{ and integer} \tag{8.8}$$

In this formulation, cap_m is the capacity of the vehicle originally scheduled for trip m. The objective function (8.1) minimizes the cost of the flow on all arcs plus an element e, explained later. Constraint (8.2) presents the standard flow conservation constraint, that is, in-flow equals out-flow for each node. Moreover, it is necessary to ensure that the logical and transportation flows are not mixed. The flow conservation constraint for the logical-arcs is presented in (8.3). Further restrictions on the transportation flow are required by Constraint (8.4), which prevent the flows of different vehicle types to be mixed. Constraint (8.3) together with Constraint (8.4) do not make the first flow conservation Constraint (8.2) obsolete because the circulation-arcs have not been assigned to any subset of arcs and also have not been considered yet. Constraint (8.5) guarantees that the capacity of the vehicles assigned to a trip is greater than or equal the originally planned capacity.

Hassold and Ceder (2014a) show also how to use alternative timetables (as per Section 5.4) and how to deal with vehicle substitution following the description in Figure 8.12. An element e is introduced in Constraint (8.6) to represent empty seats on the vehicles serving transportation-arcs for trip m. This will equal to zero in case that of no substitution of vehicles. Adding e to the objective function, with the possibility to assign it a weight, firstly prevents the use of transportation-arcs, and secondly avoids the assignment of overly large capacity to a trip if not required. The substitution of vehicles can reduce the total number of vehicles and thus reduce the cost of the vehicle schedule. However, in case that a vehicle type K_1 with capacity v_{K_1} is operated by a vehicle of type K_2 with capacity $v_{K_2} > v_{K_1}$, there will be more empty seats on the vehicle than originally calculated. Therefore, the actual penalty for empty-seats time for this timetable has to be recalculated and considered in the final evaluation of the solutions. The remaining Constraints (8.7) and (8.8) ensure the compliance with the flow boundaries l_{ij} (lower bound) and u_{ij} (upper bound) and also of the flow (representing vehicles) being an integer value.

The min-cost formulation assumes that the operator is indifferent between the given timetables and does not have any preference toward either even headway or a reduction in empty-seat-kilometer. It is shown by Hassold and Ceder (2014a) that this model can be used for the cases with of alternative timetables, and with vehicle substitution.

Finally, Hassold and Ceder (2014a) demonstrated that a substitution of vehicles is beneficial and can lead to significant cost reductions in the range of more than 27%. The methodology of this section was applied to a real-life case study in Auckland, New Zealand, with a saving of more than 15% of the cost of fleet compared with vehicle schedules that are provided by standard models. In addition, Hassold and Ceder (2014b) examined the benefits attained from an energy-efficiency perspective. Their results show that using timetables with multiple vehicle types can increase passenger occupancy of the vehicles by 19%, hence reducing total energy consumption. These results are achieved without any increase of the total passenger waiting time and fleet size.

8.6 VEHICLE SIZE DETERMINATION

There is doubt that there is a connection between different vehicle types and the size of vehicles used in public transit services. In the foregoing section, it was assumed that all vehicle types possessed a similar size so that the frequencies and timetables determined remained unchanged. However, practice often entails making a decision about vehicle size and the amount available for purchasing. That is, a bus fleet can consist of vehicles of the same size, thereby simplifying operations planning, or have a mix of sizes; that is, minibuses, standard buses, articulated/double-decker buses, etc. This section provides some preliminary analytical tools for evaluating trade-offs between different vehicle sizes and for reviewing the models developed.

8.6.1 Example: A standard bus versus two minibuses

Given a bus route operated by standard 50-seat buses, can the standard bus be exchanged more cost effectively and viably for 2 minibuses with 25 seats each? The first line of thought is directed to the question: What operational strategy should be conducted for the two minibuses? Basically, two operational alternatives can be drawn: (1) Let the two minibuses depart at the same time, with each minibus picking up and dropping off passengers at each other stop. This will result in reducing travel time and, in parallel, retaining the timetable (frequency). (2) Let the minibus headway be half that of the standard bus and let each minibus pick up and drop off at each stop. This alternative will result in reducing passenger waiting time and, in parallel, maintaining average travel time.

The following basic analysis assumes that for alternative (1) we have a long bus route (end effects ignored) that is a typical commuter collector—for example, to the CBD—and the ridership is unchanged. The assumptions for alternative (2) are that the average waiting time is half the headway (deterministic case; on headway distribution see Chapter 18), that the probability of no one being at the stop is the same for both the minibus and the standard bus, and that the ridership is fixed.

For an example of alternative (1), the following data are used: standard bus (SB) operating cost = \$15/h, minibus (MB) operating cost = \$12/h, SB travel speed denoted by μ miles/h = 3/4 MB travel speed, P = average number of passengers on board SB, and the value of a passenger's riding time is \$(1/3)/h. Hence, SB operating cost per μ miles = 15 × 1 = \$15, and 2 MB operating cost per μ miles = 2(12 × 3/4) = \$18. The problem arises: Who is going to pay the additional cost for the 2 MB (\$3 per μ miles)? A possible answer is the passenger because their riding time is reduced. A saving in travel time by MB per μ miles is 1 − 3/4 = 1/4 h. The value of the total saving is 1/4·P·1/3 = P/12. Requiring that the saving be greater than the additional cost results in P/12 > \$3 or P > 36 passengers. In other words, a two-minibus operation is preferable in alternative (1) for average loads greater than 36 on the standard bus.

For an example of alternative (2), the following data are used: SB and MB operating cost and P are the same as in the foregoing example, SB headway = 10 min, MB headway = 5 min, and the value of waiting time per passenger = \$1/h. The number of hourly on board passengers is 6P, and the saving in average waiting time when using MB is 10/2 − 5/2 = 2.5 min per passenger. Total hourly saving, therefore, is 6P·2.5/60·1 = \$(1/4)P, to be viable must be greater than the additional MB hourly operating cost. That is, (1/4)P > 2·12 − 15, which results, as in the example for alternative (1), in P > 36 passengers.

8.6.2 Vehicle size square root formula

A simple derivation of transit vehicle size accentuates a trade-off between vehicle capacity cost and passenger waiting-time cost. We assume that operating cost per vehicle-hour is the same (independent of vehicle size and load carried), average waiting time is half the headway, riding time is independent of vehicle size, and travel time and stopping time are independent.

The following notation is used:

Z is the desired average occupancy in the max load segment, which corresponds to *vehicle size*

C_b is the operating cost per vehicle-hour

P is the number of average hourly passengers carried in the max load segment

C_w is the value of hourly waiting time

The hourly frequency, utilizing Method 2 in Chapter 3, is P/Z, and the headway (in hours) is Z/P. Hence, the total operating cost is $C_b \cdot P/Z$, and the passenger waiting-time cost is $\frac{1}{2} Z/P \cdot P \cdot C_w = \frac{1}{2} Z C_w$. The overall cost, therefore, is $C_b \cdot P/Z + \frac{1}{2} Z C_w$.

The optimal vehicle size, Z_0, is attained by minimizing the overall cost using the first derivative of the overall cost with respect to $Z = 0$. This results in the following formula:

$$Z_0 = \sqrt{\frac{2 C_b P}{C_w}} \qquad\qquad (8.9)$$

Equation 8.9, called the *square root formula*, signifies, generally speaking, that the optimal vehicle size varies according to the square root of the number of passengers carried. In addition, this optimal size is sensitive to changes in C_b and C_w. The next section reviews more advanced vehicle-size modeling.

8.7 OPTIMAL TRANSIT VEHICLE SIZE: LITERATURE REVIEW

The following notation is used throughout this section (unless mentioned otherwise):

Z is the optimal vehicle size

a is the constant of operational cost function (vehicle operational cost as a function of size, $a + bZ$)

b is the slope of operational cost function

w is the value of passenger waiting time

x is the value of access/walking time

r is the value of riding/in-vehicle time

L is the route length

P is the total passenger flow

Q is the peak passenger flow

V is the vehicle speed

ϕ is the maximum allowed load factor

In some cases, parameters are given in a different notation in order to enable a convenient comparison.

An early reference to bus-size optimization was made by Jansson (1980), who paid significant attention to the cost of operating similar services at peak and off-peak during daylight. The following formula for optimal bus size was developed:

$$Z = \frac{1}{\phi} \sqrt{\frac{ahJQ}{(\beta Ew/2) + tQ(\beta Er + b/\phi)}} \qquad (8.10)$$

where
    h is the running and transitional time per km
    J is the average trip length
    β is the ratio of mean flow rate for the whole day to the peak-hour flow
    E is the extent of peak + off-peak periods per day
    t is the boarding and alighting time per passenger

Calculation of the optimal bus size, given various levels of passenger flows, leads to the conclusion that most buses in actual operation are too big.

Gwilliam et al. (1985) describe a simple formula for an optimal bus size:

$$Z = 2\sqrt{aP} \qquad (8.11)$$

Based on assigning common values of a and P to their formula, the authors claim that operating much smaller buses than those commonly used is not justified.

A more sophisticated model that aims at determining optimal bus size is provided by Oldfield and Bly (1988). Unlike most other models, this one assumes elastic demand; that is, demand that changes with passenger-trip cost. The model also takes into account the influence of changes in bus demand on road congestion; it further assumes that in-vehicle trip time increases with average bus load because of boarding times. Significant attention is given to the effect of capacity constraints on average passenger waiting time in cases in which passengers are unable to board the first bus that arrives because it is full. Complex expressions are introduced explaining the dependence of headway on bus size. The expression developed for optimal bus size is the following:

$$Z^2 = AY_i \left\{ \frac{aY_i^{1/2}W_i}{(1+\beta)Z} + \frac{nP}{K} + U + R \right\}^{-g} \qquad (8.12)$$

where
    K is the number of bus-km provided
    U is the average cost of walking time to/from bus stop
    R is the cost of time spent in the bus
    n is the time-value constant that takes into account the extra time that a big bus spends at stops owing to a large number of boardings
    g is the elasticity with respect to generalized cost (positive)
    A, β, Y_i, W_i are the constants (expressions for calculating them are developed in the paper)

Assigning common values of urban bus services in the United Kingdom, assuming no subsidies, the authors find that the optimal bus size lies between 55 and 65 seats. Under various other conditions, the model suggests that the optimum size at typical urban levels of demand will not be fewer than about 40 seats. Although this model is very sensitive to various phenomena that influence passenger demand and operator costs, it seems that the preliminary modeling and calibrating efforts required for using it are too intensive for practical use.

Jansson (1993) presents a model that simultaneously optimizes vehicle size, frequency, and journey price. All passengers having the same origin and destination along the route are referred to as a group, and each group may have its own value of time. Optimal vehicle size is computed as follows:

$$Z = \frac{\sum_i X_i \sum_{mi} \sum_j X_{j/mi}(\partial r/\partial R_m)(h_m/FNZ)}{F((\partial c_c/\partial Z)h + L(\partial c_\gamma/\partial Z))}$$

(8.13)

where
X_i is the number of passengers belonging to group i
$X_{j/mi}$ is the number of passengers in group j who travel on link m, where passengers in group i also travel
R_m is the number of passengers per seat on link m
F is the frequency of service
N is the number of cars in the train if the vehicle discussed is a train, otherwise 1
h is the round-trip time
h_m is the time on link m
c_c is the operating cost, which increases with vehicle size
c_γ is the operating cost, which increases with distance traveled

Implementation of this model may be somewhat more difficult compared with other models since it requires detailed data about the origin and destination of all passengers.

Shih and Mahmassani (1994) developed a vehicle-sizing model that assumes that the total demand matrix of the whole route system is given, not the demand for each line. The load profile of each line is determined in a transit assignment. The optimal bus size is computed as follows:

$$Z = \frac{Q_k}{\phi} \sqrt{\frac{2aL_k}{wP_k}}$$

(8.14)

where k is the route index.

An iterative process is suggested that includes reassigning the total demand matrix on the route system after the determination of optimal vehicle size and frequency for each route. Each iteration of the transit assignment yields corrected values of $(Q_k)_{max}$ and P_k so that new optimal vehicle sizes can be calculated. The iterative process does not seek to minimize system-wide cost, but to determine the optimal bus size for each separate line. Implementation of the procedure is illustrated using data from a transit network in Austin, Texas. Of 40 bus routes, results suggest that 37 have an optimal bus size of below 25 seats.

A model developed by Lee et al. (1995) attempts to find the optimal bus size not only for each route but also for each period of day so that more than one bus size can be used on one route. The model also tries to determine the conditions under which it is better to use one bus size or, alternatively, a mixed-size fleet. The bus size that gives the minimum total operator and user cost on one line is determined as follows:

$$Z = \sqrt{\frac{2aLQ^2}{wVP}} \qquad (8.15)$$

If only one bus size is operated on all routes during all periods, the system-wide optimal size is found as follows:

$$Z = \sqrt{\frac{\sum_{r=1}^{n}\sum_{t=1}^{m} 2aL_rQ_{rt}/V_{rt}}{\sum_{r=1}^{n}\sum_{t=1}^{m} wP_{rt}/Q_{rt}}} \qquad (8.16)$$

where
   r is the route index
   n is the number of routes
   t is the time-period index
   m is the number of time periods

If two bus sizes (Z_1 and Z_2) are used, the following test can determine which is better for each route:

$$\frac{Q^2L}{PV} = \frac{wZ_1Z_2}{2a} \qquad (8.17)$$

If the left-hand side is greater than the right-hand side, bigger buses should be used. If the right-hand side is greater, the operation of smaller buses is justified. For cases in which Z_1 and Z_2 are not given, the paper describes an algorithm for determining the two bus sizes that give minimum cost. The algorithm is illustrated in a simple four-route network with a two-period demand. The optimal fleet is concluded to consist of 15 buses with 33 seats and 29 buses with 20 seats. The total fleet size required for a network operation with this mix is smaller than that needed for operating the same route system with buses all of the same size. It is also shown that under the conditions of the given example, a mixed-fleet operation is preferable if the ratio of peak demand to off-peak demand is more than 1.92. If there is no significant demand variation between periods of the day, then the operation of a uniform bus size is preferable.

Gronau (2000) presents this model for determining the optimal bus size for a specific route:

$$Z = \sqrt{\frac{\alpha_0 + \beta_0 t_0}{(\lambda r/P) + t_1(\beta_1 + r)}} \qquad (8.18)$$

where
   $\alpha_0 + \alpha_1 Z$ is the distance-related operating cost
   $\beta_0 + \beta_1 Z$ is the time-related operating cost
   $t_0 + t_1 Z$ is the bus travel time
   λ is the ratio of the value of waiting time to the value of vehicle time, divided by the ratio
       of the headway to the waiting time (expected value: 1.5–2)

This model, introduced as a basis for a series of mathematical developments, aims at examining the option of using two bus sizes on one route that serves passengers with different values of time. The usefulness of this option is investigated in detail.

Tisato (2000) analyzes the variation of public transit subsidy levels among several constraint cases with respect to bus size and load factor. Four cases are analyzed: fixed-load factor with variable bus size; fixed bus size with variable load factor; both fixed; and both variable. Conditions for maximum economic surplus are determined for each of the cases. In the case in which bus size is the only variable, the expression derived for optimal bus size is

$$Z = \sqrt{\frac{aLPA^2(1.25 - 1.65\phi)}{15w\phi^2}} \tag{8.19}$$

where A is the average passenger-trip length, divided by bus-trip length.

In addition, it is shown how optimal bus size changes with a varying target-load factor.

A subgroup of vehicle-size optimization tools consists of the formulation of bus-sizing models as part of a comparison between fixed-route and flexible-route services. Chang and Schonfeld (1991), who make such a comparison, argue that a rectangular service area should be connected to an adjacent transportation terminal. If a fixed-route service is provided, the optimal bus size is the following:

$$Z = \left(\frac{8a^2gSDL^2}{xwV^2}\right)^{1/3} \tag{8.20}$$

where
   g is the access speed
   S is the length of service area
   D is the passenger-demand density in service area

If the bus route is flexible, the optimal size is

$$Z = \left(\frac{ua^3L_T^3D}{wk^2V(b + r/2)^2}\right)^{1/5} \tag{8.21}$$

where
   u is the average number of passengers per pickup point
   L_T is the equivalent-line haul distance
   k is the constant (estimated value described in the paper)

Using these formulas, the authors show that flexible routes require smaller buses than do conventional services.

Another model that compares conventional and flexible routes is presented by Chien et al. (2001). They assume a given probabilistic demand function and a nonadditive value of time; that is, the cost of one 10 min wait is higher than the total combined cost of ten 1 min waits. The model is not solved analytically, so that there is no explicit formula for optimal bus size. The optimum size in a fixed-route system is calculated by minimizing the following expression, c, with decision variables h and Z:

$$\min c = \frac{L(a+bZ)}{VZ} + r\left(\frac{M}{V}\right)^2 + w\left(\frac{Z}{2hSq(Q', \sigma)}\right)^2 + x\left(\frac{r+d}{4g}\right)^2 \tag{8.22}$$

where
    h is the route spacing
    M is the average passenger-trip distance
    S is the length of service area
    q(Q', σ) is the probabilistic demand-density function
    d is the stop spacing
    g is the average walking speed

If a flexible-route system is operated, the optimal bus size is calculated by solving the following problem, with Z and A as decision variables:

$$\min c = \frac{L_T(a+bZ)}{VS} + \frac{k\sqrt{A/(uZ)}(a+bZ)}{V} + v\left(\frac{L_T}{2V}\right)^2$$
$$+ v\frac{AZ}{u}\left(\frac{k}{2V}\right)^2 + w\left(\frac{Z}{2Aq(Q', \sigma)}\right)^2$$

where
    A is the service zone area
    L_T is the equivalent bus-trip distance
    k is the constant (estimated value described in the paper)
    u is the average number of passengers per pickup point

The models reviewed are summarized in Table 8.2 according to their characteristics.

Finally, a recent study of Tirachini et al. (2014) is worth mentioning. This study investigates the impacts of travel time variability and crowding discomfort on the optimal bus frequency and bus size for a highly congested corridor with cars, buses, and walking as travel alternatives. The goal of this research was not directly about bus size, but to study the optimal pricing structure of both cars and buses. Tirachini et al. (2014) propose a social welfare maximization framework using a known mean–variance model with the use of an empirical relationship between the mean and standard deviation of travel times. They conclude, regarding bus sizes, that both travel time variability and crowding discomfort leads to higher optimal bus sizes.

8.7.1 Literature review conclusions

The following may be concluded from the review of the literature on optimal transit vehicle size:

- According to most of the models, an increase in the value of passenger time decreases the optimal vehicle size. Vehicle size in some of the models is proportionally opposite to the square root of the time value.
- There is no consistency among models concerning which trip-time elements influence the optimal vehicle size. Each element (access, waiting in-vehicle) appears as a variable in some models, but no one element appears in all of them.

Table 8.2 Summary of the characteristics of the models reviewed

Comparison subject		Source									
	Jansson (1980)	Gwilliam et al. (1985)	Oldfield and Bly (1988)	Jansson (1993)	Shih and Mahmassani (1994)	Lee et al. (1995)	Gronau (2000)	Tisato (2000)	Chang and Schonfeld (1991)	Chien et al. (2001)
Principles and assumptions: Elastic route demand, with/without a fixed, system-wide, demand constraint	No	No	Yes, without	No	Yes, with	No	No	No	No	The model enables the use of any demand function
Demand influences road congestion	No	No	Yes	No	No	No	No	No	No	No
In-vehicle time increases when in-vehicle load increases	No	No	Yes	No	No	No	Yes	No	No	No
Waiting time is influenced by the chance of a full-vehicle arrival	No	No	Yes	No	No	No	No	No	No	No
Possibility of more than one vehicle size on one route	No	No	No	No	No	Yes	Yes (but not in the formula quoted here)	No	No	No
Differences between periods of the day	Yes	No	No	No	No	Yes	No	No	No	No

(Continued)

Table 8.2 (Continued) Summary of the characteristics of the models reviewed

Comparison subject	Jansson (1980)	Gwilliam et al. (1985)	Oldfield and Bly (1988)	Jansson (1993)	Shih and Mahmassani (1994)	Lee et al. (1995)	Gronau (2000)	Tisato (2000)	Chang and Schonfeld (1991)	Chien et al. (2001)
Total seats per hour (product of vehicle size and frequency)	Fixed	Fixed	Both fixed and not-fixed cases are examined	Fixed	Fixed	Fixed	Fixed	Both fixed and not-fixed cases are examined	Fixed	Fixed
Required effort in preparing input data	Reasonable	Easy	Intensive	Intensive Detailed demand data are needed	Reasonable Transit assignment model is needed	Reasonable	Reasonable	Reasonable	Reasonable	Reasonable
Optimal bus size depends on:										
Subsidy level	No	No	Yes	No	No	No	No	No	No	No
Fare	No	No	Yes	No	No	No	No	No	No	No
Cost of waiting	Yes	No	Yes	No	Yes	Yes	Yes	Yes	Yes	Yes
Cost of access to/from bus stop	No	No	Yes	No	Yes	No	No	No	Fixed-route: Yes Flexible: No	Fixed-route: Yes Flexible: No
Cost of in-vehicle time	Yes	No	Yes	Yes	Yes	Yes	Yes	No	Fixed: No Flexible: Yes	Yes
Maximum allowed bus occupancy	Yes	No	No	No	Yes	No	No	Yes	No	No
Constant (a) of operating cost function	No	Yes	Yes	Yes	Yes	Yes	Yes	Yes	Yes	Yes
Slope (b) of operating cost function	Yes	No	No	Yes	No	No	Yes	No	Fixed: No Flexible: Yes	Yes
Route length or time	Yes	No	Yes	Yes	Yes	Yes	No	Yes	Yes	Yes

- In the majority of the models reviewed, the constant operating cost (parameter a of operating cost function $C = a + bZ$) influences optimal vehicle size. In most of these cases, optimal vehicle size is proportional to the square root of this constant. Parameter b (slope of operating cost function) appears in some of the models, but not in a consistent manner.
- Dependency of optimal vehicle size on route length or time shows, in general, that the longer the transit route, the bigger are the vehicles required.
- Different models show opposite viewpoints on the relationship between demand and vehicle size: in some, a high passenger flow will increase the vehicles needed; in others, it will justify using smaller vehicles.
- Calibration of an operating cost function is required for each of the models discussed. In some of the models (such as the one presented by Gronau), different functions are needed for distance-related and for time-related operating costs. Calibrated values of passenger time are usually compulsory. In most of the models reviewed, no additional special effort is needed to calculate the optimal vehicle size other than using data that are readily available to most transit agencies. The model presented by Oldfield and Bly is different in that it requires much more sophisticated data. Similarly, the model developed by Jansson (1993) requires detailed information about passenger demand.

EXERCISES

8.1 Given a set of terminals, a, b, c, and d; four types of vehicles with their unit cost c_i; $i = 1, 2, 3, 4$ (in monetary units); DH travel-time matrix; and a fixed schedule of trips.

(a) Find the number of vehicles for each type required to obtain the minimum total cost; use algorithm VTSP.
(b) Find the solution to (a) after changing the costs of type II and type III from 8.5 and 5 to 10 and 4 cost units, respectively; use algorithm VTSP, but think about steps that there is no need to repeat.

Trip number	Type of vehicle	Departure time	Departure terminal	Arrival time	Arrival terminal
1	I	6:00	a	6:30	c
2	II	6:20	a	6:50	b
3	IV	6:20	b	6:50	c
4	III	6:40	d	7:10	a
5	I	7:10	a	7:40	a
6	II	7:10	b	7:40	a
7	III	7:20	d	7:50	a
8	I	7:40	c	8:10	a
9	IV	7:50	d	8:10	b
10	IV	8:00	b	8:30	c
11	IV	8:10	b	8:30	d
12	III	8:20	a	8:50	d

DH travel time (min)

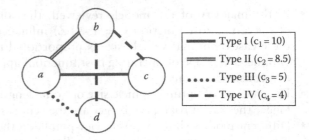

	a	b	c	d
a	0	10	10	10
b	10	0	10	10
c	10	10	0	20
d	10	10	20	0

8.2 Solve Example 1 in Section 8.4.1 using the following three new sets of input cost values; use algorithm VTSP, but think about steps that there is no need to repeat:

(a) $c_1 = 10$, $c_2 = 5$, $c_3 = 3$
(b) $c_1 = 10$, $c_2 = 11$, $c_3 = 6$
(c) $c_1 = 12$, $c_2 = 8$, $c_3 = 9$

8.3 Given a set of terminals, a, b, c; three types of vehicles with unit costs c_i; i = 1, 2, 3 (in monetary units); DH travel-time matrix; shifting tolerance of -15 min $\leq \Delta \leq 15$ min; and a fixed schedule of trips. Find the number of vehicles for each type required to obtain the minimum total cost; use algorithm VTSP.

DH travel time (min)

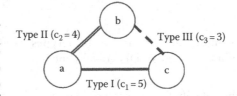

	a	b	c
a	0	30	30
b	30	0	30
c	30	30	0

Trip number	Type of vehicle	Departure time	Departure terminal	Arrival time	Arrival terminal
1	I	6:00	a	7:00	c
2	II	6:00	a	7:00	b
3	III	6:00	b	7:00	c
4	III	6:00	b	7:00	c
5	II	6:30	a	7:30	b
6	II	6:30	b	7:30	a
7	III	7:00	c	8:00	b
8	II	7:00	b	8:00	a
9	I	8:00	a	9:00	a
10	III	8:00	c	9:00	b
11	I	8:00	c	9:00	a
12	I	8:30	c	9:30	a
13	II	8:30	a	9:30	b
14	II	9:00	a	10:00	b

8.4 Examine the trade-off between operating an articulated bus (AB) with 75 seats and operating 3 minibuses (MB) with 25 seats each. The following data and information are given: AB operating cost = \$25/h, MB operating cost = \$15/h, value of passenger-riding time is \$0.5/h, frequency of AB and MB services are the same, MB travel speed is below AB travel speed

of some operational tactics for the MB service are utilized, and for P > 30, in which P is the average number of passengers on board AB, use of the MB service is preferred.

(a) Find the values of P for preferring the MB service if the MB travel speed (described above) is reduced by 10%.
(b) What are the upper and lower bounds on the travel speed of the AB service for positive P values?

REFERENCES

Bertossi, A., Carraresi, P., and Gallo, G. (1987). On some matching problems arising in vehicle scheduling. *Networks*, 17, 271–281.

Bodin, L., Golden, B., Assad, A., and Ball, M. (1983). Routing and scheduling of vehicles and crews: The state of the art. *Computers & Operations Research*, 10(2), 63–211.

Bunte, S. and Kliewer, N. (2009). An overview on vehicle scheduling models. *Public Transport*, 1(4), 299–317.

Ceder, A. (1995). Transit vehicle-type-scheduling problem. *Transportation Research Record*, 1503, 34–38.

Ceder, A. (2011). Public-transport vehicle scheduling with multi vehicle-type. *Transportation Research*, 19C(3), 485–497.

Chang, S.K. and Schonfeld, P.M. (1991). Optimization models for comparing conventional and subscription bus feeder services. *Transportation Science*, 25(4), 281–298.

Chien, S., Spasovic, L.N., Elefsiniotis, S.S., and Chhonkar, R.S. (2001). Evaluation of feeder bus systems with probabilistic time-varying demands and non-additive value of time. *Transportation Research Record*, 1760, 47–55.

Costa, A., Branco, I., and Paixao, J. (1995). Vehicle scheduling problem with multiple types of vehicles and a single depot. In *Computer-Aided Transit Scheduling*. Lecture Notes in Economics and Mathematical Systems, vol. 430 (J.R. Daduna, I. Branco, and J.M.P. Paixao, eds.), pp. 115–129, Springer-Verlag, Berlin, Germany.

El-Azm, A. (1985). The minimum fleet size problem and its applications to bus scheduling. In *Computer Scheduling of Public Transport 2* (J.M. Rousseau, ed.), pp. 493–512, North-Holland Publishing Co., Amsterdam, the Netherlands.

Forbes, M.A., Holt, J.N., and Watts, A.M. (1994). An exact algorithm for multiple depot bus scheduling. *European Journal of Operational Research*, 72(1), 115–124.

Gronau, R. (2000). Optimum diversity in the public transport market. *Journal of Transport Economics and Policy*, 34, 21–42.

Gwilliam, K.M., Nash, C.A., and Mackie, P.J. (1985). Deregulating the bus industry in Britain—(B): The case against. *Transport Reviews*, 5(2), 105–132.

Hassold, S. and Ceder, A. (2014a). Public transport vehicle scheduling featuring multiple vehicle types. *Transportation Research*, 67B, 129–143.

Hassold, S. and Ceder, A. (2014b). Improving energy efficiency of public transport bus services by using multiple vehicle types. *Transportation Research Record*, 2415, 65–71.

Jansson, J.O. (1980). A simple bus line model for optimization of service frequency and bus size. *Journal of Transport Economics and Policy*, 14(1), 53–80.

Jansson, K. (1993). Optimal public transport prices, service frequency, and transport unit size. In *Selected Proceedings of the Sixth World Conference on Transport Research*, pp. 1591–1602, Lyon, France.

Kliewer, N., Amberg, B., and Amberg, B. (2012). Multiple depot vehicle and crew scheduling with time windows for scheduled trips. *Public Transport*, 3, 213–244.

Kliewer, N., Mellouli, T., and Suhl, L. (2006). A time–space network based exact optimization model for multi-depot bus scheduling. *European Journal of Operational Research*, 175(3), 1616–1627.

Lee, K.K.T., Kuo, S.H.F., and Schonfeld, P.M. (1995). Optimal mixed bus fleet for urban operations. *Transportation Research Record*, 1503, 39–48.

Lenstra, J.K. and Kan, A.H.G.R. (1981). Complexity of vehicle routing and scheduling problems. *Networks*, **11**(2), 221–227.

Löbel, A. (1997). Optimal vehicle scheduling in public transit. PhD thesis, Technischen Universität Berlin, Berlin, Germany.

Oldfield, R.H. and Bly, P.H. (1988). An analytic investigation of optimal bus size. *Transportation Research*, **22B**, 319–337.

Shih, M.C. and Mahmassani, H.S. (1994). Vehicle sizing model for bus transit networks. *Transportation Research Record*, **1452**, 35–41.

Tirachini, A., Hensher, D.A., and Bliemer, M.C. (2014). Accounting for travel time variability in the optimal pricing of cars and buses. *Transportation*, **41**(5), 947–971.

Tisato, P. (2002). A comparison of optimization formulations in public transport subsidy. *International Journal of Transport Economics*, **27**, 199–228.

Chapter 9

Crew scheduling

Pride in what you're doing is one of the greatest sources of enthusiasm

A.C.

CHAPTER OUTLINE

PRACTITIONER'S CORNER

One of the most time-consuming and cumbersome scheduling tasks is assigning the crew (drivers) to vehicle blocks. The task requires the service of imaginative, experienced schedulers, and usually, it is performed automatically. Consequently, it is not surprising to learn that most of the commercially available transit-scheduling software packages concentrate primarily on crew-scheduling activities. After all, from the transit agency's perspective, the largest single-cost item in the budget is the driver's wage and fringe benefits. Because of the important implications of crew scheduling for providing good transit service, practitioners ought to comprehend the root of the problem and be equipped with basic tools to be able to arrive at a solution.

This chapter consists of four principal parts, following an introductory section. Section 9.2 uses the deficit-function (DF) properties for constructing vehicle chains (blocks) that take into account maximum paid idle time (swing time that is an unpaid break in a split duty). The latter consideration is aimed at helping to construct crew schedules with minimum cost. Section 9.3 presents the basic mathematical formulation used in solving crew-scheduling and crew-rostering problems. In addition, this section accentuates an approximate solution to produce low-cost duty pieces of a crew schedule. Section 9.4 describes a case study performed for the New Jersey Transit Corporation that needed a tool for analyzing the marginal cost impacts of changes in train schedules and for producing near-optimal crew assignments (duties), given a schedule and a set of work rules. Finally, Section 9.5 discusses and provides examples of crew rostering. A roster is a pattern of duties to be fulfilled for a certain number of consecutive days; commonly, the pattern repeats itself in cycles (of a week, a month, or any other period). The chapter ends with a literature review and exercises.

Practitioners are encouraged to visit all the sections while skipping only the mathematical formulation. They should, though, pay special attention to the examples and their figures.

A crew schedule is called a pick. Often the selection of picks involves conflicts. Some may say that a selection without a conflict is almost as inconceivable as a nation without crises. There are two short stories: first, about being told that the pick is excellent. A man bought a newspaper for 50 cents. He found, however, that the price marked on the paper was 35 cents. He returned to the kiosk and asked about it. The reply was, "Don't believe everything written in newspapers." Second, two rabbits arrive at the planning department of a transit agency, and one stopped to cry. "Why are you crying?" she was asked. And she answered, "Because here they cut the fifth leg of every rabbit." "But you have only four legs, so there's nothing to worry about," she was told. She replied, "Yeah, but here they cut first and then count."

9.1 INTRODUCTION

The functional diagram of a typical transit operations planning process, Figure 1.2, ends its fourth and last planning activity with crew scheduling, the aim being to assign drivers according to the outcome of vehicle scheduling. This activity is often called driver run cutting (splitting and recombining vehicle blocks into legal driver duties, shifts, runs, or assignments). Part of a vehicle block is called a duty piece or a task. This crew assignment process must comply with some constraints, which are usually dependent on a labor contract. The purpose of the assignment function is to determine a feasible set of driver duties in an optimal manner. Usually, the objective is to minimize the cost of duties so that each duty piece is included in one of the selected duties. It should be noted that vehicle-scheduling activity for railways is not cumbersome; however, it is important to consider how to build the work schedule of the train crews (drivers and conductors together) efficiently.

The criteria for crew scheduling are based on an efficient use of manpower resources while maintaining the integrity of any work rule agreements. The construction of the selected crew schedule is usually a result of the following subfunctions: (1) duty piece analysis, (2) work-rule coordination, (3) feasible duty construction, and (4) duty selection. The duty piece analysis partitions each vehicle block at selected relief points into a set of duty pieces. These duty pieces are assembled in a feasible duty construction function. Other required information are travel times between relief points and a list of relief points designated as required duty stops and start locations.

Theoretically, each relief point may be used to split the vehicle block into new duty pieces. Usually, it is more efficient to use one or more of the following criteria to select the relief points: (a) minimum duty piece length; (b) next relief point, selected as close as possible to the maximum duty piece time (maximum time before having a break); (c) only a few (say, two) relief points in each piece; and (d) operator decisions. In order to utilize any crew-scheduling method, a list of work rules to be used in the construction of feasible driver duties is required. The work rules are the result of an agreement between the drivers (or their unions) and the public transit agency (and/or public authorities).

The determination of different feasible sets of duties may be selected on the basis of, for example, one or more of the following performance measures: (1) number of duties (drivers), (2) number of split duties, (3) total number of changes, (4) total duty hours, (5) average duty length, (6) total working hours, (7) average working time, (8) number of short duties, and (9) costs. Sections 9.2 through 9.4 provide tools for handling both the preparation for and analysis of the construction of efficient crew schedules (duties).

Once the set of duties are established, it is common to group them into rosters. A roster is defined as a duty assignment over a certain amount of consecutive days, guaranteeing that all the trips are covered for a certain (usually cyclic) period. Commonly, a roster contains a subset of duties covering six consecutive days (called weeks). The length of a roster is typically between 30 and 60 days (5–10 weeks). The usual rostering problem is to find a feasible set of rosters to cover all the duties, using one or more of these four objectives: (a) minimum number of crews required, (b) minimum sum of roster costs, (c) minimum of the maximum roster duration, and (d) balancing (the equity of) workload and days-off. The rostering problem is presented in Section 9.5.

9.2 VEHICLE-CHAIN CONSTRUCTION USING A MINIMUM CREW-COST APPROACH

There are two predominant characteristics in transit operations planning: (a) different resource requirements between peak and off-peak periods and (b) working during irregular hours. These characteristics result in split duties (shifts) with unpaid periods in between. It is often called swing time. The inconvenience accompanying split duties led driver (crew) unions to negotiate for an extension of the maximum allowed driver's idle time for which the driver can still get paid. It is common, therefore, to have a constraint in a labor union agreement specifying this maximum paid idle time (swing time), to be termed T_{max}.

The crew-scheduling problem (CSP) from the agency's perspective is known to be the minimum crew-cost problem. With this minimum-cost orientation in mind, we can use the deficit-function (DF) properties to construct vehicle chains (blocks) that take into account

T_{max}, in other words, to maximize idle times (swing times) that are larger than T_{max} and hence to reduce crew costs. What follows is based on the first edition of this book and on Ceder (2011).

9.2.1 Arrival–departure joinings within hollows

The following description uses the notation and definitions associated with the DFs of Section 6.5.1. Each hollow of a DF, $d(k, t)$ at terminal k, contains the same number of departures and arrivals, except for the first and last hollow at the beginning and end of the schedule horizon. This is due to the fact that each arrival reduces $d(k, t)$ by one and each departure increases it by one, so that the hollow starts and ends at $D(k)$.

For a given hollow, H_m^k, let I_m^k be the set of all arrival epochs t_e^i in H_m^k, and let J_m^k be the set of all departure epochs t_s^j in H_m^k. The difference in time between departure and arrival is defined as $\Delta_{ij} = t_s^j - t_e^i$ for $t_s^j > t_e^i$ in H_m^k. The joining (connection) between t_e^i and t_s^j in a vehicle block is effectively the idle time between trips; hence, Δ_{ij} may represent this idle time. We may also define local peak uv within hollow H_m^k as $d(k, t_{uv})$ between t_s^u and t_e^v, in which $e_m^k < t_s^u \leq t_{uv} \leq t_e^v < s_{m+1}^k$, where H_m^k starts and ends at e_m^k and s_{m+1}^k, respectively. Note that if the start and/or end of a local peak, uv, has more than one departure or arrival, then it suffices to refer to only one of them (as u or v). Let $d_{uv}^{k,m}$ be the number of departures in H_m^k before and including t_s^u and $a_{uv}^{k,m}$ be the number of arrivals in H_m^k before t_s^u.

Lemma 9.1:

The number of arrival–departure joinings in hollow H_m^k before t_s^u must be $d_{uv}^{k,m}$.

Proof: If some departure epochs before a local peak, uv, are left without a joining, it will be impossible to connect them with arrival epochs after t_e^v. That is, each departure epoch before and including t_s^u must have a joining to an earlier arrival time within H_m^k. This can be seen in Figure 9.1a.

Lemma 9.2:

The number of arrival–departure joinings that can be constructed after t_e^v within H_m^k is $(a_{uv}^{k,m} - d_{uv}^{k,m})$.

Proof: Given hollow H_m^k and local peak uv, then based on Lemma 9.1 and the characteristics of local peaks in hollows, $d_{uv}^{k,m}$, departure epochs must and can be joined to earlier arrival epochs in H_m^k; hence, the number of arrival epochs left over without joinings is $(a_{uv}^{k,m} - d_{uv}^{k,m})$ for all local peaks. Figure 9.1b displays this explanation.

Lemma 9.3:

The sum of all idle times within any hollow is a fixed number and independent of any procedure aimed at joining arrival and departure epochs, that is, $\sum_{i,j} \Delta_{ij} = $ constant.

Proof: Let H_m^k have n arrivals and n departures. We noted previously that the number of departures and arrivals are the same within each middle hollow (i.e., excluding the first and

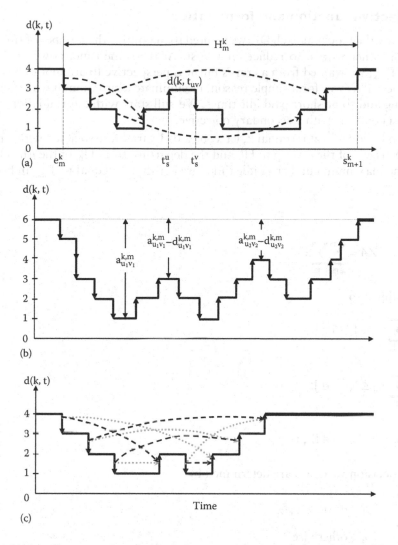

Figure 9.1 Part (a) describes examples of arrival–departure joining to support Lemma 9.1, part (b) interprets Lemma 9.2, and part (c) shows two 4-joining examples of Lemma 9.3.

last hollows). Let two different n-joining arrangements with idle times Δ_{ij}^1 and Δ_{ij}^2 for all joinings between $i \in I_m^k$ and $j \in J_m^k$ be expressed as follows:

$$\sum_{i,j} \Delta_{ij}^1 = \sum_{i,j} \left(t_{s1}^j - t_{e1}^i \right) = \sum_j t_{s1}^j - \sum_i t_{e1}^i$$

and similarly

$$\sum_{i,j} \Delta_{ij}^2 = \sum_j t_{s2}^j - \sum_i t_{e2}^i$$

We know that the sum of all departure or arrival times in a hollow is a fixed number; hence, $\sum_{i,j} \Delta_{ij}^1 = \sum_{i,j} \Delta_{ij}^2 = $ constant. Figure 9.1c further clarifies this argument.

9.2.2 Objective function and formulation

In constructing the blocks at each DF, we wanted to maximize the number of times in which $\Delta_{ij} \geq T_{max}$, in other words, to reduce crew cost. At the same time, however, for cases in which $\Delta_{ij} < T_{max}$, we wanted from a crew's fairness perspective to attempt to have equitable paid idle times. This was for a simple reason, to eliminate a situation in which some drivers will have long and some short paid idle times. We will start with a formulation of the main objective and continue with a secondary objective.

For a given hollow H_m^k at terminal k, let x_{ij} be a 0–1 variable associated with a trip joining between the arrival of the ith trip to H_m^k and the departure from H_m^k of the jth. The problem of finding the maximum number of idle times greater than or equal to T_{max} in hollow H_m^k is as follows:

Problem P4

$$\text{Max } Z4 = \sum_{i \in I_m^k} \sum_{j \in J_m^k} x_{ij} \tag{9.1}$$

Subject to

$$\sum_{j \in J_m^k} x_{ij} \leq 1, \quad i \in I_m^k \tag{9.2}$$

$$\sum_{i \in I_m^k} x_{ij} \leq 1, \quad j \in J_m^k \tag{9.3}$$

$$x_{ij} = \{0,1\}, \quad i \in I_m^k, j \in J_m^k \tag{9.4}$$

The binary decision variables are determined by

$$x_{ij} = \begin{cases} 1, & t_s^j - t_e^i \geq T_{max} \\ 0, & \text{otherwise} \end{cases}$$

A solution with $x_{ij} = 1$ indicates that joining trips i (arrival epoch) and j (departure epoch) results in an idle time larger than or equal to T_{max}. Constraints (9.2) and (9.3) ensure that each trip in H_m^k may be joined with, at most, one successor trip and one predecessor trip, respectively.

Trips that were not joined in the solution of P4 are subject to a secondary objective: equitable paid idle times. It is shown in Theorem 9.1 that joinings with this secondary objective are based on the first-in, first-out (FIFO) rule. Balancing Δ_{ij} for $\Delta_{ij} < T_{max}$ is the same as minimizing the difference between each Δ_{ij} and its average $\overline{\Delta}_{ij}$ either by absolute difference or by least-square difference. The FIFO rule used for this balancing is stated in the following theorem.

Theorem 9.1

Minimizing the least-square differences between $\overline{\Delta}_{ij}$ and each Δ_{ij} for all $i \in I_m^k$ and $j \in J_m^k$ in H_u^k is accomplished by constructing joinings using the FIFO rule.

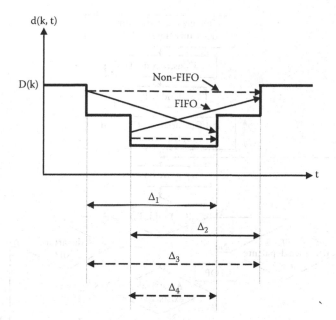

Figure 9.2 Example of a comparison of joinings based on FIFO and other rules.

Proof: It is sufficient to prove Theorem 9.1 on a simple but generalized example, as illustrated in Figure 9.2, with a hollow containing two arrivals and two departures.

We may show that

$$(\Delta_1 - \overline{\Delta})^2 + (\Delta_2 - \overline{\Delta})^2 < (\Delta_3 - \overline{\Delta})^2 + (\Delta_4 - \Delta)^2 \tag{9.5}$$

where $\overline{\Delta}$ is the average arrival–departure joining length in the example. Using an algebraic expression, then (9.5) becomes

$$\Delta_1^2 + \Delta_2^2 - \Delta_3^2 - \Delta_4^2 < 2\overline{\Delta}(\Delta_1 + \Delta_2 - \Delta_3 - \Delta_4) \tag{9.6}$$

Lemma 9.3 states that $\Delta_1 + \Delta_2 = \Delta_3 + \Delta_4$, and hence the right-hand side of expression (9.6) is zero. From Lemma 9.3, we can further obtain $(\Delta_1 + \Delta_2)^2 = (\Delta_3 + \Delta_4)^2$ or $\Delta_1^2 + \Delta_2^2 - \Delta_3^2 - \Delta_4^2 = 2\Delta_3\Delta_4 - 2\Delta_1\Delta_2$. The latter is inserted into (9.6) to yield

$$\Delta_3\Delta_4 < \Delta_1\Delta_2 \tag{9.7}$$

Based, again, on Lemma 9.3, let $\Delta_3 - \Delta_1 = \Delta_2 - \Delta_4 = B$ or $\Delta_1 = \Delta_3 - B$ and $\Delta_4 = \Delta_2 - B$; these last two equations are inserted into (9.7) to obtain $\Delta_3(\Delta_2 - B) < \Delta_2(\Delta_3 - B)$, which yields $\Delta_3 > \Delta_2$. The last result must be correct from Figure 9.2, and therefore, it agrees with expression (9.5).

9.2.3 Procedure for determining the maximum number of unpaid idle times

The mathematical programming formulation in Equations 9.1 through 9.4 is aimed at maximizing the number of idle times that are longer than or equal to T_{max}. However, this

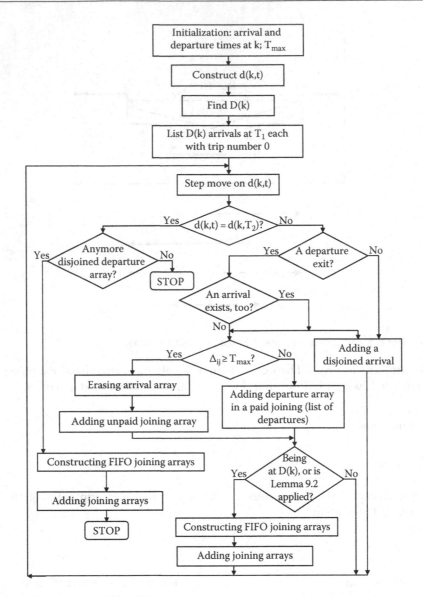

Figure 9.3 Flow diagram of algorithm T_mF.

formulation may involve a very large number of computations (nonpolynomial [NP] complete, see Section 5.5), hence entailing the use of another (more simplified) procedure. Such a procedure is described in a flow diagram in Figure 9.3 and contains both T_{max} and FIFO rule considerations; the latter is for joining arrivals and departures with paid idle times. Let us call this procedure algorithm T_mF.

The input for algorithm T_mF for each terminal *k* consists of two arrays, the arrival and departure arrays, and a given T_{max}. This input enables constructing DF at k and obtaining D(k) following the insertion of deadheading (DH) trips and the shifting of departure times for minimizing fleet size (see Chapters 6 and 7). Because D(k) vehicles are required at k, we assume their arrivals there to be at (or before) T_1 (the start of the schedule horizon). Algorithm T_mF moves by steps on d(k, t), in which each step refers to a change in d(k, t) or the detection of a dot on d(k, t); the dot means that arrival and departure epochs at k overlapped at t.

Algorithm T_mF continues with a check of the end of the schedule horizon and detects the nature of the change (or dot) in d(k, t). For each departure epoch, Δ_{ij} is examined to determine whether it is greater than or equal to T_{max}; if greater, then an unpaid joining array is added; otherwise a disjoined departure time array is added. Each arrival epoch (detected in a step move in Figure 9.3) is added as a disjoined arrival array. If a departure epoch is identified in a step move, the algorithm looks for a possible dot on d(k, t), adding its arrival epoch to the list of disjoined arrival arrays. At the end of the process, the algorithm constructs joining arrays from the disjoined arrival and departure arrays, using the FIFO rule. The complete process is shown in Figure 9.3.

An example of constructing vehicle chains (blocks), including the employment of algorithm T_mF, is shown in Figures 9.4 through 9.6. The example, consisting of three terminals and a 24-trip schedule, is exhibited in Figure 9.4, including DH travel time matrix, shifting tolerance, T_{max}, and schedule horizon. It should be noted, though, that DH travel time between terminals b and c is considered in both directions although there is only a service route between c and b. The fleet-reduction procedure, involving the shifting of departure times and DH trip insertions, is shown in Figure 7.12; here it is applied to the example in Figure 9.5. Two DH trips and two shifts are introduced into the process to reduce D(a) and D(c) from 4 to 3, resulting in a fleet size of 11 vehicles. The shifts are shown in Figure 9.5

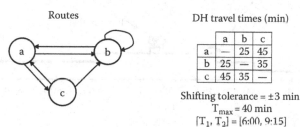

Routes

DH travel times (min)

	a	b	c
a	—	25	45
b	25	—	35
c	45	35	—

Shifting tolerance = ±3 min
T_{max} = 40 min
$[T_1, T_2]$ = [6:00, 9:15]

Trip number	Departure terminal	Departure time	Arrival terminal	Arrival time
1	a	6:00	b	6:30
2	a	6:15	c	7:20
3	a	6:20	c	7:25
4	a	6:43	c	7:38
5	a	7:10	c	8:00
6	a	7:30	b	8:25
7	a	7:50	b	8:35
8	a	8:00	b	8:50
9	a	8:10	b	8:50
10	b	6:00	a	6:30
11	b	6:00	b	6:40
12	b	6:15	b	7:00
13	b	6:15	a	7:00
14	b	6:20	a	7:30
15	b	7:30	a	8:30
16	b	7:35	a	8:40
17	b	7:45	a	8:50
18	b	8:05	a	9:00
19	c	6:30	b	7:15
20	c	7:00	a	7:40
21	c	7:15	b	7:45
22	c	7:15	a	7:50
23	c	7:20	b	8:50
24	c	7:38	b	9:03

Figure 9.4 Example consisting of 24 trips and 3 terminals for constructing vehicle chains with the T_{max} constraint.

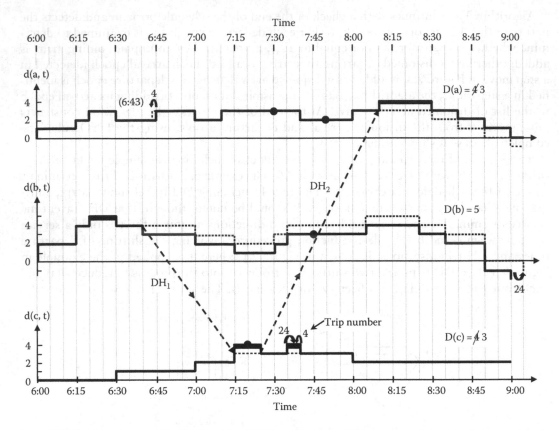

Figure 9.5 The 24-trip example (depicted by 3 DFs) undergoing a DH trip-insertion procedure, combined with shifting departure times.

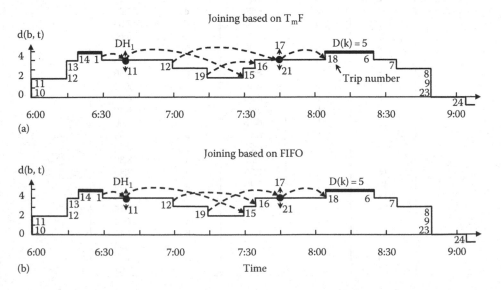

Figure 9.6 Arrival–departure joinings for constructing vehicle chains (blocks) in the middle hollow of terminal *b*, utilizing in (a) the T_mF algorithm and in (b) the FIFO rule.

by their shifting length and trip number. We can see from this figure that the only middle hollow containing more than a single departure is the second hollow of d(b, t); hence, only this hollow is subject to the process of algorithm T_mF.

Figure 9.6a describes the solution for algorithm T_mF in comparison with a solution based only on the FIFO rule in Figure 9.6b. The trip numbers of the example, appearing in Figure 9.4, are added to Figure 9.6. Algorithm T_mF results in two unpaid joinings between the arrivals of trips 11 and 12 and the departures of trips 15 and 17, respectively. In both cases, $\Delta_{ij} > T_{max} = 40$ min. The remaining joinings in Figure 9.6a are based on the FIFO rule. The use of only the FIFO rule for the entire process results in only one unpaid joining (that between trips 11 and 15) as is shown in Figure 9.6b.

The final phase of the arrival–departure joining process is to construct vehicle blocks. This will contain the joinings created and other FIFO-based joinings in order to make a complete set of blocks. The 11 blocks of the 24-trip-schedule example, based on algorithm T_mF (at terminal b), are given by their numbers in the following list: [1–DH_1–22–7], [10–4–24], [11–15], [2–23], [12–17], [13–5], [3–DH_2–9], [14–6], [19–16], [20–8], [21–18]. The process based only on the FIFO rule results in the same blocks, except for the fifth and ninth blocks, which become [12–16] and [19–17], respectively.

9.3 MATHEMATICAL SOLUTIONS

Crew scheduling and rostering are extensively treated mathematically in the books edited by Wren (1981), Rousseau (1985), Daduna and Wren (1988), Desrochers and Rousseau (1992), Daduna et al. (1995), Wilson (1999), and Voss and Daduna (2001) and in a forthcoming volume by Hickman et al. (2008). Specific detailed studies are reviewed in the succeeding text in Section 9.6.

The basic OR formulation of the CSP is a zero-one integer linear programming (ILP), called a set-partitioning problem (SPP). SPP has as its objective the selecting of a minimum cost set of feasible duties, such that each duty piece is included in exactly one of the duties. CSP is often illustrated by rows and columns, in which the rows are the duty pieces and the columns are the duties, each accompanied by a duty cost. In the latter, SPP is aimed at minimizing the cost of a set of columns, such that each row is included exactly once in one of the columns in the solution. The following is the SPP formulation:

Problem P5

$$\text{Min}\, Z5 = \sum_{q \in Q} c_q x_q \tag{9.8}$$

Subject to

$$\sum_{q \in Q(j)} x_q = 1, \quad j \in J \tag{9.9}$$

$$x_q = \{0,1\}, \quad q \in Q \tag{9.10}$$

where
    Q is the set of all feasible duties
    c_q is the cost of duty $q \in Q$
    $Q(j) \in Q$ is the set of duties covering duty piece $j \in J$

A binary zero-one variable x_q is used to indicate whether duty q is selected in the solution or not. Constraint (9.9) assures that each piece will be covered by exactly one duty. The running time of P5 belongs to the class of NP complete (see Section 5.5).

An easier way to solve the CSP is to relax constraint (9.9):

$$\sum_{q \in Q(j)} x_q \geq 1, \quad j \in J \tag{9.11}$$

Equation 9.11, together with Equations 9.8 and 9.10, represents a new problem, called a set-covering problem (SCP). SCP is usually solved first (before the SPP), and the solution is changed to handle SPP by deleting overlapping trips. This experienced-based deletion process involves changes in the duties, in the SCP solution, and in a crew member who is so assigned and will make the trip as a passenger. Freling et al. (1999) explain that such a change affects neither the feasibility nor the cost of the duties considered.

One way to speed up the running time of the SPP of CSP is by the use of the *Lagrangian-relaxation* method. In this method, the complicated set of constraints shown by Equation 9.9 is removed and put into the objective function (9.8). Each constraint removed is weighted by a given Lagrange multiplier. Another useful and commonly used method for reducing the SPP complexity is the *column-generation* (CG) technique. The number of columns (representing duties) in the SPP is usually a very large number. The CG technique, instead of examining all the columns over and over again, selects subsets of columns and either checks each subset for optimality or solves several subproblems. Some of the articles reviewed in Section 9.6 cover these methods.

A known heuristic approach to solve CSP within a framework of resolving both CSP and the vehicle-scheduling problem (VSP) is discussed by Bodin et al. (1983) and presented by Ball et al. (1981, 1983). This approach is chosen to illustrate the inherent combinatorial difficulties that exist in CSP. The basic steps of the approach are as follows: (1) generate all feasible pieces of work (to be derived from the vehicle blocks); (2) establish an internal piece cost (based on piece characteristics and past experience); (3) create an acyclic network from each block, in which nodes are the relief points and arcs represent the cost of each feasible piece in the block; (4) solve the shortest-path problem in order to establish the best (minimum cost) pieces; and (5) solve a matching problem while using a two or three legal-piece combination, and if the solution is not feasible, reiterate part of the process by updating the piece combination cost and redo the shortest path (*Step 4*) until the solution is feasible and satisfactory. Figure 9.7 presents the procedure schematically.

The example presented in Figure 9.4 is used for demonstrating the shortest-path and matching (SPM) approach exhibited in Figure 9.7. The 11-block results of the T_mF algorithm of this example appear in Table 9.1; they are extracted from the analysis shown in Figures 9.5 and 9.6a. Given that all three terminals a, b, and c are relief points, each block can be considered a driver schedule (duty) or be partitioned into alternative pieces covering all possible combinations. Figure 9.8 shows how to partition the blocks into combinations of pieces. In addition, each piece is assigned an internal piece cost based on the piece characteristics (e.g., time of day, arrival and departure locations, type of vehicle required) and past experience (the cost of a similar piece in past crew schedules). Two sets of blocks have the same pieces and costs (numbered 3 and 5, and 9 and 11). It should be noted that the first set includes unpaid idle time as a result of algorithm T_mF.

The right-hand column of Figure 9.8 contains the acyclic network of each block, representing all the possible pieces and their internal costs. This acyclic network has undergone a

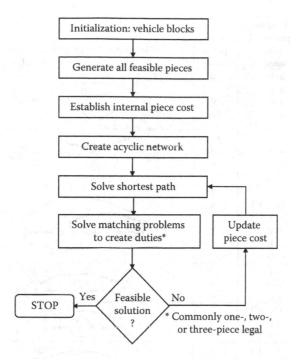

Figure 9.7 Flow diagram of the SPM heuristic approach to creating duties.

Table 9.1 Solution for vehicle blocks and routing of the example appearing in Figures 9.4 through 9.6a, using the algorithm T_mF

Block number	Trips in block (see Figure 10.4)	Block routing (see Figure 10.4) (# represents DH trip)
1	1-DH₁-22-7	a–b # c–a–b
2	10-4-24	b–a–c–b
3	11-25	b–b–a
4	2-23	a–c–b
5	12-17	b–b–a
6	13-5	b–a–c
7	3-DH₂-9	a–c#a–b
8	14-6	b–a–b
9	19-16	c–b–a
10	20-8	c–a–b
11	21-18	c–b–a

shortest-path analysis, such as the known algorithm Dijkstra (1959). For the sake of clarity, the Dijkstra algorithm is described with an example in Appendix 9.A; it will also be used in Chapters 11, 12, 15, and 16. The results of the Dijkstra procedure are emphasized in Figure 9.8, along with the minimum piece cost needed to cover the whole block. These results are illustrated in Figure 9.9 for each block. Given that each piece is eligible to be covered by 1 driver, then Figure 9.9 shows that the 11 blocks require 17 drivers at a total cost of 82 (the cost units are meaningless for the example). However, there are more possibilities for matching some pieces if they are legal (from the labor-agreement perspective) and can reduce the total cost.

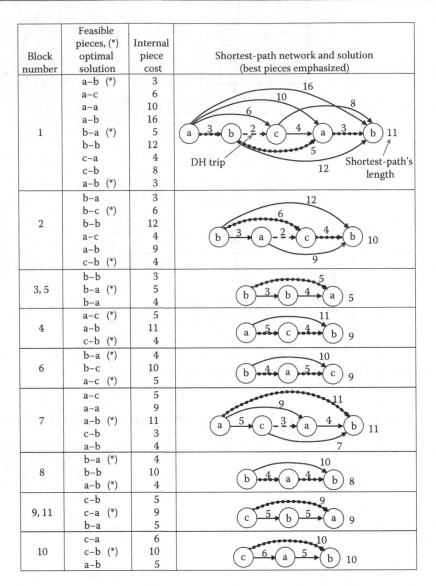

Block number	Feasible pieces, (*) optimal solution	Internal piece cost	Shortest-path network and solution (best pieces emphasized)
1	a–b (*)	3	
	a–c	6	
	a–a	10	
	a–b	16	
	b–a (*)	5	
	b–b	12	
	c–a	4	
	c–b	8	
2	a–b (*)	3	
	b–a	3	
	b–c (*)	6	
	b–b	12	
	a–c	4	
	a–b	9	
	c–b (*)	4	
3, 5	b–b	3	
	b–a (*)	5	
	b–a	4	
4	a–c (*)	5	
	a–b	11	
	c–b (*)	4	
6	b–a (*)	4	
	b–c	10	
	a–c (*)	5	
7	a–c	5	
	a–a	9	
	a–b (*)	11	
	c–b	3	
	a–b	4	
8	b–a (*)	4	
	b–b	10	
	a–b (*)	4	
9, 11	c–b	5	
	c–a (*)	9	
	b–a	5	
10	c–a	6	
	c–b (*)	10	
	a–b	5	

Figure 9.8 Partitioning of blocks into minimum-cost pieces.

Taking into account the DH travel times in Figure 9.4 as a measure of moving between a, b, and c, there are five possible matchings that need to be examined for the first piece of block 1 and one possible matching for the first piece of block 6. These possibilities are shown by the dashed lines in Figure 9.9. Usually, if the best (minimum cost) feasible matching results in a cost reduction (i.e., less than 82), then this matching is selected and the number of drivers can be reduced by one or two (to 16 or 15). The more pieces examined for a single duty, the more running time the matching process requires (Ball et al., 1983). Commonly, the check is performed on two and three legal pieces. More details on possible optimal and heuristic matching procedures can be found, for example, in Ahuja et al. (1993).

The simplified example using an SPM approach demonstrates some of the complexities involved in handling CSP, hence justifying the use of OR software. Moreover, in addition to the combinatorial problems inherent in the scheduling tasks, there are human dissatisfaction issues concerning the crews that deserve attention and make this undertaking even more cumbersome.

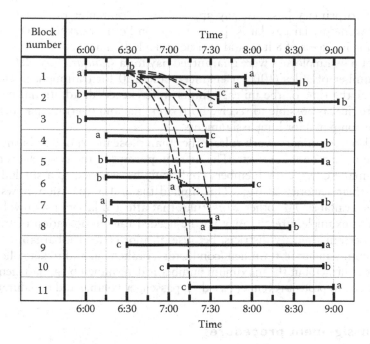

Figure 9.9 Result of partitioning the blocks into best pieces with feasible matching alternatives.

9.4 CASE STUDY: NJ COMMUTER RAIL

This section describes a case study that was performed for the New Jersey (NJ) Transit Corporation. In this study (Tykulsker et al., 1985), we will experience that all computers wait at the same speed; what counts is the flexibility of the solution and the alternatives that can be produced.

9.4.1 Background

NJ Transit provides a large commuter rail operation, carrying over 70,000 passengers and running more than 400 trains daily. At the time of the study, NJ Transit had decided to take under its wing both the service and labor force then maintained by Conrail. The collective bargaining agreements then in effect between Conrail and the various bargaining units operating in the labor force would then no longer remain in effect. Thus, NJ Transit was placed in the position of negotiating new collective bargaining agreements.

With labor-crew costs representing a major component of commuter rail costs, it was critical that NJ Transit management understand the cost ramifications of different work rules. Given the variations in these rules that already existed for the different NJ commuter rail lines and the multitude of permutations, it was important that NJ Transit have some mechanism for quickly analyzing options. Thus, the objective of the project was to provide the NJ Transit management with a tool that would enable them to quickly analyze the implications of work rule changes for labor costs. This tool would be used during the nego-tiation process, since it would provide management with the ability to quantify the impacts of various proposals and counterproposals. In addition, NJ Transit needed a tool that could be used on an ongoing basis in order to (1) analyze the marginal cost impact of changes in train schedules and (2) produce near-optimal crew assignments (duties), given a schedule and a new set of work rules.

There were two primary bases of pay applicable to commuter rail. The first pay basis was similar to an industrial pay basis: pay is based on hours worked, with a guarantee of 8 h/day. Any time in excess of 8 h is paid at time and a half. The second pay basis was called the dual basis of pay. Employees were guaranteed pay for a *standard day*, defined as 8 h and a prescribed number of miles (100 for engineers and 150 for trainmen, in the case of NJ Transit); if either the time or the mileage limits were exceeded, additional payments were made. Therefore, it was possible to receive both overtime and overmile payments, a double penalty (for the company) for long assignments.

There were also two primary work rules, other than those incorporated in the basis of pay, that had a large impact on labor costs. The more important of the two relates to interlining restrictions. In many cases, crew members could only be assigned to a single line or group of lines within their seniority district. This limited the ability to assign crews in the most efficient manner available. The other work rules that affected labor costs involved manning requirements. For example, firemen may not be needed for the operation of trains but may be required, in certain seniority districts, because of prior collective bargaining agreements. Also, the minimum number of trainmen (conductors, brakemen, and ticket collectors) is usually specified by contract, but the maximum number is determined by a management decision relating to the need to protect revenue, based on passenger volume and trip duration.

9.4.2 Crew assignment procedures

9.4.2.1 Overview of model

To address the issues described, a crew assignment/work rule model was developed. The approach consisted of three steps: (1) defining train schedules (arrival and departure times and locations, trip distances, and manpower requirements) and work rules (interlining restrictions, DH rules, minimum layovers, hours of service limits, and the locations of the beginning and ending of duties), (2) generating a range of alternative crew assignments, and finally, (3) selecting the set of *best* assignments (duties).

Crew assignments were generated, starting with the earliest departing train, by finding the first, second, third, etc., feasible departing connections at each location. From each of these trains, additional connections were explored; thus, a tree of assignments was found for each train. The size of the tree can be limited by several user-supplied parameters. If the generation is limited, the entire procedure is repeated, starting with the latest arrivals and using feasible connections. This is to ensure that a good set of morning and evening assignments are generated. Assignments were then costed by applying the appropriate bases of pay. The costing procedure allowed for hourly or mileage-pay bases; daily, weekly, and monthly guarantees; both overtime and overmile payments; spread premiums; and split duties.

The last step was to determine the actual crew assignments. Several heuristic techniques were developed to reduce the magnitude of the computational problem. An optimization package was used separately for weekdays, Saturdays, and Sundays. For each day, a relaxed (noninteger) SCP was formulated and solved, assignments in the solution were set with overlapping trains, additional assignments were made without the overlap, and this reduced SCP was solved for an integer solution. Finally, the separate daily solutions were combined, producing assignments with the required relief days. Figure 9.10 presents an overview of the model's components.

9.4.2.2 Model development

The first stage of the project was to identify an appropriate methodology. To accomplish the assignment-selection procedure, careful consideration was given to various mathematical

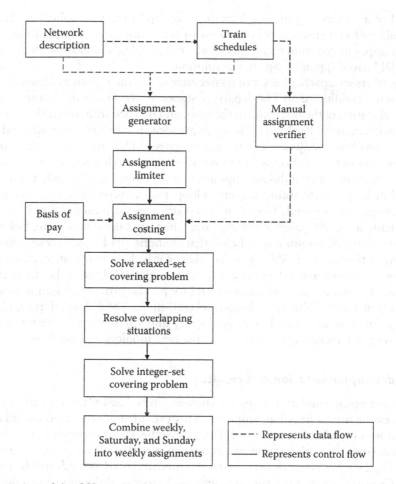

Figure 9.10 Overview of the CSP process in the NJ case study.

programming and heuristic techniques. It is worth mentioning that in a comparison of bus- and rail-crew scheduling, interlining and DH issues are more appropriate to bus-crew assignments. The combinatorial complexities of bus-crew scheduling, introduced by interlining and DH, may dictate sequential optimization techniques. For the case study, use of the first general technique of eliminating assignments prior to optimization was found to be sufficient.

We have already noticed that it is time consuming and expensive to solve large-size problems by exact integer-programming procedures. Consequently, it is common either to relax the integer restriction (solving SCP) or to use heuristics. The SCP formulation was deemed more appropriate for the rail CSP of the case study, since crews can and do deadhead on trains. The SCP is formulated in P5 in Equations 9.8, 9.10, and 9.11; in the case study, the elements in P5 have the following interpretation: the set Q is all feasible duties (assignments), c_q is the crew cost of duty (assignment) $q \in Q$, $Q(j) \in Q$ is the set of duties (assignments) covering a set of trains (or trips) $j \in J$, and x_q is used for indicating whether duty q is to be worked (selected in the solution) or not.

In commuter rail operations, crews occasionally perform DH trips while traveling on service trains. In bus operations, crews may deadhead via a route other than the service route, with or without their vehicle. Usually, bus-scheduling programs include a special DH travel-time matrix so that the DH crew/vehicle can reach the next departure point prior to the arrival of a service trip (with which they overlap in the SCP solution). Although the early availability of DH crews may create new opportunities for bus operations, this situation does not apply to commuter rail operations; the problematic issue with the latter is the existence of DH (overlapping) loops in the solution.

A DH loop refers to a particular set of consecutive trains on a given assignment that fulfills the two following conditions: (a) The departure station of the first train on the loop coincides with the arrival station of the last train on the loop, and (b) each train on the loop is covered by other assignments in the SCP solution. The loops are identified by first creating a train-coverage table, which shows how many times each train is covered. The procedure is then iterated over all the assignments in the solution set and over all trains in each assignment. When a train is covered more than once, all candidate loops are reiterated (initially for each assignment, there are no candidate loops). If the train completes a loop, the trains on the loop are eliminated and the loopless assignment is found. This train then starts a new candidate loop. Eliminating DH loops will usually lower the value of the objective function. One should note, however, that it is possible to obtain a minimum-cost solution that contains DH loops if these assignments are already costed at the minimum daily pay (in other words, loop elimination cannot lower the cost). From an operations and safety perspective, though, all DH loops should be eliminated.

To overcome the undesirable situation of DH loops, the process chosen used three main steps as shown in Figure 9.10: (1) solving a relaxed (linear) SCP formulation, (2) resolving DH loops by forming new loopless assignments, and (3) solving an exact (integer) SPP formulation with the incumbent solution plus the new loopless assignments.

9.4.2.3 Model implementation and results

The model was implemented as a set of *assignment filters*: each program was separated and acted to process or filter a file of assignments. This provided a structured modular package. In addition, it was found necessary to build extensive checks for error and inconsistency into this procedure, since the optimization procedure assumes that all assignments presented to it are feasible. The existing assignments were then costed and used as *seeds* for the optimization procedures. Figure 9.11 illustrates the computational process of the model in a flow diagram.

A major component of the model was the assignment generator. Assignments were generated without regard to crew cost. This is due to the fact that the model was explicitly designed to investigate alternative bases of pay. The generation was carried out in two passes, forward and backward, which helped to ensure that each train was covered at least once; also, this provides better-quality assignments at the start and finish of the day. Five parameters were used in order to limit the number of assignments generated: (1) limit on the connection-time window, (2) limit on the number of connections that need to be examined for each train in the search tree, (3) maximum number of assignments to be formed from each root tree, (4) minimum total time for any assignment, and (5) maximum total time for any assignment.

Split duty and part-time assignments were generated by specifying three additional parameters: the time at which a split can occur, the minimum length of the split, and the minimum length of a part-time operation. In splitting a duty, the generator simply enforces a layover of at least the minimum length at the appropriate time.

Once a set of regular, split, and part-time assignments was generated, a separate procedure was employed to reduce the number of assignments. This is done so as not to overwhelm the optimization procedure. This reduction first computes a productivity measure for each assignment. Several measures can be used: pay hours per total hour (effective hourly rate), pay dollars per train hour, pay dollars per train, total hours per train hour, and so forth. It was decided to use total hours per train hour because it was felt to be politically less sensitive to use a measure

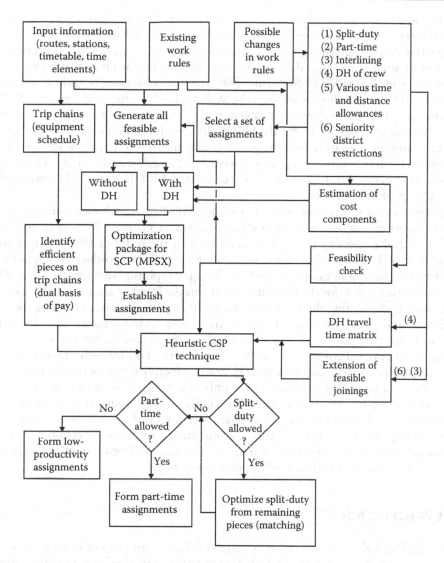

Figure 9.11 Flow diagram listing the major computational elements included in the NJ case study.

divorced from pay. The assignments were then sorted by productivity measures. The reduction was carried out by specifying a coverage limit—how many times each train must be covered before an assignment is dropped. The assignments were processed in order of productivity measure, updating a coverage count for each train on the assignment. If all the trains on an assignment are already covered to the limit, then that assignment is dropped. However, if at least one train is not at the limit, then the assignment is passed on. Thus, it is possible for some trains to be covered more than the coverage parameter. Furthermore, a train could be covered less than the coverage limit if the assignment generator did not produce that many assignments for this train. In practice, a coverage limit of 20 times per train was found to work well.

The assignment-costing procedure was designed to be as flexible as possible. It must be very simple in order to recost some or all of a set of assignments, using alternative bases of pay. More important, it must be easy to modify the procedure to incorporate completely new concepts in costing assignments. The mechanics of costing are as follows: For each assignment, select the appropriate pay basis (from the line-basis-of-pay definition), apply the appropriate rates to the miles and hours, and compute any guarantees. New pay bases are easily added by inserting the necessary code and referring to a previously unassigned parameter.

After costing, the assignments were ready to be optimized. As described previously, the process was to solve a relaxed SCP, resolve overlapping by forming new loopless assignments, recost, and then solve an SPP using both the new assignments and the relaxed SCP solution. The SPP was much smaller than the SCP; typically, it limits the relaxed SCP to 20 assignments per train and solves the problem with an average of 4 trains per assignment. Thus, the number of assignments for the SPP could be reduced by two orders of magnitude.

Another technique used to reduce the problem dimension is to partition the weekly crew-assignment problem into three components: weekday, Saturday, and Sunday. Three separate solutions were generated and solved and then combined with the three solutions to produce a full weekly set of assignments.

The optimization steps used the software package MPSX/370 with mixed-integer programming. The number of assignments (columns) ranged between 3379 and 7591 covering 250 trains (rows). The central processing unit (CPU) times were in the order of 100 s, during which it was found that the DH loop-elimination procedure was needed to produce good solutions: the objective function reductions ranged from 0% to 20% but typically were around 2% to 5%. The second linear optimization (for the relaxed solution plus the loopless assignments) usually terminated the integer, eliminating the need for branch and bound. For cases requiring the employment of branch and bound, the value of the objective function usually increases by less than 1%.

The model generally produced very good results; typical savings for manually generated crew assignments ranged from 2% to 20%. However, the DH loop-elimination procedure occasionally produced assignments with very low productivity. For example, an assignment with only two trains would uniquely cover only one train. The reason is that there are no penalties for these poor assignments: the choice among competing assignments during optimization is done solely on a cost basis and not productivity. NJ Transit used the model extensively during the transition period and during recent labor negotiation to evaluate alternative bases of pay and work rules. The Operations Department has exhaustively reviewed the assignments produced, with positive reactions.

9.5 CREW ROSTERING

The CSP establishes duties in an optimal manner with the aim of minimizing crew assignments for a given set of constraints. Usually transit agencies then arrange the duties in a set of patterns for a specified time horizon. Each pattern is defined as a *roster*, containing duties to be fulfilled over a certain number of consecutive days. Commonly, the pattern repeats itself in cycles (whether a week, a month, or any other period). The crew-rostering problem (CRP) is usually to find a feasible set of rosters to cover all duties, using one or more of these four objectives: (a) minimum number of crews required, (b) minimum sum of roster costs, (c) minimum of the maximum roster duration, and (d) balancing (equity of) workload and days-off.

9.5.1 Literature review and problem definition

A broad literature survey on CRP by Ernst et al. (2004) reveals that the number of articles on rostering in public transit is relatively small. The survey refers to a total of 193 articles on staff scheduling and rostering, with a classification of applications, problems, and models. Basically, the solution methods proffered for CRP in this survey are similar to those of CSP. That is, math-programming (e.g., SPP, SCP), heuristic, and metaheuristic approaches. The last is based on methods drawn from artificial intelligence, neural engineering, biological evolution, and more (see, e.g., Lourenco et al., 2001). Among the few studies on CRP, Carraresi and Gallo (1984) and Bianco et al. (1992) utilized network concepts for solving CRP. Catanas and Paixao (1995) used a SPP and SCP formulation combined with heuristic rules for CRP. Caprara et al. (1998) determined rosters complying with the minimum number of weeks in which each duty is carried out but once a day; mixed-integer programming was used as a base to develop an efficient heuristic algorithm. Sodhi and Norris (2004) decomposed the CRP at the London underground into two stages: The first established a roster pattern for each depot, followed by the insertion of duties into each of the patterns, and the second was formulated as an assignment problem (a known problem in OR; a special case of the minimum-cost flow problem on networks) with side constraints.

The CRP involves compliance with work and legal rules, institutional requirements, and individual preferences. The rules and requirements (some are safety oriented) commonly established in labor contracts necessitate constraints, as in CSP, that are difficult to treat analytically, for instance, the minimum number of hours required between two consecutive working days, exact (or between minimum and maximum) number of days-off per week, minimum number of weekend days in days-off per month, maximum working hours per day, equity of weekend working hours, and more.

The rosters are commonly arranged in cyclic patterns; the crew can have the same cyclic pattern or rotate between patterns, depending on the arrangement between the transit agency and their employees. The upper part of Figure 9.12 illustrates seven weekly roster types covering all possible rosters complying with two consecutive weekly days-off. Each duty is denoted by d_j^q, in which q is the day in the week and j is the index of a specific duty. Certainly some of the duties (among a total of 35) can be the same if, for example, there are fewer than five duties for a given day. Moreover, more rosters of the same type can be used when more duties are involved. A single crew member may be tied to a single roster or switched between rosters every month (or every few weeks). In the latter case, the number of days-off in the transition period can be fewer or greater than the required 2 days.

The lower part of Figure 9.12 presents a sequence of quintuplets (five consecutive working days) for a single crew member, which certainly results in one weekly day-off. The complete cycle of these quintuplets covers 36 duties in 6 weeks, some of which can be the same. Observation of the upper part of Figure 9.12 will show quintuplets if the same crew member repeats the same roster, but with two weekly days-off and not one. Rosters generally established in transit agencies may contain a combination of quintuplets, quadruplets, and triplets of working days.

9.5.2 Heuristic approach

This section provides a heuristic approach to a simplified CRP; this is done for the sake of illustrating the different planning elements involved and for illustrating the complexity of the analysis. The heuristic method is presented by an example in which the rosters are built on the basis of a 4-day week rather than the usual week of 7 days. That is, the planning horizon considered for the example is 4 days.

Table 9.2 provides the input data of the example. The objective is to find which rosters will determine the minimum number of crew members (drivers) required to cover all the duties, subject to given constraints. All together there are four different duties, d_j^q, j = 1, 2, 3, 4, in 4 days, q = 1, 2, 3, 4, to be assigned to three possible types of rosters. The lower part of Table 9.2 lists the start and end times, as well as the length of each duty. Two constraints exist: first, on the minimum number of (rest) hours required between two consecutive working days, which is *10 h*, and second, on the maximum allowed roster hours for each roster type (see Table 9.2; each roster in this table contains one duty per day for days marked with a "V").

		Mon	Tue	Wed	Thur	Fri	Sat	Sun
Roster type	R_1	d_1^M	d_2^{Tu}	d_3^W	d_4^{Th}	d_5^F	Off	Off
	R_2	Off	d_6^{Tu}	d_7^W	d_8^{Th}	d_9^F	d_{10}^{Sa}	Off
	R_3	Off	Off	d_{11}^W	d_{12}^{Th}	d_{13}^F	d_{14}^{Sa}	d_{15}^{Su}
	R_4	d_{16}^M	Off	Off	d_{17}^{Th}	d_{18}^F	d_{19}^{Sa}	d_{20}^{Su}
	R_5	d_{21}^M	d_{22}^{Tu}	Off	Off	d_{23}^F	d_{24}^{Sa}	d_{25}^{Su}
	R_6	d_{26}^M	d_{27}^{Tu}	d_{28}^W	Off	Off	d_{29}^{Sa}	d_{30}^{Su}
	R_7	d_{31}^M	d_{32}^{Tu}	d_{33}^W	d_{34}^{Th}	Off	Off	d_{30}^{Su}
Complete cycle of duties assigned to a single crew member, by week	1	d_1^M	d_2^{Tu}	d_3^W	d_4^{Th}	d_{23}^F	Off	d_6^{Su}
	2	d_7^M	d_8^{Tu}	d_9^W	d_{10}^{Th}	Off	d_{11}^{Sa}	d_{12}^{Su}
	3	d_{13}^M	d_{14}^{Tu}	d_{15}^W	Off	d_{16}^F	d_{17}^{Sa}	d_{18}^{Su}
	4	d_{19}^M	d_{20}^{Tu}	Off	d_{21}^{Th}	d_{22}^F	d_{23}^{Sa}	d_{24}^{Su}
	5	d_{25}^M	Off	d_{16}^W	d_{27}^{Th}	d_{28}^F	d_{29}^{Sa}	d_{30}^{Su}
	6	Off	d_{21}^{Tu}	d_{32}^W	d_{33}^{Th}	d_{34}^F	d_{35}^{Sa}	d_{36}^{Su}

Figure 9.12 Roster types with two weekly consecutive days-off and possible duties assigned to a single crew member with one weekly day-off for a sequence of quintuplets (five consecutive working days).

Table 9.2 Input data for the example problem

		Day 1	Day 2	Day 3	Day 4	
Duties distributed		d_1^1	d_1^2	d_1^3	—	
		d_2^1	—	d_2^3	d_2^4	
		—	d_3^2	d_3^3	d_3^4	
		—	—	d_4^3	—	*Maximum roster hours*
Type of roster	R_1	V	V	V	—	25
	R_2	—	V	V	—	18
	R_3	—	—	V	V	18

Duty for day q (q = 1, 2, 3, 4)	Start and end times	Duty length (h), L_j
d_1^q	6:00–16:00	10
d_2^q	12:00–20:00	8
d_3^q	16:00–23:00	7
d_4^q	18:00–midnight	6

The following heuristic procedure intends to find rosters with the maximum workload possible, thus fulfilling the objective of arranging the minimum number of crew members.

9.5.2.1 Procedure roster

Step 0: Initialization; determine index j of R_j by arranging all roster types in a list in decreasing order of their maximum possible lengths. For cases with the same length, assign a higher ranking in the list to roster types that contain more consecutive days (duties).

Step 1: Select the first nontreated roster type from the list in *Step 0*; construct a feasible network $G = \{N, A\}$ consisting of a set of nodes N and a set of arcs A per roster type, using a source node "s" and a target node "t"; let the arc's value be

$$v(i,j) = \begin{cases} L_j, j \in N, j \neq t \\ 0, \text{ otherwise} \end{cases}$$

where L_j is the duty length. Label the (unlabeled) nodes (say, by "V") contained in List-1 and List-2 paths; if no more rosters, go to *Step 4*.

Step 2: Solve a longest-path problem by using the rearranging node number (RNN) algorithm in Appendix 9.A (by multiplying all $v(i, j)$, $(i, j) \in A$ by (-1) and applying the shortest-path procedure); if no s–t path exists, go to *Step 1*.

Step 3: Delete and store in List-1 paths (arcs and nodes) belonging to the solution attained, except nodes s and t; if the current solution contains labeled node(s), store their path and the sum W in List–2, where $W = \sum_{j \in p^*, j \neq s,t} L_j$ and p^* is a longest path determined, and delete only the incoming arcs to the labeled nodes (in p^*); go to *Step 2*.

Step 4: Determine the best solution using the paths (rosters) with the highest W values from List-1 and List-2, such that each duty is covered only once. List all nodes (duties) that remain outside any path solution for a manual rostering-assignment process.

The example problem in Table 9.2 undergoes roster procedure. The index j assigned to the three types in Table 9.2 conforms to *Step 0* of the procedure. A preliminary stage to *Step 1* in roster is a feasibility check. Table 9.3 lists by roster type all infeasible duty connections; these infeasibilities affect *Step 1*. The three feasible networks constructed for R_1, R_2, and R_3 appear in Figure 9.13. The longest-path solution is attained in *Step 2* by algorithm RNN,

Table 9.3 Infeasible consecutive duty connections

Roster type	Infeasible connection (q = 1, 2, 3, 4)	Dominant constraint
R_1	$d_1^q-d_1^q-d_1^q$	Maximum roster hours
	$d_1^q-d_1^q-d_2^q$	
	$d_1^q-d_1^q-d_3^q$	
	$d_1^q-d_1^q-d_4^q$	
	$d_2^q-d_1^q-d_1^q$	
	$d_1^q-d_1^q-d_2^q$	
R_2	$d_1^q-d_1^q$	Maximum roster hours
	$d_3^q-d_1^q$	Minimum hours between consecutive working days
R_3	None	—

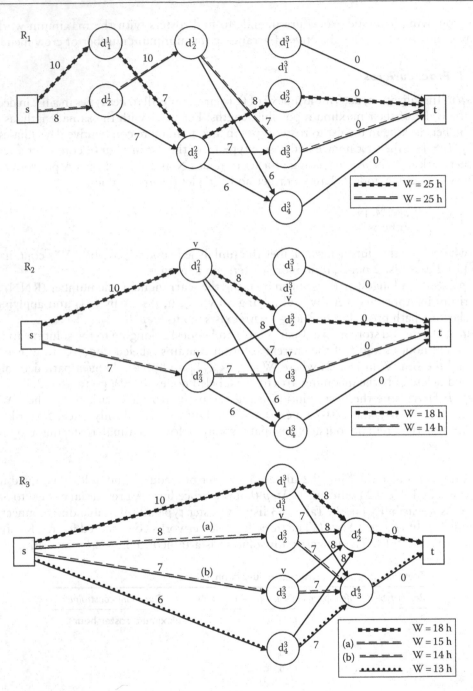

Figure 9.13 Heuristic rostering process for the example problem, using a longest-path approach for each roster type (R_1, R_2, R_3).

presented in Section 9.A.2. The RNN procedure is designed for finding the shortest path in networks with any arc value (it can also be negative) and especially without cycles (in order to avoid the possibility of an infeasible solution derived from negative cycles). The latter characteristic is compatible with the networks constructed in *Step 1*. The longest path is determined by the shortest path using RNN, in which v(i, j) is replaced by −v(i, j).

The first feasible network for R_1 in Figure 9.13 results, with the use of roster, in a two-path solution, both with W = 25 h; these paths are inserted into List-1. The second network for R_2 also results in two paths of pairs of duties (second-type rosters), but all four nodes—besides s and t—in the solution are labeled; that is, they were included in a previous solution. An analysis of R_2 provides the longest path, $s-d_1^2-d_2^3-t$ with W = 18 h. Then according to *Step 3* of roster, arc $s-d_1^2$ is deleted and the second longest path is found. Both paths of R_2 are inserted into List-2. The third and last roster-type network, R_3, begins *Step 2* with a nonlabeled path, $s-d_1^3-d_2^4-t$ with W = 18 h. The second solution, marked (a) in Figure 9.13, contains a labeled node; thus, only arc $s-d_2^3$ is deleted before continuing with the solution marked (b). In the latter case, arc $s-d_3^3$ is deleted, thereby allowing a fourth longest path with W = 13 h.

The first three steps of roster conclude with List-1 containing the solutions of R_1 and first and last solutions of R_3 and List-2 containing the solutions of R_2 and the second and third solutions of R_3. If we continue now to *Step 4* of roster, List-1 (usually having paths with the highest W values) covers all 10 duties and, therefore, presents the ultimate solution. Explicitly, four (as a minimum) crew members are required, two of roster type R_1 and two of R_3. The rosters are as follows: $[d_1^1-d_3^2-d_2^3], [d_2^1-d_1^2-d_3^3], [d_1^3-d_2^4]$, and $[d_4^3-d_3^4]$ with W = 25, 25, 18, and 13, respectively. The total hours of all duties are a given fixed number (81 h in the example). It should be noted that (1) if d_4^3, for instance, is removed from the input data, then d_3^4 will be left for manual assignment, and (2) List-2, though it may seem unnecessary, could open up opportunities for practical solutions utilizing a partial manual-assignment process.

9.6 LITERATURE REVIEW AND FURTHER READING

The CSP has been discussed abundantly both in and out of the transportation literature; relevant papers are found in journals related to mathematics, computing, operations research, and specialized scheduling resources. We will focus here mainly on the latest developments in this field. The literature for the CRP is sparse (see previous section).

CSP is often formulated as an SCP, as is explicitly shown in Equations 9.8, 9.10, and 9.11. For this purpose, a large set of driver workdays is defined, and a subset is then chosen that attempts to minimize costs, subject to constraints that make sure that all the necessary driving duties are performed. Most CSP formulations also verify that the labor-agreement rights of all drivers are maintained. A full classic CSP normally includes the generation of the feasible workdays as a first separate stage and then the choice of a subset that best satisfies all needs as a second stage. In most practical transit networks, there can be thousands of driving duties to be carried out, hence the need for millions of possible workdays. It is, therefore, common to construct in advance only a limited number of feasible workdays.

Most traditional CSP formulations concentrate on the work of bus drivers. Some of these formulations are transferable to railway use. This fact is emphasized throughout the review of papers that refer specifically to rail crews. The use of digital computers and heuristics to solve the CSP has attracted the attention of researchers since the middle 1960s (Elias, 1964). Wren and Rousseau (1995) reviewed the bus driver-scheduling problem and classified the methods used to solve the problem. Mitra and Darby-Dowman (1985) proposed a set-covering solution for CSP, using ILP. ILP was introduced to CSP-solution practice mainly during the 1980s as a result of improved computer power. A special feature of the formulation presented is its willingness to allow some uncovered duties for cases in which an optimal covering solution is not found.

Smith and Wren (1988) introduce another CSP formulation based on SCP, solved by using ILP. The elements covered are duty pieces; the covering process uses slack and surplus variables. Using pieces rather than entire duties helps to reduce the computational effort involved. The approach presented by Desrochers and Soumis (1989) also uses SCP but with a CG-solution procedure rather than ILP. The first stage of the CG process seeks a subset that best covers all duties within the range of the feasible workdays, which are already known. A second stage follows, in which new feasible workdays that improve the existing solution are generated iteratively.

Paixao (1990) formulates another SCP, using dynamic programming. The search for solutions employs a state-space relaxation method. This approach is extended in Paias and Paixao (1993), who show that this method finds the lower-bound solution. Carraresi et al. (1995) propose another CG approach, one that starts with a feasible set of workdays and iteratively replaces some workdays to obtain a better solution. The preconstructed workdays are built of duty pieces, and the solution uses a Lagrangian-relaxation method. Another somewhat similar CG method is proposed by Fores et al. (1999).

Clement and Wren (1995) introduce a solution for the CSP using a genetic algorithm (GA): a group of chromosomes, each of which represents a feasible crew schedule, is subject to repeated mutations, crossovers, and other actions, based on the idea that the search for an optimal solution can follow rules similar to a genetic survival mechanism. Several *greedy* algorithms are used for assigning duties to pieces of work. Another CSP-solution procedure based on a GA is presented by Kwan et al. (1999).

The approach demonstrated by Beasley and Cao (1996) does not follow an SCP concept; only a single type of workday is considered rather than attempting to search a broader range. A Lagrangian-relaxation method provides a lower-bound solution, which is later improved by using subgradient optimization. Next, a tree-search algorithm is used to obtain the final optimum. Beasley and Cao (1998) again use a similar approach but, instead of the Lagrangian-relaxation tool, seek the optimal lower bound by using a dynamic programming algorithm.

Another method that does not rely on SCP is suggested by Mingozzi et al. (1999). The authors describe two different duty-based heuristic solution procedures in which relaxed problems are formulated; their solutions also solve the original CSP. A third proposed solution procedure is based on SPP. The dual concept of linear relaxation programming is used to obtain a lower-bound solution. The number of variables in this problem is then reduced by using this lower bound; finally, the reduced-size problem is solved through a branch-and-bound technique. Haase et al. (2001) present an approach for the simultaneous solution of vehicle scheduling and CSPs in a single depot with a homogenous vehicle fleet. The CSP is based on an SPP formulation that incorporates side constraints for the bus itineraries. Their proposed approach consists of a CG process for the crew schedules integrated into a branch-and-bound scheme.

Lourenco et al. (2001) bring a multiobjective CSP, led by the concept that in practice there is a need to consider several conflicting objectives when determining the crew schedule. The multiobjective problem is tackled using metaheuristics, a Tabu-search technique, and GAs. Banihashemi and Haghani (2001) formulate a CSP as a duty-based network-flow problem. First, minimum cost is sought in a binary programming problem that is a relaxed version of the original, omitting labor-rights constraints. Then an iterative CG procedure using two sets of constraints is performed: *hard* constraints restricting the building of specific workdays (e.g., too long) and *soft* constraints penalizing nonefficient workdays, but not strictly forbidding their inclusion in the solution.

Freling et al. (2001a) and Huisman et al. (2005) present an integrated approach to solve a VSP and a CSP on a single bus route (i.e., they assume that different routes do not share the same drivers or vehicles). This combined methodology enables an examination of such scenarios as and when drivers are allowed to change vehicles between any two runs. First, CSP and VSP are defined separately. VSP is described as a network-flow problem, in which each path represents a single feasible vehicle schedule and each trip a node. CSP is an SCP that is solved after the VSP to form a sequential solution. An integrated approach is then presented, in which the network formulation of the VSP is combined with an SPP formulation for the CSP. The integrated model is one programming problem instead of two, but it cannot use pregenerated workdays, since there are too many options; therefore, duty pieces are used.

Shen and Kwan (2001) introduce a process that involves partitioning a predetermined vehicle schedule into a set of driver duties. The focus is on refining an existing small set of workdays; hence, the methodology does not include the common stage of generating all feasible solutions. A Tabu search is used to improve the given crew schedule. Tabu search is a class of metaheuristic that tries to avoid being trapped in a local optimum solution by basing the solution choice in each iteration on a few-iterations-back analysis; sometimes, this means that a solution is chosen even if it leads to a poorer performance than the previous iteration. Fores et al. (2001) describe a traditional ILP formulation of the CSP, with some added flexibility. The formulation accepts different objective functions (minimize the number of duties, minimize costs, or a combination), different optimization techniques (primal CG or dual-steepest edge techniques), and different criteria for reducing the number of feasible workdays. The optimization technique chosen is used to solve a relaxed noninteger problem; a branch-and-bound process then finds an integer solution.

Freling et al. (2001b) discuss differences between bus and train CSPs and propose a methodology for the scheduling of train crews. The problem, formulated as an SCP with additional constraints, is solved using a heuristic branch-and-price algorithm. Branch-and-price is a special application of branch-and-bound, in which a CG technique is used to solve linear programming relaxations with a huge number of variables. Feasible workdays are generated in a network, in which each node corresponds to a duty piece and each path through the network to a feasible duty; the optimal solution is sought through dynamic programming, that of a resource-constraint, shortest-path algorithm. Finally, Kroon and Fischetti (2001) present an SCP for railway crews, which allows some flexibility in specifying penalties for undesirable types of workdays. A dynamic CG procedure is used; hence, duties are not generated *a priori* but in the course of the solution process. Regeneration and reselection of workdays are carried out in each iteration. Generation is preformed in a network in which trips are represented by arcs. To solve the SCP, a Lagrangian-relaxation method and subgradient optimization are used instead of the common linear programming.

Huisman et al. (2004) proposed a solution approach to the dynamic single-depot and multiple-depot VSP. A *cluster-reschedule* heuristic is proposed to solve the multiple-depot VSP. By using this heuristic, the authors first assign trips to depots by solving the static problem and then solve dynamic single-depot problems. Sensitivity analysis about the deviations of the actual travel times from estimated ones is conducted.

Yunes et al. (2005) proposed a way to combine pure-integer programming and declarative constraint logic programming techniques into hybrid CG algorithms that solved, to optimality, huge instances of these real-world crew management problems. Hybrid CG algorithms are designed to solve the problems. They reported that the hybrid algorithms always performed better, when obtaining optimal solutions, than the two previous isolated approaches. In particular, the algorithms proved to be much faster for the scheduling problem.

Desaulnier (2007) developed an exact dynamic fixed cost procedure compatible with CG that starts with a relatively small fixed cost value and increases it iteratively until optimality is reached. This procedure can overcome some drawbacks of the traditional CG method.

Portugal (2009) developed new alternative models that, from a practical point of view, are better for the DSP-solving process. The models are based on set partitioning/covering models and take into account the bus operator issues and the user's standpoint and environment. Commercial ILP solver, Cplex, is used to solve the new mathematical models. It is demonstrated that the solutions generated by these models can easily be implemented in real situations.

Tian and Niu (2012) studied the modeling and algorithms of crew roster problem with given cycle on high-speed railway lines. Two feasible compilation strategies to work out the crew-rostering plan, with given cycle of high-speed railway lines, are discussed. Subsequently, an integrated compilation method is proposed to obtain a plan with relatively higher regularity in execution and lower crew members arranged. The process of plan making is divided into two subproblems, which are decomposition of crew legs and adjustment of nonmaximum crew roster scheme. The decomposition subproblem is transformed to finding a Hamilton chain with the best objective function in network, which was solved by an improved ant colony algorithm, whereas the adjustment of nonmaximum crew-rostering scheme is finally presented as an SCP and solved by a two-stage algorithm. Finally, the effectiveness of the proposed models and algorithms are tested through a numerical example. The authors conclude that although the compiling method of crew-rostering plan presented in this study has its advantages, it sacrifices both objectives to some extent. However, this algorithm can be used separately when preparing the crew-rostering plan with a single requirement.

Chen and Shen (2013) present an improved CG strategy to enhance the efficiency of the proposed crew-scheduling approach. The main idea presented is to precompile a reasonably large set of *good* potential shifts (called a shift pool) using problem-specific knowledge. Significant time savings can be achieved by simply selecting the negative reduced cost columns from the shift pool to add to the restricted master problem, which reduces the times of calling the resource-constrained shortest-path problem. Experiment results have been presented, which demonstrate that the improved CG algorithm can outperform the traditional CG algorithms. Future work has been proposed in the area of designing more sophisticated upper-bound strategies and more efficient branching rules that expedite the overall solution process.

Shen et al. (2013) present an adaptive evolutionary approach incorporating a hybrid GA for public transit CSPs. This GA is utilized to iteratively generate a near-optimal schedule. Unlike the traditional GA-based methods for crew scheduling where the chromosome length is fixed, a novel shift-based approach has been introduced where the chromosome length varies adaptively during the iterative process, and its initial value is elaborately designated as the lower bound of the number of shifts to be used in an unachievable optimal solution. Next, the hybrid GA with such a short chromosome length is employed to find a feasible schedule. During the GA process, the adaptation on chromosome lengths is achieved by genetic operations of crossover and mutation with removal and replenishment strategies aided by a simple greedy algorithm. If a feasible schedule cannot be found when the GA's termination condition is met, the GA will restart with one more gene added. The aforementioned process is repeated until a feasible solution is found. Computational experiments based on 11 real-world CSPs in China show, compared to a fuzzy GA that is known to be well performed for crew scheduling, that better solutions are found for all the testing problems. Moreover, the authors demonstrate that the algorithm works fast, has achieved results close to the lower bounds obtained by a standard linear programming solver in terms of the number of shifts, and has much potential for future developments.

The main features of the methodologies reviewed are summarized in Table 9.4.

Table 9.4 Summary of the characteristics of the methodologies reviewed

Source	Approach	Solution method	Comments
Elias (1964)	Set covering	Heuristics	
Mitra and Darby-Dowman (1985)	Set covering	ILP	Allows leaving some tasks uncovered
Smith and Wren (1988)	Set covering	ILP	
Desrochers and Soumis (1989)	Set covering	Column generation	
Paixao (1990); Paias and Paixao (1993)	Set covering (dynamic)	ILP; state-space relaxation	
Carraresi et al. (1995)	Set covering	Column generation; Lagrangian relaxation	
Clement and Wren (1995)	Genetic algorithm	Genetic algorithm	
Beasley and Cao (1996)	Single-type workdays	Lagrangian relaxation; subgradient optimization; tree search	
Beasley and Cao (1998)	Single-type workdays	Dynamic program	
Kwan et al. (1999)	Genetic algorithm	Genetic algorithm	
Mingozzi et al. (1999)	Set partitioning	Linear relaxation; branch and bound	
Haase et al. (2001)	Set partitioning	Column generation; branch-and-bound	Simultaneous crew and vehicle scheduling
Lourenco et al. (2001)	Multiobjective program	Tabu search; genetic algorithm	
Banihashemi and Haghani (2001)	Network-flow problem	Column generation	
Freling et al. (2001a); Huisman et al. (2005)	Set covering; set partitioning	Column generation	Simultaneous crew and vehicle scheduling
Huisman et al. (2004)	Zero-one integer programming	Dynamic optimization	
Shen and Kwan (2001)	Set partitioning	Tabu search	Workday generation; method not included
Fores et al. (2001)	Set covering	Primal column generation; dual-steepest edge; branch and bound	Varying objective functions accepted
Freling et al. (2001b)	Set covering; network-flow problem	Branch-and-price	Rail crews
Kroon and Fischetti (2001)	Set covering (dynamic); network-flow problem	Dynamic column generation; Lagrangian relaxation and subgradient optimization	Rail crews
Yunes et al. (2005)	Mathematical programming; constraint logic programming	Hybrid column generation	

(Continued)

Table 9.4 (Continued) Summary of the characteristics of the methodologies reviewed

Source	Approach	Solution method	Comments
Desaulnier (2007)	Lexicographic bilevel objective formulation	Column generation	
Portugal (2009)	Set covering/partitioning-based models	ILP, Cplex	
Tian and Niu (2012)	Set covering	Ant colony algorithm	
Chen and Shen (2013)	Set covering	Column generation	
Shen et al. (2013)	Set covering	Hybrid genetic algorithm	

EXERCISES

9.1 Given 29 arrival epochs and 29 departure epochs (a total of 58 trips) at terminal k between 6:00 and 8:10 a.m. (see the following table) and a maximum paid idle time of T_{max} = 35 min

(i)   Apply algorithm T_mF and find D(k) and the maximum number of joinings at k in which $\Delta_{ij} \geq T_{max}$; list all joinings found.

(ii) Use the FIFO rule for constructing the joinings at k; list all joinings and compare the results with those found in (i).

Departures		Arrivals	
Trip number	Departure time	Trip number	Arrival time
1	6:00	12	6:25
2	6:00	13	6:30
3	6:00	14	6:30
4	6:05	15	6:30
5	6:10	16	6:35
6	6:10	17	6:35
7	6:15	21	6:50
8	6:15	22	6:55
9	6:20	23	6:55
10	6:20	24	6:55
11	6:20	25	7:00
18	6:40	29	7:15
19	6:45	30	7:15
20	6:45	34	7:30
26	7:05	35	7:35
27	7:05	36	7:35
28	7:10	37	7:35
31	7:20	38	7:35
32	7:25	48	8:00
33	7:25	49	8:00
39	7:40	50	8:05
40	7:45	51	8:05
41	7:45	52	8:05
42	7:50	53	8:05

Departures		Arrivals	
Trip number	Departure time	Trip number	Arrival time
43	7:50	54	8:05
44	7:50	55	8:10
45	7:55	56	8:10
46	7:55	57	8:10
47	7:55	58	8:10

9.2 Given two vehicle blocks with their relief points, Block 1 starts at relief point a and continues through points b, c, d, c, and b, in which the piece from b to c is a DH trip. Block 2 starts at relief point *a* and continues through points b, c, a, and d. The following table lists the segments of consecutive blocks with their associated internal piece cost (ignore cost unit). Find and list the best sets of duty pieces (minimum cost) for each block.

Block 1		Block 2	
Feasible pieces	Internal piece cost	Feasible pieces	Internal piece cost
a–b	3	a–b	3
a–c	5	a–c	7
a–d	9	a–a	12
a–c	14	a–d	24
a–b	20	b–c	3
b–d	7	b–a	8
b–c	8	b–d	16
b–b	12	c–a	7
c–d	5	c–d	12
c–c	5	a–d	8
c–b	9		
d–c	2		
d–b	4		
c–b	4		

9.3 Use procedure roster to determine the minimum number of drivers required to operate the following given scheduling data: Two constraints are imposed—one is 10 h as the minimum rest period required between two consecutive working days and the second concerns the maximum allowed roster hours for each roster type (see these two constraints in the tables of this exercise). Each roster can contain one duty per day for days marked "V."

		Mon	Tue	Wed	Thur	Fri	Sat	Sun	
Duties distributed		d_1^1	d_1^2	d_1^3	—	d_1^5	d_1^6	—	
		d_2^1	d_2^2	d_2^3	d_2^4	d_2^5	—	—	
		—	d_3^2	d_3^3	d_3^4	d_3^5	d_3^6	—	
		d_4^1	—	d_4^3	d_4^4	—	—	d_4^7	Maximum roster hours
Type of roster	R_1	V	V	V	V	V	—	—	43
	R_2	V	V	—	V	V	—	—	36
	R_3	—	—	—	—	—	V	V	16

Duty for day q (q = 1, 2, 3, 4)	Start and end times	Duty length (hours), L_j
d_1^q	6:00–16:00	10
d_2^q	12:00–20:00	8
d_3^q	16:00–23:00	7
d_4^q	18:00–midnight	6

APPENDIX 9.A: SHORTEST-PATH PROBLEM

The following description is based on Ceder (1978). Further reading can be found in almost any OR book. This appendix supplements Sections 9.3 and 9.5, as well as material in Chapters 11, 12, 15, and 16.

Consider a connected network $G = \{N, A\}$ with set of nodes N, set of directed arcs A, and given *distances* (travel time, cost, etc.), $d(i, j) \ \forall \ (i, j) \in A$. The number of nodes and arcs are $|N|$ and $|A|$, respectively. In a connected network, the shortest path may have the following four categories:

(a) From one node to all other nodes with $d(i, j) \geq 0, \forall (i, j) \in A$
(b) From one node to all other nodes with $d(i, j)$ any
(c) From all nodes to all nodes with $d(i, j) \geq 0, \forall (i, j) \in A$
(d) From all nodes to all nodes with $d(i, j)$ any

The shortest u–v path, termed $\bar{P}$, is defined as a u–v path, $u, v \in N$, such that
$$d(\bar{p}) = \sum_{(i,j) \in \bar{p}} d(i, j) = \mathop{\mathrm{Min}}_{\substack{u-v \text{ paths} \\ p \text{ on } G}} \sum_{(i, j) \in p} d(i, j).$$

9.A.1 Dijkstra algorithm (one to all, d(i,j) ≥ 0)

The Dijkstra algorithm (Dijkstra, 1959) produces a spanning tree of G containing shortest paths from a given node to all other nodes in G, with nonnegative arc lengths. It is described as follows.

Step 0: Initially, assign the origin node u with a permanent label, $\pi^*(u) = 0$.
Let the set of nodes with permanent labels be $N(u) = \{u\}$. All other nodes $i, i \in N, i \neq u$, have temporary labels:

$$\pi(i) = \begin{bmatrix} d(u, i), & (u, i) \in A \\ \infty, & \text{otherwise} \end{bmatrix}$$

Note that a permanent label $\pi^*(i)$ represents the shortest distances from node u to node i.
The symbol ":=" means *is replaced by* in the following steps.
Step 1: Find the node with a temporary label, say, node k, such that $\pi(k) = \mathrm{Min}_{i \in N(u)}\pi(i)$. Set $\pi^*(k) = \pi(k)$ and mark/store the arc that leads from N(u) to k. Add k to N(u).
Step 2: if N(u) contains all nodes $i, i \in N$, terminate; otherwise, continue.
Step 3: For each arc (k, j) such that k is a node determined in *Step 1* and $j \notin N(u)$, $\pi(j) :=$ $\mathrm{Min}[\pi^*(k) + d(k, j), \pi(j)]$, go to *Step 1*.

The following example illustrates the Dijkstra algorithm with four iterations. The tree emphasized (i.e., connected paths without cycles) after the fourth iteration is the solution with all nodes permanently labeled. The arcs emphasized represent the desired spanning tree. The minimum spanning tree from node 1 to all other nodes is provided.

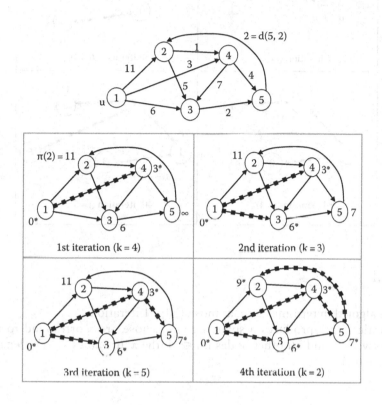

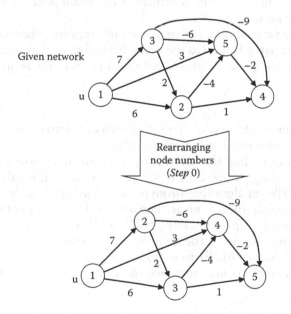

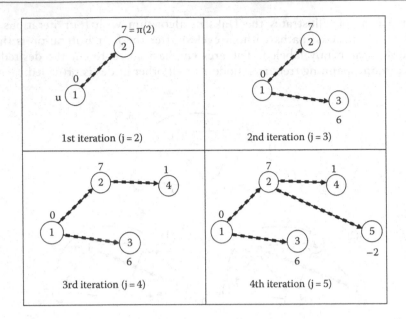

1st iteration (j = 2)

2nd iteration (j = 3)

3rd iteration (j = 4)

4th iteration (j = 5)

Theorem 9.A.1

(a) Dijkstra algorithm terminates in, at most, $|N| - 1$ iterations.
(b) Dijkstra algorithm produces a spanning tree whose arcs correspond to the shortest paths between u and all other nodes i, $i \in N$ (the $\pi^*(i)$ values are the lengths of these shortest paths).

Proof

(a) Almost trivial, because one node is permanently labeled per iteration, and the network considered is connected.
(b) By induction. The assumption is that at the mth iteration, the tree emphasized represents the shortest paths from u to the nodes belonging to that tree. From this point, it is not difficult to complete the proof, and therefore, it is left as an exercise.

Remarks

(1) If the interest is in the shortest u–v path (between only two nodes), then terminate once $\pi(v)$ becomes permanently labeled; that is, $\pi^*(v)$.
(2) If the interest is in finding the shortest paths between all nodes $i \in N$ and a node u (from all to one), simply reserve all arcs $(i, j) \in A$; that is, (i, j) will become an (j, i) arc; then apply the Dijkstra algorithm (from node u to all other nodes).
(3) For a complete graph (there is an arc between each two pairs of nodes), the Dijkstra algorithm requires approximately $|N|^2/2$ additions and $|N|^2$ comparisons, or $3|N|^2/2$ operations. By using the complexity notation in Section 5.3 (Chapter 5), the algorithm runs in $O(|N|^2)$ time for dense (in arcs) networks. For sparse networks, other algorithms may improve the running time; see, for example, Ahuja et al. (1993).

9.A.2 Algorithm rearrange-node-numbers (one to all, d(i, j) any, no cycles)

The major limitation of the Dijkstra algorithm is that it can run only on nonnegative arc lengths. There are other algorithms that handle optimally any d(i, j) that can be found, for example, in Ahuja et al. (1993). This section introduces an algorithm aimed at special cases in which G = {N, A} contains no cycles. The algorithm, which is called rearrange-node-numbers (RNN) because of its first step, is described as follows:

Step 0: Initially, let the origin node be u = 1 and number all nodes i, i ∈ N, such that arc (i, j) ∈ A if and only if I < j. Remove all nodes with assigned numbers less than 1 (since they cannot be reached from u). If this step cannot be done, a cycle is found. Set j:= 2, $\pi(1):= 0$, $A_T x = \Phi$.

Step 1: $\pi(j) = \underset{(i,j)\in A/i<j}{\text{Min}} [\pi(i) + d(i,j)] := \pi(i^*) + d(i^*, j)$. Add (i^*, j) to A_T. If there is no arc, connecting A_T with j, assume $\pi(i) = \infty$, and continue.

Step 2: Let j:= j + 1. If j = |N| terminate. The tree T = {N, A_T} is the required spanning tree, and π(i) are the shortest paths. For π(i) = ∞, there is no path between u and i on G. If j < |N|, go to Step 1.

Theorem 9.A.2

(a) Algorithm RNN terminates in |N| − 1 iterations.
(b) Algorithm RNN produces the required spanning tree; the π(i) value is the shortest path from u to node i (for π(i) = ∞, there is no path from u to i).

Proof

(a) Self-evident, because one node is added per iteration.
(b) By induction. The assumption is that at the mth iteration, the tree emphasized represents shortest paths from u to the nodes on the tree. The rest is left as an exercise.

Remarks

(1) If the interest is in the shortest u–v path (between only two nodes), then terminate once i equals the node number of v.
(2) If the interest is in finding the shortest paths between all nodes i and a node u (from all to one), simply reserve all arcs; that is, (i, j) will become (j, i) arc; then apply algorithm RNN.
(3) The RNN algorithm is basically applied to known project-scheduling methods, such as the program evaluation and review technique (PERT) and critical path method (CPM). Small changes in RNN enable one to introduce earliest-starting and latest-end times of each task (described by an arc).

The example in the succeeding text, in which some d(i, j) < 0, shows the RNN algorithm in four iterations. The solution is represented by the last tree, with shortest paths from u = 1 indicated as π(i) on the nodes.

REFERENCES

Ahuja, R.K., Magnanti, T.L., and Orlin, J.B. (1993). *Network Flows*. Prentice Hall, Englewood Cliffs, New Jersey.

Ball, M., Bodin, R., and Dial, R. (1981). Experimentation with a computerized system for scheduling mass transit vehicles and crews. In *Computer Scheduling of Public Transport: Urban Passenger Vehicle and Crew Scheduling* (A. Wren, ed.), pp. 313–334, North-Holland, Amsterdam, the Netherlands.

Ball, M., Bodin, R., and Dial, R. (1983). A matching based heuristic for scheduling mass transit crew and vehicles. *Transportation Science*, 17, 4–31.

Banihashemi, M. and Haghani, A. (2001). A new model for the mass transit crew scheduling problem. In *Computer-Aided Scheduling of Public Transport*. Lecture Notes in Economics and Mathematical Systems, vol. 505 (S. Voss and J.R. Daduna, eds.), pp. 1–15, Springer-Verlag, Berlin, Germany.

Beasley, J.E. and Cao, E.B. (1996). A tree search algorithm for the crew scheduling problem. *European Journal of Operational Research*, 94, 517–526.

Beasley, J.E. and Cao, E.B. (1998). A dynamic programming based algorithm for the crew scheduling problem. *Computers & Operations Research*, 25, 567–582.

Bianco, L., Bielli, M., Mingozzi, M.A., Ricciardelli, S., and Spadoni, M. (1992). A heuristic procedure for the crew rostering problem. *European Journal of Operational Research*, 58, 272–283.

Bodin, L., Golden, B., Assad, A., and Ball, M. (1983). Routing and scheduling of vehicles and crews: The state of the art. *Computers and Operation Research*, 10, 63–211.

Caprara, A., Toth, P., Vigo, D., and Fischetti, M. (1998). Modeling and solving the crew rostering problem. *Operations Research*, 46(6), 820–830.

Carraresi, P. and Gallo, G. (1984). A multilevel bottleneck assignment approach to the bus driver's rostering problem. *European Journal of Operational Research*, 16, 163–173.

Carraresi, P., Nonato, M., and Girard, L. (1995). Network models, lagrangean relaxation and subgradients bundle approach in crew scheduling problems. In *Computer-Aided Transit Scheduling*. Lecture Notes in Economics and Mathematical Systems, vol. 430 (J.R. Daduna, I. Branco, and J.M.P. Paixao, eds.), pp. 188–212, Springer-Verlag, Berlin, Germany.

Catanas, F. and Paixao, J.M.P. (1995). A new approach for the crew rostering problem. In *Computer-Aided Transit Scheduling*. Lecture Notes in Economics and Mathematical Systems, vol. 430 (J.R. Daduna, I. Branco, and J.M.P. Paixao, eds.), pp. 267–277, Springer-Verlag, Berlin, Germany.

Ceder, A. (1978). *Network Theory and Selected Topics in Dynamic Programming*. Dekel Academic Press (in Hebrew), Tel Aviv, Israel.

Ceder, A. (2011). Public-transit vehicle schedules using a minimum crew-cost approach. *Total Logistic Management International Journal*, 4, 21–42.

Chen, S. and Shen, Y. (2013). An improved column generation algorithm for crew scheduling problems. *Journal of Information & Computational Science*, 10(1), 175–183.

Clement, R. and Wren, A. (1995). Greedy genetic algorithms, optimizing mutations and bus driver scheduling. In *Computer-Aided Transit Scheduling*. Lecture Notes in Economics and Mathematical Systems, vol. 430 (J.R. Daduna, I. Branco, and J.M.P. Paixao, eds.), pp. 213–235, Springer-Verlag, Berlin, Germany.

Daduna, J.R., Branco, I., and Paixao, J.M.P. (eds.) (1995). *Computer-Aided Transit Scheduling*. Lecture Notes in Economics and Mathematical Systems, vol. 430, Springer-Verlag, Berlin, Germany.

Daduna, J.R. and Wren, A. (eds.) (1988). *Computer-Aided Transit Scheduling*. Lecture Notes in Economics and Mathematical Systems, vol. 308, Springer-Verlag, Berlin, Germany.

Desaulnier, G. (2007). Managing large fixed costs in vehicle routing and crew scheduling problems solved by column generation. *Computers & Operations Research*, 34(4), 1221–1239.

Desrochers, M. and Rousseau, J.M. (eds.) (1992). *Computer-Aided Transit Scheduling*. Lecture Notes in Economics and Mathematical Systems, vol. 386, Springer-Verlag, Berlin, Germany.

Desrochers, M. and Soumis, F. (1989). A column generation approach to the urban transit crew scheduling problem. *Transportation Science*, 23(1), 1–13.

Dijkstra, E. (1959). A note on two problems in connexion with graphs. *Numeriche Mathematics*, 1, 269–271.

Elias, S.E.G. (1964). The use of digital computers in the economic scheduling for both man and machine in public transportation. Kansas State University Bulletin, Manhattan, KS, Special Report, Number 49.

Ernst, A., Jiang, H., Krishnamoorthy, M., and Sier, D. (2004). Staff scheduling and rostering: A review of applications, methods and models. *European Journal of Operational Research*, 153, 3–27.

Fores, S., Proll, L., and Wren, A. (1999). An improved ILP system for driver scheduling. In *Computer-Aided Transit Scheduling*. Lecture Notes in Economics and Mathematical Systems, vol. 471 (N.H.M. Wilson, ed.), pp. 43–61, Springer-Verlag.

Fores, S., Proll, L., and Wren, A. (2001). Experiences with a flexible driver scheduler. In *Computer-Aided Scheduling of Public Transport*. Lecture Notes in Economics and Mathematical Systems, vol. 505 (S. Voss and J.R. Daduna, eds.), pp. 137–152, Springer-Verlag, Berlin, Germany.

Freling, R., Huisman, D., and Wagelmans, A.P.M. (2001a). Applying an integrated approach to vehicle and crew scheduling in practice. In *Computer-Aided Scheduling of Public Transport*. Lecture Notes in Economics and Mathematical Systems, vol. 505 (S. Voss and J.R. Daduna, eds.), pp. 73–90, Springer-Verlag, Berlin, Germany.

Freling, R., Lentink, R.M., and Odijk, M.A. (2001b). Scheduling train crews: A case study for the Dutch railways. In *Computer-Aided Scheduling of Public Transport*. Lecture Notes in Economics and Mathematical Systems, vol. 505 (S. Voss and J.R. Daduna, eds.), pp. 153–165, Springer-Verlag, Berlin, Germany.

Freling, R., Wagelman, A.P.M., and Paixao, J.M.P. (1999). An overview of models and techniques of integrating vehicle and crew scheduling. In *Computer-Aided Transit Scheduling*. Lecture Notes in Economics and Mathematical Systems, vol. 471 (N.H.M. Wilson, ed.), pp. 441–460, Springer-Verlag, Berlin, Germany.

Haase, K., Desaulniers, G., and Desrosiers, J. (2001). Simultaneous vehicle and crew scheduling in urban mass transit systems. *Transportation Science*, 35(3), 286–303.

Hickman, M., Mirchandani, P., and Voss, S. (eds.) (2008). *Computer-Aided Systems in Public Transport*. Lecture Notes in Economics and Mathematical Systems, vol. 600, Springer, Berlin, Germany.

Huisman, D., Freling, R., and Wagelmans, A.P.M. (2004). A robust solution approach to the dynamic vehicle scheduling problem. *Transportation Science*, 38(4), 447–458.

Huisman, D., Freling, R., and Wagelmans, A.P.M. (2005). Models and algorithms for integration of vehicle and crew scheduling. *Transportation Science*, 39, 491–502.

Kroon, L. and Fischetti, M. (2001). Crew scheduling for Netherlands railways "destination: customer." In *Computer-Aided Scheduling of Public Transport*. Lecture Notes in Economics and Mathematical Systems, vol. 505 (S. Voss and J.R. Daduna, eds.), pp. 181–201, Springer-Verlag, Berlin, Germany.

Kwan, A.S.K., Kwan, R.S.K., and Wren, A. (1999). Driver scheduling using genetic algorithms with embedded combinatorial traits. In *Computer-Aided Transit Scheduling*. Lecture Notes in Economics and Mathematical Systems, vol. 471 (N.H.M. Wilson, ed.), pp. 81–102, Springer-Verlag, Berlin, Germany.

Lourenco, H.R., Paixao, J.P., and Portugal, R. (2001). Multiobjective metaheuristics for the bus-driver scheduling problem. *Transportation Science*, 35(3), 331–343.

Mingozzi, A., Boschetti, M.A., Ricciardelli, S., and Bianco, L. (1999). A set partitioning approach to the crew scheduling problem. *Operations Research*, 47, 873–888.

Mitra, G. and Darby-Dowman, K. (1985). CRU-SCHED: A computer-based bus crew scheduling system using integer programming. In *Computer Scheduling of Public Transport 2* (J.M. Rousseau, ed.), pp. 223–232, North-Holland, Amsterdam, the Netherlands.

Paias, A. and Paixao, J.M.P. (1993). State space relaxation for set-covering problems related to bus driver scheduling. *European Journal of Operational Research*, 71, 303–316.

Paixao, J.M.P. (1990). Transit crew scheduling on a personal workstation (MS/DOS). In *Operational Research'90* (H. Bradley, ed.), pp. 421–432, Pergamon Press, Oxford, U.K.

Portugal, R., Lourenco, H.R., and Paixao, J.P. (2009). Driver scheduling problem modeling. *Public Transport*, 1(2), 103–120.

Rousseau, J.M. (ed.) (1985). *Computer Scheduling of Public Transport 2*. North-Holland, Amsterdam, the Netherlands.

Shen, Y. and Kwan, R.S.K. (2001). Tabu search for driver scheduling. In *Computer-Aided Scheduling of Public Transport*. Lecture Notes in Economics and Mathematical Systems, vol. 505 (S. Voss and J.R. Daduna, eds.), pp. 121–135, Springer-Verlag, Berlin, Germany.

Shen, Y., Peng, K., Chen, K., and Li, J. (2013). Evolutionary crew scheduling with adaptive chromosomes. *Transportation Research*, **56B**, 174–185.

Smith, B.M. and Wren, A. (1988). A bus crew scheduling system using a set covering formulation. *Transportation Research*, **22A**, 97–108.

Sodhi, M. and Norris, S. (2004). A flexible, fast, and optimal modeling approach applied to crew rostering at London Underground. *Annals of Operations Research*, **127**, 259–281.

Tian, Z. and Niu, H. (2012). Modeling and algorithms of the crew rostering problem with given cycle on high-speed railway lines. *Mathematical Problems in Engineering*, **2012**, Article ID 214607.

Tykulsker, R.J., O'Neill, K.K., Ceder, A., and Sheffi, Y. (1985). A computer rail crew assignment/work rules model. In *Computer Scheduling of Public Transport 2* (Rousseau, J.M., ed.), pp. 233–246, North-Holland, Amsterdam, the Netherlands.

Voss, S. and Daduna, R. (eds.) (2001). *Computer-Aided Scheduling of Public Transport*. Lecture Notes in Economics and Mathematical Systems, vol. 505, Springer-Verlag, Berlin, Germany.

Wilson, N.H.M. (ed.) (1999). *Computer-Aided Scheduling of Public Transport*. Lecture Notes in Economics and Mathematical Systems, vol. 471, Springer-Verlag, Berlin, Germany.

Wren, A. (ed.) (1981). *Computer Scheduling of Public Transport: Urban Passenger Vehicle and Crew Scheduling*, North-Holland, Amsterdam, the Netherlands.

Wren, A. and Rousseau, J.M. (1995). Bus driver scheduling: An overview. In *Computer-Aided Scheduling of Public Transport*. Lecture Notes in Economics and Mathematical Systems, vol. 430, pp. 173–187, Springer-Verlag, Berlin, Germany.

Yunes, T.H., Moura, A.V., and De Souza, C.C. (2005). Hybrid column generation approaches for urban transit crew management problems. *Transportation Science*, **39**(2), 273–288.

Chapter 10

Passenger demand

Demand will rise when service is perceived as a delicious food

Transit

A.C.

CHAPTER OUTLINE

10.1 Introduction
10.2 Transit demand, its factors, and elasticity
10.3 Example of a demand forecasting, method, and process
10.4 Multinomial logit model
10.5 Literature review and further reading (origin–destination estimation)
Exercises
References

PRACTITIONER'S CORNER

One of the main measures of the improvement of transit services is the increase in passenger demand. However, there is a trade-off between the cost of improvement and the extra benefit gained by the additional demand. It is common to find transit managers who are puzzled about the relationship (if any) between service changes and their impact on changes in demand. In fact, the following humoristic aphorism readily applies: "The more unpredictable the transit demand becomes, the more we rely on predictions." Although it is easier to try out service changes for bus services (with conclusions drawn after a learning period), the rail industry requires more assurances.

The chapter contains four main parts following the introductory section. Section 10.2 presents and discusses the basic attributes and tools used in passenger-demand analysis; among these attributes are fares, travel time, service frequency, walking time, routing and transferring, and comfort and inconvenience elements. Section 10.3 provides an example of a transit-demand forecast methodology that is adapted to predict, for a given transit service, the future patronage of a specific set of routes in certain required years. Section 10.4 exhibits a known share model that divides passengers among various travel modes according to each mode's relative desirability; such a technique has been commonly used to determine modal split, although its accuracy depends heavily on the underlying mathematical model. The last main part, Section 10.5, presents a literature review on estimating origin–destination matrices (ODMs), which constitute an essential input for most transit planning and design procedures. The chapter ends with exercises.

Practitioners are advised to read the entire chapter, rather than just Section 10.4. It should be noted, however, that demand-prediction models, no matter how good they are, should be treated carefully because of the vast number of assumptions (behavioral and others) that are inherent in them. In other words, one should not believe that the models are reality (*don't eat the menu*), but should properly weigh practical issues against modeling precision. The following story may *teach* us about life and precision (squareness).

During the French Revolution, three professionals were sentenced to be executed by guillotine. The first person had been collecting practical data, the second analyzing practical data, and the third modeling the precision of data. For the third professional, precision was above all else (a square person). The first person was asked if he wanted his head facing up or down. He requested down, and the guillotine dropped, stopping 2 cm before his neck; there was an old rule that if the guillotine stopped, you're sent free, and indeed he was released. The second person was asked about his head position; he asked face up, and the guillotine again dropped and, unbelievable, for this had never happened before, stopped 1 cm before his throat, allowing him, too, to go free. The third person, for whom precision was all important, replied to the same question: "I don't want to go under that damn machine before you fix it!"

10.1 INTRODUCTION

Chapter 9 ends a main subject of the book, transit scheduling. Chapters 12 through 17 are more design and behavioral oriented, focusing on service, network, route, and shuttle design and user behavior. Before any discussion of transit-design elements, however, it would be only natural to present an overview of the tools and models used in analyzing

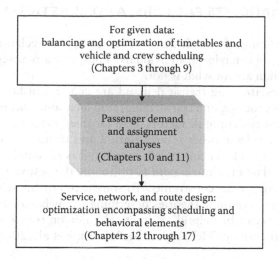

For given data:
balancing and optimization of timetables and
vehicle and crew scheduling
(Chapters 3 through 9)

Passenger demand
and assignment
analyses
(Chapters 10 and 11)

Service, network, and route design:
optimization encompassing scheduling and
behavioral elements
(Chapters 12 through 17)

Figure 10.1 Position of Chapters 10 and 11 in the sequence of main subjects of the book.

passenger demand (Chapter 10) and the related prediction of the way the demand will flow on the transit network (Chapter 11). The first question confronting us in designing a transit service concerns the size, composition, and distribution of passenger demand; it is the dominant input parameter for any new, improved, or redesigned undertaking. The sort of sandwich position of this and the next chapter between the subjects of the book is displayed in Figure 10.1.

The purpose of transit-demand studies is to estimate and evaluate passenger demand by collecting and analyzing data and models pertaining to current and future transit needs. Transit-demand studies and models form an essential part of any transit planning process. The venerable transportation-planning practice is based on a four-step sequential process: *trip generation* (number of trip ends associated with a zone of land), *trip distribution* (distribution of trips among zones), *modal split* or *mode choice* (choice among travel modes for personal trips), and *traffic assignment* (choice among available origin–destination [O-D] routes and the resulting accumulated traffic). A thorough overview of this four-step process and its variants appears in de Dios Ortuzar and Willumsen (2011). This planning practice for trip forecasting contains trips associated with transit modes, thereby calling for the utilization of some of its features in analyzing passenger demand.

The four-step sequential planning process is discussed in Boyce and Daskin (1997) in conjunction with solving optimal travel-choice models. Those researchers explained that the four-step process had to be solved with feedback in order to forecast equilibrium traffic conditions. In their iterative procedure involving the four-step process, the solution of the trip-assignment (route choice) model is required to solve the trip-distribution (O-D choice) model—that is, equilibrium travel times and costs. A variant of the four-step process (de Dios Ortuzar and Willumsen, 2011) indeed considers route choice as the step preceding the O-D choice step. Boyce and Daskin (1997) believe that the four-step process can at best be seen as a heuristic for solving a traffic-network equilibrium problem because it lacks the mechanism by which the process could converge to the desired equilibrium solution. Following this explication of the conceptual issues and limitations surrounding the four-step process, the following sections will provide examples of analyses of passenger-demand forecasting.

10.2 TRANSIT DEMAND, ITS FACTORS, AND ELASTICITY

The analysis and estimating of transit demand are complex mechanisms mainly because they involve the aggregate behavior of individuals. Thus, in estimating demand, it is better to doubt what is true than accept what is not.

The known attributes affecting transit demand are changes in fare, travel time, service frequency, walking time, routing and transferring, stop location, comfort and inconvenience elements, information, socioeconomic factors (e.g., income), external factors (e.g., land use, security), and competition from other transportation and transit services. Usually, passengers (travelers) have a travel choice among alternatives; it is reasonable to assume that their selection process is based on maximizing *utility* or finding the best way to realize their travel objectives while complying with constraints. To some extent, this selection process varies among individuals; thus, economic theories describing the behavior of a customer searching for services or goods can be helpful. A practical guide on transit demand appears in a Transport Research Laboratory (TRL) report by Balcombe et al. (2004).

10.2.1 Factors affecting demand

The choice between public and private transportation is an individual decision that is influenced by government/community decisions. These various decisions often send mixed signals to transit and potential transit passengers while failing to recognize their more system-wide and integrated implications. Generally speaking, most large cities have effectively encouraged the use of private cars through planning (dispersed land use in the suburbs), infrastructure (available parking and circulation traffic flow), pricing, and financial decisions. Consequently, there is growing confusion in many of those cities about what to do. It is therefore important to identify the factors affecting transit demand and to use them for improvements.

For the most part, increasing the quality of service (e.g., in terms of service coverage, service frequency, and fare reduction) has been found to have a significant positive outcome on transit demand. Two practical guides that cover the factors affecting demand are Balcombe et al. (2004) and TranSystems et al. (2006).

The greatest impacts from increasing the transit demand are extracted from a combination of strategy-related actions and initiatives. TranSystems et al. (2006), based on a large survey of transit agencies in the United States, categorized these actions and initiatives in the following list (in decreasing order of effectiveness):

1. *Service adjustments or improvements* (increased route coverage, route restructuring, improved schedule/route coordination, increased service frequency, increased span of service, improved reliability/on-time performance, improved travel speed and reduced stops, targeted services, passenger-facility improvements, new/improved vehicles, increased security and safety)
2. *Partnerships and coordination* (university and school passenger programs; travel demand management strategies; privately subsidized activity center service; consistent regional, and interagency, operating policies; coordination with other transportation agencies; promotion of transit-supportive design)
3. *Marketing, promotional, and information initiatives* (targeted marketing and promotions, general marketing and promotions, improved informational materials, improved customer information and assistance)
4. *Fare collection and fare structure initiatives* (improved payment convenience, regional payment integration, fare structure simplification, fare reduction)

In addition, both Balcombe et al. (2004) and TranSystems et al. (2006) indicated that there are *external factors* influencing transit demand. The factors listed as significant by TranSystems et al. (2006) are as follows:

1. *Population characteristics and changes* (general growth in the region, high and increasing immigration, high and increasing number of elderly, high and increasing tourism, high number of college students)
2. *Economic conditions* (employment/unemployment levels, per-capita income levels, household auto-ownership levels)
3. *Cost and availability of alternative modes* (fuel and toll pricing, parking pricing and availability, taxi fares, fuel taxes, auto purchase and ownership costs, availability of commuter-benefit programs for employers)
4. *Land use and development patterns and policies* (density of development; relative locations of major employers and residential areas as from, e.g., increasing suburbanization; land use and zoning controls and incentives)
5. *Travel conditions* (climate and weather patterns, traffic congestion levels and highway capacity, traffic disruptions owing to, e.g., major construction projects)
6. *Public policy and funding initiatives* (air quality mandates, auto emission standards, federal and state operating capital and transit-funding levels)

Another indirect set of factors affecting transit demand is related to *mode choice elements* extracted from various types of public policies. Examples of these elements given by TranSystems et al. (2006) are as follows:

1. *Price and availability of each mode* (cost of auto use, cost of transit, parking cost)
2. *Quality of service of each mode* (travel time, convenience, comfort, service reliability, perceived personal security and safety, perceived overall *image* of each mode)
3. *Trip characteristics for each particular trip* (trip length and purpose, number of people to be making the trip, multiple destinations)
4. *Sociodemographic characteristics of the traveler* (income, origin and destination locations, status)

Retaining a high level of passenger satisfaction while fully maintaining the protection of access to less-affluent citizens is another way to look at factors for improving transit demand. For instance, TranSystems et al. (2006) reported on large-scale surveys conducted for the Washington Metropolitan Area Transit Authority. The results of these surveys, shown in Table 10.1, indicate significant differences between the most desired transit-service improvements for riders and nonriders. It was concluded that targeting new riders warrants a different set of actions and initiatives than those used to influence existing riders.

Often, desired service improvement items can be interpreted by the willingness of riders and potential riders to pay for an increase in their satisfaction (Ceder, 2004). In one study, Kottenhoff (1998) employed the stated-preference method to ascertain Swedish train passengers' willingness to pay. The main question was how to increase the attractiveness of the rail system (the willingness to pay) and, simultaneously, to decrease its costs.

To perform these quantifications of willingness to pay, train passengers were asked to answer computer-assisted self-interviews. Table 10.2 presents the results by percentage of fare levels for which passengers are willing to pay extra in order to improve the specified item. These results were also checked (and agreed with in general) against the results from other European countries.

Table 10.1 Transit-service improvements desired by bus riders and nonriders in the Washington, DC, region

Improvement item	Riders (%)	Nonriders (%)
Better reliability	49	6
More frequent service	31	16
More convenient stops	13	18
Better information	9	30
Better shelters	10	21
Faster service	10	16
Better vehicle condition	8	13
Service to destination	12	13
Longer hours[a]	25	N/A
Less crowding[a]	22	N/A
Lower fares[a]	8	N/A
Better costumer service[a]	7	N/A

Source: TranSystems et al., Elements needed to create high ridership transit systems: Interim guidebook, TCRP Report 32, Transportation Research Board, Washington, DC, 2006.

Note: Percentage opting for each item.

[a] Not applicable for nonriders.

One example of a train service for which the conditions were improved and the payment was increased is the Danish IC/3, marketed under the name Kustpilen, which replaced simple rail–bus services. The number of travelers rose by almost 200% in the first 4 years as reported by Kottenhoff and Lindh (1996). The traveling time from Karlskrona to Malmo was reduced by 15 min (7%), and passengers traveling all the way no longer had to make a transfer. The frequency was increased from 6 to 11 trains per day in each direction. Although the fare level was reduced by about 10%–30%, the enhanced comfort offered by the new IC/3 was found to be the main attribute for the increase in patronage.

10.2.2 Demand function and elasticity

Theoretically, the (transit) demand can be mathematically expressed as a function of the attributes (explanatory variables) mentioned earlier. That is,

$$D_f = f(y_1, y_2, \ldots, y_m) \tag{10.1}$$

where
D_f is transit demand (in passengers)
y_i, $i = 1, 2, \ldots, m$, are the attributes

Basically, no preference is given to any functional form of Equation 10.1; the form is determined empirically for the type of application under question, using statistical techniques.

The responsiveness (sensitivity) of demand to changes in y_i attributes is known as elasticity. This is to say, demand elasticity concerning y_i is the ratio of the percentage change in demand to the percentage change in y_i. Assuming the existence of a demand function D_i for a given attribute y_i,

$$D_i = f(y_i), \quad i = 1, 2, \ldots, m \tag{10.2}$$

Table 10.2 Passengers' willingness to pay for train service in Sweden (by percentage of fare level to be paid as extra)

Improvement item	Fare level (%)
Timetable factors	
Change of train: one less transfer	19
Speed: 20% less travel time	15
Frequency: 2 h–1 h	5
Comfort	
Shaking and vibrations: *a little less*	11.5
Noise: *a little less noisy*	10
Climate: better ventilation	10
Seat adjustment: reclining seat backs	8
Seat orientation: face to face or face to back	8
Leg room: 10 cm less or more	6.5
Seat size: 5 cm wider	4
Onboard services	
Restaurant with hot food	13
Bistro with some food	10
Division: reading salon/quiet salons	9
Division: play areas	9
Coffee: free coffee and tea in the car	6
Entertainment: video or cinema on board	5
Entertainment: music/radio outlets	3
Office: service	2
Quality satisfaction	
On-time arrivals: from 80% to 90%	16
Lavatories: *modern and roomy*	14
Modernity: modern coach	9
Reservation: optional seat booking	9
Cleanliness: clean inside	5
Mode: going by double-decker	3.5

Source: Kottenhoff, K., *Trans. Res. Record*, 1623, 144, 1998.

then there are two main direct elasticity definitions as explicated by Balcombe et al. (2004): *point* elasticity and *arc* elasticity. The direct point elasticity of y_i is calculated as the slope of the demand curve times the ratio of the values of y_i to D_i:

$$\varepsilon_i^{point} = \left(\frac{\partial D_i}{\partial y_i} \right) \frac{y_i}{D_i} \tag{10.3}$$

The direct arc elasticity is calculated between two points on the demand curve, using a logarithmic form extracted from the differential $(dD_i/D_i)/(dy_i/y_i) = d(\ln D_i)/d(\ln y_i)$:

$$\varepsilon_i^{arc} = \frac{\log D_{i1} - \log D_{i2}}{\log y_{i1} - \log y_{i2}} = \frac{\Delta(\log D_i)}{\Delta(\log y_i)} \tag{10.4}$$

where D_{i1} and D_{i2} exhibit the values of the demand before and after the change of the attribute from y_{i1} to y_{i2}, respectively.

In a competitive environment, a change in an attribute (e.g., fare) for one transit service may affect the demand for another transit service. The sensitivity of these cross effects is indicated by the phenomenon of cross-elasticity, expressed mathematically as

$$\varepsilon_i^{uv} = \left(\frac{\partial D_i^u}{\partial y_i^v} \right) \frac{y_i^v}{D_i^u} \tag{10.5}$$

where

ε_i^{uv} is the (point) cross-elasticity of demand for service u concerning the change in attribute y_i of service v

D_i^u is the value of the demand function for attribute y_i, using service u

y_i^v is the value of attribute y_i, using service v

In practice, it may be difficult to measure cross-elasticity (since direct effects frequently outweigh cross effects). The following relationship shows how to calculate it from direct elasticities:

$$\varepsilon_i^{uv} = \left| \varepsilon_i^u \right| \frac{D_i^u}{D_i^v} \left(\frac{\partial D_i^v}{\partial D_i^u} \right) \tag{10.6}$$

where

ε_i^u (an absolute value) is the direct elasticity of demand for service u for the change in attribute y_i of (the same) service u

D_i^u / D_i^v is the ratio between the competing volumes of services, u and v

$\partial D_i^v / \partial D_i^u$ is the proportion of the demand change in services v and u as a result of the change in y_i^v

More on elasticity functions and approaches can be found in Balcombe et al. (2004).

10.2.3 Overview of elasticity results

Litman (2004) and Balcombe et al. (2004) summarized and compared numerous studies on elasticity values for several attributes, with an emphasis on transit fares. Litman (2004) concluded the research with the recommendation to use ranges rather than point values because of the inevitable uncertainty inherent in elasticity analysis.

Litman's (2004) recommendations are shown in Table 10.3 for short- and long-run predictions. The short run refers to periods of less than 2 years and the long run to periods of 5–10 years. The transit demand considered is elastic for elasticity values approaching or above 1.0 (or approaching or below −1.0). Measurable attributes influencing the quality of service are included in Table 10.3 under *attributes of service*, that is, frequency of service, access and egress times, waiting time, in-vehicle time, and transfer time. According to Litman, it is assumed in practice that demand is inelastic for the short run; a fare increase and/or service reduction will have a marginal effect on the demand, thus increasing net revenue.

Both Balcombe et al. (2004) and Litman (2004) indicated that fare elasticities were about twice that for off-peak and leisure trips as for peak and commuting trips. In addition,

Table 10.3 Recommended direct-transit elasticity values

	Market segment	Short run	Long run
Transit demand in regard to the attribute of fares	Overall	−0.5 to −0.2	−0.9 to −0.6
	Peak hours	−0.3 to −0.15	−0.6 to −0.4
	Off-peak hours	−0.6 to −0.3	−1.0 to −0.8
	Suburban commuters	−0.6 to −0.3	−1.0 to −0.8
Transit demand in regard to the attributes of service quality	Overall	0.5 to 0.7	0.7 to 1.1
Transit demand in regard to the attribute of car operating cost	Overall	0.05 to 0.15	0.2 to 0.4
Car users in regard to transit costs	Overall	0.03 to 0.1	0.15 to 0.3

Source: Litman, T., J. Public Trans., 7(2), 37, 2004, Table 11.

short-distance trips and trips made in small towns have higher fare elasticities than do longer trips and trips made in large cities.

Balcombe et al. (2004) separated the attributes influencing the quality of service into those that can be quantified and those that can be only indirectly estimated. The latter include attributes related to transit reliability, comfort, convenience, and safety. In the first group of attributes, Balcombe et al. (2004) used both elasticity values and attribute weighing in terms of equivalent in-vehicle time. They estimated that the elasticity of demand to vehicle-km for buses was between 0.1 and 0.7 for short runs and between 0.2 and 1.0 for long runs; for rail, it is roughly 0.6–0.9 for short run. The in-vehicle time elasticity for (local) bus ranges is −0.4 to −0.6 and for rail −0.6 to −0.8.

The use of attribute weighing was applied to access/egress walking time, waiting time, and service headway. Balcombe et al. (2004) estimated that access/egress walking time can be on average 1.68 times the value of in-vehicle time (depending on trip length and walking time); when transfers are involved, this average is around 1.81, which may also reflect a transfer penalty. Waiting time was valued at 1.76 times that of in-vehicle time (varying by trip length). Service headway was valued at 0.77 times in-vehicle time.

An additional attribute considered by Balcombe et al. (2004) was income, including car-ownership effects. Based on the national travel survey in the United Kingdom (1985–1997), they recommended values of −1.08 and 0.34 for long-run elasticities, for this attribute, for commuting bus and rail trips, respectively, and values of −0.33 and 0.42 for long-run elasticities for leisure trips for bus and rail, respectively. They noted that the bus-income elasticity, including car-ownership effects, was negative, compared to positive rail-income elasticity.

Hensher (2001), using mixtures of revealed-preference and stated-preference data, demonstrated an example of direct elasticity and cross-elasticity in Sydney. Revealed-preference applies to observed data (e.g., by agencies in regard to ticket sales, direct home surveys); stated-preference applies to questionnaires inquiring the relative importance of factors that are not easily appraised through revealed-preference methods. Table 10.4 presents Hensher's results for commuters; each column has one direct (in italic and bold) and six cross-elasticities. For instance, concerning the change in the single-fare bus ticket (fourth column in Table 10.4), an increase of 10% in this ticket's fare tends to reduce its sales by 3.57% and to increase the sale of single-fare train tickets by 0.57% and to increase the number of car users by 0.66%. Similarly, this fare change can apply to the other types of tickets.

Another study on fare (or trip cost) direct elasticity and cross-elasticity for commuting trips in Sydney was presented by Taplin (1997). This study, whose findings are summarized in Table 10.5, furnished adjusted fare (or cost) elasticities for trips performed by different transit modes and cars. The direct elasticities are emphasized (in italic and bold) and have

Table 10.4 Direct elasticity and cross-elasticity for Sydney commuters

Mode	Train			Bus			Car
Type of ticket	Single fare	Weekly	Pass	Single fare	Ten fare	Pass	—
Single fare (train)	−0.218	0.001	0.001	0.057	0.005	0.005	0.196
Weekly (train)	0.001	−0.093	0.001	0.001	0.001	0.006	0.092
Pass (train)	0.001	0.001	−0.196	0.001	0.012	0.001	0.335
Single fare (bus)	0.067	0.001	0.001	−0.357	0.001	0.001	0.116
Ten fare (bus)	0.020	0.004	0.002	0.001	−0.160	0.001	0.121
Pass (bus)	0.007	0.036	0.001	0.001	0.001	−0.098	0.020
Car	0.053	0.042	0.003	0.066	0.016	0.003	−0.197

Source: Hensher, D., Modal diversion, in: Hensher, D. and Button, K., eds., *Handbooks of Transport Systems and Traffic Control*, Elsevier Science, Amsterdam, the Netherlands, 2001, pp. 107–123, Table 3.

Table 10.5 Adjusted fare (or trip cost) direct elasticity and cross-elasticity for commuting trips in Sydney

Mode of travel	Train	Bus	Ferry	Car
Train	−0.156	0.032	0.003	0.037
Bus	0.063	−0.070	0.006	0.046
Ferry	0.039	0.037	−0.195	0.003
Car	0.016	0.011	0.000	−0.024

Source: Balcombe, R. et al., The demand for public transport: A practical guide, TRL Report, TRL593, TRL Limited, Crowthorne, U.K., 2004, Table 9.22.

negative values. For example, an increase of 10% in train tickets in Sydney tends to reduce train-passenger demand (sale of this ticket) by 1.56% and to increase bus and ferry riders and car users by 0.63%, 0.39%, and 0.16%, respectively.

Finally, elasticity may possibly be modeled through the functional relationship between demand changes and fare (or trip cost) changes. For example, Hong Kong transit agencies, according to Zhou et al. (2005), are more concerned with the transit-fare structure than with frequency of service in maximizing their profits. This concern derives from the relatively large sensitivity of passenger demand to fares. Zhou et al. (2005) mentioned that for a 1% reduction in transit fares, there is a consequent 1.33%–1.45% increase in transit patronage in Hong Kong ($\varepsilon < -1.0$), compared with only a 0.3% ($\varepsilon = -0.3$) increase in Canada. They illustrated the demand-elasticity effects on revenues through changing fares in the following assumed exponential function:

$$D_f = Be^{\alpha p} \tag{10.7}$$

where
   D_f is the transit demand
   p is the route fare
   B and α are constant parameters (estimated from observed data)

The *point* elasticity of Equation 10.3 for fares is

$$\varepsilon = \frac{dD_f}{dp} \cdot \frac{p}{D_f} = \alpha p \tag{10.8}$$

which is typically negative. Zhou et al. (2005) also referred to a denoted revenue R, in which $R = D_f \cdot p$. By using the aforementioned relationships, the revenue has the following derivative:

$$\frac{dR}{dp} = Be^{\alpha p}(1 + \alpha p) \tag{10.9}$$

For elasticity ε values less than –1.0 (thus, $\alpha < -1/p$), an increase in fare will result in a decrease in revenue. Consequently, Hong Kong transit agencies can increase their revenue by actually decreasing their fares.

10.3 EXAMPLE OF A DEMAND FORECASTING, METHOD AND PROCESS

This section presents and describes the rationale for a transit-demand forecast methodology adapted to predict, for a given transit service, which we will call Transit-A, the future patronage of a specific set of routes in certain required years. This example of a forecasting method also considers relevant survey results of passenger demand for competitor transit services associated with the same underlying set of routes. Taking into account all the parameters that affect transit-service demand is an exhaustive undertaking. Thus, simple but useful models addressing the major attributes of the passengers' decision process will be shown in this section. The method presented follows Ceder (2006) in a study of passenger demand for Hong Kong ferries.

10.3.1 Framework

We assume that Transit-A patronage relies not only on certain attributes but also on the changes in the levels of service of the competitive transit services. Therefore, whenever there is a change in the level offered by these other services, an alternative model, such as modal split, should be employed for a coarse first estimation.

In brief, the applied approach of the demand forecast covers the following four components: (1) identification of main attributes and their weights in affecting potential Transit-A passengers; (2) demand prediction using a calibrated growth factor from past Transit-A demand figures for any projected changes in the established Transit-A routes; (3) transit modal split between specific O-D pairs on any new Transit-A routes and/or changes in fares and/or journey times in competitor routes; and (4) development of Transit-A O-D demand matrices for each alternative network in a required year.

The forecast approach principally relates to the changes in Transit-A-relevant attributes when there are changes in passenger demand; whenever changes in competitive transit links and/or new Transit-A routes arise, a modal split is necessitated for establishing the O-D demand. The relationship between change in attribute values and the corresponding change in Transit-A demand is estimated by conducting a survey (see Chapter 2 for methods) to capture the different behavior of existing and potential Transit-A users. A demand forecast is usually network-based. The existing Transit-A network provides the base framework from which other networks can be established. Specifically, each network contains a definite set of routes and fares, an operational strategy, and different extents of competition (from other transit services). In addition to the network scenario, a route scenario can also be established to analyze individual routes when it is reasonable to assume that they are independent of one another. The latter was the case with the ferry study (Ceder, 2006).

Table 10.6 Input and output lists for forecasting models

Input	Output
1. O-D survey	1. Base-year Transit-A patronage and potential increase in Transit-A patronage from competitors
2. Opinion survey	2. Attribute weights in linear regression models
3. Estimation of transit-passenger demand from local transportation-planning studies	3. Expansion of the base-year Transit-A patronage to the design-year patronage at low, medium, and high levels
4. Local new-route service assumptions	4. Patronage estimate of new Transit-A routes
5. O-D matrices, fare matrices, travel-time matrices, and calibrated split-model coefficients taken from local transportation-planning studies	5. O-D demand forecast by modes given by MNL modal split for a design year as a base for any further changes in new Transit-A routes

In order to integrate the demand forecast for Transit-A into transportation-planning studies of the local area being considered, the ODMs for all transit services should be derived directly from the local area studies, except for those of the existing Transit-A routes, which are to be revealed by the survey. Combining all this information, the future demand for existing routes can be predicted with regard to any change in the level of service. For new routes and/or changes in other transit services, the common attributes of all relevant services between specific O-D pairs need to be identified before a multinomial logit (MNL) model (see de Dios Ortuzar and Willumsen, 2011) can be used for modal split purposes. The coefficients in the MNL model adopted the values of the local transportation-planning studies. The following section explicates this MNL model.

The possible input/output lists for a transit-demand forecast methodology are shown in Table 10.6. The framework for this forecasting methodology is presented in Figure 10.2, which summarizes the flow of the forecasting tasks with the use of two models. It is worth noting that the left-hand side of Figure 10.2 relates not only to changes in the existing Transit-A routes but also to changes in the candidate routes following the establishment of their O-D demand on the right-hand side of this figure.

Another important input in Figure 10.2 is the growth factor for a required year. This factor should be updated continuously to allow for more accurate forecasting. The growth factor is extracted from percentage changes in patronage for a given design year over the past few years. It contains all the changes incurred, including changes in travel behavior. Extrapolation (e.g., utilizing regression models) of the growth factor, which can be negative, exhibits this ongoing patronage change.

10.3.2 Opinion survey (input) interpretation

This section explains how to apply the opinion survey information to demand forecasting. Usually the main attributes affecting transit passenger demand are as follows: fare, trip time, waiting time, walking time, and comfort level. Two sets of questionnaires can be designed to obtain information on the aforementioned attributes, one set for the Transit-A survey (possibly on board) and the other for the competitor(s). The first part of the questionnaires can collect O-D information on commuters, which would help to develop the base-year Transit-A demand matrices. The second half helps in establishing the commuter's choice under different scenarios of changed services. Finally, basic questions on the trip purpose and personal details can be included for the sake of acquiring more information. For a better understanding of people's behavior and, hence, for more accurate future passenger-demand predictions, a stated-preference type of questionnaire can be designed for fare.

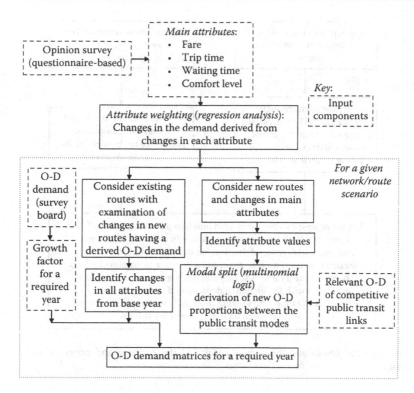

Figure 10.2 Framework of a given forecasting method for a given transit mode.

This will enable a direct deduction of the relationship between a fare level and a change in demand. A similar arrangement and rationale holds for the item of comfort level, which requires indirect transformation.

The attribute values can then be converted into equivalent monetary values in order to refine the prediction mechanism by making use of the direct relationship between the fare level and a change in demand. For developing a global model for all Transit-A routes, regardless of their O-D characteristics, tailor-made questionnaires, addressing differences in the fare and the other attributes, need to be constructed for individual routes. A list of comfort-level items, from which a respondent can choose some but not all of the most important ones, will be included in the questionnaires. By combining this information with other fare-related information from other questions, an indirect deduction of the monetary value of each comfort level can be achieved and consequently used in the prediction model.

Examples of stated-preference questions are as follows:

1. Will you still ride on a Transit-A vehicle if the fare level is increased to (1) $7, (2) $6.50, (3) $6, (4) $5.50, or (5) $5.30?
2. If the Transit-A trip travel time is 5 min longer, how much do you think the fare should be reduced? (Mark an amount.)
3. If the Transit-A service waiting time is 3 min longer, how much do you think the fare should be reduced? (Mark an amount.)
4. Please mark up to three comfort items that are most important to you for improving the service? (Choose from a given list.)
5. Given improvements in the comfort items you marked in (4), are you willing to pay for an increase in the Transit-A fare to (1) $10, (2) $7.50, or (3) $6?

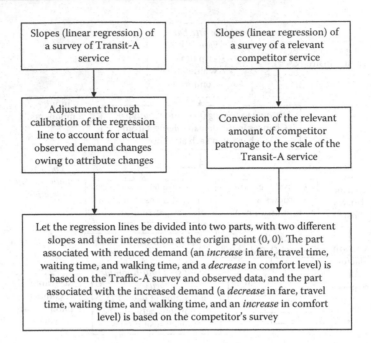

Figure 10.3 Procedure for combining the survey results of the existing and potential users of Transit-A service.

A linear regression model can be utilized as a tool to relate changes in the attribute value to corresponding changes in demand. Because of different bases in most of the attributes associated with different routes, a global method can be established by expressing the percentage change for both the attribute values and the demand. For simplicity, each attribute can be assumed to be independent of all other attributes in that they can be examined separately while developing the model. For different levels of changes in fare, a corresponding percentage change in Transit-A patronage can be deducted from the survey result. This information can then be combined, using a linear regression, to determine the relative weight of the fare attribute from the slope of the regression line. Similar procedures may be used for other attributes after they are converted to monetary values, thereby making use of the fare–demand relationship developed. A procedure to combine existing riders and potential riders from a competitor within Transit-A is shown in Figure 10.3.

The regression lines are supposed to intersect the origin (0, 0), assuming that if there is no change in the attribute, there will be no change in the demand. However, it is known that the answers of survey respondents are biased, reflecting riders' strong wishes to improve the ride without additional cost and/or the natural resistance to increase Transit-A fares and/or worsen its other attributes. As a result, the best fit for the regression line does not intersect the origin; it shows that demand is dropped even if there is no change in the attribute. One basic way to correct this bias, over and above its calibration with actual data, is to ignore the free constant element in the linear regression and to allow the regression line to intersect the origin.

10.3.3 Example of attribute derivation

A detailed example will now be provided as a useful device for understanding the underlying procedures. This example explains how to calibrate the weights of attributes from the raw survey data. It is assumed that 20 passengers were interviewed and their choices for each question randomly assigned.

First, let us consider the question: Will you still ride on a Transit-A vehicle if the fare level is increased to (1) $7, (2) $6.50, (3) $6, (4) $5.50, or (5) $5.25? These five options correspond to five levels of fare increase from $5 (by 5%, 10%, 20%, 30%, and 40%), with "1" being assigned to the option selected. In case a passenger is unwilling to pay for any fare increase, no option is assigned. The existing fares range between $2.20 and $5.30. The acceptable increase in fare equals the existing fare times the increase option chosen by the passenger interviewed. For example, the acceptable fare increase for one passenger was $5.00·10% = $0.50; that is, for $5.50, this passenger will still ride on Transit-A. The demand change for the fare is calculated from the number of passengers still willing to ride Transit-A given a fare increase. For instance, for a fare increase of 30%, five (out of twenty) still will ride on Transit-A; this means a 75% patronage decrease. Plotting the percentage change in demand versus the percentage change in fare for all routes and cases will construct a data set suitable for a regression analysis.

Second, we will now present a full analysis of the question of an increase in travel time. Similarly this analysis can be performed, for example, for waiting time, walking time/distance, and comfort items. Table 10.7 presents the raw survey data output for the question: If the Transit-A trip travel time is x min longer, how much do you think the fare should be reduced? (Interviewees need to mark an amount.)

The value of travel time is calculated by the ratio of the first two columns in Table 10.7. A set of travel-time changes is tested, whether or not passengers still ride Transit-A. For example, for the first passenger interviewed (first raw data in Table 10.7), an increase in

Table 10.7 Interpretation of survey results concerning a change in travel time

Value of x more minutes ($)	x (min)	Value of travel time ($/min)	Existing travel time (min)	Change in travel time					
				40%	30%	20%	10%	5%	1%
1	3	0.33	7	0	0	0	1	0	0
0.5	3	0.17	7	1	0	0	0	0	0
1	3	0.33	8	0	0	0	1	0	0
0.5	3	0.17	8	0	0	0	0	1	0
1	5	0.20	13	0	0	0	1	0	0
2	5	0.40	13	0	0	0	1	0	0
2.5	3	0.83	7	0	0	1	0	0	0
1.5	3	0.50	7	0	0	0	0	0	0
3	3	1.00	5	0	1	0	0	0	0
3	3	1.00	5	0	0	0	0	0	1
2	3	0.67	7	0	0	0	0	1	0
3	3	1.00	7	0	0	1	0	0	0
2.5	5	0.50	11	0	0	0	0	0	1
1.5	5	0.30	11	0	0	1	0	0	0
2	5	0.40	12	0	0	0	0	1	0
2.5	5	0.50	12	0	0	0	0	1	0
1	5	0.20	15	1	0	0	0	0	0
2	5	0.40	15	0	0	0	0	0	0
2.5	5	0.50	10	0	0	1	0	0	0
3	5	0.60	10	0	0	0	0	0	1
Demand change for travel-time change				−90%	−85%	−65%	−45%	−25%	−10%

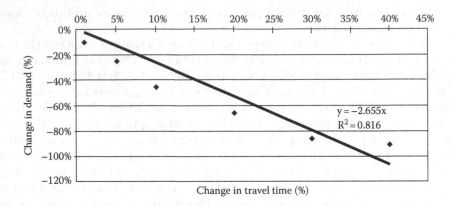

Figure 10.4 Regression line of change in travel time against change in demand.

travel time of 40% is equivalent to a fare increase of 7·40%·0.33 (existing travel time *times* travel-time change *times* the value of travel time) = $0.924. Because the acceptable increase in fare for the first passenger was calculated (in the first question) to be $0.44, this passenger will not ride on Transit-A when the travel time increases 40% or more, and thus "0" is assigned. If, however, the travel time should increase by only 10%, this would be equivalent to a fare increase of 7·10%·0.33 = $0.231, which is less than the maximum acceptable increase, and therefore "1" is assigned in Table 10.7. Thus, the demand change for the travel-time change can be calculated with this conversion; this is plotted in Figure 10.4 with the resultant fitted regression line. The R-squared value (0.816) is the fraction of the variance in the data, which is explained by the regression.

10.4 MULTINOMIAL LOGIT MODEL

The MNL model (see, e.g., de Dios Ortuzar and Willumsen, 2011) is a share model that divides the individuals among various travel modes according to each mode's relative desirability for any given trip. Such a technique has been commonly used to determine modal split in a number of studies, though its accuracy depends heavily on the given data and underlying mathematical format of the logit model.

This modal split model is usually used when there are some factors that the linear regression model cannot fully address, for example, the introduction of new Transit-A routes and changes in the degree of competition with other public transit modes (e.g., via improvements, fare changes). Four public transit modes are commonly included and examined in the modal split process along with the private vehicle. These modes are train, bus, ferry, and taxi.

The input attributes for the MNL model should be consistent and valid for all the available transit modes and services. We will consider here the two main attributes affecting modal split—fare and travel time. Because of the limited number of attributes employed in the model, it may be necessary to fine-tune Transit-A demand.

The MNL model takes the following mathematical form:

$$P_m = \frac{e^{u(m)}}{\sum_{i=1}^{n} e^{u(i)}} \quad \text{for } 1 \leq m \leq n \tag{10.10}$$

where
  P_m is the probability of an individual's choosing mode m or the proportion of trips using mode (or service) m
  u(i) is the utility of the ith mode, which includes the modal parameters and coefficients for a certain trip among the n available modes

For instance, the utility function with reference to buses can take this form:

$$u(i) = T(t_i - t_{bus}) + C(c_i - c_{bus}) + B_1 + B_2 \qquad (10.11)$$

where
  t_i is the travel time incurred by mode i
  c_i is the cost (fare) by mode i
  T and C are the calibrated coefficients
  B_1 and B_2 are biases for the mode and movement limitations, respectively, varying according to whether the mode is accessed by walking or by another transit mode and according to the degree of access difficulty

The following are typical MNL assumptions used in the forecast: (1) given zoning system; (2) fares and travel times for railway, bus, ferry, and taxi taken from given matrices for peak and nonpeak periods, as well as the cost and travel time of private vehicles; (3) total transit patronage taken from given local transportation-planning studies (aimed at capturing the right mixture of population and economy growth affecting transit demand) and based on self-zones selection; (4) given number of peak and off-peak hours for the local area considered; and (5) specific zone selection depending on the proportion of transit users.

Following is an example of the use of the MNL model. Suppose a new railway line is introduced in a transit network consisting of railway, bus, and passenger ferry. In order to arrive at an estimate of daily patronage for this new line, the corresponding zones of the line's end points should initially be identified; say, Z1 and Z2. The corresponding transportation-planning study figures for this particular O-D pair for a designated year are as follows:

- Total daily transit demand: 164,000 trips (Z1 to Z2) and 161,000 trips (Z2 to Z1)
- Travel times (includes the trip, waiting, walking, and transfer times) and fares for each of the three transit modes:

Mode	Travel time (min)		Fare ($)
	Z1 to Z2	Z2 to Z1	
Railway	27.1	27.1	7.22
Bus	51.9	51.5	7.91
Ferry	49.2	48.2	5.37

The travel times and fares are inserted into the utility function (10.11) with the following coefficients:

Mode	T	C	B_1	B_2
Railway	−0.00373	−0.00153	0.202	0.267
Ferry	−0.00373	−0.00153	0.491	−1.1

The utility value of each of the transit modes is then calculated by Equation 10.11:

$$u(railway) = 0.567409, \quad u(bus) = 0, \quad u(ferry) = -0.595789$$

Utilizing the MNL model in Equation 10.10 yields the following modal split results:

$$\text{Modal split by railway} = \frac{e^{0.567409}}{e^{-0.595789} + e^{0} + e^{0.567409}} = 53.2\%$$

$$\text{Modal split by bus} = \frac{e^{0}}{e^{-0.595789} + e^{0} + e^{0.567409}} = 30.2\%$$

$$\text{Modal split by ferry} = \frac{e^{-0.595789}}{e^{-0.595789} + e^{0} + e^{0.567409}} = 16.6\%$$

Applying these results to the new railway line between Z1 and Z2 will determine the estimated patronage level: $(164,000 + 161,000) \cdot 53.2\% = 172,900$ daily trips.

10.5 LITERATURE REVIEW AND FURTHER READING (ORIGIN–DESTINATION ESTIMATION)

ODMs constitute an essential input for most transit planning and design procedure. The literature describes many methods of matrix estimation, but most studies refer to the generation of car-trip matrices based on data from traffic counts. This section will review methods of estimating transit-trip ODMs based on passenger counts or surveys.

Simon and Furth (1985) describe the method developed by Tsygalnitzky in 1977 for creating a route O-D without O-D survey data. A good representation of transit demand can usually be achieved if the trip matrix is based on a survey, in which passengers are asked directly about their precise origin and destination; however, such surveys are expensive and time-consuming; therefore, several methods deal with situations in which data from surveys are not available. Tsygalnitzky's method uses only passenger on–off counts; the model's passenger flow from one stop to another along the route is given in terms taken from fluid mechanics. The matrix is generated using a recursive algorithm. Simon and Furth analyze results statistically from the implementation of Tsygalnitzky's method and compare them to actual trip demand. They conclude that this method yields reasonable results in cases in which the route structure is not too complicated.

Ben Akiva et al. (1985) review and test several techniques for generating route-level trip matrices. The common practice of expanding onboard survey results by means of total boardings is compared with expanding results by iterative proportional fitting, constrained generalized least squares, and constrained maximum likelihood methods. An intervening-opportunity model, which does not use the onboard survey, is also tested. The more complex methods achieve better accuracy and reduced bias by combining the survey data with passenger counts. An empirical case study demonstrates that under the assumption of error-free ride-check data, the proportional fitting technique is preferred because of its computational ease without any loss of accuracy. Ben Akiva (1987), in a different study discusses methods of combining transit data from different sources of partial information to create a transit-trip matrix. The same matrix-estimation techniques described by Ben Akiva et al. (1985) are examined. The use of alternative data sources, such as ticket-sales information, is illustrated. Aspects concerning the level of accuracy of combined data, biases, and computational difficulties are discussed.

Nguyen et al. (1988) propose a method for the dynamic estimation of transit-trip matrices, using passenger counts. The matrix period is divided into several time intervals, and formulations that connect the link flow at a given time to link flows over time segments are developed. The paper also presents an assignment model that is used for the estimation. The optimization is based on maximum entropy and maximum likelihood approaches.

Furth and Navick (1992) analyze the relationship between Tsygalnitzky's recursive method, which is commonly used for creating trip matrices from counts, and the bioproportional method, which combines data from counts and from an O-D survey. In the bioproportional method, the survey data are used to create an initial seed matrix, which is then adjusted iteratively by balancing row and columns to match on–off totals. The authors show that the Tsygalnitzky method is, in fact, a special case of the bioproportional method. As a seed matrix, it implicitly uses a null matrix, which contains information about trip directionality and minimum length. The paper illustrates why the recursive method is inappropriate when there is significant competition among routes. In addition, the authors offer a correction for cases in which the on–off data have been aggregated to the segment level.

Navick and Furth (1994) discuss possible sources of the seed matrix, which is often used as an initial basis for adjustments, and propose a model for generating this matrix. According to this model, the number of trips in each matrix cell is influenced by two factors: popularity, which is a feature of the destination zone and is not explained by models, and propensity, which is a feature of each O-D pair and depends on the distance between them. The researchers develop a propensity function that is the product of a power term and an exponential term and is equivalent to a gamma distribution. The power-function exponent is estimated by maximum likelihood, based on survey data. The gamma seed combined with the bioproportional method to match O-D totals is shown to be effective in generating ODMs.

Gur and Ben-Shabat (1997) develop another model that uses passenger counts for building or improving trip matrices. Their model is formulated as a nonlinear programming problem; for each pair i–j of stations, the model calculates the probability that a passenger who boards at i will alight at j. The problem takes advantage of information that is embedded in counts of individual vehicles and in the differences between different counts. The least-squares method is used for adjusting boarding totals to counts, and a minimum information model (Fratar) then generates a detailed trip table.

Wong and Tong (1998, 2003) propose a methodology for creating transit-trip matrices from passenger counts, using a maximum entropy programming formulation. The matrix-generation process depends on more detailed time data than do most previous methods; it uses a schedule-based dynamic assignment model to determine least-cost paths between all O-D pairs and clock-arrival times. Cases with and without a given seed matrix are examined. The authors also present a solution algorithm in which the matrix reduces computer-memory requirements in an efficient way.

Friedrich et al. (2000) present a technique for the continuous updating of demand matrices that is based on fuzzy set theory. According to this theory, counts are treated as intervals with lower and upper bounds, not as precise values. Matrix values are estimated such that all totals fall within the proper bandwidths. The method, formulated as a programming problem, may be implemented for estimating either car or transit-passenger matrices.

Nuzzolo and Crisalli (2001) and Nuzzolo et al. (2003) develop another method for estimating time-varying matrices from time-varying counts. The reference period is divided into several subperiods, and detailed timetable information is considered. A dynamic schedule-based model of a stochastic type is used to predict passengers' path choices. The method is formulated as a maximization problem without constraints. The match between matrix and counts is achieved by minimizing generalized least squares.

Cui (2006) documents the development of an algorithm to estimate a bus passenger trip ODM based on automatic data collection system (ADCS) archived data including automated fare collection (AFC) data, automatic passenger count (APC) data, and automatic vehicle location (AVL) data. The proposed algorithm consists of three steps: data preparation, estimation of single-route ODMs for all routes, and estimation of a network level ODM using transfer flow information. The proposed single-route ODM estimation utilizes AFC and APC data to develop the *seed* matrix and marginal values, respectively. Using iterative proportional fitting and maximum likelihood estimation (MLE) techniques, the single-route ODMs based on seed matrices and marginal values are estimated and results compared. The transfer flows for network level ODM estimation are derived by considering the consecutive transactions from AFC data. The resulting network level ODM is provided at a route segment level of detail. This O-D estimation algorithm has been applied to a selected corridor of the Chicago Transit Authority bus network. Finally, the author recommends that the MLE method be used to estimate the single-route ODMs and the proportional distribution method be used to estimate the transfer flow ODM.

Farzin (2008) describes the process used to create an ODM in São Paulo, Brazil, with data available from ADC systems. The approach used in this paper illustrates how these large quantities of ADC systems data can be utilized to take advantage of the constant improvements in computer processing speed to derive new planning and measurement applications. The proposed methodology includes development of ODM using three data sources: bus stops, AVL data, and AFC data. The results indicate reasonable patterns of ODMs. Furthermore, using AFC system provides a larger sample set in comparison to traditional survey data, thus allowing more comprehensive analysis and generating more specific results. Future improvements suggested by the author include addition of transfer-level information to the ODM and verification of available bus stop data.

Chapleau et al. (2008) argue that with a well-defined methodology, the transaction boarding data, collected via smart car AFC system, can be used to monitor the supply and consumption of the transit system, the mobility of its users, and the activity pattern at trip generators. The authors provide descriptions of the validation and enrichment procedures as well as their logic engaged, so it could be adopted for other similar systems. Furthermore, examples of characterization of various objects in the transit system and a multiday analysis are presented and discussed. The authors emphasize that more transit data will become available as more transit agencies adopt the smart card AFC technology, thus creating opportunities to develop innovative ways to extract and analyze data for transit planning and modeling purposes.

Wang (2010) explores the application of archived data from ADCS to transportation planning with a focus on bus passenger O-D inferences at the bus-route level and on travel behavior, using London as an example. The feasibility and convenience of the trip-changing methods' application, to obtain bus passengers' boarding and alighting locations, has been demonstrated. These results have subsequently been validated by comparing against the bus passenger O-D survey data in London. The inferred ODMs and the AVL data enable estimation of alighting times for bus passengers, therefore allowing bus journey stages to be conveniently linked to form complete journeys based on the difference between the subsequent trip's boarding time and the previous trip's alighting time for each bus passenger. Comparison of interchange time and the connecting bus route's headway can provide a methodology to evaluate the connecting bus services and bus passengers' interchange patterns. Finally, the author suggests that this research can be expanded to a full bus network and other travel modes, thus enabling effective intermodal network planning through the development of comprehensive databases.

Horváth (2012) investigated methods for obtaining reliable passenger data to construct time-dependent ODMs, in order to facilitate effective planning of public transit systems. A forecasting model, which consists of three stages, has been developed to derive time-dependent passenger data. First, full-scope cross-section data are collected with personnel or with an automatic counting system. Then, boarding and alighting data are linked to obtain the ODM for each run. Finally, ODMs of the run-through transfers are combined by assuming that the probability of a transfer between two runs in a given stop is proportional to the travel possibilities in this relation. The overall methodology has been tested through an application in a Hungarian city (Dunaújváros). The author concludes that these results were reliable, so they could be used in the planning process.

Table 10.8 Summary of characteristics of the origin–destination methods reviewed

Sources	Method requires seed matrix?	Mathematical approaches used	Time segmentation (static/dynamic model)
Simon and Furth (1985) (method by Tsygalnitzky)	No		Static
Ben-Akiva et al. (1985)	Yes	Iterative proportional fitting, constrained generalized least squares, constrained maximum likelihood	Static
Ben-Akiva (1987)	Yes		Static
Nguyen et al. (1988)	No	Maximum entropy, maximum likelihood	Dynamic
Furth and Navick (1992)	Methods that do/don't require a seed matrix are compared	Bioproportional method	Static
Navick and Furth (1994)	A method is presented for creating a seed matrix	Bioproportional method, maximum entropy	Static
Gur and Ben-Shabat (1998)	Optional	Least squares, minimum information model (Fratar)	Static
Wong and Tong (1998, 2003)	Cases with/without a seed matrix are examined	Maximum entropy	Dynamic
Friedrich et al. (2000)	Yes	Fuzzy set theory	Static
Nuzzolo and Crisalli (2001) and Nuzzolo et al. (2003)	No	Generalized least squares	Dynamic
Cui (2006)	Yes	Maximum likelihood estimation method is used to estimate the single-route ODMs, and the proportional distribution method is used to estimate the transfer flow ODM	Static
Farzin (2008)	No		Static
Chapleau et al. (2008)	No		Static
Wang (2010)	Yes		Static
Horváth (2012)	No		Static
Tirachini et al. (2013)	No	Multinomial logit model	Static

Tirachini et al. (2013) investigated the impact of crowding on the passenger-demand estimation. They argued that people's perception of crowding should be considered as a source of disutility for passengers when estimating passenger demand for public transport. They found that (1) a model that assumes users as indifferent to the occupancy levels of vehicles overestimates the value of travel-time savings (VTTS) for low load factors and underestimates VTTS for high load factors, and likewise (2) a model that is insensitive to crowding levels underestimates demand if vehicles are uncrowded and overestimates demand if vehicles are crowded. These findings were not proven using empirical studies. However, the authors analytically found that when crowding was omitted, there was a load factor threshold that marks the underestimation or overestimation of passenger demand by using an MNL model for mode choice.

The main characteristics of the methods reviewed previously are summarized in Table 10.8.

EXERCISES

10.1 Assuming that a rail company has the following linear demand relationship: $p = 20 - 0.04q$, where p is the route fare and q is the number of hourly sold tickets. Find the total revenue associated with each (p, q) pair and indicate when its maximum value is reached. In addition, specify the p–q zones under which the point elasticity is elastic and inelastic.

10.2 Given the following power demand function, $N = p^{-0.3} \cdot t^{-0.3} \cdot a^{0.2} \cdot c^{-0.3}$, where N is the number of transit trips, p is the fare (in dollars), t is the travel time (in hours), a is the cost of automobile trip (in dollars), and c is the average income (in dollars). (a) Given 20,000 passengers per hour who currently use the transit system at a flat fare of $1.20 per trip, what will be the change in N for a flat fare of $0.80 and what will be the transit agency gain? (b) Given the cost of an automobile trip (including parking fee) of $4.00, what will be the change in N if the parking charges increases by $0.60?

10.3 Given a calibrated utility function, $u = b - 0.04C - 0.02t$, where C is the cost of travel (in cents), t is the travel time (in minutes), and the MNL model is as described in the chapter. (a) What will be the modal split (percentage demand traveling by each mode) for the following data given? (b) Given rising gasoline price so that C for an automobile increases by $1.20, what impact will this change have on the modal split?

Mode	b	C	t
Bus	−0.30	85	30
Light rail	−0.35	100	50
Automobile	−0.25	110	35

REFERENCES

Balcombe, R., Mackett, R., Paulley, N., Preston, J., Shires, J., Titheridge, H., Wardman, M., and White, P. (2004). The demand for public transport: A practical guide. TRL Report, TRL593, TRL Limited, Crowthorne, U.K.

Ben Akiva, M. (1987). Methods to combine different data sources and estimate origin-destination matrices. In *Transportation and Traffic Theory* (N.H. Gartner and N.H.M. Wilson, eds.), pp. 459–481, Elsevier Science, New York.

Ben Akiva, M., Macke, P.P., and Hsu, P.S. (1985). Alternative methods to estimate route-level trip tables and expand on-board surveys. *Transportation Research Record*, 1037, 1–11.

Boyce, D.E. and Daskin, M.S. (1997). Urban transportation. In *Design and Operation of Civil and Environmental Engineering Systems* (C. ReVelle and A.E. McGarity, eds.), pp. 277–341, John Wiley & Sons, New York.

Ceder A. (2004). New Urban Public Transportation Systems: Initiatives, effectiveness and challenges. *ASCE Journal of Urban Planning and Development*, 130(1), 56–65.

Ceder A. (2006). Planning and policy of ferry passenger service in Hong Kong. *Transportation Journal*, 33, 133–152.

Chapleau, R., Trépanier, M., and Chu, K.K. (2008). The ultimate survey for transit planning: Complete information with smart card data and GIS. In CD-ROM of the *Eighth International Conference on International Steering Committee for Travel Survey Conferences*, Lac d'Annecy, France.

Cui, A. (2006). Bus passenger origin-destination matrix estimation using automated data collection systems. Doctoral dissertation, Massachusetts Institute of Technology, Boston, MA.

de Dios Ortuzar, J. and Willumsen, L.G. (2011). *Modelling Transport*, 4th ed., John Wiley & Sons, Chichester, U.K.

Farzin, J.M. (2008). Constructing an automated bus origin-destination matrix using farecard and global positioning system data in Sao Paulo, Brazil. *Transportation Research Record*, 2072, 30–37.

Friedrich, M., Mott, P., and Noekel, K. (2000). Keeping passenger surveys up to date: A fuzzy approach. *Transportation Research Record*, 1735, 35–42.

Furth, P.G. and Navick, D.S. (1992). Bus route O-D matrix generation: Relationships between biopro-portional and recursive methods. *Transportation Research Record*, 1338, 14–21.

Gur, Y.J. and Ben-Shabat, E. (1997). Estimating bus boarding matrix using boarding counts in individual vehicles. *Transportation Research Record*, 1607, 81–86.

Hensher, D. (2001). Modal diversion. In *Handbooks of Transport Systems and Traffic Control* (D. Hensher and K. Button, eds.), pp. 107–123, Elsevier Science, Amsterdam, the Netherlands.

Horváth, B. (2012). A simple method to forecast travel demand in Urban Public Transport. *Acta Polytechnica Hungarica*, 9(4), 165–176.

Kottenhoff, K. (1998). Passenger train design for increased competitiveness. *Transportation Research Record*, 1623, 144–151.

Kottenhoff, K. and Lindh, C. (1996). The value and effects of introducing high standard train and bus concepts in Blekinge, Sweden. *Transport Policy*, 2(4), 235–241.

Litman, T. (2004). Transit price elasticities and cross-elasticities. *Journal of Public Transportation*, 7(2), 37–58.

Navick, D.S. and Furth, P.G. (1994). Distance-based model for estimating a bus route origin-destination matrix. *Transportation Research Record*, 1433, 16–23.

Nguyen, S., Morello, E., and Pallottino, S. (1988). Discrete time dynamic estimation model for passenger origin/destination matrices on transit networks. *Transportation Research*, 22B, 251–260.

Nuzzolo, A. and Crisalli, U. (2001). Estimation of transit origin/destination matrices from traffic counts using a schedule-based approach. In *Proceedings of the AET European Transport Conference*, Homerton College, PTRC, Cambridge, U.K.

Nuzzolo, A., Russo, F., and Crisalli, U. (2003). *Transit Network Modelling—The Schedule-Based Dynamic Approach*, Franco Angeli, Milano, Italy.

Simon, J. and Furth, P.G. (1985). Generating a bus route O-D matrix from on-off data. *Journal of Transportation Engineering*, 111, 583–593.

Taplin, M. (1997). A world of trams and urban transit. *Light Rail and Modern Tramway*, 60(718), 1–8.

Tirachini, A., Hensher, D.A., and Rose, J.M. (2013). Crowding in public transport systems: Effects on users, operation and implications for the estimation of demand. *Transportation Research*, 53A, 36–52.

TranSystems Corp., Planner Coll., Inc., and Crikelair, T. Assoc. (2006). Elements needed to create high ridership transit systems: Interim guidebook. TCRP Report 32, Transportation Research Board, Washington, DC.

Wang, W. (2010). Bus passenger origin-destination estimation and travel behavior using automated data collection systems in London, U.K. Doctoral dissertation, Massachusetts Institute of Technology, Boston, MA.

Wong, S.C. and Tong, C.O. (1998). Estimation of time-dependent origin-destination matrices for transit networks. *Transportation Research*, **32B**, 35–48.

Wong, S.C. and Tong, C.O. (2003). The estimation of origin-destination matrices in transit networks. In *Advanced Modeling for Transit Operations and Service Planning* (W.H.K. Lam and M.G.H. Bell, eds.), pp. 287–315, Elsevier Science, Amsterdam, the Netherlands.

Zhou, J., Lam, W.H.K., and Heydecker, B. G. (2005). The generalized Nash equilibrium model for an oligopolistic transit market with elastic demand. *Transportation Research*, **39B**, 519–544.

Route choice and assignment

CHAPTER OUTLINE

PRACTITIONER'S CORNER

Once we have an idea of how many people want to travel between two points, by time of day and transit mode, the next phase is to determine the routes (direct or via transfers) they should take to reach their destination. This determination relies on passenger behavior in regard to choosing a route if there are alternatives. What is the typical (average) choice strategy if there is a slower or less direct route, but one that does not involve long waiting times? This and other route choice issues are discussed in this chapter in order to prepare the ground for assigning demand on the transit network. Utilizing an assignment procedure will allow for assessing/predicting changes in transit service design at the network level.

The following riddle may indirectly indicate the complexity of the analysis exhibited in this chapter. A passenger arrives at an intersection with two bus stops in opposite directions of a desired route. He doesn't know which direction to take, since there is no indication of any destination. The passenger meets two people who know the required direction, but one tells only the truth and the other always lies (both these people, who are friends, know that). The passenger has no idea who is telling the truth and who is lying. How can the passenger, by receiving an answer from one of these two people to but one question, obtain the information required? (See the end of this practitioner's corner for the answer, but try first to think of the answer.)

The chapter consists of five main parts, following an introductory section. Section 11.2, using probabilities, analyzes the alternatives facing a transit passenger at a stop who is deciding whether to board an arrived vehicle or to wait for a later vehicle that will have a shorter in-vehicle time. This route choice dilemma is viewed as a decision between two categories of routes, slow and fast, in Section 11.3, from which the proportion of each category is derived. Section 11.4 crystallizes the proportions between the two categories for the case of regular vehicle arrivals. Section 11.5 presents some of the transit assignment features that are related to route choice modeling; emphasis is placed on estimating the accumulated demand on each segment of the network. The last part, Section 11.6, provides a literature review on transit assignment studies, with guidance for further reading.

Practitioners may find this chapter too technical, especially the mathematical handling of probability formulae. However, it is recommended that the introduction to each section be read in order to capture the essence of the subject. For instance, knowing the rationale behind passengers' route choices may help overcome planning and operational problems. One humorous example concerns passenger behavior in situations characterized by a high inflation rate; then it will be better to ride a taxi than a bus because in the bus you pay upfront and in the taxi at the end of the trip.

As for the question to be put to one of the two friends (it doesn't matter which one): If I'll tell your friend that this direction (points to a direction) is the correct one, what will be his answer? The passenger will then choose the opposite direction of the answer given.

11.1 INTRODUCTION

In constructing or revising a transit network of routes, it is not sufficient to know only the demand, its elasticity and factors affected it. What is needed in addition is a comprehension of and insight into passenger behavior when traffic flows disperse along the network of routes. What, if any, are the passenger paths (use of more than one route with transfers) through the network? What factors influence a passenger's choice of routes? The answers to these questions represent route choice behavior (modeling), which is then used in assigning passenger demand for prediction purposes (of traffic volume, capacity analysis, fleet size required, design and control elements needed, and more). This chapter will review the subject of passenger route choice and assignment, placing some emphasis on passenger waiting-time strategies.

One of the most crucial characteristics of a transit network is the existence of overlapping routes that share the same transit stops while running on common segments. Emanating from this characteristic is the fact that more than one transit routes can serve the demand between a certain origin–destination (O-D) pair. We will focus in this chapter on the decision-making process of the individual passenger who has to select the most efficient route serving a given stop. This process is one of the main obstacles in developing transit assignment procedures.

An additional complication of transit assignment methods in real-world problems relates to the possibility that several transit modes (bus, train, light-rail, metro, etc.) and a walking mode exist, each with its own characteristics. Moreover, each mode is perceived differently by passengers. To simplify the presentation of transit assignment approaches, this chapter includes the simple case of a single transit mode. The analysis follows Marguier and Ceder (1984) and Israeli and Ceder (1996). The literature review of this theme appears in Section 11.6. Summaries of the main contributions of this (transit assignment) theme can be found in Bell (2003), Bell and Schmöcker (2004), Nuzzolo (2003a,b), and Nuzzolo and Crisalli (2004).

11.2 ROUTE CHOICE USING WAITING-TIME STRATEGY

The decision problem facing a transit passenger at a stop with alternative choices is whether to board an arrived vehicle or to wait for a later vehicle that will have less in-vehicle time. This issue reflects real choice situations in which the passenger can distinguish between two categories of overlapping routes: slow and fast. The objective is to minimize total travel time (waiting time and in-vehicle time), which would result from adopting an optimal strategy in the choice process. This action depends on certain parameters: time between each O-D pair, vehicle regularity, distribution of passenger arrivals at the stop, and the structure of possible paths constituting the network.

The analysis shown in this section deals with the problem of passenger waiting strategies by using mathematical formulations in a probabilistic fashion in order to derive the proportion of passengers for each category (slow and fast). This calculation depends on the evaluation of a probabilistic function that considers such affecting parameters as vehicle-frequency share, in-vehicle time difference, passenger arrivals, and vehicle headway distribution. In addition, theoretical evaluations are presented in order to achieve correct implementation of (a) mean waiting-time assumptions and (b) the intuitive rule circumstances (according to which, passengers select routes in proportion to their vehicle frequencies). The outcome of the analysis can be integrated into transit assignment models to approach realistic situations.

A practical passenger-assignment procedure on transit networks consists of the following characteristics: (1) the exact structure of the network between each origin and destination (direct routes, parallel routes, sequential routes, and transfer paths); (2) in-vehicle time (direct routes and transfer-path times); (3) passenger waiting time at a stop; (4) passenger waiting strategies at a stop served by overlapping routes; (5) the *circular problem* (because frequencies are not always known at this stage) in which the passenger flows depend on route frequencies and, conversely, the frequencies rely on the load profiles of the routes, which are the outcome of the flows; (6) network size (treating each O-D pair separately and, subsequently, considering the entire network simultaneously can help to combat computational complexities of large networks); and (7) failure-to-board probabilities. This section treats the second, third, and fourth characteristics and their relationship to the first, fifth, and sixth; the seventh characteristic, as well as all the others, are discussed in the literature review in Section 11.6.

11.2.1 Passenger waiting-time dilemma

The objective of this section is to study the problem of route choice encountered by a passenger who is able to select one of several routes (alternatives) in order to travel from stop A to stop B. The common characteristic of all the transit routes considered is that they stop at both points A and B. The central idea is that a passenger wishing to travel from A to B may disregard some of the vehicles arriving at A because of their association with relatively long travel times.

The analysis, following Marguier and Ceder (1984), is performed in a probabilistic context, in which the passengers, based on their experience, have a fairly good sense of the characteristics of each route. That is, they have some information about the headway distribution (route frequency) and expected in-vehicle time from A to B. The passengers are also influenced in the route selection process by the amount of time they have already waited since their arrival at the transit stop, which can be considered additional information available to the passengers. Total travel times, consisting of waiting time and in-vehicle travel time for each route, are compared in the analysis. The waiting time, with its formulation shown in the following text and in Chapter 18, clearly plays an important role in total travel time because of its relationship to and dependency on vehicle-interval reliability and the passenger-arrival process at stop A. Before proceeding with the main thrust of the analysis, let us introduce some related formulations and assumptions.

11.2.2 Basic relationships

Chapter 18 (on transit reliability) will show how to arrive to the following commonly used formula for mean passenger waiting time under the assumption of random passenger arrivals:

$$E(w) = \frac{E(H^2)}{2E(H)} = \frac{E(H)}{2}\left[1 + \frac{VarH}{E^2(H)}\right] \tag{11.1}$$

where
  w is the waiting time
  E(w) is the mean waiting time
  E(H) and VarH are, respectively, the mean and variance of the time headway H between vehicles

Underlying this relationship is the following assumption: the average waiting time of passengers arriving at an interval of length t is ½t, and the average number of passengers arriving during such an interval is proportional to t. In addition, from data on off-peak bus intervals, the following relationship was found:

$$\mathrm{VarH} = \frac{AE^2(H)}{A + E^2(H)} \tag{11.2}$$

where A is a constant (with time square dimension) between 0 and infinity.

A = 0 corresponds to the deterministic headway case (regular vehicle arrivals). $A \to \infty$ corresponds to the completely random case (Poisson vehicle arrivals, i.e., exponential headways). Actual values found for A are between 15 and 35.

The transit vehicle interval irregularity is actually characterized by the headway distribution coefficient of variation C, which is defined as the ratio of the headway standard deviation to the headway mean, that is, its square is

$$C^2 = \frac{\mathrm{VarH}}{E^2(H)} \tag{11.3}$$

For transit vehicle (especially buses) headway distributions, C^2 ranges between 0 and 1, where $C^2 = 0$ corresponds to perfectly regular vehicle arrivals (deterministic headway) and $C^2 = 1$ to the completely random case (exponential headway).

Combining Equation 11.2 and the intrinsic definition of C^2 given by Equation 11.3 yields

$$C^2 = \frac{A}{A + E^2(H)} \tag{11.4}$$

It may be noted that the parameter A lies between 0 (for which $C^2 = 0$) and infinity ($C^2 \to 1$). Using, respectively, the notations $f_H(t)$ and $F_H(t)$ for the probability density function and the cumulative distribution function of the headway H, defining $\overline{F}_H(t) = 1 - F_H(t)$, and denoting the vehicle frequency as $F = 1/E(H)$, yields the following:

1. For $A = 0, H(\text{deterministic}) = \dfrac{1}{F}$ and $F_H(t) = \begin{cases} 1, & t \le (1/F) \\ 0, & t \ge (1/F) \end{cases}$

2. For $A = 1$, $f_H(t) = F \cdot e^{-Ft}$ and $\overline{F}_H(t) = e^{-Ft}$

Equation 11.4 may also be rewritten as $A = C^2/F^2(1 - C^2)$, which shows how, for a given frequency F, the parameter A and the coefficient of variation are uniquely related.

The idea, then, is that for a given A (or C^2 when using the aforementioned relationship), the headway distribution belongs to a family of functions that can approach the two extremes, the deterministic and the exponential cases, whereas Equations 11.2 and 11.4 are imposed. Two such families are considered here. The first one, developed by

Marguier and Ceder (1984) and referred to as *power* distributions, verifies the following relationships:

$$f_H(t) = \begin{cases} \dfrac{2F\cdot C^2}{1+C^2}\left(1 - \dfrac{1-C^2}{1+C^2}\cdot F\cdot t\right)^{3C^2-1/(1-C^2)}, & t \le \dfrac{1+C^2}{1-C^2}\cdot\dfrac{1}{F} \\[2ex] 0, & t \ge \dfrac{1+C^2}{1-C^2}\cdot\dfrac{1}{F} \end{cases}$$

(11.5)

$$\bar{F}_H(t) = \begin{cases} \left(1 - \dfrac{1-C^2}{1+C^2}\cdot F\cdot t\right)^{2C^2/(1-C^2)}, & t \le \dfrac{1+C^2}{1-C^2}\cdot\dfrac{1}{F} \\[2ex] 0, & t \ge \dfrac{1+C^2}{1-C^2}\cdot\dfrac{1}{F} \end{cases}$$

Underlying the definition of this family of distributions is the fact that for such distributions at the limit for A = 1 (see the relationship between A and C^2 in the preceding text), the distributions approach the exponential case in a fashion corresponding to the way the binomial processes approach the Poisson process.

The second family is made up of gamma distributions and verifies the following relationship:

$$f_H(t) = \frac{(F/C^2)^{1/C^2}}{\Gamma(1/C^2)}\cdot t^{1-C^2/C^2}\cdot e^{-Ft/C^2}$$

(11.6)

where Γ is the gamma function, that is, $\Gamma(C^{-2}) = \int_{x=0}^{\infty} x^{C^{-2}-1}\cdot e^{-x}dx$.

The idea of selecting gamma distributions is that the exponential is a particular case of gamma and the generic gamma distribution has two independent parameters related to the mean and variance, which allows the constraint given by Equation 11.2 to be met. Expressions (11.5) and (11.6) are used here in the route choice problem in order to illustrate the generality of the results.

Certainly, both families give identical curves for $C^2 = 0$ and $C^2 = 1$, since these values correspond to the deterministic and the exponential cases, respectively. Real-life data show headway histograms with increasing irregularity (C^2) along a transit route; the histograms have a maximum point, as observed for the gamma situation, and an intersection with the ordinate axis (different from the origin), as observed for the *power* situation distributions. Therefore, realistic situations fall somewhere between the *power* and the gamma situations.

The general assumptions of this analysis are the following:

1. The (between days) vehicle-arrival variability is high enough so that passengers cannot identify any minimum wait in order to time their arrival at the transit stop. This assumption is particularly appropriate for urban routes without published timetables in which the values of A and C^2, or F, are relatively large. This assumption can be relaxed in accordance with the study by Jolliffe and Hutchinson (1975).
2. The proportion of passengers arriving coincidentally with a vehicle (*see and rush*) is fixed and independent of other parameters. Observations in urban areas (in London and Paris) found this proportion to be up to 16% (Marguier and Ceder, 1984). Together with assumption (1), we assume then that a major proportion (e.g., 84%–100%) of the passengers arrives randomly.

Assumptions related to the route choice problem are the following:

1. Passengers are assumed to know the routes and to have a general knowledge of the headways and waiting time from experience. Thus, they select a strategy of choice among routes by attempting to minimize the sum of expected waiting and in-vehicle times. This assumption could be extended by considering, in a similar fashion to Jolliffe and Hutchinson's (1975) methodology, that only a certain proportion of the passengers follow the principle of minimization of total waiting and in-vehicle times, whereas the others will take the first bus that arrives and then continues on to point B.
2. At point A, the routes are statistically independent of each other; this is true if the routes do not share another common section, upstream of point A.

11.3 PROPORTION OF PASSENGERS BOARDING EACH ROUTE

The route choice dilemma can be viewed as a decision between two categories of routes. When there are more than two routes to travel from point A to point B, it is not irrational to assume that passengers will not consider each route individually, but rather will tend to group them into categories. Two likely categories are fast routes and slower ones. Passengers will then not choose between routes, but only between categories, and thus take the first vehicle to arrive in the category selected. The routes belonging to one category can then be viewed as forming one *equivalent* route. For simplicity sake, we will refer in the following text only to the faster route, Route 1, and to the slower route, Route 2. The following analysis is suitable for individual Routes 1 and 2, but may be extended for the concept of two-route categories.

Using the aforementioned formulations, we can then derive the estimated proportion of passengers that travel from points A to B by Route 1 and the complementary proportion that uses Route 2. This proportion represents the quantity of interest in the travel-assignment process for different transit routes. The estimated proportion depends on two factors: the frequency of each route and the headway coefficients of variation. The latter can be replaced by Equation 11.4.

Following Marguier and Ceder (1984), use of the optimal strategy will result in Route 1 users being either those who board the first vehicle, which is a Route 1 vehicle, or those who allow Route 2 vehicles to pass and wait for the Route 1 vehicle. Conversely, Route 2 users are those for whom the first-arriving vehicle is from Route 2 and board that vehicle. With the subscripts 1 and 2 used for Route 1 and Route 2, respectively, for all quantities, the latter are those for which

$$w_2 \leq w_1 \quad \text{and} \quad \tau_2 - \tau_1 \leq RW_1(w_2) \tag{11.7}$$

where
  w is the waiting time
  τ is the in-vehicle time between points A and B
  $RW(t^*)$ is the expected time that remains before the arrival of a Route 1 vehicle, given that the passenger has already waited time t^*

The first inequality in (11.7) corresponds to the fact that a Route 2 vehicle arrives first (the time W_2 for the first vehicle on Route 2 to arrive is smaller than the time W_1 for Route 1). The second inequality corresponds to the criterion for boarding the Route 2 vehicle instead of waiting for a faster vehicle (Route 1). The proportion P_2 of Route 2 passengers can thus be written as the probability that both inequalities in (11.7) exist:

$$P_2 = \text{prob}\left[w_2 \leq w_1 \text{ and } \tau_2 - \tau_1 \leq RW_1(w_2)\right] \tag{11.8}$$

Equation 11.8 can be expressed in a function of the distributions of the waiting times W_1 and W_2 (see Marguier and Ceder, 1984) as

$$P_2 = \begin{cases} 0, & \tau_2 - \tau_1 > \dfrac{1+C^2}{2F_1} \\[2ex] \displaystyle\int_0^{RW_1^{-1}(\tau_2-\tau_1)} \overline{F}_{w_1}(t)f_{w_2}(t)dt, & rw_1 < \tau_2 - \tau_1 \leq \dfrac{1+C^2}{2F_1} \\[2ex] \displaystyle\int_0^{\infty} \overline{F}_{w_1}(t)f_{w_2}(t)dt, & \tau_2 - \tau_1 \leq rw_1 \end{cases} \tag{11.9}$$

where
   RW_1^{-1} is the reciprocate of the RW_1 function (RW_1 can be reciprocated, since it decreases monotonically)
   rw_1 is the asymptotical value of the function RW_1

The value of rw_1 depends on the type of distribution on C_1^2 and on F, for example, $rw_1 = 0$ for the *power* distribution family and $rw_1 = C_1^2/F_1$ for the gamma distribution family. In the limit case of *exponential headways*, $C_1^2 = 1$, $RW_1(t) = 1/F_1 = rw_1$, and

$$\int_0^{\infty} \overline{F}_{w_1}(t) \cdot f_{w_2}(t)dt = \int_0^{\infty} e^{-F_1 t} \cdot F_2 \cdot e^{-F_2 t}dt = \frac{F_2}{F_1 + F_2}$$

Equation 11.9 in this case is reduced to

$$P_2 = \begin{cases} 0, & \tau_2 - \tau_1 > \dfrac{1}{F_1} \\[2ex] \dfrac{F_2}{F_1 + F_2}, & \tau_2 - \tau_1 \leq \dfrac{1}{F_1} \end{cases} \tag{11.10}$$

If in this exponential case $\tau_2 - \tau_1$ is large, the proportion of Route 2 users is 0. Otherwise, the share of each route's users is equal to its frequency share.

   The proportion P_2 certainly does not depend on the timescale (i.e., the time unit selected). Therefore, the quantities F_1, F_2, and $\tau_2 - \tau_1$ can be grouped within the following two variables: (a) the Route 2 frequency share, $FS_2 = F_2/(F_1 + F_2)$, and (b) the ratio of the in-vehicle time difference to the headway of Route 1, $TH_1 = \tau_2-\tau_1/H_1$. The proportion P_2 is plotted against FS_2 and TH_1 in Figures 11.1 and 11.2, respectively, for each type of headway distribution (*power* or gamma) and several values of C_1^2 and C_2^2. The curves in Figure 11.1 are based on $TH_1 = 0.54 = (31/57)$, and in Figure 11.2 on $FS_2 = 1/2$ (equal frequency share).

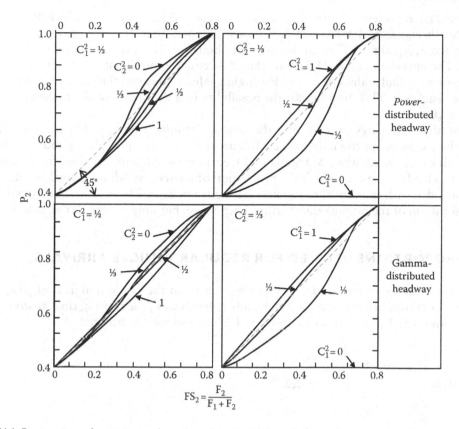

Figure 11.1 Proportion of passengers boarding the slow Route 2 (complementary to the fast Route 1 proportion) as a function of the slow Route 2 frequency share of $(\tau_2 - \tau_1)/H_1 = 0.54$.

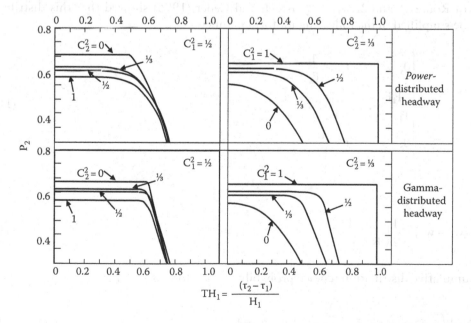

Figure 11.2 Proportion of passengers boarding the slow Route 2 (complementary to the fast Route 1 proportion) as a function of $(\tau_2 - \tau_1)/H_1$ for $FS_2 = 0.5$.

Figure 11.1 represents the market share–frequency share curves, in which P_2 increases with FS_2. In Figure 11.1 for $C_2^2 = 1/3$, it should be noted that the proportion P_2 for $C_1^2 = 0$ remains fixed, equal to 0. This can be seen from Equation 11.9 because $TH_1 = 0.54 > 0.5$ for $C_1^2 = 0$. The curves in Figure 11.2 show that P_2 is either approximately or exactly constant with respect to small values of $\tau_2 - \tau_1$. For higher values, however, the value of P_2 drops rapidly and reaches 0. This drop signals the possibility of a threshold for the passengers' route choice decision.

All in all, this analysis shows that the general intuitive rule $P_2 = FS_2$ (i.e., passengers board the routes proportionally to their frequencies) does not provide a good approximation in all cases. In addition, Marguier and Ceder (1984) present a revised formulation of Equation 11.9 for a case in which the proportion of passengers whose arrival at the transit stop coincides with the arrival of the first vehicle is not zero. The next section constructs a full description of the proportion boarding each route, but only for regular vehicle arrivals.

11.4 PROPORTIONS DERIVED FOR REGULAR VEHICLE ARRIVALS

This section follows Israeli and Ceder (1996). Based on the random arrival of passengers and regular transit-vehicle arrivals (deterministic headways), the waiting time at stops associated with i = 1, 2 (referring to Routes 1 and 2) is uniformly distributed:

$$f_{w_i}(t) = \begin{cases} F_i, & 0 \le t \le \dfrac{1}{F_i} \\[2mm] 0, & t \ge \dfrac{1}{F_i} \end{cases}, \quad i = 1, 2 \tag{11.11}$$

The relevant component of interest is the distribution of the difference between the waiting times of Routes 1 and 2, $w_1 - w_2$. Israeli and Ceder (1996) showed that this distribution could be simplified in the following uniform manner:

$$f_{w_1 - w_2}(t) = \begin{cases} 0, & t \le -\dfrac{1}{F_2} \\[2mm] \dfrac{F_1 F_2}{F_1 + F_2}, & -\dfrac{1}{F_2} \le t \le \dfrac{1}{F_1} \\[2mm] 0, & t \ge \dfrac{1}{F_1} \end{cases} \tag{11.12}$$

with

$$E(w_1 - w_2) = \frac{1}{2}\left(\frac{1}{F_1} - \frac{1}{F_2}\right) \tag{11.13}$$

The cumulative distribution (choice probability), therefore, is given by

$$\mathrm{Prob}(w_1 - w_2 \le t) = F_{w_1 - w_2}(t) = \int_{-\infty}^{t} f_{w_1 - w_2}(t)dt \tag{11.14}$$

to obtain

$$F_{w_1-w_2}(t) = \begin{cases} 0, & t \le -\dfrac{1}{F_2} \\ \dfrac{F_1}{F_1+F_2}(1+F_2 \cdot t), & -\dfrac{1}{F_2} \le t \le \dfrac{1}{F_1} \\ 0, & t > \dfrac{1}{F_1} \end{cases} \tag{11.15}$$

The complementary cumulative distribution $F_{w_2-w_1}(t)$ can be obtained by $1 - F_{w_1-w_2}(t)$.

The probability (proportion) P_1 of selecting Route 1 (fast) vehicle can be obtained by assigning $t = \tau_2 - \tau_1$ to Equation 11.15. Because $\tau_2 > \tau_1$, then $t > 0$ and Equation 11.15 becomes

$$P_1 = F_{w_1-w_2}(t) = \begin{cases} \dfrac{F_1}{F_1+F_2}(1+F_2 \cdot t), & 0 \le t \le \dfrac{1}{F_1} \\ 1, & t \ge \dfrac{1}{F_1} \end{cases}, \quad t = \tau_2 - \tau_1 \tag{11.16}$$

Based on Equation 11.16, the complementary probability $P_2 = 1 - P_1$ can be derived with the following interesting property: P_2 can never bear the entire passenger demand. That is,

$$P_2 = F_{w_2-w_1}(t) = \begin{cases} \dfrac{F_2}{F_1+F_2}(1-F_1 \cdot t), & 0 \le t \le \dfrac{1}{F_1} \\ 0, & t \ge \dfrac{1}{F_1} \end{cases}, \quad t = \tau_2 - \tau_1 \tag{11.17}$$

Because $t = \tau_2 - \tau_1 \ge 0$ and $F_1, F_2 \ge 0$, the result is that $P_2 \ne 1$.

Figure 11.3 presents P_1 as a function of the frequency proportion FS_1, $FS_1 = F_1/(F_1 + F_2)$, where $F_1 + F_2 = 10$ veh/h, and $\tau_2 - \tau_1$ is in hours. For those cases in which the passenger

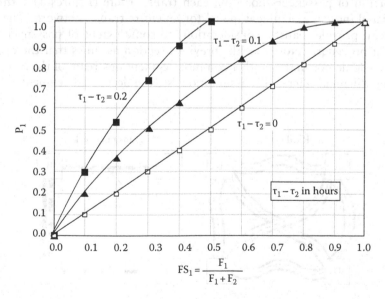

Figure 11.3 Proportion of passengers boarding Route 1 (fast) vehicle as a function of the frequency share.

is unaware of the in-vehicle travel-time difference ($\tau_2 - \tau_1 = 0$), the proportion rule can be applied. Otherwise, Equations 11.16 and 11.17 apply. It should be noted that for $\tau_2 - \tau_1 = 0$, passengers select the first transit vehicle that arrives.

11.5 PASSENGER ASSIGNMENT BASED ON ROUTE CHOICE

The beginning of this chapter indicated that vehicle and traveler assignment on a network of choices (paths or routes) is an essential part of estimating (predicting) the accumulated demand on each segment of the network. It is, therefore, an integral tool of transportation planning, in which route choice modeling forms the base of any assignment algorithm. In transit planning, the assignment algorithms for passengers warrant (at the planning stage) changes in routes, location of stops, selection of operational strategies, priority schemes for transit vehicles, traffic and parking arrangements, environmental impact studies, and more. Section 11.6 provides a literature review of transit assignment studies. The section presents some of the transit assignment features that are related to route choice modeling.

Passenger path may involve transfers and, hence, the use of more than one transit route. An arc in the transit network is a road segment that can have more than one crossing route, thus creating the possibility of overlapping transit services. Figure 11.4 illustrates the differences between a road network and a transit network. A passenger traveling from node A to node D on the transit network in Figure 11.4 has two possible paths: Routes I and II with a transfer at B, or routes I and II with a transfer at C. For instance, in the network in Figure 11.4, we can examine routes A–C–D and A–B–C, in which route A–C–D replaces route B–C–D, thus making node C a single transfer point from B to D. This examination (as well as other sets of alternative routes) can use passenger-assignment procedures for given frequencies; the latter will be derived from the O-D demand. Based on given criteria or objectives, different routing recommendations can be made by the assignment algorithms. This simple description is further analyzed by Guan et al. (2006) in combining transit assignment and route configuration.

The proportion of passengers boarding each transit route (Figures 11.1 through 11.3) can be integrated into an assignment model for an entire transit network. The use of such a procedure can provide realistic considerations to some extent of passenger behavior in the assignment procedure. Note that the previous section assumes that the routes operate independently and do not share another common segment upstream of the transit stop with a waiting-time dilemma. Here, this assumption is extended to all segments of the transit network.

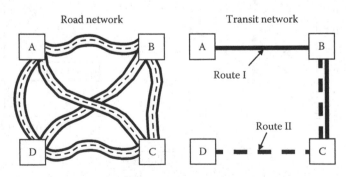

Figure 11.4 Road and transit network illustrations of four locations.

Passenger assignment on the transit network is interpreted as intermitted flow of passengers: waiting to board or transferring at stops and riding on the vehicles. At node u, the total flow of arriving passengers who want to obtain service (boarding or transferring) and the frequency parameters of the service will determine the total waiting time W_u at that node. The total expected waiting time, using Equation 11.1, can be formulated as

$$E(W_u) = \sum_r \sum_v p_{uv}^r \frac{E(H_r)}{2}\left[1 + \frac{VarH_r}{E^2(H_r)}\right] \tag{11.18}$$

where

p_{uv}^r is the number of arriving passengers at u from node v who seek to board/transfer to a route (or group of routes) r

H_r is the headway (or combined headway for a group) of r

One objective (e.g., Spiess and Florian, 1989) can be to minimize the sum of the cost of passengers and their expected waiting times, subject to constraints for all nodes in the network. The constraints ensure conservation-of-flow conditions (what goes into a node equates with what comes out) and that the flow on each arc is greater than or equal to 0 and less than or equal to a function of the frequencies and waiting times on that arc.

Passenger-choice strategy was defined as being between fast and slow route categories, but in a network context. As the proportion boarding a vehicle in each category is dependent on the frequency share and the time difference between the two categories, it is necessary to define a *combined* frequency and *equivalent* in-vehicle time for each category. The network structure contains different possible paths between each origin and destination, that is, direct routes and transfer paths. The latter can consist of overlapping (parallel) routes, successive routes, or both. Although calculating the *equivalent* in-vehicle time for each category is simple enough (passengers tend to make an average of all possible paths connecting the origin with the destination), the *combined* frequency is more complicated, as will be explained. The complex network structure accounts for why researchers have tried to avoid this problem by presenting nonrealistic simplifications, such as using only the frequencies of routes that connect directly to the transit stop or employing different frequencies for each arc of the network. The latter requires *smoothing* techniques in order to find the correct frequency even for direct routes, a process that leads to unrealistic results.

At the outset of the process, as has been seen, passengers have to choose a category (slow or fast routes); then, the route within this category needs to be chosen. Those passengers who do not distinguish among route lengths (in-vehicle times) within each category board the first vehicle to arrive. That is, passengers are assigned to the routes in proportion to the routes' frequencies.

Each category, which refers to a subnetwork between the corresponding origin and destination, is part of a *modified* transit network: the nodes are transit stops with overlapping routes (i.e., enabling transfers), and the arcs are direct route segments between these nodes. The number of arcs in a possible path between an origin and a destination denotes the number of vehicle changes, minus one.

Figure 11.5 shows the *combined* frequency calculation and, consequently, the demand assignment within a category, or the routes between i and j. The *combined* frequency for parallel routes is their sum, and for successive routes, it is their minimum. In such a manner, passengers between k and j in Figure 11.5a experience a *combined* frequency of 3 + 1 = 4 veh/h, while their effective frequency along i–k–j is min(6, 4) = 4 veh/h. The combined frequency

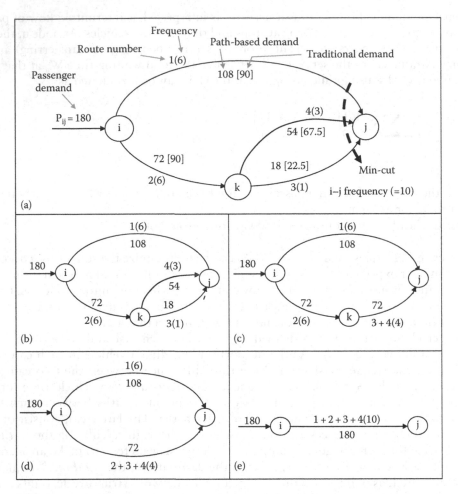

Figure 11.5 Passenger assignment on an hourly basis, in which part (a) shows that the traditional assignment is a path-based assignment and parts (b) to (e) describe in steps the network synthesis process.

of all categories between i and j is 6 + 4 = 10 veh/h, which basically is the minimum cut of the small network in Figure 11.5a (for the definition of minimum-cut frequency, see Section 6.4). This result (which can be either F_1 or F_2, depending on the type of category) is then utilized in Equations 11.9, 11.16, and 11.17 in order to predict the proportion of passengers between i and j for the two categories.

Further analysis of the network shown in Figure 11.5a is illustrated in Figure 11.5b through e and called network synthesis. It is based on recursive calculations of the combined frequencies in a subnetwork, until a single *combined* frequency arc is reached. Network synthesis is analogous to an analysis made for capacitor (electric) networks, with parallel and successive summations. A parallel summation of frequencies is equal to the sum of the frequency associated with the cut of the parallel arcs; a successive summation of frequencies equates with the minimum among the frequencies considered. The calculation of the divided demand can be viewed in a reversed process. The demand of $p_{ij} = 180$ passengers in Figure 11.5e, which is a single-frequency arc, is divided (going backward) in Figure 11.5d on the basis of the proportional frequency share rule, for example, see Equation 11.10, that is,

180·(6/10) and 180·(4/10). The latter demand of 72 is kept along i–k–j in Figure 11.5c and again (going backward) is divided proportionally in Figure 11.5b to 54 and 18.

The separation into two route categories for each i–j pair (if passengers notice a time difference between them) creates two separate sets (subnetworks) for each route; this means that a route or path serving the demand between i and j (directly or via transfer) will be included in the subnetwork of either the fast or slow category in this pair, but not in both. Thus, assignment equations can be formulated in such a manner that *combined* frequencies and demand shares will be formulated separately for the two route categories, while the relation between variables in each category will be given in the frequency calculation constraints.

Considering the interrelationship of all O-D pairs in the network enables more manageable handling of large, complex networks. This interrelationship is affected by the mutual dependence of the unknown frequencies and demand flows (serving as variables) in an assignment procedure. While actual frequencies are based on the methods of Chapter 3 (consisting, via ride and point-check counts, of all O-D demands for a route), the demand flows are usually based on the frequency share (with or without considering route travel times). More on transit assignment studies and considerations appears in Section 11.6.

11.6 LITERATURE REVIEW AND FURTHER READING

One of the vital ingredients in transit planning is the prediction of the paths that passengers choose for transit routes to take them from origin to destination. This prediction relies on the use of transit assignment models that have appeared in the welter of professional papers on the subject in the past 40 years. Much progress was achieved in past research in adding more realistic features of passenger behavior and operational planning elements to these models. Among the added features are waiting times at stops, transfer times between routes, preferred set of routes for an O-D pair from a passenger's perspective, preferred passengers' strategies, distributed travel and waiting times, crowding level on vehicles, and failure-to-board probabilities. A summary of some of the papers can be found in Bell (2003), Bell and Schmöcker (2004), Nuzzolo (2003a,b), and Nuzzolo and Crisalli (2004). This section reviews briefly and chronologically the main contributions to transit assignment modeling.

Early methods of transit route choice and assignment are works by Dial (1967) and Le Clercq (1972), who used heuristic rules in considering both waiting and travel times when computing the shortest paths on a network. Passengers take the first vehicle to arrive, and a transfer penalty is equal to the expected waiting time. Chriqui and Robillard (1975) introduced, for the first time, a choice-behavior feature by which passengers can select, from a set of alternative routes, a subset of routes from which they will board the first vehicle to arrive. This subset of routes, called *attractive routes*, represents routes that have a better chance to offer shorter travel times (based on prior knowledge). The researchers presented their ideas with the use of a simple network and a single origin and destination.

Chriqui and Robillard's direction was extended in a doctoral study by Spiess in 1984. Nguyen and Pallotino (1988) introduced a graph interpretation for a strategy of selecting a set of attractive routes at a boarding point; this graph representation was denoted a *hyperpath*. Part of Spiess's original study was incorporated into a work that became known (Spiess and Florian, 1989). The latter developed a two-part algorithm to assign passenger demand from the user's perspective. Spiess and Florian assumed that passenger behavior reflected a minimization of the expected value of access, that is, of waiting and in-vehicle times or a weighted sum of these time elements. The first part of their algorithm computes the total expected time elements between origin and destination, including transfers; the second part assigns demand according to the strategy of choosing a set of attractive routes. Both Spiess and Florian (1989) and Nguyen and Pallotino (1988) considered the effect of the inconvenience of crowded vehicles through discomfort functions in their equilibrium-assignment models.

De Cea and Fernàndez (1993), inspired by Spiess and Florian, introduced a transit-equilibrium assignment model in which waiting times on access links depended on passenger flows, that is, they applied congestion functions to passengers requesting to board the first-arrived vehicle. De Cea and Fernàndez (1993) incorporated heuristically the discomfort effect of crowded vehicles and crowded stops. Their model was solved by Jacobi's method of using a similar diagonalization algorithm to that used by Florian (1977).

Wu et al. (1994), using the concept of hyperpaths (strategies) proposed by Nguyen and Pallotino (1988), introduced a network consisting of road-based and transit route–based arcs. Their passenger-related arcs included the time elements of walking, waiting, boarding, in vehicle, transferring, and alighting. Their assignment model considered that the time required to board a vehicle increased with flow (of transit vehicles); the distribution of flows across the set of attractive routes is proportional to the minimum frequency share.

The foregoing articles may be grouped under frequency-based models, in which within-day dynamics (in transit vehicle headways) are not taken into account. In the mid-1990s, some route choice and transit assignment models started explicitly to consider different headways (timetable) as an input. These studies were grouped under schedule-based models. In most such studies, passenger-choice behavior is interpreted by a *utility* function. That is, passengers are assumed to assign a utility value to each alternative route from a given choice set (of routes) and to select the one with maximum utility. Usually, the utility values of the alternatives are treated as random variables, thus converting the choice process (among routes) into a search for maximum (utility) probability. Each utility function is dependent on a set of passenger attributes. Hickman and Bernstein (1997) used a sequential deterministic path-choice approach with a utility function consisting of stochastic travel-time attributes and passenger information. They studied high-frequency service, including congested networks, using equilibrium and dynamic processes.

Tong and Wong (1999) showed the differences between frequency-based and schedule-based approaches and constructed a stochastic schedule-based dynamic model using simulation. In their model, passengers either move on a walking segment, wait (queue) on a network segment, or travel on a transit vehicle along a selected route. They employ a time-dependent branch-and-bound method for the least-cost path (shortest path); the path's costs contain the time elements of walking, waiting, in vehicle, and transferring penalty.

Lam et al. (1999) introduced a frequency-based, stochastic user-equilibrium model for the transit assignment process. Their model considers capacity constraints on each transit route; the route on which the capacity constraints are not fulfilled is excluded from the set of attractive routes. For an overcrowded service, some passengers (who could already be on board) may choose alternative services, which in practice is the case only with boarding passengers who see a loaded vehicle. Bell and Schmöcker (2004) concluded that the model by Lam et al. (1999) could fit situations with spare capacity, such as seat reservation or high-fare systems.

Cominetti and Correa (2001) and Bouzaïene-Ayari et al. (2001) advanced the consideration of limited vehicle capacities with more realistic waiting-time functions at stops. Their frequency-based models used the effect of changes in frequencies (because of overcrowded vehicles) rather than minimum-frequency flow share. Their equilibrium-assignment models take into account queuing processes at transit stops to reflect more realistic changes in establishing the set of attractive routes. However, no specific algorithm was provided for computing their models.

In a schedule-based study, Nuzzolo et al. (2001) used a random utility function with passenger information for a transit assignment procedure. They included departure-time choice and stop choice in their stochastic model. Stop choice was defined as the probability of selecting the boarding point from a set of stops located within a given access distance. They developed the model for high-frequency service with a sequential choice process for both within-day and day-to-day dynamics.

Kurauchi et al. (2003) introduced a frequency-based model in which a cost is assigned to the probability of failing to board the transit vehicle. They considered the separation between passengers on board a crowded vehicle and passengers at stops seeking to board. The latter have a reduced priority to flow on the network. Kurauchi et al. used Markov chains, in which the boarding probability is dependent on the leftover capacity of the transit vehicle. They also discussed the use of the Markovian approach in stochastic and deterministic user-equilibrium route choice processes.

Cepeda et al. (2006) provided a frequency-based formulation built on the model by Cominetti and Correa (2001). The former introduced a transit assignment algorithm intended for large-scale networks and tested it on real-size networks. Cepeda et al. (2006) discussed the assumptions made, the difficulties in reaching a satisfactory solution, and some possible future improvement directions.

Cats et al. (2011) present a dynamic transit analysis and evaluation tool that represents timetables, operation strategies, real-time information (RTI), adaptive passenger choices, and traffic dynamics at the network level. A mesoscopic transit and traffic simulation model has been developed to represent traffic dynamics, dwell times, timetables and vehicle scheduling, control strategies, and passenger path-choice processes. This model incorporates passengers' progress within the transit network as a sequence of discrete choice decisions on boarding, walking, and alighting. At each decision point, consideration has been given to the time-dependent expectations of passengers about travel-time components. This model has been applied to the metro network area of Stockholm, Sweden, under various operating conditions and information provision scenarios, as a proof of concept. Analysis of the results indicate substantial path-choice shifts and potential time savings associated with more comprehensive RTI provision and transfer coordination improvements. The authors suggest that transit operators should provide RTI at station levels in order to enable more informed travel decisions, which will ultimately result in time savings. Future work on this model has been proposed to allow the development of a mixed frequency and schedule-based passenger-arrival patterns to represent more realistic scenarios.

Szeto et al. (2011) propose a nonlinear complementary problem (NCP) formulation for the risk-aversive stochastic transit assignment problem in which in-vehicle travel time, waiting time, capacity, and the effect of congestion are considered as stochastic variables simultaneously and both their means and variances are incorporated into the formulation. This model has indicated a number of advantages compared to the frequency-based transit assignment model. A new congestion model is developed and captured in the proposed NCP formulation to account for different effects of onboard passengers and passengers waiting at stops. Surveys and numerical studies have been performed to validate the risk-aversion behavior of passengers and to illustrate the properties of the problem and effectiveness of the proposed solution method. The results indicate that underestimating the congestion effect and ignoring the risk-aversion behavior can overestimate the patronage of transit service. As such, the proposed model can be incorporated in the transit network design model to determine the optimal service frequency and fare structure, thus allowing the transit operators to maximize profitability without compromising the service quality. Future work would include extension of this model to account for multiple passenger classes, and also, consideration can be given to the perception errors on travel time and waiting time.

Wu (2013) aims to find an efficient solution algorithm for the reliable a priori shortest path (RASP) problem. The subject problem is analyzed from two aspects: the convolution method and the stochastic dominance (SD) scheme. An alternative to the convolution method is developed on the basis of the adaptive discretization approach, which was originally proposed to solve the RASP problem in static networks. In comparison, the higher-order SD rule can be engaged to reduce the number of nondominated paths, thus improving the overall computational efficiency. The experiments undertaken illustrate that the second-order SD rule provides approximation of reasonable quality while reducing the central processing unit time by at least 50%.

Trozzi et al. (2013) present a dynamic user equilibrium (DUE) for bus networks where recurrent overcrowding results in queues at transit stops. The route choice model embedded in the dynamic assignment explicitly considers common lines and strategies with alternative routes. As such, the shortest hyperpath problem is extended to a dynamic scenario with capacity constraints where the diversion probabilities depend on the time the stop is reached and on the expected congestion level at that time. The bottleneck-queue model with time-varying exit capacity is extended to reproduce congestion for all the lines sharing a stop. This model is applied to separate queues for each line in order to satisfy the first-in, first-out principle within every attractive set while allowing overtaking among passengers having different attractive sets but queuing single file. The resulting DUE can reproduce the results from the static model in the uncongested scenario with time-independent (constant) travel variables; however, it clearly illustrates the changes in the hyperpath, total travel cost, and flow pattern when these become dynamic. Future work has been proposed to extend this approach to other transport modes. Furthermore, this new model can be tested on large transit networks to verify the value of heuristic solution methods and algorithm speedup techniques.

Tirachini et al. (2013) examine the multiple dimensions of passenger crowding related to public transit demand, supply, and operations, including effects on operating speed, waiting time, travel-time reliability, passengers' well-being, valuation of waiting and in-vehicle time savings, route and bus choice, and optimal levels of frequency, vehicle size, and fare. The authors illustrate the impact of crowding on the estimated value of in-vehicle time savings and demand prediction by estimating the externalities for rail and bus services in Sydney. Furthermore, using multinomial logit (MNL) and error component models, the authors show that alternative assumptions concerning the threshold load factor that triggers a crowding externality effect do have an influence on the value of travel time savings (VTTS) for low-occupancy levels (all passengers sitting); however, for high-occupancy levels, alternative crowding models estimate similar VTTS. Importantly, if demand for a public transport service is estimated without explicit consideration of crowding as a source of disutility for passengers, demand will be overestimated if the service is designed to have a number of standees beyond a threshold, as analytically shown using an MNL choice model. The authors conclude by stating that more research is needed to explore if these findings hold with more complex choice models and in other contexts. For instance, the negative impacts of crowding on the reliability of public transport services can be analyzed along with engineering and design factors in vehicles and stations that may help to reduce the effects of crowding on increasing travel time and worsening the experience of traveling in public transport.

Ceder et al. (2013) present a study with two main objectives. The first objective is to determine the effects of uncertainty, in out-of-vehicle times during transfers, on transit users' willingness to use transfer routes. The second objective is to determine the influence of out-of-vehicle facilities, offered by public-transport operators, on transit users' perception of trip attributes related to transfers. A user preference survey has been conducted at two major transit terminals in Auckland, New Zealand. The survey data were modeled using cumulative prospect theory (CPT) and fuzzy logic. The results indicate that for all trip attributes, except for comfort, transit users' exhibited risk-averse behavior; users revealed greater preference for the transfer route with less uncertainty in the out-of-vehicle times. For comfort, transit users were found to display risk-taking characteristics when the waiting time for a seat was less than 5 min. Such findings suggest that increasing the consistency in out-of-vehicle times will increase attractiveness of transfer routes, thus enabling a more efficient and integrated network of transit routes to result in enlargement of ridership. The authors assert that the policymakers and transit planners must focus on methods of reducing uncertainty in out-of-vehicle times during transfers. Analysis of transit users' perception of trip attributes, given their current station, revealed statistical evidence of differences for two trip attributes, transfer waiting time and vehicle delay. Such findings indicate that transit users who are accustomed to better out-of-vehicle facilities have a lower tolerance for uncertainty in transfer waiting times and delay times. To the authors' knowledge, this study provides for the first time in literature a comparison between the two cognitive models. The comparison revealed that CPT and fuzzy logic models are both capable of representing transit users' decision-making process. However, while CPT provides an indication of transit users' preference for a transfer route, fuzzy logic is capable of providing a closer approximation of the proportion of transit users preferring a transfer route.

Hamdouch et al. (2014) present a schedule-based transit assignment model considering supply uncertainties and assuming that users adopt strategies to travel from their origins to destinations. Unlike previous transit assignment studies with fixed timetables, the proposed model is capturing the stochastic nature of the transit schedules and in-vehicle travel times due to road conditions, incidents, or adverse weather. The solution method developed does not rely on any path enumeration or column generation technique; the transit network loading procedure relies on the usage of Bellman's recursion principle (dynamic program). The model developed ensures that onboard passengers, continuing to the next stop, have priority, and waiting passengers are loaded on a first-come-first-serve basis considering vehicle capacities. The authors use a mean–variance approach considering the covariance of travel time between links in a space–time graph; the equilibrium conditions are stated as a variational inequality involving a vector-valued function of effective strategy costs. The equilibrium solution is found by method of successive averages in which the optimal strategy of each iteration is generated by solving a dynamic program. The numerical examples presented in this study produce a few interesting results, namely, that for an increase of travel-time variance, passengers may decide to leave later and that vehicle capacity and early/late arrival penalties have no effect on departure time choices.

A summary of the articles reviewed appears in Table 11.1, in which separate attention is given to frequency- and schedule-based models.

Table 11.1 Summary of features concerning the articles reviewed

Advance in transit assignment research	Frequency-based features	Schedule (timetable)-based features	Route choice features
Dial (1967), Le Clercq (1972)	Heuristic consideration of waiting time (and travel time) in shortest-path approach.		Boarding the first transit vehicle to arrive.
Chriqui and Robillard (1975)	Probabilistic selection of a subset of routes to minimize the expected sum of (wait + travel) time.		Boarding the first vehicle from a set of attractive routes.
Nguyen and Pallotino (1988)	Origin to destination is interpreted on an acyclic directed graph (called *hyperpath*).		Boarding the first vehicle (from a set of routes) with a planned strategy of path movement.
Spiess and Florian (1989)	Choosing a set of routes to minimize expected sum of (access + wait + travel) time; using equilibrium model with linear programming.		Boarding the first vehicle using a strategy of choosing only among attractive routes.
De Cea and Fernàndez (1993)	Incorporating a limited capacity for each route (of an attractive set) at stops in which waiting time depends on passenger flow; using asymmetric equilibrium model with Jacobi method.		Boarding the first vehicle (from a set of routes), given that passenger flows do not exceed a route's capacity.
Wu et al. (1994)	Hyperpaths are used for walk, wait, board, in-vehicle, transfer, and alight time elements; boarding time increases with flow, using equilibrium model with Jacobi method.		Boarding the first vehicle using a strategy of choosing only among attractive routes.
Hickman and Bernstein (1997)		Model for high-frequency service using sequential choice approach.	Deterministic utility path-choice model with passenger information.
Tong and Wong (1999)		Simulation model consisting of a network of routes with a given number of departure times; using shortest-path of weighted (walk + wait + travel) time and route change penalty.	Random utility path-choice model for frequent service.

(Continued)

Table 11.1 (Continued) Summary of features concerning the articles reviewed

Advance in transit assignment research	Frequency-based features	Schedule (timetable)-based features	Route choice features
Lam et al. (1999)	Stochastic user equilibrium with explicit route capacity constraints.		Boarding the first vehicle; for overcrowded service, some passengers may choose alternative services.
Cominetti and Correa (2001), Bouzaïene-Ayari et al. (2001)	Congestion functions at stops obtained from queuing theory to increase waiting time and affect passenger-flow share.		Boarding the first vehicle (from a set of routes) with available capacity.
Nuzzolo et al. (2001)		Introducing a set of departure-time choices and a set of stop choices for a given access distance.	Random utility path-choice model for frequent service with possible passenger information.
Kurauchi et al. (2003)	Introducing failure-to-board probability in which the demand exceeding capacity remains on the platform; using Markov chains and user-equilibrium approach.		Boarding the first vehicle of a chosen single route in which route choice depends on risk of failing to board.
Cepeda et al. (2006)	Congestion functions at stops with formulation for large-scale networks.		Boarding the first vehicle (from a set of routes) with available capacity.
Cats et al. (2011)		Represents timetables, operation strategies, real-time information, and traffic dynamics at the network level.	Transit path choices are modeled as a sequence of boarding, walking, and alighting decisions that passengers undertake when carrying out their journey.
Szeto et al. (2011)	A reliability-based user-equilibrium condition is defined based on the proposed generalized concept of travel-time budget referred to as effective travel cost in the formulation.		A nonlinear complementarity problem formulation for the risk-aversive stochastic transit assignment problem in which in-vehicle travel time, waiting time, capacity, and the effect of congestion are considered simultaneously.
Wu (2013)			An alternative convolution method is developed on the basis of the adaptive discretization approach to solve the reliable a priori shortest path problem.

(Continued)

Table 11.1 (Continued) Summary of features concerning the articles reviewed

Advance in transit assignment research	Frequency-based features	Schedule (timetable)-based features	Route choice features
Trozzi et al. (2013)			The shortest hyperpath problem is extended to a dynamic scenario with capacity constraints where the diversion probabilities depend on the time the stop is reached and on the expected congestion level.
Tirachini et al. (2013)		Multinomial logit and error component models are used to test whether the threshold load factor that triggers a crowding externality effect does have an influence on the value of travel time.	
Ceder et al. (2013)			A user preference survey has been conducted at two major public transport terminals in Auckland, New Zealand. The survey data were modeled using cumulative prospect theory and fuzzy logic.
Hamdouch et al. (2014)		Both travel strategies and supply uncertainties are considered to capture the stochastic nature of the transit schedules and in-vehicle travel times; dynamic program is used.	Onboard passengers, continuing to the next stop, have priority, and waiting passengers are loaded on a first-come-first-serve basis considering vehicle capacities.

EXERCISES

11.1 Given the following four-route schematic transit network adapted from the EMME/2 user's manual (reference: INRO Consultants Inc. EMME/2 User's Manual, Release 9.6, May 2005), this network contains the following headways and average travel times.

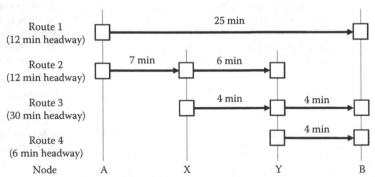

Assume that walking times and transfer times are zero:

(a) Describe the possible travel options from A to B; include the possibility that a passenger may choose among different combinations of routes.
(b) Calculate expected travel times for each option.
(c) Which set of options gives the minimum expected travel time?

REFERENCES

Bell, M.G.H. (2003). Capacity constrained transit assignment models and reliability analysis. In *Advanced Modeling for Transit Operations and Service Planning* (H.K. Lam and M.G.H. Bell, eds.), pp. 181–199, Pergamon, Elsevier Science, Oxford, U.K.

Bell, M.G.H. and Schmöcker, J.-D. (2004). A solution to the transit assignment problem. In *Schedule-Based Dynamic Transit Modeling. Theory and Applications* (N.H.M. Wilson and A. Nuzzolo, eds.), pp. 263–279, Kluwer Academic, Norwell, Massachusetts, USA.

Bouzaïene-Ayari, B., Gendreau, M., and Nguyen, S. (2001). Modeling bus stops in transit networks: A survey and new formulations. *Transportation Science*, 35, 304–321.

Cats, O., Koutsopoulos, H.N., Burghout, W., and Toledo, T. (2011). Effect of real-time transit information on dynamic path choice of passengers. *Transportation Research Record*, 2217, 46–54.

Ceder, A., Chowdhury, S., Taghipouran, N., and Olsen, J. (2013). Modelling public-transport users' behaviour at connection point. *Transport Policy*, 27, 112–122.

Cepeda, M., Cominetti, R., and Florian, M. (2006). A frequency-based assignment model for congested transit networks with strict capacity constraints: Characterization and computation of equilibria. *Transportation Research*, 40B, 437–459.

Chriqui, C. and Robillard, P. (1975). Common bus lines. *Transportation Science*, 9, 115–121.

Cominetti, R. and Correa, J. (2001). Common-lines and passenger assignment in congested transit networks. *Transportation Science*, 35, 250–267.

De Cea, J. and Fernàndez, J.E. (1993). Transit assignment for congested public transport systems: An equilibrium model. *Transportation Science*, 27, 133–147.

Dial, R.B. (1967). Transit pathfinder algorithm. *Highway Research Record*, 205, 67–85.

Florian, M. (1977). A traffic equilibrium model of travel by car and public transit modes. *Transportation Science*, 11, 166–179.

Guan, J.F., Yang, H., and Wirasinghe, S.C. (2006). Simultaneous optimization of transit line configuration and passenger line assignment. *Transportation Research* B, 40, 885–902.

Hamdouch, Y., Szeto, W.Y., and Jiang, Y. (2014). A new schedule-based transit assignment model with travel strategies and supply uncertainties. *Transportation Research*, 67B, 35–67.

Hickman, M.D. and Bernstein, D.H. (1997). Transit service and path choice models in stochastic and time-dependent networks. *Transportation Science*, 31, 129–146.

Israeli, Y. and Ceder, A. (1996). Public transportation assignment with passenger strategies for overlapping route choice. In *Transportation and Traffic Theory* (J.-B. Lesort, ed.), pp. 561–588, Elsevier Science, Oxford, U.K.

Jolliffe, J.K. and Hutchinson, T.P. (1975). A behavioural explanation of the association between bus and passenger arrival at a bus stop. *Transportation Science*, 9, 248–282.

Kurauchi, F., Bell, M.G.H., and Schmöcker, J.-D. (2003). Capacity constrained transit assignment with common lines. *Journal of Mathematical Modeling and Algorithms*, 2(4), 309–327.

Lam, W.H.K., Gao, Z.Y., Chan, K.S., and Yang, H. (1999). A stochastic user equilibrium assignment model. *Transportation Research*, 33B, 351–368.

Le Clercq, F. (1972). A public transport assignment method. *Traffic Engineering and Control*, 14(2), 91–96.

Marguier, P.H.J. and Ceder, A. (1984). Passenger waiting strategies for overlapping bus routes. *Transportation Science*, 18, 207–230.

Nguyen, S. and Pallotino, S. (1988). Equilibrium traffic assignment for large-scale transit networks. *European Journal of Operational Research*, 37, 176–186.

Nuzzolo, A. (2003a). Transit path choice and assignment models. In *Advanced Modeling for Transit Operations and Service Planning* (H.K. Lam and M.G.H. Bell, eds.), pp. 93–124, Pergamon, Elsevier Science, Oxford, U.K.

Nuzzolo, A. (2003b). Schedule-based transit assignment models. In *Advanced Modeling for Transit Operations and Service Planning* (H.K. Lam and M.G.H. Bell, eds.), pp. 125–163, Pergamon, Elsevier Science, Oxford, U.K.

Nuzzolo, A. and Crisalli, U. (2004). The schedule-based approach in dynamic transit modeling: A general overview. In *Schedule-Based Dynamic Transit Modeling. Theory and Applications* (N.H.M Wilson and A. Nuzzolo, eds.), pp. 1–24, Kluwer Academic.

Nuzzolo, A., Russo, F., and Crisalli, U. (2001). A doubly dynamic schedule-based assignment model for transit networks. *Transportation Science*, 35, 268–285.

Spiess, H. and Florian, M. (1989). Optimal strategies: A new assignment model for transit networks. *Transportation Research*, 23B, 83–102.

Szeto, W.Y., Solayappan, M., and Jiang, Y. (2011). Reliability-based transit assignment for congested stochastic transit networks. *Computer-Aided Civil and Infrastructure Engineering*, 26(4), 311–326.

Tirachini, A., Hensher, D. A., and Rose, J. M. (2013). Crowding in public transport systems: Effects on users, operation and implications for the estimation of demand. *Transportation Research*, 53A, 36–52.

Tong, C.O. and Wong, S.C. (1999). A stochastic transit assignment model using dynamic schedule-based network. *Transportation Research*, 33B, 107–121.

Trozzi, V., Gentile, G., Bell, M., and Kaparias, I. (2013). Dynamic user equilibrium in public transport networks with passenger congestion and hyperpaths. *Procedia-Social and Behavioral Sciences*, 80, 427–454.

Wu, J.H., Florian, M., and Marcotte, P. (1994). Transit equilibrium assignment: A model and solution algorithm. *Transportation Science*, 28, 193–203.

Wu, X. (2013). Finding reliable shortest paths in dynamic stochastic networks. *Transportation Research Record*, 2333, 80–90.

Chapter 12

Service design

Elements, operational parking, and stop location

To think too long
about amending a
service often becomes
its undoing

A.C.

CHAPTER OUTLINE

12.1 Introduction
12.2 Service-design elements
12.3 Scheduling-based solution for operational parking conflicts
12.4 Optimum stop location: Network-based theoretical approach
12.5 Stop placement considering uneven topography
12.6 Literature review and further reading
Exercises
References

PRACTITIONER'S CORNER

After reviewing and investigating transit-scheduling components and passenger demand and assignment, we open this chapter with a description of transit service and network and route-design components, subjects that will also be covered in the three following chapters. Service-design and stop placement issues, in the subjects of this chapter, facilitate an understanding of the importance of transit planning in enhancing existing or new transit services. The planning table is the mechanism with which to start changing the prevailing opinion that transit service and problems are tied to each other as in the adage: urban areas with transit facilities are exciting places where something is almost always happening, mostly unsolved.

The chapter consists of four principle parts, following an introductory section. Section 12.2 outlines the predominant design elements, service strategies, and possible actions for both existing and new transit services. In addition, this section provides standards and measures for system and performance. Section 12.3 pinpoints and solves a specific design problem in which transit vehicles need to park at route departure/controlled/holding stops located on street; because of the lack of parking spaces, they block the traffic lane adjacent to the parking bay/lane. The solution approach employs a surplus function model representing the surplus of parking vehicles required at a particular departure point in a multiterminal transit system. Section 12.4 furnishes a framework and theoretical analysis for finding the smallest number of public transit stops and their locations so that no passenger is further away from a stop than a preselected distance. An optimal algorithm with an example is provided. Section 12.5 provides formulations and framework of analyzing the location of public transit stops considering uneven topography environment. This is to enable easy and convenience access and egress to and from the stops.

Practitioners are encouraged to visit thoroughly all the sections besides Sections 12.3 through 12.5, except the introductions of these sections. They should, though, inspect the practical problems portrayed in these three sections and the examples with their solution.

Finally, among the important transit service and connectivity attributes are waiting time and passenger information. When designing the service, it is worthwhile to put yourself into the shoes of waiting passengers, who wish the wait had a fast-forward button. This may lead the planner to recommend hanging the following sign at all stops: "Waiting times perceived are larger than they are." In addition, useless information should be avoided, like that given to a passenger who asked where to get off the bus: "See where I get off, and get off two stops before."

12.1 INTRODUCTION

Profound service-design approaches, including well-coordinated service elements, comprise the foundation of any successful public transit service. In fact, an adequate, well-designed service is a present the transit agency gives to itself. The importance of service-design elements may be better comprehended after reading the following examples of two randomly selected reviews.

12.1.1 Transit service frustration

Examples of passengers' reviews are from the *Washington Metropolitan Area Transit Authority* (WMATA, 2014), website: http://www.yelp.com/biz/washington-metropolitan-area-transit-authority-washington-3.

January 15, 2014

Being a native Washingtonian and having experienced the Metro since its opening in 1976, it pains me to be critical … however, with thousands of trips over the years, I feel I can give an honest, unvarnished review. I've lived in or traveled to many cities with far more dependable mass transit systems.

Frequent service delays and maintenance issues have become the norm. Horrible sound system in the cars making station announcements either inaudible or garbled. Seating arrangement encourages "hogging" of two seats, compelling many to stand, mostly crowding around the doors as the aisles are too narrow for the oversized passengers. The tiered fare system causes confusion, particularly for tourists, newbies, and infrequent users. Don't get me started on people standing on the left side of the escalators. The only good thing about broken escalators is that it forces people to actually move up/down the steps.

I give it 2-stars, but there are days when I completely lose faith in Metro. No system is perfect, but the Metro needs some serious help to get it up to a reasonable standard of reliability …!

December 21, 2013

Welcome to the land of frustration! Where the escalators are always, ALWAYS broken, the train is ALWAYS delayed, and you never know where the hell you are because the driver does not know how to speak into a mike. Next stop … jfeiwroweir. What did you just say? L'Enfant?

Perhaps I am spoiled. I mean, I did sit in my nice clean car for hours … in traffic, but at least it was clean, rarely ever broke down and was on time! Imagine that? Even with the Metro D.C. areas horrific traffic … why the hell can't metro be the same?

Why is it so dirty? Why is it always late? Why is everything always broken? Where does all the money go? It's not like it's cheap … riding the Metro is expensive. It cost me $3.00 each way and to get my car out of the parking garage it's another $4.75. I imagine the large crowds of people moving through the station like wild animals are probably spending about the same if not more.

And what's with the weird smell? Why do the breaks on the train smell like a 10 day old fish that has sat out in the sun? Why? I think metro is in desperate need of new management, repairs and updates. Metrorail needs to get it together! Let's start over.

12.2 SERVICE-DESIGN ELEMENTS

A discussion of service and evaluation standards and guidelines appears in Section 1.3. These standards and guidelines establish service needs as seen by the local authority/government and the transit agency, but they do not cover the entire spectrum of required service-design elements. This section outlines the predominant design elements for both existing and new transit services.

The main service-design elements are as follows:

- Potential markets
- Network size and coverage
- Network structure, followed by route structure
- Route coordination (intra- and interagency)
- Route classification
- Span of service
- Service frequency, followed by public timetables

- Schedule coordination (intra- and interagency)
- Vehicle scheduling, followed by vehicle types and fleet size
- Crew scheduling, followed by rostering
- Fare policy
- Passenger amenities and information systems
- Data collection systems
- Determination of measures of performance
- Setting service and evaluation standards
- Ridership, cost, and revenue estimation

12.2.1 Description of service-design elements

The elements in the foregoing list will now be described briefly.

According to TranSystems et al. (2006), each *potential market* element requires an analysis of demographics and travel patterns as well as market research. The analysis is used to identify areas with the potential to support transit services and to locate current and projected travel markets. The market research is intended to identify market segments, along with passengers' degree of satisfaction from the transit service; the research focuses on service preferences and the inclination to use/increase the use of the service, given specific improvements.

The *network size and coverage* element applies to the set of all routes—that is, the entire system—and serves to fix the recommended spacing (distance) between routes for varying residential area densities. TranSystems et al. (2006) provide a few examples of a coverage measure. For instance, it suggests the provision of transit service within walking distance (defined as 400 m) of all residents living in areas with population densities greater than 2000 people per square kilometer.

The *network structure followed by route structure* element can take one or a combination of the four common forms: multimodal, radial, grid and time transfer, and pulse. A multimodal network of routes coordinates short and long trips through different transit modes, for example, short trips by buses that feed long trips by rail. The network of radial routes aims at providing a considerable amount of service to central points, for example, the central business district (CBD). A grid structure network of routes, on the one hand, allows easy access to the transit system but, on the other hand, requires many transfers; thus, the transfers in this network are timed and preferably synchronized online. The last network structure is based on a pulse system, in which routes are initiated at the same (central) point, which becomes a transfer point; usually, this suits small urban areas.

The *route coordination (intra- and interagency)* element applies at the intra-agency front to a route-design system with coordinated meeting points in terms of convenient passenger transfer facilities. At the interagency front, coordination is manifested in terms of operating policies and transit support promotion programs.

The *route classification* element exhibits various route-type needs for different geographical areas. Route classification schemes usually comprise one or a combination of the following: line haul, local, express, feeder/distributor, branching, radial line haul, zonal, commuter, circulator, crosstown, short turn, and limited stop.

The *span of service* element indicates the length of time that a service should be provided, by time of day and day of week. For example, a span of service guideline can suggest that the first trip should arrive no later than, and the last trip should depart no earlier than, the (specified) times shown for weekdays, Saturdays, and Sundays.

The *service frequency followed by public timetable* element is described and discussed explicitly in Chapters 3 and 4. Essentially, this element sets minimum service-frequency and maximum vehicle-occupancy thresholds to guarantee a basic level of service for different geographical areas. For the timetable, this design element sets the type of vehicle headway to be utilized, for example, even headway, clock headway, even-load headway, and by time of day and day of week.

The *schedule coordination (intra- and interagency)* element applies at the intra-agency front to a design of timetables that will maximize simultaneous arrivals at transfer points (see Chapter 5). At the interagency front, schedule coordination is expressed in terms of jointly designed timetables by different agencies and the use of different transit modes to allow for easy transfers between transit routes, with minimum waiting time.

The *vehicle scheduling followed by vehicle type and fleet size* element is described and discussed in Chapters 6 through 8. Basically, this element ensures design efficiency in terms of the fleet size required, balancing deadheading (DH) trips and shifts in departure times and minimizing the cost involved in purchasing vehicles.

The *crew scheduling followed by rostering* element is described and discussed in Chapter 9. Fundamentally, this element focuses on minimizing the crew wages involved and provides a satisfactory working schedule from the crew's perspective.

The *fare policy* element aims at improving payment convenience and integration between different transit services. For instance, a fare policy can be established with free transfers, discount options, and the elimination of fare zones and special surcharges.

The *passenger amenities and information systems* element aims at improving passenger facilities and vehicle amenities and improving passenger information for pretrip planning, en route riding, and waiting at terminals and stops. This element is also concerned with improving passenger safety and security.

The *data collection system* element is described and discussed in Chapters 2, 3, and 18. This vital element constructs the foundation for effective and efficient transit operations planning. The key components of this element are suitability, accuracy, and an adequate amount of data.

The *determination of measures of performance* element aims at quantifying tools for measuring the quality of service. The measures determined show whether the service is appropriate, convenient, and reliable, especially from the passenger's perspective. Two basic types of measures of performance are described later in this section: system performance and connectivity performance.

The *setting service and evaluation standards* element is described and discussed in Chapter 1. On one hand, standards maintain and improve the service; on the other hand, they present a source of fiscal pressure on the transit agencies. Evaluation standards aim at improving the efficiency, effectiveness, and productivity of the transit service.

The last design element on the list is *ridership, cost, and revenue estimation*. This element provides the tools for forecasting the three linked components of passenger demand, cost of service, and revenue. Some of these tools are described in Chapter 10, which emphasizes ridership prediction.

12.2.2 Service strategies and possible actions

A good summary of transit service strategies appears in the U.S. Transit Cooperative Research Program (TCRP) H-32 report (TranSystems et al., 2006). The following five lists provide new design and design-adjustment examples by purpose and possible implementation actions (in parenthesis):

1. New forms of service
   - Improve travel speed (introduction of express/zonal/rail/bus rapid transit service)
   - Attract new passengers (circulator shuttles, dial-a-ride service)

2. Area coverage service
   - Increased route coverage (service expansion; integrated circulator and line-haul services)
   - Increased span of service (late night, weekends, holidays)
3. New and adjusted routing
   - New routing (linking routes, splitting routes, feeder based, crosstown based)
   - Routing adjustment (route extension/shortening/realignment, express/zonal/local)
   - Coordination (intra- and interagency transfer centers/points)
4. New and adjusted scheduling
   - Introduce interlining (area based, trip based, vehicle type based)
   - Introduce/improve coordination (intra- and interagency synchronized/timed transfers)
   - Change frequency (increase/decrease, even headways, clock headways)
   - Change departure times (shifts within tolerances, even-load departure times)
   - Modify time elements (average travel time, layover time, recovery time)
   - Improve reliability (see Chapter 18 for lists of possible actions)
5. Improved amenities
   - Introduce/improve passenger facilities (stops/station, transfer points, transit centers, park-and-ride amenities)
   - Introduce/improve vehicles (amenities, new vehicle type)
   - Increase safety (introduce/improve safety features on vehicles and at stops, increase awareness and preparedness)
   - Increase security (introduce/improve security agents and features on vehicles and at stops, increase awareness and preparedness)

12.2.3 Standards and measures of system and performance

This section presents service standards and measures of performance. The service standards, accompanied by the required data, follow the list given in Table 1.3; these standards are shown in Table 12.1. The measures are divided into measures of system performance (MOSP) and measures of connectivity performance (MOCP).

The MOSP and MOCP, which should be selected at the design stage, quantify how well transit routes are used. Table 12.2 lists 10 MOSP and 4 MOCP, along with their notation, interpretation, and the data required. Measures of reliability of service are not included in Table 12.2; they will be discussed in Chapter 18.

The measure of emissions E_{rt} in Table 12.2 can be extracted from real data. For instance, Table 12.3 shows calculated data from the U.S. Environmental Protection Agency's (EPA) MOBILE5 and PART5 software-based models. MOBILE5 is a vehicle-emission modeling software, and PART5 is a software model for estimating particulate emissions from highway vehicles. More information can be found in the EPA's national emission inventories air pollutant trend website: www.epa.gov/ttn/chief/eiinformation.html. A more updated model by EPA is MOBILE6, which can be found at www.epa.gov/otaq/m6.htm. Table 12.3 (using MOBILE5 and PART5 software) provides data by vehicle type, per single vehicle, and on a per-passenger basis; the last is based on an occupancy of 35 passengers/bus and 150 passengers/train.

12.3 SCHEDULING-BASED SOLUTION FOR OPERATIONAL PARKING CONFLICTS

One of the operational problems for urban transit, especially with buses, occurs in a situation in which vehicles need to park at route departure/controlled/holding stops located on street;

Table 12.1 Service standards and data required

Standard item	Data required
Route length	Average running time; distance
Stop spacing	Population density; type of service (e.g., local, express); number of passengers using a given stop
Route directness[a]	Average running time; distance; passenger origin–destination count
Short turn[a]	Average passenger counts by stop
Route coverage	Population data; land-use data; public view
Route overlapping	Scale maps of the network of routes
Route structure[a]	Average passenger origin–destination counts; population data; public opinion
Route connectivity	Maps of network of routes and feasible transfer points
Span of service	Timetables by route and zone of operation
Load (crowding) level	Average passenger counts by stop or at max load points
Standees[a]	Passenger counts by stop
Headway upper limit[a]	Average passenger counts; permits between operator and authority
Headway lower limit	Average passenger counts; number of vehicles available
Transfers	Average passenger counts; transfer counts; waiting time
Passenger shelters[a]	Average passenger boardings by stop; number of elderly and handicapped
Schedule adherence[a]	Onboard counts of departure and arrival times (manually or automatic device)
Timed transfer	Passenger origin–destination counts by time of day; trip timetables
Missed trips[a]	Dispatcher log and maintenance records (agency data)
Passenger safety[a]	Accident reports, combined with average passenger counts and km driven
Public complaints	Ordinary mail, e-mails, fax, telephone, visits (by category of complaint)

[a] Standards commonly in use.

however, because of a lack of parking spaces, the buses block the traffic lane adjacent to the parking bay/lane. Such operational scenarios are commonly observed at school dismissal times. Although a special arrangement can be made for school/factory dismissal times—for example, using side streets—the blocking of a traffic lane in other situations can result in severe traffic congestion. One way to solve the problem is to construct more parking spaces at these conflict points; however, this is a costly and time-consuming solution. This section presents a scheduled-based design solution to this problem in which lane-blocked situations are caused at route departure points, that is, eliminating or reducing the impact of lane-blocked situations through the use of shifting departure times and/or inserting DH trips in timetables.

12.3.1 Surplus function model

The model constructed to solve the operational parking conflicts is basically a mirror image of the deficit function (DF) model described extensively in Chapters 6 through 8. This model is called a *surplus function* (SF) model because it represents the surplus of parking vehicles required at a particular departure point in a multiterminal transit system. SF is a positive step function that basically increases by one at the time of each trip arrival and decreases by one at the time of each trip departure; its starting value depends on the starting and maximum DF values. This definition shows a strong mutual dependency between DF and SF.

To construct the SF model, the only information needed is a timetable of required trips and the number of available parking spaces at each departure point. The main advantage of SF, as with DF, is its visual nature.

Let K be the set of all departure points, $p(k, t, S)$ denote the SF for departure point k for all $k \in K$ at time t for schedule S, and PS_k be the number of available parking spaces at k.

Table 12.2 Notations and interpretation of transit measures of performance

Type of measure	Measure	Designated notation	Interpretation and data required
Measure of system performance	Number of passengers	P_{rt}	Number of passengers by route r and time of day t
	Revenue	R_{rt}	Agency revenue by route r and time of day t ($)
	Vehicle trips	V_{rt}	Number of vehicle trips, timetable based, by route r and time of day t
	Vehicle-kilometers of travel	VKT_{rt}	Vehicle-kilometers of travel, based on V_{rt}, by route r and time of day t (veh-km)
	Passenger travel times	PTT_{rt}	Passenger travel times, based on trip times and load profiles, by route r and time of day t (h)
	Emissions	E_{rt}	Vector of average emission of four pollutants (CO, NO_x, VOC, PM_{10}), based on VKT_{rt}, by route r and time of day t (kg)
	Vehicle-hours of travel	VHT_{rt}	Vehicle-hours of travel, based on V_{rt}, by route r and time of day t (veh-hr)
	Passengers per veh-h	PVH_{rt}	Ratio of passengers per vehicle-hour, based on P_{rt} and VHT_{rt}, by route r and time of day t (pass/veh-h)
	Passengers per veh-km	PVK_{rt}	Ratio of passengers per vehicle-kilometer, based on P_{rt} and VKT_{rt}, by route r and time of day t (pass/veh-km)
	Revenue per passenger	RP_{rt}	Agency revenue per passenger, based on R_{rt} and P_{rt}, by route r and time of day t ($)
Measure of connectivity performance	Passenger waiting times	PWT_{rt}	Passenger waiting times, based on P_{rt} and service frequency, by route r and time of day t (h)
	Passenger transfers	PTR_{rt}	Number of passenger transfers, within and between modes of travel, by route r and time of day t
	Connectivity-production cost	CPC_{rt}	Cost of passenger waiting time and transfer penalty plus the cost of vehicle-hour, based on VHT_{rt}, PWT_{rt}, and PTR_{rt}, by route r and time of day t ($)

Table 12.3 Average emission of pollutant per vehicle (in parentheses, per passenger) in gram/mile

	CO	NO_x	VOC	PM_{10}
Diesel bus	**23.2** (0.66)	**22.1** (0.63)	**4.2** (0.12)	**0.63** (0.02)
Automobile	**23.0** (19.17)	**3.9** (3.25)	**3.7** (3.08)	**0.09** (0.075)
Rail	**0.03** (0.0002)	**0.47** (0.003)	**0.02** (0.0001)	**0.009** (0.0001)

The maximum value of $p(k, t, S)$ over the schedule horizon $[T_1, T_2]$ is designated $P(k, S)$. For simplicity, S will be deleted when it becomes clear which underlying schedule is being considered. The DF notations defined in Sections 6.5.1, 6.5.2, 7.3.1, and 7.3.3 will also be utilized for the SF analyses.

SF, unlike DF, cannot be negative, because each departure must be preceded by an arrival, thereby creating the need for a parking space between the arrival and the departure epochs. DF is defined as the total number of departures minus the total number of trip arrivals at terminal k, up to and including time t. SF is defined as the total number of arrivals minus the total number of trip departures at terminal k, up to and including time t, in which the

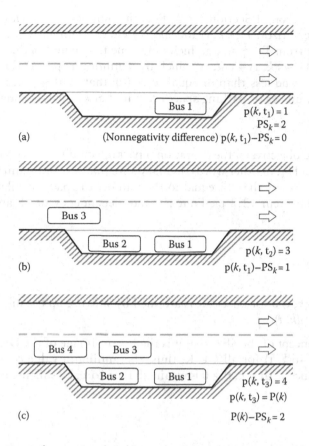

Figure 12.1 Three situations of operational parking at *k* = route departure stop (a) without lane-blocked, (b) lane-blocked by one bus, and (c) lane-blocked by two buses (the value of the surplus function and its difference from the two available spaces are shown).

number of arrivals at T_1 is $D(k)$ minus the number of departures at T_1. The next section further describes these relationships and definitions in a more formal way.

Figure 12.1 depicts three operational parking situations with $PS_k = 2$, at t_1, t_2, and t_3. Figure 12.1a shows a single parked bus; hence, $p(k, t_1) = 1$; because of the excess number of buses, $p(k, t) - PS_k$ is zero (not –1). Figure 12.1b and c exhibit lane-blocked situations; in Figure 12.1c, $p(k, t_1) = P(k) = 4$, representing the maximum number of parking vehicles over $[T_1, T_2]$.

12.3.2 Minimum parking spaces required

Following the definition of SF, we can establish a few formulae and rules that will constitute the basis of an algorithmic-based solution. The first of these is Theorem 12.1.

Theorem 12.1:

The minimum number of parking spaces required at *k*, $N_p(k)$ is equal to the maximum SF at *k*.

$$N_p(k) = P(k) = \max p(k,t) \quad \forall k \in K$$

$$t \in [T_1, T_2]$$

(12.1)

Proof: Using the notation of Section 6.5.2, F_k is the number of vehicles present at k at the start of the schedule horizon T_1; $s(k, t)$ and $e(k, t)$ yield the cumulative number of trips starting and ending at k from T_1 up to and including time t. The number of vehicles parking at k at time $t \geq T_1$ is $F_k - s(k, t) + e(k, t)$, which by definition is $p(k, t)$. This expression must be both nonnegative and less than or equal to $N_p(k)$; that is, $0 \leq p(k, t) \leq N_p(k)$, $T_1 \leq t \leq T_2$. The minimum number of parking spaces required at k, then, is equal to the maximum SF $P(k)$.

Consequently, the objective of the operational parking conflict is to eliminate or minimize cases in which $P(k) > PS_k$, in other words, to make sure that the minimum number of parking spaces required at k is less than or equal to the number of spaces available. The following Lemma 12.1 and Theorem 12.2 facilitate the interdependency of SF and DF through the maximum SF $P(k)$.

Lemma 12.1:

Compliance with $p(k, t)$ 0 for all $k \in K$ is attained by shifting up the mirror image of $d(k, t)$ by $+D(k)$ to obtain $p(k, t)$.

Proof: The basic concept of the SF is that it is the mirror image of the DF. The mirror image of $d(k, t)$ is simply $-d(k, t)$ for all $k \in K$; thus, the minimum value of the mirror image is $-D(k)$. The compliance with $p(k, t) \geq 0$ is attained, therefore, by shifting up $-d(k, t)$ by $+D(k)$.

Theorem 12.2:

$$P(k) = D(k) - \text{Min } d(k, t), \quad \forall k \in K$$

$$t \in [T_1, T_2]$$

(12.2)

Proof: Based on Lemma 12.1, $p(k, t) = -d(k, t) + D(k)$ for all $k \in K$; thus, the following holds: max $p(k, t) = P(k) = D(k) - \text{Min } d(k, t)$ for all $k \in K$.

Figure 12.2 illustrates a nine-trip fixed schedule with two terminals/departure points. Both $d(k)$ and $p(k)$ for $k = a, b$ are shown, with an emphasis on $PS_a = 2$ and $PS_b = 1$ available parking spaces. The value of $p(a, T_1) = 2$, $T_1 = 5{:}00$ is determined by shifting the mirror image of $d(a, t)$ three units up because of $D(a) = 3$; thus, because the mirror image of the DF, $-d(a, t)$, starts with $-d(a, T_1) = -1$, its shift will result in $p(a, T_1) = -1 + 3 = 2$. The same applies to $p(b, T_1) = 3$.

   Note that in order to have a sufficient number of parking spaces, $P(k)$ must be less than or equal to PS_k. If it is not, attempts could be made to attain this constraint. The following section utilizes the scheduled-based tools to reduce $P(k)$ for all $k \in K$, in terms of shifting departure times and inserting DH trips.

12.3.3 Scheduled-based reduction of P(k) procedure

The DF theory explicated in Chapters 6 and 7 is used for a heuristic procedure to reduce required parking spaces for situations in which $P(k) > PS_k$. Ostensibly, this procedure involves

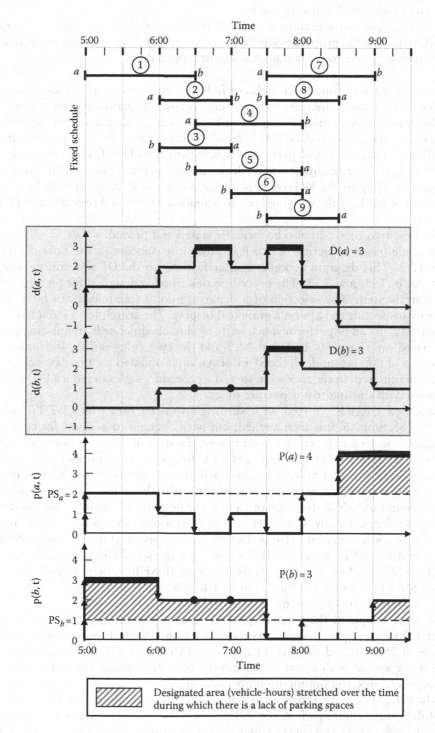

Figure 12.2 Construction of two surplus functions, dependent on their deficit functions (gray background), for a nine-trip example.

shifting trip-departure times and DH trip insertion so as to reduce P(k), if necessary, without increasing D(k) for all relevant $k \in$ K.

The mirror-image configuration of the DF, including shifting up this configuration by D(k) to attain p(k, t), has properties similar to the original DF. As a result, the two following rules are applied for the construction of a heuristic procedure:

1. Shifting departure times (left, right, or in both directions), within their tolerances, for reducing P(k) cannot increase D(k), mainly because the maximum intervals of P(k) and D(k) do not overlap; this rule applies to both a single shift and a chain of shifts.
2. Inserting DH trip to reduce P(k) is feasible only if the following two conditions are fulfilled: (a) the DH trip departs at or before the start of the first maximum interval of P(k), and, at the same time, this DH departure must come at or after the last maximum interval of D(k); (b) the DH trip must arrive, from k to k', at or after the last maximum interval of k', this rule applying both to a single DH trip and to a chain of DH trips.

Note that these two rules can also be formally stated and proved.

The schedule-based reduction of the P(k) procedure appears in the flow diagram form in Figure 12.3. This diagram contains similar features to the DF procedures described in Figures 6.9a, b, 7.11, and 7.12. The procedure described is designed for an interactive person–computer system. The selection of a departure point/terminal u can be made by the scheduler by inspecting the SFs on a graphical display. The search for a reduction of P(k) at u (see Figure 12.3) can be performed manually by the scheduler or by requesting procedures that are based on Theorems 12.1 and 12.2 and the two rules earlier. If a unit reduction shifting chain of P(k) is found, all the SF information is updated for this new schedule and a new iteration initiated. In the following step, the procedure seeks to reduce P(k) by DH trips incorporated with shifting trip-departure times.

The modified URDHC (mixed with shifting trip-departure times [SDT]) subroutine, described in Section 7.5.3, is then applied, but with changes to account for rule (2) previously, that is, inserting a DH trip from (or before) the start of the maximum interval of P(k), compared to bringing a DH trip to the start (or before) of the maximum D(k) interval in the DF procedures. In this DH procedure, the feasibility requirement for DH trip insertion is based on Equation 7.3. Another point to be mentioned is that DH trips that are added to the schedule S should include shifting tolerances for the next iterations of the SDT and modified URDHC procedures. Finally, if a URDHC has been found, the DH chain cost involved is compared with the saving of the cost of a single parking space. If the DH cost is higher than the saving cost, the URDHC is cancelled. Otherwise, the set of DH trips and required shifts (if any) {DH+SH} is added to the previous schedule; all SF information, including shifting, is updated for a new schedule. The modified URDHC subroutine continues, with updated S, until all terminals are examined.

A complete example of the required parking-space reduction procedure is illustrated in Figure 12.4. This example consists of a nine-trip schedule with three terminals, a single DH travel time of 75 min, and shifting tolerances of 15 min (both directions). The three DFs of the example appear with a gray background after going through the minimum fleet-size reduction procedures; the optimum number of vehicles required is four. The SFs are illustrated below the DFs, and the only complete mirror-image configuration (multiplied by –1) is that of departure point c, where the shifting up is not performed, because D(c) = 0. The procedure to reduce P(k) first checks which departure points are characterized by P(k) > PS$_k$; in fact, this is the case for all k = a, b, c.

Departure point/terminal a is then selected in Figure 12.4 to search for possible shifting departure-time opportunities; indeed, for this search, two shifts of trips 6 and 8 are

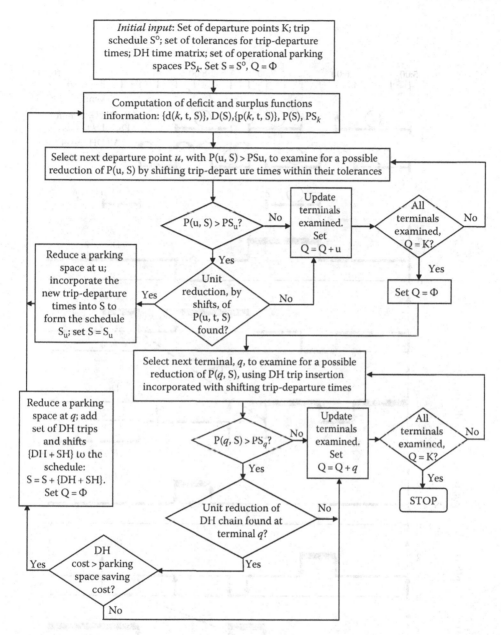

Figure 12.3 Flow diagram of the required parking-space reduction procedure, involving the shifting of departure times and deadheading insertions (modified URDHC subroutine of Chapter 7).

found feasible to reduce $P(a)$ from 2 to 1, thus making it equal to PS_a. These 15 min shifts are shown in the upper part of Figure 12.4. The second step is the attempt to reduce $P(b)$ by shifting trips 1 and 5 (15 min each) in opposite directions, but this cannot be realized. The reason is that by shifting trip 1 to the right, the value of $p(b, T_1)$ is changed from 1 to 2, which is the value of $D(b)$; the value reduces to 1 only at $p(b, 5:15)$. Thus, the two shifts cannot reduce $P(b) = 2$. The next step examines the reduction of $P(c)$ by shifts, but here, too, it cannot be performed. The procedure in Figure 12.3 proceeds to look for a DH trip insertion while fulfilling the two conditions of rule (2). Although it is impossible to insert a DH

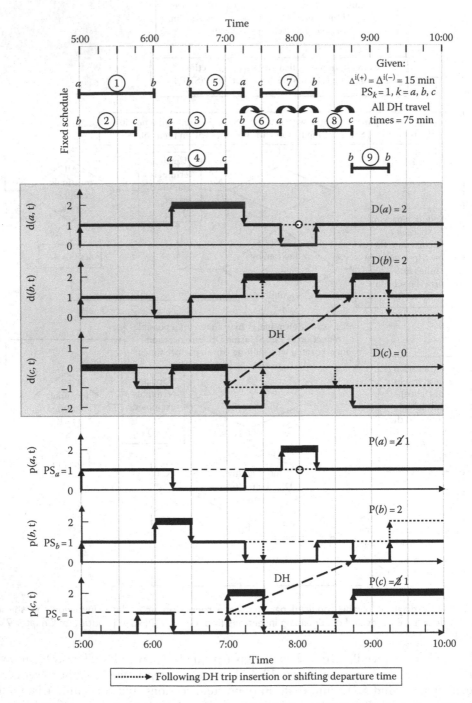

Figure 12.4 Nine-trip example of maximum reduction of the difference P(k) − PS$_k$, k = a, b, c of the surplus functions, utilizing shifting and deadheading insertion (the impact of changes on the deficit functions [gray background] is also shown).

trip from departure points a and b, it is found that it is feasible to do so from c to b; this results in reducing $P(c)$ from 2 to 1, making it equal to PS_c. The example ends up with only one excess-parking situation in b, between 5:45 and 6:30, but with the elimination of this undesirable circumstance in a and c.

12.4 OPTIMUM STOP LOCATION: NETWORK-BASED THEORETICAL APPROACH

Network coverage is among the service-design elements presented in Section 12.2, which contains the criterion for providing transit service within a defined walking distance of all residents living in a certain area. The present section furnishes a framework and theoretical analysis for finding the smallest number and the locations of public transit stops so that no passenger is further away from a stop than a preselected distance. An optimum algorithm is described and discussed for a general road network in which the nodes are community locations and the stops are to be located along the arcs (streets) or on nodes. The algorithm follows the method described in Ceder et al. (1983).

In order to ensure a high level of service for public transit users, walking distances to stops should be as short as possible. Farewell and Marx (1996) state that people consider walk time to be much less convenient than in-vehicle travel time and proposed a maximum walking distance of 400 m to a transit stop. Ceder et al. (1983) exhibit convenient ratios between in-vehicle travel time and walking time: 2.0 in the Netherlands, 3.5 in Chicago, and 6.2 in San Francisco. For instance, people in San Francisco consider 1 min walking to be 6.2 times less convenient than 1 min of in-vehicle travel time; this high value may be attributed to the difficult terrain in that city. Therefore, by reducing the length of the walking distance to stops in that city, public transit agencies will make their service more attractive.

12.4.1 Framework of analysis

The problem under consideration is to find the smallest number and the location of transit stops in a general network, so that no passenger is further away than a preselected distance, assuming that demand is generated at specific locations/nodes along the arcs/streets/roadway segments. The stops could be located either on the nodes or on any point along each arc. That is, the points lying on the arc are admissible as is each node. This problem, called the "m-center" problem in the operations research (OR) field, will be reviewed briefly here.

Minieka (1970) first proposed the m-center problem as an optimal algorithm; this was followed by an independent study by Christofides and Viola (1971), which also presented an iterative optimal algorithm. Handler (1973) developed an improved algorithm for Minieka's approach and showed that his method was preferable to that suggested by Christofides and Viola, particularly as the problem size increases. Summaries of the algorithms developed appear in Christofides (1975) and Minieka (1978). Several network examples serve as a stimulus for further examination of this m-center problem, since both the Handler and the Christofides and Viola approaches might require too many steps (computational time) for the optimal solution. It should be noted, however, that the available procedures were constructed with the intention of finding the optimal locations of m-centers in a given network, rather than locating centers according to a critical distance constraint. Both problems can be solved by the same method, though more computational time is usually required for a *given m-center* than for a *given-distance* problem.

Consider a connected network $G = \{N, A\}$ with a set of nodes N, a set of directed arcs A, and given distances $d(i, j)$, $\forall (i, j) \in A$. The numbers of nodes and arcs are $|N|$ and $|A|$, respectively. More notations are as follows:

$D(k, q)$ = shortest distance between k and q; k, q are two points anywhere on G (shortest-path algorithms appear in Appendix 9.A).

$SDM(i, j)$ = shortest-distance matrix for all pairs i and j; $i, j \in N$.

SDM_d = $SDM(i, j)$ matrix information indicating the shortest paths by directions from node i to node j, $\forall i, j \in N$.

ℓ = critical distance (i.e., for each node i, $i \in N$ must be no further away than ℓ units from its closest stop).

$S_{u,v} = \{u, v \in N : D(u, v) \leq 2\ell\}$, that is, set of all node pairs u, v that have a value equal to or less than 2ℓ in $SDM(i, j)$.

$D_i(u, v) = Min[D(u, i), D(v, i)]$; $i \in N$; $u, v \in S_{u,v}$.

s = candidate stop.

$(i, j)_{u,v}$ = arc on which a candidate stop related to u, v can be located; $(i, j)_{u,v} \in A$.

$ST = \{s \in G : (i, j)_{u,v} \text{ exists}\} \cup \{s = i, \forall i \in N\}$, that is, a set of all candidate stops, including each node as a candidate stop.

ST^* = set of stops in the optimal solution.

$d(i, s)$ = distance between node i and a candidate stop point s lying on $(i, i)_{u,v}$.

$P_{u,v}$ = (potential active path: $(i, j)_{u,v}$ exists, and the critical distance criterion is satisfied for a given pair u, v); that is, for the set of all possible paths in G between u and v, u, v $\in S_{u,v}$, so that (1) if the candidate stop is on either i or j, the shortest distance between u and v through i or j, respectively, is equal to or less than 2ℓ and (2) if the candidate stop lies on $(i, j)_{u,v}$, excluding i and j, the shortest distance (simple path) between u and v through $(i, i)_{u,v}$ is equal to or less than 2ℓ.

$P_a = \bigcup_{u,v \in S_{u,v}} P_{u,v}$, that is, the set of all potential active paths.

SCP = Set covering problem; the corresponding problem is related to a matrix in which each row represents a node (there are n rows) and each column a candidate stop. If the distance between node i and the candidate stop is less than or equal to ℓ, the entry of 1 is indicated. The set covering problem (SCP) is to find the least number of columns such that every row contains an entry 1 under at least one of the selected columns, that is, finding the minimum number of columns to *cover* all the rows.

The objective function is to find the minimum number of stops, $s \in ST^*$, so that $D(i, ST^*) \leq \ell$ for all $i \in N$, where $D(i, ST^*)$ = the shortest distance between node i and its closest stop, s, s $\in ST^*$. Certainly, the SCP solution, where ST is represented by all the columns, is the solution required for locating the transit stops. Note that there is always a feasible solution, since each node is also a candidate stop; that is, for large ℓ values, each node will be its own stop.

12.4.2 Set-ST algorithm and explanation

The algorithm to find the set ST is constructed as follows:

Step 1: Construct $SDM(i, j)$ and its related matrix information, SDM_d.

Step 2: Identify $S_{u,v}$ from $SDM(i, j)$.

Step 3: Select u, v $\in S_{u,v}$ and search for $(i, j)_{u,v} \forall (i, j) \in A$ through the procedure to determine ST (see the flow diagram in Figure 12.6); complete the procedure for ST for all u, v $\in S_{u,v}$.

Step 4: Store $SDM(i, j)$ and call SDM_d.

Step 5: Identify P_a (based on SDM_d).

Step 6: Construct SCP matrix for all s $\in$ ST (known from *Step 3*), based on P_a.

Step 7: Solve the SCP.

For *Step 1* of the Set-ST algorithm, various methods can be implemented; one of them is the Dijkstra's algorithm described in Appendix 9.A. More algorithms can be found, for example, in Ahuja et al. (1993). *Steps 2, 4, 5,* and *6* in the algorithm are self-explanatory by utilizing the aforementioned notations. *Step 3* deserves clarification and *Step 7* a comment.

In *Step 3*, the search for $(i, j)_{u,v}$ is performed on any $(i, j) \in A$ for each u, v pair. This search enables one to check whether a candidate stop (i.e., whether the shortest distance to u and to v is less than or equal to ℓ) can be located at any point on G. The interpretations of $(i, j)_{u,v}$ point to the existence of a simple path on G (with no repetitions of either nodes or arcs along this path) between u and v, so that (1) its length is less than or equal to 2ℓ; (2) its midlength point, p^*, lies on $(i, j)_{u,v}$; and (3) the shortest distance between u and p^* and between v and p^* is that along the simple path considered. In that last case, $p^* = s \in$ ST. The search for $(i, j)_{u,v}$, is based on Lemma 12.2.

Before specifying Lemma 12.2, let us refer to a *special diagram* that was first demonstrated by Hakimi (1964) and further discussed by Handler (1973) and Minieka (1978). The x-axis in this *special diagram* represents the distance from node i to node j along (i, j), and the y-axis represents the shortest distance from u and v to each point along (i, j). The distances are represented by 45° lines that can be monotonically decreased or increased or can have a unique broken point of 90° along (i, j).

Lemma 12.2:

If k is a point lying on (i, j) and $D(u, k)$ and $D(v, k)$ are described by the *special diagram*, there are only are four possible cases for $(i, j)_{u,v}$:

(a) If the lines for $D(u, k)$ and $D(v, k)$ intersect, a candidate stop is located on (i, j).

(b) If the lines for $D(u, k)$ and $D(v, k)$ do not intersect, a candidate stop cannot be located on (i, j).

(c) If the lines for $D(u, k)$ and $D(v, k)$ are the same at one node, say i, and split toward the other node j, then i becomes a candidate stop.

(d) If the lines for $D(u, k)$ and $D(v, k)$ are the same along each point k, $k \in (i, j)$, then both i and j are candidate stops.

Proof: The proof of this lemma is almost straightforward with the use of the *special diagram* configurations, such as the one illustrated in Figure 12.5. If we follow the line (triangle shape) in Figure 12.5a for $D(u, k)$, it can be seen that the shortest path from u to a point between i and k_1 is through i, whereas the shortest path from u to a point between j and k_1 is through j. It is also clear that $D(u, i) + d(i, k_1) = D(u, j) + d(i, j) - d(i, k_1)$ because this is the peak point (equal to the shortest distance from i and from j) of $D(u, k)$. Based on the interpretation for $(i, j)_{u,v}$ it becomes evident that point s satisfies all the requirements for a candidate stop. Figure 12.5b shows schematically that the solid line is the desirable simple path, whereas the dashed lines exhibit the direction of the shortest distance from u and v to other points than s that lie on (i, j). The critical distance criterion can be checked by the length of the heavy solid line in Figure 12.5a. It is postulated that the length of this line should be less than or equal to 2ℓ (note that point s is the midpoint of the heavy solid line). The explanation for Figure 12.5 can be applied to all the other *special diagram* configurations, that is, all the possible combinations

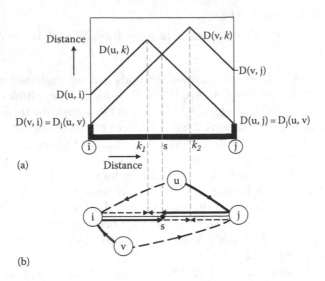

Figure 12.5 Example of a configuration for case (a) of Lemma 12.2 in order to search for $(i, j)_{u,v}$ and its corresponding candidate stop where in (a) is the triangle shape, and in (b) is schematic paths.

of the lines for $D(u, k)$ and $D(v, k)$ categorized by the cases of Lemma 12.2. For cases (c) and (d), these configurations will include equal lengths of $D(u, i)$ and $d(v, i)$ or $d(u, j)$ and $d(v, j)$.

The basic rules concerning the search for candidate stops (Set ST) constitute Lemma 12.2. These rules are integrated into a procedure to determine the set ST in Figure 12.6; this procedure exhibits *Step 3* of the Set-ST algorithm and examines each (i, j), $(i, j) \in A$ for all u, v pairs. The set of arcs $(i, j) \in A$ should be arranged such that each node (except those connected by a single arc) is considered at least once as node i of one arc (i, j). Note that if a node j is connected by a single arc (i, j) and can serve as a candidate stop for i (and perhaps for other nodes), then it can be replaced as a candidate stop by node i. This efficient arrangement allows the omission of a special check of node j as a potential candidate stop, reduces the computation time, and ensures a determination of all s, $s \in ST$, on G. Finally, in Figure 12.6, the critical distance is examined before the determination of $(i, j)_{u,v}$, an order of steps that can be changed if the value of ℓ is relatively small; for example, $(i, j)_{u,v}$ can be determined first and then the critical distance examined.

Following are remarks pertain to *Step 7* of the Set-ST algorithm. The solution to large-scale SCPs often involved a considerable amount of computation time (Cristofides, 1975). SCPs have attracted intensive research attention, particularly because of their wide applicability, for example, for airline crew scheduling. Rubin (1973) developed an effective heuristic algorithm for SCPs, and Balas and Padberg (1972) identified special properties of an SCP matrix that improve its solution through the simplex method of linear programming (by avoiding degeneracy difficulties). Thus, *Step 7* can be based on known heuristics procedures in the OR field. At this stage, it is possible to set the following theorem:

Theorem 12.3:

The Set-ST algorithm achieves an optimal solution for the problem of *minimum stops for a given critical distance* in a finite number of steps.

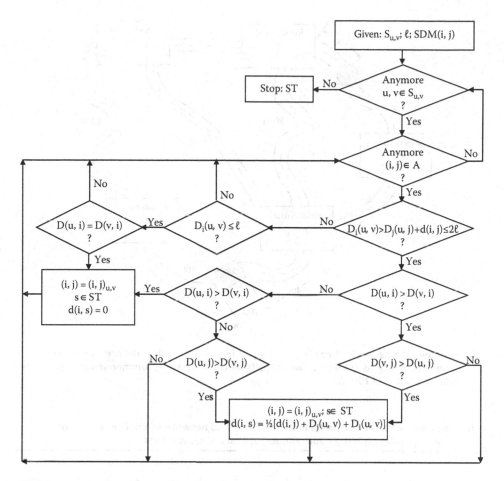

Figure 12.6 Determination of a set of candidate-stop locations.

Proof:

1. *Optimality* is attained by construction; the SCP solution guarantees the selection of the minimum number of stops satisfying the distance constraint. The procedure in Figure 12.6, through Lemma 12.2, establishes that all candidate stops are to be considered for the analysis of the SCP. Handler (1973) showed that the optimal solution for the problem combines only candidate sources. These simple arguments suffice.

2. *Finite convergence* is attained because an iteration of the Set-ST algorithm, which is performed only through the procedure in Figure 12.6, clearly indicates a finite operation ($S_{u,v}$—see Figure 12.6—is based on a finite number $|A|$).

12.4.3 Numerical example

Let us consider a small network of roads for further explanations of the Set-ST algorithm. This example is illustrated in Figure 12.7. The critical distance is $\ell = 500$ m. The shortest-distance matrix is given in Table 12.4, with the corresponding path (list of sequential nodes) in parentheses; those distances forming $S_{u,v}$ are in bold.

The procedure (*Step 3* of the algorithm) for Figure 12.6 is shown in Table 12.5. For each u, v pair, there is an X-shaded symbol for those $(i, j) \neq (i, j)_{u,v}$; for each $(i, j)_{u,v}$, the

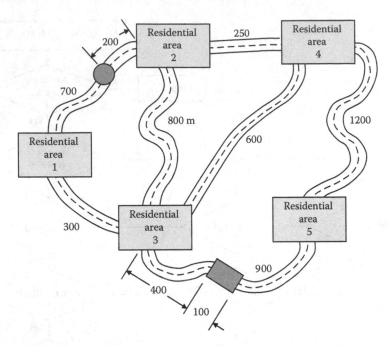

Figure 12.7 Example of a network of roads consisting of five nodes (residential areas) and seven road segments, with lengths in meters; two stops (gray colored) are recommended at a specific location along a stretch of 100 m.

Table 12.4 Shortest-distance matrix (in meters) of the road network example in which the corresponding path (by node number) appears in parentheses

i \ j	1	2	3	4	5
1	0	700 (1–2)	300 (1–3)	900 (1–3–4)	1200 (1–3–9)
2		0	800 (2–3)	300 (2–4)	1500 (2–4–5)
3	Symmetrical		0	600 (3–4)	900 (3–5)
4				0	1200 (4–5)
5					0

Table 12.5 Results of the procedure in Figure 12.6 for the road network example

$u, v \in S_{u,v}$	\multicolumn{7}{c}{$(i, j) \in A$}
	(1, 2)	(1, 3)	(2, 3)	(3, 4)	(3, 5)	(4, 2)	(5, 4)
1,2	700 (1–2)	X	X	X	X	X	X
1,3	X	300 (1–3)	X	X	X	X	X
1,4	1000 (1–2–4)[a]	X	X	900 (1–3–4)	X	X	X
2,3	1000 (2–1–3)[a]	X	800 (2–3)	900 (3–4–2)	X	X	X
2,4	X	X	X	X	X	300 (4–2)	X
3,4	X	X	X	600 (3–4)	X	X	X
3,5	X	X	X	X	900 (3–5)	X	X

[a] This $P_{u,v}$ is not the shortest-distance path between u and v.

Table 12.6 Information and results required for the set covering problem analysis of the road network example (first four rows) and the set covering problem matrix (last five rows)

u, v	1,2	1,3	1,4	1,4	2,3	2,3	2,3	2,4	3,4	3,5	1,1	2,2	3,3	4,4	5,5
DL[a]	700	300	1000	900	1000	800	900	300	600	900	0	0	0	0	0
$(i,j)_{u,v}$	(1,2)	(1,3)	(1,2)	(3,4)	(1,2)	(2,3)	(3,4)	(2,4)	(3,4)	(3,5)	0	0	0	0	0
d(i, s)	350	150	500	150	200	400	450	150	300	450	0	0	0	0	0
1[b]	1	1	1	1		1					1				
2	1			1		1	1	1	1			1			
3		1		1	1	1	1		1	1			1		
4			1	1				1	1	1				1	
5											1				1

[a] DL = $D_i(u, v) + D_j(u, v) + d(i, j)$, in meters; if s is on node I, then DL = $2D_i(u, v)$.
[b] Node number for the rows of the SCP matrix.

value of DL = $D_i(u, v) + D_j(u, v) + d(i, j)$, which should be less than or equal to $2\ell = 1000$ m, is indicated in the upper part of each filled cell in Table 12.5. The path $P_{u,v}$ is shown in the lower cell part for $(i, j)_{u,v}$. In this example, none of the candidate stops is located on a node.

The information and intermediate results related to $s \in ST$ are shown in Table 12.6, including the SCP matrix. The optimal solution combines the two circle columns in Table 12.6, in which $(i, j)_{u,v} = (3,5)_{3,5}$ and $(1,2)_{1,4}$. It is interesting, and important for planning purposes, to note that the difference between 2ℓ and the DL distance indicated in Table 12.6 allows for flexibility of stop location. That is, the optimal stop lying on arc (3,5) can be located 50 m closer toward either node 3 or node 5 without violating the optimal solution. On the other hand, the optimal stop on (1,2) can be located only at one point (200 m away from node 2 on that arc).

In addition, it is possible to observe—or to obtain from the SCP matrix in Table 12.6—the optimal solution for $\ell < 500$. For example, for $150 \leq \ell < 500$, three stops will be required. Practically speaking, it makes sense to have the objective of maximum location flexibility; such an objective yields the solution $(i, j)_{u,v} = (1,3)_{1,3}$, $(2,4)_{2,4}$, and $(5,5)_{5,5}$, where the last is node 5.

This section approached the problem of transit stop location in a network in which the nodes are located in concentrated population areas. A more challenging problem distributes the origins of passenger demand along the arcs (the continuous case) rather than in specific nodes. Nonetheless, the Set-ST algorithm can also be used as a preliminary tool to assess the location of transit stops in such a continuous case. For instance, artificial nodes (creating a new set of arcs) could be added, such that the distance between two adjacent nodes on the same arc would not be more than a given value.

12.5 STOP PLACEMENT CONSIDERING UNEVEN TOPOGRAPHY

Route coverage and stop spacing are among the service-design elements presented in Section 12.2. This section provides a formulation and analysis for finding the location of public transit stops to enable easy and convenient access and egress to and from the stops. One of the major parameters considered in this section, to attain convenient access and egress, is the effect of uneven topography on walking; this has barely been considered in the literature, particularly due to a lack of data and the complexity involved with its analysis.

It is estimated that the human heart beats approximately twice as fast while walking uphill at a grade of 15%, as compared to walking on level ground. This means that a person walking uphill consumes twice the amount of oxygen as a person walking on level ground (Terrier and Schutz, 2010). Slope not only affects walking speed but also reduces the attractiveness of using the transit service as shown by Rodriguez and Joo (2004) in their paper studying the factors of pedestrian and cyclist mode choice.

Locating bus stops on steep slopes is also disadvantageous to bus operators. The effect of gravity on already weak engines can lead to a considerable additional delay if a bus has to accelerate from a stop located at the bottom of a hill. The marginal impact of grade causes an additional stopping delay ranging from 5 to 11 s for grades greater than 3% (Furth and SanClemente, 2006). The bus stop infrastructure design guidelines (Bus Stop, 2009) of the Auckland Regional Transport Authority include topography as one of 10 primary factors to consider when locating new bus stops, noting that "in areas where the topography is hilly or very steep, closer spacing of bus stops may be required. Grade of road should not impede accessibility." In addition to this, a recent survey published as part of the TCRP (2013) shows that 10 out of 42 agencies surveyed across the United States consider steep hills to be a significant factor affecting the operation of their fleet.

This section covers the methodology developed recently by Ceder et al. (2015) for optimal stop placement considering uneven topography. This methodology was tested on the CBD area of the city Auckland, New Zealand. The problem is formulated as follows: for a single bus route r, select from a predetermined set of potential stop locations, the optimal set of stops to serve demand accessing the route in both service directions considering the effects of topography on access time and operational cost. The headway is assumed as a fixed input to the model.

12.5.1 Framework and modeling

The underlying street network is represented by a digraph $G = \{N, A\}$ where A is a set of arcs representing streets, roads, footpaths, etc., and N is a set of nodes representing intersections, potential stop locations, demand points, origin points, and terminal points. G can represent the street network at varying levels of complexity as appropriate for a particular application of the model. For example, a street network exhibiting large changes in grade over a relatively small area would be more appropriately modeled at a finer resolution. We define D as the set of demand points, a set of theoretical points at which transit demand, generated by some facility (e.g., building, land-use area, zone, or census block), enters or leaves the street network. The total hourly demand between two demand points i and j is denoted d_{ij}, $\forall i, j \in D$ (pass/h).

A bus route, r, is assumed to be a simple path in G beginning at origin point o^r and terminating at terminal point t^r. Potential stops are located along the route, on either side of the carriageway, in order to serve the route in both service directions. $S = \{s_1, s_2, ..., s_n\}$ is defined as the set of potential stop locations along r, ordered by the distance from o^r. S is assumed as an input to the model that in practice could be generated either manually by an experienced planner or at fixed increments along the route as per Ibeas et al. (2010) and Ceder et al. (2015). The optimal stop configuration will be a subset of S, selected to minimize operational and accessibility costs. The decision variables $x_m^{(1)}$ and $x_m^{(-1)}$ represent the selection of a potential stop $m \in S$ in the forward service direction (from o^r to t^r) and backward service direction (from t^r to o^r), respectively. We define l_{mn} as the distance in kilometers between two stops m, n $\in$ S, l^r as the total length of the route in kilometers, h as the route headway in minutes, and α_m as the slope (%) of the carriageway at stop $m \in S$, in the forward service direction.

The headway h is assumed to be a fixed input to the model, appropriately selected, to ensure that the demand along the route is satisfied at a reasonable level of bus occupancy.

It is assumed that a trip from demand point i to j consists of three components: (1) walking from i to a stop m, (2) traveling in-vehicle from stop m to n, and (3) walking from stop n to j. It is further assumed that passengers always select the paths from i to m and n to j that have the least *access cost* denoted by ac_{im} and ac_{nj}, respectively. The cost of an access path is equivalent to its total travel time in minutes (denoted by at_{im} and at_{nj}) weighted by an attractiveness factor, ε. The factor ε represents the degree to which hilly terrain is perceived as unattractive to a passenger (for reasons including increased energy consumption and difficulty for the elderly and disabled). We note that in reality passenger route choice may be influenced by a variety of other factors including bus stop proximity to an intersection, safety issues, and land use along the access path. At this stage, such factors are not included in the model for simplicity.

The attractiveness factor ε is based on an elasticity value relating pedestrian mode share to estimated delay due to sloping terrain, derived by Rodriguez and Joo (2004) and defined as

$$E_{X_{slope}}^{P_{ped}} = \frac{\delta P_{ped}/P_{ped}}{\delta X_{slope}/X_{slope}}$$

(12.3)

where

$\delta P_{ped}/P_{ed}$ represents a percentage change in pedestrian mode share
$\delta X_{slope}/X_{slope}$ represents a percentage change in delay due to slope

Rodriguez and Joo (2004) supply (in Table 3 of their paper) three different values for $E_{X_{slope}}^{P_{ped}}$, relating to three modes of pedestrian, bicycle, and bus. We use the result of the pedestrian mode for the one-level logit model with $E_{X_{slope}}^{P_{ped}} = -0.162$, significant at a 99% level of confidence. This value describes the percentage change in pedestrian mode share expected for a 1% increase in delay due to sloping terrain. Thus, for example, a path with hilly topography that causes a 20% increase in travel time is 3.2% less attractive than a level path of equal length. Conversely, the perceived cost of the hilly path is 3.2% greater than the level path. This is based on the assumption that the relationship between access path hilliness and demand, established by Rodriguez and Joo (2004), implies a similar, but inverse, relationship between route hilliness and access path cost. Defining $\varepsilon = -E_{X_{slope}}^{P_{ped}} = 0.162$, we define total access cost as

$$ac_{im} = at_{im}\left[1 + \left(\frac{at_{im} - l_{im}/w(0)}{l_{im}/w(0)}\right)\varepsilon\right]$$

(12.4)

where

l_{im} is the total length of the access path between demand point i and stop m
$w(0)$ is the walking speed on level terrain ($\alpha = 0\%$)

The ac_{nj} is similarly defined.

Let us first consider access time. The total access time at_{im} between demand point i and stop m is calculated based on a function of walking speed, $w(\alpha)$ in m/s that varies with slope, α. The definition of $w(\alpha)$ is based on an approximation of data collected by Terrier and Schutz (2010) and shown in Figure 12.8.

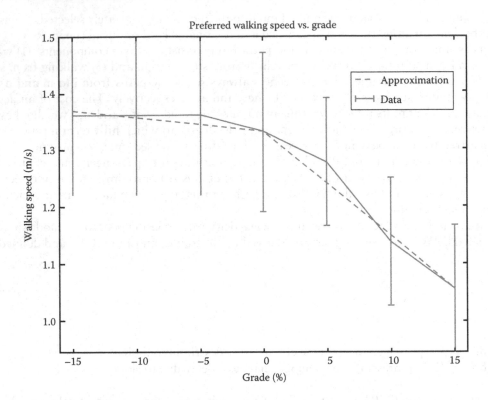

Figure 12.8 Relationship between slope and walking speed.

A linear piecewise function, obtained using least-squares regression with fixed intercept (i.e., fixing $w(0\%) = 1$), is used to approximate the data. This approximation is only appropriate for grades between –15% and 15%. We assume that outside of this range, the walking speed remains constant. This assumption is unlikely to have a considerable impact on the analysis as road grades typically lie within the bounds shown in the following equation:

$$w(\alpha) = \begin{cases} 1.058 & \alpha > 15 \\ (-0.066\alpha + 4.8)/3.6 & 15 \geq \alpha \geq 0 \\ (-0.0086\alpha + 4.8)/3.6 & 0 > \alpha \geq -15 \\ 1.37 & \alpha < -15 \end{cases} \tag{12.5}$$

Slope is not constant, but varies along the length of the access path, producing a slope profile. The slope profile is calculated as the derivative of the elevation profile of the access path, $E_{im}(x)$, where x is the distance along the access path from i. The total access time between i and m is therefore calculated in Equation 12.6 as the integral of distance along the access path divided by the walking speed (which is itself a function of distance along the access path), over the length of the access path. That is,

$$at_{im} = \int_0^{l_{im}} \frac{x}{w(E'_{im}(x))} dx \tag{12.6}$$

The at_{nj} access time between n and j is similarly defined. In reality, $E_{im}(x)$ is a smooth continuous profile; however, in practice, it is likely represented as a discrete set of elevation points $E_{im} = \{(x_1, e_1),(x_2, e_2), ..., (x_n, e_n)\}$, where x_k is the distance along the access path, from i to point k, e_k is the elevation at point k, and n is the total number of elevation points. Assuming that gradient is constant between points in E_{im}, at_{im} can be rewritten as follows:

$$at_{im} = \sum_{k=1}^{n-1} \frac{x_{k+1} - x_k}{w\big((e_{k+1} - e_k)/(x_{k+1} - x_k)\big)} \tag{12.7}$$

Equation 12.7 gives the correct total access time if $x_1 = 0$ and $x_n = l_{im}$, in other words, where the first and last elevation points are located at the start and end points of the access path. Where the elevation of start and end points is unknown, these may be calculated using interpolation as follows:

$$e_1 = e_2 - \frac{e_3 - e_2}{x_3 - x_2} x_2 \tag{12.8}$$

$$e_n = e_{n-1} + \frac{e_{n-1} - e_{n-2}}{x_{n-1} + x_{n-2}}(x_n - x_{n-1}) \tag{12.9}$$

Furthermore, if the elevation points are determined at fixed intervals of elevation, δe, then interpolated start and end points should be constrained to within $\pm\delta e$ (in units of m above sea level) of the nearest elevation point (i.e., $e_2 - \delta e \leq e_1 \leq e_2 - \delta e$, and $e_2 - \delta e \leq e_1 \leq e_2 - \delta e$).

An example of an elevation profile from the case study of Ceder et al. (2015) is shown in Figure 12.9. The figure shows the elevation profile of the access path and cumulative travel time in both directions, with and without accounting for the effects of slope on walking speed. As can be seen, in the forward direction, the effect of slope causes approximately 20 s of delay over a distance of approximately 650 m. In the backward direction, the effect of slope is minimal. This is due to the fact that preferred walking speed on downhill grades is not significantly faster than on level ground as shown in Figure 12.8.

The last point of this section is the introduction of the concept of reachable stops and reachable demand points. A stop m is reachable from a demand point i if $ac_{im} \leq \gamma$, where γ is a predefined upper limit on access time in seconds. Conversely, i is reachable from m if $ac_{mi} \leq \gamma$. Note that $ac_{im} \neq ac_{mi}$ since the elevation profiles E_{im} and E_{mi} are not equivalent (and are, in fact, opposite). We thus define in Equations 12.10 and 12.1 the set of binary variables β_i^m and β_m^i where $\beta_i^m = 1$ if stop m is reachable from demand point i and $\beta_m^i = 1$ if i is reachable from m.

$$\beta_i^m = \begin{cases} 1 & ac_{im} \leq \gamma \\ 0 & \text{otherwise} \end{cases} \quad \forall i \in D, m \in S \tag{12.10}$$

$$\beta_m^i = \begin{cases} 1 & ac_{mi} \leq \gamma \\ 0 & \text{otherwise} \end{cases} \quad \forall i \in D, m \in S \tag{12.11}$$

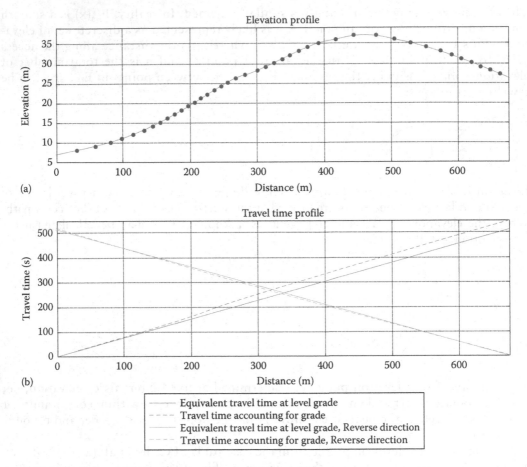

Figure 12.9 Examples of (a) elevation and (b) travel time profile.

12.5.2 Optimal selection of bus stop locations

The selection of bus stop location involves a hierarchical process in which at one level planners and operators make decisions in order to improve the overall performance of the route, while at another level individuals make choices with regard to their selection of stops in order to minimize their individual objectives, for example, travel time. Therefore, in order to evaluate the cost of a certain bus stop configuration, user behavior in response to this configuration must simultaneously be modeled. This property of the stop selection problem naturally lends itself to formulation as a bi-level optimization problem.

A bi-level optimization problem is a class of optimization problem that has an optimization problem embedded in its constraints. It has the following general formulation (Colson et al., 2007):

$$\min_{x\in X, y\in Y} F(x,y)$$

$$\text{s.t.} \quad G(x,y) \le 0 \tag{12.12}$$

$$\min_{y} f(x,y)$$

$$\text{s.t.} \quad g(x,y) \le 0$$

The problem is divided into two levels, termed the upper and lower levels, where x and y are the upper- and lower-level variables, F(x, y) and f(x, y) are the upper- and lower-level objective functions, and G(x, y) and g(x, y) are the upper- and lower-level constraints. The advantage of using a bi-level optimization formulation for this problem is to find a demand distribution that is optimal according to some objective at the lower level, for a given stop selection at the upper level, which is in turn evaluated according to a different objective.

In Ceder et al. (2015), a bi-level optimization problem has been developed in which the upper level of the model minimizes a cost function that represents the perspectives of the passenger, the agency, and the community (see more regarding these perspectives in Chapter 15). This function consists of agency costs relating to the cost of operating the bus route and user costs relating to accessing and using the route. The operational costs considered are the total operating cost per kilometer of route covered (C_{km}), the total idling cost incurred during dwell time at stops (C_{idle}), the total personnel costs over the duration of the analysis period (C_{pers}), and the total fixed costs relating to the required fleet size (C_{fix}). The user costs considered relate to the total access time to and from stops (T_{access}) and the total in-vehicle time (T_{veh}). The upper-level variables represent the selection of stops along the route in forward and backward service directions and are denoted $x_m^{(1)}$, $\forall m \in S$ and $x_m^{(-1)}$, $\forall m \in S$. The upper-level constraints ensure that at least one reachable stop is supplied for each demand point in both service directions and that each demand point is reachable from at least one stop in both service directions.

The lower-level model serves to assign demand between demand points to the route via selected stops. Two alternative lower-level objective functions are discussed by Ceder et al. (2015), the first seeking to minimize the total passenger travel time and the second seeking to minimize only the total passenger access time. Lower-level decision variables represent the number of passengers traveling between any pair of demand points i, j, via a pair of stops (m, n) at which they board and alight. The lower-level constraints ensure that all demand d_{ij} is assigned to the route. In order to define the set of lower-level decision variables, the concept of *reachable stop pairs* has been introduced. A pair of stops (m, n) is reachable for a pair of demand points (i, j), if m is reachable from i and j is reachable from n, or in other words $\beta_i^m \beta_m^i = 1$. For an arbitrary pair of demand points (i, j), the set of reachable stop pairs, S^{ij}, defined as

$$S^{ij} = \left\{ (m, n) : \beta_m^i \beta_n^i = 1, \forall (m, n) \in S \times S, m \neq n \right\}, \quad \forall (i, j) \in D \times D, i \neq j \tag{12.13}$$

The notation $(i, j) \in D \times D$, $i \neq j$, refers to all possible ordered pairs of demand points where $i \neq j$, and $(m, n) \in S \times S$, $m \neq n$, refers to all possible ordered pairs of potential stops where $m \neq n$. Then the set of lower-level decision variables can be defined as

$$\delta = \left\{ \delta_{ij}^{mn}, \forall (i, j) \in D, (m, n) \in S^{ij} \right\} \tag{12.14}$$

Where only passenger access time is considered in the lower-level objective function, the model outlined previously can be classified as a mixed integer, nonlinear–linear, bi-level programming problem where *nonlinear–linear* refers to the linearity of the upper- and lower-level problems, respectively. Note that if in-vehicle time is included in the lower-level objective function, then the problem becomes a mixed integer, nonlinear–nonlinear bi-level problem. There is little treatment of such problems in literature, with most of the research effort being invested in the study of linear bi-level problems (Li and Jiao, 2008). Because of their nonconvex, nondifferentiable nature, bi-level problems are inherently difficult to solve, with even the simplest instance, the linear–linear bi-level problems, having been shown (Colson et al., 2007) to be of NP-hard complexity (regarding complexity levels, see Section 5.5). Thus, it is necessary to use a heuristic approach to approximate an optimal bus stop configuration. The solution method used by Ceder et al. (2015) is an evolutionary algorithm based partially on the work of Li and Jiao (2008) and Beyer and Schwefel (2002).

The approach outlined in this section is detailed by Ceder et al. (2015) with the complete mathematical formulation of the bi-level optimization model, the solution algorithm and a case study (in Auckland, New Zealand). The main benefits of this approach are as follows: (1) the explicit modeling of access paths allows for a high degree of accuracy in relation to route accessibility; (2) the technique developed for determining elevation profiles is reasonably simple to implement using geographical information systems (GIS) software and enables topography to be considered in bus network design; (3) the bi-level type model realistically models the hierarchical nature of the transit planning process; (4) the GIS-based software framework developed as part of the case study allows the street network to be represented at a high level of detail, with the ability to include pedestrian-only paths; and (5) the model also makes use of (mostly) freely available GIS data covering a very large geographic area; thus, there is potential for the method to be applied to bus routes in almost every major city.

12.6 LITERATURE REVIEW AND FURTHER READING

This section reviews research works that discuss various indicators of the level of transit service (LOS). Such indicators enable a measurement of the service quality of both existing transit systems and proposed schemes and, therefore, serve as important planning tools. Note that transit coordination/connectivity studies are reviewed in Chapter 5 and reliability studies in Chapter 18.

Alter (1976) introduces a measure of transit LOS whose value is affected by accessibility, travel time, reliability, service directness, service frequency, and onboard density criteria. A given set of tables and conversion tables assigns grades, from A to F, to each criterion. The accessibility grade is based on the walking time or distance to and from a stop. The reliability grade is based solely on frequency. Travel time is evaluated according to the ratio of transit travel time to car travel time on the same route. The grade for directness depends on a combination of the number of required transfers and waiting time. Grades for peak and off-peak frequencies are weighted by population density in the area served. Grades for onboard density depend on the average area for a single passenger. All six measures are aggregated into an overall LOS indicator. Some of the criteria use qualitative considerations, but the whole process is generally quantitative. Although sensitive to various criteria, the proposed evaluation methodology seems fairly simplistic.

Horowitz (1981) proposes an indicator for bus-service quality that incorporates the main performance variables as perceived by transit riders: in-vehicle time, transfer time, walking time, and waiting time. The need to wait and the need to transfer, which are known to have a significant effect on passengers' perception of LOS, are accounted for by both a numerical variable and a dummy variable. To take equity considerations into account, LOS measures are calculated separately for different ridership sectors and then weighted and summarized. The weights used represent the equivalent travel time of a private car. The final LOS rating depends on the number of people within each population sector who can travel within a certain travel-time standard.

Polus and Shefer (1984) measure bus LOS by the ratio of average bus travel time to the average car travel time on the shortest distance between the same origin and destination. The model they propose for making forecasts of this LOS indicator is based on physical or traffic-related variables: route length, stop spacing, number of intersections on route, etc. It can, therefore, be used when detailed information about bus performance or demand is unavailable, which makes their model useful for analyzing suggested route changes.

Madanat et al. (1994) use an ordered probit model to correlate discrete responses from a bus-passenger survey with a quantitative variable representing onboard crowding levels. The ordered probit technique assists in identifying the thresholds between successive ratings presented to the respondents. The authors compare their analysis to the different onboard crowding levels that were used in the highway capacity manual (HCM) manual at that time to measure transit LOS. The results are very different from the HCM values.

Henk and Hubbard (1996) propose a measure for the availability of bus or rail services that take into account service considerations: coverage, frequency, and capacity. Various alternative measures of each of these considerations are discussed, and then preferred indicators are chosen: coverage is calculated by the number of directional route-kilometers per square kilometer; frequency is indicated by the amount of vehicle-kilometers per route-kilometer; and capacity is measured by seat-kilometers per capita. The values of the three components in each zone of the network studied are normalized by subtracting the mean and dividing by the standard deviation; an average of the three components is then computed.

Murugesan and Rama Moorthy (1996) develop an index for bus LOS. Survey results are processed using the theory of fuzzy sets, which enables a quantitative analysis based on qualitative rating responses. Twenty LOS attributes are included in the analysis, but the authors focus on a description of the fuzzy set methodology, not the LOS indicators.

Friman et al. (1998) study a database of transit riders' complaints and incidents in order to determine which LOS attributes are perceived as the most important. The analysis is qualitative, and the authors conclude that the service features most important to passengers are a driver's behavior and service reliability. The simplicity of the transit network and the provision of sufficient information are found to be important as well.

A TCRP report (1999) includes a detailed discussion of LOS and capacity issues for any transit mode. The calculation of LOS takes account of passenger loads and the accessibility of the service. In addition to the analysis of a system-wide LOS, there is a discussion of LOS considerations for a specific stop or for a specific route segment.

Prioni and Hensher (2000) develop an LOS model based on stated-preference survey data that depict the influence of various factors on the LOS as perceived by bus users. Factors found to have the highest impact on LOS perception are the tariff, travel time, and access time to the stop. Driver friendliness and the smoothness of the bus ride are also found to be quite significant. The authors examine the differences in the LOS perception of buses operated by different companies and find that there is some bias toward certain operators; they show how the index developed could be used in the contracting process to monitor bus operators and performance.

Ryus et al. (2000) introduce a measure for transit availability based on the concept that transit routes serve a small group of potential users. The measure is defined as the percentage of time in which transit service is available and accessible to an average person within a reasonable walking distance. Computation of this indicator requires the use of a GIS. Frequency, service coverage, stop location, and operation hours influence the value of the indicator; population density affects it, too, but does not influence the demand for transit services.

Guttenplan et al. (2001) propose a methodology for determining LOS in a combined bus, cycling, and walking network. The bus LOS is influenced not only by common factors, such as frequency, but also by intermodal factors that are unique to this method, such as difficulty in crossing the road or the connection between the bus stop and the sidewalk.

Yang and Wang (2001) evaluate LOS on the traffic-zone level, based on fuzzy c-means method. Using GIS functions, the authors compute each zone's characteristics, such as its route network density and population density. The fuzzy c-means method is used to identify clustering patterns and to aggregate the detailed measurements into composite, zone-level LOS indicators.

Tyrinopoulos and Aifadopoulou (2008) propose a methodology for the quality control of passenger services in the public transit business. An overview of the methodology developed by the Hellenic Institute of Transport to assess the levels of quality and performance of public transport services is presented and discussed. Overall, a total of 39 indicators are analyzed, classified in the following seven basic categories: safety, comfort, cleanliness; information, communication with the passengers; accessibility; terminals and stop points' performance; lines performance; general elements of the public transport system; and compound indicators based on the results of the indicators of the previous categories. Among the compound indicators, a customer satisfaction measure is considered in order to take into account customer perceptions. In fact, the authors suggest using factor analysis and multinomial logistic regression for investigating the influence of the operational performance indicators of the transportation system on customer satisfaction. This proposed methodology requires adaptation to the local characteristics of the transportation system under consideration. Furthermore, the indicator, the sample size, the surveys, and other elements of the methodology must be customized based on the size of the network under analysis. The adoption of the proposed methodology by a public transit operator will enable assessment of the performance of transportation services, development of service improvement strategies, and monitoring the progress of service quality and more importantly provide a better understanding of the needs and priorities of the passengers.

Lai and Chen (2011) explore the relationships between passenger behavior and the various factors that affect them by constructing a comprehensive model considering public transit involvement, service quality, perceived value, satisfaction, and behavioral intentions. Notwithstanding the factors recognized by past studies, such as service quality, perceived value, and satisfaction, this study addresses the importance of the involvement of public transit services in passenger behavioral intentions. A real-life application has been presented using passenger survey data from the Kaohsiung Mass Rapid Transit, a newly operating public transit system in Taiwan. The structural equation modeling technique is applied to analyze the conceptualized relationship model. The findings reveal that all causal relationships are statistically significant. The relevant managerial implications arising from the findings have subsequently been discussed.

Eboli and Mazzulla (2011) present a methodology for measuring transit service quality by considering both passenger perceptions and transit agency performance measures. These two types of service quality measurement, in conjunction, account for both perceived and measured service quality, respectively. The basic formulation of this procedure entails that when the variance of the objective/subjective indicator is close or equal to zero, the new indicator tends to or coincides with the objective/subjective indicator. The proposed procedure is subsequently applied to a real-life case study of a suburban bus line; a series of subjective and objective indicators are calculated on the basis of users' perception about the service and measurements provided by the transit agency. Variance of the objective indicators was found to be lower than the variance of the subjective ones—thus leading to move the value of the new indicator toward the value of the objective one. Therefore, it is imperative to carefully consider the most appropriate objective indicators for measuring the quality of the service attributes.

Hadas (2013) presents a unified methodology for extraction of transit data, data storage, and analysis, which enable relatively easy spatial analysis with GIS techniques based solely on Google transit feeds and any available road layer with no need for additional data.

Four connectivity indicators are introduced for the analysis of transit systems: (1) road coverage level indicator, (2) intersection coverage level indicator, (3) stop transfer potential indicator, and (4) routes overlap indicator. The model that has been developed is a simple-to-use tool enabling the decision makers to analyze the connectivity of transit networks and to perform benchmarking. Automatic calculation of the connectivity indicators within the GIS package enables the large public transit networks to be analyzed and compared effectively. Finally, this study demonstrates the application of the proposed methodology with the analysis for the transit systems in Auckland (New Zealand), Vancouver (Canada), and Portland (Oregon, USA). The possibility of undertaking what-if analysis, such as altering the transit network, changing frequencies, relocating stops, and merging stops for better accessibility, has also been demonstrated. Furthermore, this tool allows checking the effect of different stop transfer potential parameters in terms of policy making (maximum walking distance, maximum transfer time, etc.). Subsequently, the analysis can be carried out based on the dynamic properties of network, specifically based on the time of day and day of week.

The major features of the LOS indicators reviewed are presented in Table 12.7.

Table 12.7   Summary of the characteristics of the models reviewed

Source	Transit modes	Quantitative?	Factors influencing LOS
Alter (1976)	All modes	No	Accessibility, travel time, reliability, service directness, service frequency, onboard density
Horowitz (1981)	Bus	Yes	In-vehicle time, transfer time, walking time, waiting time
Polus and Shefer (1984)	Bus	Yes	External factors (route length, stop spacing, number of intersections on route, etc.)
Madanat et al. (1994)	Bus	Partially	Onboard density
Henk and Hubbard (1996)	All modes	Yes	Coverage, frequency, capacity
Murugesan and Rama Moorthy (1996)	Bus	Partially	(Twenty factors are discussed)
Friman et al. (1998)	All modes	No	Driver behavior, reliability, network simplicity, information
TCRP 35 (1999)	All modes	Yes	Accessibility, passenger load
Prioni and Hensher (2000)	Bus	Yes	Tariff, travel time, access time, driver friendliness, smoothness
Ryus et al. (2000)	All modes	Yes	Frequency, service coverage, stop location, operation hour, population density
Guttenplan et al. (2001)	Bus	Yes	Factors related to the pedestrian environment (e.g. difficulty in crossing road, etc.)
Yang and Wang (2001)	All modes	Yes	Network density, population density
Tyrinopoulos and Aifadopoulou (2008)	All modes	Partially	39 LOS indicators presented
Lai and Chen (2011)	All modes	Yes	Several core service and psychical environment factors discussed
Eboli and Mazzulla (2011)	Bus	Yes	11 service aspects and 26 service attributes discussed
Hadas (2013)	Bus	Yes	Four connectivity indicators introduced for the analysis of public transit (PT) systems

EXERCISES

12.1 Given the following timetable with seven bus trips, three departure points, and shifting tolerances of ±5 min and DH travel time of 25 min for all trips, the number of available parking spaces in each of the three departure points is one space. Note that the DH time is larger than the travel time (buses need to detour the service route when empty).

(a) Find the minimum fleet size required by using DF modeling.
(b) While maintaining the result of (a), find the best bus scheduling solution so as to minimize the number of departure points with maximum parking spaces required greater than the given number of available spaces; utilize SF modeling to arrive at the solution.

Trip number	Departure terminal	Departure time	Arrival terminal	Arrival time
1	a	6:00	b	6:15
2	b	6:25	a	6:40
3	b	5:40	c	5:55
4	c	5:45	a	6:00
5	a	6:30	b	6:50
6	b	6:35	c	6:50
7	a	6:20	c	6:42

12.2 Use the same technical data and queries as in Exercise 12.1, but with 10 trips, 4 departure points, and the following timetable:

Trip number	Departure terminal	Departure time	Arrival terminal	Arrival time
1	a	7:00	b	7:20
2	b	7:30	a	7:45
3	c	8:15	d	8:30
4	b	7:00	c	7:15
5	a	7:25	c	7:40
6	b	7:45	a	7:55
7	a	8:05	c	8:15
8	c	7:25	c	7:40
9	d	7:30	b	7:45
10	d	7:25	b	7:45

12.3 Employ the same technical data and queries as in Exercise 12.1, but with nine trips and this timetable:

Trip number	Departure terminal	Departure time	Arrival terminal	Arrival time
1	a	8:00	b	8:20
2	b	8:30	a	8:45
3	c	9:15	b	9:30
4	b	8:00	c	8:15
5	a	8:25	c	8:40
6	b	8:45	a	8:55
7	a	9:05	c	9:15
8	c	8:25	c	8:40
9	a	9:15	b	9:35

12.4 Given a network of 9 two-way street segments and 6 residential-area nodes in which the distances between nodes, in meters, appear in the following table and are symmetrical (same distances between nodes i and j and between j and i),

Node	2	3	4	5	6
1	900	300			
2		1100	1300		
3			1000	400	
4				700	500
5					600

(a) Find the minimum number of stops in the network, such that the critical distance between each node and its closet stop is less than or equal to 900 m.

(b) Find the optimal locations of four stops in the street network so as to minimize the maximum distance between a node and its closet stop; what is this distance?

(c) Find the optimal location of a single stop in the network using the same criterion as in (b); what is the minimax distance obtained, and from what node?

REFERENCES

Ahuja, R.K., Magnanti, T.L., and Orlin, J.B. (1993). *Network Flows*, Prentice Hall, Englewood Cliffs, New Jersey.

Alter, C.H. (1976). Evaluation of public transit services: The level-of-service concept. *Transportation Research Record*, **606**, 37–40.

Balas, E. and Padberg, M.W. (1972). On the set covering problem. *Operations Research*, **20**, 1152–1161.

Beyer, H.-G. and Schwefel, H.-P. (2002). Evolution strategies: A comprehensive introduction. *Natural Computing*, **1**(1), 3–52.

Bus Stop Infrastructure Design Guidelines. (2009). Private Bag 92 236, Auckland Regional Transport Authority, Auckland, New Zealand.

Ceder, A., Butcher, M., and Wang, L. (2015). Optimization of bus stop placement for routes on uneven topography. Submitted to *Transportation Research Part B* (already received favourable comments of the reviewers).

Ceder, A., Prashker, J., and Stern, H.I. (1983). An algorithm to evaluate public transportation stops for minimizing passenger walking distance. *Applied Mathematical Modeling*, **7**, 19–24.

Christofides, N. (1975). *Graph Theory: An Algorithmic Approach*, Academic Press, Orlando, FL, USA.

Christofides, N. and Viola, P. (1971). The optimum location of multicenters on a graph. *Operations Research Quarterly*, **22**, 45–54.

Colson, B., Marcotte, P., and Savard, G. (2007). An overview of bilevel optimization. *Annals of Operation Research*, **153**, 235–256.

Eboli, L. and Mazzulla, G. (2011). A methodology for evaluating transit service quality based on subjective and objective measures from the passenger's point of view. *Transport Policy*, **18**(1), 172–181.

Farewell, R.G. and Marx, E. (1996). Planning, implementation, and evaluation of OmniRide demand-driven transit operations: Feeder and flex-route services. *Transportation Research Record*, **1557**, 1–9.

Friman, M., Edvardson, B., and Garling, T. (1998). Perceived service quality attributes in public transport: Inferences from complaints and negative critical incidents. *Journal of Public Transportation*, **2**(1), 67–89.

Furth, P.G. and SanClemente, J. (2006). Near side, far side, uphill, downhill: Impact of bus stop location on bus delay. *Transportation Research Record*, **1971**, 66–73.

Guttenplan, M., Landis, B.W., Crider, L., and McLeod, D.S. (2001). Multimodal level-of-service analysis at the planning level. *Transportation Research Record*, **1776**, 151–158.

Hadas, Y. (2013). Assessing public transport systems connectivity based on Google Transit data. *Journal of Transport Geography*, 33, 105–116.

Hakimi, S.L. (1964). Optimum locations of switching centers and the absolute centers and medians of a graph. *Operations Research*, 12, 450–459.

Handler, G.Y. (1973). Minimax location of a facility in an undirected graph. *Transportation Science*, 7, 287–293.

Henk, R.H. and Hubbard, S.M. (1996). Developing an index of transit service availability. *Transportation Research Record*, 1521, 12–19.

Highlights of the Transit Capacity and Quality of Service Manual. (1999). TCRP Research Results Digest, 35, Project A-15, Publication of the Transportation Research Board. Transportation Research Board.

Horowitz, A.J. (1981). Service-sensitive indicators for short-term bus-route planning. *Transportation Research Record*, 798, 36–39.

Ibeas, A., Dell'Olio, L., Alonso, B., and Sainz, O. (2010). Optimizing bus stop spacing in urban areas. *Transportation Research*, 46E, 446–458.

Lai, W.T. and Chen, C.F. (2011). Behavioral intentions of public transit passengers—The roles of service quality, perceived value, satisfaction and involvement. *Transport Policy*, 18(2), 318–325.

Li, H. and Jiao, Y-C. (2008). A hybrid evolutionary algorithm for mixed-integer nonlinear bilevel programming problems. In *Proceedings of the Second International Conference on Genetic and Evolutionary Computing*, WGEC (Du, Y. and Zhao, M., eds.), pp. 549–553, IEEE Computer Society, Washington, DC.

Madanat, S.M., Cassidy M.J., and Ibrahim, W.H.W. (1994). Methodology for determining level of service. *Transportation Research Record*, 1457, 59–62.

Minieka, E. (1970). The *m*-center problem. *SIAM Review*, 12, 138–139.

Minieka, E. (1978). *Optimization Algorithms for Networks and Graphs*, Marcel Dekker, New York, NY.

Murugesan, R. and Rama Moorthy, N.A. (1998). Level of public transport service evaluation: A fuzzy set approach. *Journal of Advanced Transportation*, 32(2), 216–240.

Polus, A. and Shefer, D. (1984). Evaluation of a public transportation level of service concept. *Journal of Advanced Transportation*, 18(2), 135–144.

Prioni, P. and Hensher, D. (2000). Measuring service quality in scheduled bus services. *Journal of Public Transportation*, 3(2), 51–74.

Rodríguez, D.A. and Joo, J. (2004). The relationship between non-motorized mode choice and the local physical environment. *Transportation Research Part D*, 9, 151–173.

Rubin, J. (1973). A technique for the solution of massive set covering problems, with application to airline crew scheduling. *Transportation Science*, 7, 34–48.

Ryus, P., Ausman, J., Teaf, D., Cooper, M., and Knoblauch, M. (2000). Development of Florida's transit level-of-service indicator. *Transportation Research Record*, 1731, 123–129.

Terrier, K.A.P. and Schutz, Y. (2010). Can accelerometry accurately predict the energy cost of uphill/downhill walking? *Ergonomics*, 44(1), 48–62.

Transport Cooperative Research Program (TCRP). (2013). System-specific spare bus ratios update—Appendix D, Synthesis 109, Transportation Research Board, Washington, DC.

TranSystems Corp., Planner Coll., Inc., and Crikelair, T. Assoc. (2006). Elements needed to create high ridership transit systems: Interim guidebook. TCRP Report 32, Transportation Research Board, Washington, DC.

Tyrinopoulos, Y. and Aifadopoulou, G. (2008). A complete methodology for the quality control of passenger services in the public transport business. *European Transport*, 38, 1–16.

Yang, X. and Wang, W. (2001). GIS-based Fuzzy C-means clustering analysis of urban public transit network service: The Nanjing city case study. *Road and Transport Research, ARRB Transport Research*, 10(2), 56–65.

Washington Metropolitan Area Transit Authority (WMATA). (2014). Recommended Reviews. In: http://www.yelp.com/biz/washington-metropolitan-area-transit-authority-washington. (Accessed June 18, 2014).

Chapter 13

Coordination and connectivity

Measures and analysis

A large umbrella
is for heavy rain;
connected vessels
are for heavy
passenger loads

A.C.

CHAPTER OUTLINE

PRACTITIONER'S CORNER

With changes in lifestyles, trip-making behavior has also changed. Commuters nowadays make multiple stops between home and work. This has resulted in an increased demand for more complex trip chains. In order to be competitive with private vehicles, transit operators need to provide services that can handle such complex trip chains. Well-connected routes offer commuters to access the complete network. Commuters are not confined to using direct routes but are able to make transfers efficiently to go where they need to go. It allows them to take control of their journey and their own destinations instead of having to choose predetermined destinations by transit operators. This is similar to what we observe in transit-metro systems.

To assess the connectivity of routes, we need a tool. This chapter outlines the measures that can be used to assess the attractiveness of a route or multiple routes. The versatility of the connectivity tool has been demonstrated by studies on comparison of routes and networks.

The chapter consists of four main parts, following an introductory section. Section 13.2 shows how to measure transit connectivity based on distinguished quantitative and qualitative attributes including the interpretation, significance, and application of each of the measures. Section 13.3 describes directions and tools for detecting weak segments in interroute and intermodal transit chains (paths), for possible revisions/changes. Section 13.4 provides a method, based on the max-flow concept in networks, to determine weak routes in the transit networks through an analysis to detect the bottlenecks existing in current transit networks. Section 13.5 presents an application of the transit connectivity measures in the area of social equity. The chapter ends with a literature review and exercises.

Practitioners are basically encouraged to visit thoroughly all the sections. The writer James Collins said: "an integrated life is one where you're able to fit the different pieces of your life together in seamless fashion." Same will be with an integrated transit trip: just to move seamlessly. This chapter interprets *seamlessly*.

13.1 INTRODUCTION

Service-design criteria always contain postulates to improve routing and scheduling coordination (intra- and interagency transfer centers/points and synchronized/timed transfers). Ostensibly, the lack of well-defined connectivity measures precludes the weighing and quantifying of the result of any coordination effort. This section provides an initial methodological framework and concepts for (1) quantifying transit connectivity measures and (2) directions and tools for detecting weak segments in interroute and intermodal chains (paths) for possible revisions/changes.

The importance of service-connectivity elements may be better comprehended after reading the following selections of a newspaper article, written by its editorial writer. Although the writer colors his personal experience with metaphors, it is hard to argue with the basic facts of the article.

13.1.1 Transit connectivity frustration

From the *Los Angeles Times*, February 5, 2006 ("Taking the rapid out of transit" by Dan Turner):

> Like many epic journeys of exploration, mine began not out of necessity but out of curiosity—the ancestral human urge to test the boundaries of endurance and knowledge.

My quest: to get from my house in the Hollywood Hills to LAX, using only public transportation. "I had not anticipated that the work would present any great difficulties," said Sir Ernest Shackleton after surviving his harrowing, failed attempt to reach the South Pole in 1915, his icebound ship by that time at the bottom of the sea ...

Trains ferry passengers in and out of most big airports across the country, including Atlanta's Hartsfield, Chicago's O'Hare and even San Francisco International. But not at Los Angeles International Airport. It is the fifth-busiest airport in the world, with more than 60 million passengers a year, and more people start their flights there than anywhere else—yet it is not served by any rail line. Like reaching the Pole, getting to the airport using only public transit is a feat requiring courage, fortitude and very bad judgment ...

Twenty-four minutes later the bus arrives. Knowingly, I put $1.25 in the slot and take a seat. Leaving the bus at Hollywood Boulevard, I ask the driver for a transfer. He fixes me with a fishy stare ... buses do not issue transfers. You have to buy a day pass, which is $3. I hold out a $5 bill. The driver looks at it as if it's a used tissue. He does not give change. So begins the 1.5-mile trek to the Hollywood and Highland Red Line station, with not a sled dog or Sherpa to lead the way. Yes, I could take another bus, but I'm still steamed about the day-pass snub ...

At least I didn't have to contend with bus drivers anymore. Riding the escalator into the bowels of Hollywood, I enter the Mercedes of L.A. public transit, the $4.5-billion Red Line subway. The 17.4-mile system is fast, semi-clean, quiet—a wonder of efficiency with nearly 120,000 boardings a day. It would attract many thousands more if only it went somewhere. Originally planned to run all the way down Wilshire Boulevard, the city's densest corridor, it instead ends with a whimper at Wilshire and Western Avenue, its spine hacked off by community opposition and weak-kneed politicians ...

From here it's 17 minutes to the 7th Street station in downtown L.A., where I transfer to the Blue Line. Eight minutes later, I'm flashing through downtown at 25 mph ...

Several days later—or maybe it's 24 minutes—I'm at the Imperial/Wilmington station in Lynwood, prepared to transfer to the Green Line. Thirteen minutes later, I'm on the train heading toward LAX. At last I can see it up ahead—the LAX/Aviation Boulevard station. But my adventure isn't over.

The Green Line from Norwalk was originally planned to end inside the airport, but in 1995, after the money ran short, so did the line. An $11-billion plan to remodel LAX, approved in 2004, called for a people mover that would carry passengers to the terminals from a big transportation center connected to the Green Line, but when most of the plan was recently scrapped to settle a lawsuit with airport neighbors, so was the people mover. Instead, there is a shuttle bus from the Aviation station ...

My expedition from home to LAX takes two hours and 47 minutes, yet I am flushed with the thrill of accomplishment when the shuttle finally arrives. I am footloose and free, untied to a vehicle in a long-term parking lot. Records are sketchy, but I'm confident that no one else from the Hollywood Hills has ever attempted this journey. After all, they could drive or take a cab to LAX in about 40 minutes. Unlike Shackleton, I have reached my Pole.

Along the line of the article's metaphors, we can conclude that when it is dark enough, one can see the stars; when the transit service is poor, the transit agency can see how good it can be and proceed to attain this objective. The need to transfer between routes generates a major cause of discomfort for transit users. Improving public transit connectivity is one of the most vital tasks in transit operations planning. A poor connection can cause some passengers to stop using the transit service.

13.2 CONNECTIVITY MEASURES

Before proceeding with the main themes of this section, let us review a perceived concept of transit connectivity. One possible definition of a prudent, well-connected transit path is this: *an advanced, attractive transit system that operates reliably and relatively rapidly, with smooth (ease of) synchronized transfers, part of the door-to-door passenger chain.*
Interpretation of each component is defined as follows:

Attractiveness: Available information (by telephone, Internet, newspaper, radio, TV, mail leaflets), simple communication (abbreviated telephone number, automatic storage of users' telephone number and address), clear service meeting-point characteristics (clear stop sign, vehicle color and logo), boarding/alighting/riding comfort (low floor, extra space next to driver, comfortable seats, features for physically challenged people, low noise), onboard service (newspapers, magazines, free coffee/tea, TV/video display of timetable, weather, etc.), simple payment (electronic ticketing, prepaid, transfers, smart-card ticketing).

Reliability: Small variance of measures of concern to passengers (total travel time, waiting time, in-vehicle time, seat availability), small variance of measures of concern to transit vehicles (schedule adherence, headways, on-time pullouts, missed trips, breakdowns, load counts, late reports), small variance of measures related to pretrip information using telephone communication (online timetable, travel time to caller); for more on reliability features, see Chapter 18.

Rapidly: Easy access/egress and comfortable stops (fixed stops with shelters and information, transit vehicle bays at time points, with an extra approach lane at signalized intersection), transit vehicle preference at unsignalized intersections (*yield* or *stop* not according to traffic procedures, special bypass arrangements at strategic points), transit vehicle preference at signalized intersections passive priority by extending or shortening green, active priority using actuated transit vehicle signals (e.g., radio, inductive loop), purchase and validation of tickets (electronically, ordinary) on transit vehicles (one way, round trip, transfer, daily, weekly, monthly).

Smoothness (ease): Comfortable routing (maximum criterion for walking distance, round-trip deviation from designated route in bad weather, evolution of flexible routing and scheduling), special train-station entrance (transit vehicle special gate, passengers' special entrance door with comfortable stairs/escalator to/from the train platform), special train exit (exit door next to the train platform for transit vehicle ticket holders, transit vehicle waiting at exit or under shelter with vehicle-arrival announcement on variable message signs [VMS]).

Synchronized: Online communication between all modes of transit vehicles (vehicle equipped with arrival information for the relevant stations and time difference, positive or negative, for synchronization), passenger subscriptions with serial numbers (adding a variable scheduling element to suit subscribers, planning the fixed-scheduling component with subscriber information), short-turn and shortcut routing strategies (computerized suggestions for the transit vehicle driver on short turn and shortcut, VMS onboard information on meeting time with another or the same transit mode); for more on routing strategies and timed transfers, refer to Chapters 17 and 20, respectively.

13.2.1 Developing connectivity measures

The common denominator for all transit services are the following quality-of-connectivity attributes:

e_1 = average walking time (for a connection)
e_2 = variance of walking time
e_3 = average waiting time (for a connection)
e_4 = variance of waiting time
e_5 = average travel time (on a given transit mode and path)
e_6 = variance of travel time
e_7 = average scheduled headway
e_8 = variance of scheduled headway

These eight attributes, which can be measured, will be termed *quantitative attributes*. However, there are other important attributes that cannot easily be quantified and measured. Three of these are

e_9 = smoothness (ease) of transfer (on a given discrete scale)
e_{10} = availability of easy-to-observe and easy-to-use information channels (on a given discrete scale)
e_{11} = overall intra- and interagency connectivity satisfaction (on a given discrete scale)

These hard-to-quantify attributes will be termed *qualitative attributes*. Nonetheless, it is true that the value of all 11 attributes may be perceived differently by different passengers or even by the same passenger in different situations. These different perceptions are captured in the average weighting of each attribute. The weight of each attribute is survey based and/or based on the results of a mode (path)-choice model. The analysis framework proposed in this section distinguishes between quantitative and qualitative attributes, though both can be combined with some agreeable weighting factors. The reason for this separation is to make it easy for decision makers to evaluate improvements and changes in the transit connectivity chains.

As noted earlier, measuring transit connectivity involves various parameters and components. Therefore, the following notations are introduced in order to ease the explicit construction of connectivity measures:

Notations
For a given time window (e.g., peak hour, average weekday),

O = (Koichi and Kambayashi) is the set of origins O_i
D = $\{D_u\}$ is the set of destinations D_u
P_{D_k} = $\{P\}$ is the set of interroute and intermodal paths to D_k
P_{O_k} = $\{P_i\}$ is the set of interroute and intermodal paths from O_k
M_p = $\{m\}$ is the set of transit routes and modes included in path p
E_t = $\{e^t\}$ is the set of quantitative attributes suitable for connectivity measures
E_ℓ = $\{e^\ell\}$ is the set of qualitative attributes suitable for connectivity measures
e^j_{mp} is the value of attribute e^j, j = t, ℓ, related to mode m on path p
α_e is the weight/coefficient for each attribute e^j, j = t, ℓ
c^j_p is the quantitative and qualitative (j = t, ℓ) connectivity measure of path p
F_p is the average number of passengers using path p
$c_p(i, j)$ is the capacity (flow of passengers) of arc (i, j) between route and mode i and between route and mode j; each i can also be an origin O_i or destination D_i; (i, j) is contained in path p and is part of a network-flow model

Based on the notations earlier, the following equation-based notations are established:

$$c_p^j = \sum_{m \in M_p} \sum_{e^j \in E_j} \alpha_e e_{mp}^j, \quad j = t, \ell \tag{13.1}$$

$$C_{D_k}^j = \sum_{p \in P_{D_k}} c_p^j, \quad j = t, \ell \tag{13.2}$$

$$C_{O_k}^j = \sum_{p \in P_{O_k}} c_p^j, \quad j = t, \ell \tag{13.3}$$

$$C_D^j = \sum_{D_k \in D} c_{D_k}^j, \quad j = t, \ell \tag{13.4}$$

$$C_O^j = \sum_{O_k \in O} c_{O_k}^j, \quad j = t, \ell \tag{13.5}$$

$$c_p^{jF} = c_p^j \cdot F_p, \quad j = t, \ell \tag{13.6}$$

$$C_{D_k}^{jF} = \sum_{p \in P_{D_k}} c_p^{jF}, \quad j = t, \ell \tag{13.7}$$

$$C_{O_k}^{jF} = \sum_{p \in P_{O_k}} c_p^{jF}, \quad j = t, \ell \tag{13.8}$$

$$C_D^F = \sum_{D_k \in D} C_{D_k}^{jF}, \quad j = t, \ell \tag{13.9}$$

$$C_O^F = \sum_{O_k \in O} C_{O_k}^{jF}, \quad j = t, \ell \tag{13.10}$$

Note that the required weight/coefficient α_e in Equation 13.1 for measuring the level/quality/goodness of connectivity in Equations 13.1 through 13.10 can be estimated by the results of both passenger surveys and the path-/mode-choice model (some of which may need to be constructed, being site specific).

The interpretation, significance, and application of each of the proposed connectivity measures appear in Figure 13.1. This figure also presents graphical examples of the relevant intermodal path for each measure and, under the applications, the purpose of measure.

The purpose of the first connectivity measure in Figure 13.1, Equation 13.1, is to compare paths, that is, chains of trips, each from an origin to a destination, including transfers. Usually, this comparison takes place following an improvement or change in one or more paths. Otherwise, this comparison, using the evaluation tool proposed, categorizes the different paths by their access/egress connectivity quality. The purpose of the second connectivity measure, Equation 13.2, is to compare destinations. The purposes of the third and

Connectivity measure	Interpretation	Significance	Graphical example	Application
$c_p^j = \sum_{m \in M_p} \sum_{e^j \in E_j} \alpha_e^* \, e_{mp}^j, \; j = t, \ell$	Sum of all connectivity-component measures along a given path k	Overall connectivity value or quality for a given path k	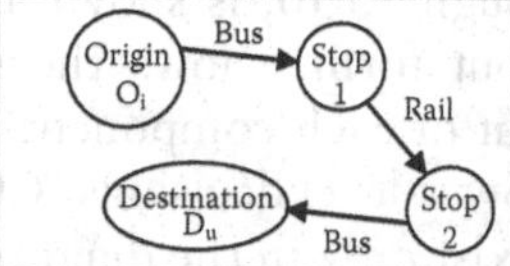	Evaluation of path k access-connectivity value. *Purpose*: comparison among paths
$c_{D_k}^j = \sum_{p \in P_{D_k}} C_p^j, \; j = t, \ell$	Sum of all connectivity-component measures along all access paths to a given destination D_u	Overall connectivity values for all paths related to D_u	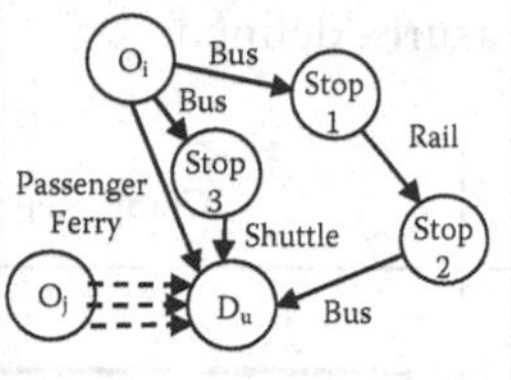	Evaluation of destination D_u access-connectivity value considering only existing paths (for new paths, C_{FD}^j are applicable). *Purpose*: comparison among destinations
$c_p^{jF} = c_p^j \cdot F_p, \; j = t, \ell$	Sum of all connectivity-component measures along a given path k, weighted by the average amount of passengers using this path	Overall exposure-connectivity measure (for all passengers) for a given path k	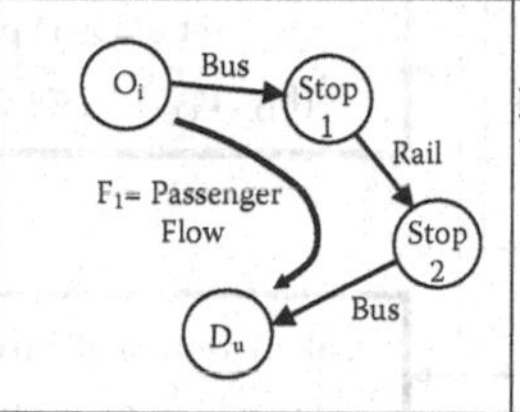	Evaluation of path k people-access-connectivity value (considering amount of passengers exposed). *Purpose*: comparison among paths, considering passenger flow
$C_{D_k}^{jF} = \sum_{p \in P_{D_k}} c_p^{jF}, \; j = t, \ell$	Sum of all connectivity-component measures along all access paths to D_u, weighted by the average amount of passengers using these paths	Overall exposure-connectivity measure for all paths related to terminal Du	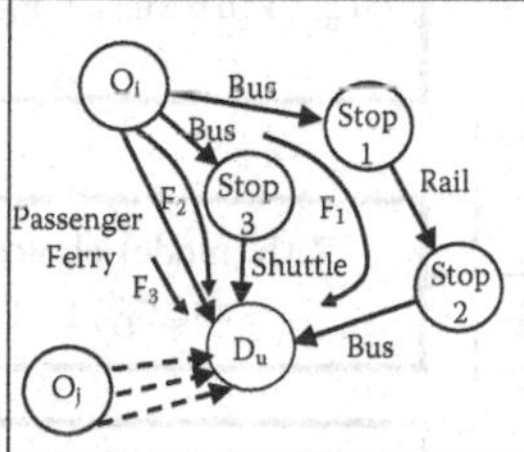	Evaluation of terminal D_u people-access-connectivity value, considering exiting and new paths. *Purpose*: comparison among destinations, considering passenger flow
$C_D^F = \sum_{D_k \in D} C_{D_k}^{jF}, \; j = t, \ell$	Sum of all connectivity-component measures along all access paths to a set of destinations, weighted by the average amount of passengers using these paths	Overall exposure-connectivity measure for all paths in a given set of destinations	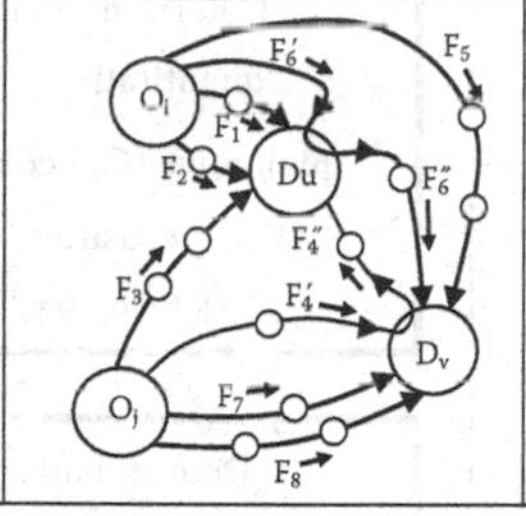	Evaluation of a set of destinations, considering existing and new paths. *Purpose*: comparison among destinations in a given set (e.g., zone-based, purpose-based)

Figure 13.1 Interpretation, significance, and application of quantitative O-D connectivity measures* at different levels and for a given time period.

fourth measures, Equations 13.6 and 13.7, are the same as for the first and second measures, respectively, but with the consideration of passenger flow, that is, a determination of the average number of passengers exposed to the calculated level of connectivity. The purpose of the fifth measure in Figure 13.1, Equation 13.9, is to compare groups of destinations in regard to overall existing connectivity quality.

All connectivity measures that consider passenger flows should be updated, following routing and/or scheduling and/or service improvements or changes. When referring to a group of destinations (zonal based, purpose based) in the last measure in Figure 13.1, paths can have a stop at one destination and continue to others. For example, in the last-row graphical representation in Figure 13.1, the access path with F_4' and F_4'' flows has a stop at D_v for F_4' flow between O_j and D_v and for F_4'' flow between D_v and D_u.

---

* These measures also apply to qualitative attributes.

The process for the determination of transit connectivity measures, using the notations in Equations 13.1 through 13.10, is shown in Figure 13.2. This figure, constructed by an input–component–output form, shows the systematic decision sequence and process of the analysis. The output of each component positioned higher in the sequence becomes an important input into lower-level decisions. Clearly, the independence and orderliness of the separate components exist only in the diagram. In what follows are two case studies utilizing the connectivity measures defined.

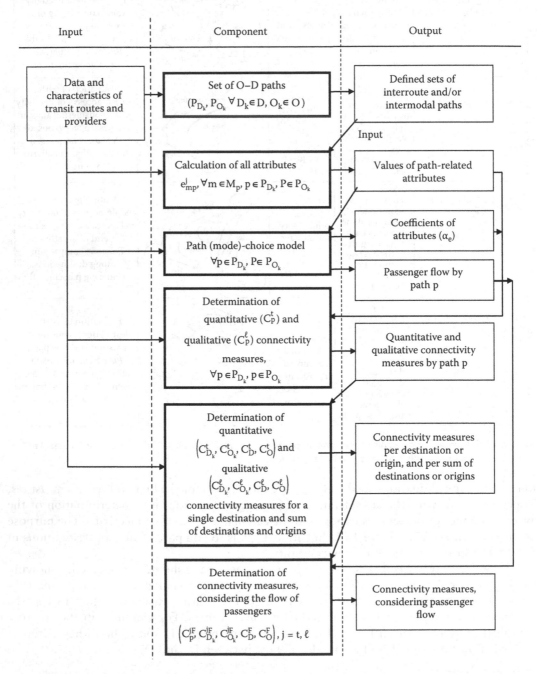

Figure 13.2 Analysis framework for the determination of connectivity measures.

13.2.2 Case study I: Measuring transit connectivity performance of a city

This first case study by Ceder et al. (2009) provides a methodological framework and concepts for quantifying transit connectivity measures and uses these new measures in a case study of Auckland's public transit network for possible revisions/changes. In February 2008, a busway was opened in the North Shore area of Auckland, New Zealand. This busway goes from Albany to Britomart in the central business district (CBD), through four new stations in the North Shore area. Because of a dedicated road and comfortable and frequent buses, it improves journey speeds and reliability and provides an attractive alternative to private vehicle use.

For this busway project, local and express bus routes were designed to link into the busway through five new stations, so that people can easily transfer and join the CBD from everywhere in the North Shore. Figure 13.3 illustrates the North Shore bus network with all the routes (A_{ij}), destinations (D_i), origins (O_i), and hubs (H_i) considered. Transferable bus tickets were introduced, allowing passengers to change buses using only one ticket. In addition, two park-and-ride stations were built at the Albany and Constellation stations, next to the motorway. People can park their car the whole day for free and travel to and from their destination on the bus. All these improvements make transfers easier and enhance people's willingness to use transit. In this section, the transit connectivity measures refer to morning peak-hour period (7–9 a.m.) and to the flows of passengers to the CBD (Britomart station) and Takapuna station—the two main employment areas of Auckland.

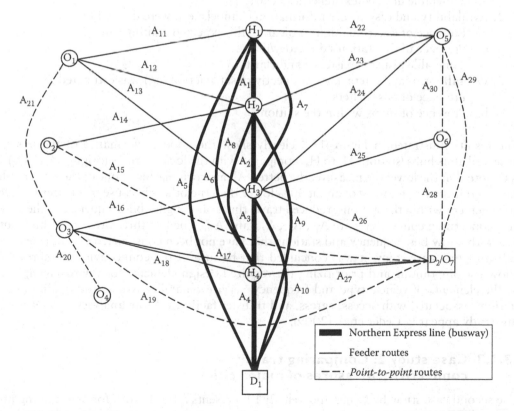

Figure 13.3 The selected bus network in Case study 1.

The computation of the attributes, of the network in Figure 13.3, was performed (Ceder et al., 2009) for each transit-network arc (part of a path) as follows:

1. The average scheduled headway and its variance were derived by the bus timetables during the morning peak hour and actual observations.
2. The average waiting time was calculated by Equation 11.1, namely,

$$\bar{W} = \bar{H} \cdot \frac{1}{2} \cdot \left[1 + \frac{var(H)}{\bar{H}^2} \right] \quad \text{H: scheduled headway.}$$

3. The average travel time, its variance, and the average number of passengers per arc were estimated by the data collected of a survey and some data given by the bus companies.
4. The average walking time and its variance were estimated according to the configuration of the stations and the distances between stops.

The qualitative attributes were also included in this study. Qualitative attributes are more of passenger-perception attributes; thus, it was decided to break down each of the attributes and specify their related elements in the passenger survey as follows:

1. Smoothness (ease) of transfer (e_9) was estimated by
   a. Distance between the stops (meters)
   b. Presence of stairs, sloping ground, etc. (scale of 1–5)
   c. Length of the transfer (minutes)
   d. Timetable display (existence and clarity)
2. Availability and easy-to-use information channels (e_{10}) were derived by
   a. Presence of electronic signs indicating the expected waiting time
   b. Presence of the station's directory/map
   c. Timetable display (existence and clarity)
3. Overall intra- and interagency connectivity satisfaction (e_{11}) was estimated by
   a. Presence of bus shelters
   b. Presence of shops within the station

The results illustrated in Figure 13.4 clearly show the good performance of the busway observed at its hubs (stations H_1 to H_4) compared with the local stations (origins O_1 to O_6); it is to note that the lower the measure, the better. All arcs originating at one of these four hubs have a connectivity measure of about 30% lower than the others because of a better transfer arrangement within the stations, reduced travel times, and higher bus frequencies. The weak arcs concerning connectivity are A_{13}, A_{28}, A_{29}, and A_{30}. The last three arcs are of bus route 858 with a low bus frequency and stations that have not been constructed for transfers.

In this first case study, it was concluded that the use of the connectivity measures will allow for pinpointing and prioritizing the required design elements for improvement, that is, the elements of vehicle type and frequency, service-reliability issues, and quality of the stations associated with access, egress, and transfer facilities. More analyses and results of this study appear in Ceder et al. (2009).

13.2.3 Case study 2: Comparing transit connectivity measures of major cities

The second case study by Ceder and Teh (2010) presents a framework for determining and comparing between the transit connectivity measures of the CBDs of Christchurch City,

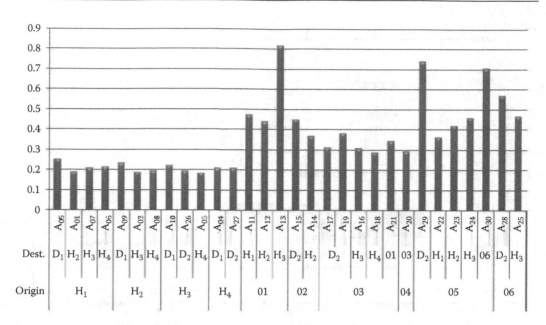

Figure 13.4 Arc connectivity across Auckland's North Shore routes.

Wellington City, and Auckland City in New Zealand. The morning time period 7–9 a.m. was selected. According to statistics New Zealand's 2008 census population data, the population of Christchurch City CBD was 369,000 compared to Wellington CBD's population of 192,800 and Auckland CBD's population of 438,100. The study areas for these three CBDs are defined based on the relative population of these cities. Auckland City's study area is the largest among the three cities at 6.6 km², followed by 5.7 km² study area at Christchurch City. Wellington City has the smallest study area of roughly 2.8 km² because of its smaller population.

The network nodes selected are existing key timing points or key stops as indicated in the public timetable. A directional arc was inserted into the bus network map where there is an existing bus service operating. Among these network nodes, a list of origins/destinations, which are the most frequently visited nodes, was determined. The arc connectivity values were determined after establishing the bus network map. Because of not conducting actual survey in the three study areas, the base data consist of only the existing bus frequencies and timetables and of available data and information provided by the local transport authorities. Connectivity values of paths were calculated by summing up connectivity values of individual arcs to form the path from a defined origin to a defined destination. A total of 115 paths have been selected for Auckland City, followed by 105 paths in Christchurch City. Wellington City has the smallest sample size of 96 paths, which coincides with the relatively smaller study area and population.

The comparison of path connectivity, shown in Figure 13.5, indicates that Auckland provides the best path connectivity by having the lowest median and interquartile values and the smallest range of data. Having the lowest interquartile value means that 50% of the paths selected for Auckland has smaller variance and hence lesser uncertainties. It is further noted that more than 75% of Auckland's selected path has a connectivity value of less than Wellington's 25th percentile connectivity value. Further to the comparison of origins/destinations, it is desirable to compare the connectivity between central transit terminals because of its importance to provide good connectivity for the public to enhance future patronage. The qualitative attributes, of this second case study, were excluded from the calculation.

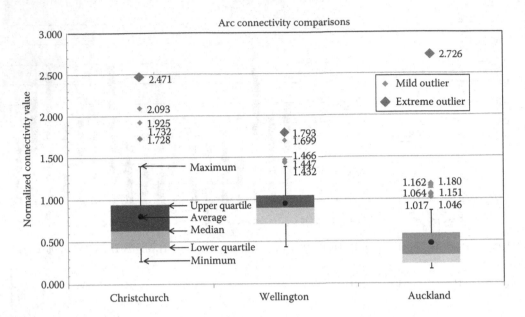

Figure 13.5 Arc connectivity ranges of three cities.

   The central transit terminals in both Christchurch and Auckland are located on the main street. However, Christchurch's terminal is located approximately 500 m south of the central location, while Auckland's terminal is located at the northern end of the main street. Wellington's central transit terminal is not located along the main street but at the northwestern end of the existing bus network. Analysis of these three terminals shows that a bus terminal should be located at a more centralized location along the main street to provide the best connectivity. This should also be supplemented by providing frequent (lesser headway) and more reliable (lesser variance in headways and travel times) transit services along the main street.

   The conclusions of this second case study were that Auckland City's transit outperforms both Christchurch and Wellington cities with lower connectivity measure overall. The location of the existing central city transit terminal in all three cities should be improved to provide better transit connectivity as both an origin and a destination. In general, more centrally located transit terminal and more paths between each origin–destination (O-D) pair could improve the transit connectivity. More analyses and results of this study appear in Ceder and Teh (2010).

13.3 DETECTING WEAKNESSES ON INTERROUTE AND INTERMODAL PATHS

An important aspect of a tool for transit connectivity improvements is the ability to detect current or anticipated weaknesses in interroute and/or intermodal chains/paths. Figure 13.6 portrays a possible process identifying these weaknesses, based on the defined notations. Whereas the first two lists (output in Figure 13.6) do not consider passenger flow, the third and fourth lists do take this flow into account. The fourth list attempts to identify passenger-flow bottlenecks at the network level. The process of attaining the fourth list is explicated in the next section.

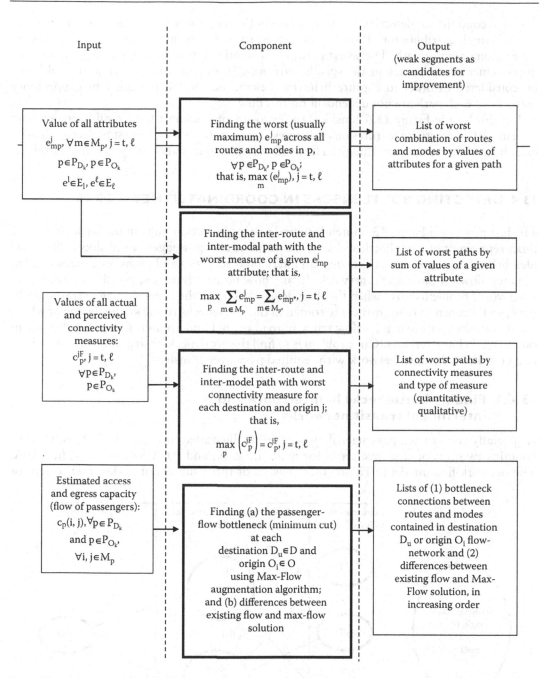

Figure 13.6 Analysis framework for detecting weak segments on interroute and intermodal paths.

The first list of weak segments in Figure 13.6 is constructed by means of the value of each attribute across all routes and modes on a given path. That is, the worst route (operated by a certain mode) for each attribute on a given path is identified and can automatically be considered a candidate for improvement. If improvement cannot be made, the second worst route is examined, and so on. This process does not consider/weigh the amount of passenger flows on each path. It simply identifies weak segments (owing to their inherent features), regardless of how many passengers use them.

The second list for detecting weak segments in Figure 13.6 considers the sum of the value of each single attribute for all routes on a given path, for example, the sum of all waiting times along a given path. The worst path, p*, is identified and can serve as a candidate for improvements or changes in the specific attribute. In practice, passenger flow could also be considered in order to capture behavioral elements, though the inherent connectivity features of each path are not dependent on this flow.

The third list in Figure 13.6 considers the exposure of passenger flow to all interroute and intermodal connectivity features on each access/egress path. A list of worst paths can then be established and treated accordingly. The next section explains and interprets this third list.

13.4 DETECTING BOTTLENECKS IN COORDINATED NETWORKS

The last process in Figure 13.6 intends to find weak connection links in the sense that those links represent potential bottlenecks for passenger-access/passenger-egress flows. The basic idea behind this approach is to review all intermodal activities and paths on a specially constructed flow (of passengers) network. In any flow process (vehicles, people, or freight) on a network, problems arise when the flow demand exceeds the capacity. These traffic problems, well known in road traffic, are rooted in the bottlenecks of the system, some of which are dynamic in nature (e.g., a slow truck in road traffic) and are not necessarily dependent on physical characteristics. Our problem is to find the bottleneck obstructing passenger flow in a coordinated transit network with required passenger transfers.

13.4.1 Finding bottlenecks in interroute and intermodal transit networks

A specially constructed passenger-flow network is illustrated in Figure 13.7. This network contains two sets of elements: set N for nodes i, i ∈ N, and set A for arcs (i, j), (i, j) ∈ A. The network-flow model in Figure 13.7 consists of three subsets of nodes: first, origin or

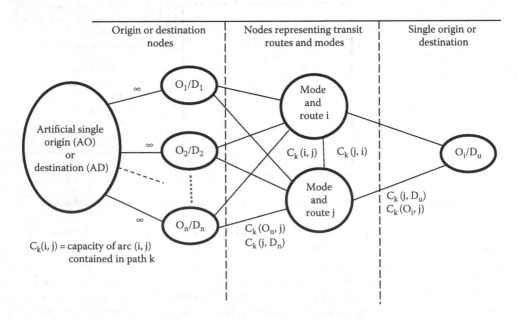

Figure 13.7 Conceptual O-D connectivity network-flow model.

destination nodes; second, nodes related to transit routes and modes; and third, a node related to a single origin or destination. Each arc of this network has a capacity value for either access or egress flow; the arc is associated with a certain interroute and intermodal path. In the first subset of nodes, there is a source (origin) or sink (destination) node connected artificially to all the O-D nodes. The process is to load this special network with either access or egress passenger flows while having a capacity constraint on each arc.

Passenger-access flows originate in the artificial node AO in Figure 13.7 and move to reach the destination D_u; passenger-egress flows move from AD to O_i. Finding bottlenecks requires maximizing the flow created at AO/AD and absorbed in D_u/O_i. In the maximum-possible-flow solution, those arcs having a flow equal to capacity naturally form the bottlenecks of the network.

A vital element of the network-flow model is the capacity assigned to each arc, in other words, the maximum amount of passengers, for a given time period (peak hour, average day), that can traverse each arc. Generally speaking, there are a finite number of arcs in an actual transit network of routes and modes. For each path, there is a need to search the capacity (passengers/hour) that can be offered by each route and mode, for instance, the maximum feasible rail/bus/passenger ferry frequency times the vehicle's passenger capacity.

The solution for the maximum-flow (max-flow) problem corresponding to the network-flow model in Figure 13.7 appears in Appendix 6.A; it is based on the max-flow augmentation algorithm. An example of passenger-flow transit network is shown in Figure 13.8a. This is an example of access intermodal paths from a given origin to two destinations, D_1 and D_2. There are five transit modes, and the capacities (passengers/hour) are the numbers on the arcs. Based on the max-flow min-cut theorem in Appendix 6.A, the bottlenecks of the network-flow model are the arcs along the minimum cut. That is, the solution is a maximum produced flow of 1500 passengers/h. The max-flow augmentation algorithm is based on the seven following paths (in each, more pass/h flow is added):

1. Origin—bus: D_1–AD,   Augmented (aug.) flow = 400
2. Origin—bus: D_2–AD,   Aug. flow = 200
3. Origin—bus: passenger ferry—D_2–AD,   Aug. flow = 100
4. Origin—shuttle bus–rail: D_1–AD,   Aug. flow = 300
5. Origin—shuttle bus–cable car–rail: D_2–AD,   Aug. flow = 250
6. Origin—shuttle bus–cable car–passenger ferry: D_1–AD, Aug. flow = 150
7. Origin—shuttle bus–cable car–passenger ferry: D_2–AD, Aug. flow = 100

Total flow = 1500

The last augmented network flow, based on the procedure in Appendix 6.A, appears in Figure 13.8b. The sum of augmented flows on (i, j) in Figure 13.8b is represented by a reverse arc (j, i). The difference in flow between the capacity on (i, j) and this sum of flows is represented by a forward arc. Only those arcs with a flow equal to the capacity do not have forward arcs in Figure 13.8b. Those arcs that are emphasized are the bottleneck segments.

The following is an additional procedure for the best exploitation of the results of the max-flow algorithm:

1. Solve the max-flow algorithm for maximum capacity ranges.
2. Find differences between the max-flow solution(s) and existing flow (representing existing demand).
3. Check for any current bottleneck, in which existing demand is higher than the optimal max-flow solution; if such exists, suggest immediate improvement.

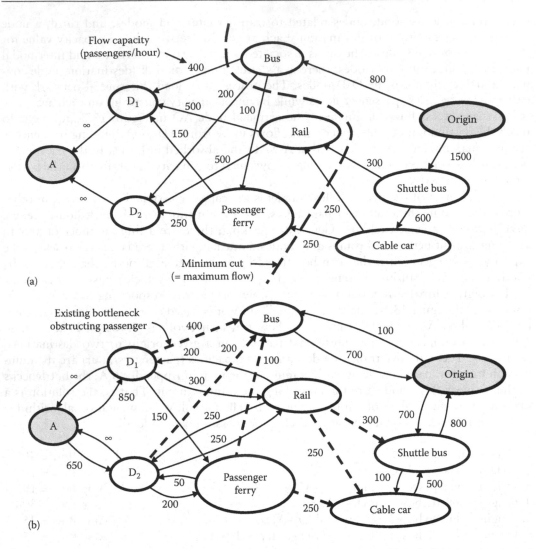

Figure 13.8 Example of the network-flow model and its minimum cut in part (a) and of the last augmented network flow in part (b).

4. Examine weak segments in the existing interroute and intermodal paths (see Figure 13.7).
5. Propose service modifications for actual and perceived attributes.
6. Update relevant attributes using path (mode)-choice model.
7. If no changes are made, continue; otherwise, go to 1.
8. Set final conclusions.

Practically speaking, it is recommended that the following data be collected from each relevant transit agency. These data constitute the base for capacity calculation, estimation of attributes, and the creation of interroute and intermodal connectivity proposals:

1. *Frequencies*—existing, minimum (estimated), maximum (estimated)
2. *Vehicle sizes*—existing (number of seats, allowed standees); which (and when, if not continuously) sizes are used

Table 13.1 Access/egress transit modes and paths to airport terminal

Path	Access/egress mode	Path description	Notes
k_1	Train service	Auto–*wait*–train–*wait*–shuttle–terminal	
k_2	Public transit bus	Walk–*wait*[a]–public bus–terminal	
k_3	Scheduled airport bus	Auto–*wait*–airport bus–terminal	
k_4	Shared-ride door-to-door van[b]	Van–terminal	No waiting, but not a straight ride
k_5	Charter bus	Auto–*wait*–charter bus–terminal	Special ride

Note: Private vehicle access/egress to and from train, airport bus, and charter bus is assumed; passengers are assumed to walk to and from a public bus.

[a]   There is an additional wait for each transfer between two buses.
[b]   Additional distance and time for passengers using van, by time of day.

3. *Passenger load*—by time of day
4. *Average travel time*—between each two nodes connected by time of day
5. *Routing*—existing; flexibility for routing adjustments
6. *Routing strategies*—existing; flexibility for adding strategies (e.g., skip stop, short turn, shortcut)

A good example of transit connectivity paths is an airport terminal. Table 13.1 demonstrates access and egress paths (k_1 to k_5) of transit service to an airport terminal.

13.4.2 Case study of two world sport events

This section is based on a study by Ceder and Perera (2014) who examined the transit networks of two known world sport events from the coordination/connectivity perspective. It is known that weak segments of the transit connectivity network will increase passenger's congestion and will result in delays and also in passengers' frustration. The objective of the two case studies was to use the aforementioned defined connectivity measures to analyze and determine the weak routes in the transit networks.

The case studies present two major cities undergoing two major sporting events: the city of London in the United Kingdom hosting the XXX Olympic Games in 2012 and the city of Auckland in New Zealand that hosted the Seventh Rugby World Cup in 2011. These cities are used to offer a comparison of the bottlenecks in their transit networks.

The steps and their corresponding data requirements for identifying weak segments of a transit network are outlined in Table 13.2.

In London, two origins were chosen, both from Central London, because most of the tourists accommodated these locations during the Olympics season. One of the origin points is the Charing Cross station and the other is the London Bridge station that is located on the other side of the Thames River. The Wembley Stadium (with a seating capacity of 90,000 spectators) was considered as the destination because this is the venue staging the London 2012 Olympic Games Football competition. Frequencies and passenger capacities of each of the transit modes selected were obtained from the information provided in Transport for London website.

In Auckland, two origins were chosen, one from Central Auckland (Britomart Transport Centre) and the other from the North Shore part of Auckland City (Devonport). These two origins reasonably represent the majority of the spectators, both locals and tourists. The Eden Park Stadium was chosen as the destination, because this was the venue for the 2011

Table 13.2 Data requirements for identifying weak segments of a transit network

Step	Data requirements for the case studies
Establish origin(s) and destination.	There must be more than one feasible PT route between the origin(s) and destination. Create, first, origin nodes; second, nodes related to transit routes and modes; third, a node related to a single destination.
Consider all possible transit routes from origin(s) to destination.	Each arc in the network is associated with a certain interroute and intermodal path. All the routes considered entail a maximum walking distance of 1 km by the passengers, from origin to the destination including the walking involved during transfers.
Obtain the frequency of each transit mode through published timetables.	The hourly frequencies of each transit mode are obtained from the timetables by calculating the mean frequency between 8:00 a.m. and 8:00 p.m. on weekdays.
Establish the passenger capacities of each transit vehicle.	For simplicity, on each PT vehicle, the full passenger capacity (seated + standees) has been used.
Assign flow capacities (passengers/hour) to each arc of the network.	This is calculated as equal to the hourly frequency of transit vehicle times the passenger capacity of the vehicle.
Identify the minimum cut (= maximum flow).	The bottlenecks of the network-flow model are the arcs along the minimum cut.

Rugby World Cup. The Eden Park Stadium is New Zealand's largest rugby stadium with a seating capacity of 60,000 spectators. Frequencies and passenger capacities of each of the transit modes selected were obtained from the information provided in Maxx Auckland Transport website.

The passenger-flow transit network for London is shown in Figure 13.9a. This demonstrates the access intermodal paths from two given origins (Charing Cross station and London Bridge station) to a single destination (Wembley Stadium). There are four transit modes, and the capacities (passengers/hour) are the numbers indicated on the

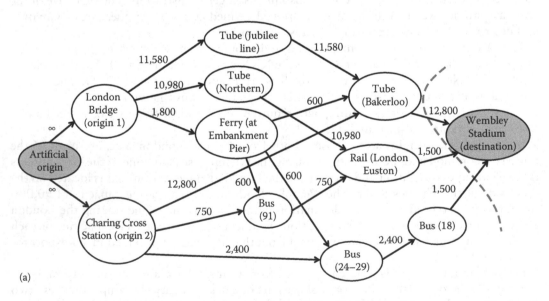

(a)

Figure 13.9 London's network with capacities and min cut (dashed line) in part (a).   (*Continued*)

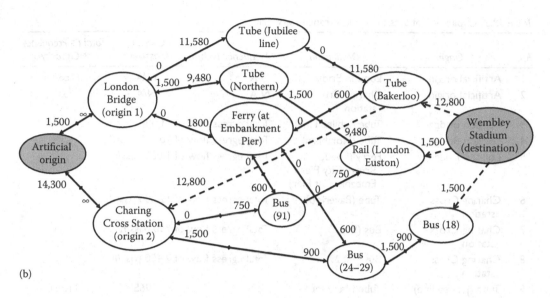

(b)

Figure 13.9 (Continued) London's network with capacities and min cut (dashed line) its augmented flows in part (b).

arcs. Table 13.3 summarizes, as an example, the capacity calculations for each arc on the London network.

The sum of augmented (i, j) flows is shown in Figure 13.9b, and it is represented by a reverse arc (j, i). The difference in flow between the capacity on (i, j) and this sum of flows is represented by a forward arc. Only those arcs with a flow equal to the capacity do not have forward arcs in Figure 13.9b. Those arcs that are emphasized with a dashed arrow are the bottleneck segments.

It is to note that Table 13.4 contains, as an example, all the possible paths from an artificial origin to the destination. However, the max-flow augmentation algorithm is based on only three of these paths as is shown in the succeeding text with an augmented flow greater than zero.

Similar to the analysis explicated earlier for the London's case study, the passenger-flow transit network for Auckland's case study and its solution are shown in Figure 13.10a and b, respectively. This demonstrates the access intermodal paths from two given origins (Britomart Transport Centre and Devonport) to a single destination (Eden Park Stadium). There are three transit modes, and the capacities (passengers/hour) are the numbers indicated on the arcs of Figure 13.10a. We note that in Figure 13.10b, like in Figure 13.9b, only arcs that are emphasized with a dashed arrow are the bottleneck segments.

It is evident from the network-flow diagrams of Figures 13.9 and 13.10 that the Auckland network is not capable of handling as many passengers as the London network. However, these two networks are similar wherein the bottlenecks of both networks lie on the arcs adjacent to the destination. Therefore, a major passenger accumulation at these specific stations can be expected, and indeed this was the case in Auckland in 2011. That is, these bottlenecks, if not addressed properly, as was in the case in Auckland, will cause major congestion problems at specific transit stations or will force the passengers to seek alternative transportation modes and paths to the destination (including some flow along walk arcs). One possible solution would entail introducing a shuttle service

Table 13.3 Capacities of arcs in London transit network

Arc	Origin	Destination	Frequency (veh/h)	Capacity (pass/veh)	Total (= Frequency × Capacity)
1	Artificial origin	London Bridge	N/A	N/A	∞
2	Artificial origin	Charing Cross station	N/A	N/A	∞
3	London Bridge	Tube (Jubilee Line)	Total egress flow of 11,580 pass/h		
4	London Bridge	Tube (Northern)	Total egress flow of 10,980 pass/h		
5	London Bridge	Ferry (London Bridge City Pier to Embankment Pier)	Total egress flow of 1,800 pass/h		
6	Charing Cross station	Tube (Bakerloo)	Total egress flow of 12,800 pass/h		
7	Charing Cross station	Bus (91)	Total egress flow of 750 pass/h		
8	Charing Cross station	Bus (24, 29)	Total egress flow of 2,400 pass/h		
9	Tube (Jubilee line)	Tube (Bakerloo)	12	965	11,580
10	Tube (Northern)	Rail (London Euston)	12	915	10,980
11	Ferry (London Bridge City Pier to Embankment Pier)	Tube (Bakerloo)	2.4	250	600
12	Ferry (London Bridge City Pier to Embankment Pier)	Bus (24, 29)	2.4	250	600
13	Ferry (London Bridge City Pier to Embankment Pier)	Bus (91)	2.4	250	600
14	Bus (91)	Rail (London Euston)	7.5	100	750
15	Bus (24, 29)	Bus (18)	24	100	2,400
16	Tube (Bakerloo)	Wembley Stadium	10	1280	12,800
17	Rail (London Euston)	Wembley Stadium	3	500	1,500
18	Bus (18)	Wembley Stadium	15	100	1,500

from these points to the respective destinations to increase the capacity of the final arc heading toward the destination.

In order to improve the capacity of the transit network (adapting the infrastructure to the event), specific plans are made for the bus, train, metro, tram, and passenger ferry services required for the event. These plans mainly consist of adapting the frequency of the service, introducing larger vehicles (if available), extending the operating time, reducing the crew rest period, closing the nearest overfilled stations, transferring stops and rerouting lines, and issuing special day tickets. The planning process attempts to ascertain the maximum capacity of the available transit networks (buses, railway lines, stations, stairways, passageways). Such a review of capacity takes account of the expected access, egress, and onboard demands. More analyses and results of this study appear in Ceder and Perera (2014).

Table 13.4 Possible origin–destination paths and their augmented flow

Route number	Route	Augmented flow (pass/h)
1	Artificial origin (AO)–London Bridge station–tube (Jubilee line)–tube (Bakerloo)–Wembley Stadium	0
2	AO–London Bridge station–tube (Northern)–rail (London Euston)–Wembley Stadium	1,500
3	AO–London Bridge station–ferry (London Bridge City Pier to Embankment Pier)–tube (Bakerloo)–Wembley Stadium	0
4	AO–London Bridge station–ferry (London Bridge City Pier to Embankment Pier)–bus (24–29)–bus (18)–Wembley Stadium	0
5	AO–London Bridge station–ferry (London Bridge City Pier to Embankment Pier)–bus (91)–rail (London Euston)–Wembley Stadium	0
6	AO–Charing Cross station–tube (Bakerloo)–Wembley Stadium	12,800
7	AO–Charing Cross station–bus (91)–rail (London Euston)–Wembley Stadium	0
8	AO–Charing Cross station–bus (24–29)–bus (18)–Wembley Stadium	1,500
	Total flow	**15,800**

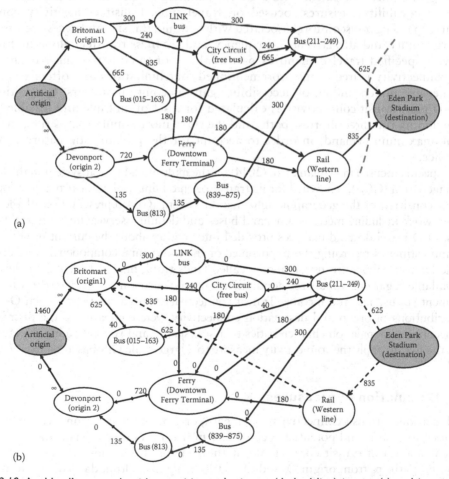

(a)

(b)

Figure 13.10 Auckland's network with capacities and min cut (dashed line) in part (a) and its augmented flows in part (b).

13.5 EQUITY IN TRANSIT PROVISION

Another application of the previously defined transit connectivity measures lands in the area of social equity with a connotation somehow to the word *public* in public transit. This section is based on a recent work by Kaplan et al. (2014) using data from Denmark.

The last decades are witnessing a slow but steady paradigm shift from planning *mass transit* to considering equity and social inclusion as an integral part of the transit planning process. While equity and social inclusion have been initially discussed with respect to fare policies, concessionary fares, and transit subsidies, the perspective has been widened to include population groups with mobility limitations. The interest in considering equity and social inclusion was first manifested during the 1990s by discussing the need to integrate equity as a policy goal in transport provision. Accessibility is broadly defined as the ability and ease of reaching activities, opportunities, services, and goods, and accessibility gaps are defined as the differences in accessibility across geographical areas, population groups, and time. These accessibility gaps serve as indicators for identifying spatial, vertical, temporal, and intergenerational inequities.

Kaplan et al. (2014) proposed to use transit connectivity as a comprehensive impedance measure for the calculation of both location-based and potential-accessibility measures that relate to equity assessment within transit planning and evaluation processes. While previous accessibility measures focused on travel time, transit connectivity considers travel time, passenger discomfort associated with waiting, transfer and access/egress times, service reliability and attractiveness, frequency, and *seamless* transfers along multimodal paths with specified travel demand as part of the door-to-door passenger chain. Thus, transit connectivity is free of the aforementioned four limitations and offers a deep and comprehensive understanding of accessibility gaps as equity indicators. In addition, for each O-D pair, transit connectivity is calculated for a set of multiple and feasible transit paths including the three shortest paths and the three most popular paths (i.e., the paths with the maximum demand) in order to account for the probabilistic nature of transit path choice.

This aspect is focused in Kaplan et al. (2014) on the multimodal transit system in the Greater Copenhagen area (GCA), renowned for its transit-oriented finger plan for urban development. The data consisted of the geographic information system (GIS) representation of the multimodal network including metro, trains, and buses and detailing service lines, timetables, and stations. O-D travel demand matrices provided information about the current use of the network, and estimates regarding the importance of the travel time components were obtained from the Danish national transport model called lands trafik model (LTM). Zone-level data were available regarding zone size, population, and socioeconomic characteristics including employment status, age, and income. Transit connectivity was calculated for each O-D pair, and distributions of origin and destination connectivity were compared to the distributions of socioeconomic population characteristics to assess spatial and vertical equity. Notably, this study is the first to apply the connectivity measure to a large-scale GIS-based metropolitan size network.

13.5.1 Calculation of measures

The calculation of transit connectivity at the path, origin, and destination level allows computing location-based and potential-accessibility measures. Location-based measures relate to the calculation of transit connectivity at the zone level. Consider the set $P_{O_i} = \{P_i\}$ of multimodal paths p_{ij} from origin O_i and the set $P_{D_j} = \{P_j\}$ of multimodal paths p_{ij} to destination D_j. Having computed the connectivity c_p of each path p, it is possible to calculate the connectivity C_i for each zone as the origin of transit trips and the connectivity C_j for each zone as the destination of transit trips:

$$C_i = \sum_{p \in P_{O_i}} c_p = \sum_{p \in P_{O_i}} \left(\sum_{e^q \in E} \left(\omega_{e^q} \cdot e_p^q \right) \right) \tag{13.11}$$

$$C_j = \sum_{p \in P_{D_j}} c_p = \sum_{p \in P_{D_j}} \left(\sum_{e^q \in E} \left(\omega_{e^q} \cdot e_p^q \right) \right) \tag{13.12}$$

The origin connectivity C_i measures the possibility of each zone i to reach every other zone in the study area and hence the prospect for the population of zone i to reach opportunities (e.g., jobs, universities, hospitals) in every other zone. The destination connectivity C_j measures the possibility of each zone j to be reached from every other zone in the study area and hence the prospect of the opportunities in zone j to be accessible by the population of every other zone. Notably, the connectivity C_{ij} for each O_i–D_j pair expresses the possibility of the population in zone i to reach the opportunities in zone j. When comparing origin and destination connectivity across zones, a classification of the zones according to connectivity distribution percentiles provides insight into the spatial equity in the study area. When adding to the comparison of a layer of socioeconomic characteristics of each zone, the comparison of the connectivity distribution percentiles with the socioeconomic characteristic percentiles provides insight into the vertical equity in a study area.

Potential-accessibility measures relate to the calculation of the connectivity at the zone level and the size of the zone. It is possible to calculate the transit provision PTP_i of each zone i by modifying a formulation that offers appropriate balance between complexity and interpretability

$$PTP_i = \sum_{D_j \in D} \frac{P_{gi}}{C_i} = \sum_{D_j \in D} \frac{P_{gi}}{\sum_{p \in P_{O_i}} \left(\sum_{e^q \in E} \left(\omega_{e^q} \cdot e_p^q \right) \right)} \tag{13.13}$$

The transit provision PTP_i weighs the origin connectivity C_i according to the size that may be expressed as a portion P_{gi} of the population (e.g., number of jobs, number of students), in order to express the facility to reach a locus of opportunity. When comparing PTP_i across zones, and adding a socioeconomic layer, especially if relating specific population groups to specific opportunities (e.g., students and universities), the analysis provides valuable insight into the vertical equity in a study area.

13.5.2 Case study in Denmark

This case study reported by Kaplan et al. (2014) focuses on the GCA, which comprises 18 municipalities extending for about 3,000 km^2 with a population of about 2 million people. The planning and the development of the transit system in the GCA follows the finger-plan directives, which indicate five cities in Denmark (Køge, Roskilde, Frederikssund, Hillerød, and Helsingør) as the direction from Copenhagen for the fingers to be served by transit and road connections.

The transit network of the GCA consists of seven major modes: (1) metro, (2) local trains, (3) suburban trains (S-trains), (4) regional and intercity trains (Reg-trains and IC-trains), (5) regular buses, (6) high-frequency buses (A-buses), and (7) suburban and express buses (S-buses and E-buses). The metro serves central Copenhagen and the airport. The Reg- and IC-trains lead north and west of Copenhagen, while the S-trains follow the radial finger lines from central Copenhagen to the mentioned five cities. E-buses and S-buses serve the S-train stations (primarily in rings), A-buses operate in central Copenhagen, and the remaining buses run in Copenhagen, the suburbs, and the rural areas at a lower frequency.

Figure 13.11 presents the distributions of origin and destination connectivity of the GCA zones. The origin connectivity is presented also as a graph of connectivity values alongside the corresponding map, in order to provide an example of the construction of the maps from the calculated connectivity values. The values in the graph are presented at an aggregate zone level because of the large number of zones in the study area, while the maps are presented at the traffic zone level.

From the origin connectivity perspective emerges the general advantage of the transit-oriented development in terms of spatial equity. Zones with higher population density in the metropolitan core and along the train lines enjoy better connectivity. Nevertheless, the finger plan is only partially associated with spatial equity because the same origin connectivity is not observed for all the *fingers*. From the destination connectivity perspective, Kaplan et al. (2014) found that while most jobs are located in the metropolitan core and along the train lines of the finger plan, not all enjoy the same connectivity. In terms of vertical equity, while the zones with the highest average income enjoy better connectivity, the zones that are less better off in terms of connectivity are not much worse off in terms of income. Nevertheless, there are exceptions in zones that are highly populated, have a low-average income, and have a relatively low-connectivity level, albeit their proximity to the metropolitan core. The connectivity cavities associated with these zones were found in all the investigated aspects of the equity assessment. More on this theme appears in Kaplan et al. (2014).

13.6 LITERATURE REVIEW AND FURTHER READING

This section reviews research works that discuss various indicators of the level of transit service (LOS). Such indicators enable a measurement of the service quality of both existing transit systems and proposed schemes and, therefore, serve as important planning tools. Note that transit coordination/connectivity studies are reviewed in Chapter 6 and reliability studies in Chapter 17.

Alter (1976) introduces a measure of transit LOS whose value is affected by accessibility, travel time, reliability, service directness, service frequency, and onboard density criteria. A given set of tables and conversion tables assigns grades, from A to F, to each criterion. The accessibility grade is based on the walking time or distance to and from a stop. The reliability grade is based solely on frequency. Travel time is evaluated according to the ratio of transit travel time to car travel time on the same route. The grade for directness depends on a combination of the number of required transfers and waiting time. Grades for peak and off-peak frequencies are weighted by population density in the area served. Grades for onboard density depend on the average area for a single passenger. All six measures are aggregated into an overall LOS indicator. Some of the criteria use qualitative considerations,

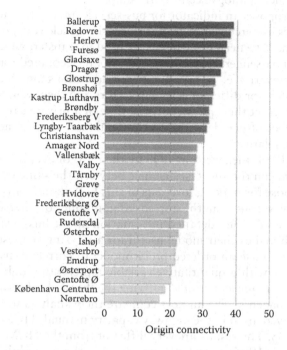

Origin connectivity values

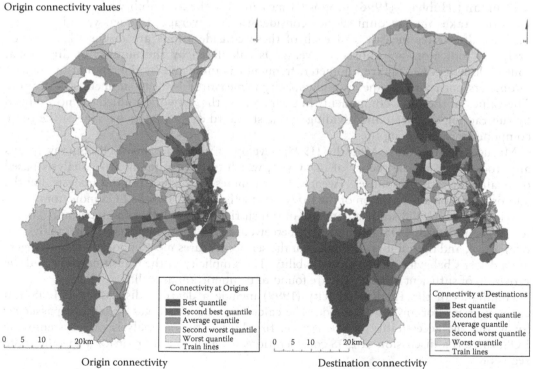

Figure 13.11 Connectivity to employment opportunities.

but the whole process is generally quantitative. Although sensitive to various criteria, the proposed evaluation methodology seems fairly simplistic.

Horowitz (1981) proposes an indicator for bus-service quality that incorporates the main performance variables as perceived by transit riders: in-vehicle time, transfer time, walking time, and waiting time. The need to wait and the need to transfer, which are known to have a significant effect on passengers' perception of LOS, are accounted for by both a numerical variable and a dummy variable. To take equity considerations into account, LOS measures are calculated separately for different ridership sectors and then weighted and summarized. The weights used represent the equivalent travel time of a private car. The final LOS rating depends on the number of people within each population sector who can travel within a certain travel-time standard.

Polus and Shefer (1984) measure bus LOS by the ratio of average bus travel time to the average car travel time on the shortest distance between the same origin and destination. The model they propose for making forecasts of this LOS indicator is based on physical or traffic-related variables: route length, stop spacing, number of intersections on route, etc. It can, therefore, be used when detailed information about bus performance or demand is unavailable, which makes their model useful for analyzing suggested route changes.

Madanat et al. (1994) use an ordered probit model to correlate discrete responses from a bus-passenger survey with a quantitative variable representing onboard crowding levels. The ordered probit technique assists in identifying the thresholds between successive ratings presented to the respondents. The authors compare their analysis to the different onboard crowding levels that were used in the highway capacity manual (HCM) manual at that time to measure transit LOS. The results are very different from the HCM values.

Henk and Hubbard (1996) propose a measure for the availability of bus or rail services that takes into account service considerations: coverage, frequency, and capacity. Various alternative measures of each of these considerations are discussed, and then preferred indicators are chosen: Coverage is calculated by the number of directional route-kilometers per square kilometer, frequency is indicated by the amount of vehicle-kilometers per route-kilometer, and capacity is measured by seat-kilometers per capita. The values of the three components in each zone of the network studied are normalized by subtracting the mean and dividing by the standard deviation; an average of the three components is then computed.

Murugesan and Rama Moorthy (1998) develop an index for bus LOS. Survey results are processed using the theory of fuzzy sets, which enables a quantitative analysis based on qualitative rating responses. Twenty LOS attributes are included in the analysis, but the authors focus on a description of the fuzzy-sets methodology, not the LOS indicators.

Friman et al. (1998) study a database of transit riders' complaints and incidents in order to determine which LOS attributes are perceived as the most important. The analysis is qualitative, and the authors conclude that the service features most important to passengers are a driver's behavior and service reliability. The simplicity of the transit network and the provision of sufficient information are found to be important as well.

Highlights of the Transit Capacity (1999) include a detailed discussion of LOS and capacity issues for any transit mode. The calculation of LOS takes account of passenger loads and the accessibility of the service. In addition to the analysis of a system-wide LOS, there is a discussion of LOS considerations for a specific stop or for a specific route segment.

Prioni and Hensher (2000) develop an LOS model based on stated-preference survey data that depict the influence of various factors on the LOS as perceived by bus users. Factors found to have the highest impact on LOS perception are the tariff, travel time, and access time to the stop. Driver friendliness and the smoothness of the bus ride are also found to be quite significant. The authors examine the differences in the LOS perception of buses operated by different companies and find that there is some bias toward certain operators; they show how the index developed could be used in the contracting process to monitor bus operators and performance.

Ryus et al. (2000) introduce a measure for transit availability based on the concept that transit routes serve a small group of potential users. The measure is defined as the percentage of time in which transit service is available and accessible to an average person within a reasonable walking distance. Computation of this indicator requires the use of a GIS. Frequency, service coverage, stop location, and operation hours influence the value of the indicator; population density affects it, too, but does not influence the demand for transit services.

Guttenplan et al. (2001) propose a methodology for determining LOS in a combined bus, cycling, and walking network. The bus LOS is influenced by common factors, such as frequency, but also by intermodal factors that are unique to this method, such as difficulty in crossing the road or the connection between the bus stop and the sidewalk.

Yang and Wang (2001) evaluate LOS on the traffic zone level, based on fuzzy-c means method. Using GIS functions, the authors compute each zone's characteristics, such as its route-network density and population density. The fuzzy-c means method is used to identify clustering patterns and to aggregate the detailed measurements into composite, zone-level LOS indicators.

Tyrinopoulos and Aifadopoulou (2008) propose a methodology for the quality control of passenger services in the transit business. Essentially, the work assesses the levels of quality and performance of transit services. In this work, 39 indicators are analyzed, classified in the following seven categories: safety–comfort–cleanliness, information–communication with the passengers, accessibility, terminals and stop points' performance, lines performance, general elements of the transit system, and compound indicators based on the results of the indicators of the previous categories. Among the compound indicators, a customer satisfaction measure is considered in order to take into account customer perceptions. In fact, the authors suggest to use factor analysis and multinomial logistic regression for investigating the influence of the operational performance indicators of the transportation system on customer satisfaction.

Lai and Chen (2011) study the passenger behavioral intentions and explore the relationships between passenger behavioral intentions and the various factors that affect them. Apart from the factors recognized by past studies, such as service quality, perceived value, and satisfaction, this study addresses the importance of the involvement of public transit services in passenger behavioral intentions. By using passenger survey data from the Kaohsiung Mass Rapid Transit (KMRT), a newly operating transit system in Taiwan, the structural equation modeling technique is applied to analyze the conceptualized relationship model. The findings reveal that all causal relationships are statistically significant. Finally, the managerial implications are discussed.

Eboli and Mazzulla (2011) present a methodology for measuring transit service quality. This methodology is based on the use of both passenger perceptions and transit agency performance measures involving the main aspects characterizing a transit service. The combination of these two types of service quality measurement fulfills the need to provide a reliable as possible measurement tool of the transit performance. The proposed procedure is then applied to a real case study of a suburban bus line; a series of subjective and objective indicators are calculated on the basis of users' perception about the service and measurements provided by the transit agency.

Mishra et al. (2012) propose a graph theoretical approach to determine connectivity measures for multimodal transit systems. The purpose of this work is to construct a methodology enabling important information on transit system performance, especially multimodal transit systems, with the lowest possible data requirements to support the decision-making process of both transit agencies and users. Their methodology consists of performance indicators of connectivity index from four different levels: node connectivity, line connectivity, transfer center connectivity, and region (large area) connectivity. An example problem and a real-life case study in Washington–Baltimore region are employed to illustrate how the proposed methodology can be used to quantify and evaluate the network connectivity of multimodal transit systems.

Hadas (2013) presents a unified methodology for extraction transit data, data storage, and analysis, to enable spatial analysis with GIS techniques. The techniques are based solely

Table 13.5 Summary of the characteristics of the models reviewed

Source	Modes	Quantitative?	Factors influencing LOS
Alter (1976)	All transit modes	No	Accessibility, travel time, reliability, service directness, service frequency, onboard density.
Horowitz (1981)	Bus	Yes	In-vehicle time, transfer time, walking time, waiting time.
Polus and Shefer (1984)	Bus	Yes	External factors (route length, stop spacing, number of intersections on route, etc.).
Madanat et al. (1994)	Bus	Partially	Onboard density.
Henk and Hubbard (1996)	All transit modes	Yes	Coverage, frequency, capacity.
Murugesan and Rama Moorthy (1998)	Bus	Partially	(Twenty factors are discussed.)
Friman et al. (1998)	All transit	No	Driver behavior, reliability, network simplicity, information.
TCRP 35 (1999)	All transit modes	Yes	Accessibility, passenger load.
Prioni and Hensher (2000)	Bus	Yes	Tariff, travel time, access time, driver friendliness, smoothness.
Ryus et al. (2000)	All transit modes	Yes	Frequency, service coverage, stop location, operation hour, population density.
Guttenplan et al. (2001)	Bus	Yes	Factors related to the pedestrian environment (difficulty in crossing road, etc.).
Yang and Wang (2001)	All transit modes	Yes	Network density, population density.
Tyrinopoulos and Aifadopoulou (2008)	All transit modes	Partially	Thirty-nine LOS indicators are presented.
Lai and Chen (2011)	All transit modes	Yes	Several core service and psychical environment factors are discussed.
Eboli and Mazzulla (2011)	Bus	Yes	Eleven service aspects and twenty-six service attributes are discussed.
Mishra et al. (2012)	Multimodal transit	Yes	Four connectivity indicators from the node, line, transfer center and region levels.
Hadas (2013)	Bus	Yes	Four connectivity indicators are introduced for the analysis of PT systems.

on Google transit feeds and any available road layer with no need for additional data. Four connectivity indicators are introduced for the analysis of transit systems: (a) road coverage level indicator, (b) intersection coverage level indicator, (c) stop transfer potential indicator, and (d) routes overlap indicator. Finally, this study demonstrates the proposed methodology with the analysis for the transit systems in Auckland (New Zealand), Vancouver (Canada), and Portland (Oregon, United States).

The major features of the LOS indicators reviewed are presented in Table 13.5.

EXERCISES

13.1 Given the following network (the numbers on the arcs represent capacities in thousand passengers/hour),

(a) Find the maximum value *1–8* (s-t or O-D) flow, its distribution and min cut.
(b) Find the maximum value *4–5* (s-t or O-D) flow, its distribution and min cut.

Use for both (a) and (b) the augmentation algorithm shown in Appendix 6.A.

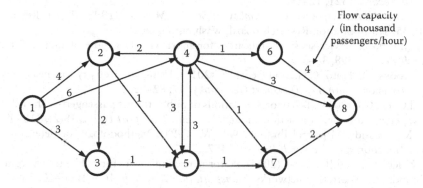

13.2 Given the following network (the numbers on the arcs represent capacities in thousand passengers/hour),

(a) Find all the minimum cuts between nodes 1 (origin) and 6 (destination).
(b) Assume that the cost to increase the capacity of each arc by one unit (each unit is in terms of 1,000 passengers per hour) is $1,000 (e.g., increase frequency of service). What is the minimum-cost procedure to increase the maximum value of the 1–6 (O-D) flow by one unit?

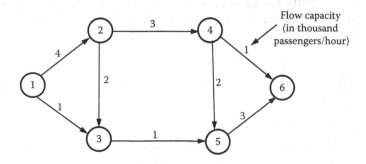

REFERENCES

Alter, C.H. (1976). Evaluation of public transit services: The level-of-service concept. *Transportation Research Record*, **606**, 37–40.

Ceder, A. and Perera, S. (2014). Detecting and improving public-transit connectivity with case studies of two world sport events. *Transport Policy*, **33**, 96–109.

Ceder, A. and Teh, C.S. (2010). Comparing public transport connectivity measures of major New Zealand cities. *Transportation Research Record*, **2143**, 24–33.

Ceder, A., Yann, L.-A., and Coriat, C. (2009). Measuring public-transport connectivity performance applied in Auckland. *Transportation Research Record*, **2111**, 138–147.

Eboli, L. and Mazzulla, G. (2011). A methodology for evaluating transit service quality based on subjective and objective measures from the passenger's point of view. *Transport Policy*, **18**(1), 172–181.

Friman, M., Edvardson, B., and Garling, T. (1998). Perceived service quality attributes in public transport: Inferences from complaints and negative critical incidents. *Journal of Public Transportation*, **2**(1), 67–89.

Guttenplan, M., Landis, B.W., Crider, L., and McLeod, D.S. (2001). Multimodal level-of-service analysis at the planning level. *Transportation Research Record*, **1776**, 151–158.

Hadas, Y. (2013). Assessing public transport systems connectivity based on Google Transit data. *Journal of Transport Geography*, **33**, 105–116.

Henk, R.H. and Hubbard, S.M. (1996). Developing an index of transit service availability. *Transportation Research Record*, **1521**, 12–19.

Highlights of the Transit Capacity and Quality of Service Manual. (1999). TCRP Research Results Digest, **35**, Transportation Research Board, Washington, DC.

Horowitz, A.J. (1981). Service-sensitive indicators for short-term bus-route planning. *Transportation Research Record*, **798**, 36–39.

Kaplan, S., Popoks, D., Prato, C.G., and Ceder, A. (2014). Using connectivity for measuring equity in transit provision. *Journal of Transport Geography*, **37**, 82–92.

Lai, W.T. and Chen, C.F. (2011). Behavioral intentions of public transit passengers—The roles of service quality, perceived value, satisfaction and involvement. *Transport Policy*, **18**(2), 318–325.

Madanat, S.M., Cassidy, M.J., and Ibrahim, W.H.W. (1994). Methodology for determining level of service. *Transportation Research Record*, **1457**, 59–62.

Mishra, S., Welch, T.F., and Jha, K.M. (2012). Performance indicators for public transit connectivity in multi-modal transportation networks. *Transportation Research*, **46A**, 1066–1085.

Murugesan, R. and Rama Moorthy, N.A. (1998). Level of public transport service evaluation: A fuzzy set approach. *Journal of Advanced Transportation*, **32**(2), 216–240.

Polus, A. and Shefer, D.(1984). Evaluation of a public transportation level of service concept. *Journal of Advanced Transportation*, **18**(2), 135–144.

Prioni, P. and Hensher, D. (2000). Measuring service quality in scheduled bus services. *Journal of Public Transportation*, **3**(2), 51–74.

Ryus, P., Ausman, J., Teaf, D., Cooper, M., and Knoblauch, M. (2000). Development of Florida's transit level-of-service indicator. *Transportation Research Record*, **1731**, 123–129.

Tyrinopoulos, Y. and Aifadopoulou, G. (2008). A complete methodology for the quality control of passenger services in the public transport business. *European Transport*, **38**, 1–16.

Yang, X. and Wang, W. (2001). GIS-based Fuzzy C-means clustering analysis of urban public transit network service: The Nanjing city case study. *Road and Transport Research, ARRB Transport Research*, **10**(2), 56–65.

Chapter 14

Coordination and connectivity
Behavioral aspects

An easy transfer
will get you there
quicker than
a car

A.C.

CHAPTER OUTLINE

14.1   Introduction
14.2   Psychological factors influencing passengers' intentions to make transfers
14.3   Effects of variation in out-of-vehicle times on passengers' decisions to make transfers
14.4   Just noticeable difference of travel time and cost savings for route choice
14.5   Planned and unplanned transfers: A guideline for policymakers
14.6   Literature review and further reading
Exercises
References

PRACTITIONER'S CORNER

There are always two sides to a story. The same can be said about perception of public transit services. While operators are focused on minimizing cost of service, passengers are concerned about maximizing ease of use. Travelers demand an efficient and reliable door-to-door service from transit systems to consider it a viable alternative to private vehicles. Transport agencies have responded by implementing integrated multimodal systems with effective interconnectivity as a strategy for attracting and retaining patronage. Integration supports service providers in reducing operational costs while increasing network coverage. It also means more routes involving transfers for users. Many studies have shown that passengers consider transfers to be an inconvenience. This chapter discusses strategies to increase public users' willingness to make transfers. The goal is for passengers to perceive transfer as a beneficial component of their trip rather than a disturbance.

This chapter consists of five principle parts, following an introductory section. Section 14.2 discusses the psychological factors that create intentions to make transfers. Section 14.3 outlines the two cognitive models that can be used to assess transfer behavior. Section 14.4 describes the results of a study that adopted Weber's law *just noticeable difference* to determine the travel-time and cost savings desired by public transit users. Section 14.5 provides a guideline for policymakers and planners for designing well-connected transfers in an integrated transport network. Each principle section ends with recommendations for practitioners. The chapter ends with a literature review and exercises.

Methods and solutions presented in this chapter are expected to assist network planners and operators to include *seamless* transfers in an integrated multimodal transit system as a service to passengers. This inclusion will offer travelers with a viable alternative to private cars and thus increase patronage for service providers. Three-part notes for practitioners appear at the end of Sections 14.2.2, 14.4.4, and 14.5.2. Finally, the complications, though solvable, involved with understanding transit users can, to some extent, be described by the saying of Friedrich Nietzsche: "You have your way. I have my way. As for the right way, the correct way, and the only way, it does not exist."

14.1 INTRODUCTION

In today's society, for reasons ranging from the increase in demand for employees' flexibility in the labor market to the decline of traditional household travel patterns, private vehicles play a dominant role in travel behavior. In 2004, 23% of the world's energy-related greenhouse gas emissions were from the transport sector and private vehicles contributed 44% to that share. The need for user-friendly transit systems has become crucial with road transport contributing significantly toward climate change. The intergovernmental panel on climate change recommended modal shift from private vehicles to sustainable transport to be a key climate change mitigation strategy. In March 14, 2008, the Transportation Research Board released a research problem statement titled *What Do Passengers Want, What Do Passengers Need?* The statement discussed that little is known regarding what passengers want, need, or expect from intermodal facilities. Both the motivation to change and the means to facilitate such change, by the individual overcoming the perceived barriers, are required for altering travel behavior. The following article discusses sustainable travel behavior as a new era for passenger transport.

From *The Economist*, September 22, 2012 (*The Road Less Travelled*).

Modern life is unimaginable without the car. The automobile has powered the growth of cities and steered their sprawl. Its manufacture has created millions of jobs and eased the development of many millions more. In rich countries, 70% of journeys are now by car. More than a billion cars now roll on the world's roads.

Measured globally, car use will go on rising, for as people in emerging markets get rich, they want the mobility and status that car-ownership offers. But in the rich world the decades-long link between rising incomes and car use has been severed, and miles driven per person have been falling. That is partly because of recession and high oil prices, but the trend pre-dates 2007. Other, longer-term, factors are at work. One is generational: car-ownership is reaching saturation.

....

Congested roads, smog and fears about global warming have led many cities to try to change the way people move around. Tokyo has shown that mass-transit systems need not be poor or dirty. Portland, which grew with the car, has since the 1980s developed its light rail. London has devoted more space exclusively to buses and cycles; cars pay to enter the centre. Singapore has congestion-pricing too. For the past 30 years Copenhagen has cut the number of parking bays by 3% a year. By contrast, in places where petrol is under-taxed so the motorists are shielded from the costs of the pollution (America) or where urban design has included public transport as an afterthought (Los Angeles), policy has supported the car.

....

Governments in emerging markets, where hundreds of cities are taking shape, should learn from mistakes and successes elsewhere. Policies that encourage people to drive into urban centres— by, for instance, requiring businesses to offer parking spaces for employees and customers— condemn metropolitan areas to heavy car use and congestion. Planning that provides mass transit systems and good pedestrian and cycle ways can make them more efficient. That is happening in some places. China is building rail networks in more than 80 cities. Eighteen Indian cities are developing metro systems. Yet many cities continue to drive themselves round the same bend as in the developed world. Bangkok, Dhaka and Jakarta are building more freeways in response to already clogged ones.

The car will bring freedom and fun to millions in emerging markets, just as it has done in the rich world. But if technology and policy are enabling people to find cheaper and cleaner ways to work and enjoy themselves, that is all to the good.

To make public transit a viable alternative to private vehicles, globally, transport agencies have been developing strategies to produce an effective integrated multimodal system. The objective of such integrated systems has been to provide travelers with a wide spectrum of destination choices through a convenient, accessible, comfortable, safe, speedy, and affordable transport system. This can be achieved by strategically positioning transfer points in the network to optimize resources. However, travel behavior specialists have established that passengers find transfers to be an inconvenience to their journey. As such, transport modelers give routes involving transfers a penalty termed *transfer penalty*. With this in mind, this chapter attempts to answer the question: *What can be done to increase passengers' willingness to use routes that involve transfers?*

Leaving one's comfort zone is never easy. Change is always undertaken with caution by every individual in matters of concern. The idea of *sustainability* is no longer new, but the concept of sustainable transport is. As such, transport practitioners must assist in the development of creating more *attractive* integrated transit systems, and the first step in doing so is to understand what is desired by the passengers.

14.2 PSYCHOLOGICAL FACTORS INFLUENCING PASSENGERS' INTENTIONS TO MAKE TRANSFERS

Many researchers have confirmed that travel behavior does not depend directly on the objective service level. Instead, it depends on psychological factors, such as beliefs, that can be influenced by the service level of the transport system. The *theory of planned behavior* (TPB), a well-established social psychological model, has been commonly adopted in travel behavior studies to gain such knowledge.

14.2.1 Theory of planned behavior

The TPB was first proposed by Ajzen (1985). As shown in Figure 14.1, one central assumption of the theory is that intention is the main direct determinant of behavior. Intention is viewed as the resultant of three antecedents: attitude, social norm, and perceived behavioral control (PBC). The strength of intention provides an indication of how hard people are willing to try, of how much effort they are willing to exert, in order to conduct the behavior. In other words, intention captures the motivational factors that influence behavior. According to the TPB, intention is viewed as the resultant of three antecedents: attitude, social norm, and PBC. These intention antecedents are based on behavioral beliefs, normative beliefs, and control beliefs, respectively. Ajzen (2005) claims that these beliefs are the foundation of behavior and changes in these beliefs should lead to change in behavior. The rationale for a direct link between PBC and behavior is that given a sufficient degree of actual control over the behavior, people are expected to carry out their intentions when requisite opportunities and resources (e.g., time, money, skills) are available (Ajzen, 2002).

Individuals create beliefs about behavior by associating it with certain attributes. These beliefs are termed *behavioral beliefs* and form the basis of attitude. Attitude is defined as the individual's positive or negative evaluation of performing the intended action. For example, Iseki and Taylor (2009) have shown that passengers perceive transfers to be burdensome due to the extra effort required, that is, generally, there is a negative attitude toward making transfers. Normative beliefs are created from approval or disapproval of important reference

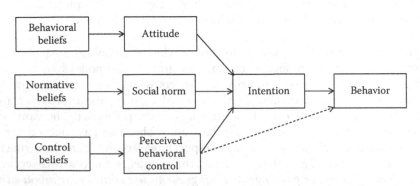

Figure 14.1 Theory of planned behavior. (From Ajzen, I., *Attitudes, Personality and Behavior*, McGraw-Hill, New York, 2005.)

group about performing behavior. These beliefs form the basis of *social norm* defined as the individual's perception of social obligation to perform or not perform the intended action. For example, Bamberg et al. (2007) revealed that proenvironmental reasons, through social norm, have a greater effect on the intention to use public transit if travelers are in a society that is aware of the negative environmental consequences caused by cars and supports transit use.

Control beliefs are created based on barriers that are perceived by the individual to undertake an action. When individuals perceive a higher level of control, the control factor is no longer important and the emphasis is placed on whether the individual has the intention to perform the behavior, thus strengthening the intention–behavior relationship. When control beliefs are weak, the individual is unlikely to have the intention to perform the behavior, thus weakening the intention–behavior relationship. PBC refers to an individual's perception of ease or difficulty in performing the intended behavior. For example, Bamberg and Schmidt (2001) have shown that PBC is more positive when ease of using public transport has been improved through better service quality.

14.2.2 Effects of PBC on intention and behavior

Two recent studies were undertaken to investigate the effects of PBC on passengers' intentions to use routes that involve making transfers.

14.2.2.1 Effects of improved trip attributes on passengers' intention

The first study by Chowdhury and Ceder (2013a) investigated the effects of improvements in trip transfer-based attributes on passengers' intention to use routes with transfers. The trip attributes selected are reliability of the connection, personal safety, transfer time (walking and waiting), journey time, and information on making transfers. Figure 14.2 shows the conceptual framework investigated by this study.

A survey was undertaken in two major transport hubs: New Lynn Transport Center and Northern Busway, in Auckland, New Zealand. A five-point Likert scale (strongly disagree to strongly agree) was used in the questionnaire. A neutral point (*no opinion*) was included in the scale to acquire responses for all items in the questionnaire and to keep participants interested in completing the questionnaire. The questionnaire was designed to be completed within 10 min. To determine statistical difference in the responses for each of the items, the data were fitted into generalized linear models (GLMs) of the Poisson family. Table 14.1 gives the results of the GLM models.

Findings of this study support the proposed theoretical framework, illustrated in Figure 14.2. Statistical analysis showed that as transfer walking and waiting times increased, passengers were less willing to use transfers. For delay in connecting vehicle, users were negatively disposed to any delay greater than 5 min, even if occurrence was rare. During the survey, participants verbally expressed that they are more comfortable with the direct route service because it is inconvenient to make transfers due to past experience with unreliable services. Reliability of the transfer duration and reduction in travel time were shown to have the most influence on users' intention to use transfer routes. This result is in line with findings of Beirao and Sarsfield-Cabral (2007).

Findings confirm the importance of the selected trip attributes. Results suggest that the service provided by operators must match or exceed passengers' expectation for users to have the intention to make transfers. According to TPB, users are required to possess strong underlying control beliefs to produce a positive PBC, which, in turn, creates a higher intention to use transfer routes. More details are in Chowdhury and Ceder (2013a).

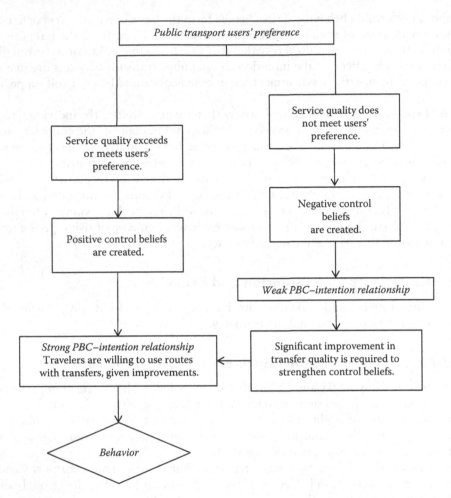

Figure 14.2 Effect of public transit users' preference for quality of interchange attributes on their perceived behavioral control–intention relationship.

14.2.2.2 Effect of PBC on passengers' intention

The second study by Chowdhury and Ceder (2013b) was conducted to further investigate the effect of PBC on passengers' intention to make transfers. PBC was broken down to its constituents, self-efficacy, and perceived controllability (PC) as shown in Figure 14.3.

Self-efficacy is defined as confidence in one's ability to perform the behavior. PC refers to one's control over performance or nonperformance of the behavior. There exists sufficient clear and consistent evidence for the distinction between self-efficacy and PC. Findings by Armitage and Conner (2001) have shown that the control beliefs used to measure self-efficacy and PC overlap. A certain level of correlation also exists between the two components (Kraft et al., 2005). Ajzen (2002) discussed that past studies commonly used different control beliefs to measure self-efficacy and controllability.

PC has been suggested to be predominantly dependent on external factors such as available resources and opportunities. Similarly, for the study of Chowdhury and Ceder (2013b), control beliefs of PC were measured using operational trip attributes: reliability of connection, transfer walking time, and transfer waiting time. Self-efficacy has been suggested to be predominantly dependent on internal factors such as ability, perceived inconvenience,

Table 14.1 Results (Chowdhury and Ceder, 2013a) from generalized linear models

Likert item (variable)	1 (Strongly disagree) Intercept	2 (Disagree)	3 (Neutral)	4 (Agree)	5 (Strongly agree)
Delay in connecting vehicle arrival					
5 min	0.14	0.17	0.22	0.25	0.22
	Reference	(0.400)	(0.058)	(0.008)	(0.036)
5–10 min	0.26	0.26	0.23	0.16	0.08
	Reference	(1.000)	(0.446)	(0.019)	(0.000)
10–15 min	0.45	0.19	0.18	0.12	0.06
	Reference	(0.000)	(0.000)	(0.000)	(0.000)
Transfer walking time					
5 min	0.04	0.05	0.14	0.24	0.53
	Reference	(0.827)	(0.002)	(0.000)	(0.000)
5–10 min	0.17	0.12	0.32	0.25	0.14
	Reference	(0.213)	(0.001)	(0.050)	(0.548)
10–15 min	0.39	0.19	0.22	0.11	0.09
	Reference	(0.000)	(0.000)	(0.000)	(0.000)
Transfer waiting time					
5 min	0.06	0.04	0.13	0.25	0.51
	Reference	(0.534)	(0.001)	(0.000)	(0.000)
5–10 min	0.14	0.15	0.30	0.30	0.11
	Reference	(0.901)	(0.000)	(0.000)	(0.287)
10–15 min	0.40	0.22	0.22	0.12	0.04
	Reference	(0.001)	(0.000)	(0.000)	(0.000)
Willingness to pay for security					
Better lighting	0.25	0.09	0.25	0.14	0.27
	Reference	(0.000)	(0.924)	(0.012)	(0.710)
CCTV	0.23	0.13	0.27	0.21	0.15
	Reference	(0.016)	(0.398)	(0.545)	(0.054)
Security personnel	0.34	0.17	0.25	0.10	0.13
	Reference	(0.000)	(0.0829)	(0.000)	(0.000)
Transfer information					
Real-time display	0.04	0.03	0.10	0.17	0.66
	Reference	(0.796)	(0.014)	(0.000)	(0.000)
Advice on the Internet	0.05	0.06	0.15	0.24	0.50
	Reference	(0.549)	(0.002)	(0.000)	(0.000)
More information about tickets and routes	0.07	0.03	0.17	0.30	0.43
	Reference	(0.041)	(0.004)	(0.000)	(0.000)
Intention to use transfer given improvement to service					
Reliability	0.06	0.06	0.23	0.31	0.35
	Reference	(1.000)	(0.000)	(0.000)	(0.000)
Connection	0.04	0.07	0.32	0.29	0.28
	Reference	(0.151)	(0.000)	(0.000)	(0.000)

(Continued)

Table 14.1 (Continued) Results (Chowdhury and Ceder, 2013a) from generalized linear models

Likert item (variable)	*1 (Strongly disagree)* Intercept	*2 (Disagree)*	*3 (Neutral)*	*4 (Agree)*	*5 (Strongly agree)*
Information	0.06	0.13	0.41	0.19	0.20
	Reference	(0.0252)	(0.000)	(0.000)	(0.000)
Travel-time savings	0.05	0.05	0.27	0.30	0.33
	Reference	(0.67)	(0.000)	(0.000)	(0.000)
Choosing to use transfer route if all four improvements were made					
New route with transfer	0.15	0.07	0.16	0.30	0.32
	Reference	(0.018)	(0.718)	(0.002)	(0.000)
Old direct route	0.22	0.13	0.21	0.18	0.26
	Reference	(0.028)	(0.739)	(0.359)	(0.527)

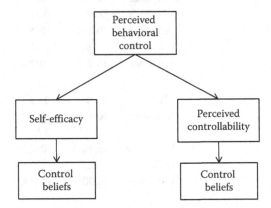

Figure 14.3 Perceived behavioral control.

and willpower. For the study, the control beliefs of self-efficacy were measured using trip attributes that are commonly evaluated by user perception: personal safety at terminals (Atkins, 1990) and need for information to make transfers (Molin and Chorus, 2009). Three hypotheses were formed and tested. Figure 14.4 shows the hypothetical relationship among the variables assessed for significance in passengers' willingness.

> *Hypothesis 1*: The stronger is passengers' PC and self-efficacy, the greater will be their intention to use transfer routes.
> *Hypothesis 2*: The stronger is the passengers' intention of use, the greater will be the probability of actually using a transfer route.
> *Hypothesis 3*: Passengers' sociodemographics and trip characteristics have a direct effect on their self-efficacy and PC and an indirect effect on their intention.

A user preference survey was undertaken at two key transport hubs: New Lynn Transport Centre and Newmarket Train Station in Auckland, New Zealand. These two, chosen from the Auckland public transit network as survey locations, support a high percentage of passengers who make either an intermodal or an intramodal transfer. The questionnaire consisted of two sections: Sections A and B. Section A contained sociodemographic and trip characteristic questions. Section B consisted of measurement items for PBC, intention,

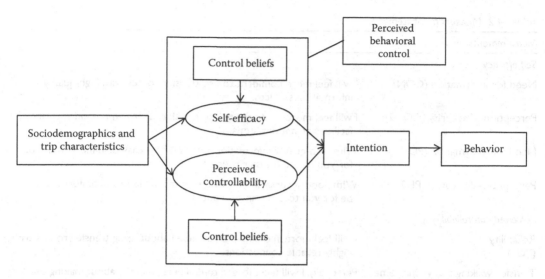

Figure 14.4 Hypothetical model explaining the causal relationship among the latent and manifest variables.

and behavior. The questionnaire was designed to be completed within 10 min Table 14.2 gives the measurement items and their notations. The data were analyzed using structural equation modeling (SEM). As the TPB contains a chain of mediating causal variables, SEM provides an appropriate statistical tool to determine the effects and relationships among the variables. Figure 14.5 gives the final SEM model based on the definitions of Table 14.2. Fit indices—such as the goodness-of-fit-index (GFI) value, 0.96; adjusted GFI, 0.90; normed fit index, 0.91; comparative fit index, 0.97; and the root mean square error of approximation, 0.04—have indicated the model to be of good fit.

Taken together, results from the final model support the adoption of the TPB to explain passengers' intention to use routes requiring transfers. In line with past findings, the two constructs of self-efficacy were seen to have discriminant validity and similarity. Similar to the findings by Ajzen (2002) and other studies (Armitage, 1999; Armitage and Conner, 2001), it is evident from the analysis that self-efficacy is more closely associated with intention and behavior than PC. Only two observed variables of PC had a significant effect on intention: *next vehicle in 10 min plus a 5–10 min walk* (PC4) and *next vehicle in 15 min plus a 5–10 min walk* (PC5). The *next vehicle in 10 min plus a 5–10 min walk* (PC4) variable had a positive effect, whereas other significant variable (PC5) had a negative effect. This finding is in accordance with the design of the measurement items. The *next vehicle in 15 min plus a 5–10 min walk* (PC5) variable was designed to be the worst connection scenario of the four items related to transfer waiting and walking time. Overall, the results seem to support *Hypothesis 1*: the stronger the public transit users' PC and self-efficacy, the greater will be their intention to use transfer routes. The significant positive coefficient (0.29), concerning the path from intention to decision, supports *Hypothesis 2* (Chowdhury and Ceder, 2013b).

Contrary to *Hypothesis 3*, it was found that sociodemographics and trip characteristics have direct effects on intention. Only gender, frequency of transit use, and current use of transfer routes had significant effects on intention. The positive effect of frequency and current use of transfers on intention indicates that positive control beliefs are related to regular use of public transit. Such finding is consistent with past studies that have shown that the transfer penalties given by infrequent passengers tend to be higher than those given by regular passengers (Guo and Wilson, 2007; Blainey et al., 2012). Regular passengers tend to be more familiar with the system and therefore are less likely to face uncertainties when transferring.

Table 14.2 Measurement items

Measurement items	Statements
Self-efficacy	
Need for information (CON1)	I will feel more confident to use transfer routes with high-quality information services.
Perception of security (CON2)	I will feel more confident to use transfer routes with good security facilities at stations/stops.
Need for information (PD1)	With high-quality information services, how easy or difficult will it be for you to use transfer routes?
Perception of security (PD2)	With good security facilities, how comfortable or uncomfortable will it be for you to use transfer routes?
Perceived controllability	
Reliability (PC1)	I will feel more *in control* (less anxious) about using transfer routes with highly reliable connections.
Transfer walking and waiting time	Personally, I will feel more *in control* (less anxious) about making the transfer if
(PC2)	The next vehicle is scheduled to arrive in 5 min and walking to the next vehicle takes 2–5 min.
(PC3)	The next vehicle is scheduled to arrive in 10 min and walking to the next vehicle takes 2–5 min.
(PC4)	The next vehicle is scheduled to arrive in 10 min and walking to the next vehicle takes 5–10 min.
(PC5)	The next vehicle is scheduled to arrive in 15 min and walking to the next vehicle takes 5–10 min.
Intention	
(Inten1)	Given this opportunity, I am willing to use transfer routes for my daily trips (e.g., work, recreation).
(Inten2)	I intend to use transfer routes to combine most of my daily trips (e.g., work, recreation).
Behavior	
(Decision)	I will choose to use transfer routes for most of my daily trips (e.g., work, recreation).

Source: Chowdhury, S. and Ceder, A., *Int. J. Trans.*, 1(1), 3, 2013b.

Male participants were seen to be less willing to use transfer routes than female participants. A probable reason for the gender effect is the different travel patterns between men and women, due to household responsibilities, child care, and employment status.

14.2.2.3 Notes for practitioners

The importance of well-connected public transit routes has been highlighted by the findings of the two psychological studies. Passengers must possess strong and positive underlying control beliefs, particularly those associated with confidence in their ability to make transfers, for them to have the intention to use routes involving transfers. The perceived benefits of using routes involving transfers need to be greater than their perceived barriers of making the connections. This section recommends that improvements in service quality of transfers need to be targeted at supporting the users' ability to make transfers successfully

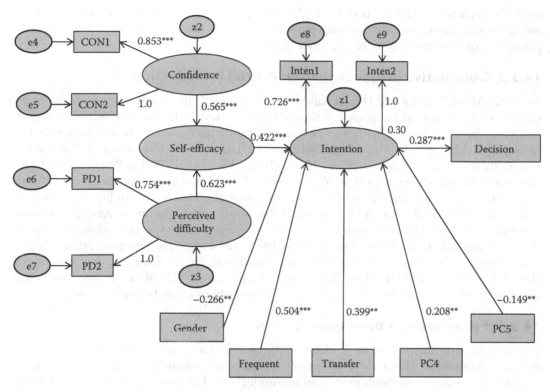

Figure 14.5 Final structural equation modeling model (**p ≤ 0.01, ***p ≤ 0.001).

and with ease. Therefore, authorities need to focus on developing attractive transfer routes (travel-time savings) with comfortable transfers (easy switch between vehicles), from a user perspective, to encourage ridership and increase patronage of public transit.

14.3 EFFECTS OF VARIATION IN OUT-OF-VEHICLE TIMES ON PASSENGERS' DECISIONS TO MAKE TRANSFERS

Out-of-vehicle times are perceived as being more onerous than in-vehicle times by passengers when making transfers. In this section, two cognitive models will be discussed to assess the effects of variation in out-of-vehicle times on passengers' decisions to make transfers. Out-of-vehicle times are defined as the duration of time when beginning a trip (waiting times) and when making a transfer (walking and waiting times). Iseki and Taylor (2009) show that perceived waiting time is more onerous than the actual waiting time and is dependent on waiting conditions such as personal safety, reliability of connection, and comfort. These authors further show that reducing perceived walking and waiting times, for transfers, substantially increases the attractiveness of public transit.

14.3.1 Route choice problem

When modeling route choice in travel behavior studies (see Chapter 11), it has been a common practice to assume that passengers have perfect knowledge about their choices and make rational decisions based on utility maximization. Discrete choice models derived using expected utility theory (EUT) and random utility theory (RUT) have been commonly adopted to analyze choice in travel behavior studies. An increasing number of route choice studies have been challenging the assumption of absolute rationality of passengers by showing evidence of violations of EUT (Avineri, 2004; Xu et al., 2011). Statistical models using the assumption of utility maximization overlook the fact that human decision-making process is approximate rather than precise (Ceder et al., 2013).

14.3.2 Cumulative prospect theory: A brief introduction

In 1992, Amos Tversky and Daniel Kahneman introduced the *cumulative prospect theory* (CPT), an extension of the prospect theory (Tversky and Kahneman, 1992). The theory conceives human choice behavior to be predominantly an *intuitive* process rather than a conscious deliberate one. The central feature of CPT is that it is able to model diminishing sensitivity in human decision-making process (Brandstatter et al., 2002). Pesendorfer (2006) discussed that one of the main shortcomings of CPT is the dependency of the outcome framing on the reference point. The reference point is often obtained by researchers from experimental settings and thus essentially unobservable outside the experiment. Researchers have often dealt with this issue by treating the reference point as a free variable chosen to match the observed behavior. Others have proposed the use of multiple reference points (Senbil and Kitamura, 2004). Van de Kaa (2010) suggested that, for the moment, the estimates of the CPT parameters obtained by Tversky and Kahneman (1992) offer the best functional description of choice under uncertainty in a wide variety of travel behavior context.

14.3.3 Fuzzy logic: A brief introduction

Fuzzy logic, first introduced by Zadeh (1965), enables formation of logical statements to compute vagueness. Using the concept of *approximate reasoning*, fuzzy logic makes it possible to model imprecision in human reasoning and thus decision making. Each fuzzy logic system can be divided into three stages: fuzzification, fuzzy inference, and defuzzification. Figure 14.6 shows the link among the stages and the input and output of each stage. A disadvantage of fuzzy logic is the task of fine-tuning the membership functions and adjusting the fuzzy rules. Construction of the membership functions and the fuzzy rules is a trial-and-error process until an appropriate fit of the input–output set is achieved. It is evident from a number of route and mode choice studies (Murugesan and Rama Moorthy, 1998; Teodorovic, 1999) that fuzzy logic is capable of modeling ambiguity in perception and appraisal of trip attributes.

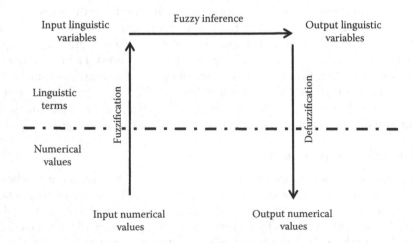

Figure 14.6 Fuzzy logic.

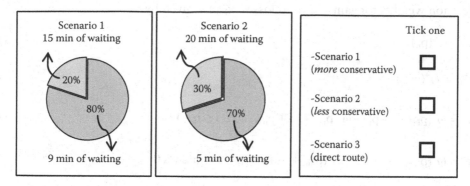

Figure 14.7 Case A for transfer waiting time.

14.3.4 Assessment of modeling

Ceder et al. (2013) conducted a study that utilized the two models for assessing the effects of variation in out-of-vehicle times. A user preference survey was undertaken at two major terminals: Britomart Transport Center and Newmarket Train Station, in Auckland, New Zealand. Respondents of the survey were presented with a total of 15 hypothetical cases, three cases (A, B, C) for each of the out-of-vehicle time variables. The variables chosen are transfer walking time, transfer waiting time, delay in connection, security at stops, and seating availability for comfort. Each case consisted of a direct route (Scenario 3) and two transfer routes (Scenarios 1 and 2) with an equal travel-time savings of 15–20 min in comparison to the direct route. For the transfer routes, one of the scenarios was designed to be perceived as being *less conservative* (risky route) and the other to be *more conservative* (less risky route). Three cases with varying probabilities of uncertainty in trip attributes were presented to better understand the variations in users' risk-taking behavior with variations in risk probabilities. Participants were instructed to select one of the three scenarios for each case. Figure 14.7 shows one of the cases (Case A) for transfer waiting time.

14.3.4.1 Results of assessment

14.3.4.1.1 Cumulative prospect theory modeling

According to Tversky and Kahneman (1992), the main formulae used in calculating the cumulative prospect values are

$$V(f) = V(f^+) + V(f^-) \tag{14.1}$$

$$V(f^+) = \sum_{i=0}^{n} \pi i^+ v(xi) \tag{14.2}$$

$$V(f^-) = \sum_{i=-m}^{0} \pi i^- v(xi) \tag{14.3}$$

where
   $V(f)$ is the cumulative prospect value
   $V(f^+)$ is the prospect gain value
   $V(f^-)$ is the prospect loss value

The decision weights for gains (π_i^+) and losses (π_i^-) are given by

$$\pi_n^+ = \omega^+(p_n) \tag{14.4}$$

$$\pi_{-m}^- = \omega^-(p_{-m}) \tag{14.5}$$

$$\pi_i^+ = \omega^+(p_i + \cdots + p_n) - \omega^+(p_{i+1} + \cdots + p_n), \quad 0 \le i \le n-1 \tag{14.6}$$

$$\pi_i^- = \omega^-(p_{i-m} + \cdots + p_i) - \omega^-(p_{-m} + \cdots + p_{i-1}), \quad 1-m \le i \le 0 \tag{14.7}$$

The functional form is used to calculate value functions for gains ($x \ge 0$) and losses ($x < 0$) as follows:

$$v(x) = \begin{cases} x^\alpha, & x \ge 0 \\ -\lambda(-x)^\beta, & x < 0 \end{cases} \tag{14.8}$$

The weighed functions for gains and losses and weighted sum are

$$w^+(p) = \frac{p^\gamma}{[(p^\gamma) + (1-p)^\gamma]^{1/\gamma}} \tag{14.9}$$

$$w^-(p) = \frac{p^\delta}{[(p^\delta) + (1-p)^\delta]^{1/\delta}}$$

$$u(x_i, p_i) = \sum_{i=0}^{n} v(\Delta x_i)\pi^+(p_i) + \sum_{j=-m}^{-1} v(\Delta x_i)\pi^-(p_{-j}) \tag{14.10}$$

Tversky and Kahneman (1992) estimated the values for CPT parameters as follows:

$\alpha = \beta = 0.88; \lambda = 2.25; \gamma = 0.61; \delta = 0.69$

The weighted sum approach was undertaken to derive the reference point values for each trip attribute from the survey data.

Cumulative prospect values were calculated using two reference point values (± 5 min for operational attributes and ± 2 min for comfort and safety) to determine the degree of fluctuation in the values with respect to the participants' preference. Table 14.3 gives the cumulative prospect values for each reference point according to each variable. The scenarios that were presented as being *more conservative* are highlighted in gray.

14.3.4.1.2 Fuzzy logic modeling

Step 1: Fuzzification

Each trip attribute (input), the difference in weighted times (WTs), and the difference in transfer delay time (DT) of the two transfer route scenarios (input) were classified into three groups: low, moderate, and high. Passengers' preference for a transfer route scenario (output) was grouped into eight ridership categories: A, B, C, D, E, F, G, and H. The ridership categories represent the proportion of passengers willing to use the transfers.

Table 14.3 Cumulative prospect values

Transfer waiting time	Case A		Case B		Case C	
	S1	S2	S1	S2	S1	S2
Reference point (min)	*Cumulative prospect value*					
4	−11.66	−9.97	−4.16	−6.11	−5.55	−6.11
9	−2.80	−4.27	0.14	−0.71	−0.98	−0.71
14	1.93	0.12	4.43	4.55	3.67	4.55
	Proportion of respondents from survey data (%)					
	67.7	23.7	38.3	53.7	34.3	56.3

Transfer walking time	Case A		Case B		Case C	
	S1	S2	S1	S2	S1	S2
Reference points (min)	*Cumulative prospect value*					
1.5	−7.01	−7.22	−5.67	−7.02	−7.22	−5.27
6.5	0.09	0.14	−0.21	0.58	−1.18	1.71
11.5	3.72	4.28	4.36	4.55	4.55	5.47
	Proportion of respondents from survey data (%)					
	37.0	55.0	32.0	60.0	22.7	69.3

Delay time	Case A		Case B		Case C	
	S1	S2	S1	S2	S1	S2
Reference point (min)	*Cumulative prospect value*					
2.5	−5.10	−6.69	−6.25	−10.33	−6.17	−7.15
7.5	0.11	0.80	−1.12	−1.30	−0.76	−0.55
12.5	4.62	4.90	3.11	3.64	4.12	4.49
	Proportion of respondents from survey data (%)					
	22.0	70.0	31.7	59.7	30.7	61.0

Comfort (seating while waiting)	Case A		Case B		Case C	
	S1	S2	S1	S2	S1	S2
Reference point (min)	*Cumulative prospect value*					
2.5	−14.02	−12.54	−6.25	−10.88	−17.02	−9.46
4.5	−10.86	−9.04	−3.71	−5.93	−14.83	−5.93
6.5	−7.55	−14.48	−1.97	−3.27	−8.23	−3.27
	Proportion of respondents from survey data (%)					
	53.3	38.3	51.0	40.3	21.0	71.0

Personal security (at terminal)	Case A		Case B		Case C	
	S1	S2	S1	S2	S1	S2
Reference point (min)	*Cumulative prospect value*					
1	−7.57	−9.07	−7.40	−8.08	−9.03	−8.48
3	−4.54	−5.60	−4.54	−3.63	−6.03	−5.00
5	−2.79	−2.50	−2.79	−1.45	−4.19	−1.84
	Proportion of respondents from survey data (%)					
	37.7	54.3	24.7	67.3	21.0	71.0

Source: Ceder, A. et al., *Trans. Policy*, 27, 112, 2013.

Step 2: Fuzzy inference

Fuzzy inference contains a set of *if–then* logic statements. The fuzzy rules were in the following form.

If [trip attribute] is $[X_{TA}]$ and $[\Delta WT]$ is $[X_{WT}]$, *then* ridership is $[Y_R]$.

Based on Figure 14.8, the input data are fuzzified, and then based on the fuzzified input data, the corresponding fuzzy rules were used. The max–min composition method is applied for making the fuzzy inference. This procedure is shown in Table 14.4.

Step 3: Defuzzification

Defuzzification is the final stage of the fuzzy system. A common approach is the center of gravity method (Zhang and Prevedouros, 2011). The process involves converting the fuzzy inference outputs into an aggregated (*crisp*) value. The aggregated value, denoted y*, will change continuously with continuous change in the input values, and its expression is shown in Equation 14.7.

$$y^* = \frac{\int \mu(y) y \, dy}{\int \mu(y) \, dy} \tag{14.11}$$

An example from Ceder et al. (2013) shows the resulting ridership categories from the fuzzy inference; these results were F (0.38) and G (0.62). The center of gravity method of this example is illustrated in Figure 14.9. The shaded area represents the ridership categories (F and G) for which the y* calculation was undertaken (Zhang and Prevedouros, 2011). Based on the result shown in Figure 14.9, 24% of transit users who were willing to use transfer routes preferred Scenario 1 and 76% preferred Scenario 2 (see Figure 14.7 for the two scenarios of Case A). Table 14.5 presents the comparison between the fuzzy model output and the survey data.

Analysis of the survey data revealed that for all trip attributes, except for comfort, users exhibited greater preference for the scenario that was perceived to be *more conservative* despite a higher probability of shorter out-of-vehicle time in the *less conservative* scenario. For comfort, users displayed risk-taking characteristics when the waiting time for an available seat was less than 5 min. Results in Tables 14.3 and 14.5 from the CPT and fuzzy logic model, respectively, illustrate that the two models are capable of representing users' out-of-vehicle behavior when making transfers. The cumulative prospect values were able to accurately reflect public transit users' preferences for the various transfer route scenarios. In most of the cases, the difference between the cumulative prospect values for the two scenarios increased, and the difference in the proportion of participants' preference also increased, with participants favoring the higher cumulative prospect value scenario. Analysis also revealed no statistical evidence of significant difference between outputs from the fuzzy system developed and the survey data. Outputs of the fuzzy system were within 5% of the actual response. Therefore, while the cumulative prospect value is able to provide an indication of users' preference, fuzzy logic is capable of providing the proportion of users preferring a transfer route. More on this study is in Ceder et al. (2013).

14.4 JUST NOTICEABLE DIFFERENCE OF TRAVEL AND COST SAVINGS FOR ROUTE CHOICE

Two trip attributes commonly considered by passengers during route choice are travel time and travel cost. A recent study was undertaken by Chowdhury et al. (2015) to determine the minimum travel-time and cost savings required from passengers' perception to use routes with transfers. The study adopted Weber's law *just noticeable difference* (JND).

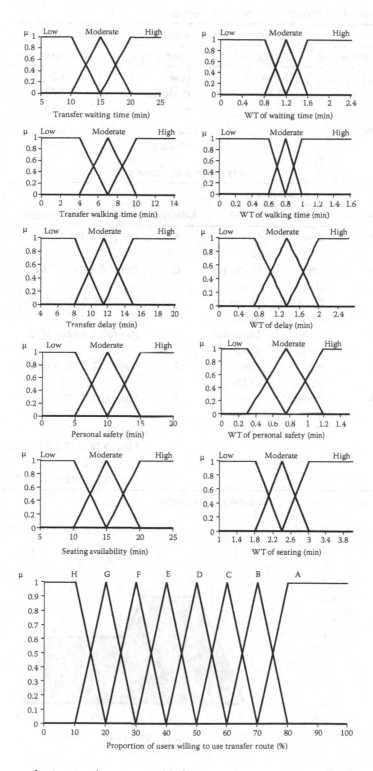

Figure 14.8 Fuzzy sets for input and output variable form.

Table 14.4 Max–min composition method

Fuzzification of input data for Scenario 1			
Input variable	Input data	Fuzzified category	Membership grade
DT	3 min	Low	1.0
	12 min	Moderate	0.88
		High	0.12
WT	1.1	Low	0.38
		Moderate	0.62

Fuzzy inference for Scenario 1				
	Input data			
Rule no.	DT	WT	Ridership	Max–min composition
9	Low (1.0)	Low (0.38)	F	Min (1.0, 0.38) = 0.38
3	Low (1.0)	Moderate (0.62)	G	Min (1.0, 0.62) = 0.62
1	Moderate (0.88)	Moderate (0.62)	G	Min (0.88, 0.62) = 0.62
2	High (0.12)	Moderate (0.62)	G	Min (0.12, 0.62) = 0.12
8	High (0.12)	Low (0.38)	F	Min (0.12, 0.38) = 0.12
7	Moderate (0.88)	Low (0.38)	F	Min (0.88, 0.38) = 0.38
				Ridership F: Max (0.38, 0.12, 0.38) = 0.38
				Ridership G: Max (0.62, 0.12, 0.62) = 0.62

Source: Ceder, A. et al., *Trans. Policy*, 27, 112, 2013.

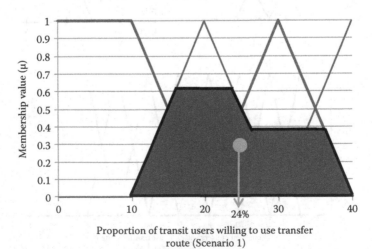

Figure 14.9 Defuzzification for ridership G and F using center of gravity method for transfer delay time of Case A, Scenario 1.

Table 14.5 Comparison between fuzzy model and survey data

Trip attribute	Case	Scenario	Fuzzy output (set B)	Survey data (set B)	Chi-squared/p-value
Transfer waiting time	A	S1	97	107	1.773/0.183
		S2	42	32	
	B	S1	56	54	
		S2	84	86	
	C	S1	56	49	
		S2	83	90	
Transfer walking time	A	S1	56	59	0.191/0.662
		S2	84	81	
	B	S1	56	54	
		S2	84	86	
	C	S1	28	33	
		S2	112	107	
Transfer delay time	A	S1	28	37	1.041/0.307
		S2	112	103	
	B	S1	42	47	
		S2	97	92	
	C	S1	42	41	
		S2	98	99	
Comfort	A	S1	82	83	0.175/0.675
		S2	55	54	
	B	S1	68	73	
		S2	68	63	
	C	S1	27	29	
		S2	110	108	
Safety	A	S1	55	50	0.071/0.789
		S2	82	87	
	B	S1	27	34	
		S2	110	103	
	C	S1	27	29	
		S2	110	108	

Source: Ceder, A. et al., *Trans. Policy*, 27, 112, 2013.

14.4.1 Just noticeable difference

JND, also known as the difference threshold, originates in the field of experimental psychology (psychophysics). The concept of JND was first discovered by the experimental investigations conducted by Ernst Weber (1795–1878), an anatomist and psychologist (Baird and Noma, 1978). Weber's law states that when two stimuli are compared, rather than simply perceiving the difference between the stimuli being compared, human beings perceive the ratio of difference to the magnitude of the stimuli. Weber's law has been mathematically formulated as

$$\frac{\Delta I}{I} = k \qquad (14.12)$$

where
    ΔI is the change required for a JND in the stimulus magnitude
    I is the stimulus magnitude
    k is the constant for the particular sense

The values of k have been seen to vary over low, medium, and high magnitude ranges of the stimuli. As the k value decreases, the perceptual sensitivity improves (Baird and Noma, 1978).

14.4.2 Survey as a case study

A user preference survey was conducted (Chowdhury et al., 2015) at the University of Auckland's city campus to test the JND concept. The study location is shown in Figure 14.10. The variables in Equation 14.12 were defined as I representing the current total travel time or travel cost experienced by the individual and ΔI defined as the change in travel-time or cost savings that motivate the passenger to switch to the alternative route with a transfer. Two surveys were conducted, with the main focus of one survey being travel-time savings (Survey I) and the other being travel-cost savings (Survey II). The influence of comfort at

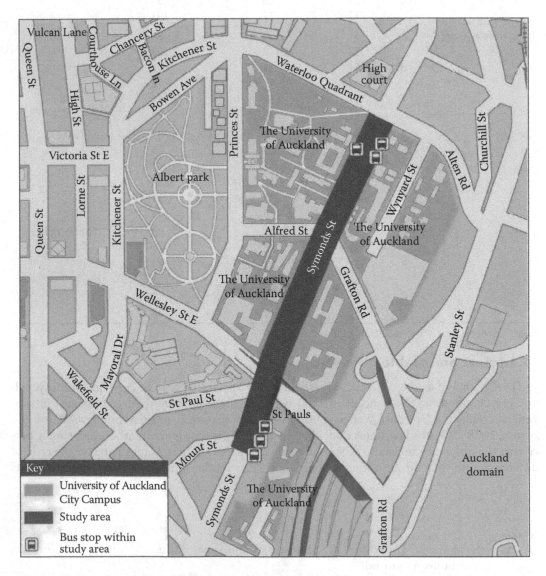

Figure 14.10 The study area of the survey conducted by Chowdhury et al. (2015).

the connection point was included in both surveys to determine its effect on the JND constants for travel-time and cost saving.

14.4.3 Results of the case study

To determine the average k values, the trimmed means of the two survey data were calculated. Wilcox (2009) recommended a 20% trimming of the data to account for outliers. The 20% Winsorized mean ($\overline{W}$) represents the average k for users' travel-time and cost savings. The ranges of the average k were determined by calculating the 95% Winsorized confidence interval (C.I.). Table 14.6 includes the 20% Winsorized variance (s_w^2) and $\overline{W}$ of the two data set with the range for both comfort levels at the interchange.

A curve fitting regression analysis was undertaken using the Statistical Package for the Social Sciences (SPSS) with 300 data points for each comfort level. Gender of the participants did not have a statistically significant effect on the k values. To determine the relationship between the k values and participants' current travel time, the best fitted curve model was chosen through observation and the R^2 value. The selected curve illustrated an inverse relationship between the k values and participants' current travel time. The following equations give the relationship for each comfort level at the connection point:

$$k_{(\text{basic interchange comfort})} = 0.249 + \frac{3.08}{\text{current travel time (min)}} \tag{14.13}$$

$$k_{(\text{high-quality interchange comfort})} = 0.172 + \frac{3.77}{\text{current travel time (min)}} \tag{14.14}$$

Distinct patterns were identified among the k values. It is shown in Figure 14.11 that 94% of the participants selected a ΔI following inverse curves at intervals of 5 min. The diagram illustrates that majority (70%) of the participants desired a travel-time savings between 10 and 20 min for basic comfort at the interchange. The change from a basic to a high-quality comfort at the connection point increased the number of participants following the ΔI curve of 5 min (41–77). The number of participants on the ΔI curve of 10 min also increased (87–103). Figure 14.12 illustrates that majority (77%) of the participants desired a time saving between 5 and 15 min for a better-quality connection point.

For travel-cost savings, gender of the participants did not have a statistically significant effect on the k values. The k values were grouped in accordance with participants' current travel cost. Table 14.7 gives the 20% Winsorized mean ($\overline{W}$) and variance (s_w^2) of the k values for each group.

Table 14.6 JND for travel-time and cost savings

Trip attribute		Basic interchange facilities	High-quality interchange facilities
Travel-time savings	$\overline{W}$	0.33 ± 0.014	0.25 ± 0.012
	s_w^2	0.0057	0.0043
Travel-cost savings	$\overline{W}$	0.16 ± 0.011	0.10 ± 0.010
	s_w^2	0.0029	0.0025

Source: Chowdhury, S. et al., J. Trans. Geogr. 43, 151, 2015.

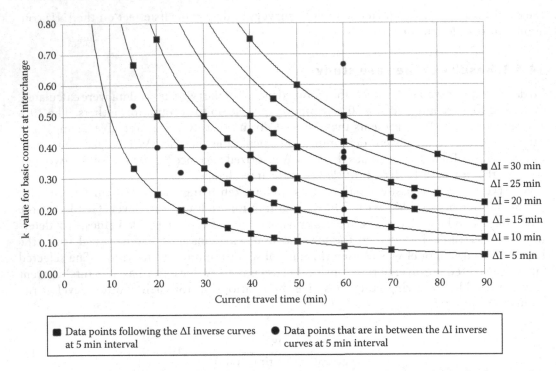

Figure 14.11 Inverse relationship between k values and transit users' current travel time for basic quality comfort at interchange.

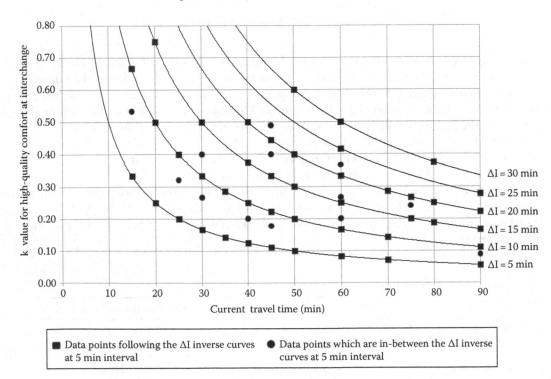

Figure 14.12 Inverse relationship between k values and transit users' current travel time for high-quality comfort at interchange.

Table 14.7 Average JND k constant for current travel cost

Current travel cost (NZ$)		$21	$28	$35	$42
$k_{\text{(basic interchange comfort)}}$	$\overline{W}$	0.160	0.178	0.149	0.162
	s_w^2	0.003	0.003	0.002	0.003
$k_{\text{(high-quality interchange comfort)}}$	$\overline{W}$	0.102	0.121	0.089	0.106
	s_w^2	0.003	0.004	0.002	0.003

14.4.4 Discussion of the JND analysis

Results of the analysis, given in Table 14.6, have shown that the average JND constant for travel-cost savings is less than the average JND constant for travel-time savings in both interchange comfort scenarios. An explanation for this finding is that the majority of the survey participants were university students. Comfort at the connection point was seen to have a positive effect in reducing the average JND constant for both travel-time and travel-cost savings. On average, passengers desired at least a 33% reduction in their current travel time and at least a 16% reduction in their current travel cost given basic comfort at the interchange. For an interchange with more comfort, on average, passengers desired at least a 25% reduction in their current travel time and at least a 10% reduction in their current travel cost. This finding suggests that high-quality interchanges are more likely to encourage the use of more efficient routes involving transfers.

The statistically significant inverse relationship between passengers' current travel time, in minutes, and the k values suggests that travelers with a longer journey time have more willingness to use routes involving transfers given some travel-time savings. For travelers with a shorter journey time, the intention to make transfers is lower possibly because of the reason that these travelers are satisfied with their current travel time. Figures 14.11 and 14.12 illustrated that majority of the participants followed ΔI (minimum travel-time savings required to use route with transfer) inverse curves at intervals of 5 min. This finding supports prior research (Van der Henst et al., 2002) that discussed the psychological disposition of rounding during verbal communications related to time in multiples of 5 min. Analysis of the travel cost-savings data, given in Table 14.7, has shown that the average k was of a similar value for each of the current travel cost groups. The finding suggests that fare discounts need to be given in proportion to users' current travel costs to increase their willingness to make transfers.

14.4.4.1 Notes for practitioners

The importance of high-quality services for public transit users' satisfaction is no secret. It has been claimed by researchers that users are unwilling to make transfers. The studies discussed in Sections 14.3 and 14.4 have shown that users are willing to make transfers given that the connection is reliable and saves time and cost. Such attractive connections can be achieved through integration among transit services.

Rachel Kyte, World Bank's vice president of sustainable development, stated the importance of integrated transport system to insure quality. The following extract is from the article presented in the magazine *Global*, "An international briefing" (Kyte, 2011):

> Having a large number of small operators allows for low-cost services, but the quality is poor due to severe competition. Other disadvantages include dangerous driving practices, pollution and a tendency to have too much service on profitable routes and virtually no service on non-profitable routes. Meanwhile, single publicly owned entities

may offer higher quality of service but costs tend to be high and the quantity of service is often inadequate. There is increasing recognition that the best industry structure falls somewhere between the two. Having a single public entity that plans the network and determines the quality of service, with a small number of private operators providing services under structured contracts, allows a balancing of public good needs with the operational efficiency of the private sector.

14.5 PLANNED AND UNPLANNED TRANSFERS: A GUIDELINE FOR POLICYMAKERS

14.5.1 Definition of planned and unplanned transfers

Chowdhury and Ceder (2013c) proposed a definition of *planned* and *unplanned* transfers. A *planned* transfer is a connection that has been intentionally designed by policymakers and network planners in the planning stage of the multimodal transport network to improve service efficiency and convenience for passengers. An *unplanned* transfer is defined as a connection that has been created by the passengers from available services without any additional guidance on how to make the connection. As given in Table 14.8, planned transfer consists of five attributes: network integration, integrated timed transfer, integrated physical connection of transfers, information integration, and fare and ticketing integration.

14.5.2 Effects of planned and unplanned transfers on passengers' perception

This section covers two studies focused on the effects of planned and unplanned transfers on passengers' perception of transfers.

14.5.2.1 Effects of attributes

The first study by Chowdhury and Ceder (2013c) is based on a survey conducted in Auckland, New Zealand. The Northern Busway and Britomart terminals were chosen as the survey locations. Local feeder routes are connected to five designated stations along the Northern Expressway. Transfers occur between the local routes and the main line. Majority of the buses entering and leaving Auckland central begin and end their trip at Britomart, thus creating the opportunity to make transfers. Assessment of the facilities at each station revealed that the transfer occurring on Northern Expressway is more closely aligned to being planned transfer than those in Britomart.

The questionnaire was composed of general sociodemographics, trip characteristics, and hypothetical scenario questions. The bandwidth for the sociodemographic questions, age and income, were adopted from the New Zealand census questionnaires. To participants who are currently using a transfer route, questions on details of the transfer connection and usage satisfaction were asked. To participants who are not using a transfer route, a multi-choice question was asked on improvements to which trip attributes will increase their willingness. A five-point Likert scale (very poor/very good) was used as the response measure.

Of the participants from Northern Busway, 61% rated their transfer route to be 4 (good). Of the participants from Britomart, 54% rated their transfer route to be 3 (neutral) and none rated their routes to be greater than 3. The average transfer waiting time and walking time for users' of Northern Busway was found to be 12 and 8 min, respectively. Similarly, for Britomart, the average transfer waiting and walking time was found to be 6 and 5 min, respectively.

Table 14.8 Attributes defining a planned transfer

Attributes	Definition	Characteristics
Network integration	Routes are required to be connected from a network perspective to allow passengers to access a wider range of destinations. Proper integration of multimodal transport system will reduce wasteful duplication of route services, thus improving the utilization of resources.	• Physical overlap of service lines. • Combination of high-frequency routes and low-frequency routes (feeder services). • Network coverage. • Easy accessibility to public transit network.
Integrated timed transfer	The aim of integrated timed transfers is to interconnect the multimodal transport network such that the transfer times are minimized. This is achieved by operators synchronizing their scheduled routes to develop a pulsed-hub network. Since the performance of timed transfer is dependent on schedule reliability, implementation improves the overall reliability of the transport system. Routes and scheduling are required to be designed simultaneously.	• Minimize transfer waiting time. • Synchronize scheduled routes.
Integrated physical connection for transfers	Terminals are required to be physically connected for the transfers among them to be considered as being planned. Integration between terminals has been defined as being sheltered walkways between terminals, security measures at connected walkways, and information provisions such as signage providing guidance between the connected terminals, map of the local street area, and location of connected walkways.	• Sheltered walkways. • Security measures to protect users between separate stations/stops. • Types of information, such as directional signage and maps, to link stations/stops at separate locations.
Information integration	An integrated information system is essential to facilitate urban and interurban multimodal trip planning. With many advanced information systems available, real-time information can be made accessible directly to the user en route. Such an information system assists users to preplan transfers and then provides guidance en route, thereby reducing the chances of missed connection and providing travel support (Grotenhuis et al., 2007; Zhang et al., 2011).	• Journey planner to assist users in planning their transfers among all public transit services. • Real-time information (arrival/departure/DTs) at stations/stops. • En route guidance providing real-time information. • Maps and timetables for all public transit services at stations/stops.
Fare and ticketing integration	Fare system integration of a multimodal transport network has been shown to facilitate seamless transfers, thus encouraging the use of transfer routes (Sharaby and Shiftan, 2012). A simple user-friendly integration, such as smart cards, can improve the efficiency of boarding and egressing.	• Smart cards used for all services. • No additional cost for transfers.

Source: Chowdhury, S. and Ceder, A., J. Public Trans., 16(2), 1, 2013c.

Figure 14.13 shows the proportion of passengers willing to use transfer routes given improvements made to the trip attributes. Users of Britomart were willing to use transfer routes given more connected routes, more information on transfers, and the total transfer time being less than 20 min. Network integration was shown to have the most influence on willingness. Northern Busway passengers' willingness depended on better seating area, sheltered walking area, and integrated ticketing system, which offers no additional cost for transfers. Cost of transfer was shown to have a greater influence than the trip attributes related to comfort for users from both locations.

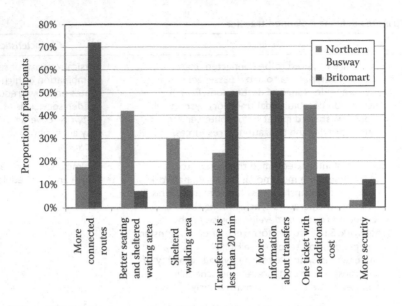

Figure 14.13 Proportion of participants willing to use transfer routes given improvements in trip attributes.

14.5.2.2 Effects of explanatory variables

The second study by Chowdhury et al. (2014) was undertaken to determine the effects of planned and unplanned transfers on captive users' (passengers who do not currently make transfers) decision to use transfer routes. Two survey locations in Auckland, New Zealand, were selected—Northern Busway and Britomart. All participants were asked a question on whether they would choose an alternative hypothetical transfer route when comparing the route to a direct route with travel time of 40 min. The hypothetical transfer route scenarios had travel-time savings of 10, 15, and 20 min with varying types of connection. This question was designed to determine passengers' perception of planned and unplanned transfers. Table 14.9 gives one of the scenarios. The survey was conducted for 10 weekdays (5 days for each location) during commuter morning peak period (7–10 a.m.). Target participants were limited to only commuters because this group represents the highest proportion of users.

For the hypothetical scenarios, as the response is dichotomous, a binary logistic regression model was developed to predict the dichotomous dependent variable as a function of the predictor variables. Equation 14.15 gives its common form. A statistical package, SPSS, was used to fit the two data sets into binary logistic regression models.

$$\text{logit}\,P(X) = \alpha + \sum \beta_i X_i \tag{14.15}$$

Table 14.10 gives the model for the Northern Busway data. The Hosmer–Lemeshow test is a goodness-of-fit statistic used to determine whether the developed model reasonably approximates the data (Kleinbaum and Klein, 2010). Results of the analysis indicated that the model adequately fit the data because the significance value for Hosmer–Lemeshow test was 0.27.

Compared with the travel-time savings of 10 min, only travel-time savings of 20 min was significant. Holding all other factors constant, current nontransfer route users were 2.92 times more likely to use a transfer route with travel-time savings of 20 min savings (odds ratio = 2.92) when compared to a travel-time savings of 10 min. In addition, users were

Table 14.9 Hypothetical transfer scenarios

Ten minutes travel-time savings, given the following service		
Option 1	• 5 min waiting time for the next vehicle	Yes
	• 5 min walking time to the next vehicle	No
	• Sheltered waiting and walking area	
	• High-quality information about transfer	
Option 2	• 5 min waiting time for the next vehicle	Yes
	• 5 min walking time to the next vehicle	No
	• Sheltered waiting and walking area	
	• Have to search and find your own transfer point	
Option 3	• 5 min waiting time for the next vehicle	Yes
	• 5 min walking time to the next vehicle	No
	• No sheltered waiting and walking area	
	• Have to search and find your own transfer point	

Source: Chowdhury, S. et al., Trans. Plan. Technol., 37(2), 154, 2014.

Table 14.10 Explanatory variables for Northern Busway

						95% C.I. of odds ratio	
Explanatory variables	Coefficient	Wald	df	p-value	Odds ratio	Lower	Upper
Northern Busway (planned transfer)							
Travel		21.350	2	0.000			
Travel (15)	0.322	2.082	1	0.149	1.380	0.891	2.137
Travel (20)	1.072	20.674	1	0.000	2.922	1.840	4.638
Shelter (yes)	2.522	112.250	1	0.000	12.458	7.813	19.865
Information (yes)	0.693	9.068	1	0.003	1.999	1.274	3.138
Gender (female)	1.770	63.216	1	0.000	5.869	3.794	9.079
Frequent (yes)	1.203	26.119	1	0.000	3.331	2.100	5.284
Constant	−3.511	100.875	1	0.000	0.030		

Source: Chowdhury, S. et al., Trans. Plan. Technol., 37(2), 154, 2014.
Test of model fit (Hosmer–Lemeshow test) chi-squared = 9.9, df = 8.
Nagelkerke R^2 = 0.46, log likelihood (final) = 710.066.
Omnibus test (intercept public transport only and final model) chi-squared = 314.81, df = 6.

12.46 times more likely to use a transfer route with the presence of shelters when compared to a transfer connection without shelters.

The following gives the logit for the fitted model for the Northern Busway data set:

$$\text{logit}(\text{Transfer}) = -3.51 + 0.32 * \text{Travel}_{15\,min}$$

$$+1.07 * \text{Travel}_{20\,min} + 2.52 * \text{Shelter} + 0.69 * \text{Information}$$

$$+1.77 * \text{Gender} + 1.20 * \text{Frequency} \qquad (14.16)$$

Table 14.11 gives the model fit for Britomart. Holding all other factors constant, current nontransfer route users were 1.52 times more likely to use a transfer route with travel-time savings of 15 min savings (odds ratio = 1.52) when compared to a travel-time savings of 10 min. Public transit users were 2.94 times more likely to use a transfer route with high-quality information. When comparing connections with and without shelter, users were 5.19 times more likely to make a transfer with shelter (odds ratio = 5.19).

Table 14.11 Explanatory variables for Britomart

Explanatory variables	Coefficient	Wald	df	p-value	Odds ratio	95% C.I. of odds ratio	
						Lower	Upper
Britomart (unplanned transfer)							
Travel (15)	0.420	4.770	1	0.029	1.522	1.044	2.218
Shelter (yes)	1.648	52.315	1	0.000	5.196	3.325	8.121
Information (yes)	1.079	21.075	1	0.000	2.943	1.856	4.665
Constant	−1.478	52.496	1	0.000	0.228		

Source: Chowdhury, S. et al., Trans. Plan. Technol., 37(2), 154, 2014.
Test of model fit (Hosmer–Lemeshow test) chi-squared = 1.634, df = 4.
Nagelkerke R^2 = 0.306, log likelihood (final) = 651.72.
Omnibus test (intercepublic transport only and final model) chi-squared = 151.14, df = 3.

The following equation gives the logit for the final logistic regression model for the Britomart data:

$$\text{logit}(\text{Transfer}) = -1.48 + 0.42 * \text{Travel}_{15\,\text{min}} + 1.65 * \text{Shelter} + 1.08 * \text{Information} \quad (14.17)$$

14.5.2.3 Discussion of both studies

Overall, results from these two studies suggested that users of public transit network with planned connections are more likely to use transfer routes than users of network with unplanned connections. Physical integration was seen to have a greater effect than information integration on passengers' willingness to make transfers. Participants from both transport terminals were willing to use the transfer routes given shelter provided during transfer walking and waiting times (Northern Busway odds ratio = 12.46 and Britomart odds ratio = 5.20) and high-quality information services (Northern Busway odds ratio = 2.0 and Britomart odds ratio = 2.94) while holding transfer walking and waiting times to be constant at 5 min each. Therefore, it can be recommended that for transfers between physically separated terminals, planning must focus on design measures to create good connections. At the very least, connections between terminals need to be weather-protected and sufficient way-finding signage must be present. Findings from the two studies have highlighted the importance of including transfers as a service component in the main planning stage of an integrated multimodal transport network development to provide users with *seamless* transfers.

14.5.2.4 Notes for practitioners

The idea behind providing planned transfers is to make public transit services *more efficient and attractive* by reducing travel times and increasing reliability and connectivity. During the development of an integrated transit network, planners are encouraged to address *all five attributes* of planned transfers for the connections to be perceived as being *seamless* by passengers. The five attributes that need to be achieved for transfers to be categorized as being *planned* are (a) network integration, (b) integrated timed transfer, (c) integrated physical connection of transfers, (d) information integration, and (e) fare and ticketing integration. It should be noted that if all five attributes are not achieved, then the connection cannot be categorized as being a planned transfer.

14.6 LITERATURE REVIEW AND FURTHER READING

The literature review of transit behavioral-based studies is quite extensive. In what follows are three themes with a review linked directly and indirectly with user's behavior related to making transfers on transit routes.

14.6.1 Mode choice between public transport and private vehicles

Literature analyzing the general attitude of travelers toward public transit use has shown that although improvement in service quality is likely to increase ridership, the level of increase can be limited if travelers hold prejudices toward the image of transit service and of becoming users (Murray et al., 2010). One of the strategies used by a government to amend travelers' misconception of transit is mass media information campaigns on the benefits of use (Tertoolen et al., 1998). The aim of the campaigns is to induce positive attitudes about transit use in society. Beale and Bonsall (2007) revealed that persuasive marketing messages on the benefits of using transit increased the negative attitude of car users. On the other hand, for habitual users, the marketing messages encouraged more use. Guo and Wilson (2007) stated that frequent users are more familiar with the network and services and therefore have fewer negative perceptions than infrequent users and nonusers. Pedersen et al. (2012) discussed that car users frequently underestimate their potential satisfaction with transit use and inaccurately recall the satisfaction associated with any previous use, which leads to persistent negative attitude toward mode switch. Taniguchi and Fujii (2007) investigated the psychological effect of an intervention, which consisted of newsletters (a marketing technique) and free bus tickets, on travelers' decisions to use transit. The results showed that including marketing techniques in interventions is an effective method of increasing ridership of transit services.

Several other studies have focused on the proenvironmental aspect of travel behavior. Nilsson and Kuller (2000) concluded that the intention to behave in a proenvironment way depends considerably on the individual's attitude related to environmental issues. Factual knowledge was seen to play a subordinate role. In terms of the attitudinal factors, environmental concern and attachment to one's car was suggested to be most influential in determining proenvironmental behaviors. Individuals living in a heavy traffic area showed more interest in proenvironmental behaviors than individuals living in a low-pollution area. Nordlund and Garvill (2003) stated that values and problem awareness significantly affect personal norms, which in turn influences the individual's willingness to incorporate proenvironmental behaviors. Collins and Chambers (2005) discussed that individual's proenvironmental behavior is a result of the interaction among their social beliefs, biospheric beliefs, egoistic beliefs, and situational factors such as travel by public transport cost, travel time, and access.

14.6.2 Trip attributes

Redman et al. (2013) stated that supply-oriented quality of transit is only part of the whole picture. It is also important to consider the quality of service as perceived by passengers to meet a consistent satisfactory level over time. This section focuses on the trip attributes that have been identified to be significant in passengers' perceived ease of making transfers. The importance of personal safety at terminals has been echoed in several travel behavior studies (Atkins, 1990; Volinski and Page, 2006). Currie and Delbosc (2013) discussed that users are concerned about traveling with unknown passengers. Transit users feel unsafe particularly during night travels (Blainey et al., 2012; Currie and Delbosc, 2013). A study in the United Kingdom has shown that an additional 10.5% of rail trips would be generated if users' fears were addressed (Currie and Delbosc, 2013). Reliability is one of the most important operational attributes of transit services. A number of studies on timetable scheduling have been done to determine methods of improving reliability. Muller and Furth (2009) examined how better planning can minimize the inconvenience to users who are making transfers. Results emphasized the importance of optimal offset in schedule planning to minimize transfer waiting times as well as to reduce missed connections. Methods such as *holding* have been suggested to improve reliability by reducing additional journey times from schedule deviance (Van Oort and van Nes, 2009; Van Oort et al., 2012).

Other studies have shown that better information systems are required to reduce users' perceived inconvenience of making multimodal trips (Zografos et al., 2008). Integration between various operators is required for an information system to facilitate urban and interurban multimodal trip planning (Zografos et al., 2008). Molin and Chorus (2009) showed that transferring to a lower-frequency service induces a need for more information and thus a higher willingness to pay for information. Pricing is another key attribute that attracts ridership of transit (Redman et al., 2013). Integration of fare systems across operators has a positive effect on the perceived ease of making transfers (Buehler, 2011). Sharaby and Shiftan (2012) demonstrated that a well-integrated fare system, which allows no additional cost for transfers, has a positive impact on ridership by improving travelers' willingness to use routes with transfers.

14.6.3 Policy and infrastructure

Two distinctions can be made among the different policy measures: *push* and *pull*. The objective of *push*-type measures is to reduce the attractiveness of private vehicles, while *pull*-type measures aim to increase the attractiveness of sustainable transport. There are four types of policies: legal policies, information and educational policies, economic policies, and physical changes (e.g., infrastructure) (Cools et al., 2009). Some examples of legal policy measures are speed restrictions, parking restrictions, and prohibiting cars in city centers (Cools et al., 2009). Types of parking restrictions include availability of parking spaces, parking costs, parking time limits, provision of employee parking, and residential parking permits. As for economic policy, some of the measures developed are congestion charges (already implemented in cities like London, Stockholm, and Durham), toll roads, and taxation of cars and fuel (Stopher, 2004; Cools et al., 2009).

In regard to information and educational policy measures, primarily in Europe and Australia, governments have identified the need to encourage voluntary travel behavior change toward sustainable modes of transport (Pramberg, 2004). Personalized travel planning initiatives are targeted marketing approaches that empower individuals to change their travel behaviors for their personal benefit (Ampt, 2004). Physical policies aim at increasing the attractiveness of transit services by improving service quality, providing infrastructures such as separate bus lanes and better land-use planning for shorter travel times (Cools et al., 2009). Such *pull*-type measures are seen to be more easily accepted by the public than *push*-type measures as they are considered to more effective and fairer (Eriksson et al., 2008). Integration of multimodal transport systems has received particular interest in recent years to promote mode switch. Hull (2008) emphasized the importance of maintaining interoperability and interconnectivity across service providers to optimize an intermodal transport system.

Table 14.12 provides a summary of the factors that influence transit users' willingness to make transfers.

Table 14.12 Summary of factors influencing public transport users' willingness

Factors influencing willingness to use transit	Source	Users	Quantitative?
Education	Beale and Bonsall (2007)	Bus users	No
	Murray et al. (2010)	Car users	Partially
	Tertoolen et al. (1998)	Car users	Partially
Frequency of use	Guo and Wilson (2007)	Train users	Yes
	Pedersen et al. (2012)	Car users	No
Proenvironmental behavior	Nilsson and Kuller (2000)	Car users	Partially
	Nordlund and Garvill (2003)	Car users	Partially
	Collins and Chambers (2005)	Students	Partially
Integrated information systems	Zografos et al. (2008)	Train users	No
	Molin and Chorus (2009)	Train users	Yes
Integrated ticketing system	Sharaby and Shiftan (2012)	Bus users	Yes
	Buehler (2011)	Car and transit users	No
Safety	Currie and Delbosc (2013)	Students	Partially
	Blainey et al. (2012)	Car users	No
	Atkins (1990)	Transit users	No
Rellability	Muller and Furth (2009)	N/A	Yes
	Van Oort et al. (2012)	N/A	Yes

EXERCISES

14.1 The TPB has been used as a common psychological model to determine travel behavior.

(1) Define the three antecedents: attitude, social norm, and PBC.
(2) Describe the two links, from intention and PBC to behavior.
(3) State the three most influential trip attributes that will improve public transit users' control beliefs to make transfers.

14.2 Given as follows is the choice of two alternative routes that involve making a transfer. The case shows the variation in transfer waiting time for Scenarios 1 and 2. The assumption is that the *scheduled* travel time of both alternatives is equal.

(1) Determine what alternative users are most likely to choose using the CPT. Assume that the *reference point* is equal to 9 min.
(2) Describe the steps that will be undertaken to determine users' route choice using fuzzy logic.

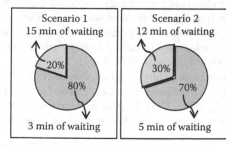

Scenario 1
15 min of waiting
20%
80%
3 min of waiting

Scenario 2
12 min of waiting
30%
70%
5 min of waiting

-Scenario 1 (*less* conservative)

-Scenario 2 (*more* conservative)

14.3　A government transport agency is developing a new guideline for designing their transit network. The current network mainly comprises train and bus services for the high- and low-frequency routes, respectively. Describe a *planned* transfer between a high-frequency train service (every 20 min) and a low-frequency bus service (every 60 min).

REFERENCES

Ajzen, I. (1985). From intentions to actions: A theory of planned behavior. In *Action Control: From Cognition to Behavior.* SSSP Springer Series in Social Psychology (J. Khul and J. Beckmann, eds.), pp. 11–39, Springer, New York.

Ajzen, I. (2002). Perceived behavioral control, self-efficacy, locus of control, and the theory of planned behavior. *Journal of Applied Social Psychology,* 32(4), 665–683.

Ajzen, I. (2005). *Attitudes, Personality and Behavior,* McGraw-Hill, New York.

Ampt, E. (2004). Understanding voluntary travel behaviour change. *Transport Engineering in Australia,* 9(2), 53–66.

Armitage, C.J. (1999). The theory of planned behaviour: Assessment of predictive validity and 'perceived control'. *British Journal of Social Psychology,* 38(1), 35–54.

Armitage, C.J. and Conner, M. (2001). Efficacy of the theory of planned behaviour: A meta-analytic review. *British Journal of Social Psychology,* 40(4), 471–499.

Atkins, S.T. (1990). Personal security as a transport issue: A state-of-the-art review. *Transportation Reviews,* 10(2), 111–125.

Avineri, E. (2004). A cumulative prospect theory approach to passengers behavior modeling: Waiting time paradox revisited. *Journal of Intelligent Transportation Systems: Technology, Planning, and Operations,* 8(4), 195–204.

Baird, J.C. and Noma, E. (1978). *Fundamentals of Scaling and Psychophysics,* John Wiley & Sons, Inc., New York.

Bamberg, S., Hunecke, M., and Blobaum, A. (2007). Social context, personal norms and the use of public transportation: Two field studies. *Journal of Environmental Psychology,* 27(3), 190–203.

Bamberg, S. and Schmidt, P. (2001). Theory-driven subgroup-specific evaluation of an intervention to reduce private car use. *Journal of Applied Social Psychology,* 31(6), 1300–1329.

Beale, J.R. and Bonsall, P.W. (2007). Marketing in the bus industry: A psychological interpretation of some attitudinal and behavioural outcomes. *Transportation Research,* 10F, 271–287.

Beirao, G. and Sarsfield-Cabral, J.A. (2007). Understanding attitudes towards public transport and private car: A qualitative study. *Transport Policy,* 14(6), 478–489.

Blainey, S., Hickford, A., and Preston, J. (2012). Barriers to passenger rail use: A review of the evidence. *Transport Reviews,* 32(6), 675–696.

Brandstatter, E., Kuhberger, A., and Schneider, F. (2002). A cognitive-emotional account of the shape of the probability weighting function. *Journal of Behavioral and Decision Making,* 15(2), 79–100.

Buehler, R. (2011). Determinants of transport mode choice: A comparison of Germany and the USA. *Journal of Transport Geography,* 19(4), 644–657.

Ceder, A., Chowdhury, S., Taghipouran, N., and Olsen, J. (2013). Modelling public-transport users' behaviour at connection point. *Transport Policy,* 27, 112–122.

Chowdhury, S. and Ceder, A. (2013a). The effects of interchange attributes on public-transport users' intention to use routes involving transfers. *Psychology and Behavioral Sciences,* 27(1), 5–13.

Chowdhury, S. and Ceder, A. (2013b). A psychological investigation on public transport users' intention to use routes with transfers. *International Journal of Transportation, Special Issue: Urban Mobility and Modal Shift towards Public Transport,* 1(1), 3–20.

Chowdhury, S. and Ceder, A. (2013c). Definition of planned and unplanned transfer of public-transport service and users' decision to use routes with transfers. *Journal of Public Transportation,* 16(2), 1–20.

Chowdhury, S., Ceder, A., and Sachdeva, R. (2014). The effects of planned and unplanned transfers on public-transport users' perception of transfer routes. *Transportation Planning and Technology,* 37(2), 154–168.

Chowdhury, S., Ceder, A., and Schwalger, B. (2015). The effects of travel time and cost savings on commuters' perception of public transport routes involving transfers using Just Noticeable Difference. *Journal of Transport Geography* **43**, 151–159.

Collins, C.M. and Chambers, S.M. (2005). Psychological and situational influences on commuter-transport-mode choice. *Environment and Behavior*, 37(5), 640–661.

Cools, M., Moons, E., Janssens, B., and Wets, G. (2009). Shifting towards environment-friendly modes: Profiling travelers using Q-methodology. *Transportation*, 36(4), 437–453.

Currie, G. and Delbosc, A. (2013). Factors influencing young peoples' perceptions of personal safety on public transport. *Journal of Public Transportation*, 16(1), 1–19.

Eriksson, L., Garvill, J., and Nordlund, A.M. (2008). Acceptability of single and combined transport policy measures: The importance of environmental and policy specific beliefs. *Transportation Research*, **42A**, 1117–1128.

Grotenhuis, J.W., Wiegmans, B.W., and Rietveld, P. (2007). The desired quality of integrated multimodal travel information in public transport: Customer needs for time and effort savings. *Transport Policy*, 14(1), 27–38.

Guo, Z. and Wilson, N.H.M. (2007). Modeling the effect of transit system transfer on travel behaviour. *Transportation Research Record*, **2006**, 11–20.

Hull, A. (2008). Policy integration: What will it take to achieve more sustainable transport solutions in cities? *Transport Policy*, 15(2), 94–103.

Iseki, H. and Taylor, B.D. (2009). Not all transfers are created equal: Towards a framework relating transfer connectivity to travel behaviour. *Transport Reviews*, 29(6), 777–800.

Kleinbaum, D.G. and Klein, M. (2010). *Logistic Regression: A Self-Learning Text*, Springer, New York.

Kraft, P., Rise, J., Sutton, S., and Roysamb, E. (2005). Perceived difficulty in the theory of planned behaviour: Perceived behavioural control or affective attitude? *The British Psychological Society*, 4(3), 479–496.

Kyte, R. (2011). A good public transport system must be easy and convenient to use, fast, safe, clean and affordable. *Global: The International Briefing*. http://www.global-briefing.org/2012/07/a-good-public-transport-system-must-be-easy-and-convenient-to-use-fast-safe-clean-and-affordable/#auth. Accessed March 27, 2015.

Molin, E. and Chorus, C. (2009). The need for advanced public transport information services when making transfers. *European Journal of Transport and Infrastructure Research*, 4(9), 397–410.

Muller, T. and Furth, P. (2009). Transfer scheduling and control to reduce passenger waiting time. *Transport Research Record*, **2112**, 111–118.

Murray, S.J., Walton, D., and Thomas, J.A. (2010). Attitudes towards public transport in New Zealand. *Transportation*, 37(6), 915–929.

Murugesan, R. and Rama Moorthy, N.V. (1998). Level of public transport service evaluation: A fuzzy set approach. *Journal of Advanced Transportation*, 32(2), 216–240.

Nilsson, M. and Kuller, R. (2000). Travel behaviour and environmental concern. *Transportation Research*, **5D**, 211–234.

Nordlund, A.M. and Garvill, J. (2003). Effects of values, problem awareness, and personal norm on willingness to reduce personal car use. *Journal of Environmental Psychology*, 23(4), 339–347.

Pedersen, T., Kristensson, P., and Friman, M. (2012). Counteracting the focusing illusion: Effects of defocusing on car users' predicted satisfaction with public transport. *Journal of Environmental Psychology*, 32(1), 30–36.

Pesendorfer, W. (2006). Behavioral economics comes of age: A review essay on advances in behavioral economics. *Journal of Economic Literature*, 44(3), 712–721.

Pramberg, P. (2004). A national move to change travel behaviour. *Transport Engineering in Australia*, 9(2), 49–52.

Redman, L., Friman, M., Garling, T., and Hartig, T. (2013). Quality attributes of public transport that attract car users: A research review. *Transport Policy*, 25, 119–127.

Senbil, M. and Kitamura, R. (2004). Reference points in commuter departure time choice: A prospect theoretic test of alternative decision frames. *Journal of Intelligent Transportation Systems*, 8(1), 19–31.

Sharaby, N. and Shiftan, Y. (2012). The impact of fare integration on travel behavior and transit ridership. *Transport Policy*, 21, 63–70.

Stopher, P.R. (2004). Reducing road congestion: A reality check. *Transport Policy*, 11(2), 117–131.

Taniguchi, A. and Fujii, S. (2007). Promoting public transport using marketing techniques in mobility management and verifying their quantitative effects. *Transportation*, 34(1), 37–49.

Teodorovic, D. (1999). Fuzzy logic systems for transportation engineering: The state of the art. *Transportation Research*, 33A, 337–364.

Tertoolen, G., van Kreveld, D., and Verstraten, B. (1998). Psychological resistance against attempts to reduce private car use. *Transportation Research*, 32A, 171–181.

Tversky, A. and Kahneman, D. (1992). Advances in prospect theory: Cumulative representation of uncertainty. *Journal of Risk and Uncertainty*, 5(4), 297–323.

Van de Kaa, E.J. (2010). Applicability of an extended prospect theory to travel behaviour research: A meta-analysis. *Transport Reviews*, 30(6), 771–804.

Van der Henst, J.B., Carles, L., and Sperber, D. (2002). Truthfulness and relevance in telling the time. *Mind and Language*, 17(5), 457–466.

Van Oort, N., Boterman, J.W., and van Nes, R. (2012). The impact of scheduling on service reliability: Trip-time determination and holding points in long-headway services. *Public Transport*, 4(1), 39–56.

Van Oort, N. and van Nes, R. (2009). Control of public transportation operations to improve reliability: Theory and practice. *Transportation Research Record*, 2112, 70–76.

Volinski, J. and Page, O. (2006). Developing bus transfer facilities for maximum transit agency and community benefit. *Transportation Research Record*, 1955, 3–7.

Wilcox, R.R. (2009). *Basic Statistics: Understanding Conventional Methods and Modern insights*. Oxford University Press Inc., Oxford, New York.

Xu, H., Zhou, J., and Xu, W. (2011). A decision making rule for modeling travelers' route choice behaviour based on cumulative prospect theory. *Transport Research*, 19C, 218–228.

Zadeh, L.A. (1965). Fuzzy sets. *Information and Control*, 8(3), 338–353.

Zhang, L., Li, J.Q., Zhou, K., Gupta, S.D., Li, M., Zhang, W.B., Miller, M.A., and Misener, J.A. (2011). Traveler information tool with integrated real-time transit information and multimodal trip planning. *Transportation Research Record*, 2215, 1–10.

Zhang, L. and Prevedouros, P.D. (2011). User perceptions of signalised intersection level of service using fuzzy logic. *Transportmetrica*, 7(4), 279–296.

Zografos, K., Spitadakis, V., and Androutsopoulos, K. (2008). Integrated passenger information system for multimodal trip planning. *Transportation Research Record*, 2072, 20–29.

Chapter 15

Network (routes) design

CHAPTER OUTLINE

PRACTITIONER'S CORNER

This chapter focuses on the first activity of the operational planning process described in Chapter 1: public transit route design and evaluation at the network level. The problem addressed is that of how to design a new transit network or to redesign an existing network. Naturally, this activity would be appropriate only very infrequently because of the disruption that would be imposed on passengers if wholesale changes were made to the transit network. In addition, many transit agencies have not gone through such a reappraisal, mainly because of the lack of clear-cut, practical, and measurable design criteria for evaluating the *goodness* of transit routes and comparing sets of routes. Theodore Roosevelt said: "Nine tenths of wisdom is being wise in time." Thus, a design and evaluation tool may provide a timely boost for reconstructing transit routes, though one should bear in mind the required inter-connectivity of a transit network.

Following the introductory section of this chapter, Section 15.2 proposes a framework for the construction of the operational objective functions of the transit network design problem. This framework takes into account the passengers as well as agency and community/government interests—the three perspectives emanating from the broad spectrum of transit activities. Section 15.3 utilizes the objective functions described for facilitating the construction of a methodology for the practical design of an efficient transit network of routes. This methodology combines the philosophy of mathematical programming approaches with decision-making techniques; it allows the transit planner to select a set of routes from a number of alternatives. Section 15.4 introduces a model and a solution algorithm to take into account seasonal demand variation, which is a somewhat neglected aspect of transit demand. The solution algorithm developed is a hybrid method that optimizes the design of a bus network at the route and network levels. Section 15.5 introduces a method for solving bus network design problem using genetic algorithm (GA), and Section 15.6 shows how to consider connected exclusive lanes within optimal solution of bus networks.

Practitioners are encouraged to visit all the sections except Section 15.5. In particular, it is recommended that they look at the introductory parts and examples illustrated in all of the sections.

It is worth noting that a new design or a redesign of a transit network needs time to be fully comprehended; this is the *same* time defined as nature's way of keeping everything from happening at once. Last, this Danish proverb seems most apt: "Better to ask twice than to lose your way once." In our case, it is better to present the new routes in at least two clear ways than to have the user misinterpret them even once.

15.1 INTRODUCTION

There are two main approaches to restructuring transit routes: (1) at the route level or for a small group of routes and (2) at the network level. For the first approach, Pratt and Evans (2004), in TCRP Report 95, suggested that restructuring was to simplify routes, accommodate new travel patterns, ease or eliminate transfers, reduce route circuitry, or otherwise alter route configuration. This approach is dealt with to some extent in Chapters 12, 16, and 17. The present chapter proffers practical solutions utilizing the second approach.

Only a few researchers have studied the interrelationship of the scheduling and the network-design planning activities shown in Figure 1.2. The interrelationship has two categories: (a) each set of routes, based on demand, yields a different set of frequencies and timetables and, ultimately, the required fleet size; (b) the operational cost derived from the scheduling activities and passenger level of service affects the search for the optimal route design while relying on a compromise solution between the agency and the passengers.

For many public transit agencies whose network of routes has not been reappraised for anywhere from 20 to 50 years, it is high time to consider such an undertaking. This should provide sufficient motivation to seek an efficient network route-design method, based on certain objective functions and a set of constraints. The main purpose of the methodologies presented in this chapter is to transport a given origin–destination (O-D) demand through the transit network in the most cost-effective way.

The special characteristics of route-design problems are as follows: (1) passenger demand is spread throughout the entire network, where it is generated and terminated at many points along the network's links and can be grouped in terms of an O-D matrix; (2) the demand must be transported simultaneously (usually during peak hours); and (3) over a given planning horizon, it is impossible to reconstruct the routes (i.e., once the route network is designed, it will remain as is over an entire planning period).

Prior approaches to the public transport network-design problem can be grouped into those that simulate passenger flows, those which deal with ideal networks, and those based on mathematical programming. *Simulation models* are presented in Dial and Bunyan (1968), Heathington et al. (1968), and Vandebona and Richardson (1985). These models require a considerable amount of data, and their proximity to optimality is uncertain. *Ideal network methods* are based on a broad range of design parameters and a choice of objectives reflecting user and agency interest. Such methods appear in Kocur and Henrickson (1982), Tsao and Shonfeld (1984), and Kuah and Perl (1988). These methods are adequate for screening or policy analyses, in which approximate design parameters rather than a complete design are determined; thus, these methods cannot represent real situations. *Mathematical programming models* are divided into generalized network-design models and transit-specific networks models. Known generalized network models are well summarized and reviewed in Kim and Barnhart (1999); also, see as an example the heuristics developed by Farvolden and Powell (1994). The transit-specific network-design models are inevitably heuristic because of the extremely high computational effort required. These partial optimization approaches appear in Lampkin and Saalmans (1967), Silman et al. (1974), Dubois et al. (1979), Mandl (1979), and Keudel (1988). None of the models and methods mentioned in this section has actually been applied. However, two survey papers by Guihaire and Hao (2008) and Kepaptsoglou and Karlaftis (2009) review methods to solve the problem of bus network design. The latter performs a three-part systematic review of the problem: design objectives, operating environment parameters, and solution approaches. This three-part classification can be used in defining further research. Section 15.7 provides a further literature review.

An overview of prior work done on this theme calls for a method that, given the availability of typical data, will be more practical and less complex than other methods and models. Such a method, described in Section 15.3, may increase the chances of its acceptance by most transit agencies. The method developed and its related methods, appearing in this chapter, seek to have transit routes as close as possible to the shortest paths.

Finally, it is interesting to observe Figure 15.1, which is currently a distributed info-page at the taxi stand at the Washington (DC) National Airport. This page includes the following note: *Fares are based on shortest route; however, the shortest route may not be the quickest route.* (Which is also a nice way of them to remind us that laughter is the shortest distance between two people.) In our case, the shortest path intends to be the quickest one because

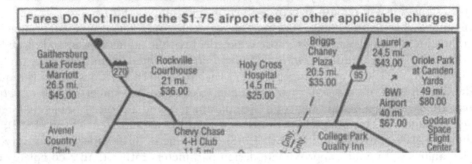

WELCOME TO
RONALD REAGAN
WASHINGTON NATIONAL AIRPORT

Once you have been comfortably seated in the taxicab, please note the Airport Taxi Operator's Permit hanging from the rear view mirror. For future reference, you should record the Permit Number #_____.

(NOTE: Driver is allowed to remove the Permit after exiting the Airport.)

APPROXIMATE FARES FOR THESE AREAS

Fares are based on shortest route, however, the shortest route may not be the quickest route.
Fares are not flat rates, they are APPROXIMATE fares and do not include the $1.75 airport fee.
Your actual fare may be different than the amount shown on the map.

Fares Do Not Include the $1.75 airport fee or other applicable charges

Gaithersburg Lake Forest Marriott 26.5 mi. $45.00

Rockville Courthouse 21 mi. $36.00

Holy Cross Hospital 14.5 mi. $25.00

Briggs Chaney Plaza 20.5 mi. $35.00

Laurel 24.5 mi. $43.00

Oriole Park at Camden Yards 49 mi. $80.00

BWI Airport 40 mi. $67.00

Goddard Space Flight Center

Avenel Country Club

Chevy Chase 4-H Club 11.5 mi.

College Park Quality Inn

Figure 15.1 Current use of the term *shortest route* by the taxi service at the Washington (DC) National Airport (see the marked section under APPROXIMATE FARES ...).

of its time units; in real-time, however, this shortest path may not hold because of traffic congestion (e.g., caused by road accidents or road repairs, etc.).

15.2 OBJECTIVE FUNCTIONS

This section proposes a framework for the construction of the operational objective functions of the public transit network-design problem. This framework takes into account the passengers, the agency, and community interests. It follows the studies by Ceder and Israeli (1992), Israeli and Ceder (1995), Ceder (2001), Ceder et al. (2002), and Yin et al. (2005).

15.2.1 Three perspectives and four criteria

From the literature review presented earlier and that at the end of the chapter, there are no clear-cut, practical, or measurable criteria for evaluating the *goodness* of transit routes and comparing sets of routes at the network level. The only comprehensible matter is that the design of transit routes should be looked at simultaneously from three perspectives: that of the passenger, the agency, and the community/government. These three perspectives emanate from the broad spectrum of transit activities.

Four criteria can be considered when measuring the quality of a transit route: (1) minimum passenger waiting time; (2) minimum empty seat/space time; (3) minimum time difference from shortest path; and (4) minimum fleet size. The first three criteria are measured in passenger-hours, and the last one in number of vehicles. Clearly, criterion (1) represents the passengers' perspective; criteria (2) and (4), the agency's perspective; and criterion (3), both the passengers' and the community's perspectives.

When the purpose of measuring is to compare sets of transit routes or different transit modes, monetary weights can be introduced to the four criteria. Optionally, criterion (3) can be replaced by the total monetary loss (or saving, if it is negative) if all the transit passengers are switched to the shortest path. For instance, when comparing a set of bus routes and a set of metro lines, in which the latter is the shortest path (in time), criterion (3) will provide the total monetary loss/saving if all the bus passengers switch to the metro lines.

The next section constructs a quantitative framework for the four criteria in order to devise tools for the design of an optimal set of transit routes. This quantitative framework furnishes a clearer picture of the four criteria, with two optional versions of criterion (3).

15.2.2 Formulation

The following established notations will be used throughout the analyses of this chapter.

15.2.2.1 Notations

Consider a connected network composed of a directed graph $G = \{N, A\}$ with a finite number of nodes $|N|$ connected by $|A|$ arcs. Define:

Route	Progressive path initiated at a given transit terminal and terminated at a certain node while traversing given arcs in sequence.
Transfer path	Progressive path that uses more than one route.
$R = \{r\}$	Set of transit routes.
$TR = \{tr\}$	Set of all transfer paths.
$S = \{sp\}$	Set of all shortest paths (minimum average travel times).
N_r	Set of nodes located on route r.
N_{tr}	Set of nodes located on transfer path tr.
N_{sp}	Set of nodes located on the shortest path sp.
d_{ij}^r	Passenger demand between i and j, i, j, $\in$ N, riding on route r.
d_{ij}^{tr}	Passenger demand between i and j along the transfer path t_r.
d_{ij}^{sp}	Passenger demand between i and j along its shortest path.
F_r	Vehicle frequency associated with route r.
F_{min}	Minimum frequency (reciprocal of policy headway) required.
t_{ij}^r	Average travel time between i and j on route r.
t_{ij}^{tr}	Average travel time between i and j on transfer path t_r (can include transfer penalties).
t_{ij}^{sp}	Average travel time between i and j on its shortest path.
t_r	Overall travel time on route r between its start and end.
L_r	Maximum passenger load on route r.
w_r	Passenger waiting time on route r.
d_o	Desired occupancy on each vehicle (load standard).

$$a_{tr}^r \qquad \begin{cases} 1, & \text{transfer tr moves through route r} \\ 0, & \text{otherwise.} \end{cases}$$

α Maximum allowed deviation from shortest path for any O-D pair on a transit path (including transfers).

k_{tr} Maximum degree of transfer path t_r (number of vehicle changing).

15.2.2.2 Two principal objective functions

The transit network-design problem is based on two principal objective functions, minimum Z_1 and minimum Z_2, across the different sets of transit routes:

$$Z_1 = \begin{bmatrix} a_1 \sum_{i,j \in N} WT(i,j) + a_2 \sum_r EH_r + a_4 \sum_{i,j \in N} DPH(i,j), & \text{for single set} \\ a_1 \sum_{i,j \in N} WT(i,j) + a_2 \sum_r EH_r + \sum_{i,j \in N} [a_3 PH(i,j) - a_4 DPH(i,j)], & \text{for comparison} \end{bmatrix} \qquad (15.1)$$

$$Z_2 = FS \qquad (15.2)$$

where

PH (i, j) is the passenger hours between nodes i and j, i, j $\in$ N (defined as passengers' riding time in a transit vehicle on an hourly basis; it measures the time spent by passengers in vehicles between the two nodes)

DPH (i, j) is the difference in passenger hours between PH (i, j) and total passenger hours from i to j when using only the shortest path, i, j $\in$ N

WT (i, j) is the waiting time between nodes i and j, i, j $\in$ N (defined as the amount of time passengers spend at the transit stops between the two nodes)

EH_r is the empty seat/space-hours on route r (defined as the unused seats/spaces in a transit vehicle on an hourly basis; empty seat/space-hours measures the unused capacity on vehicles)

FS is the fleet size (number of transit vehicles needed to provide all trips along a chosen set of routes)

a_k is a monetary or other weights, k = 1, 2, 3, 4

Equation 15.1 contains the two options for the Z_1 objective function, which can be interpreted as minimum waiting time and maximum utilization; for given weights of 1 or without units, this equation results in units of passenger hours. Equation 15.2 is simply the required minimum fleet size.

15.2.2.3 Objective function components

Equations 15.1 and 15.2 essentially combine five objective function components. The first straightforward objective is to minimize passengers' total waiting time. This is strictly the perspective of the transit user. The formulation of this objective takes the following form:

$$\text{Min } a_1 \sum_{i,j \in N} WT(i,j) \qquad (15.3)$$

where a_1 is the monetary value of 1 h's waiting time.

The second objective is to minimize the total unused seat capacity so as to allow more viable transit service. This is strictly the perspective of the agency, which wishes to see more occupation of the available seats. The following is the formulation of this objective:

$$\text{Min } a_2 \sum_r EH_r \tag{15.4}$$

where a_2 is the equivalent of 1 h's average monetary revenue divided by the average number of hourly boarding passengers. This objective is to minimize the total monetary value of unused seat capacity.

The third and fourth objectives are two versions of the same objective: (a) to minimize the total time loss (in monetary value) between riding the transit vehicle and traveling by automobile (assumed to be the shortest path) and (b) to minimize the total loss (in monetary value) if all the passengers are switched to the shortest path. Versions (a) and (b) take the following forms, respectively:

$$\text{Min } a_4 \sum_{i,j \in N} DPH(i,j) \tag{15.5}$$

$$\text{Min } \sum_{i,j \in N} [a_3 PH(i,j) - a_4 DPH(i,j)] \tag{15.6}$$

where
   a_3 is the equivalent of 1 h's difference in average cost/fare between riding the shortest path (automobile or a transit competitor) and the transit route
   a_4 is the monetary value of 1 h's in-vehicle time

The value of Equation 15.6 is the total monetary loss (or saving, if it is negative) if all the passengers are switched to the shortest path, where $a_3 PH$ = total monetary loss, with respect to cost/fare only, if all the passengers are switched to the shortest path; and $a_4 DPH$ = total monetary value of the time saved if all the passengers are switched to the shortest path. The latter fits Equation 15.5. These objectives represent the perspectives of the community/government and the passengers.

The fifth objective is to minimize the number of vehicles required for a given set of routes and frequencies (timetables). This is strictly the agency perspective, which wishes to perform all transit trips using a minimal number of vehicles. This objective takes the form:

$$\text{Min } FS \tag{15.7}$$

Objectives (15.3) through (15.6) are all in terms of passenger-hour cost; for the sake of simplicity, therefore, this can be summed up to Min Z_1 as shown in Equation 15.1. Objective (15.7) stands alone to some extent and is termed Min Z_2 in Equation 15.2.

15.2.2.4 Calculation of Z_1 functions

The objective function Z_1 is based on the passenger-load profile. Figure 15.2a presents the example of a small transit network. The input data during a given time period (usually peak hours) consist of average travel times, an estimated O-D demand in Figure 15.2b, and $a_k = 1$ for all k = 1, 2, 3, 4. Figure 15.2c displays the feasible routes complying with the following five practical constraints: (i) maximum route length of 30 min; (ii) maximum allowed

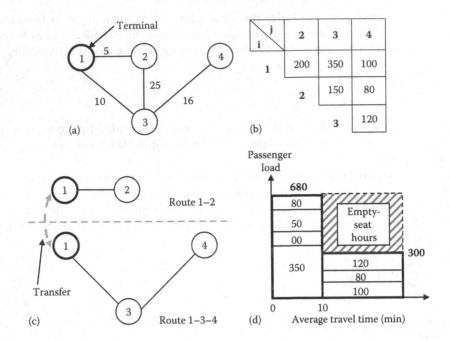

Figure 15.2 Example network, its given data, and the construction of a load profile: (a) sample network with average travel time in minutes (given); (b) symmetric O-D matrix in passenger/h; (c) feasible routes in both directions, complying with the four criteria; and (d) the load profile of route 1–3–4 (derived).

deviation of 40% from the shortest path; (iii) maximum of one transfer for all O-D pairs; (iv) a chosen route should not be included in another feasible route; and (v) no circular routes considered.

Because routes can start only at terminals (Node 1 in the example), a mapping process is applied, with constraint (i) considered first. This results in six routes: [1–2], [1–2–3], [1–2–3–4], [1–3], [1–3–2], and [1–3–4], all of which comply with the maximum of 30 min average travel time. In parallel, a shortest-path algorithm (see Appendix 9.A) is applied. Constraint (ii), maximum deviation from shortest path, can then be checked using the shortest-path information. Table 15.1 provides this check while describing the procedure for a determination of feasible routes, that is, making sure that a feasible route will not be included in an another feasible route in order to avoid overlapping.

Once the feasible routes are established, their load profiles are constructed, based on the input data. The load profile of route [1–3–4] is presented in Figure 15.2d. The load on 1–3 is the maximum load (L_{1-3-4} = 680 passengers) of the O-D demand: 1–3, 1–4, 2–3, and 2–4 (see Figure 15.2b). The load on 3–4 is 300, consisting of the O-D demand: 1–4, 2–4, 3–4. According to the 680 maximum load, the frequency of route [1–3–4] would be (680/50), F_{1-3-4} = 13.6 veh/h. The shaded area of the load profile between 680 and 300 passengers along the 16 min segment represents 101.3 empty seat-hours for d_o = 50 (number of seats on the vehicle) and F_{min} = 4 during the associated time period of the example. In addition, the load profile of the other feasible route, 1–2, is simply rectangular with a maximum load of L_{1-2} = 200 + 150 + 80 = 430 passengers and F_{1-2} = 8.6.

The first function of Z_1 in Equation 15.1 is the total wait-time hours both at the transit stops and during transfers. Different formulations of the expected passenger waiting time appear in Sections 11.2 and 18.4. Utilizing Equation 11.1, in which passengers arrive

Table 15.1 Determination of feasible routes in the example problem

Route end points	$1 \to 2$	$1 \to 3$	$1 \to 4$
Shortest-path (min)	5	10	26
Shortest-path route	1–2	1–3	1–3–4
Symmetric (both directions), O-D served optimally	1–2	1–3	1–3, 1–4
Other routes not complying with the maximum length criterion (40%)	1–3–2	1–2–3	1–2–3–4
Symmetric (both directions), O-D served by other routes, and their deviation (%) from the shortest path	1–2 (600%)	1–2 (0%)	1–2 (0%)
	1–3 (0%)	1–3 (200%)	1–3 (200%)
	2–3 (67%)	2–3 (67%)	1–4 (77%)
			2–3 (67%)
			2–4 (32%)
			3–4 (0%)
Feasible route	1–2	1–3–4	
Symmetric (both directions), O-D served optimally via one transfer	2–3, 2–4		

randomly at the transit stop and headways are deterministically distributed, we obtain the expected waiting time on route r, which is half the headway:

$$w_r = \frac{1}{2F_r}, \quad \text{for all } r \in R \tag{15.8}$$

Thus,

$$\sum_{i,j \in N} WT(i,j) = \sum_{r \in R} \frac{1}{2F_r} \left(\sum_{i,j \in N_r} d_{ij}^r + \sum_{ij \in N_{tr}} d_{ij}^{tr} a_{tr}^r \right) \tag{15.9}$$

Applying Equation 15.9 to the example of Figure 15.2 yields (for both directions of routes):

$$\sum_{i,j \in N} WT(i,j) = 2 \left[\begin{array}{c} \frac{1}{2 \times 13.6}[(350 + 100 + 120) + (150 + 80)] \\ + \frac{1}{2 \times 8.6}[(200) + (150 + 80)] \end{array} \right] = 108.8 \, \text{pass.-h}$$

where $d_o = 50$. Note that any demand between i and j, i, j $\in$ N, can split (e.g., some via a direct route and the remaining via transfers); in this case, an assignment procedure (see Chapter 11) can be applied. However, this is not the case of the example.

The second function of Z_1 in Equation 15.1 describes the total empty-space hours or empty-seat hours (when d_o equals the number of seats on the vehicle). This function represents an unproductive measure for the agency (e.g., unused seat capacity). Its formulation is

$$\sum_r EH_r = \sum_{r \in R} [\max (L_r, F_{min} \cdot d_o)] t_r - \sum_{i,j \in N} PH(i,j) \tag{15.10}$$

In the example of Figure 15.2, this equation yields $EH_r = 2(101.3 + 0) = 202.6$ passenger-hours for both directions of the feasible routes.

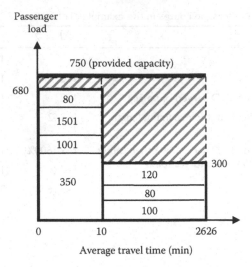

Figure 15.3 The load profile of route [1–3–4] in case of minimum frequency provided.

It is to note that the empty-seat hour calculation is based on the provided frequency usually max load based; however, if this frequency is the minimum required, the empty-seat hour would be calculated based only on the frequency. For instance, if in route [1–3–4] we want to have a clock-headway of 4 min (frequency of 15 veh/h) then the load profile is as shown in Figure 15.3, and

$$EH_r = 2\left[(750-680)\left(\frac{10}{60}\right)+(750-300)\left(\frac{16}{60}\right)\right] = 240.4 \, \text{pass.-h}$$

The first version of the third function in Equation 15.1, which appears in Equation 15.5, is the total passenger-hour difference between PH(i, j) on r and PH(i, j) on the shortest path sp:

$$\sum_{i,j\in N} DPH\,(i,j) = \sum_{i,j\in N} PH\,(i,j) - \sum_{sp\in S}\sum_{i,j\in N_{sp}} d_{ij}^{sp}\, t_{ij}^{sp} \tag{15.11}$$

in which, for the example in Figure 15.2, $\Sigma DPH = 0$; the reason for this is that routes [1–2] and [1–3–4] include only the shortest paths.

The second version of the third function in Equation 15.1, which appears in Equation 15.6, has two parts, the second part being Equation 15.11. The first part is total passenger-hours in the routing system:

$$\sum_{i,j\in N} PH\,(i,j) = \sum_{r\in R}\sum_{i,j\in N_r} d_{ij}^{r} t_{ij}^{r} + \sum_{tr\in TR}\sum_{ij\in N_{tr}} d_{ij}^{tr} t_{ij}^{tr} \tag{15.12}$$

For the example in Figure 15.2, Equation 15.12 yields

$$\sum_{i,j\in N} PH = \frac{2\left[(680\times10)+(300\times16)+(430\times5)\right]}{60} = 458.3 \, \text{pass.-h}$$

in which the transfers (both directions of 2–3, 2–4) do not include penalties (estimated extra-effort cost).

15.2.2.5 Estimation of Z_2

Estimation of the minimum fleet size can utilize the simple calculation of Section 6.2 or the deficit function modeling described in Chapters 6 and 7. Note that it may be sufficient to use the procedure to determine the stronger fleet size lower bound (see Section 7.2.2) for this estimation. Practically speaking, the design of optimal transit routes involves a vast amount of computations of sets of routes. Thus, the lower-bound-based Z_2 calculation can ease the computation effort for each route considered.

15.2.3 Applications

The framework of the objective functions described is believed to be a useful tool set for more than designing a new transit network of routes. For instance, this framework can apply to

- Optimal design for expansion or curtailment of an existing transit network of routes
- Assessment of the performance of an existing transit network from the aspects of: (1) agency efficiency (fleet size, empty-seat hours, length of routes, number of transfers) and (2) passenger level of service (average waiting time, deviation from shortest path, crowding level)
- Sensitivity analysis of transit network performance for a variety of system parameters (such as different fleet sizes, different levels of service, changes in passenger demand, changes in frequencies, changes in travel time, and more)

One application that utilized the objective functions described was presented by Yin et al. (2005). This application proposes a deployment-planning framework that provides, in a sequence of steps, a general structure for the optimal deployment of buses in a rapid transit (BRT) system. The following BRT elements were considered for system enhancement:

1. Bus signal priority (extended over existing deployment)
2. Exclusive lanes
3. Articulated buses
4. Multiple door boarding and alighting
5. Stop enhancements
6. Electronic fare payment
7. Precision docking

Given these BRT elements, the deployment-planning framework was used to determine cost-efficient combinations for system enhancement. Equations 15.1 through 15.6 were then utilized to calculate performance measures for each combination of BRT elements. From these calculations, an optimal combination was selected and recommended for deployment. Table 15.2 presents seven alternative combinations of BRT elements considered in the study, together with their estimated cost (in millions of $US).

In the Yin et al. (2005) study, the budget limit for implementing the BRT elements was considered as given in the amount of $90 million. If the number was high enough, there would be no trade-off between elements. Therefore, the financially feasible alternatives are C, D, and G. Moreover, by refining Alternative E through excluding element e, stop enhancement, and making it financially feasible, a new Alternative H is created that includes the elements a, c, d, and f. The total cost is $83.9 million.

Table 15.2 Cost estimates for combinations of BRT elements

Alternative	BRT elements	Cost ($M)
A	a, b, c, d, e, f	99.9
B	a, b, c, d, e, f, g	101.4
C	a, b, e, f	19.9
D	a, b, e, f, g	21.4
E	a, c, d, e, f	93.9
F	a, c, d, e, f, g	95.4
G	a, b, c, d	86.6

Source: Yin, Y. et al., Transport. Res. Rec., 1903, 11, 2005.

Table 15.3 Evaluation results for combinations of BRT elements

	Passenger travel time (pass.-h)	Passenger waiting times (pass.-h)	Empty seat-hours (pass.-h)	Z1	Z2
Alternative D	1999	94	1614	23,966	138
Alternative G	2025	139	1639	28,648	84
Alternative H	2075	139	1665	29,065	86

Source: Yin, Y. et al., Transport. Res. Rec., 1903, 11, 2005.

The results of the Yin et al. analysis are shown in Table 15.3, in which the objective is to find the best alternative with minimum Z_1 and minimum Z_2. It can be seen from Table 15.3 that Alternative H is dominated by Alternative G and that both G and D are non-dominated. Although the capital cost of Alternative D is much less than that of Alternative G, the former requires a much larger fleet size, which may lead to higher operating and maintenance costs. Therefore, transit agencies should look carefully at the trade-off between these two objectives and, based on their preferences and non-quantitative considerations, decide which of the two recommended alternatives (G and D) should be deployed.

15.3 METHODOLOGY AND EXAMPLE

The objective functions described facilitate the construction of a methodology for the practical design of an efficient transit network of routes. This methodology, outlined in this section together with an example, combines the philosophy of mathematical programming approaches with decision-making techniques; it allows the transit planner to select a set of routes from a number of alternatives. The methodology and the example follow Ceder and Israeli (1992) and Ceder (2003).

15.3.1 Six-element methodology

The transit route-design methodology consists of six elements, shown in Figure 15.3. The *first element* generates every feasible route and transfer (throughout the entire network) from all terminals, including the shortest-path computation. Initially, the network contains average travel times covering a time window, which is usually the peak period. These measured average travel times are then used as input for the calculation of the shortest path between each origin–destination (O-D) pair. Each candidate route determined meets the route length–factor constraint; that is, one procedure of this element screens out routes according to given boundaries of a route length. In addition, there is a limit on the route's average travel time between each O-D pair. That is to say, a given passenger demand, usually during peak hours, cannot be assigned to a candidate route if its average travel time exceeds the shortest-path travel time by more than a given percentage. Feasible transfers (between O-D pairs without direct routes) are based on establishing additional direct routes between O-D pairs characterized by high O-D demands (predetermined O-D); feasibility is determined by the travel-time limit in comparison with the shortest path. These extra direct routes are initiated and/or terminated at non-terminal nodes, and consequently deadheading trips are responsible for their connection to the terminals. The feasible transfers are created using a mapping algorithm (branching of routing possibilities along with a check of constraints). Finally, low O-D demand, without a direct route, may not be considered for implementing a transit service. The following example further explains this element.

The *second element* in Figure 15.4 creates a minimum set of routes and related transfers, such that connectivity between nodes is maintained and their total deviation from the shortest path is minimized. This problem is defined as a set-covering problem (SCP) similar to the one in Section 12.4.3. The SCP determines the minimum set of routes from the matrix of feasible routes, in which each column represents either a feasible route or a feasible transfer.

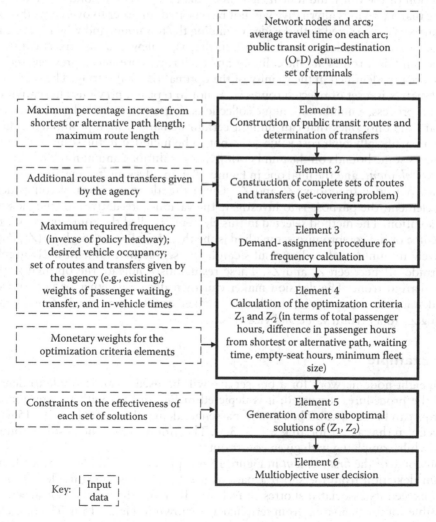

Figure 15.4 A methodology for designing public transit routes.

The procedures included in this element are described in the first edition of this book, in Section 14.4.

The *third element* assigns the entire O-D demand to the chosen set of routes. The assignment algorithm follows the procedure described in Section 11.5 and includes steps that are related to a route choice decision investigation. That is, the algorithm includes a probabilistic function for passengers who are able to select the transit vehicle that arrives first or, alternatively, who wait for a faster vehicle. The passengers' strategy is to minimize the total weight of wait, transfer, and in-vehicle times.

The *fourth element* represents the optimization criteria from the passengers', agencies', and community/governments' perspectives. It is detailed in the previous section and based on Equations 15.1 through 15.6.

The *fifth element* is responsible for constructing alternative sets of routes in order to search for additional (Z_1, Z_2) values in the vicinity of their optimal setting. The procedure for this search is based on incremental changes in the set of routes, much like the known reduced-gradient methods. Given the set of routes associated with the minimum Z_1 value, the single route that is the worst contributor to Z_1 is deleted and then the SCP is resolved, followed by the execution of the third and fourth elements. This process could continue, but there is no guarantee that a previous alternative will not be repeated. In order to overcome this problem, a new matrix is constructed with the idea of finding the minimum and worst set of candidate routes for possible deletion in each iteration, that is, a new SCP matrix is constructed in which the candidate routes are the columns and each row represents a previous set of routes that was already identified in the vicinity of the optimal (Z_1, Z_2) setting. The solution to this new SCP matrix is a set of rejected routes so as not to repeat a previous alternative solution. During this process, a number of unique collections of routes are termed *prohibited columns*, as they are the only ones that can fulfill a certain demand. These prohibited columns are assigned an artificially high cost value so as not to be included in the solution. This process also involves some bounds on the number of (Z_1, Z_2) solutions and number of iterations.

The *sixth element,* and the final one in Figure 15.4, involves multi-objective programming of the two objective functions, Z_1 and Z_2. Given the alternative sets of routes derived in the fifth element, the purpose is to investigate the various alternatives for the most efficient (Z_1, Z_2) solution. The method selected in this element is called the compromise-set method. The outcome of the compromise-set method is the theoretical point at which (Z_1, Z_2) attains its relatively minimum value. The results can be presented in a table or a 2D graph showing the trade-off between Z_1 and Z_2. These results also indicate the optimal zone or the so-called Pareto front. The decision maker can then decide whether or not to accept the proposed solution; for example, the decision maker can see how much Z_1 is increased by decreasing Z_2 to a certain value and vice versa.

15.3.2 Example

A simple eight-node network for a bus service will be used as an example to demonstrate some of the procedures discussed; it is depicted in Figure 15.5 with two terminals (from which trips can be initiated). The input of passenger demand appears in Table 15.4; in addition, it is given that $a_k = 1$ for all $k = 1, 2, 3, 4$. The aim is to find the best bus routes in the network while complying with given constraints.

The outcome of the *first element* in Figure 15.4 is presented in Table 15.5, while using the maximum deviation from the shortest path as $\alpha = 0.4$, that is, no route length or portion of it can exceed its associated shortest travel time by more than 40%. Construction of the nine feasible routes emanating from terminal 1 is shown in Figure 15.6. The first element is

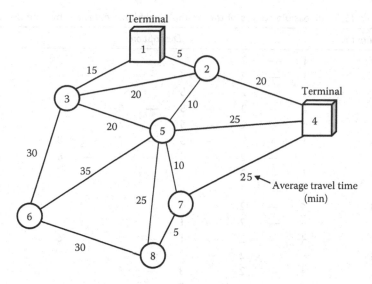

Input : Maximum increase (from shortest/alternative path) factor, $\alpha = 0.4$
Maximum number of transfers allowed, $k_{tr} = 2$
Desired vehicle occupancy, $d_o = 50$ passengers

Figure 15.5 Example of an eight-node network with its basic input.

Table 15.4 Passenger demand between nodes (assumed to be symmetrical) for the example problem

Nodes	2	3	4	5	6	7	8
1	80	70	160	50	200	120	60
2		120	90	100	70	250	70
3			180	150	120	30	250
4				80	210	170	230
5		Symmetrical			250	40	130
6						130	120
7							70

based on an algorithm that mainly produces feasible transfers throughout the entire network. The transfers that are created using a mapping algorithm are shown in Table 15.6 for the example problem, the numbers in parentheses being the route numbers of Table 15.5 that comprise the transfers. In the transfer-path description, the numbers outside the parentheses represent nodes while those inside the parentheses represent routes.

The *second element* of the methodology creates a minimum set(s) of routes and related transfers, defined as an SCP matrix. Each row in this matrix represents either a feasible route or a transfer. The "1" and "2" in the matrix are inserted whenever an O-D demand can be, respectively, feasibly and optimally (shortest path) transported by the route or transfer; or "0" otherwise. The word *covering* in the SCP refers to at least one column with "1" in each row. Section 14.4 of the first edition of this book provides further description of the SCP analysis. In the example problem, about 100 sets of routes were generated. An example of one set is {4, 6, 9, 11, 25, 28, 32, 46}, in which the numbers refer to those in Tables 15.5 and 15.6; five routes and two transfer paths cover all the O-D pairs in the defined feasible manner. The results of the SCP analysis for this set of routes appear in Table 15.7.

Table 15.5 All feasible routes of the example problem generated by Element I

Route no.	Description						
1	1 — 2						
2	1 — 2 — 4						
3	1 — 2 — 5						
4	1 — 2 — 5 — 6						
5	1 — 2 — 5 — 7						
6	1 — 2 — 5 — 7 — 8						
7	1 — 2 — 5 — 7 — 8 — 6						
8	1 — 3						
9	1 — 3 — 6						
10	4 — 2						
11	4 — 2 — 1 — 3						
12	4 — 2 — 1 — 3 — 6						
13	4 — 2 — 3						
14	4 — 2 — 3 — 6						
15	4 — 2 — 5						
16	4 — 2 — 5 — 6						
17	4 — 5						
18	4 — 5 — 3						
19	4 — 5 — 6						
20	4 — 5 — 7						
21	4 — 5 — 7 — 8						
22	4 — 5 — 7 — 8 — 6						
23	4 — 7						
24	4 — 7 — 5						
25	4 — 7 — 5 — 3						
26	4 — 7 — 5 — 6						
27	4 — 7 — 8						
28	4 — 7 — 8 — 6						

The *third element* assigns the entire O-D demand to the chosen set of routes. Vehicle frequency in this element is derived from the passenger-load profile of each route; the load on each route is determined by the demand and assignment method. The *fourth element* calculates the optimization parameters PH_r, DPH_r, WH_r, EH_r on a selected route basis for computing Z_1, and determines the minimum fleet size Z_2 required to meet the passenger demand. For instance, the calculated values of these optimization parameters of the set of routes, of Table 15.7, are shown in Figure 15.7. The step-by-step derivation of these parameters is shown in Figure 15.7 for Route 4 (the first route in the set); that is, given the assigned passenger demand for each O-D pair (the result of the third element), a load profile is constructed. Then, the max load, frequency (from which the waiting time is derived), and empty-space hours can be calculated. The values of PH_r and DPH_r depend on the average travel time for each O-D pair as well as on its assigned demand.

The *fifth element* is responsible for constructing alternative sets of routes in order to search for additional (Z_1, Z_2) values in the vicinity of their optimal setting. In the example problem, nine sets were produced by this element; these sets are as shown in Table 15.8, including the resultant Z_1 and Z_2 values.

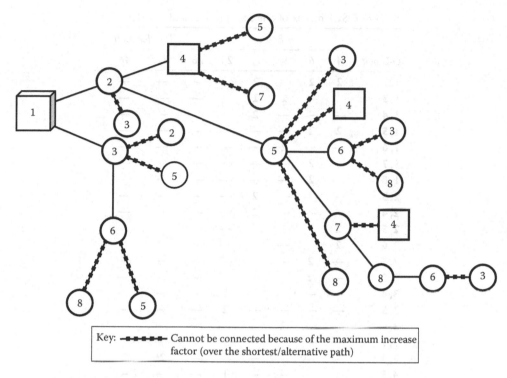

Key: ▬■▬■▬ Cannot be connected because of the maximum increase factor (over the shortest/alternative path)

Figure 15.6 Example solution steps in generating all (nine) feasible routes from Terminal 1.

Table 15.6 All transfers for the example problem connecting nodes 3 and 8 during their 35 min shortest path

Transfer no.			Description			
29(5, 18, 27)	3(18)	—	5(5)	—	7(27)	—	8
30(5, 18, 28)	3(18)	—	5(5)	—	7(28)	—	8
31(6, 18)	3(18)	—	5(6)	—	7(6)	—	8
32(6, 25)	3(25)	—	5(25, 6)[a]	—	7(6)	—	8
33(7, 18)	3(18)	—	5(7)	—	7(7)	—	8
34(7, 25)	3(25)	—	5(25, 7)	—	7(7)	—	8
35(18, 20, 27)	3(18)	—	5(20)	—	7(27)	—	8
36(18, 20, 27)	3(18)	—	5(20)	—	7(28)	—	8
37(18, 21)	3(18)	—	5(21)	—	7(21)	—	8
38(18, 22)	3(18)	—	5(22)	—	7(22)	—	8
39(18, 24, 27)	3(18)	—	5(24)	—	7(27)	—	8
40(18, 24, 28)	3(18)	—	5(24)	—	7(28)	—	8
41(18, 26, 27)	3(18)	—	5(26)	—	7(27)	—	8
42(18, 26, 28)	3(18)	—	5(26)	—	7(28)	—	8
43(21, 25)	3(25)	—	5(25, 21)	—	7(21)	—	8
44(22, 25)	3(25)	—	5(25, 22)	—	7(22)	—	8
45(25, 27)	3(25)	—	5(25)	—	7(27)	—	8
46(25, 28)	3(25)	—	5(25)	—	7(28)	—	8

Note: Route numbers in parenthesis.

[a] When there is a possibility of more than one transfer, the node in which the transfer is considered indicates the two routes in the parentheses.

Table 15.7 SCP matrix of one set of routes and transfers

O-D pair	Routes						Transfer path	
	4	6	9	11	25	28	32	46
1, 2	2	2	—	2	—	—	—	—
1, 3	—	—	2	2	—	—	—	—
1, 4	—	—	—	2	—	—	—	—
1, 5	2	2	—	—	—	—	—	—
1, 6	—	—	2	—	—	—	—	—
1, 7	—	2	—	—	—	—	—	—
1, 8	—	2	—	—	—	—	—	—
2, 3	—	—	—	2	—	—	—	—
2, 4	—	—	—	2	—	—	—	—
2, 5	2	2	—	—	—	—	—	—
2, 6	2	—	—	—	—	—	—	—
2, 7	—	2	—	—	—	—	—	—
2, 8	—	2	—	—	—	—	—	—
3, 4	—	—	—	2	1	—	—	—
3, 5	—	—	—	—	2	—	—	—
3, 6	—	—	2	—	—	—	—	—
3, 7	—	—	—	—	2	—	—	—
3, 8	—	—	—	—	—	—	2[a]	2[a]
4, 5	—	—	—	—	1	—	—	—
4, 6	—	—	—	—	—	2	—	—
4, 7	—	—	—	—	2	2	—	—
4, 8	—	—	—	—	—	2	—	—
5, 6	2	—	—	—	—	—	—	—
5, 7	—	2	—	—	2	—	—	—
5, 8	—	2	—	—	—	—	—	—
6, 7	—	—	—	—	—	2	—	—
6, 8	—	—	—	—	—	2	—	—
7, 8	—	2	—	—	—	2	—	—

—, The O-D pair is not covered by the route.

1, Covered but not optimally.

2, Optimally covered.

[a] Optimum path: contains a single transfer.

The *sixth element* of the methodology involves multiobjective programming of the two objective functions, Z_1 and Z_2. Given the alternative sets of routes derived in the fifth element, the purpose here is to investigate which set provides the more efficient solution. The trade-off situation regarding the example problem is depicted in Figure 15.8. The lower left corner of the envelope contour of the nine solutions represents the best sets; for a bi-objective problem, this boundary is called the Pareto front (Coello et al., 2002). The user is then able to choose a desired solution with this information in hand. In the example case, the choice is between [$Z_1 = 787$, $Z_2 = 106$], [$Z_1 = 866$, $Z_2 = 102$], and [$Z_1 = 997$, $Z_2 = 101$] for sets 1, 4, and 6. More details of the multi-objective analysis appear in the first edition of this book.

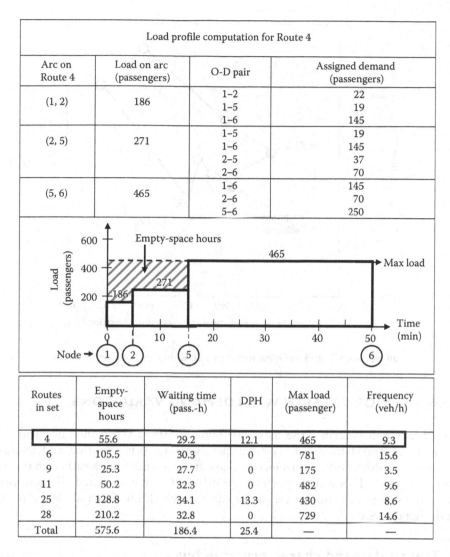

	Load profile computation for Route 4		
Arc on Route 4	Load on arc (passengers)	O-D pair	Assigned demand (passengers)
(1, 2)	186	1–2 1–5 1–6	22 19 145
(2, 5)	271	1–5 1–6 2–5 2–6	19 145 37 70
(5, 6)	465	1–6 2–6 5–6	145 70 250

Routes in set	Empty-space hours	Waiting time (pass.-h)	DPH	Max load (passenger)	Frequency (veh/h)
4	55.6	29.2	12.1	465	9.3
6	105.5	30.3	0	781	15.6
9	25.3	27.7	0	175	3.5
11	50.2	32.3	0	482	9.6
25	128.8	34.1	13.3	430	8.6
28	210.2	32.8	0	729	14.6
Total	575.6	186.4	25.4	—	—

Figure 15.7 Computation of the load profile of Route 4 and the analysis required for Set 1 of routes (#4, 6, 9, 11, 25, 28) and transfers (#32, 46).

Table 15.8 All selected sets of the example problem and their objective functions values

Set	Description	Z_1	Z_2
1	{4, 6, 9, 11, 25, 28, 32, 46}	787	106
2	{7, 9, 11, 19, 25, 27, 34, 45}	900	109
3	{7, 9, 11, 25, 28, 34, 46}	1105	117
4	{6, 9, 11, 16, 25, 28, 32, 46}	866	102
5	{6, 12, 19, 25, 28, 32, 46}	937	103
6	{7, 12, 25, 27, 34, 45}	997	101
7	{7, 12, 25, 28, 34, 46}	1213	113
8	{4, 6, 12, 25, 28, 32, 46}	869	103
9	{6, 12, 16, 25, 28, 32, 46}	961	105

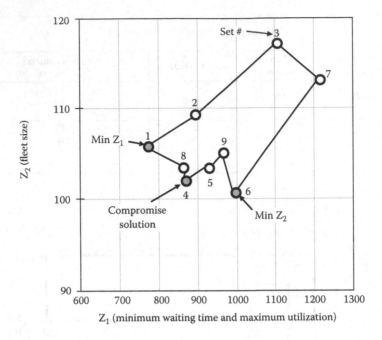

Figure 15.8 Trade-off between Z_1 and the minimum fleet size Z_2 in the example problem.

15.4 BUS NETWORK DESIGN WITH DEMAND VARIATIONS

Chapter 10 lists and describes the main attributes to affect the transit demand. That is, changes in fares, travel time, service frequency, walking time, routing and transferring, stop location, comfort and inconvenience elements, information, weather changes, tourism flows, and more. This section portrays the different demand-related characteristics and focuses on the design of transit network for cases in which the demand varies in a certain pattern during the year.

15.4.1 Transit demand characteristic in bus network design problem

Three types of demand variations attribute-related can be identified as follows.

15.4.1.1 Service dependency

Transit demand is an output of the transportation-demand analysis process. Modal-split procedure defines the share of transit demand, out of the total demand in a city, by using the utility of transportations modes. The utility of a transit mode is derived from current transit system attributes, such as travel time, number of transfers, and convenience. It can be easily seen that transit network design would result in a new set of utilities and consequently with a new transit demand figure. This variation of transit demand is called service dependency. Lee and Vuchic (2005), Fan and Machemehl (2006), and Van Nes et al. (1988) have all considered service dependency of demand in their attempt to solve a transit network design problem. Figure 15.9 shows the concept of service dependency with respect to the bus network design problem (BNDP).

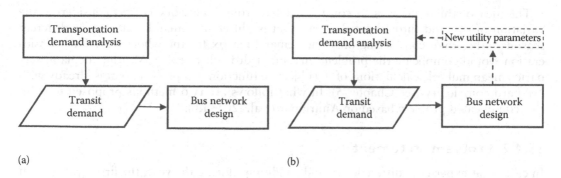

Figure 15.9 Service dependency (fixed vs. elastic demand) related to the BNDP. (a) Fixed demand consideration for bus network design and (b) elastic demand consideration for bus network design.

15.4.1.2 Temporal variation

Passenger demand constantly changes during the days of a week and the weeks of a month according to people's activities and seasonal requirements. These temporal variations can produce a continuous demand function with the possibility to characterize it by peak and off-peak demand. One important aspect of temporal variations is related to seasonal demand with its annual variation pattern. For instance, summer tourism can significantly increase the transit demand. Generally speaking, in many cities recreational trips increase the demand during summer with a reduction of educational and work trips; it results in a complete different demand pattern than in other seasons. This multi-period demand consideration has been studied on one occasion by Chang and Schonfeld (1991). They developed an analytic model for optimizing bus service, considering the time dependency and elasticity in demand characteristics. Their analytic model has been developed for feeder bus systems and cannot be applied for the entire network. Although their method was a step forward in this area, the simplifications made of network description and decision variables do not allow for utilizing it with real-life transit networks. For instance, dividing the city into triangular zones and optimizing line spacing for the feeder network is not a realistic assumption, especially in irregular city networks.

15.4.1.3 Randomness

Random demand variation is another aspect of changes in transit demand. This aspect was studied by Yan et al. (2006) in their investigation of intercity bus routing and scheduling. However, the random fluctuations of demand are negligible compared with the significant changes attribute based. For example, the peak (or off-peak) hours demand fluctuations between two days are significantly smaller than the differences between the observed peak and off-peak demands for a given day. Accordingly, and also because of the complexity involved with the consideration of demand distribution functions, this type of variation is assumed to be insignificant in the analysis of the network (routes) design problem.

A good choice related to transit-demand characteristics can make the problem closer to reality. There are a number of cities all over the world that experience demand variations as a result of tourist arrivals and other demand-related activities. Currently, these situations are the main reasons for the change in schedule of the existing transit service. Naturally, there is a need to investigate the extent to which changes in the transit routing can assist in attaining a better network-design solution when considering the variation of demand.

The idea of this section is to construct a base-robust network (BRN) considering the variation of demand throughout the year, that is, different demand matrices for different seasons rather than considering only one demand matrix for the whole year. This consideration would complicate the problem, making it difficult to solve. Multiple-demand scenarios mean multiple calculations of the objective function in a problem that is already with NP-hard complexity (see Chapter 5). In what follows are two methods proposed to solve this complicated problem based on Amiripour et al. (2014a,b).

15.4.2 Problem statement

In cities that experience different demand conditions during the year, the first strategy that transit agencies choose is to reschedule the routes and provide extra services where needed. If this strategy does not satisfy the raised demand, then some new routes are introduced to cover new origin–destination demand. The need to introduce new routes implies that the regular network cannot provide the required service when a new demand arises. In other words, the base network that has been designed for one demand scenario is not optimum in a new demand scenario. Thus, the question is whether it is possible to have a base-robust network (BRN) that keeps its optimality in all demand scenarios occurring during the year and remains intact.

For clarification, a four-node network is illustrated in Figure 15.10 with two demand matrices, D1 and D2, representing the first and the second half year (6-month) demand, respectively. The D1 and D2 represent two seasons. The idea of the BRN is illustrated on this network. There are 12 possible routes (R1 to R12) covering the four nodes of the network of Figure 15.10, given that routes can start and end at any of the four nodes. These routes are shown in Figure 15.10. The initial objective is to minimize the total travel time (TTT) in passenger-minutes such that all the node-to-node demand is satisfied.

Basically, there exist two possible strategies to handle the variation of the demand. The first is to use/implement the same network of routes for the entire year, and the second is to use a different (best suited) network for each season. However, the second strategy is not convenient and confusing from the passenger's perspective, thus it is not really applicable. The main question, therefore, is which set of routes to apply for the entire year so as to best

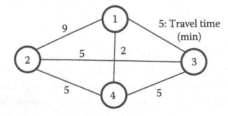

R1: 1–2–3–4	R7: 2–1–3–4
R2: 1–3–2–4	R8: 2–1–4–3
R3: 1–4–2–3	R9: 2–3–1–4
R4: 1–4–3–2	R10: 2–4–1–3
R5: 1–3–4–2	R11: 3–1–2–4
R6: 1–2–4–3	R12: 3–2–1–4

Demand matrix 1 of first season (D1)

Nodes	1	2	3	4
1	0	10	12	15
2		0	15	20
3	SYM		0	30
4				0

Demand matrix 2 of second season (D2)

Nodes	1	2	3	4
1	0	10	30	12
2		0	40	10
3	SYM		0	15
4				0

Figure 15.10 Example of four-node network with 12 possible routes.

Table 15.9 Total travel time of all possible routes in the sample network

Route no.	Path	TTT-D1 (passenger-min)	TTT-D2 (passenger-min)	Sum of TTT-D1 and TTT-D2 (passenger-min)
1	1–2–3–4	968	1113	2081
2	1–3–2–4	860	830	1690
3	1–4–2–3	719	854	1573
4	1–4–3–2	659	729	1388
5	1–3–4–2	760	945	1705
6	1–2–4–3	928	1353	2281
7	2–1–3–4	1040	1185	2225
8	2–1–4–3	814	1149	1963
9	2–3–1–4	715	699	1414
10	2–4–1–3	650	879	1529
11	3–1–2–4	1240	1303	2543
12	3–2–1–4	1063	1084	2147

accommodate the variable demand. An optimization approach is to select a network of routes that minimizes the total travel time with the fulfillment of all pairs of the demand. Table 15.9 shows the TTT for satisfying D1, D2, and the sum of both representing the TTT for the entire year.

Fundamentally, there are four different scenarios to consider. That is, Scenario 1, best case (applying each half a year a different bus network); Scenario 2, applying the best network of the first half year throughout the entire year; Scenario 3, applying the best network of the second half year throughout the entire year; and Scenario 4, applying the network with the minimum TTT for the entire year (BRN). The solution for Scenario 1 is to apply Route 10 and Route 9 for the first and second half year, respectively, with a TTT of 1349 (650 + 699) passenger minutes. The solution for Scenario 2 is to apply only Route 10 throughout the year with a TTT of 1529 (650 + 879) passenger minutes. The solution for Scenario 3 is to apply only Route 9 throughout the year with a TTT of 1414 (699 + 715) passenger minutes. Finally, the BRN solution is to apply Route 4 throughout the year with a TTT of 1388 (659 + 729) passenger minutes.

Figure 15.11 presents a comparison between the highlighted results of Table 15.9 for the yearly TTT results; this is the last column of Table 15.9 except for Scenario 1 in which the best results are combined of each half year demand (650 + 699). Although Scenario 1 is the best result, as told, it is not convenient and will confuse the user who will have to deal with different routes in the first and second half year for the same origin–destination (OD) pair. Thus, the BRN (Route 4 and Scenario 4) is the best to represent both the first and second half year demands using a single route.

In this simple example, also shown in Amiripour et al. (2014b), a single route covers the demand. However, in the real problem of finding the BRN the complicated configurations of routes should be considered and the objective function would not be the simple TTT. That is, the objective function is usually a multi-objective function covering the three perspectives of the users, operators, and community.

15.4.2.1 Problem formulation

All possible transit networks of all of the demand scenarios must be examined under an optimization framework. This is in contrast to cases where the network has been designed for a single demand scenario. The general formulation of the problem is described as follows.

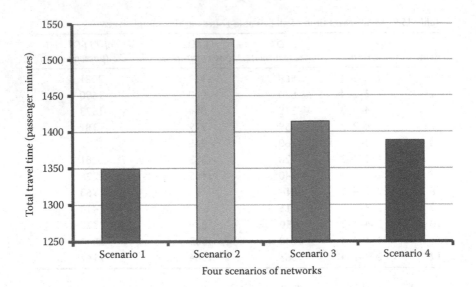

Figure 15.11 The TTT yearly results of the four scenarios related to the sample four-node.

Assume n demand scenarios (D_q), $q = 1, 2, \dots, n$, $D_q \in DS$, where DS is a demand-scenario set, and each D_q occurs at a specific time window during the year. In addition, θ_q indicates the percentage of year that D_q occurs for each q, thus $\sum_{q \in DS} \theta_q = 1$. Let $Z_{BN_p}^{D_q}$ be the objective function of transit network BN_p on demand scenario D_q. $BN_p \in BNS$, where BNS is the set of different networks designed for different demand scenarios. Thus, the objective in a year implementing BN_p would be $\sum_{q \in DS} \theta_q Z_{BN_p}^{D_q}$.

For each demand scenario D_q one optimum transit network BN_q can be designed. One possibility is to have n transit networks each fitting best to its corresponding D_q with an objective function value of $\sum_{q \in DS} \theta_q Z_{BN_q}^{D_q}$. Another possibility is to consider a single network BN_p for all demand scenarios D_q, and attain $\sum_{q \in DS} \theta_q Z_{BN_p}^{D_q} \ \forall BN_p \in BNS$. Between these two possibilities, it is therefore reasonable to construct the following formulation of the BRN to be termed BN_r with the objective function $\sum_{q \in DS} \theta_q Z_{BN_r}^{D_q}$ for demand scenario D_q. This yields the following problem:

$$\text{Min} \sum_{q \in DS} \theta_q Z_{BN_r}^{D_q} \tag{15.13}$$

subject to

$$\sum_{q \in DS} \theta_q Z_{BN_r}^{D_q} < \sum_{q \in DS} \theta_q Z_{BN_p}^{D_q} \quad \forall BN_p \in BNS \tag{15.14}$$

The goal is to minimize $\sum_{q \in DS} \theta_q Z_{BN_r}^{D_q}$ with the result being a base-robust network for all demand changes.

Constraint (15.14) implies that the robust network will give better results than cases where only one (most frequently used) demand matrix is used for the transit network design problem; in these cases, the variation of demand is considered only by changes of schedules, not in the routing.

Equation 15.15 shows that certainly the objective function value of the sum of all best networks for each scenario is less than or equal to the one searched for (BRN). The left-hand-side of Equation 15.15, therefore, represents the best-case scenario. However, in practice, this is unrealistic because transit agencies and the authorities would not agree to frequent routing changes (a few times per year) of the entire network system; this would obviously create confusion among the passengers:

$$\sum_{q \in DS} \theta_q Z_{BN_q}^{D_q} \leq \sum_{q \in DS} \theta_q Z_{BN_r}^{D_q} \tag{15.15}$$

The objective function used is the summation of Equations 15.1 and 15.2 plus the unsatisfied demand on the given network G = (N, A), where N and A are sets of nodes and links, respectively. In addition, and in order to avoid the void solution with nil routes, we add a penalty on any unsatisfied demand. The whole optimization formulation with the interpretation of its elements and parameters appears as follows with Equation 15.16 being the objective function:

$$\text{Min} \sum_{q \in DS} \theta_q Z_{BN_r}^{D_q}$$

$$= \sum_{q \in DS} \theta_q \left[\begin{array}{c} a_1 \sum_{i,j \in N} (WT_{i,j})^{D_q} + a_2 \sum_{k \in BN_r} (EH_k)^{D_q} \\ + a_3 \sum_{i,j \in N} (DPH_{i,j})^{D_q} + a_4 \sum_{k \in BN_r} (FS_k)^{D_q} + a_5 (UD)^{D_q} \end{array} \right] \tag{15.16}$$

subject to

$$\sum_{q \in DS} \theta_q Z_{BN_r}^{D_q} < \sum_{q \in DS} \theta_q Z_{BN_p}^{D_q} \quad \forall BN_p \in BNS \tag{15.17}$$

$$f_{min} < f_k \quad \forall k \in BN_p, \quad \forall BN_p \in BNS, \quad \forall D_q \in DS \tag{15.18}$$

$$L_{min} < L_k < L_{max} \quad \forall k \in BN_p, \forall BN_p \in BNS \tag{15.19}$$

$$T_{i,j}^k < (1 + \alpha) T_{i,j}^{sp} \quad \forall k \in BN_p, \forall BN_p \in BNS, \forall i,j \in N_k \tag{15.20}$$

$$Tr_{i,j}^{covered} \leq Tr_{max} \quad i,j \in N \tag{15.21}$$

$$(UD_T)^{D_q} < UD_{max} \quad \forall D_q \in DS \tag{15.22}$$

$$\sum_{k \in BN_p} (FS_k)^{D_q} < FS_{max} \quad \forall k \in BN_p, \forall BN_p \in BNS, \forall D_q \in DS \tag{15.23}$$

where

$\sum_{q \in DS} Z_{BN_p}^{D_q}$ is the objective function associated with the network BN_p under demand scenario D_q

$\sum_{q \in DS} Z_{BN_r}^{D_q}$ is the objective function associated with the robust network BN_r under demand scenario D_q

$(WT_{ij})^E$ is passengers' waiting time under demand scenario D_q between i and j

$(EH_k)^{D_q}$ is the unused seat capacity of route k under demand scenario D_q

$(DPH_{ij})^{D_q}$ is the total time loss between riding the transit vehicle and travelling by car under demand scenario D_q between i and j

$(UD)^{D_q}$ is the total unsatisfied demand under demand scenario D_q

$(FS_k)^{D_q}$ is the fleet size of route k under demand scenario D_q

The following list explicates the parameters used:

a_1: Monetary value of 1 h's waiting time.
a_2: The equivalent of 1 h's average monetary revenue divided by the average number of hourly boarding passengers.
a_3: Monetary value of 1 h in-vehicle time.
a_4: Monetary value of 1 h's operation of one transit vehicle.
a_5: The time penalty associated with an unsatisfied transit user.
f_k: Corresponding headway of route k.
f_{min}: Its minimum allowable value (policy-headway based).
L_k: Length of route k, L_{Min}, and L_{Max} are its minimum and maximum allowable values.
T_{ij}^k: Travel time between i and j on route k.
T_{ij}^{sp}: Travel time between i and j using shortest path.
α: Maximum allowable deviation of one route from shortest path.
N_k: Set of nodes in route k.
$Tr_{ij}^{covered}$: Number of transfers for a trip between i and j if the OD pair is covered.
Tr_{max}: Maximum allowable number of transfers.
UD_{max}: Maximum allowable unsatisfied demand.
FS_{max}: Maximum allowable fleet size.

Constraint (15.17) guarantees that the robust network would be better compared with other networks designed for a single demand scenario. Constraint (15.18) limits the model to define frequencies for all routes within an acceptable range. Constraint (15.19) is the upper and lower bounds of the route length, and Constraint (15.20) assures that the route length is below an acceptable deviation from the shortest path. Constraint (15.21) limits the number of transfers below a given maximum number. Constraint (15.22) is the maximum allowable unsatisfied demand, and Constraint (15.23) represents budget limitation associated with the operator's perspective; it keeps the fleet size below a given acceptable level.

Solving this problem would take a great deal of computational effort (NP-hard) because of the calculations involved with the transit assignment process and of the objective function for n(DS) times. The next section, therefore, proposes two methodologies enabling solutions for large-scale transit network sizes.

15.4.3 Two solution methodologies

Based on Newell (1979), the transit network design problem is a problem of astronomical portions. This quote is another way of showing the NP-hard nature of the transit network design problem. Among the three categories of the problem solution methods discussed earlier (math solutions, heuristic methods, and meta-heuristic methods), the math methods cannot be used for a real-life transit network design problem because of the level of simplifications it requires for the problem. Developing a heuristic application to different large-scale transit networks is also a complicated procedure that requires different types of evaluations and definitions within the algorithm. Thus, there are not many heuristics available for the transit network design problem in contrast to meta-heuristic methods.

Considering seasonal demand variations in the transit network design problem would make it more complicated; the required time depends on how many demand scenarios exist. The more demand scenarios the more required computational effort to solve the problem. The use of meta-heuristic methods, such as genetic algorithm, is one way to attain a satisfied solution. Let us call this solution method DMH (direct meta-heuristic). In a DMH method, the fitness function would be the weighted objective function for different demand scenarios.

Although DMH is a straightforward solution, it cannot be the ultimate procedure because of the required computational time. Hence, two-solution methodologies, enabling to solve the problem in a reasonable amount of time, are presented in the following text.

15.4.3.1 Solution methodology I(DMH-F)

Following Amiripour et al. (2014a), a hybrid method, called DMH-F, is constructed. This method is developed in two major steps: *route-level optimization* and *network-level optimization*. The *route-level optimization* limits the set of feasible routes to routes that have greater potential to enter BRN. This heuristic process ignores routes that do not satisfy the BRN requirements and creates a smaller feasible set for the next step. The *network-level optimization* is a meta-heuristic method where the possible BRNs are created to check whether or not they can be considered as robust using genetic algorithm. Afterward, a heuristic algorithm checks the created BRNs for possible improvements. Figure 15.12 illustrates this procedure schematically.

In order to solve the transit network design problem with seasonal variable demand amid the proposed algorithm, the sample network with eight nodes (see Figure 15.5) is used as an explanatory device to describe the new method developed. The method is based on the following four steps.

15.4.3.1.1 Step 1: Data collection

In this step, demand scenarios and their corresponding demand matrices are identified for each time period during the year. Other data related to policies and levels of service are also inserted for creating the constraints.

Sample-related: Four demand scenarios are considered for the sample network of Figure 15.5; it is depicted in Table 15.10 (one for each season) with *SYM*, meaning symmetrical.

15.4.3.1.2 Step 2: Route-level optimization

15.4.3.1.2.1 STEP 2.1: CREATING AN INITIAL SET OF FEASIBLE ROUTES

In this step, all possible routes are identified using the shortest path or the K-shortest path algorithm (Yen, 1971) between all nodes or terminal points. This requires the input of the

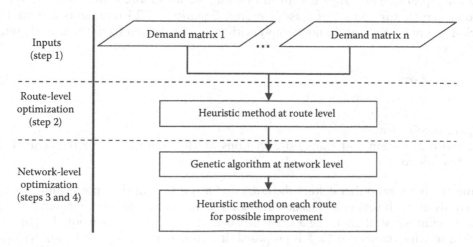

Figure 15.12 Procedure of DMH-F for transit network design with seasonal variable demand.

Table 15.10 Four demand matrices of the sample network in Figure 15.5

	0	1	2	3	4	5	6	7
0	0	80	70	160	50	200	120	60
1		0	120	90	100	70	250	70
2			0	180	150	120	30	250
3				0	80	210	170	230
4					0	250	40	130
5	SYM					0	130	120
6							0	70
7								0

	0	1	2	3	4	5	6	7
0	0	80	70	40	50	120	120	60
1		0	120	40	100	70	250	70
2			0	40	150	120	30	250
3				0	80	210	170	230
4					0	250	120	130
5	SYM					0	65	120
6							0	70
7								0

	0	1	2	3	4	5	6	7
0	0	80	120	160	90	200	120	60
1		0	120	90	40	120	20	120
2			0	180	150	120	30	20
3				0	80	210	1120	10
4					0	20	40	130
5	SYM					0	130	120
6							0	120
7								0

	0	1	2	3	4	5	6	7
0	0	20	70	160	50	200	120	60
1		0	120	90	250	70	250	70
2			0	120	250	120	30	250
3				0	20	210	170	230
4					0	250	40	130
5	SYM					0	130	120
6							0	70
7								0

maximum allowable deviation from the shortest path (α) and the parameter K. That is, the second shortest path will be based on the shortest path (see Appendix 9.A), the third shortest path will be based on the second shortest path, and so on. Yen (1971) provided an algorithm to solve it efficiently.

Sample-related: For the sample network, of Figure 15.5, K = 10 and the allowable deviation from shortest path (α) is 50%; this results in the creation of 113 feasible routes.

15.4.3.1.2.2 STEP 2.2: ELIMINATING SHORT OR LONG ROUTES

Excessively short or long routes are not economic and operable. Thus, a range of acceptable route length (in terms of time or distance) is defined such that routes outside this range are not acceptable. This range is defined by using the mean and standard deviation of all the routes created in the set of feasible routes. Equation 15.24 represents a route length criterion where routes that do not comply with it will be deleted from the feasible set. The criterion is:

$$|l_k - \mu| \leq \xi \cdot \sigma \tag{15.24}$$

where
   l_k is the route length
   μ and σ are mean and standard deviations of all route lengths of the created set of feasible routes

Parameter ξ is the ratio that defines the range and it is determined by performing a sensitivity analysis on each network considered. If a large number is chosen for ξ only few routes would be eliminated; if choosing a small number, most of the routes would be eliminated. Hence, a starting number of $\xi = 2$ is proposed. It is to note that a sensitivity analysis can be useful for determining the amount of ξ for different transit networks.

Sample-related: In the sample network, the mean and variance length of the 113 routes, initially selected, are 46.9 and 19.3 (in min) with $\xi = 2$; it results with an acceptable route length within the time (in min) range [8.2–85.6]. Four routes were not within this range and thus removed to make 109 routes in the feasible set.

15.4.3.1.2.3 STEP 2.3: IMPORTANT NODE COVERAGE

Covering important demand-related attraction/production (A/P) points within a transit network is a target to guarantee reaching meaningful demand points. This step identifies important A/P points and the routes are ranked by the number of important nodes they cover; then routes with the lowest number of important nodes covered are eliminated from the set of feasible routes. That is, using the following procedure:

1. For all nodes ($i \in N$) in the network:

$$\text{Set ODI}_i = \sum_{j \in N} (d_{ij} + d_{ji}).$$

2. Sort all the nodes by their ODI_i.
3. Mark top $\lambda\%$ of the nodes as important A/P points.
4. For all the routes in the set of feasible routes, count the number of important A/P points.
5. Sort all the routes by their important A/P points count.
6. Eliminate $\beta\%$ of the routes with the lowest important A/P point count.

where ODI_i is the total amount of demand that node i attracts or produces; d_{ij} is the amount of demand between nodes i and j.

Determination of λ and β is a decision to take place after a sensitivity analysis.

Sample-related: On the sample network, the ODIs of nodes 4 and 6 (see Figure 15.5) are larger than others, thus chosen to be important A/P points. Routes that do not cover nodes 4 or 6 are removed. The procedure, with $\lambda = 0.3$ and $\beta = 0.2$, results with a removal of 21 routes, bringing it to 88 routes in the feasible set.

15.4.3.1.2.4 STEP 2.4: DEMAND VARIABILITY INDEX

A good route of the process portrayed is perceived also in terms of having sufficient passenger loads. In other words, a good route will attract passengers continually; this type of route is expected to be part of the BRN. Thus, in this step, the covered demand by each route in different demand scenarios is calculated including the mean and standard deviation of the demand covered. Then the remaining routes are sorted by their coefficient of variation (CoV) = σ/μ. The $\gamma\%$ of the routes having the largest CoV are dropped (γ is selected based on sensitivity analysis). This filter assures that all routes of the set of feasible routes do not have large variations of covered demand for the different demand scenarios.

Sample-related: In the sample network, the CoV of each of the 88 routes is considered using the four demand scenarios. An initial value of $\gamma = 0.3$ is considered. By doing so, 26 more routes were deleted by this filter bringing it to 62 routes in the set of feasible routes, from which the best solution will be found.

15.4.3.1.2.5 STEP 2.5: PRACTITIONERS' ROUTES

It is to note that following the three filtering steps, there is another step allowing practitioners to insert existing routes chosen on top of the routes in the feasible set. This will form the final set of feasible routes. This additional step is important in practice, but was not considered for the sample network.

15.4.3.1.3 STEP 3: GENETIC ALGORITHM OPTIMIZATION

Step 3 employs the approach of genetic algorithm (GA) to optimize the specified objective function. GA is a stochastic optimization algorithm, inspired by body-evolution process, searching for a good solution of problems in which the feasible set is excessively large. This known algorithm (Amiripour, 2014b) has been widely used for complex problems. Operators of crossover, mutation, and elitism are the major components of GA. First, a random initial population of solutions, called chromosomes, is made; in this section, chromosomes are transit networks where the routes come from a set of feasible routes. Then the fitness function or the objective function value of each chromosome is calculated as is shown schematically in Figure 15.12. By using a weighted random selection, two chromosomes are chosen to enter the mating pool for the crossover operator, called parents. Crossover operator combines two parents and makes two new children (termed offspring) as is depicted in Figure 15.13. If the offspring have a better fitness value than their parents, then the parents are replaced by the offspring; otherwise the parents stay in the population. This process, described in Figure 15.12, continues until the new generation is born. To avoid stagnation in terms of local optima, the mutation operator randomly changes some solutions in each population as is shown in Figure 15.13. Elitism operator gets the

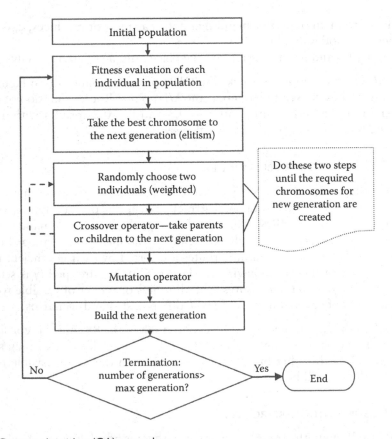

Figure 15.13 Genetic algorithm (GA) procedure.

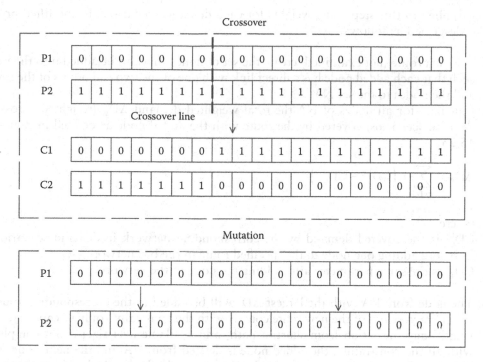

Figure 15.14 Crossover and mutation operators in GA.

best chromosome in each generation and passes it to the next. This ensures that the iterative process of GA always gets better or at least remains the same.

There exist many variations of GA in the literature applicable to a variety of problems. The classic form of GA is illustrated in Figure 15.13. Crossover operator, shown in Figure 15.14, combines two parent chromosomes (P1, P2) and produces two children chromosomes (C1, C2); the mutation operator changes one random gene in one chromosome of P1 (Figure 15.14). Narrowing the set of feasible routes affects the convergence speed of the algorithm thus reducing the computational effort for solving the problem.

The chromosome structure is a binary vector with the length of *the set of feasible routes*. If a route is in the network, the corresponding gene would be 1 and otherwise 0. The amount of 1 s in the binary vector is the number of routes in the network. With the use of GA, this method seeks to optimize the objective function value considering a large number of passenger-demand possibilities. Following the selection of the routes, Step 4 deals with the assurance of having a large number of these possibilities.

15.4.3.1.4 STEP 4: HEURISTIC METHOD FOR INCREASING THE DEMAND COVERAGE OF THE NETWORK

By using GA, Step 3 creates a transit network that covers a certain demand level for each demand scenario. The stochastic nature of GA and the complications brought to the problem by introducing multidemand scenarios are the main reasons the GA cannot attain a large number of demand coverage. Thus, after finding the base network, each route is treated by Step 4 to increase the demand coverage for all of the demand scenarios. The procedure used is a modification of node selection and insertion strategies developed by Baaj and Mahmassani (1995). The original algorithm has four different strategies that can be applied in this step using weights for each demand scenario. The modified procedure is described as follows:

1. For each route (not examined) create the set of nodes that can be added into the route such that each added node has a direct link with one of the two end points of the route; add the nodes to the set NA.
2. Calculate for all nodes of NA the total weighted demand ΔD_w (weighted across all demand scenarios) covered by the route with the added node as defined in Equation 15.25:

$$\Delta D_w = \sum_{q \in DS} \theta_q \left(CD_1^{D_q} - CD_0^{D_q} \right)$$

(15.25)

where
$CD_0^{D_q}$ is the covered demand by the corresponding network in demand scenario D_q
 before adding one node to the specified route in the bus network
$CD_1^{D_q}$ is the same parameter after adding the node to the network

3. The node from NA with the largest ΔD_w will be added to the corresponding route if the characteristics of the new route comply with the constraints of max capacity, max length, max deviation from shortest path, and available fleet size; if not complying with all the constraints, the added node is deleted from NA and the next node from NA with the largest ΔD_w is examined; otherwise, add the checked node to the route, delete it from NA, and create a new route (with the added node); go to (a) with the new route; if no more nodes exist in NA then STOP.

The ΔD_w is the total demand covered, by the route, with the consideration of the added node(s). This total demand is the weighted sum of passenger demand across all scenarios. Figure 15.15 illustrates the elements of Step 4 using two routes [1 to 5] and [6 to 9]. That is, the set NA comprises nodes 10–15 of the [1 to 5] route and of nodes 14–19 of the [6 to 9] route. The ΔD_w is then calculated for each possible added node by figuring out the total new demand (weighted across all scenarios) of its corresponding route. In this example, there will be 12 such calculations from which the node associated with max ΔD_w will be selected to be added. Then the process continues with additional nodes to be added to the set NA coming from the new added node that will become a new end point. This process follows major parts of the route expansion procedure of Baaj and Mahmassani (1995).

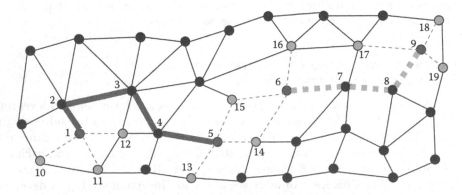

Figure 15.15 Interpretation of Step 4.

Sample-related: In the sample network, shown in Figure 15.5, no node could be added because of the constraints given.

Step 4 allows taking into account the maximization of the annual, not season-based, passenger demand. The total weighted demand ΔD_w can be calculated at the network level considering also trips with transfers. At the end of this four-step procedure, the demand considered for the BRN is increased and ready for processing.

15.4.4 Analysis of the sample network

Table 15.11 presents a summary of results and processes of Steps 1 and 2 of the developed procedure for the sample network (Figure 15.5).

In Step 3, the objective is to minimize the weighted summation of waiting time (in passenger-hours), deviation from the shortest path (in passenger-hours), empty-seat hours, the minimum fleet size, and unsatisfied demand as presented in Equation 15.16.

The population size for the GA is 1000 chromosomes for each generation; the chromosome structure is a binary vector with the length of the number of feasible routes (62: the reduced number for the 8-node network). Crossover rate is set to 0.6 and mutation is set to 0.003 in a population. In addition and in order to provide more diversity of solutions and avoid stagnation at local optima, a new population is randomly generated on each 50 generations. The next generation will be formed by selecting the top chromosomes from both the current and the new population. The elitism operator takes the best chromosomes to the next generation. Figure 15.16 refers to the sample network and shows the performance of the robust network by comparing its objective function value (Equation 15.16) with other possible networks. The best objective function values belong to a set of networks termed *Best Networks* that consider the best possible network for each demand scenario—in other words, to set a different transit network for each season, thus being unacceptable in practice. This is used as the best-case scenario for comparison purpose. The regular networks in Figure 15.16 are those that have been designed using only one demand matrix. Figure 15.16 shows that the BRN is closer to the best-case scenario than for the regular (usually existing) network results, that is, the BRN has the potential to significantly improve existing networks.

The advantage of the BRN is one aspect of the problem studied. The other aspect of the problem is the solution methodology. In order to solve the previously stated problem, either a DMH or the DMH-F algorithm can be used.

Figure 15.17 shows the results, of the sample network, using the DMH-F and DMH methodologies. Both methods are iterative as is seen in Figure 15.17; the convergence to a better solution is an important factor which shows that the DMH-F converges much faster than the common DMH method. The initial filtering of the set of feasible routes narrowed the

Table 15.11 Summary of implementing the route-level process on the sample network in Figure 15.5

Set of feasible routes	113 routes	Running K-shortest path algorithm between all node pairs with K = 10 and allowable deviation from shortest path = 50%.
Eliminating short or long routes	4 routes deleted	Mean and variance of routes are 46.9 and 19.3 min. For $\xi = 2$ the acceptable route length is in the range [8.2, 85.6] in min.
Important node coverage	21 routes deleted	Values of $\lambda = 0.3$ and $\beta = 0.2$ are used. Two nodes identified as important nodes; 21 routes are removed because of zero coverage of A/P points.
Demand variability index	26 routes deleted	Initial value of $\gamma = 0.3$ considered, and leading to the removal of 26 more routes.
Remaining routes	62 routes	

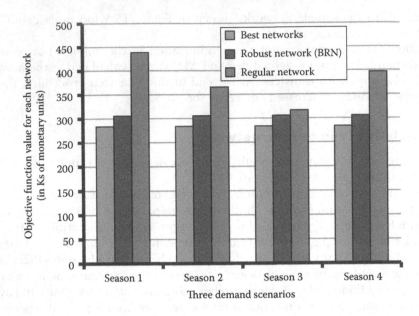

Figure 15.16 Objective function values using three scenario networks of the sample network.

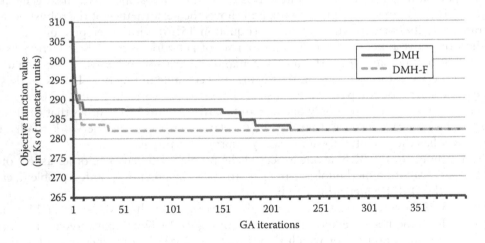

Figure 15.17 Comparing the proposed algorithm and direct GA method for the sample network.

area and consequently the GA may start from a better solution. It is to note that according to the stochastic nature of the GA the last statement is not always true, but it is expected that the initial population will be better in terms of objective function.

As represented in Figure 15.17, the DMH-F converges to a network with an objective-function value of 281,976 (in monetary units) by choosing routes out of the set with 62 routes. The DMH chose routes among 113 routes and resulted in a network with the same objective function value of 281,976. This means that the DMH results are the same as DMH-F, but the GA reaches this result in iteration 222 (DMH) compared with iteration 36 for the DMH-F method. In other words, the DMH-F converges to the sample-network solution six times faster. In addition, the chromosome length is shorter in the DMH-F method than in the DMH to reduce the runtime for each generation. It is to note that in small networks the risk of narrowing the set of feasible routes is much higher than for large networks.

That is, by narrowing the set of feasible routes some of the good solutions are removed and it would affect the final results. More of this methodology is in Amiripour et al. (2014a).

15.4.5 Solution methodology 2 (HBRD-I)

The second solution methodology is based on Amiripour et al. (2014b) in which a hybrid method, called HBRD-I, for solving the transit network design with seasonal demand variation (finding BRN) is preferred. The objective function and constraints are those shown in Equations 15.16 through 15.23. This HBRD-I procedure utilizes GA efficiently in creating the BRN. Figure 15.18 illustrates the HBRD-I procedure in general terms, based on eight specific steps described as follows:

Step 0: Network set of BRN starts with no routes.
Step 1: Design one corresponding network ($BN_p \in BNS$) for each demand scenario considering the routes of the network set of BRN as fixed. These networks would be the best possible networks for the corresponding demand scenarios (using GA).
Step 2: Calculate the overlapping score (OS) for all the routes using the following process:

For all the routes ($\hat{k} \in BN_p$) in all networks ($BN_p \in BNS$)
Calculate overlapping index of route $\hat{k}$ with network $BN(OI_{\hat{k},BN})$
Compare $\hat{k}$ with all routes (k) of the other networks ($BN_q | BN_q \in BNS, q \neq p$) using

$$OI_{\hat{k},BN} = Max\{OI_{\hat{k},k} : \forall k \in BN\} \qquad (15.26)$$

with

$$OI_{\hat{k},k} = \frac{L_{common}}{L_{\hat{k}} + L_k} \qquad (15.27)$$

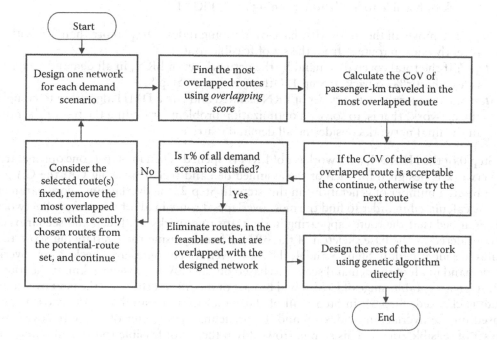

Figure 15.18 Flowchart of the HBRD-I method.

The overlapping score of route $\hat{k}$ would be

$$OS_{\hat{k}} = \sum_{BN_q \in BS, q \neq p} OI_{\hat{k}, BN_q} \tag{15.28}$$

where

L_{common} is the common length of route $\hat{k}$ and k
$L_{\hat{k}}$ and L_k are length of routes $\hat{k}$ and k

A route with a higher $OS_{\hat{k}}$ is more likely to enter the network set of BRN because it will be selected in different demand scenarios.

Step 3: Sort all the routes in all the networks according to their $OS_{\hat{k}}$.
Step 4: Calculate the total passengers-km of the route with the largest $OS_{\hat{k}}$ in all demand scenarios within the designed network. Calculate the coefficient of variation (CoV) (σ/μ) of passenger-km for the chosen route.
Step 5: If the coefficient of variation of passenger-km for the chosen route is within the acceptable range, this route would enter the designed-network (BRN) set and the procedure proceeds from Step 6. Otherwise, the next route with largest $OS_{\hat{k}}$ with the desired condition is chosen to enter the designed network and the process goes back to Step 4.

This step results with routes having an acceptable amount of passengers in all of the demand scenarios. The coefficient of variation helps to find routes that are more robust in terms of demand, thus it leads to a network with less frequency changes required during the year.

Note: The acceptable range of CoV should be defined by performing a sensitivity analysis of the network of feasible routes before running the HBRD-I.

Step 6: Remove all the routes that have overlapping index ($OI_{\hat{k},k}$) more than δ% with the recently chosen route(s) from the set of feasible routes.
Step 7: If the total covered demand by the network set of BRN, in all demand scenarios, is less than τ% go back to Step 1. Otherwise, go to Step 8.
Step 8: Consider the network set of BRN as fixed and run a DMH algorithm to complete the network, that is, to solve an optimization problem for finding the rest of the routes in the final network considering all demand matrices.

At Step 0 of the HBRD-I, a network set of BRN is opened. Then in Step 1, one optimal transit network is designed for each of the assumed demand scenarios using GA; the GA goes for a limited number of generations in this step. In Step 2, the overlapping score parameter (OS) is calculated in order to find the most overlapped route in all of the designed networks. It is assumed that the routes appearing in the transit networks of all demand scenarios can be considered as the base skeleton of the BRN. Thus, by using the overlapping score these routes are identified. In Steps 3 and 4, the most overlapped routes are checked for covering the demand in all of the demand scenarios based on the CoV of passenger-km. If the route is within an acceptable range of CoV, it will be part of the BRN, otherwise the next overlapped route is checked (Step 5). In Step 6 all of the routes that are similar to the chosen routes (based on the overlapping index OI and the predefined parameter of δ) are removed from the set of feasible routes. This step narrows down the set of feasible routes and reduces the chromosome length and feasible solution area for a better convergence of GA. At this stage if τ% of the demand, for each demand scenario, is covered by the BRN the process proceeds with Step 8, which is running DMH to find the other routes of the BRN. Otherwise, the process goes back to Step 1 to design a new transit network for each demand scenario by considering the routes in the BRN as pre-chosen routes.

The HBRD-I procedure attempts to find routes to help construct the BRN incrementally. At each iteration, one or more routes are added to the BRN and all the similar routes are removed from the set of feasible routes. Then another iteration starts with previously selected routes considered fixed. After some iterations, the BRN covers a certain portion of the demand, so that a DMH method can complete the process and finds the rest of the routes to increase the demand coverage. The decision of the time to switch to a DMH method is determined by the τ parameter; this parameter should be determined by performing a sensitivity analysis of different sets within the network.

As mentioned earlier, the HBRD-I is a hybrid method using GA efficiently. Because of the stochastic nature of the GA, the complexity of this method is unknown; however, searching for a solution in a smaller set of feasible routes is much faster than in a larger space (it converges to the solution faster). Thus, the HBRD-I method is more efficient than the DMH method because it makes the set of feasible routes smaller at each iteration. Besides, searching for a smaller amount of routes at each iteration makes this procedure faster than the DMH method. More of this methodology is in Amiripour et al. (2014b).

15.5 METHODOLOGY FOR ACTUAL-SIZE TRANSIT NETWORK DESIGN

Designing transit network for actual-size road networks is described by Bagloee and Ceder (2011). This study focuses on providing systematic transit-planning tools for actual-size networks considering the following issues collectively:

Practicality—being able to tackle actual-size problems.
Integrated planning—synchronizing routes and frequency setting.
Multiclass—design of a system containing different transit modes (e.g., metro, bus, ferry) and different route hierarchies (e.g., collector, feeder, mass (or dominant)-route) in a systematic manner.
Transit assignment—a transit assignment method is required for a given solution; it is noted that in transit network design most researchers, because of computational complexities, use simplifications.
Stop positioning—despite the importance of positioning the stops on the routes, only a handful of research papers consider it and these are mostly concerned with the issue of optimal stop spacing by the use of rules-of-thumb.

This method of Bagloee and Ceder (2011) undertakes actual-size cases with a process of transit network design synchronized with frequency setting. The algorithm developed considers the concept of clustering to identify the best set of candidate stops and of different transit modes and hierarchies. Two transit assignment methods have been employed in the planning process: capacity-free and capacity-constraint transit assignment. The principals of Bagloee and Ceder (2011) methodology to design optimal transit network are as follows.

Given a road network, an exogenous fixed transit demand, and a limited budget, the problem is to derive a comprehensive transit plan (including routes and assigned vehicles) so that the total passengers' discomfort is minimized. A route comprises two end points (terminals), a sequence of links, a timetable based on frequency setting, and is 2-way (bidirectional).

The algorithm allows different and even distant paths (depending upon demand concentration) for either direction, thus loop routes can also be considered. In any transit plan (scenario), the discomfort index of passengers from origin i to destination j is designated as generalized time $GT_{i,j}$:

$$GT_{i,j} = In\,Veh\,Tim_{i,j} + \omega_1 \cdot Walk\,Tim_{i,j} + \omega_2 \cdot Wait\,Tim_{i,j} + \omega_3 \cdot Transfer_{i,j} \qquad (15.29)$$

where
$In\,Veh\,Tim_{i,j}, Walk\,Tim_{i,j}, Wait\,Tim_{i,j}, Transfer_{i,j}$ are in-vehicle time, access/egress-walking time, waiting time, and number of transfers or boarding, respectively, to get from i to j
$\omega_1, \omega_2, \omega_3 > 0$ are weights for waiting time, walking time, and transfer/boarding penalty while in-vehicle's weight is assumed to be 1

The objective function (Φ) is established as the total saved generalized time with respect to no-transit-plan scenario:

$$\max \Phi^x = \sum_{i,j} GT_{i,j}^0 - \sum_{i,j} GT_{i,j}^x \qquad (15.30)$$

where 0 and x denote no-transit-plan and scenario number, respectively. This formulation makes the definition of the objective function clearer in terms of discomfort index interpreted as generalized times, and emphasizes the catering to captive users who have no other travel choice than transit.

The solution algorithm consists of three components illustrated in Figure 15.19. In Component 1, the location of stops is identified based on closeness to highly concentrated transit demand places and clustering factor. Component 2 is inspired by the Newton gravity theory, to generate a set of candidate routes. In Component 3, a search algorithm is run through the generated candidate routes to seek a good solution; this search procedure is encoded by means of a genetic algorithm (GA) and equipped with ant-system (AS) collective points as the search engine.

The algorithm was tested on the actual-size transit network of the city of Winnipeg; the results show that under the same conditions (budget and constraints) the set of routes resulted in a reduction of 14% of total travel time compared to the existing transit network. In addition, the methodology developed is compared favorably with other studies using the transit network of Mandl benchmark. The generality of the methodology was tested on the real dataset of the larger city of Chicago, in which a more efficient and optimized scheme is proposed for the existing rail system. More on this study appears in Bagloee and Ceder (2011).

15.6 CONNECTED PRIORITY-LANE CONSIDERATION IN BUS NETWORKS

This section is somewhat different from the previous sections of this chapter. It is not about the general problem of transit network design; it is about a special transit network that consists of only exclusive or dedicated or priority bus lanes. One of the major drawbacks in constructing priority bus lanes is the lack of a well-connected network of these lanes so as to maintain the prioritized movement of its patronage.

Bus priority schemes and techniques on urban roads and highways have proven effective for almost a half century. Many bus-priority strategies have been demonstrated worldwide.

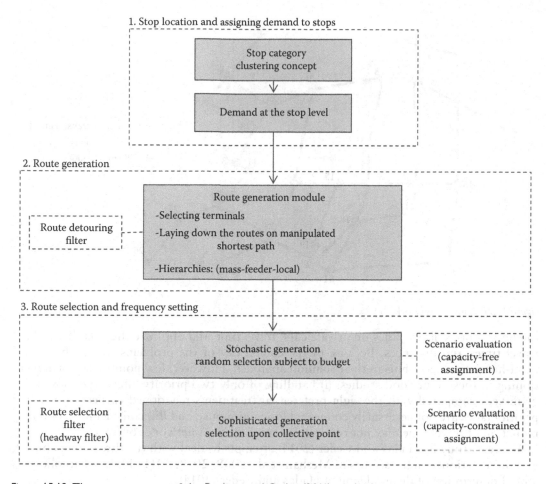

1. Stop location and assigning demand to stops

Stop category
clustering concept

Demand at the stop level

2. Route generation

Route generation module

-Selecting terminals

-Laying down the routes on manipulated
shortest path

-Hierarchies: (mass-feeder-local)

Route detouring
filter

3. Route selection and frequency setting

Stochastic generation
random selection subject to budget

Scenario evaluation
(capacity-free
assignment)

Route selection
filter
(headway filter)

Sophisticated generation
selection upon collective point

Scenario evaluation
(capacity-constrained
assignment)

Figure 15.19 Three components of the Bagloee and Ceder (2011) method.

Traditionally, priority is granted for bus operation at stops, at intersections, and by preferential/exclusive lanes. It is known that bus travel times, reliability of service, and vehicle productivity are improved when buses are able to use higher-speed, uncongested lanes. These improvements make the bus systems more attractive and thus increase the potential to gain new riders.

Eight preferential treatments to buses on street lanes are known as follows: exclusive curb lane, semi-exclusive curb lane (shared only with cars about to turn), exclusive median lane (with stop island), exclusive lane in the center of a street, bus malls (limited to pedestrians and buses), exclusive freeway/highway lanes, ramp bypass (for entering a freeway/highway during traffic congestion), and congestion bypass (exclusive lanes to bypass traffic bottlenecks).

Network-level consideration of priority/exclusive lanes is one way to improve transit performance. Mees (2010) introduced the term *network effect* where the transit network has short waiting times, easy transfers, good coverage, and high reliability to allow it to compete with the private cars. Thus, planning system-wide priority lanes share the same philosophy.

The first to introduce a system-wide approach for designing priority lanes were Mesbah et al. (2008, 2010, 2011a,b). In these research studies, a bi-level optimization model is proposed to combine priority lanes selection with traffic assignment. The model assesses

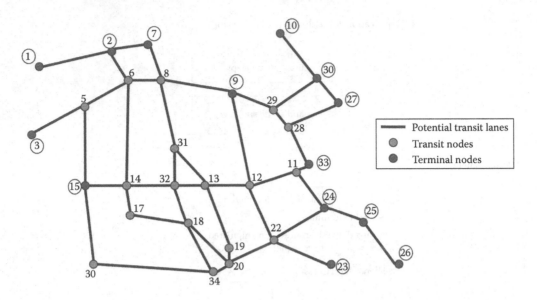

Figure 15.20 Sample network.

the impact of exclusive lanes on private cars' travel time and optimize the overall weighted travel times and distances. Because of the complexity of the problems, the authors used meta-heuristics algorithms in their solution approach. However, few points are worth mentioning in these four good studies: (i) handling of only two priority alternatives (exclusive or mixed) compared with the eight preferential treatments mentioned earlier; (ii) resulted priority lanes are not necessarily connected (or continuous); and (iii) consideration of only a reduction of travel time may not result in the most efficient network coverage.

A recent study by Hadas and Ceder (2014) attempts to deal with the three points (i), (ii), and (iii) mentioned in the studies of Mesbah et al. (2008, 2010, 2011a,b). In what follows is a brief description of the modeling of Hadas and Ceder (2014).

Each arc in the sample network presented in Figure 15.20 (between two numbered nodes) is a road section that can be constructed as part of a possible priority lane (exclusive or semi-exclusive). Each priority-lane alternative will be examined in terms of its cost and benefits (time saving). All circled nodes are a set of possible origins and destinations for the priority lanes. The goal is to construct a set of priority lanes that connects transit stations, transfer hubs, routes' start/end stops, and link one priority lane to other priority lanes. By doing so, the transit network will be characterized by uninterrupted routes (such as 15–14–32–13–12–11–33 in Figure 15.20), as opposed to the construction of isolated priority lanes, which often experience traffic bottlenecks in the form of non-prioritized sections. Balancing origin and destination nodes is crucial for a well-connected network. For instance, optimizing time saving alone can result with two North–South priority lanes compared with a balanced result of one North–South priority lane and one East–West priority lane; the latter slightly decrease the time saving, but maintain a balanced priority lanes network.

15.6.1 Model formulation

Let $G(N, A)$ be a directed network that comprises all road sections traversed by transit routes. Let $x_{i,j}^k$ be a binary decision variable, such as "1" represents the selection of priority lane alternative k for road section (i, j), and "0" otherwise. Furthermore, let $c_{i,j}^k$ be the construction cost, $v_{i,j}^k$ the travel time saving per passenger, and $f_{i,j}$ the total passengers' flow of all routes passing through road section (i,j). Let $I \subseteq N$ be a set of all nodes from which a priority lane starts or ends. For constructing paths, let $p_{i,j}^{m,s,t}$ be an indicator whether road section (i, j) is part of path m that starts from node $s \in I$ and terminates at node $t \in I$. For clarity the index m will be omitted henceforth. Let $px^{m,s,t}$ be a decision variable, such as "1" represents the selection of path m that starts from node $s \in I$ and terminates at node $t \in I$. Let matrix P represent the paths; P can be easily calculated, as is described in the following text. Again, for clarity, the index m will be omitted. Furthermore, let B be the budget available, and D_l, D_u be the lower and upper bounds for nodes' degree. The optimization formulation is

$$\max \sum_i \sum_j \sum_k x_{i,j}^k \cdot v_{i,j}^k \cdot f_{i,j} \tag{15.31}$$

$$\max \min_{i \in SL} \left\{ \min_j \left(\sum_t px^{j,t}, \sum_s px^{s,j} \right) \right\} \tag{15.32}$$

subject to

$$\sum_i \sum_j \sum_k x_{i,j}^k \cdot c_{i,j}^k \leq B \quad \forall i, j \in n \tag{15.33}$$

$$\sum_k x_{i,j}^E \leq 1 \quad \forall i, j \in N \tag{15.34}$$

$$\sum_k x_{i,j}^k - \left[\sum_s \sum_{t \neq s} \left(p_{i,j}^{s,t} \cdot px^{s,t} \right) \geq 1 \right] = 0 \quad \forall i, j \in N \tag{15.35}$$

$$x_{i,j}^k = \{0, 1\} \tag{15.36}$$

$$px^{s,t} = \{0, 1\} \tag{15.37}$$

Equation 15.31 maximizes total time saving resulted from using the selected transit priority lanes. Equation 15.32 maintains a balanced connectivity between the selected terminal nodes. This balance is maintained by maximizing the minimal in-degrees and out-degrees (the number of nodes directly connected to/from a given node) of all terminal nodes among all feasible solutions (SL). An unbalanced priority lane set will impact the overall reliability of the transit network and reduce the level of service. Constraint (15.33) enforces budget availability and constraint (15.34) maintains the selection of only one alternative. Constraint (15.35) enforces the following: if at least one path $\left(\left[\sum_s \sum_{t \neq s} \left(p_{i,j}^{s,t} \cdot px^{s,t} \right) \geq 1 \right] \right)$ from s to t is selected ($px^{s,t} = 1$), then one alternative ($x_{i,j}^k$) for road section (i,j) must be selected given that the road section is part of the path from s to t ($p_{i,j}^{s,t} = 1$). This constraint also maintains the continuity of each selected priority lane.

The complex multi-objective problem can be converted to a single-objective problem. That is, Equation 15.32 can be substituted by the following constraints:

$$\sum_t px^{s,t} \geq Dl \quad \forall s \in I, s \neq t \tag{15.38}$$

$$\sum_t px^{s,t} \leq Du \quad \forall s \in I, s \neq t \tag{15.39}$$

$$\sum_s px^{s,t} \geq Dl \quad \forall t \in I, s \neq t \tag{15.40}$$

$$\sum_s px^{s,t} \leq Du \quad \forall t \in I, s \neq t \tag{15.41}$$

Equations 15.38 and 15.39 enforce that the out-degree is between upper and lower bounds. Equations 15.40 and 15.41 enforce the in-degree bounds. Hence, it is possible to explore different solutions with the adjustments of D_u and D_l.

The input to this model is based on matrix P; this matrix is a set of all possible paths (shortest, quickest, etc.) starting at node s and terminating at node t. Pre-calculating P will result with an efficient optimal algorithm, because it is not required to calculate on-the-fly paths from G. Constructing matrix P is a straightforward process of implementing an all-pairs shortest path algorithm (see Appendix 9.A), K-shortest path (Yen, 1971), and multiple reasonable routes (Park et al., 2002). Finally, a case study is presented in Hadas and Ceder (2014) to demonstrate this approach.

15.7 LITERATURE REVIEW AND FURTHER READING

This section contains a review of papers that propose methods of optimizing the configuration of transit-route systems. The output of such methods is a route itinerary that usually includes headways or frequencies. Papers that focus on headway determination, without a route-itinerary design, are discussed in Chapter 3. Papers in which special attention is given to unique objectives, such as the minimization of transfer times, are concentrated in Chapter 5.

Hobeika and Cho (1979) present a method for determining the structure of a bus-route system. A heuristic algorithm partitions the existing bus stops in an urban area into sectors and seeks a way to link the stops while trying to minimize the total distance traveled by all buses. In the optimization process, each bus is subject to capacity and distance constraints. The routes developed are improved iteratively; in each iteration, a disaggregate choice model is used to examine passengers' behavior. Equilibrium between supply and demand is reached when the proportion of passengers using buses cannot be increased by improving the bus network.

Marwah et al. (1984) develop a method for the simultaneous design of routes and frequencies. First, passenger flows are assigned on the road network. Then, a large set of possible bus routes that satisfy certain constraints is generated. Finally, routes that minimize the number of system-wide transfers are selected. Heuristics are used for the concentration of flows on the road network and for the initial generation of routes. Linear programming is used for the selection of optimal routes and for assigning frequencies.

Van Nes et al. (1988) formulate a programming problem that sets route itineraries and frequencies with the objective of maximizing the number of trips in which a transfer is not needed. The constraints include a given fleet size and budget limitations. The authors discuss the advantages of formulating the route-design problem as a programming problem, such as the ability to add further constraints. With additional constraints, it is possible to use the model in systems in which some existing routes may not be changed or there is a given limit to the number of routes, etc.

List (1990) describes a methodology for preparing optimal sketch-level service plans. The sketch-level plan does not include precise routes, but it does determine passenger-flow values and frequencies on the road network. The model includes constraints for demand satisfaction, fleet size, minimal frequency, load factor, junction capacity, train length, and crew requirements. Several sub-models are developed that help to formulate those parts of the model that stem from limited resources, such as energy consumption. The model makes sure that network flows are balanced and takes into account multi-period passenger demand.

Baaj and Mahmassani (1991, 1992, 1995) develop transit network design methods based on artificial intelligence (AI). The methods discussed are developed by a typical formulation of the network-design problem as a programming problem with minimum frequency, load-factor, and fleet size constraints. The first paper (1991) uses flowcharts to present a quantitative description of a three-stage design process for a route network. In the first stage, a large set of routes is generated; the second stage involves network analysis and a determination of frequencies; the third stage is network improvement. The second paper (1992) focuses on a method of representing the transportation network by using lists and arrays to make the solution procedure efficient. The third paper (1995) concentrates on the stage of creating the initial set of routes, which are supposed to be modified now and improved later on. In order to generate this initial route set, a set of basic skeletons is created along the shortest paths between nodes with high passenger demand; the skeletons are expanded, using a set of node-insertion manipulations.

Spasovic and Schonfeld (1993) introduce a method for determining optimal route lengths, route spacing, headways, and stop spacing in a radial network with one central business district (CBD). Cost functions are developed for both the operators and the users by minimizing their sum; equations are derived for the optimal values of all decision variables. A many-to-one demand pattern is assumed, and passenger density can either be uniform or decrease linearly with distance from the CBD. In addition, a solution algorithm is developed that incorporates realistic vehicle-capacity constraints.

Spasovic et al. (1994) extend the model described in the previous paragraph. The decision variables in this version include route lengths, route spacing, headways, and fares. Two alternative optimality criteria are examined—operator profit and social welfare, the latter being the sum of user and operator surpluses. Social welfare is optimized with both unconstrained subsidies and break-even constraints. Analytical solutions are sought for a rectangular transit corridor with elastic demand, uniformly distributed passenger-trip density, and many-to-one travel patterns.

Ramirez and Seneviratne (1996), using GIS, propose two methods for route-network design with multiple objectives. Both methods involve ascribing an impedance factor to each possible route and then choosing those routes that have the minimum impedance. In the first method, the impedance factor depends on passenger flow and on the road length traveled. This method requires the use of an assignment model. In the second method, the impedance factor depends on the number of employees who have a reasonable walking distance from the route.

Pattanik et al. (1998) present a methodology for determining route configuration and associated frequencies, using a genetic algorithm. Solutions are chosen in an iterative process from a large set of possibilities in which the chances of a solution's surviving through the iterations are higher if it yields a high value for a given fitness function. The method presented here adopts the typical programming formulation of the route-network-design problem with the objective of minimizing a weighted combination of passenger-time costs and operator-time costs; the objective function provides the basis for the calculation of the fitness-function values. A methodology is also presented for the coding of variables as strings with a fixed or variable length.

Soehodo and Koshi (1999) formulate a programming problem for designing transit routes and frequencies. Similar to other models, the problem is solved by first creating all feasible routes and then choosing an optimal subset. In addition to some traditional components, such as minimum frequency and fleet size constraints, the problem adds some unique elements, such as the inclusion of private car user costs, transit-passenger crowding costs, and transfer costs, to the minimized objective function. A sub-model is developed for each of these cost types. Equilibrium of network flows is another constraint. The model assumes that demand is elastic, and therefore the shift of passengers between different modes of transport plays a major role. Both transit and non-transit demand-assignment models are used.

Bielli et al. (2002) describe another method for designing a bus network, using a genetic algorithm. As in other genetic algorithms, each population of solutions goes through reproduction, crossover, and mutation manipulations, whose output is a new generation of solutions. In the proposed model, each iteration involves demand assignment on each network of the current set of solutions and a calculation of performance indicators based on the assignment results. These indicators supply input to a multi-criteria analysis of each network, leading to the calculation of its fitness-function value.

Yan and Chen (2002) present a method for designing routes and timetables that aims at optimizing the correlation between bus-service supply and passenger demand. The method is based on the construction of two time-space networks: a fleet-flow network and a passenger-flow network. Both networks are depicted in bi-dimensional diagrams in which the horizontal dimension represents bus stops and the vertical dimension represents time. While the fleet-flow network shows the potential activities of the bus fleet, the passenger-flow network illustrates trip demand. The objective of the model is to feed buses and passengers at minimum cost in both networks simultaneously. A mixed-integer, multiple-commodity, network-flow problem and a solution algorithm based on Lagrangian relaxation are presented.

Van Nes and Bovy (2002) and Van Nes (2003) investigate the influence of the definition of objective functions on the design of stop spacing and line spacing, and the preferences of different traveler groups, respectively. Van Nes and Bovy develop an analytical model, in which the objective functions are defined for the traveler, the agency, and the authorities. Two alternative objectives for the traveler and six objectives for the authorities are examined, and the results of the interactions of different combinations are discussed. In addition, various alternative assumptions regarding demand elasticity are accounted for, and the difference in outcomes between different city sizes is analyzed. Van Nes (2003) investigates the same model, but with different weights for different traveler groups. Van Nes found that the optimal result of the network of routes was similar to the traditional single-user-class approach, thus concluding that a more realistic description of passenger groups was not necessary for the network-design theme.

Tom and Mohan (2003) continue the development of genetic methods for route-network design. In the current model, frequency is the variable; thus, it differs from earlier models in terms of the coding scheme adopted. Whereas fixed-string length coding and variable-string length coding were used in previous models, a combined route and frequency-coding model is proposed here.

Zhao and Ubaka (2004) present a mathematical methodology for transit route network optimization. The goal of this study is to provide an effective computational tool for the optimization of large-scale transit route networks. The objectives are to minimize transfers and optimize route directness while maximizing service coverage. The formulation of the methodology consists of three parts: (1) representation of transit route network solution space; (2) representation of transit route and network constraints; and (3) solution search schemes. The methodology has been implemented as a computer program and has been tested using previously published results. Finally, the results of these tests and results from the application of the methodology to a large-scale realistic network optimization problem in Miami-Dade County, Florida, are presented.

Zhao and Zeng (2008) present a meta-heuristic method for optimizing transit networks, including route network design, vehicle headway, and timetable assignment. Given information on transit demand, the street network of the transit service area, and total fleet size, the goal is to identify a transit network that minimizes a passenger cost function. The methodology described in this paper includes a representation of transit network variable search spaces (route network, headway, and timetable); a user cost function based on passenger random arrival times, route network, vehicle headways, and timetables; and a meta-heuristic search scheme that combines simulated annealing, tabu, and greedy search methods. Finally, this methodology has been tested with problems reported in the existing literature, and applied to a large-scale realistic network optimization problem. The results show that the methodology is capable of producing improved solutions to large-scale transit network design problems in reasonable amounts of time and computing resources.

Mauttone and Urquhart (2009a) proposed a heuristic method, called pair insertion algorithm (PIA), for designing bus network exhibiting the viewpoints of both users and operators. They modified the heuristic route generation algorithm of Baaj and Mahmassani (1995), where its original expansion of routes is replaced by a strategy of insertion of pairs of vertices. Using a case of the city of Rivera, Uruguay, they show that their model has an advantage over the Baaj and Mahmassani model from the operator perspective in terms of number of routes, and total travel time.

Mauttone and Urquhart (2009b) developed a method to solve the multi-objective transit network design problem using greedy randomized adaptive search procedure (GRASP) meta-heuristic. They implemented their method on Mandl's benchmark network and in a case study of the city of Rivera, Uruguay. Their results show that the proposed algorithm produces more non-dominated solutions than the weighted-sum method with the same computational effort.

Wan and Lo (2009) developed a framework to design transit services taking into account other existing transit services, travelers' preferences for combined-mode choices, and a financial viability requirement. Inter-route and inter-modal transfers are modeled through the state-augmented multimodal (SAM) network approach. They developed a two-phase methodology; the first phase seeks to maximize the social welfare by providing the largest set of bus service connections, subject to the financial viability constraints, and the second phase seeks to minimize the operation cost by detailing the bus routings and frequencies to fulfill the service connections determined in the first phase. They implemented their method on an illustrative sample network.

Fan and Mumford (2010) presented a framework for solving the urban transit routing problem, dealing with the following components: (1) a suitable representation of the problem, (2) an initialization procedure to construct randomly feasible sets of routes, and (3) a make-small-change routine to generate neighborhood moves. They implemented hill-climbing and simulated annealing for testing their technique and showed that on Mandl's network their method gives better results than those of previously published papers. In addition their method was examined for use on large-scale networks.

Feng et al. (2010) employ fuzzy programming and use four indicators representing sustainable transportation, such as accessibility, operator profit, user cost, and reduction of external cost, to optimize urban bus route allocation. The proposed model is applied to Taoyuan County, the most rapidly growing city in Taiwan, and managerial implication are identified to help decision-makers align bus routes based on sustainability criteria. Consideration is also given to extending the model to more comprehensive evaluation functions.

Szeto and Wu (2011) designed the bus network for Tin Shui Wai, a suburban residential area in Hong Kong, to improve the existing bus services by reducing the number of transfers and the total passenger travel time. They proposed an integrated solution method to solve the route design and frequency setting problems simultaneously. In their proposed solution method, a genetic algorithm for the route design problem is hybridized with a neighborhood search heuristic for the frequency setting problem. In order to illustrate the robustness and quality of their solutions, computational experiments are performed based on 1000 perturbed demand matrices. The t-test results showed that the design obtained by the proposed solution method is robust under demand uncertainty.

Poorzahedy and Safari (2011) utilized ant colony optimization algorithm for bus network design problem and minimized total passenger travel time on the transit network taking into account fleet requirements. They applied and calibrated the model based on the network of Sioux Falls. They also implemented their method on the case study of the city of Mashhad, Iran. Finally, this study compared the developed method with a genetic-algorithm method on two case studies and showed that their results are superior to those obtained by the genetic algorithm.

Yu et al. (2012) created a traveler density model with the maximization of demand density of routes considering resource constraints. Their transit network design problem is divided into three stages: (1) skeleton route design, (2) main route design, and (3) branch route design. They used an ant colony optimization technique for solving the model on a case study from Dalian city in China. Their results showed that the approach can improve the solution quality if the transfer coefficient of their objective function is reasonably set.

Cipriani et al. (2012) describe a procedure for solving the bus network design problem and its application in a large urban area (the city of Rome), characterized by (a) a complex road network topology, (b) a multimodal transit system (rapid rail transit system, buses, and tramways lines), and (c) a many-to-many transit demand. The solving procedure consists of a set of heuristics, which includes a first routine for the route generation based on the flow concentration process and a parallel genetic algorithm for finding a suboptimal set of routes with the associated frequencies. The final goal of the research is to develop an operative tool to support the mobility agency of Rome for the bus network design phase.

Nikolic and Teodorovic (2013) developed a model for solving the transit network design problem based on the swarm intelligence (SI) concepts of artificial intelligence. The objectives were to maximize the number of served passengers, minimize the total in-vehicle time of all served passengers, and to minimize the total number of transfers in the network. Their proposed bee colony optimization (BCO) algorithm can generate high-quality solutions within reasonable CPU times. They used Mandl's benchmark network and a larger test network to examine their development.

Nayeem et al. (2014) developed a population-based model for the transit network design problem to maximize the number of satisfied passengers, the total number of transfers, and to minimize the total travel time of all served passengers. They used genetic algorithm with elitism (GAWE) for solving the problem. The study claims to have the best results on Mandl's network compared with all previous methods.

Szeto and Jiang (2014) developed a bi-level transit network design problem, where the transit routes and frequency settings are determined simultaneously. They also developed a hybrid artificial bee colony (ABC) algorithm to solve the bi-level problem. The upper-level problem is formulated as a mixed integer nonlinear program with the objective of minimizing the number of passenger transfers, and the lower-level problem is the transit assignment problem with capacity constraints.

Table 15.12 summarizes the main features reviewed and discussed in this section.

Table 15.12 Summary of the characteristics of the methods reviewed

Source	Special features	Unique objectives/constraints	Intermodal considerations
Hobeika and Cho (1979)	Stops are divided into sectors.		Yes
Marwah and Umrigar (1984)		Minimizing the system-wide number of transfers.	No
Van Nes et al. (1988)		Maximizing the number of direct trips.	No
List (1990)	Sketch-level design only; multi-period demand.	Junction capacity, train length, and crew-requirement constraints.	Yes
Baaj and Mahmassani (1991, 1992, 1995)	Artificial intelligence.		No
Spasovic and Schonfeld (1993)		Minimizing the sum of user and operator costs; capacity constraint.	No
Spasovic et al. (1994)		Maximizing either operator profit or social welfare, with unconstrained subsidies and break-even constraints.	No
Ramirez and Seneviratne (1996)	GIS-based; the second proposed model takes into account potential demand (based on employment density), and not just existing demand.		No
Pattanik et al. (1998)	Genetic algorithm.		No
Soehodo and Koshi (1999)	Elastic demand.	Private car user costs, transit passenger crowding costs, and transfer costs are included in the minimized objective function.	Yes
Bielli et al. (2002)	Genetic algorithm.	The fitness function is based on multi-criteria analysis.	No
Yan and Chen (2002)	Network-flow problem formulation.		No
Van Nes and Bovy (2002)	Elastic and fix demand examined.	Multiple objectives of the traveler, the agency, and the authorities are defined and compared.	No
Van Nes (2003)	Different traveler groups.	Multiple objectives in which travel time and relationship between supply and demand depend on traveler groups.	No
Tom and Mohan (2003)	Genetic algorithm.		No
Zhao and Ubaka (2004)		Minimizing transfers and optimizing route directness while maximizing service coverage.	Yes

(Continued)

Table 15.12 (Continued) Summary of the characteristics of the methods reviewed

Source	Special features	Unique objectives/constraints	Intermodal considerations
Zhao and Zeng (2008)	The formulation includes definitions of solution search spaces (route network, vehicle headway, and timetable) and a stochastic solution search scheme.	Minimizing the total user cost of the transit service area subject to the operator budgetary constraint and a set of route, headway, and timetables feasibility constraints.	No
Mauttone and Urquhart (2009a)	Heuristic algorithm.	The interests of users (total in-vehicle time) and operators (total route mileage).	No
Mauttone and Urquhart (2009b)	Multi-objective consideration in bus network design problem.	Minimizing total travel time as the first objective and fleet size as the second.	No
Wan and Lo (2009)	Taking into consideration travelers' preferences for the combined-mode choices, as well as a financial viability requirement.	Social welfare maximization.	Yes
Fan and Mumford (2010)	Heuristic method.	Minimizing travel time and number of transfers.	No
Feng et al. (2010)	Employ fuzzy programming, and use four indicators representing sustainable transportation.	Four objectives, including accessibility, operator profit, user cost and the reduction of external cost, referring to criteria of sustainable transportation are formulated.	No
Szeto and Wu (2011)	Solving the route design and frequency setting problems simultaneously.	Sum of the number of transfers and total passengers' travel time.	No
Poorzahedy and Safari (2011)	Use of ant colony optimization for bus network design.	Minimize the total travel time of the users of the network, while being feasible in fleet requirements.	No
Yu et al. (2012)	Used ant colony optimization for bus network design.	Maximize the total demand density of the route (the number of transit demands divided by the length of the route).	No
Cipriani et al. (2012)	Genetic algorithm.	The objective function is defined as the sum of operator's costs and users' costs plus an additional penalty related to the level of unsatisfied demand.	Yes
Nikolić and Teodorovic (2013)	Based on the bee colony optimization (BCO) meta-heuristics.	Maximize the number of satisfied passengers, minimize the total number of transfers, and the total travel time of all served passengers.	No

(Continued)

Table 15.12 (Continued) Summary of the characteristics of the methods reviewed

Source	Special features	Unique objectives/constraints	Intermodal considerations
Nayeem et al. (2014)	Genetic algorithm with elitism.	Minimizing number of satisfied passengers, the total number of transfers, and the total travel time of all served passengers.	No
Szeto and Jiang (2014)	Artificial bee colony (ABC) algorithm.	Minimizing the number of passenger transfers.	No

EXERCISES

15.1 Given: (1) A bus service network consisting of four nodes (one terminal) and five arcs with average travel times (minutes) for both directions of travel.

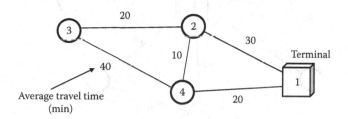

(2) An O-D symmetrical demand matrix for a 2 h peak period, the arrival pattern of passengers is assumed to be homogeneously distributed for each cell.

		To	
From	2	3	4
1	960	380	160
2		220	200
3	Symmetrical		240

(3) The desired occupancy is 60 passengers per bus.

(4) Bus trips are generated (start and end) only at the terminal (Node 1) and deadheading trips are not allowed.

(5) Two routes (round trips) serve the given network demand, Route A: 1 → 2 → 3 and backward to Node 1; and Route B: 1 → 2 → 4 and backward to Node 1.

(6) Boarding, alighting, and transfer times are neglected.

    (a) For the two existing routes, derive the required headways based on the maximum load section method; if more than one alternative exists, select the most appropriate one and explain your decision.

    (b) What is the minimum fleet size required for Routes A and B, based on the derived schedule (headways)?

(c)   Your task is to evaluate all possible single routes (round trips—same route forward and backward from the terminal) in the given network, visiting all nodes during the 2 h peak period. Note that the round trips may end after the peak hours.

(c-1) Define all possible combinations of single routes (visiting all nodes).

(c-2) Set an appropriate criterion (criteria) for your evaluation.

(c-3) Construct the load profile with respect to travel time for each possible single route.

(c-4) What is the minimum fleet size required for each single route?

(c-5) Suggest a single route and compare it with the existing two routes.

15.2 Given: (1) The following bus-service network:

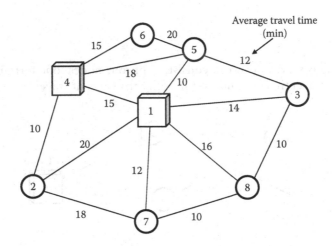

(2) In accordance with their respective sequence of nodes, the following three routes exist: the first employs an articulated bus, 1–2–4–6, with a desired occupancy (d_o) of 75 passengers; the second, 4–5–3, uses a standard bus with $d_o = 50$; and the third, 1–8–7, employs a minibus with $d_o = 25$.

(3) Symmetrical O-D matrix of a peak period, by number of passengers.

From	To						
	2	3	4	5	6	7	8
1	80	100	120	100	180	30	40
2		100	100	140	200	50	30
3			100	80	120	20	30
4				60	140	30	20
5	Symmetrical				100	0	40
6						10	20
7							50

(a)   For each route r (one direction), calculate: difference in passenger-hours, PHr; empty-space hours, EHr; difference between passenger-hours and shortest path, DPHr (without transfers); and passenger waiting time, wr.

(b)   Calculate total DPH for all three routes, including transfers consideration.

15.3 One of two independent transit modes is considered for implementation, either fast ferry or fast train. The route examined is A → B → C. The following data are given:

Data item	Mode	Transit segment A → B	A → C	B → C
Average travel time (min)	Fast train	40	50	10
	Fast ferry	70	90	20
Expected fare ($)	Fast train	35	45	12
	Fast ferry	30	35	9
O-D demand (passengers)	Both modes	280	1100	700

Additional given data and information: (1) desired occupancy, in terms of number of seats, on the fast train and fast ferry is 600 and 450 passengers, respectively; (2) value of passenger waiting time is $9/h; (3) value of passenger travel time is $3/h; (4) passengers arrive randomly at stops A and B; and (5) timetables in both modes are based on even headways and Method 2 (see Section 3.2).

(a) Calculate the hourly waiting-time cost of each alternative mode.
(b) Calculate the loss cost of hourly empty seats and travel times.
(c) Calculate the hourly income and profit of each alternative mode.
(d) Based only on the given data, suggest the preferred mode.
(e) List briefly actual cost elements that were neglected in this exercise.

REFERENCES

Amiripour, M., Ceder, A., and Mohaymany, A.S. (2014a). A hybrid method for bus network design with high seasonal demand variation. *Journal of Transportation Engineering—ASCE*, **140**(6), June 2014. doi: http://dx.doi.org/10.1061/(ASCE)TE.1943-5436.0000669.

Amiripour, M., Ceder, A., and Mohaymany, A.S. (2014b). Method for designing large-scale bus network with seasonal variations of demand. *Transportation Research*, 48C, 322–338.

Baaj, M.H. and Mahmassani, H.S. (1991). An AI-based approach for transit route system planning and design. *Journal of Advanced Transportation*, 25, 187–210.

Baaj, M.H. and Mahmassani, H.S. (1992). Artificial intelligence-based system representation and search procedures for transit route network design. *Transportation Research Record*, 1358, 67–70.

Baaj, M.H. and Mahmassani, H.S. (1995). Hybrid route generation heuristic algorithm for the design of a transit network. *Transportation Research*, 3C, 31–50.

Bagloee, S.A. and Ceder, A. (2011). Transit-network design methodology for actual-size road networks. *Transportation Research*, 45B, 1787–1804.

Bielli, M., Caramia, M., and Carotenuto, P. (2002). Genetic algorithms in bus network optimization. *Transportation Research*, 10C, 19–34.

Ceder, A. (2001). Operational objective functions in designing public transport routes. *Journal of Advanced Transportation*, 35, 125–144.

Ceder, A. (2003). Designing public transport network and routes. In *Advanced Modeling for Transit Operations and Service Planning* (W. Lam and M. Bell, eds.), pp. 59–91, Pergamon Imprint, Elsevier Science, New York.

Ceder, A., Gonzalez, O., and Gonzalez, H. (2002). Design of bus routes: Methodology and the Santo Domingo case. *Transportation Research Record*, 1791, 35–43.

Ceder, A. and Israeli, Y. (1992). Scheduling consideration in designing transit routes at the network level. In *Computer-Aided Transit Scheduling*. Lecture Notes in Economics and Mathematical Systems, vol. 386 (M. Desrochers and J.M. Rousseau, eds.), pp. 113–136, Springer-Verlag, Berlin, Germany.

Chang, S.K. and Schonfeld, P.M. (1991). Multiple period optimization of bus transit systems. *Transportation Research*, 25B, 453–478.

Cipriani, E., Gori, S., and Petrelli, M. (2012). Transit network design: A procedure and an application to a large urban area. *Transportation Research*, 20C, 3–14.

Coello, C.A., Van Veldhuizen, D.A., and Lamont, G.B. (2002). *Evolutionary Algorithms for Solving Multi-Objective Problems*, Kluwar Academic/Plenum Publishers, New York.

Dial, R.B. and Bunyan, R.E. (1968). Public transit planning system. *Socio-Economic Planning Sciences*, 1, 345–362.

Dubois, D., Bel, G., and Libre, M. (1979). A set of methods in transportation network synthesis and analysis. *Operations Research*, 30, 797–808.

Fan, L. and Mumford, C. (2010). A metaheuristic approach to the urban transit routing problem. *Journal of Heuristics*, 16(3), 353–372.

Fan, W. and Machemehl, R.B. (2006). Optimal transit route network design problem with variable transit demand: Genetic algorithm approach. *Journal of Transportation Engineering*, 132(1), 40–51.

Farvolden, J.M. and Powell, W.B. (1994). Subgradient methods for the service network design problem. *Transportation Science*, 28, 256–272.

Feng, C.M., Hsieh, C.H., and Peng, S.C. (2010). Optimization of urban bus routes based on principles of sustainable transportation. *Journal of the Eastern Asia Society for Transportation Studies*, 8, 1137–1149.

Guihaire, V. and Hao, J.-K. (2008). Transit network design and scheduling: A global review. *Transportation Research*, 42A, 1251–1273.

Hadas, Y. and Ceder, A. (2014). Optimal connected urban bus network of priority lanes. *Transportation Research Record*, 2418, 49–57.

Heathington, K.W., Miller, J., Knox, R.R., Hoff, G.C., and Bruggman, J. (1968). Computer simulation of a demand scheduled bus system offering door to door service. *Highway Research Record*, 91, 26–40.

Hobeika, A.G. and Cho, C. (1979). Equilibration of supply and demand in designing bus routes for small urban areas. *Transportation Research Record*, 730, 7–13.

Israeli, Y. and Ceder, A. (1995). Transit route design using scheduling and multi-objective programming techniques. In *Computer-Aided Transit Scheduling*. Lecture Notes in Economics and Mathematical Systems, vol. 430 (J.R. Daduna, I. Branco, and J.M.P. Paixao, eds.), pp. 56–75, Springer-Verlag, Berlin, Germany.

Kepaptsoglou, K. and Karlaftis, M. (2009). Transit route network design problem: Review. *Journal of Transportation Engineering*, 135(8), 491–505.

Keudel, W. (1988). Computer-aided line network design (DIANA) and minimization of transfer times in networks (FABIAN). In *Computer-Aided Transit Scheduling*. Lecture Notes in Economics and Mathematical Systems, vol, 308 (J.R. Daduna and A. Wren, eds.), pp. 315–326, Springer-Verlag, Berlin, Germany.

Kim, D. and Barnhart, C. (1999). Transportation service network design: Models and algorithms. In *Computer-Aided Scheduling of Public Transport*. Lectures Notes in Economics and Mathematical Systems, vol. 471 (N.H.M. Wilson, ed.), pp. 259–283, Springer-Verlag, Berlin, Germany.

Kocur, G. and Hendrickson, C. (1982). Design of local bus service with demand equilibration. *Transportation Science*, 16, 149–170.

Kuah, G.K. and Perl, J. (1988). Optimization of feeder bus routes and bus-stop spacing. *Journal of Transportation Engineering*, 114, 341–454.

Lampkin, W. and Saalmans, P.D. (1967). The design of routes, service frequencies, and schedules for a municipal bus undertaking: A case study. *Operations Research*, 18, 375–397.

List, G.F. (1990). Towards optimal sketch level transit service plans. *Transportation Research*, 24B, 324–344.

Lee, Y.J. and Vuchic, V.R. (2005). Transit network design with variable demand. *Journal of Transportation Engineering*, 131(1), 1–10.

Mandl, C.E. (1979). Evaluation and optimization of urban public transportation networks. *European Journal of Operation Research*, 5, 396–404.

Marwah, B.R., Umrigar, F.S., and Patnaik, S.B. (1984). Optimal design of bus routes and frequencies for Ahmedabad. *Transportation Research Record*, 994, 41–47.

Mauttone, A. and Urquhart, M.E. (2009a). A route set construction algorithm for the transit network design problem. *Computers and Operations Research*, 36(8), 2440–2449.

Mauttone, A. and Urquhart, M.E. (2009b). A multi-objective metaheuristic approach for the transit network design problem. *Public Transport*, 1(4), 253–273.

Mees, P. (2010). *Transport for Suburbia: Beyond the Automobile Age*, Earthscan, London, U.K..

Mesbah, M., Sarvi, M., and Currie, G.(2008). New methodology for optimizing transit priority at the network level. *Transportation Research Record*, 2089, 93–100.

Mesbah, M., Sarvi, M., and Currie, G. (2011a). Optimization of transit priority in the transportation network using a genetic algorithm. *IEEE Transactions on Intelligent Transportation Systems*, 12(3), 908–919.

Mesbah, M., Sarvi, M., Currie, G., and Saffarzadeh, M. (2010). Policy-making tool for optimization of transit priority lanes in urban network. *Transportation Research Record*, 2197, 54–62.

Mesbah, M., Sarvi, M., Ouveysi, I., and Currie, G. (2011b). Optimization of transit priority in the transportation network using a decomposition methodology. *Transportation Research*, 19C, 363–373.

Nayeem, M.A., Rahman, M.K., and Rahman, M.S. (2014). Transit network design by genetic algorithm with elitism. *Transportation Research*, 46C, 30–45.

Newell, G.F. (1979). Some issues relating to the optimal design of bus routes. *Transportation Science*, 13(1), 20–35.

Nikolić, M. and Teodorović, D. (2013). Transit network design by Bee Colony Optimization. *Expert Systems with Applications*, 40(15), 5945–5955.

Park, D., Sharma, S.L., Rilett, L.R., and Chang, M. (2002). Identifying multiple reasonable alternative routes: Efficient vector labeling approach. *Transportation Research Record*, 1783, 111–118.

Pattnaik, S.B., Mohan, S., and Tom, V.M. (1998). Urban bus transit route network design using genetic algorithm. *Journal of Transportation Engineering*, 124, 368–375.

Poorzahedy, H. and Safari, F. (2011). An ant system application to the bus network design problem: An algorithm and a case study. *Public Transport*, 3(2), 165–187.

Pratt, R. and Evans, J. (2004). Traveler response to transportation system changes: Chapter 10-Bus routing and coverage, *TCRP Report 95*. Transportation Research Board, Washington, DC.

Ramirez, A.I. and Seneviratne, P.N. (1996). Transit route design applications using geographic information systems. *Transportation Research Record*, 1557, 10–14.

Silman, L.A., Barzily, Z., and Passy, U. (1974). Planning the route system for urban buses. *Computers and Operations Research*, 1, 201–211.

Soehodo, S. and Koshi, M. (1999). Design of public transit network in urban areas with elastic demand. *Journal of Advanced Transportation*, 33, 335–369.

Spasovic, L.N., Boile, M.P., and Bladikas, A.K. (1994). Bus transit service coverage for maximum profit and social welfare. *Transportation Research Record*, 1451, 12–22.

Spasovic, L.N. and Schonfeld, P.M. (1993). Method for optimizing transit service coverage. *Transportation Research Record*, 1402, 28–39.

Szeto, W.Y. and Jiang, Y. (2014). Transit route and frequency design: Bi-level modeling and hybrid artificial bee colony algorithm approach. *Transportation Research*, 67B, 235–263.

Szeto, W.Y. and Wu, Y. (2011). A simultaneous bus route design and frequency setting problem for Tin Shui Wai, Hong Kong. *European Journal of Operational Research*, 209(2), 141–155.

Tom, V.M. and Mohan, S. (2003). Transit route network design using a frequency coded genetic algorithm. *Journal of Transportation Engineering*, 129, 186–195.

Tsao, S. and Schonfeld, P. (1984). Branched transit services: An analysis. *Journal of Transportation Engineering*, 110, 112–128.

Vandebona, U. and Richardson, A.J. (1985). Simulation of transit route operations. *Transportation Research Record*, 1036, 36–40.

Van Nes, R. (2003). Multiuser-class urban transit network design. *Transportation Research Record*, 1835, 25–33.

Van Nes, R. and Bovy, P.H.L. (2002). Importance of objectives in urban transit-network design. *Transportation Research Record*, 1735, 25–34.

Van Nes, R., Hamerslag, R., and Immers, B.H. (1988). Design of public transport networks. *Transportation Research Record*, 1202, 74–83.

Wan, Q. and Lo, H. (2009). Congested multimodal transit network design. *Public Transport*, **1**(3), 233–251.

Yan, H., Chi, C.J., and Tang, C.H. (2006). Inter-city bus routing and timetable setting under stochastic demands. *Transportation Research*, **40A**, 572–586.

Yen, J.Y. (1971). Finding the K shortest loopless paths in a network. *Management Science*, **17**(11), 712–716.

Yin, Y., Miller, M., and Ceder, A. (2005). Framework for deployment planning of bus rapid transit systems. *Transportation Research Record*, **1903**, 11–19.

Yu, B., Yang, Z.-Z., Jin, P.-H., Wu, S.-H., and Yao, B.-Z. (2012). Transit route network design-maximizing direct and transfer demand density. *Transportation Research*, **22C**, 58–75.

Zhao, F. and Ubaka, I. (2004). Transit network optimization: Minimizing transfers and optimizing route directness. *Journal of Public Transportation*, **7**(1), 63–82.

Zhao, F. and Zeng, X. (2008). Optimization of transit route network, vehicle headways and timetables for large-scale transit networks. *European Journal of Operational Research*, **186**(2), 841–855.

Chapter 16

Designing short-turn trips

If you want to change the routing across the board, start with a small board

A.C.

CHAPTER OUTLINE

PRACTITIONER'S CORNER

Subsequent to the overview and methods for designing routes at the network level, the next two chapters deal with specific design features at the route level. This present chapter presents a set of procedures to design transit timetables efficiently with trips that are initiated beyond the route departure point and/or terminated before the route arrival point. Such trips are called short-turn trips. In practice (see Chapter 4), transit frequency is determined at the heaviest load-route segment, whereas the operation at other segments may be inefficient because of situations, characterized by empty seats. Transit planners attempt to overcome this problem by manually constructing short-turn trips with the objective of reducing the number of vehicles required to carry out the transit timetable. The purpose of this chapter is to improve and automate this task.

The following riddle may serve as a stimulus for the need to handle a design element only after comprehending the process in which this element appears. The riddle: given a microorganism-biological process in which a microbe that is put into a glass splits into two after I s, the two then split into four after another second, the four into eight after another second, and so on; after I min, the glass is full of microbes. How many seconds are required to fill half of the same glass? (Answer is given at the end of this corner.)

The chapter contains four main parts, following an introductory section. Section 16.2 outlines the framework of the methodology for the efficient design of short-turn trips. Section 16.3 identifies a minimum number of candidate short-turn points and presents the example that is subsequently used throughout the other sections of the chapter. Section 16.4 provides a procedure to adjust the number of departures at each short-turn point to that required by the load data, provided that the maximum headway (associated with passenger waiting time) attained is minimized. Section 16.5 constructs another procedure to minimize the number of short-turn trips while ensuring that the minimum fleet size is preserved. The chapter ends with a literature review and exercises.

Practitioners are encouraged to visit Sections 16.1 through 16.3. In addition, they should follow all the figures, tables, and corresponding paragraphs of the chapter that are related to the example problem.

It is known that an error in the premise will appear in the conclusion. Often, transit agencies avoid a profound design because of their premise that it would be costly, time consuming, and unrewarding. It is analogous to a common saying, and perhaps even belief, that good things in life are illegal, immoral, or fattening. It will be shown in this chapter that a profound design (of short-turn trips) can, on the contrary, reduce cost, reduce time, and be rewarding. Last, the answer to the riddle is 59 s (in the 60th second, the half will multiply itself); obviously this answer is strongly dependent on the process described.

16.1 INTRODUCTION

Chapter 15 discussed the attainment of an efficient network of transit routes. This chapter focuses on improving the cost-effectiveness of each single route. Transit planners certainly understand the need to accommodate the observed passenger demand as well as possible. At the same time, however, their efforts are also directed at the minimization of vehicle and driver costs. The trade-off between increasing the passenger's comfort and reducing the cost of the service makes the planner's task cumbersome and complex. The design of short-turn trips is one such trade-off. A short-turn trip is initiated beyond the route's departure terminal and/or terminated before its arrival terminal. The possibility of generating short lines opens up an opportunity to further save on vehicles while ensuring that the passenger load on each route segment does not exceed the desired occupancy (load factor).

Planners in most transit agencies usually include a short-turn operating strategy in their attempts to reduce the cost of the service. The procedures commonly used are based only on the visual observation of the load profile, that is, a potential turn point is determined at the time point (major stop) nearest to a stop at which a sharp decrease or increase of passengers is observed. Although this procedure is intuitively correct, planners do not know whether all the short-turn trips are actually needed to reduce the fleet size. Unfortunately, each short-turn trip limits the service and, hence, tends to reduce the passenger level of service.

The major objectives of a short-turn operating strategy, and therefore of this chapter, are as follows:

- To identify minimum candidate short-turn points based on passenger load profile data
- To adjust the number of departures at each short-turn point to that required by the load data, provided that the maximum headway obtained is minimized; this objective results in the maximum possible short-turn trips and the minimum required fleet size (including shifting departure times and deadheading (DH) trips)
- To minimize the number of short-turn trips, provided that the minimum fleet size attained (with short-turn trips) is maintained; for a given timetable, this objective results in increasing the level of service as seen by the passengers

Several methods will be presented to meet the objectives. These methods are based on procedures and algorithms that use data commonly inventoried or collected by most transit agencies. The chapter is based on the methodology and modeling presented by Ceder (1990, 1991), with some improved and corrected elements.

16.2 METHODOLOGY

The proposed methodology relies on the following input data: (1) a complete timetable of all route time points, (2) passenger load profiles, by load and distance, for each time period, (3) minimum frequency or policy headway, and (4) a set of candidate short-turn points. These data are given for both directions of the route (each direction with its own data). Candidate short-turn points are usually all the major route stops (time points) at which the public timetable exists. In some cases, it may be limited to only those time points at which vehicles can actually turn back.

The comprehensive tasks needed to accomplish the objectives of the methodology are described in flow-diagram form in Figure 16.1; this methodology also appeared in Ceder (2003). It starts with a procedure to determine the set of feasible short-turn points R_j among the candidate points. Then the deficit function (DF) theory (Chapters 6 and 7) is used to derive the minimum number of vehicles required to carry out all the trips in the complete, two-direction timetable, N_{min}. The required number of departures is determined at each of the feasible short-turn points, and then the so-called Minimax H algorithm is applied. The basis of the algorithm is the elimination of some departures from the complete timetable in order to obtain the number of departures required. In this procedure, the algorithm minimizes the maximum difference between two adjacent departure times (headway). At this stage, as shown in Figure 16.1, the DF method derives the minimum fleet size required with

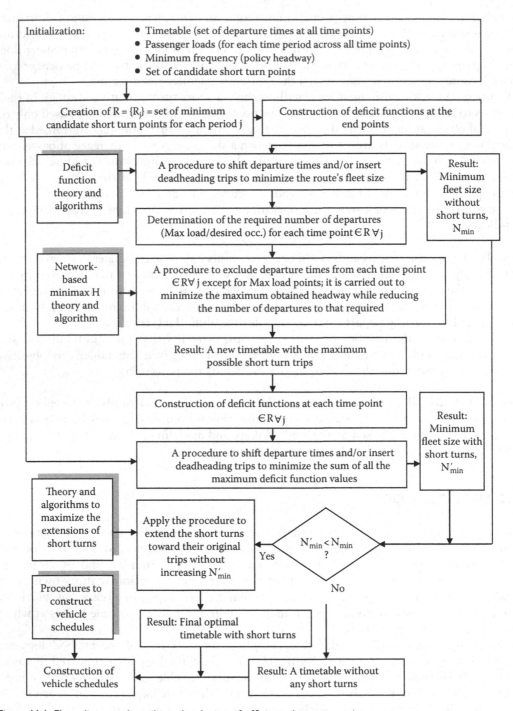

Figure 16.1 Flow diagram describing the design of efficient short-turn trips.

short turns, N'_{min}. If this minimum is less than the size required without short turns, then another procedure is applied. This second procedure inserts (back) the maximum possible departures among those previously eliminated, provided that the minimum fleet size, N'_{min}, is maintained. The final step of the overall program is to create vehicle blocks to cover all the trips that appear in the last version of the two-direction timetable.

16.3 CANDIDATE POINTS AND EXAMPLE

The short-turn points are usually route time points at which the vehicle can turn back without interfering with the traffic flow. It is therefore anticipated that for each route, the initial set of candidate short-turn points will be given by the transit planner.

16.3.1 Minimum candidate short-turn points

Let the set of candidate short-turn points be designated as set R_1 for one direction and R_2 for the opposite route direction. Note that R_1 does not necessarily coincide with R_2. More specifically,

$$R_1 = \left\{ r_{11}, r_{12}, \ldots, r_{1n} \right\}$$
$$R_2 = \left\{ r_{21}, r_{22}, \ldots, r_{2q} \right\} \tag{16.1}$$

where
 r_{ij} is the jth candidate short-turn point in the jth direction ($j = 1, 2$)
 n and q are such points for directions 1 and 2, respectively

For a given time period, the fluctuation of a passenger load along the entire route (load profile) may reveal that some short-turn points are actually redundant; thus, we can establish a set of a minimum number of candidate points. For example, say a load profile consists of 13 stops and 5 candidate short-turn points as shown in Figure 16.2. Theoretically, each segment between two adjacent short-turn points can be treated independently with respect to its required frequency. This frequency is determined by the maximum observed load on the segment, which is marked by the hatched area in Figure 16.2. In short-turn strategy, however, all the trips must serve the heaviest load segment of the route (in the example, all trips must cross the r_4–r_5 segment). Another observation is that fewer trips are required between r_3 and r_4 than between r_2 and r_3 while both groups of trips must cross the r_4–r_5 segment. Consequently, the point r_3 is redundant.

The exclusion of the redundant points at each time period j results in a set of minimum candidate short-turn points, R_j; this analysis is important from the viewpoint of computational time. The formal description of the algorithm to determine the set R_j includes an additional analysis of the difference between the required frequencies at the short-turn point

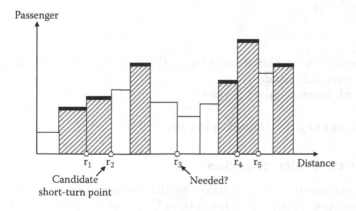

Figure 16.2 Construction of a set R of feasible short-turn points.

associated with the max load segment and a considered short-turn point. If this difference is small, the considered short-turn point can be deleted. Note that the difference in the frequencies is equivalent to the difference in the load, and there are always stochastic variations of that load. Hence, if this difference is small, it is not reasonable to consider short-turn trips from the associated short-turn point. This is actually similar to the manual procedure performed in current practice in which the planner selects short-turn points only on the basis of an observed sharp increase or decrease in the load profile. For subsequent analyses, the union of all R_j for all (time periods) j is denoted as the set R.

16.3.2 Example

A simple example is used as an expository device to illustrate the DF approach and the procedures developed. This example appears in Figure 16.3 for a 2 h schedule of departure times at the max load points. The route includes three time points (a, b, c), which is the set R. The average travel times for service and DH trips are given in the following timetable in Figure 16.3; no shifting in departure times is allowed.

Construction of d(a,t) and d(c,t) can then take place, and the minimum number of vehicles required without DH trips is D(a) + D(c) = 11. However, a DH trip can be inserted from *a* to c, departing after the last maximum interval of d(a, t) and arriving just before the start of the first maximum interval of d(c, t). Both d(a, t) and d(c, t) are then changed, as seen by the dashed line in Figure 16.3; thus, D(c) is reduced from 6 to 5, and the overall fleet size is reduced from 11 to 10. After that, it is impossible to further reduce the fleet size through DH trip insertion; hence, N_{min} = 10. This condition can also be detected automatically by the lower-bound test. That is, the maximum of the combined DFs is 10, and therefore, N_{min} reaches its lower bound.

16.4 EXCLUDING DEPARTURE TIMES

The basic information required to consider short turns is the route's load profile. Based on this load-profile information, each route segment between two adjacent short-turn points can be treated separately. That is, the required number of trips between the (k − l)th and *k*th short-turn points for a given direction and given time periods is similar to Equation 3.2 in Chapter 3:

$$F_k = max\left(\frac{P_k}{d_o}, F_{min} \right)$$ (16.2)

where
  P_k is the maximum load observed between the two adjacent short-turn points
  d_o is the desired occupancy
  F_{min} is the minimum required frequency

F_r is the route frequency, determined as in Chapter 15.

16.4.1 Level-of-service criterion

The manual procedure usually undertaken by the planner to create short-turn trips simply involves the exclusion of departure times in order to set the frequency at each short-turn point k to F_k instead of to F_r. This exclusion of departure times is performed without any

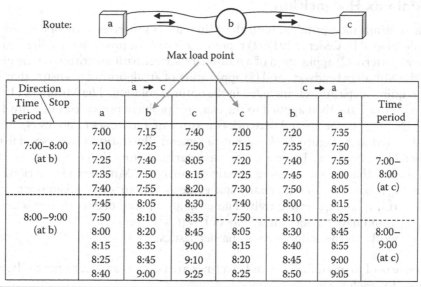

Route:

Max load point

Direction		a → c			c → a			
Time period \ Stop	a	b	c	c	b	a	Time period
7:00–8:00 (at b)	7:00	7:15	7:40	7:00	7:20	7:35	
	7:10	7:25	7:50	7:15	7:35	7:50	7:00–
	7:25	7:40	8:05	7:20	7:40	7:55	8:00
	7:35	7:50	8:15	7:25	7:45	8:00	(at c)
	7:40	7:55	8:20	7:30	7:50	8:05	
8:00–9:00 (at b)	7:45	8:05	8:30	7:40	8:00	8:15	
	7:50	8:10	8:35	7:50	8:10	8:25	
	8:00	8:20	8:45	8:05	8:30	8:45	8:00–
	8:15	8:35	9:00	8:15	8:40	8:55	9:00
	8:25	8:45	9:10	8:20	8:45	9:00	(at c)
	8:40	9:00	9:25	8:25	8:50	9:05	

Hours at the max load point	Travel times (min)				DH times (min)		
	a–b	b–c	c–b	b–a	a–c	a–b	b–c
7:00 –8:00	15	25	20	15	25	15	5
8:00 –9:00	20	25	25	15			

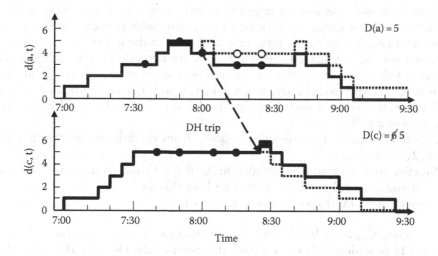

Figure 16.3 Example of a 2-way route with a 2 h schedule for which 10 vehicles are required (based on the graphical deficit function method).

systematic instructions in the (scheduler) belief that it is possible thereby to reduce the number of vehicles required to carry out the timetable.

The result of excluding certain departure times is that some passengers will have to extend their wait at the short-turn points. To minimize this adverse effect, it is possible to set the following (Minimax H) criterion: *Delete $F_r–F_k$ departure times at* k *with the objective of minimizing the maximum headway obtained.*

The Minimax H criterion attempts to achieve the minimization of maximum passenger waiting time; the result may represent an adequate passenger level of service whenever short-turn strategy is employed.

16.4.2 Minimax H algorithm

To solve the optimization problem with the Minimax H criterion, a theory based on three stages was developed by Ceder (1991): (1) representation of the problem on a directed network with a special pattern; (2) application of a modified shortest-path algorithm on the network to determine the Minimax headway; and (3) application of an algorithm to ensure that the exact number of required departures is included in the optimal solution. The Minimax H algorithm will now be outlined and then applied to the example problem presented in Figure 16.3.

Let $G_m = \{N_m, A_m\}$ be the special network consisting of a finite set of nodes N_m and a finite set A_m of directed arcs. Figure 16.4 presents a general illustration of the special network, accompanied by an example. There are n given departures from the complete timetable, and the requirement is that only m < n need remain to satisfy the Minimax H criterion. The construction of G_m is based on m – 2 equally spaced departure times between the first and last given departures, t_1, and t_n, respectively. These equally spaced departure times are denoted by $t'_2, t'_3, \ldots, t'_{m-1}$ and have an equal headway of $t_e = (t_n - t_1)/(m - 1)$.

The G_m network has the following six characteristics:

1. G_m consists of m rows; the first and last rows are nodes t_1 and t_n, respectively, and there is a row for each t_j, j = 2, 3, …, m – 1.
2. Each node in N_m represents a departure time in the given set of departures; however, it is not necessary that all given departures be included in N_m (see 7:10, 8:50 in the example of Figure 16.4), and also the same departures may be represented by several nodes (see 7:45, 8:00, 8:20 in Figure 16.4).
3. The nodes in each row are organized from left to right in increasing time order with respect to their associated t'_j. That is, all the given nodes t_i, such that $t'_k \leq t_i < t'_{k+1}$, are positioned twice, once to the right of t'_k and once to the left of t'_{k+1}, where t'_k, t'_{k+1} are two adjacent, equally spaced departure times. An exception occurs in the second and the (m – 1)th rows, where only one node is positioned to the left of t'_2 and one to the right to t'_{m-1}, respectively. These single nodes, t_3 and t_{n-2} in Figure 16.4, are selected such that t_3 is the closest node to t'_2, provided that $t_3 < t'_2$, and t_{n-2} is the closest node to t'_{m-1}, provided that $t'_{m-1} \leq t_{n-2}$.
4. The directed arcs in A_m connect only nodes from the kth row to the (k + 1)th row, k = 1, 2, …, m – 1.
5. A directed arc from t_i to t_j is included in A_m if $t_j > t_i$, and from t_i to t_i (in subsequent rows) if and only if G_m is disconnected without this arc.
6. The length of an arc from t_i to t_j is exactly $t_j - t_i$.

After constructing G_m, a modified shortest-path algorithm is applied as the second stage of the Minimax H procedure. This is a modified version of the Dijkstra algorithm described in Appendix 9.A. The Dijkstra algorithm is based on assigning temporary labels to nodes, the label being an upper bound on the path length from the origin node to each node. These labels are then updated (reduced) by an iterative procedure. At each iteration, exactly one of the temporary labels becomes permanent, implying that it is no longer the upper bound but rather the exact length of the shortest path from the origin to the considered node.

The modification of the Dijkstra method takes place in the computation step, in which the labels are updated. It is modified from $\pi(t_i) = \text{Min}[\pi^*(t_k) + (t_i - t_k), \pi(t_i)]$ (see Step 3 of Dijkstra algorithm in Appendix 9.A), to

$$\pi(t_i) = \text{Min}\{\max[\pi^*(t_k), (t_i - t_k)], \pi(t_i)\} \tag{16.3}$$

where $\pi(t_i)$, $\pi^*(t_k)$ are temporary and permanent labels of nodes t_i and t_k, respectively.

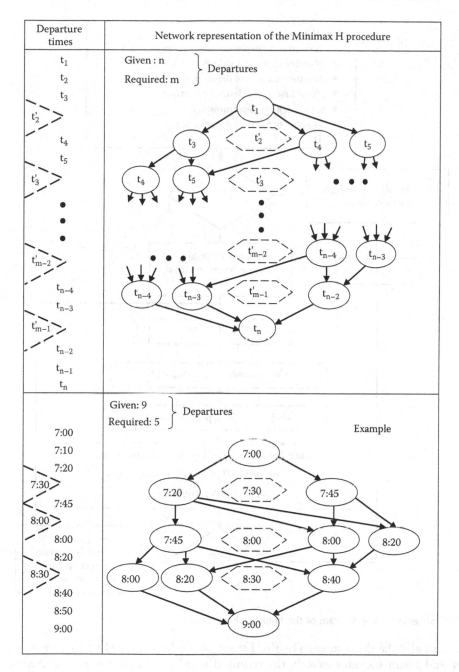

Figure 16.4 General network representation of the Minimax H procedure at one terminal, accompanied by an example.

This modified-Dijkstra algorithm, instead of minimizing the sum of arc values, searches for the minimum of the maximum arc value. The algorithm is applied to G_m when the origin node is t_i; the algorithm terminates when the temporary label on node t_n becomes permanent.

Figure 16.5 exhibits the flow diagram of the three stages of the Minimax H algorithm. When the modified-Dijkstra procedure ends with $p < m$ departures, a secondary criterion, for the second largest headway, is introduced to find the difference of $m-p$ departure times;

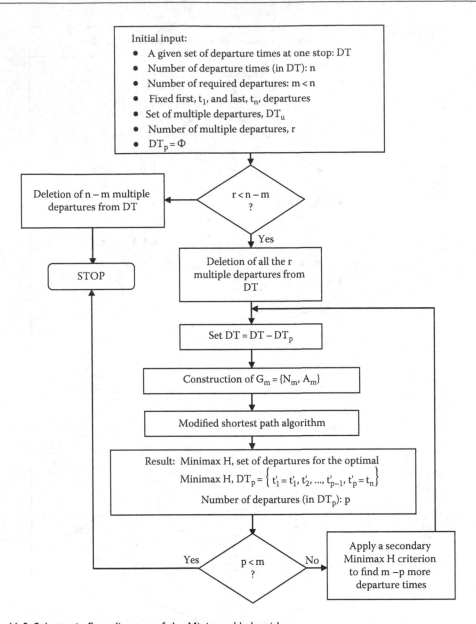

Figure 16.5 Schematic flow diagram of the Minimax H algorithm.

this is, basically the third stage. The third stage of the Minimax H procedure ensures that the optimal result includes exactly the required number of departures (m). Although the modified shortest-path algorithm for G_m determines the value of the Minimax headway, it does not ensure that the result will include all the m required departures.

The third stage is interpreted utilizing the 10 different network configurations illustrated in Figure 16.6. Each example in Figure 16.6 is based on a different initial DT set of five departures, from which one of the middle three is unnecessary. Cases (e) and (f) consist of a single path with three departures, whereas four departures are required. The additional departure required will be selected between t_2 and t_3 for case (e) and between t_3 and t_4 for case (f). The Minimax H = 45 min in both cases. If we select the t_2 departure for case (e),

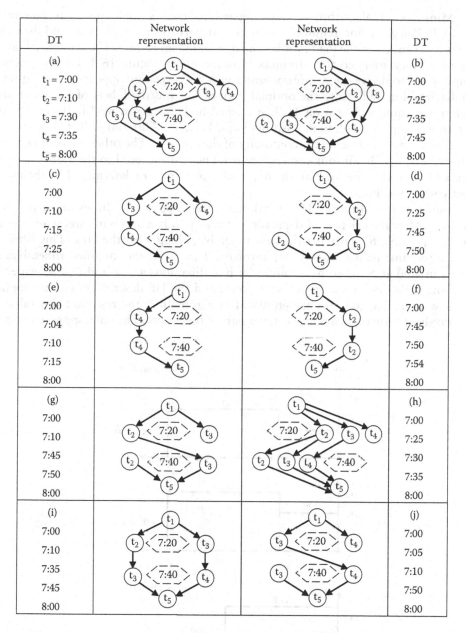

Figure 16.6 Different networks (based on a given five-departure timetable at one stop) for a required schedule of four departures from 7:00 to 8:00.

then the second largest headway will be 11 min, as opposed to 10 min for t_3. A similar situation exists when selecting t_3 instead of t_4 for case (f).

Cases like (e) and (f) call for a secondary criterion of minimizing the maximum headway of a subset of DT. That is, for case (e), t_1 remains the same, t_n becomes $t_4 = 7:15$, $n = 4$, $m = 3$, and a new G_m is constructed and solved. For case (f), $t_n = t_5$ remains the same, t_1 becomes $t_2 = 7:45$, $n = 4$, and $m = 3$ of the new G_m. The detailed description of the third stage of the Minimax H algorithm appears in Ceder (1991), along with a procedure for treating multiple departures (with the same or more than one departure time in the given timetable).

The Minimax H algorithm is now applied to the example problem presented in Figure 16.3. The given and required numbers of departures for each hour and direction of travel are determined by the load profiles and $d_o = 50$ in Figure 16.7. These required numbers (values of m) then undergo the Minimax H procedure in Figure 16.8. Four G_m networks are constructed to derive the Minimax headway. This derivation appears in the figure with an emphasized line indicating the optimal path of G_m and the labels of the shortest-path algorithm according to Equation 16.3. A dashed line and arrowhead indicates the direction of another optimal solution. Also note that between 7:00 and 8:00 t_n becomes t_1 for 8:00–9:00 in order to preserve the continuity of the analysis. The other parts of Figure 16.8 are self-explanatory. In all four cases, there is no need to proceed to the third stage of the Minimax H algorithm, because all the required departures are determined by the modified shortest-path algorithm.

The results of the Minimax H algorithm are then applied to the example problem in Figure 16.3. After the deletion of departures at time points a and b in directions a → c and c → a, it is possible to construct the new timetable, along with the DFs. This time, however, all three time points (a, b, c) are involved. That is, in the modified timetable, some trips are initiated at b and some terminate at b in directions a → c and c → a, respectively. Thus, point b also becomes an end/start point, and the DF description can be applied to it. The new timetable and DFs are presented in Figure 16.9; the resultant timetable (with maximum short turns) appears in the upper part of this figure. The corresponding DFs show

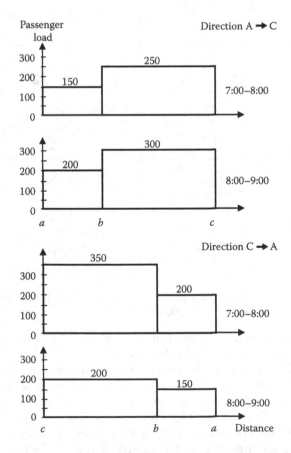

Figure 16.7 Load profiles of the example problem with a desired occupancy of 50 passengers per vehicle.

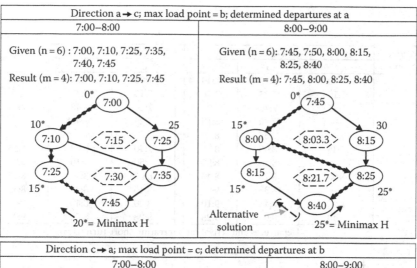

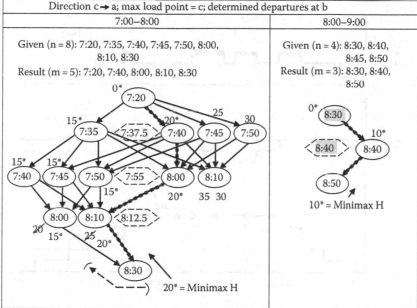

Figure 16.8 Minimax H algorithm and results for the example problem where * refer to Equation 16.3.

that 10 vehicles are required to carry out the timetable without DH trips and that 9 vehicles (N'_{min}) are required with a single DH trip from d(c, t) to arrive at d(b, t) before or at 8:35. Following the condition of Figure 16.1 for $N'_{min} < N_{min}$, where N_{min} = 10 vehicles, the next step is to attempt to reduce the number and impact of short-turn trips, provided that N'_{min} is maintained.

16.5 MAXIMUM EXTENSIONS OF SHORT-TURN TRIPS

There are two similar undertakings for possible extensions of short-turn trips: (1) convert trips to points $r_i \in R$ from DH to service trips and (2) extend arrival and departure trips to points $r_i \in R$ toward their original schedule.

Direction	a → c			c → a		
Stop	a	b	c	c	b	a
	7:00	7:15	7:40	7:00	7:20	7:35
	7:10	7:25	7:50	7:15	7:35*	—
	7:25	7:40	8:05	7:20	7:40	7:55
	—	*7:50	8:15	7:25	7:45*	—
	—	*7:55	8:20	7:30	7:50*	—
	7:45	8:05	8:30	7:40	8:00	8:15
	—	*8:10	8:35	7:50	8:10	8:25
	8:00	8:20	8:45	8:05	8:30	8:45
	—	*8:35	9:00	8:15	8:40	8:55
	8:25	8:45	9:10	8:20	8:45*	—
	8:40	9:00	9:25	8:25	8:50	9:05

Max load point

* Departures and arrivals at b (creating short-turn trips)

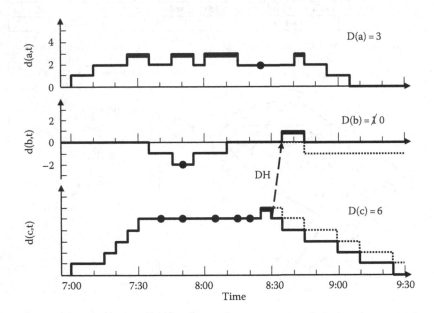

Figure 16.9 New timetable with maximum excluded departure times (following the Minimax H procedure) and associated deficit functions.

16.5.1 Converting DH to service trips

To convert DH to service trips, start by denoting the modified timetable with maximum short turns by DT′; the route's end points by e_q, q = 1, 2; and the intermediate short-turn points by $r_i \in R$, i = 1, 2, ..., V, where there are V short-turn points. To attain N'_{min}, the overall schedule carrying out DT′ might also include DH trips. This overall schedule is designated S. In this section, the DF properties will be exploited to check whether a DH trip can be interpreted as an extension of a short-turn trip in DT′.

By using DF theory, a DH trip can be inserted into a certain time window in order to reduce the fleet size by one. To simplify this possibility, a DH trip is inserted from one terminal to terminal k so that its arrival time always coincides with the first time that d(k, t) attains its maximum. The following steps attempt to describe the procedure to convert DH trips in S into service trips, used in the original timetable:

Step 1: Select a DH trip in S and call it $\overline{DH}$; if there is no DH trip in S, stop.

Step 2: If the $\overline{DH}$ is from r_i to $e_{q'}$ ($q' = 1$ or 2, $r_i \in R$), go to *Step 3*; if the $\overline{DH}$ is from $e_{q'}$ to r_i, go to *Step 4*; and if the $\overline{DH}$ is from r_i to r_k ($r_i, r_k \in R$), go to *Step 5*.

Step 3: Examine the arrival in $d(r_i, t)$ to the left of the departure time of $\overline{DH}$ (start with the one closest to that departure time) to see whether it can be extended to $e_{q'}$ (by replacing $\overline{DH}$). If the arrival considered is associated with trip p_1, the extension can be executed but if and only if the following three conditions are met: (a) the arrival time of p_1 at r_i is within the hollow that contains the $\overline{DH}$ departure time; (b) p_1 was originally planned to continue toward $e_{q'}$; and (c) the originally planned arrival time of p_1 at $e_{q'}$ is equal to or less than the arrival time of $\overline{DH}$. If all the three conditions are fulfilled, delete $\overline{DH}$ from S, update DT', $d(r_i, t)$, and $d(e_{q'}, t)$, and go to *Step 1*; otherwise, $\overline{DH}$ remains in S, go to *Step 1*.

Step 4: Examine the departure in $d(r_i, t)$ to the right of the arrival time of $\overline{DH}$ (start with the one closest to that arrival time) to see whether it can be extended to $e_{q'}$ (by replacing $\overline{DH}$). If the departure considered is associated with trip p_2, the extension can be executed but if and only if the following three conditions are met: (a) the departure time of p_2 at r_i is less than or equal to the arrival time of $\overline{DH}$; (b) p_2 was originally planned to start at $e_{q'}$; and (c) the originally planned departure time of p_2 at $e_{q'}$ is within the hollow that contains the $\overline{DH}$ departure time. If all three conditions are fulfilled, delete $\overline{DH}$ from S, update DT', $d(e_{q'}, t)$, and $d(r_i, t)$, and go to *Step 1*; otherwise, $\overline{DH}$ remains in S, go to *Step 1*.

Step 5: Set $r_k = e_{q'}$ and use the procedure in *Step 3*; if it is terminated successfully ($\overline{DH}$ is converted to a service trip), execute Adjustment A and go to *Step 1*. Otherwise, set $r_i = e_{q'}$ and use the procedure in *Step 4*; if it is terminated successfully, execute Adjustment A and go to *Step 1*. Otherwise, $\overline{DH}$ remains in S, go to *Step 1*. Adjustment A: delete $\overline{DH}$ from S, update DT', $d(r_i, t)$, and $d(r_k, t)$.

16.5.2 Extensions at short-turn points

Following the use of DH trips from/to short-turn points, this section will now describe the principles and procedures for possible extensions of short-turn trips toward their original schedule without increasing N'_{min}. To this end, let the updated timetable DT' be denoted by DT'_1, including the conversion of DH trips to service trips by extending their associated short-turn trips. An extension of a short-turn trip can be viewed as stretching the trip toward the route's end points, $e_{q'}$. An extension does not necessarily mean that the short-turn trip is converted to a full trip along the entire route, because it can be only partially extended; that is, an extension can be performed only from r_i to r_k ($r_i, r_k \in R$). The three stages at which the extensions at $r_i \in R$ can be analyzed and executed are as follows: (a) zeroing the maximum DF; (b) stretching the maximum interval; and (c) treating the DF hollows.

16.5.2.1 *Zeroing the maximum deficit function*

On the basis of the DF properties, it is possible to prove that while N'_{min} is preserved, the number of extensions in each $r_i \in R$ from r_i to $e_{q'}$ is greater than or equal to $D(r_i)$. This rule is based on the observation that in each $r_i \in R$, exactly $D(r_i)$ departures can be extended to their original departure point without increasing N'_{min}. This procedure will eventually lead to $D(r_i) = 0$ for all $r_i \in R$. These extensions are obtained through the following basic steps. Note, for subsequent steps, as well, that t_s and t_e denote the beginning and the end of the DF maximum interval.

Step 1 (Initialization): Set R = $\overline{R}$.

Step 2: Select $r_i \in \overline{R}$; if $\overline{R} = \varphi$ (empty), stop.

Step 3: Check to see whether $D(r_i) = 0$; if so, delete r_i from $\overline{R}$ and go to *Step 2*; otherwise continue.

Step 4: Identify a trip (there might be more than one) whose departure time is at t_s of $d(r_i, t)$ and extend this departure to its original time at $e_{q'}$; update DT_1', $d(r_i, t)$, and $d(e_{q'}, t)$ and go to *Step 3*.

Figure 16.10 illustrates an example of three extensions to $d(r_i, t)$. The first two (numbered 1 and 2) induce $D(r_i)$ to decrease from two to zero. Each extension in Figure 16.10 is followed by an update of $d(r_i, t)$. After the three extensions, only two (of five) short-turn trips remain at r_i. Note that these two trips could also be extended if both are associated with the same terminal k and that their arrival time in the original schedule is at or before the departure time at k.

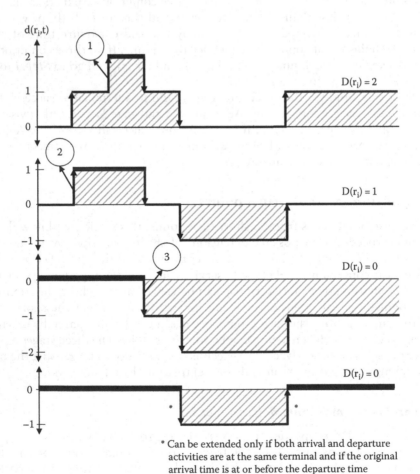

* Can be extended only if both arrival and departure
activities are at the same terminal and if the original
arrival time is at or before the departure time

Figure 16.10 Schematic example of updated deficit functions at an intermediate short-turn point after each of the three indicated extensions.

16.5.2.2 Stretching the maximum interval

After reducing all $D(r_i)$ to zero, it is possible to prove the following rule: while preserving N'_{min}, further extensions can be performed from $r_i \in R$ to $e_{q'}$, up to the point at which the maximum interval is stretched over the whole span of the schedule horizon. This rule is based on the observation that certain arrivals can be extended without increasing $D(r_i)$ above zero. The span of the schedule horizon is determined by the earliest departure and latest arrival in the original timetable.

These additional extensions to the route's end points are executed using the following steps, where DT'_2 denotes the updated timetable after the stage described previously (zeroing the DF of r_i):

Step 1 (Initialization): Set $R = R'$.
Step 2: Select $r_i \in R'$; if $R' = \varphi$ (empty), stop.
Step 3: Check whether the r_i maximum interval (from t_s to t_e) coincides with the span of the schedule horizon; if so, delete r_i from R' and go to *Step 2*; otherwise, continue.
Step 4: Identify a trip (there might be more than one) whose arrival is at t_e of $d(r_i, t)$ and extend this arrival to its original time at $e_{q'}$; update DT'_2, $d(r_i, t)$, and $d(e_{q'}, t)$ and go to *Step 3*.

The preceding procedure is demonstrated by Extension 3 in Figure 16.10.

16.5.2.3 Treating the deficit function hollows

At this third stage, a search is made to determine more extensions at $r_i \in R$ regarding departures and arrivals in hollows. Each hollow in $d(r_i, t)$ contains the same number of arrivals as departures. The procedure developed does not treat hollows consisting of only one point. In Figure 16.10, for example, the third DF with Extension 3 has one hollow that consists of two arrivals followed by one departure. DF theory, as outlined in Chapters 6 and 7, permits the construction of the following extension search procedure, in which DT'_3 denotes the updated timetable after the stages just discussed:

Step 1 (Initialization): set $R = \overline{R'}$.
Step 2: Select $r_i \in R$; if $\overline{R'} = \varphi'$ (empty), stop.
Step 3: Check the next (with respect to time) trip in $d(r_i, t)$; if it is the last departure, go to *Step 2*; if it is an arrival, go to *Step 5*; otherwise, continue.
Step 4: Examine this departure by extending it to its original time at $e_{q'}$; execute this extension if $D(e_{q'})$ is unchanged or if $D(e_{q'})$ is increased, but it can also be reduced (back) through the unit reduction DH chain (URDHC) procedure; update DT'_3 and all the DFs involved. Then, if t_e of $d(r_i, t)$ does not coincide with the right boundary of the schedule horizon, go to the extension procedure described in the previous stage (stretching the maximum interval); otherwise go to *Step 3*. If the extension cannot be made, repeat this extension examination toward a different short-turn point r_k (instead of $e_{q'}$) each time, selecting them backward from $e_{q'}$ to r_i.
Step 5: Examine this arrival by extending it to its original time at $e_{q'}$ and use the URDHC procedure to check whether $D(r_i)$ can remain the same; if so, execute the extension; update DT'_3 and all the DFs involved; otherwise, repeat this extension examination the same way as in *Step 4*.

Finally, if this procedure leads to the introduction of a new DH trip, the procedure for extensions of DH trips needs to be repeated.

16.5.3 Extensions of the example problem

The example problem, after the Minimax H procedure, is shown in Figure 16.9; this result will now be subjected to possible extensions related to the arrivals and departures at d(b, t). First, the extensions of DH trips procedure will be used to determine whether or not the DH trip can be converted into a service trip. In the example it is found that it cannot be converted, and hence the single DH trip remains in the schedule.

Second, the procedures described for extensions of short-turn points are applied; their execution is shown in Figures 16.11 and 16.12. Because D(b) = 0 (after inserting the DH trip), the algorithm in the first stage cannot be utilized. However, because of the algorithm

Max load point

Direction	a → c			c → a		
Stop	a	b	c	c	b	a
Timetable with maximum short turns,* including DH trip from c to b (8:30–8:35)	7:00	7:15	7:40	7:00	7:20	7:35
	7:10	7:25	7:50	7:15	7:35*	—
	7:25	7:40	8:05	7:20	7:40	7:55
	—	*7:50	8:15	7:25	7:45*	—
	—	*7:55	8:20	7:30	7:50*	—
	7:45	8:05	8:30	7:40	8:00	8:15
	—	*8:10	8:35	7:50	8:10	8:25
	8:00	8:20	8:45	8:05	8:30	8:45
	—	*8:35	9:00	8:15	8:40	8:55
	8:25	8:45	9:10	8:20	8:45*	—
	8:40	9:00	9:25	8:25	8:50	9:05

* Departures and arrivals at b.

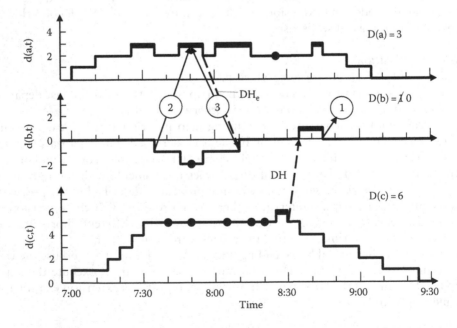

Figure 16.11 Modified timetable and deficit functions following the Minimax H procedure, along with an indication of three short-turn trip extensions.

Direction	a → c			c → a		
Stop	a	b	c	c	b	a
	7:00	7:15	7:40	7:00	7:20	7:35
	7:10	7:25	7:50	7:15	7:35	(7:50)*
Final timetable	7:25	7:40	8:05	7:20	7:40	7:55
with minimum	—	7:50	8:15	7:25	7:45	—
number of	—	7:55	8:20	7:30	7:50	—
short turns, but	7:45	8:05	8:30	7:40	8:00	8:15
with the same	(7:50)*	8:10	8:35	7:50	8:10	8:25
minimum	8:00	8:20	8:45	8:05	8:30	8:45
number of	—	8:35	9:00	8:15	8:40	8:55
vehicles	8:25	8:45	9:10	8:20	8:45	(9:00)*
	8:40	9:00	9:25	8:25	8:50	9:05

(...)* is an extension (see the following).

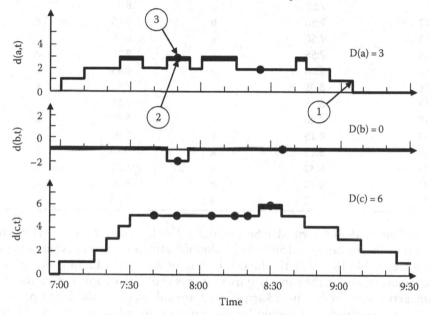

Figure 16.12 Optimal timetable for the problem and associated deficit functions and extensions.

in the second stage, Extension 1 can be performed, increasing d(b, t) at 8:45 from –1 to 0. Then, the algorithm in the third stage is used. It can be observed that Extension 2 alone affects D(b), increasing it by one at 8:10; the URDHC procedure, therefore, searches for a DH trip that can arrive at b at 8:10. Such a DH trip, designated DH_e in Figure 16.11, is inserted from a. Note that the insertion of DH_e is possible because d(a, t) drops to 2 at 7:50 after resuming the original arrival time associated with Extension 2; thus, D(a) remains 3.

The final step in the third stage is to check DH_e with the procedure of extensions of DH trips. This permits Extension 3 to be performed. Consequently, three of the eight short-turn trips in the timetable of Figure 16.10 were extended to their original schedule, while N'_{min} remains 9. In other words, the procedures developed identify only the minimum (crucial) short-turn trips that are required to reduce fleet size. Figure 16.12 illustrates the updated DFs after the three extensions. It can be observed that no more extensions can be made.

The timetable in Figure 16.12 achieved, for the example problem, both the minimum fleet size required (of nine vehicles) and the minimum required five short-turn trips to reach this minimum fleet size. The next step is to construct the nine blocks using either the FIFO method or

Table 16.1 List of trips of the example problem in order of increasing departure times

Trip number	Departure time	Departure location	Arrival time	Arrival location
1	7:00	a	7:40	c
2	7:00	c	7:35	a
3	7:10	a	7:50	c
4	7:15	c	7:50	a
5	7:20	c	7:55	a
6	7:25	a	8:05	c
7	7:25	c	7:45	b
8	7:30	c	7:50	b
9	7:40	c	8:15	a
10	7:45	a	8:30	c
11	7:50	c	8:25	a
12	7:50	b	8:15	c
13	7:50	a	8:35	c
14	7:55	b	8:20	c
15	8:00	a	8:45	c
16	8:05	c	8:45	a
17	8:15	c	8:55	a
18	8:20	c	9:00	a
19	8:25	c	9:05	a
20	8:25	a	9:10	c
21(DH)	8:30	c	8:35	b
22	8:35	b	9:00	c
23	8:40	a	9:25	c

the *within hollow* method described in Section 6.5.5. Finally, a summary of performance measures of the nine blocks can be calculated for administrative and further evaluation purposes.

For instance, Table 16.1 lists the final schedule of Figure 16.12 in order of increasing departure times, when for the same departure times those at a come before those at c and the last departures are at b. The 23 trips listed are subjected to the FIFO procedure and nine blocks are constructed, grouped by trip number, as follows: [1,9,20], [2,10,21,22], [3,11,23], [4,13], [5,15], [6,16], [7,12,17], [8,14,18], [19]. A summary of the performance measures of each block is shown in Table 16.2.

Table 16.2 Performance measures of the nine-block result of the example problem

Block number	Service time (min)	DH time (min)	Idle time (min)	Service (km)	DH (km)	Block time (min)
1	120	0	10	72	0	130
2	100	10	10	76	14	120
3	120	0	15	72	0	135
4	75	0	0	48	0	75
5	80	0	5	48	0	85
6	80	0	0	48	0	80
7	80	0	5	52	0	90
8	85	0	0	52	0	85
9	40	0	0	24	0	40
Total	785	10	45	492	14	840

16.6 LITERATURE REVIEW AND FURTHER READING

This section reviews primarily the research on the design of short-turn trips. It ends with several studies on using short turning as a real-time control strategy.

Furth (1987) presents a methodology for schedule coordination between short-turn and full-length trips in a system consisting of several transit modes. The objective is to find an optimal trade-off between the minimum possible fleet size and the minimum passenger waiting time. The schedule offset between full and short-service patterns is determined as are the turning points of the short-turn service. Optimal service headways and vehicle size are also calculated. The frequency of the short-turn route is set to a multiple of that of the full-length route, and the ratio of frequencies is called scheduling mode; different models are developed for different scheduling modes. Vehicle capacity is given significant attention. The researcher shows that a problem of overcrowding may occur even when overall capacity exceeds the volume on every road link. The timetable-design problem includes an integer-minute constraint.

Miller and Bunt (1987) describe a computer program that simulates light-rail operations on a specific route in Toronto. This program is designed to analyze the impact of a range of operating policies on the regularity of the streetcar service and to compare alternate means of improving regularity. One of the strategies compared is the introduction of short-turn service. The model assists in determining where and when a vehicle will be directed to turn around before the end of the route. This determination is based on a set of decision rules. The basic rules recommend turning a vehicle when a gap is found that exceeds a given threshold between successive vehicles traveling in the opposite direction of the vehicle to be turned. Other rules make sure, for example, that the gap in the original direction, which occurs as a result of the turning, does not exceed a given threshold.

Vijayaraghavan and Anantharamaiah (1995) examine the possibility of reducing the required number of buses operated on a single route by using two strategies: partial service (i.e., short turning) and express service. Fleet-assignment options and their effect on the efficiency of fleet use are analyzed graphically.

Dell Site and Filippi (1998) present a model for bus-service optimization under an elastic demand. The model is formulated as a programming problem in which decision variables are the locations of turning points, the time offsets between departures of the full-length and the short-turn routes, frequencies, and fares. Vehicle size is also considered as a variable, but is represented indirectly. The model enables taking into account different demand patterns at different periods. A different frequency and offset are ascribed to each period; for other variables, a single optimal value is determined. A numerical procedure is presented to solve the problem.

Ulusoy et al. (2010) develop a model for cost-efficient operation that optimizes all-stop, short-turn, and express transit services and the associated frequencies by considering heterogeneous demand. For a transit line with given stations and origin–destination demand, the objective function of total cost is optimized by considering a set of constraints that ensure frequency conservation and sufficient capacity subject to operable fleet size. A numerical example is designed that is based on a real-world rail transit line to demonstrate the applicability and effectiveness of the developed model. Results show that optimized integrated service patterns and associated service frequencies significantly reduce total cost.

Cortés et al. (2011) develop a model that combines short turning and deadheading in an integrated strategy for a single transit line, where the optimization variables are both of a continuous and discrete nature: frequencies within and outside the high demand zone, vehicle capacities, and those stations where the strategy begins and ends. The authors show that closed solutions can be obtained for frequencies in some cases, which resembles the classical *square root rule*. Unlike the existing literature that compares different strategies with a given normal operation (no strategy—single frequency), the authors use an optimized base case, in order to assess the potential benefits of the integrated strategy on a fair basis. It was revealed that the integrated strategy can be justified in many cases with mixed load patterns, where unbalances within and between directions are observed. In general, the short turning strategy may yield large benefits in terms of total cost reductions, while low benefits are associated with deadheading, due to the extra cost of running empty vehicles in some sections.

The main characteristics of the models reviewed to this point are summarized in Table 16.3.

The aforementioned papers relate to the design of short-turn trips. Some other papers also discuss short turning as a real-time control strategy.

Huddart (1973) discusses short-turn trips as a strategy to avoid bus bunching. The author explains that in order to use this strategy efficiently, a bus should be taken out of a bunch in one direction and be entered into a schedule gap in the opposite direction. In addition, difficulties in real-time decision-making regarding short-turn trips are described.

Strathman et al. (2001) mention that a bus chosen to turn around should ideally be one with a small number of passengers, a small gap from the preceding bus, and a small gap from the following bus. The authors mention that damage is caused for passengers whose destination is further from the turning point, since they are forced to transfer from a short-turn trip to a full-length trip.

Shen and Wilson (2001) develop a disruption-control model for rail transit systems. The model, formulated as a deterministic mixed-integer program, is used to examine the introduction of several real-time control strategies, including short turning. The main conclusion is that the best system performance is achieved when holding and short-turn strategies are combined. The paper does not include a methodology for short-turn strategy design.

Table 16.3 Summary of features of the models reviewed

Source	Decision variables	Required demand input	Elastic/constant demand	Operation period
Furth (1987)	Offset between full and short service, turning points, headways, and vehicle size	O-D matrix	Constant	Single
Miller and Bunt (1987)	Turning points and times	None	Not considered	Single (afternoon 5 h peak)
Vijayaraghavan and Anantharamaiah (1995)	Turning points, number of stops	Total demand	Constant	Single
Dell Site and Filippi (1998)	Offset between full and short service, turning points, headways, fares, and vehicle size	O-D matrix and elasticities	Elastic	Multiperiod
Ulusoy et al. (2010)	The decision variables of the objective function of total cost include integrated service patterns and associated service frequencies	O-D matrix	Elastic	Multiperiod
Cortés et al. (2011)	The impact on waiting time cost, in-vehicle time cost and operator cost	O-D matrix	Constant	Single

EXERCISES

16.1 A bus route is given between terminals *a* and *c* (both directions of travel) with a single short-turn point *b* along the route. The input data appear in the following table for the period 6:00–8:00.

Direction	$a \to c$		$c \to a$	
	$a \to b$	$b \to c$	$c \to b$	$b \to a$
Original departure times	6:00; 6:12; 6:17; 6:25; 6:32; 6:40; 6:46; 6:53; 7:00; 7:10; 7:18; 7:25; 7:35; 7:40; 7:48	—	6:00; 6:10; 6:18; 6:25; 6:32; 6:40; 6:50; 6:55; 7:10; 7:18; 7:25; 7:32; 7:40; 7:48; 7:53	—
Service travel time (min)	15	20	25	10
DH travel time (min)	10	17	20	8
Passenger load 6:00–7:00	400	250	150	200
7:00–8:00	400	300	450	550

The desired occupancy of a bus is 50 passengers; there is a minimum frequency of 3 veh/h, and no shifting in departure time is allowed.

(a) Apply the method for designing short-turn trips so as to minimize both the number of vehicles required and the number of short-turns; provide the final timetable.
(b) Establish schedules/blocks for all vehicles using the FIFO procedure.

16.2 The following two-route network is given, with stop b as a single candidate short-turn point for both routes in both directions of travel.

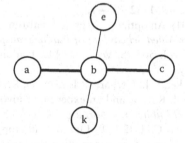

The following table contains the data required.

Route and direction	$a \to c$		$c \to a$		$e \to k$		$k \to e$	
	$a \to b$	$b \to c$	$c \to b$	$b \to a$	$e \to b$	$b \to k$	$k \to b$	$b \to e$
Service travel time (min)	20	30	25	15	10	15	20	15
DH travel time (min)	18	25	20	12	7	10	15	10
Hmin$_i$ (min)[a]	5		10		8		13	
Hmax$_i$ (min)[a]	6		15		7		20	
Passenger load 6:00–7:00	150	250	250	400				
7:00–8:00	250	350	150	200				

[a] Hmin$_i$ and Hmax$_i$ are the minimum and maximum headways permitted, respectively, between two adjacent departures on route i (see Chapter 5 for more detailed description and examples).

In addition, it is given that the determined frequency for the 2 h period considered (departures are between 6:00 and 8:00) is 12 vehicles/period for each route and direction of travel; no shifting in departure time is allowed, and the derived timetable starts at 6:00 (the first fixed departure).

(a) Construct the timetable for only routes a → c and c → a, using the method of synchronization described in Chapter 5.
(b) Apply the method to create short-turn trips for routes a → c and c → a so as to minimize both the number of vehicles required and the number of short-turns; provide a final timetable.

REFERENCES

Ceder, A. (1990). Optimal design of transit short-turn trips. *Transportation Research Record*, **1221**, 8–22.

Ceder, A. (1991). A procedure to adjust transit trip departure times through minimizing the maximum headway. *Computers and Operations Research Journal*, **18**, 417–431.

Ceder, A. (2003). Designing public transport networks and routes. In *Advanced Modeling for Transit Operations and Service Planning* (W. Lam and M. Bell, eds.), pp. 59–91, Pergamon Imprint, Elsevier Science, New York.

Cortés, C.E., Jara-Díaz, S., and Tirachini, A. (2011). Integrating short turning and deadheading in the optimization of transit services. *Transportation Research*, **45A**, 419–434.

Dell Site, P. and Filippi, F. (1998). Service optimization for bus corridors with short-turn strategies and variable vehicle size. *Transportation Research*, **32A**, 19–38, 1998.

Furth, P.G. (1987). Short turning on transit routes. *Transportation Research Record*, **1108**, 42–52.

Huddart, K.W. (1973). Bus priority in greater London: Bus bunching and regularity of service. *Traffic Engineering and Control*, **14**, 592–594.

Miller, E.J. and Bunt, P.D. (1987). Simulation model of shared right-of-way streetcar operations. *Transportation Research Record*, **1152**, 31–41.

Shen, S. and Wilson, N.H.M. (2001). An optimal integrated real-time disruption control model for rail transit systems. In *Computer-Aided Scheduling of Public Transport*. Lecture Notes in Economics and Mathematical Systems, vol. 505 (S. Voss and J.R. Daduna, eds.), pp. 335–364, Springer-Verlag, Heidelberg, Germany.

Strathman, J.G., Kimpel, T.J., Ducker, K.J., Gerhart, R.L., Turner, K., Griffin, D., and Callas, S. (2001). Bus transit operations control: Review and an experiment involving TRI-MET's automated bus dispatching system. *Journal of Public Transportation*, **4**, 1–26.

Ulusoy, Y.Y., Chien, S.I.J., and Wei, C.H. (2010). Optimal all-stop, short-turn, and express transit services under heterogeneous demand. *Transportation Research Record*, **2197**, 8–18.

Vijayaraghavan, T.A.S. and Anantharamaiah, K.M. (1995). Fleet assignment strategies in urban transportation using express and partial services. *Transportation Research*, **29A**, 157–171.

Chapter 17

Smart shuttle and feeder service

Don't worry about not being
on time; the sooner we fall behind,
the more time we'll have to catch up

Shuttle

A.C.

CHAPTER OUTLINE

PRACTITIONER'S CORNER

The choice between public and private transportation is an individual decision that is influenced by government/community decisions. These decisions often send mixed signals (e.g., "use transit"; "we're trying to reduce traffic congestion to ease your automobile driving") to transit and potential transit passengers while failing to recognize their more system-wide, integrated implications. Chapter 13 facilitates an understanding of the importance of service-connectivity elements in enhancing existing or new transit services. One such element is the design of an integrated, smart feeder/shuttle service. Such a service may stem, for instance, from the need to overcome the problem of an excessive number of automobiles arriving and parking at a train station, resulting in high parking demand around the station. The purpose of this chapter is to examine an innovative feeder/shuttle system that will comply with (1) passengers' needs and desires, (2) intelligent transportation technologies, and (3) an agency's viability. After all, a solution for being transported by a door-to-door service is the mother of attraction.

This chapter contains eight main parts, following an introductory section. Section 17.2 continues the analysis of the minimum fleet size required that was the subject of Chapter 6, but for a radial/shuttle route having a single departure and arrival point. Section 17.3 proposes 10 different feeder/shuttle routing strategies with various combinations of fixed/flexible routes, fixed/flexible schedules, a uni- or bidirectional concept, and shortcut (shortest path) and/or short-turn (turn-around) concepts. Section 17.4 investigates these strategies by employing a simulation model specifically developed and constructed for this purpose. This simulation model is used in Section 17.5 for a case study of Castro Valley in California, where the feeder/shuttle service is coordinated with the metro (BART) service; the 10 routing strategies are compared for 4 fleet size scenarios. Section 17.6 reports on a survey conducted in the case-study area concerning (a) willingness to use the smart-shuttle service, (b) willingness to pay for the service, and (c) the attributes that would enhance the shuttle service. Section 17.7 introduces a framework for the optimal design of circular feeder/shuttle fixed routes, including a method for estimating potential passenger demand. Section 17.8 shows the modeling of an optimal circular route that can handle any size road network. The chapter ends with a literature review and exercises.

Practitioners are advised to visit all sections of the chapter, except perhaps Section 17.8, which is more of a mathematical nature. They are especially encouraged to follow the examples and the case study.

Finally, an idea that has been introduced is that passengers who want to take the shuttle service should be able to use an online, intelligent information system through a telephone/cellular call. The system will announce the arrival time to the point closest to the caller unless this estimated time is not reliable enough. In the latter case, the system will call back once an estimation becomes reliable. It may even have a special ring for easy recognition. Does the name Pavlov ring a bell?

17.1 INTRODUCTION

When exploring the design possibility of a transit feeder/shuttle service, we can refer to Section 13.2, in which an ideal coordinated service is portrayed. For simplicity, we will include the transit-feeder service within a shuttle service in the remaining parts of the chapter. Ideally, a smart transit shuttle system will provide an advanced, attractive service that operates reliably and relatively rapidly, and has smooth, synchronized transfers as part of the door-to-door transit-passenger chain. In order to arrive at the design of such a system, new integration and routing concepts had to be developed. This chapter integrates work done by Ceder and Yim (2002), Yim and Ceder (2006), Jerby and Ceder (2006), Ceder (2009), and Ceder (2013).

A growing concern for public transit is its inability to encourage people to switch their mode of transportation from solo to shared driving. The majority of large cities have encouraged private car use through planning (dispersed land-use in the suburbs), infrastructure (available parking and circulation traffic flow), pricing, and financial decisions. Consequently, there is growing confusion among drivers in many of those cities about what to do. One way to handle the known decline in transit use is to retain a high level of satisfaction among transit users while fully maintaining the protection of accessibility for less-affluent travelers.

What follows is a description of the motivation that was employed by research reported in this chapter. Although overall transit ridership is declining in cities, an encouraging trend is increased ridership in long-haul express bus or rail transit. When long-haul express transit systems were built in the 1970s and 1980s, parking facilities were also provided for the riders, under the rubric of *park and ride*. The concept was readily accepted by the public, and a large number of commuters preferred to take an express bus or train to avoid rush-hour traffic and high parking costs. However, most of the arrivals at the train stations are made by private cars, hence creating traffic congestion and parking overloading in the station area.

Such a case was observed in the Bay area around San Francisco in connection with the use of the local metro system, called BART (Bay Area Rapid Transit). This observation, which served as the motivation for this chapter, appeared on the front page of the *San Francisco Chronicle* on three consecutive days in 2001. Figure 17.1 shows the relevant sections of the newspaper, providing news about the problem and commentary in the form of suggested parking solutions (bottom right of figure might be hard to read). The main parking solution considered by BART was this: "Building reserved parking lots, where users would be charged a fee for a guaranteed space." An obvious question arises: Why not employ a smart bus shuttle service instead?

Mark Twain said: "You cannot depend on your eyes when your imagination is out of focus." Our eyes see what transit services are currently providing for high- and low-density communities. Our eyes can read reports covering urban transportation characteristics, the influence of transportation investment, ground transportation strategy, and passenger transportation action plans. However, we cannot depend on our eyes alone to trigger our imagination. As Einstein said, "Imagination is more important than knowledge." Consequently, in order to design an imaginative, *ideal* shuttle service (see definition in the first paragraph of this section), it will be worthwhile to start almost from scratch in order to attain such a smart service.

The purpose of this chapter is to describe the conceptual construction of an innovative shuttle system that will (1) meet the needs and desires of end users, (2) utilize intelligent-transportation technologies, and (3) increase operational efficiency. A simulation model, a passenger survey, and a case study will help us take a practical approach to the optimal design of the shuttle route(s).

Figure 17.1 Motivation for the study effectively appeared on the front page of the *San Francisco Chronicle* on 3 consecutive days.

17.2 MINIMUM FLEET SIZE REQUIRED FOR A CIRCULAR (SHUTTLE) ROUTE

Section 6.2 discussed the case in which interlinings are not allowed and each route is operated separately. Let T_r be the average round-trip time, including layover and turnaround times, of a circular route r (whose departure and arrival points are the same). The minimum fleet size is then equal to the largest number of vehicles that depart within T_r (Salzborn, 1972).

Although Salzborn's modeling provides the basis for fleet size calculation, it relies on two assumptions that, in practice, do not hold for a circular (shuttle) route: (1) vehicle departure rate is a continuous function of time and (2) T_r is the same throughout the period under consideration. In practice, departure times are discrete (see Chapters 4 through 6), and average trip time is usually dependent on time-of-day. The analysis of Section 6.2 is used in the present section for the case of a circular (shuttle) route.

Let route r start and end at b. Let T_{rj} be the average trip time on route r for a vehicle departing at t_j from b, including its layover time at b. Also, let n_j be the number of departures from b including departure j at t_j until $t_{j'}$, but excluding departure j'. We further define that j arrives at b and continues with departure j' at $t_{j'}$, which is the *first* feasible departure from b at a time greater than or equal to $t_j + T_{rj}$. Theorem 6.1 (see Chapter 6) without n_{ia} but with the setting $n_{jb} = n_j$ yields the following:

$$N^r_{min} = \max_j n_j \tag{17.1}$$

where N^r_{min} is the minimum fleet size required for the circular route r.

The timetable for b in the example shown in Figure 6.3 is used for the shuttle case in Figure 17.2. Both a single T_{rj} value of 30 min throughout the timetable and a case in which T_{rj} varies are used in Figure 17.2. On the left-hand side of Figure 17.2, the time windows are all with the same length (30 min); on the right-hand side, they are T_{rj} dependent. The maximum number of departures within T_{rj} is emphasized for each case, leading in both instances to a determination of the minimum fleet size, N^r_{min}, according to Equation 17.1.

Following the determination of N^r_{min}, the construction of vehicle chains (blocks) can be carried out by using the FIFO (first-in, first-out) rule of first-feasible connection. For example,

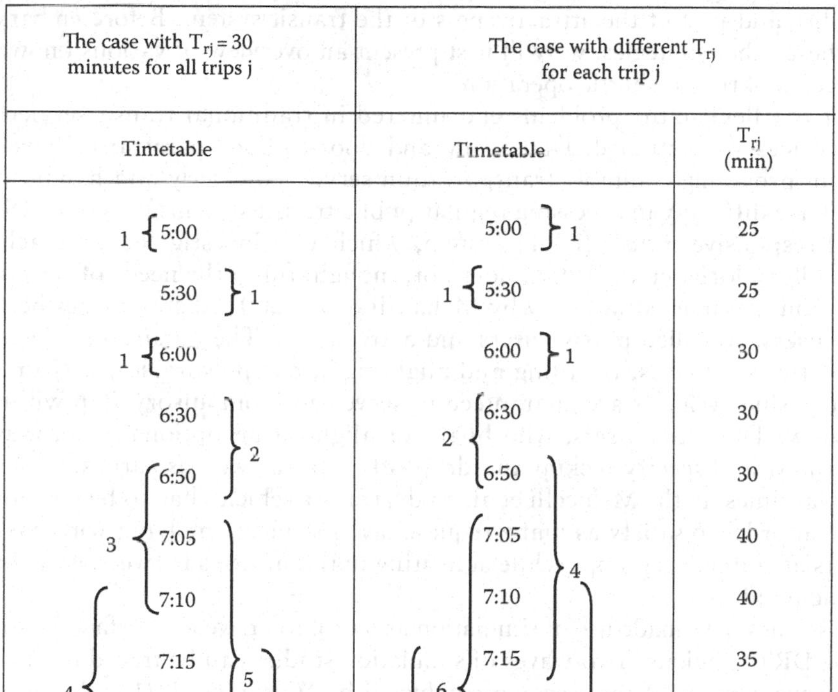

The case with $T_{rj} = 30$ minutes for all trips j	The case with different T_{rj} for each trip j	
Timetable	Timetable	T_{rj} (min)
1 { 5:00	5:00 } 1	25
5:30 } 1	1 { 5:30	25
1 { 6:00	6:00 } 1	30
6:30	6:30	30
6:50 } 2	6:50 } 2	30
7:05	7:05	40
7:10 } 3	7:10 } 4	40
7:15	7:15	35
7:20 } 5	7:20 } 6	30
7:30 } 4	7:30 } 5	25
7:40	7:40 } 3	25
8:00 } 2	8:00 } 2	–
$N^r_{min} = 5$	$N^r_{min} = 6$	

Figure 17.2 Example of the derivation of radial-route fleet size when the round-trip time, including layover time, can either be the same (left side) or vary (right side) for each departure time.

for the case with varied T_{rj} in Figure 17.2, the first block is [5:00–5:30–6:00–6:30–7:05–8:00], in which the 7:05 departure will be ready for another departure at 7:45 (T_{r6} = 40 min). The second block, using the FIFO rule, is [6:50–7:20]. Because there is need for a minimum of six vehicles, some trips of the first block can certainly be performed by the second block, thereby opening some idle time for the vehicles associated with these blocks. For instance, the first two blocks can be [5:00–6:00–6:50–7:20] and [5:30–6:30–7:05–8:00].

Finally, two points are worth noting: (1) When shifting in departure times is allowed, a further reduction in the fleet size can be achieved. (2) When more than one circular (shuttle) route is operated from the same point, it is recommended that the deficit-function method (of Chapters 6 and 7) be used to arrive at the minimum fleet size required for all routes.

17.3 ROUTING STRATEGIES

The previous section dealt with a fixed route and a fixed shuttle-service schedule; however, other routing strategies can be imagined and should be explored. These strategies represent the flexibility and part of the attractiveness of the transit system. Before embarking on a study of these other strategies, we will first present an overview of various known and relevant concepts of transit-shuttle operation.

In order to alleviate the problems encountered in traditional transit services, several flexible services were studied. Dial-a-ride and door-to-door paratransit have played a vital role in providing equitable transportation service to elderly and handicapped persons who have difficulty in accessing regular public transit systems (Cervero, 1998). Such a demand-responsive transit (DRT) system, which was investigated by Ioachim et al. (1995) and Borndorfer et al. (1999), does not, though, fulfill the needs of the entire transit population. An interesting study by Melucelli et al. (2001) distinguishes between two classes of users, so-called passive users and active users. The *passive users* make use of traditional transit, that is, boarding and alighting at compulsory stops. No reservation is necessary, since vehicles are guaranteed to serve each compulsory stop within a given time window. The *active users*, who board or alight at an optional stop, must issue a *service request* and specify pickup and drop-off stops, as well as earliest departure and latest arrival times. In the Melucelli et al. study, transit vehicles had to be rerouted and rescheduled in order to satisfy as many requests as possible, complying with passage-time constraints at compulsory stops while activating optional stops between two compulsory stops on demand.

Several studies have made use of simulation as a tool to arrive at satisfactory routing and scheduling DRT solutions. Two waves of simulation studies can be traced in the literature. The first wave consists of the research conducted by Wilson et al. (1970), who evaluated various heuristic routing rules and algorithms used in a computer-aided routing system. These rules and algorithms, developed for mainframe computers, have limitations in their handling of large-size road networks with different routing strategies. The second wave of research was conducted by Fu (1999, 2001) and Fu and Xu (2001), who considered the use of advanced technologies. Their studies present a simulation model, Sim-Paratransit, which was developed for evaluating advanced paratransit systems, such as AVL (automatic vehicle location) and CAD (computer-aided dispatch) systems. The simulation model is described by Fu (1999, 2001), and the evaluation of AVL and CAD systems by Fu and Xu (2001). The ability to track the location of a transit vehicle continuously allows for the introduction of intelligent paratransit systems, which will naturally lead to operating such systems at a significantly improved level of productivity and reliability.

This chapter, among other themes, investigates 10 routing strategies (Ceder, 2013);

1. Fixed route with a fixed schedule (timetable) and fixed direction (of travel)
2. Fixed route; flexible (demand-driven) schedule; fixed direction
3. Fixed route; flexible schedule; bidirectional
4. Fixed route; flexible schedule; fixed direction; possible short-turns
5. Fixed route; flexible schedule; bidirectional; possible short-turns
6. Fixed route; flexible schedule; fixed direction; possible shortcuts
7. Fixed route; flexible schedule; bidirectional; possible shortcuts
8. Fixed route; flexible schedule; fixed direction; possible short-turns and shortcuts
9. Fixed route; flexible schedule; bidirectional; possible short-turns and shortcuts
10. Flexible (demand-responsive) route with a flexible schedule

Fixed direction means that the shuttle will always maintain the same direction of travel (same sequence of stops), whereas bidirectional allows for flexibility in selecting a direction based on real-time demand information. The term *shortcut* means that, based on certain loading-threshold and synchronization criteria, the shuttle will not continue its fixed route and, instead, will use the shortest path (minimum travel time) to arrive at the required point, for example, the train station. For convenience, we will use the shuttle terminal as a train station throughout this chapter, though this connection point can also serve other transit modes.

The loading threshold is a given (input) number of passengers on board the shuttle. The synchronization criterion means matching the shuttle's new (via shortcut) arrival time with an earlier train than that originally planned if the entire route were completed. The term *short-turn* means that, based on certain loading-threshold and synchronization criteria, the shuttle will not continue on its fixed route. Instead, it will turn around and arrive at the train station in the opposite direction in order to try to pick up passengers who were too late to catch this shuttle when it first passed their stops. The loading-threshold and synchronization criteria for the short-turn strategy (including the consideration of more pickups) and the shortcut strategy are similar. Each strategy allows for the flexibility of the other. In other words, the loading threshold of the shortcut strategy is initially higher than the loading threshold of the short-turn strategy. If the latter is reached and there is a possibility of picking up x passengers (after turning around), where x is equal to or greater than the difference between the two loading thresholds, then the short-turn strategy is recommended.

Figure 17.3 represents the 10 strategies on a small network with two shuttle routes, one marked with a dashed line and the other with a dotted line. Strategies 2–10 are characterized by a flexible schedule, and it is emphasized by *no timetable* in Strategy 2. Arrows in both directions of the route means a bidirectional situation. The shortcut strategy has lines with arrows in Figure 17.3 that indicate deviations from the fixed route. The short-turn strategy has arrows indicating a turnaround at a certain point in the network. Both representations appear in Strategies 8 and 9, involving a possible combination of the two. The last strategy is for a DRT-type of service, allowing for the creation of a new route every time, based on trip bookings.

The idea of covering almost every possible practical routing strategy stems from the need to satisfy user desires and understandings. Certainly, there is no intention that all strategies be used at the same time; rather, it is to examine which strategy is best for a given demand pattern and magnitude while taking into consideration the real-time traffic situation in the area of the shuttle's trips. A simulation model was devised for that purpose.

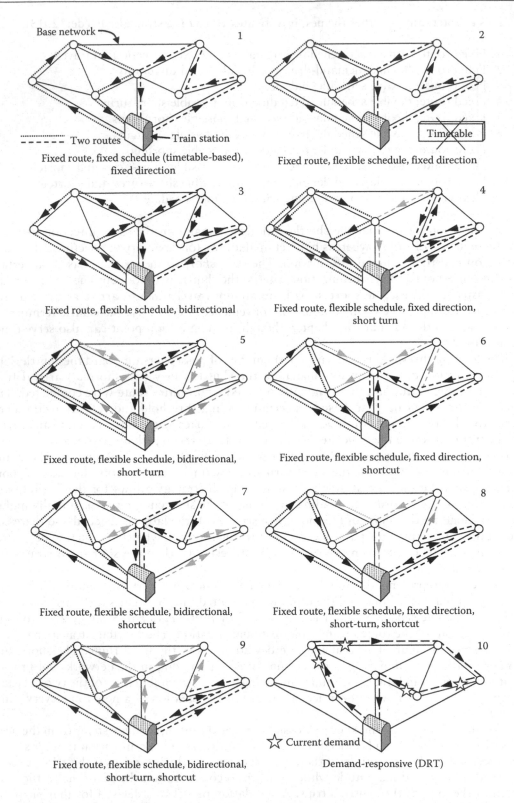

Figure 17.3 Ten routing strategies.

This simulation tool, explained in the next section, enables a comparison of the various strategies, based on the following measures:

- Sum of total time (in passenger-hours) from passenger pickup to train-departure time
- Sum of total time (in passenger-hours) riding the shuttle vehicle
- Sum of total waiting time (in passenger-hours) for the train
- Sum of total waiting time (in passenger-hours) for the shuttle vehicle
- Total number of transit vehicles (by number of seats) required to meet the demand

These measures of travel and waiting times and number of vehicles characterize the effectiveness and efficiency of each strategy. The strategy selected for a given passenger demand is the one with the minimum weighted travel and waiting times (user perspective) and the minimum number of vehicles (agency perspective).

17.4 SIMULATION

The purpose of the simulation model (Ceder, 2013) is to examine the 10 different routing strategies. Passenger demand can either be inserted as part of the input or be generated randomly on the network.

17.4.1 Simulation input variables

Following are the input variables of the simulation program, which was written in C++ language. Each variable is presented by its simulation name, as well as an explanation and interpretation of its substance. What is referred to here as *bus* can be applied to any shuttle vehicle.

Bus2Train = Time in seconds that a bus must be at the station before a train arrives in order to ensure an efficient meeting

Train2Bus = Time (seconds) that a bus must wait after a train arrival to ensure pickup

SizeType = Number of seats in this bus type

Quantity = Number of vehicles of this SizeType

FixPick = Fixed time (seconds) for one passenger pickup, including bus slow down

FixDrop = Fixed time (seconds) for one passenger drop-off, including bus slow down

FixBoard = Fixed (additional) boarding time (seconds) per passenger

FixAlight = Fixed (additional) alighting time (seconds) per passenger

NodeNo = Node index at section end points

SectionNo = Section number between two nodes

StopNo = Stop-number starts with SectionNo, representing an intersection, not a node

MeanDemand = Mean number of potential travel requests per given hour and SectionNo, to the train station

MeanDestin = Mean number of potential travel requests per given hour and SectionNo, from the train station

MeanTime = Mean section travel time (seconds)

StDevTime = Standard deviation of section travel time (seconds)

Min4Turn = Minimum number of on board passengers to allow a short-turn

Min4Cut = Minimum number of on board passengers to allow a shortcut

Min4Trip = Minimum number of travel requests by calls to allow a non-scheduled trip

Min4Dep = Minimum of number of waiting passengers to allow a non-scheduled trip

RouteNo = Unique route index
RouteDir = Direction of RouteNo (start westbound or eastbound)
TTimeTable = Fixed train timetable in hhmmss form, hh is from 00 to 24
BTimeTable = Fixed bus timetable in hhmmss form, hh is from 00 to 24
Layover = Fixed time for driver rest at the end of each trip

All these variables interact in each simulation iteration as part of the different strategies and other internal features.

17.4.2 Simulation procedures

The simulation model is based on events. The simulation starts with reading the input data and proceeds by arranging train-arrival events and passenger-arrival events. Figure 17.4 presents the basic event-oriented simulation logic.

There are eight main events, classified in Figure 17.5. Event 1 represents passengers walking to the stop to wait for the next shuttle in order to arrive at the train station. Event 2 represents passengers who have arrived on the train and are now waiting for the next shuttle. Event 3 occurs when a vehicle becomes available for the next trip. Event 4 is when the shuttle arrives at a node (intersection) on the road network being considered. Event 5 represents the arrival of passengers who want to ride the shuttle from its stop to the train station. Event 6 represents passengers who are about to arrive at the train station and will seek to ride the shuttle. Event 7 is the arrival of the train at the station, including the time for the passengers to arrive at Event 2. Last, Event 8 is the time when the shuttle departs, in accordance with the timetable.

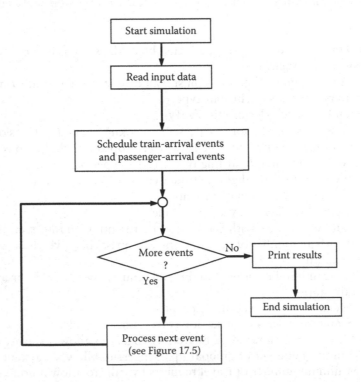

Figure 17.4 Basic simulation logic.

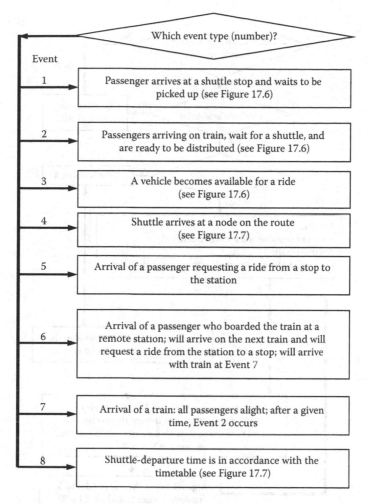

Figure 17.5 Event classification.

The actions taken for Events 1, 2, and 3 appear in Figure 17.6. They start with enquiring whether the number of passengers who want the service has reached the minimum required for dispatching a vehicle. For the DRT strategy (no fixed route), the procedure in Figure 17.6 identifies the section with current (booked) demand and the application of a shortest-path algorithm (e.g., see Appendix 9.A). The dynamic-routing procedure involves moving the shuttle from the train station across all demand points to the first demand point that is within the shortest path from the station. From the last point, the shortest-path algorithm is used again for all the other demand points that were not visited until all the points have been included in the dynamic route. This DRT routing procedure has been found to be effective and convenient to use in the simulation.

Once a vehicle is available (see Figure 17.6), the next event is of type 4, described in Figure 17.7. Thus Event 4, in Figure 17.7, starts either with a station node (train station), whether at the end or at the beginning of the shuttle ride, or at an intermediate node. For any intermediate node, the procedure checks whether the minimum number of passengers on board the shuttle has reached the threshold for either a short-turn or a shortcut strategy. The procedure then checks to see whether creating a short-turn or a shortcut will enable passengers to arrive at an earlier train than that which would be met by completing the entire route.

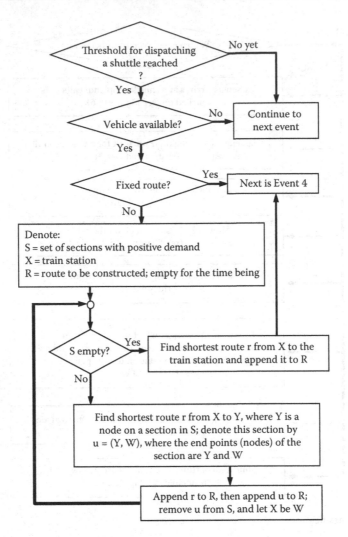

Figure 17.6 Actions taken for Events 1, 2, and 3.

Finally, there are the actions taken in the simulation for Event 5 following the process of informing the passenger through the available online service. There are two possible actions: (a) the passenger will be able to reach the shuttle stop on time after learning of the expected arrival time and (b) the passenger will not be able to arrive on time and will ask to be notified of the next arrival time by a callback. It is assumed that passengers will either call an automated system or look at the shuttle website to ascertain the arrival time. The passenger will then be asked to click, for instance, "1" (for wanting to use the service) or "2" (for not wanting to use it following the announcement of the expected arrival time). Only those who click "1" (OK) are taken into account in the simulation process. The simulation model can either consider a given demand figure or be used to generate a random demand based on the residential density of each section of the network. In the fixed-route case, passengers reach their closest stop on the network. The travel time is a random variable with a normal distribution, and the simulation model calculates the probability of being on time. If this probability is below 90%, the user is notified to wait for a callback. In this way, the system uses the philosophy of advanced technologies and maintains a highly reliable service.

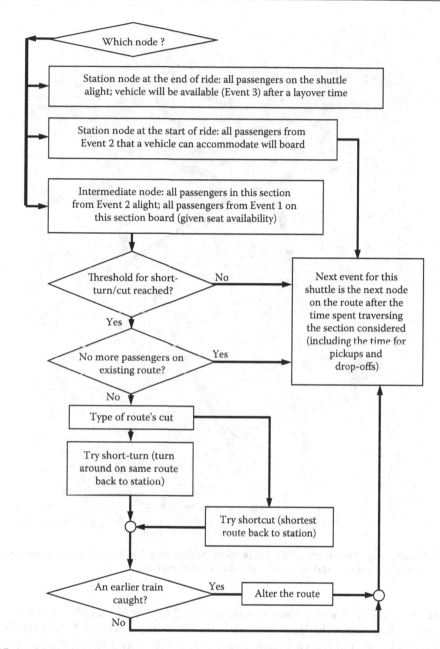

Figure 17.7 Actions for Events 4 and 8.

17.5 CASE STUDY

In order to test a real-life situation, the area of Castro Valley in California was selected for data collection and simulation runs (Ceder, 2013). The BART station in Castro Valley is on the *blue* line, Dublin/Pleasanton to Daly City. A site observation was conducted in the Castro Valley area, from which the base network and stops were created. The optimal routing procedures given in Sections 17.7 and 17.8 were not considered in this case study; instead, two routing scenarios starting at the BART station were introduced, one with a

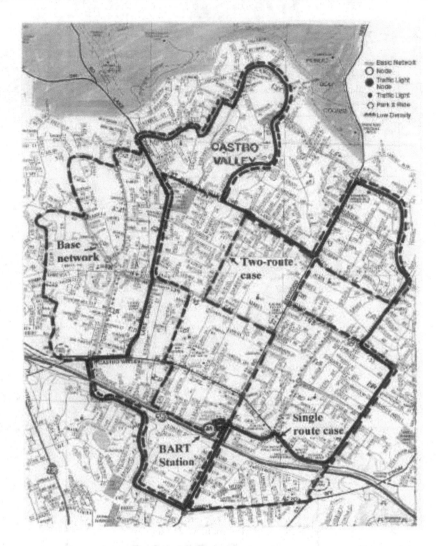

Figure 17.8 Routing map of the Castro Valley case study: base network is defined by gray lines, single-route case by a solid line, and the two-route case by dashed lines.

single route and one with a two-route shuttle system. The routing decided on followed a site visit. Figure 17.8 presents the base network (described in Section 17.7.1), both the single-route case (solid line) and the two-route case (dashed lines), and shows the location of the BART station at which the routes start and end. The two circular routes (dashed lines) are extended from the station—one to the left and one to the right.

A large number of simulation runs were executed across the 10 routing strategies described for four different groups of (available) shuttle buses: 1, 2, 3, and 4 buses. The estimated demand was 400 passengers daily, generated randomly. Table 17.1 summarizes the results obtained for the waiting time per passenger in 40 cases related to the single-route scenario. The minimum (best) passenger-waiting-time results are indicated in this table by an asterisk for each group of available buses; the second-best results are marked by "✓." The waiting time in Table 17.1 is the average time per passenger, in minutes, that elapses from the time of a phone call to the shuttle bus-information center until pickup occurs. This time period

Table 17.1 Simulation results for waiting time per passenger (in min), using different combinations of strategies and number of buses for the Castro Valley case study

| Strategy | Number of buses | | | |
	1 Bus	2 Buses	3 Buses	4 Buses
1	51	22	20	20
2	25	22	17	15
3	24 ✓	23	15 *	14 ✓
4	25	17 *	16 ✓	15
5	24 ✓	18 ✓	15 *	12 *
6	24 ✓	17 *	16 ✓	15
7	24 ✓	18 ✓	15 *	12 *
8	24 ✓	23	16 ✓	15
9	24 ✓	18 ✓	15 *	1 *
10	22 *	18 ✓	15 *	15

Note: Given demand = 400 buses daily.
* = Best result.
✓ = Second-best result.

includes the walking time from the place of the call to the bus stop (assuming the shortest walking distance) and the waiting time until the bus arrives.

The results presented in Table 17.1 show that the fixed route–fixed schedule strategy (#1) results in the highest (longest) waiting times. At the same time, the flexible route–flexible schedule (demand-responsive) strategy (#10) does not always provide the best results; hence, it cannot a priori be superior to the other strategies. In fact, the best routing strategies observed in this real-life test are those with two asterisks and two "✓": fixed route–flexible schedule and bidirectional, with possible short–turn (#5); fixed route–flexible schedule and bidirectional, with possible shortcut (#7); and fixed route–flexible schedule and bidirectional, with possible short-turn and shortcut (#9). The short-turn, shortcut, and bidirectional-based routing strategies indeed prove worthwhile to consider. These three uncommon strategies reflect the current availability of online information and communication.

In addition, six more simulation runs were performed for the two-route scenario, using strategies #1, 8, and 10 with a group of four buses for both picking up passengers for the train station and distributing them from the station. An additional criterion, attention to which had to be paid at the phone location, was established for maximum waiting time on these six runs: 20 min (could be changed in the simulation runs). This criterion reflected the fact that the caller would not actually wait if the announced waiting time for the shuttle bus was more than 20 min; in such a case, the caller would cancel the request. Table 17.2 summarizes the

Table 17.2 Simulation analysis for the two-route, four-bus case, with a 20 min criterion (maximum waiting after call)

Strategy ⇒	Fixed route, Fixed schedule (#1)	Fixed route, Flex schedule (#8)	Flex route, Flex schedule (#10)
Average wait from phone call to bus arrival (min)	13.3	8.9	6.8
Standard deviation (min)	3.0	3.7	2.5
Average wait at train station (min)	13.6	1.6	1.4
Standard deviation (min)	15.3	8.5	5.0

average waiting time per passenger for these two pickup and drop-off cases, including the standard deviation determined for each simulation run.

The cumulative curves for the waiting times in these six situations appear in Figures 17.9 through 17.11. Table 17.2 shows that the average waiting time for distributing passengers in the fixed schedule case is much higher than for the flexible schedule cases. Furthermore, the standard deviations are lower in the pickup cases than in the drop-off cases. More precise configurations for these results are shown in Figures 17.9 through 17.11.

In these figures, the upper cumulative curve refers to the pickup case, and the lower curve to the drop-off case. Obviously the waiting time in the drop-off case depends on the bus-departure time from the train station. This is the reason that the cumulative curves for waiting at the train station have the shape of a large step function. It should also be mentioned that the x-axis scale is not the same in all cases; it simply reflects the resultant waiting-time range. A comparison between Strategies 1 and 8 (Figures 17.9 and 17.10) for

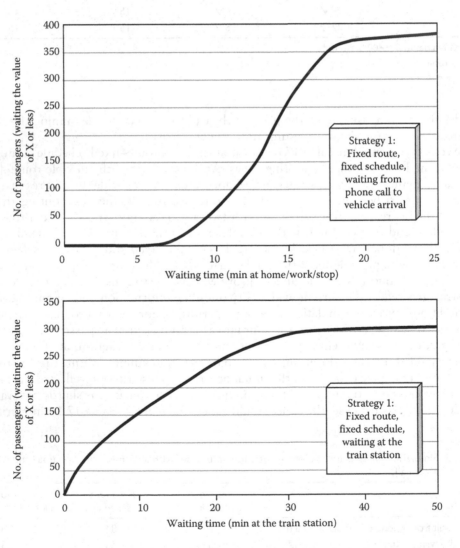

Figure 17.9 Waiting time (phone call-to-bus arrival in upper curve, and at the train station in lower curve), in minutes, for Strategy 1 (fixed route–fixed schedule); note that waiting-time scales are not the same.

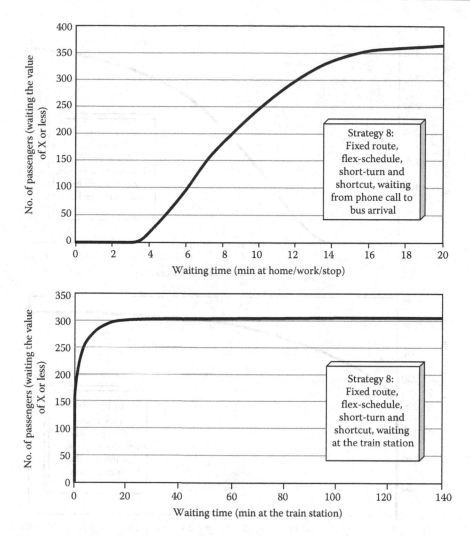

Figure 17.10 Waiting time (phone call-to-bus arrival in upper curve, and at the train station in lower curve), in minutes, for Strategy 8 (fixed route–flex schedule, with short-turn and shortcut); note that waiting-time scales are not the same.

the pickup case reveals that whereas the wait ranges from 5 to 20 min in the fixed schedule, it ranges from 3 to 18 min in the flexible schedule. In the demand-responsive case (Strategy 10, Figure 17.11), the waiting time for the pickup case ranges from 3 to 13 min.

These simulation runs are only preliminary steps toward the examination of a smart-shuttle operation. More simulation runs are required for different numbers of fixed routes and various demand levels, along with further sensitivity analyses of the input parameters.

17.6 CUSTOMER SURVEY

To obtain the needed consumer information, the case study used the survey-research method described by Yim and Ceder (2006). The test market was identified as being within a 2-mile radius of the Castro Valley BART station. Four hundred telephone interviews were conducted

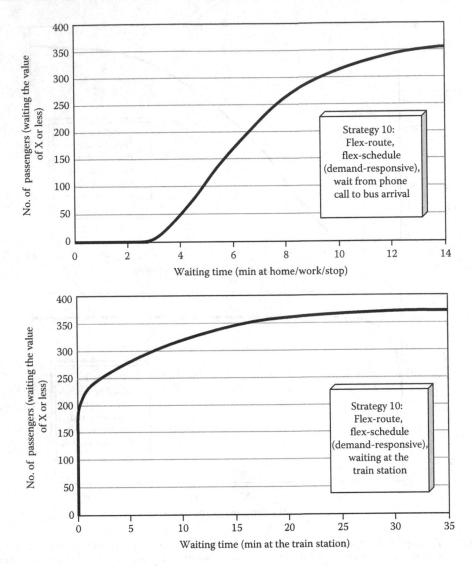

Figure 17.11 Waiting time (phone call-to-bus arrival in upper curve, and at the train station in lower curve), in minutes, for Strategy 10 (demand-responsive); note that waiting-time scales are not the same.

in this market area, using a random-digit-dial sample and the computer-aided telephone-interview (CATI) technique. The following criteria were used for screening survey participants:

- Eighteen years old or older
- Permanent resident of the household surveyed
- Said that BART was a possible means of transportation for them
- Commute or make their most frequent trip away from home by some means other than walking or bicycling

The margin of error for a 400-respondent sample is ±5.0% at the 95% level of confidence.

The survey questions covered the following topics: (1) trip characteristics, (2) mode of access transportation to BART, (3) willingness to use a smart shuttle, (4) willingness to pay for the service, (5) desired attributes of the shuttle service, and (6) demographic characteristics of the survey respondents. The study identified the features that would attract consumers in terms of routing characteristics (i.e., intermediate stop options, express service), travel time, waiting time, number of stops, and willingness to pay for the shuttle service.

17.6.1 Survey results

Several attributes were investigated with respect to the design of the shuttle service. Among them were the number of pickups and drop-offs, the size of a shuttle vehicle, acceptable number of riders, travel time, and waiting time. The results for these attributes and for other questions are as follows.

17.6.1.1 Important attributes for shuttle design

The respondents said that the most important attribute was the cost of the shuttle service. The second most important was overall travel time, including waiting time for the shuttle either at BART or at the pickup location. The third most important was the on-time reliability of the service at the pickup location or at the BART station.

17.6.1.2 Pickups and drop-offs

Most people expected four to five pickups (median of five pickups) on the way to the BART station. Similarly, they expected four to five drop-offs on the way home (median of five drop-offs).

17.6.1.3 Travel time

The question was, if the average travel time to the BART station was slower than it currently is, do you think you might use the BART shuttle service? Nearly one-third (29%) said they would take the shuttle, whereas half (53%) said they would take the shuttle if it took about the same time. Only 12% said they would take it only if it were faster. The survey suggests that people are willing to accept a longer travel time when using a shuttle because of whatever benefits they perceive it affords.

17.6.1.4 Arrival time and schedule information

One of the reasons that people are hesitant to go by public transit is the uncertainty associated with vehicle-arrival times. Advanced transit information system (ATIS) can disseminate real-time vehicle-schedule information to those who are regular transit users, as well as to the occasional transit rider. ATIS vehicle schedules can also attract those who seldom or never used public transit in the past.

17.6.1.5 Cost

The cost question for riding a shuttle was constructed to ask about the higher price first and then lower prices. Suppose the cost of the shuttle service were $1 for a one-way trip, how likely would you be to use this service? According to the survey, half the respondents said they would be interested if the cost were $1; furthermore, 41% said that they would still be interested in taking a shuttle at the price of $2.00 for a one-way trip. As expected, consumer interest in using the shuttle service is highly elastic with respect to the cost as the results in

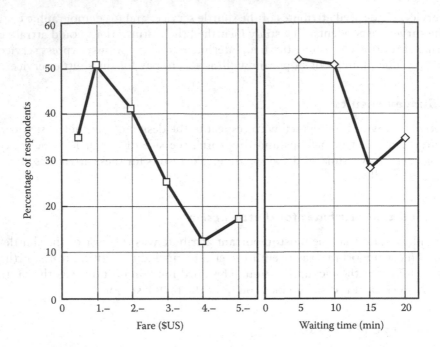

Figure 17.12 Willingness to pay and wait for the shuttle service.

Figure 17.12 illustrate. However, the survey found that price elasticity was not directly proportional to the cost of the shuttle service. The willingness to use the service is significantly (95% level of confidence) different between the low and the high cost of the shuttle service.

17.6.1.6 Waiting time

Questions about waiting time were also asked at the same time as the cost of the shuttle service. If the waiting time was 20, 15, 10, and 5 min, how likely would respondents be to use the shuttle service? The results are shown in Figure 17.12. The survey showed that the longer the waiting time, the less willing people are to take the shuttle; however, there was not a significant difference between a 10 min waiting time and a 5 min waiting time. This suggests that half of the shuttle users are willing to wait 5 min and up to 10 min for a vehicle.

17.6.1.7 Frequency of using the shuttle service

Most respondents said that they would use the service two to three times a week.

17.6.1.7.1 Payment method

Over half (55.4%) of the respondents were interested in paying for the service on a per-use basis. Only 11.2% responded favorably to a weekly fee basis, and 31.2% said they could work with a monthly subscription arrangement.

17.6.1.7.2 Preferred means of receiving information about the shuttle

The survey showed that 62% would like to receive their information from a pamphlet. Approximately one-third (30.9%) preferred to retrieve it through the Internet, and only 5.9% would like to receive it by telephone.

17.6.1.7.3 Benefits of the shuttle

A question pertained to the biggest benefits derived personally from using the shuttle service. It was an open-ended question and accepted up to three responses; therefore, the percentages shown in this section are not mutually exclusive. Respondents mentioned a variety of personal benefits. Among them were these: (1) convenience, including no need to park (25%), avoid walking in bad weather (2%), avoid wear on vehicle (21%), and others (30%); (2) safety, including reduced stress and anxiety (8%), less chance of an accident (2%), and the avoidance of traffic fights (18%); (3) travel time savings (14%); (4) less cost (18%); (5) reduced pollution (7%); and (6) the chance to meet people and to socialize (2%).

17.7 OPTIMAL ROUTING DESIGN: BASE NETWORK

This and the following sections present, respectively, a methodology and a procedure for designing optimal shuttle (circular) routes, following Ceder (2009) and Jerby and Ceder (2006). The objective set forth for an urban road network is maximal coverage of potential passenger demand to a transportation center (e.g., train station), with travel time along the entire route not exceeding a certain threshold value. Generally, an objective function that aspires for maximum coverage is supposed to create a long, winding route, which in this case is blocked by the route-time constraint to ensure an adequate level of service. The problem input consists of an urban network with trip-generation nodes and a single destination (main) node, average travel time between each segment on the network to the main node, and a constraint defining maximum travel time along the route.

The methodology described is based on a modular approach, enabling the partitioning of a complex problem into a series of sub-problems. Hence, each sub-problem can be referred to as an independent component. The methodology consists of eight main stages as shown in Figure 17.13. The *first stage* is the characterization of urban network attributes for deriving a base network for route design. The *second stage* deals with the insertion of average travel times into the links of the road network, while the *third stage* defines the service area. The *fourth stage* focuses on a method to estimate potential demand for trips on a designed shuttle route. This method includes the development of a potential demand measure for each link on the road network, based on urban and spatial criteria and using density and walking-distance parameters. This measure is used as an input to an optimization model developed in the *fifth stage*. The model enables the automated design of an optimal circular route complying with a given total travel-time constraint. Test runs using the optimization model are performed in the *sixth stage*, showing that its complexity is high and cannot be efficient for medium and large-size networks. Accordingly in the *seventh stage*, an alternative heuristic algorithm is proposed and developed. The algorithm, which enables the automatic design of circular routes, ensures good, mostly but not always optimal, results. The *eighth stage* examines the heuristic algorithm on different networks.

17.7.1 Base road network

Theoretically there could be a situation in which all city streets are used for the passage of transit vehicles; in actuality, this is not the case. For various reasons, some of the network links will never be used for the passage of transit vehicles; therefore, these links can a priori be excluded from the planning process. Such an action will define a more reduced transit-road network (the base network). Working with a base network allows the planner to reduce the complexity of the problem and enables the transit agency and the passengers to

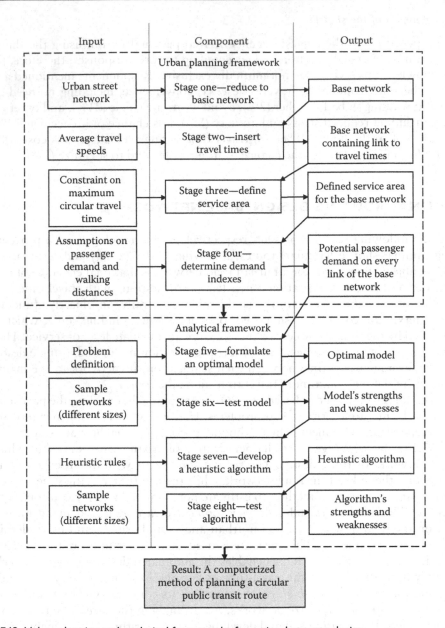

Figure 17.13 Urban planning and analytical frameworks for a circular route design.

rely on a simpler network that is easier to understand and operate. Determination of a base road network consists of the following elements: street characteristics (width, slope, parking arrangements); spacing between parallel streets; and safety considerations or any other criterion set by the transit agency's constraints on a case-by-case basis.

According to the stages in Figure 17.13, the travel time in each network link is determined after creating the base network. This value can be either measured directly or calculated according to the length of the link and the average travel speed. The next stage in Figure 17.13 is to define the *service area* that can be served reasonably by a single circular route. The size of the service area depends on the route's travel-time constraint, that is, links that cannot be covered in the framework of the travel time allocated should be eliminated from the base network. The proposed process includes testing of each of the base network links in order to determine whether a circular route can be created from the link to the main node and back while complying with the time constraint. For convenience, the main node will continue to be referred to as the train station. Should the duration of the shortest route (in travel-time values) from the link to the train station and back exceed the time constraint, such a link should be eliminated from the network. Applying this procedure on all links of the base network will result in a further reduction of the network and the elimination of distant links that should not have been taken into consideration in the first place.

17.7.2 Potential passenger demand

The demand potential is assessed on the basis of spatial data from the urban network, and is subject to a number of assumptions. Methods for assessing transit demand, which were presented in Chapter 10, include the counting of passengers, conducting surveys, using known databases of residential addresses and work places, and weighting of socioeconomic and demographic data of the region. In addition to these methods, following is a two-phase alternative method that uses spatial and demographic data.

17.7.2.1 Phase A: Calculation of average walking distance per link

This phase considers the maximum acceptable walking distance to the shuttle route for potential passengers. This variable is known to be part of a *catchment area*. It is customary to assume that it encompasses an area of up to 400 m from every side of the road link although some passengers will be willing to walk a longer distance under certain conditions. Figure 17.14 illustrates

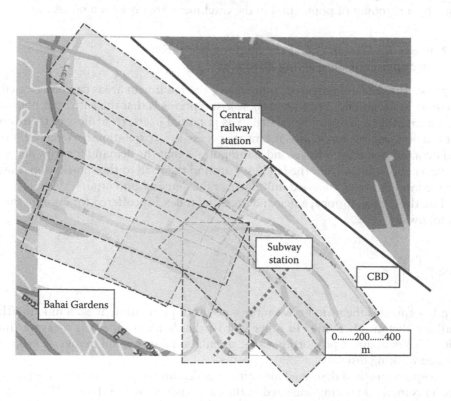

Figure 17.14 Catchment areas for the base network of the case study (City of Haifa).

the catchment areas for pedestrians in a case study area of downtown Haifa (Israel) as polygons on the base network. In this case, the catchment area amounts to 300 m from every side of the route. In Figure 17.14, one can identify areas where overlapping is created between two or more polygons; the assumptions discussed in the following text are used for such overlapping cases.

The first assumption is that passengers choose the road link nearest to the point of origin as the waiting point for the shuttle vehicle. The second assumption is that passengers will walk to the nearest possible waiting point by taking the shortest *bee line* distance (a reasonable assumption, since the passenger in many cases can walk to the stop in an almost straight line through parks or alleys). Therefore, in cases in which catchment areas overlap, every point in the overlapping area can be attributed to the nearest bee-line link. A value assessing the average walking distance in the area will now be defined.

The average walking distance $\overline{wd}_{i,j}$ on link (i, j) is influenced by land uses within the catchment area, and is calculated as follows:

$$\overline{wd}_{i,j} = \sum_{b=1}^{m} \frac{pop_b \cdot wd_b}{\overleftarrow{pop}_{i,j}} \tag{17.2}$$

where

m represents the total number of buildings in the catchment area
b is a certain building in the catchment area
pop_b represents the amount of population (residents, employees) in building b
wd_b is the walking distance from building b to the nearest road link
i and j represent network nodes
$\overleftarrow{pop}_{i,j}$ is the amount of population in the catchment area between nodes i and j

17.7.2.2 Phase B: Assessment of the demand potential as dependent on walking distances

In this phase, the demand potential from each of the catchment areas can be assessed, based on two more assumptions. The first of these assumptions is that the entire population in the pedestrian catchment area can be related to as potential users, regardless of their motorization and socioeconomic levels; this means that there is a correlation between building density and demand potential. The second assumption is that the demand for trips also depends on the walking distances from the land uses around the road to the shuttle stop, that is, the further the land use from the road link, the less the demand potential.

Based on these assumptions, a demand potential index is proffered for every network link (i, j) as follows:

$$pd_{i,j} = \frac{\overleftarrow{pop}_{i,j}}{\overline{wd}_{i,j}} \tag{17.3}$$

where $pd_{i,j}$ expresses the ratio between the amount of population at each link and the average walking distance per link. In other words, the demand potential measure (index) in Equation 17.3 is in direct proportion to the population density and in inverse proportion to the average walking distance.

The two-phase method described for estimating demand requires quality geographic information. This method was implemented in the case study shown in Figure 17.14, with bands around each network link. When no exact data existed on the number of residents in each

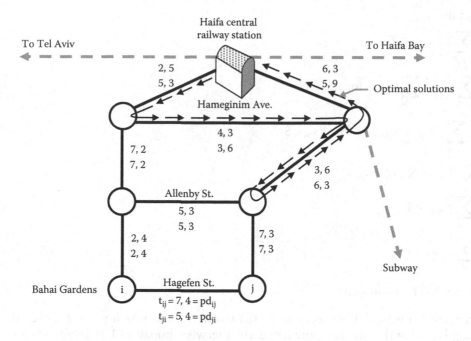

Figure 17.15 Schematic description of the case-study network and its optimal circular route solution for T = 25 min.

block, assessments were carried out based on a multiplication of the number of building floors (using the number of apartments per floor) and the approximate size of households. The outcome for the case study in Figure 17.14 is presented in the schematic network of Figure 17.15.

17.8 OPTIMAL ROUTING DESIGN: ALGORITHM

This section contains two main parts. First, an optimization model formulation is described, using operations research (OR) terms (see Section 5.5) and following Ceder (2009). Second, a heuristic algorithm, following Ceder (2009) and Jerby and Ceder (2006), is presented for handling large-size networks.

17.8.1 Optimization model

Let $G(N, A)$ be defined as the graph representing a network of streets in an urban area, where network nodes are represented by the set $N = \{1, 2, ..., n\}$ and network streets (graph links) as $A = \{(i, j): i, j \in N; i \neq j\}$. For each of the links $(i, j) \in A$ an average travel time of $t_{(i,j)}$ minutes by time-of-day is defined, and also a demand potential index $pd_{(i,j)}$. The potential demand value is assumed to be homogeneously distributed at the link. The model described produces a circular route r, starting and ending at node n_1. The model maximizes the demand potential of passengers while complying with the constraints of maximum circulating travel time T. The case study of Figure 17.15 includes the optimal circular solution for T = 25 min and shows the values of $t_{(i,j)}$ and $pd_{(i,j)}$ for each link.

Model formulation is as follows:

$$\text{Max} \sum_{(i,j)\in A} pd_{(i,j)} y_{(i,j)} \tag{17.4}$$

Subject to:

$$\sum_{(i,j)\in A} t_{(i,j)} y_{(i,j)} \leq T \tag{17.5}$$

$$\sum_{j\in N} y_{(i,j)} = \sum_{k\in N} y_{(k,i)} \quad \text{for all } i \in N \tag{17.6}$$

$$\sum_{i\in s_m, j\in s_m} y_{(i,j)} \leq |s_m| - 1 \quad \text{for all } s_m \in S \tag{17.7}$$

$$S = \left\{ s_m \subset N : n_1 \notin s_m, |s_m| \geq 2 \right\}$$

$$\sum_{i\in N} y_{(n_1,i)} = 1 \tag{17.8}$$

$$\sum_{i\in N} y_{(i,n_1)} = 1 \tag{17.9}$$

$$y_{(i,j)} \in \{0,1\} \quad \text{for all } (i,j) \in A \tag{17.10}$$

The objective function (17.4) contains $\{0,1\}$ variables $y_{(i,j)}$ in which $y_{(i,j)} = 1$ means that the link (i, j) is part of the circular route r, or zero otherwise. Equation 17.5 represents the maximum travel-time constraint. Equation 17.6 implements a condition for creating a closed circular route in the network; it follows the known Euler theorem (e.g., see Eiselt et al., 1995), in which the number of incoming and outgoing links of every node is equal.

Equation 17.7 ensures that only circular routes originated at node n_1 will be considered, where S is a set of all the network-node combinations, which include at least two nodes without node n_1. Equation 17.7 ensures that the number of route links connecting the nodes of any sub-set s_m of S will be smaller than the number of nodes in the set. In other words, Equation 17.7 requires that all chosen links must be connected to each other, thus avoiding the formation of more than one continuous circular route. Equations 17.8 and 17.9 ensure that node n_1 is included in any solution of circular routes, and finally constraint (17.10) defines $y_{(i,j)}$.

Note that the preceding model formulation cannot create circular routes in which two (or more) routes overlap in some of the nodes (e.g., eight-shaped routes). To attain such routes, it is possible to identify potential overlapped nodes and to split them each to two nodes with the same incoming and outgoing links (artificially), that is, a potential node i will be split into i′ and i″. Another important note is that only Equation 17.7 increases in a manner that cannot be described by a polynomial equation, hence results in increasing the complexity of the problem with the increase of the network size. The conclusion, therefore, is that solving the problem of real-life road networks through the integer-programming formulation described is impractical. This complexity in the number of calculations warrants the examination of heuristic solutions that is given in the next section.

17.8.2 Heuristic algorithm

The algorithm described is based on an efficiency criterion. For that purpose, an impedance ratio, z_{ij}, is defined as follows:

$$z_{ij} = \frac{\omega \cdot t_{ij}}{\lambda \cdot pd_{ij}} \tag{17.11}$$

where ω, λ are coefficients.

As mentioned, the potential demand is a value that is in direct proportion to inhabitant density and in inverse proportion to walking distances. The efficiency criterion utilized in the algorithm is that of the shortest path between each two links of a selected group of links in accordance with Equation 17.11.

The heuristic algorithm is shown in a flow diagram in Figure 17.16. The algorithm constructs an efficient circular route from an initial group p of links (p ≤ P) characterized by relatively high demand potential; it utilizes Equation 17.11 and the shortest path between

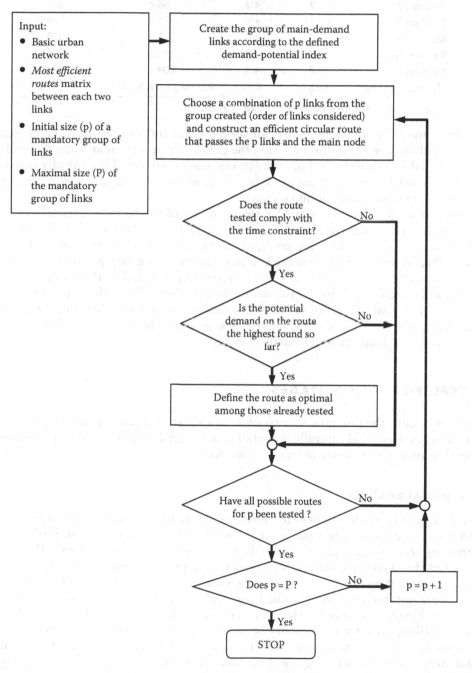

Figure 17.16 Flow diagram of the heuristic procedure.

Table 17.3 Summary of the heuristic algorithm runs, given different networks

Network tested	No. of nodes	P/A ratio	Objective function value[a] (optimal)	Objective function value[a] (heuristic)
From Figure 17.15	7	4/9	26.6	26.6
Random 1	7	4/9	191	191
Random 2	7	4/7	208.8	208.8
Random 3	7	4/7	206.4	206.4
Random 4	7	4/8	79.2	79.2
Random 5	7	4/9	338	338
Random 6	7	4/9	243.6	243.6
Random 7	8	4/9	204	204
vRandom 8 (large)	25	4/42	Too long running time	142.2

[a] In potential units of demand, following Equations 17.2 and 17.3.

the links in the selected group p. A small sub-group of links is chosen in every iteration of the algorithm. The algorithm examines the possibility of a circular route that will pass through the sub-group links and meet the travel-time constraint. Then the algorithm compares the potential demand of the candidate route to each of the routes examined so far. Lastly, the best circular route found is chosen.

Several test runs were made enabling a comparison of the heuristic algorithm with the optimal model. The results are shown in Table 17.3. The first network is that of Figure 17.15; the seven networks that follow are small, randomly selected networks; last is a larger-size network. One important feature of the heuristic algorithm is group p, which is the number of initial mandatory links chosen for route r. Figure 17.6 defines P as the maximum number of such mandatory links. Table 17.3 shows the ratio between P and the number links in the network, A. P = 4 was found to provide reasonable computer running time, including the 42-link last tested network. Finally, the results support the heuristic algorithm, as they are the same as the optimal across all small networks.

17.9 IMPLEMENTATION STAGES

In order to secure the potential success of a new shuttle service, certain steps should be undertaken gradually and carefully. Essentially, these steps, shown in Ceder (2013), involve an initial quantitative analysis and then a pilot plan.

17.9.1 Initial analysis

Figure 17.17 schematically outlines five components that are necessary to complete the initial analysis: (1) constructing a base street network; (2) creating groups of fixed routes; (3) constructing short-turn, shortcut, and bidirectional strategies; (4) creating a DRT-type of service; and (5) comparing different strategies with a given passenger demand.

Component 1 of Figure 17.17 uses site observations and measurements for a base road-network configuration, including traffic-light locations and an 85 percentile of time to and from the train-station platform. This 85 percentile of the different times observed ensures adequate walking time for the majority of people. The determination of the base road network considers the following elements: approximate (low, average, high) density of residential area, street characteristics (width, slope, parking arrangements), spacing between parallel streets, and the road-network shape of each zone in the area.

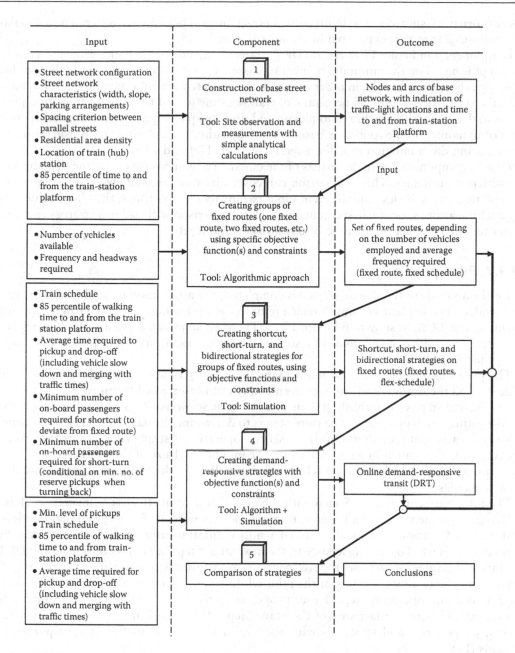

Figure 17.17 Practical methodology for constructing a feeder/shuttle service.

Component 2 of the initial analysis creates the fixed routes to be considered. These circular routes can be constructed by the heuristic algorithm described in Section 17.8, including a measure of its demand potential. The maximum length of the routes to be selected is influenced by the number of shuttle vehicles available and, if given, the minimum frequency required.

Component 3 constructs the operational strategies that can continuously ensure an adequate level of service. These strategies, outlined and explained in Section 17.3, basically cover short-turn, shortcut, and bidirectional possibilities. They can be analyzed by a simulation tool similar to that explicated in Sections 17.4 and 17.5.

Component 4 in Figure 17.17 creates DRT strategies, given certain input elements. That is, the input is based on the minimum number of passengers for whom a vehicle may be sent for a pickup; the train schedule, in order to match the expected arrival time of the DRT vehicle with the train's arrival; the 85 percentile of walking time to and from the train-station platform; and average times required to pickup and drop-off passengers. The tool to create this type of dynamic, online routing is based on an algorithm (e.g., shortest path from point to point), using the simulation tool discussed in Sections 17.4 and 17.5.

Finally, component 5, which consists of the outcome of components 2, 3, and 4, compares the different strategies. This comparison can cover different demand levels, different numbers of available vehicles, and different input parameters (travel times, threshold values for dispatching a trip, short-turn, shortcut, etc.). The comparison will lead to a strategy that can better fit a given situation (by time-of-day and demand level).

17.9.2 Pilot plan

Once the analysis of the shuttle service is completed, the application of a pilot study is recommended. The implementation of such a pilot in the area being considered can follow, for example, the 12 steps shown in Figure 17.18. These 12 steps can serve as a framework for a master plan, with each outcome of a step becoming an additional input for the next step, except for Step 6.

The pilot master plan starts with a demand analysis by time-of-day and day-of-week in order to find the origin–destination pattern and consumer-oriented features.

The second step is to establish (if done previously in Sections 17.7 and 17.8) or design the fixed-routing and stop system. The third step is to determine the base frequencies and timetables for each route or, alternatively, to set the operational strategies in accordance with Section 17.3. The fourth step is to determine the number and size of the shuttle vehicles and to create the chain of trips (vehicle schedules) that will serve the fifth step, constructing the crew schedules.

The pilot plan continues in Step 6 with the establishment of effective information channels and instruments (e.g., call center, Internet, newspapers, radio, TV, and mail leaflets) that will lead to the development of user-friendly communication procedures between the users and a selected operating agency in the next step (Step 7). Step 8 constructs the DRT operational strategies without the use of the fixed routing/stop/schedule system. Step 9 determines the testing scenarios of the pilot while Step 10 presents the process for selecting an adequate operator. Step 11 uses proper advertising tools to approach an operable pilot study. Finally, the last step of the plan (Step 12) is aimed at improving the instruments, procedures, and strategies with the use of innovative intelligent transportation systems (ITS).

17.10 LITERATURE REVIEW AND FURTHER READING

This section describes models for the design of shuttle or feeder routes. Only models corresponding with a fixed route fall within the scope of this review.

Wirasinghe (1980) formulates an analytical model for a system in which bus routes feed a railway line during peak hours. It is assumed that the street network is a perfect rectangular grid and that there is no need for synchronization between the feeders and the

Input	Step	Outcome
Survey	(1) Demand analysis	O-D, user features
Road map	(2) Routing + stops	Routes, stops
Standards, travel time	(3) Frequency + timetables	Timetables
Vehicle sizes, travel times	(4) Vehicles (No., sizes) + scheduling	Vehicle schedules
Operator work rules	(5) Crew scheduling	Crew schedules
Survey, routes, timetables	(6) Information	Information channels
Travel times, call back	(7) Communication	Caller information
Short-turns, shortcut; and bidirectional options	(8) Operational strategies	Deviation elements
Survey, candidate scenarios	(9) Testing scenarios	Operational scenarios
Operations process	(10) Bidding	Operator
Determined procedures	(11) Advertisement + execution	Pilot
Innovative ITS	(12) Update	Pilot changes

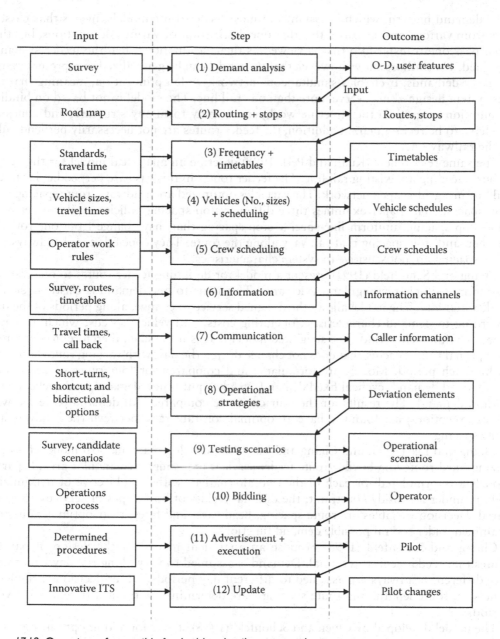

Figure 17.18 Overview of a possible feeder/shuttle pilot master plan.

rail line. Two cases are presented. In the first case, the locations of railway stations are given, and the optimal density and headways of the buses determined. In the second case, the spacing of feed points is not given, and therefore all three parameters (station spacing, route spacing, and headways) are optimized.

Kuah and Perl (1987) present an optimization methodology for the design of an integrated bus–rail system. A mathematical programming problem is formulated in which the itinerary and frequency of each route are determined under a fleet size constraint. In addition, a heuristic solving method is developed. For the basic problem, a many-to-one demand pattern is assumed.

This demand pattern, which is assumed in most feeder-route models, means that passengers from various origins travel to either one destination or various destinations, but they must pass one transfer hub that is viewed as their destination when the feeder system is designed. The authors show, however, that the model can be generalized to represent many-to-many demands. It enables simultaneous network-wide optimization, meaning that the bus system designed may serve more than one rail line. The model is not based on binding assumptions regarding the street network; streets may form any structure, and zones do not have to be rectangular. In addition, the feeder routes are not necessarily perpendicular to the railway line.

The same researchers (Kuah and Perl, 1988) describe an analytical model for the design of buses feeding an existing rail line. The feeder routes in this case are perpendicular to the rail line in a grid-shaped network. The variables optimized are headways, route spacing, and stop spacing. The paper examines three different stop-spacing policies: uniform network-wide stop-spacing, uniform line-specific stop-spacing that may change from one route to another, and stop-spacing that can vary along the route. They conclude that headways and route spacing are not sensitive to system characteristics.

Chang and Schonfeld (1991) present a model for designing feeder routes to one transfer station in a simple rectangular service area. The many-to-one demand is elastic to service quality and fare. The model allows the designed service to change, along periods of the day, according to demand characteristic, operating costs, and vehicle speeds, which are given separately for each period. The suggested model makes it possible, therefore, to determine the optimal route spacing that does not change during the day, with varying route headways within each period. Models are formulated and compared for four types of conditions: steady fixed demand, cyclical fixed demand, steady equilibrium demand, and cyclical equilibrium demand. The results for the four cases are compared. All demand patterns (with minor exceptions) are found to have an optimal constant ratio between the headway and route spacing.

Chang and Lee (1993) and Chang and Yu (1996) develop mathematical models for optimizing fixed-route and flexible-route feeder services in a simple rectangular area. Separate models are formulated for each of the two alternatives with the objective of maximizing welfare under a subsidy constraint; the optimal results of both types of services are compared. Decision variables are route spacing, headways, and fares; their optimal values are examined under several possible demand functions.

Chang and Schonfeld (1993) propose an analytical model for the design of parallel routes that feed a central terminal. Demand is assumed to vary along the day, and therefore different headways are assigned to different day periods. Other decision variables—zone size, route length, and route spacing—are determined without differences between periods.

The model developed by Chien and Schonfeld (1998) strives for a joint optimization of a rail route and its feeder buses. The variables optimized are rail-line length, feeding-station spacing, bus headways, bus stop-spacing, and bus route-spacing. Street network is assumed to be a simple grid, with the feeders perpendicular to a single rail line. On the other hand, the model is also based on the realistic assumption of many-to-many demand.

Chien (2000) presents a model for defining bus-feeder routes and their headways. One central route is fed into one major station, and the demand is many-to-one. The service area is shaped as a partial grid, that is, a grid with some of the links missing. The form of each zone must be rectangle-like in a simple grid, but the zone size may be heterogeneous. The model enables the user to decide the number of feeder routes to be defined.

Chien et al. (2001a) introduce a methodology for the design of a feeder-bus route feeding a major intermodal transfer station in a suburban area. The model is formulated as a linear programming problem with geographic, capacity, and budget constraints. Decision variables are route locations and headways. The street network has a partial grid pattern, and the demand pattern is many-to-one. Two solution algorithms are suggested. The first is an exhaustive search algorithm that promises, in principle, an optimal solution; however, it is very time-consuming in most realistic networks. The second suggested algorithm is a genetic algorithm (GA), that is, an algorithm in which the search for optimal solutions is based on the natural evolution process. In each iteration of a GA, a set of solutions is selected based on the previous set; solutions that show good performance in each iteration have a higher probability of being selected for the next iteration. Genetic algorithms are efficient in terms of calculation time, but may result in only a near-optimal solution.

Chien et al. (2001b) compare fixed and flexible route systems serving as feeders for a transportation center close to a rectangular service area. Optimal vehicle size, route spacing, and headways are determined for both types of service. The models assume a probabilistic demand, which varies over periods of the day. Passengers' value of time is assumed to be non-additive, that is, the value that a passenger ascribes to a 2 min period is higher than twice the value ascribed to a 1 min period. The researchers find it best to operate conventional buses at peak hours and flexible routes in off-peak periods. The threshold value of demand at which one system is more cost effective is calculated.

Shrivastava and Dhingra (2001) present a heuristic algorithm for the development of bus routes feeding railway stations. The algorithm uses different node-selection and insertion strategies to determine route itineraries under many-to-many demand. Routes are developed for two types of time criteria. The first is a maximum-demand-deviated, shorter, time-path criterion; this means that when nodes are inserted into the middle of a route, the length of the deviated route should not exceed a given acceptable limit. The second criterion is a path-extension time criterion, meaning that if nodes are inserted at the end of a route, then the maximum route length in terms of travel time should not exceed a certain maximum value. Use of the suggested method is possible in any network, no matter the street-network structure. A case study is presented, in which the use of this method enabled serving the same demand as previously with a smaller number of routes, a smaller fleet, and better schedule adherence because of the shorter routes.

Another paper by Shrivastava and Dhingra (2002) presents a methodology for designing a coordinated schedule for existing bus routes feeding a railway station. The model presented aims at minimizing total operating costs and total transfer times under the constraint that each transfer time should be executed within given boundaries. The model is formulated as a nonlinear, non-convex programming problem with many variables, and it is therefore difficult to solve. A GA is suggested to provide near-optimal results. The results show that the designed schedule is most sensitive to the parameters (such as the transfer penalty). The conclusion is that the value of this penalty should be carefully considered.

Zhao and Dessouky (2008) study the service capacity design problem for a mobility allowance shuttle transit system—a transportation concept that merges the flexibility of DRT systems with the low-cost operability of fixed-route bus systems. The authors analyze the relationship between service cycle time, and the length and width of the service area under a desired service level constraint. The analysis can be used as a guide for designing service area parameters. Given that a shuttle travels in a rectilinear pattern and utilizes a non-backtracking nearest-insertion algorithm, closed-form approximate solutions are derived for the service capacity design problem. Finally, the authors demonstrate that setting the length of the service area to half the travel speed of the shuttle multiplied by the cycle time is an effective approximation.

Feeder lines often switch operations between a demand-responsive and a fixed-route policy. In designing and running such systems, the identification of the condition justifying the operating switch is often hard to properly evaluate. In their study, Quadrifoglio and Li (2009) propose an analytical model and solution of the problem to assist decision makers and operators in their choice. By employing continuous approximations, the authors derive handy but powerful closed-form expressions to estimate the critical demand densities, representing the switching point between the competing operating policies. Based on the results of one-vehicle and two-vehicle operations for various scenarios, in comparison to values generated from simulation, the authors verify the validity of the analytical modeling approach developed.

Ciaffi et al. (2012) investigate the transit network design problem related to the feeder routes defined as transit services for the connection of local areas, where the demand has to be gathered, with the stops of the main transit network, usually railway or underground station. The objective of this study is the development of a procedure that simultaneously generates routes and frequencies of the feeder bus network in a real-size large urban area. The solving procedure is articulated in two phases: in the first one, a heuristic algorithm generates two different and complementary sets of feasible routes, in order to provide a good balance between maximization of the service coverage area and minimization of the overall travel time. The set of all feasible routes, generated taking into account only the main skeleton of the road network, is then the input data for the second phase where a GA is utilized for finding a sub-optimal set of routes with the associated frequencies. In the first phase, the first set of feasible routes generated is composed by circular routes using the solution of the travelling salesman problem (TSP), and the second feasible set aims at developing feeder routes more direct than the others using the k-shortest path algorithm. The proposed procedure has been implemented on two real-life size networks, Winnipeg and Rome, in order to compare its effectiveness with the performances of the existing transit networks. The results of the applications of the design procedure show that the feeder routes imply a more integrated transit network with a reduction of the total travel time, despite an increase of the number of transfers, in a more efficient way as demonstrated by the reduction of the operating costs and the increase of the average load factor.

Kim and Schonfeld (2014) present formulation and numerical examples of integration between flexible- and fixed-route services. Their decision variables are service type, vehicle size, number of zones per region, headway, and fleet size. The authors propose a probabilistic optimization model for integrating fixed and flexible bus services with timed transfers. For the coordination of passenger transfers, same or multiple headways are used including optimization of the fleet size. Kim and Schonfeld (2014) show that the integration of fixed and flexible services is especially desirable when demand densities vary considerably among the served regions. The proposed models quantify the delay and cost savings achievable from optimized coordination. Their results confirm that timed transfers are desirable for increasing the probability of vehicle connections at transfer terminals, and thus minimizing passenger wait times compared to uncoordinated operations. A GA is used in their study as a solution method.

Dikas and Minis (2014) studied the possibility to change the optimal routing of a flexible transit service so as to pickup passengers with limited mobility and drop them off at their destination. The analytical problem is to maximize the number of served paratransit (of people with limited mobility) requests and reducing the total riding time of the paratransit clients with the compliance of covering the optimal routing. The solution of this problem uses an exact branch and price method including a labeling algorithm with appropriate dominance rules in order to reach the optimal solution within reasonable computational times. The authors reported that their proposed algorithm improves the computational time by up to 50% compared with existing approaches for similar problems.

Braekers et al. (2014) are extending previous dial-a-ride transit studies by introducing multi-depot heterogeneous dial-a-ride problem (MD-H-DARP). In other words, their formulation considers different types of users and vehicles. Two algorithms are proposed to solve the MD-H-DARP: an exact branch-and-cut algorithm and a deterministic annealing meta-heuristic. Their branch-and-cut algorithm is based on a similar algorithm for the standard dial-a ride problem (DARP) and applied on a compact two-index formulation using binary variables. Results on several sets of benchmark instances indicate that the algorithm is characterized by a small computational time. The authors concluded that based on their findings the branch-and-cut algorithm may be applied to solve small-size problems and the meta-heuristic algorithm is an efficient tool for solving large-size problems.

Pan et al. (2014) develop modeling to handle simultaneously the decision of both the service area of flexible feeder transit service and best routing. They propose a mixed integer linear programming (MILP) to optimize both the service area and routing. The transit service is of the type of demand-responsive service dealing with passengers located in typical Chinese residential areas. The proposed model is based on a bi-level math programming with an upper level to maximize the number of served passengers by the feeder transit system and a lower level to minimize the operational cost for transit operators. The authors use a case study to examine their modeling and selected the network of the hi-tech district in Jinan in China. The analyses conducted use the tools of the CPLEX software and then a meta-heuristic method to show that the total cost of the objective-function differs by 2% between the optimization and heuristic treatments.

Finally Yu et al. (2014) in their recent report attempted to optimize flexible feeder dedicated transit from bus stops to rail stations and vice versa. They call their designed system *circular service*. That is, the authors formulated an optimization problem to determine the feeder-service routing and the stop sequence to collect passengers, based on real-time passenger demand. The objective function is to minimize the total tour cost incurred by passengers and operators including the minimization of the walk time of each passenger. A bi-level nonlinear mixed integer programming model was constructed and a tabu search method was implemented with different local search strategies and neighborhood evaluation methods. They introduced a sort of case study in Austin, Texas, comparing the cost performance of three different situations: utilizing two vehicles, three vehicles, and considering fixed stops. It is concluded that a penalty value should be set high for passengers who may miss the train, and that forcing the route to go via more bus stops is linked directly with the budget available to run the service.

Characteristics of the models reviewed are summarized in Table 17.4.

Table 17.4 Summary of models reviewed, with emphasis on fixed-route models

Sources	Model formulation	Decision variables	Street network structure	Feeding system	System being fed	Special demand features
Wirasinghe (1980)	Analytical model	Feeder spacing and headways. Locations of feeding points—optional	Simplistic	Parallel routes	One rail route	
Kuah and Perl (1987)	Programming problem	Feeder itineraries and headways	Realistic	Any routes	Several rail routes	Many-to-one and many-to-many
Kuah and Perl (1988)	Analytical model	Feeder spacing, headways, and stop spacing	Simplistic	Parallel routes	One rail route	
Chang and Schonfeld (1991)	Analytical model	Feeder spacing and day-period-specific headways	Simplistic	Parallel routes	One transfer station	Multiple-period elastic demand
Chang and Lee (1993); Chang and Yu (1996)	Analytical model	Feeder spacing, headways, and fares	Simplistic	Parallel routes	One rail route	Elastic demand
Chang and Schonfeld (1993)	Analytical model	Feeder spacing, length, and day-period-specific headways	Simplistic	Parallel routes	One transfer station	Multiple-period
Chien and Schonfeld (1998)	Iterative algorithm	Rail line length, location of feeding points, feeder spacing and headways, feeder stop spacing	Simplistic	Parallel routes	One rail route	Many-to-many
Chien (2000)	Iterative algorithm	Feeder itineraries and headways (the user determines the number of routes)	Semi-realistic (partial grid)	Any routes	One transfer station	
Chien et al. (2001a)	Programming problem	Feeder itinerary and headway	Semi-realistic (partial grid)	One feeder route	One transfer station	
Chien et al. (2001b)	Analytical model	Feeder spacing, headways, and vehicle size	Simplistic	Parallel routes	One transfer station	Multiple-period probabilistic demand
Shrivastava and Dhingra (2001)	Iterative algorithm	Feeder itineraries	Realistic	Any routes	Several transfer stations	Many-to-many

(Continued)

Table 17.4 (Continued) Summary of models reviewed, with emphasis on fixed-route models

Sources	Model formulation	Decision variables	Street network structure	Feeding system	System being fed	Special demand features
Shrivastava and Dhingra (2002)	Programming problem	Railway and feeder schedule	Realistic	Any routes	One/ several transfer stations	
Zhao and Dessouky (2008)	Optimization problem	Maximizing potential demand and minimizing walking distances	Realistic	Any routes	One transfer station	
Quadrifoglio and Li (2009)	Analytical model	Demand responsive policy that might be converted to a traditional fixed-route policy for higher demand	Realistic	Any routes	One transfer station	Critical demand based on utility function
Ciaffi et al. (2012)	Heuristic and genetic algorithms	Transit routes	Realistic	Any routes	Several transfer stations	Fixed demand
Kim and Schonfeld (2014)	Analytical model	Service type, vehicle size, number of zones per region, headway, and fleet size	Realistic	Any routes	One transfer station	Timed transfer
Dikas and Minis (2014)	Optimization problem	Riding requests, travel time, routing	Realistic	Any routes	One transfer station	
Braekers et al. (2014)	Optimization problem	Routing, response time, type of vehicle	Realistic	Any routes	Several transfer stations	
Pan et al. (2014)	Optimization and meta-heuristic	Service area, feeder-transit routing	Typical Chinese residential areas	Any routes	One transfer station	
Yu et al. (2014)	Optimization and meta-heuristic	Locations of rail stations, fixed bus stops, fixed cost for routing	Realistic	Any routes	One transfer rail station	

EXERCISE

17.1 The following street network connected to a train station is given. Each link direction (there are two directions of travel per link, marked by arrows) is characterized by a pair of numbers: average travel time (in minutes, on the left) and demand potential ratio (in units of Equation 16.3, on the right). The maximum circulating travel time is given as T = 30 min; in addition, $\omega = 1.0$, $\lambda = 1.0$, $p = 2$, and $P = 2$.

(a) Create the matrix of the impedance ratio z_{ij}.
(b) Assume that the initial group of the highest potential demand units contains 5 links (out of 22 links, both directions) of the given network. Find the best single route that will maximize potential demand while complying with the constraint T.

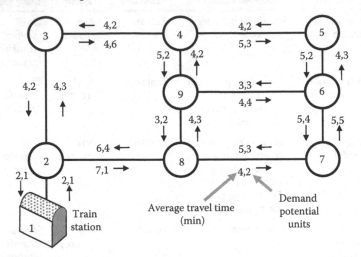

REFERENCES

Borndorfer, R., Grotschel, M., Klostemeier, F., and Kuttner, C. (1999). Telebus Berlin: Vehicle scheduling in a dial-a-ride system. In *Computer-Aided Transit Scheduling*. Lectures Notes in Economics and Mathematical Systems, vol. 471 (N.H.W. Wilson, ed.), pp. 391–422, Springer-Verlag, Berlin, Germany.

Braekers, K., Caris, A., and Janssens, G.K. (2014). Exact and meta heuristic approach for a general heterogeneous dial-a-ride problem with multiple depots. *Transportation Research*, **67B**, 166–186.

Ceder, A. (2009). Stepwise multi-criteria and multi-strategy design of public transit shuttles. *Journal of Multi Criteria Decision Analysis*, **16**(1–2), 21–38.

Ceder, A. (2013). Integrated smart feeder/shuttle transit service: Simulation of new routing strategies. *Journal of Advanced Transportation*, **47**(6), 595–618.

Ceder, A. and Yim, Y.B. (2002). Integrated smart feeder/shuttle bus service. (California PATH Report). Institute of Transportation Studies, University of California at Berkeley, Berkeley, CA.

Cervero, R. (1998). *Paratransit in America: Redefining Mass Transportation*. Praeger, Westport, Connecticut.

Chang, S.K. and Schonfeld, P.M. (1991). Multiple period optimization of bus transit systems. *Transportation Research*, **25B**, 453–478.

Chang, S.K. and Schonfeld, P.M. (1993). Optimal dimensions of bus service zones. *Journal of Transportation Engineering*, **119**, 567–585.

Chang, S.K. and Yu, W.J. (1996). Comparison of subsidized fixed- and flexible-route bus system. *Transportation Research Record*, **1557**, 15–20.

Chang, S.K.J. and Lee, C.J. (1993). Welfare comparison of fixed- and flexible-route bus systems. *Transportation Research Record*, **1390**, 16–22.

Chien, S. (2000). Optimal feeder bus routes on irregular street networks. *Journal of Advanced Transportation*, **34**, 213–248.

Chien, S. and Schonfeld, P.M. (1998). Joint optimization of a rail transit line and its feeder bus system. *Journal of Advanced Transportation*, **32**, 253–282.

Chien, S., Yang, Z., and Hou, E. (2001a). Genetic algorithm approach for transit route planning and design. *Journal of Transportation Engineering*, **127**, 200–207.

Chien, S.I., Spasovic, L.N., Elefsiniotis, S.S., and Chhonkar, R.S. (2001b). Evaluation of feeder bus systems with probabilistic time-varying demands and nonadditive time costs. *Transportation Research Record*, **1760**, 47–55.

Ciaffi, F., Cipriani, E., and Petrelli, M. (2012). Feeder bus network design problem: A new meta-heuristic procedure and real size applications. *Procedia-Social and Behavioral Sciences*, **54**, 798–807.

Dikas, G. and Minis, I. (2014). Scheduled paratransit transport systems. *Transportation Research*, **67B**, 18–34.

Eiselt, H.A., Gendreau, M., and Laporte, G. (1995). Link routing problems, part I: The Chinese postman problem. *Operations Research*, **43**, 231–242.

Fu, L. (1999). Improving paratransit scheduling by accounting for dynamic and stochastic variations in travel time. *Transportation Research Record*, **1666**, 74–81.

Fu, L. (2001). Simulation model for evaluating intelligent paratransit systems. *Transportation Research Record*, **1760**, 93–99.

Fu, L. and Xu, Y. (2001). Potential effects of automatic vehicle location and computer-aided dispatch technology on paratransit performance. *Transportation Research Record*, **1760**, 107–113.

Ioachim, I., Derosiers, J., Dumas, Y., Solomon, M., and Villeneuve, D. (1995). A request clustering algorithm for door-to-door handicapped transportation. *Transportation Science*, **29**, 35–139.

Jerby, S. and Ceder, A. (2006). Optimal routing design for shuttle bus service. *Transportation Research Record*, **1971**, 14–22.

Kim, M. and Schonfeld, P. (2014). Integration of conventional and flexible bus services with timed transfers. *Transportation Research*, **68B**, 76–97.

Kuah, G.K. and Perl, J. (1987). A methodology for feeder-bus network design. *Transportation Research Record*, **1120**, 40–51.

Kuah, G.K. and Perl, J. (1988). Optimization of feeder bus routes and bus-stop spacing. *Journal of Transportation Engineering*, **114**, 341–354.

Melucelli, F., Nonato, M., Crainic, T.G., and Guertin, F. (2001). Adaptive memory programming for a class of demand responsive transit systems. In *Computer-Aided Scheduling of Public Transport*. Lectures Notes in Economics and Mathematical Systems, vol. 505 (S. Voss and J.R. Danuna, eds.), pp. 253–273, Springer-Verlag, Berlin, Germany.

Pan, S., Yu, J., Yang, X., Liu, Y., and Zou, N. (2014). Designing a flexible feeder transit system serving irregularly shaped and gated communities: Determining service area and feeder route planning. *Journal of Urban Planning and Development*, Published online June 23, 2014. doi: 10.1061/(ASCE)UP.1943-5444.0000224, 04014028.

Quadrifoglio, L. and Li, X. (2009). A methodology to derive the critical demand density for designing and operating feeder transit services. *Transportation Research*, **43B**, 922–935.

Salzborn, F.J.M. (1972). Optimum bus scheduling. *Transportation Science*, **6**, 137–148.

Shrivastava, P. and Dhingra, S.L. (2001). Development of feeder routes for suburban railway stations using a heuristic approach. *Journal of Transportation Engineering*, **127**, 334–341.

Shrivastava, P. and Dhingra, S.L. (2002). Development of coordinated schedules using genetic algorithms. *Journal of Transportation Engineering*, **128**, 89–96.

Wilson, N.H.M., Sussman, J.M., Higonnet, B.T., and Goodman, L.A. (1970). Simulation of a computer-aided routing system (CARS). *Highway Research Record*, **318**, 66–76.

Wirasinghe, S. (1980). Nearly-optimal parameters for a rail-feeder bus system on a rectangular grid. *Transportation Research*, **14A**, 33–40.

Yim, Y.B. and Ceder, A. (2006). Smart feeder/shuttle bus service: Consumer research and design. *Journal of Public Transportation*, **9**, 97–121.

Yu, Y., Machemehl, R., and Hakimi, S. (2014). Real-time optimization of passenger collection for commuter rail systems. Technical report 600451-00082-1, Center of Transportation Research, University of Texas at Austin, National Technical Information Service, Springfield, VA.

Zhao, J. and Dessouky, M. (2008). Service capacity design problems for mobility allowance shuttle transit systems. *Transportation Research*, **42B**, 135–146.

Chapter 18

Service reliability

Don't meet reliability problems half way; they are quite capable of making the whole trip

A.C.

CHAPTER OUTLINE

18.1 Introduction
18.2 Measures of reliability and sources of unreliable service
18.3 Modeling of reliability variables
18.4 Passenger waiting time at a stop
18.5 Advanced reliability-based data and control
18.6 Techniques to resolve reliability problems
18.7 Literature review and further reading
Exercises
References

PRACTITIONER'S CORNER

Transit service reliability problems are very real and have had significant impact on transit efficiency and productivity. Transit reliability can be defined as dependability in terms of time (waiting and riding), passenger load, vehicle quality, safety, amenities, and information. There is never a good uncertainty or a bad certainty in transit operations. The question arises as to whether remedies exist to service reliability problems and, if so, whether they are implementable and exhaustive. This chapter will attempt to answer this question through a disaggregate perspective of individual elements causing unreliable service.

The chapter contains six main parts, following an introductory section with examples of passenger complaints as reported in the media. Section 18.2 discusses measures of reliability, sources, and indicators of unreliable service and the bus-bunching phenomenon. Section 18.3 provides the variables affecting service reliability and models for estimating travel time and dwell time. Section 18.4 describes and analyzes statistical models for passenger waiting time at transit stops. Section 18.5 focuses on introducing innovations into an advanced public transit system— for example, automatic vehicle locators and automatic passenger counters—their objective being to collect and control reliability-based data. Section 18.6 furnishes an overview of techniques to improve reliability problems; it covers the subjects of better planning, online control strategies, and vehicle-priority schemes. Section 18.7 reviews work done on three themes: measures of reliability, passenger waiting time, and control strategies. The chapter ends with exercises.

Practitioners may skip Section 18.4 as they read through the chapter. We argue that transit agencies usually assess measures of system performance that do not reflect what individual passenger perceives, thus warranting measurements at the disaggregate level in order to capture passengers' perceptions. We make the following suggestion to passengers; the story is illustrative:

Never let a bus driver know you're in a hurry; it will work against you. This brings us to the story. A priest and a bus driver arrive in heaven. The angel Gabriel approaches them and asks them to follow him so he can show them their new houses. They come to a nice neighborhood with villas. Gabriel pulls out a key and gives it to the bus driver: "This is your home," the angel says, pointing to one of the villas. Gabriel continues to walk with the priest and, after a while, they arrive at a poor neighborhood, with small houses that are almost tottering. The angel pulls out a key and gives it to the priest: "This is your home," he says, pointing to one of the small houses. The priest, looking completely puzzled, asks: "How come that is for me? After serving God all my life, you give me such a house and you give the bus driver a villa?" Gabriel replies: "Because when you're praying, everyone falls asleep, when he is driving, everyone prays."

18.1 INTRODUCTION

Service reliability in transit operations has been receiving increasing attention as agencies become faced more and more with the immediate problem of providing credible service while attempting to reduce operating costs. Unreliable service has been cited as the major deterrent to existing and potential passengers. For example, Balcombe et al. (2004), in a U.K. practical-transit guide, report that passengers' perception of local bus services is interpreted in the following ranking of importance in weights (given here in parenthesis) that add up to 100: reliability (34), frequency (17), vehicles (14), driver behavior (12), routes (11), fares (7), and information (5). In other words, for instance, it is twice as important from a passenger's perspective to improve reliability as to increase the frequency of service.

There are many humoristic sayings about transit reliability that represent part of the passengers' accumulated frustration on this subject. Some of these sayings are as follows:

1. Passengers waiting long give a look you could have poured on a waffle.
2. Bus and men are alike—both aren't there when you need them.
3. To go nowhere, follow the crowd or stay on the bus (stuck in a traffic jam).
4. Ad on a bus: "Love is like a bus—if you missed one, another will come along," to which one can add: like in love, you don't know when it will come and whether it will have room for you.
5. When the bus you are on is late, the bus you want to transfer to is on time (one of Murphy's laws).

Along the line of comprehending the magnitude of service reliability problems in regard to transit demand, it is also interesting to read passengers' complaints. The following are typical and have appeared in the media; they remind us what Alfred North Whitehead said: "We think in generalities, we live in details."

Misleading frequency claim

From the *New Zealand Herald*, May 16, 2006, in an article by Claire Trevett ("Link bus adverts 'misleading'").

Stagecoach's claim that its Link bus service runs every 10 min has been ruled misleading after a passenger complained of sometimes waiting for up to 40 min ...

Of the claim that the buses ran "every 10 min," the board said: "There was a passenger expectation that a bus would be available every 10 min on the route and, as that had not been the case in the consumer's experience, the Complaints Board ruled that the website advertisement was misleading. ..."

Jean McGeorge, a law student in her 30s: "They are really erratic. You'll get two buses at the some time and then nothing for 20 min and that's incredibly frustrating."

Missed connections

From the *San Diego Union-Tribune*, March 2, 2006, in a Letter to the Editor ("Transit schedules are a nightmare").

In addition to the common complaints about public transit not being on time and going where people want to go, the biggest problem I have had is in making connections.

Example 1: I am in class at Cajamarca College until 8:50 p.m., but the No. 858 bus departs at 8:44 p.m., leaving me to wait until 10:02 p.m. As a result, I don't get home until 11:30 p.m., then have to leave again at 6:30 a.m. to be at my office in time for work.

Example 2: I often have meetings in University Heights and transfer from the Green Line trolley to the No. 11 bus at San Diego State University. The No. 11 leaves at about the same minute that the trolley arrives, with no time to get upstairs to catch it.

Example 3: I can't begin to count how many times I've seen the No. 855 bus leaving Grossmont Transit Center right as the trolley arrives. The point is that the schedules don't seem to take each other into consideration. And just for fun, try to get from Santee Town Center to Scripps La Jolla.

Scott Weselis, *Santee*

Lack of amenities

From *BBC.co.uk* on October 28, 2005, by a participant Gerry Chatham.

Train announcements are rare. Customer services are a joke. Trains always packed – lucky to get a seat on some of the newer trains and if you do – you're at risk of catching DVT because of lack of leg room.

Three out of my five morning trains this week were either delayed or cancelled. Some (not all) staff are rude/unhelpful; lack of staff at train stations. Inability to purchase a ticket on a Monday morning as ticket offices are now like ghost towns. No visible security presence at stations late at night. Dirty stations; expensive fares. ...

It sounds like I am being very negative – I have good reason. Kent is a lovely place to live. It would be excellent if *we, the customers*, were offered value for money, clean trains that run on time, and overall a train company that puts the customers first on their list of priorities. Just look at the comments left by everyone else here, then make up your own mind.

Bus bunching

From the book by Larson and Odoni (1981), taken from a Sunday edition of the *Washington Post*.

I have long been trying to discover why Metro buses on Wisconsin Avenue are regularly bunched during rush hours instead of being evenly spaced. It is a common experience to wait for 10 or 15 min in rain or snow, and then find three or four buses coming along nose to tail.

As Metro refuses to reply to letters on this subject, I can only assume that it has something to hide.

From the passenger's point of view, the advantages of even spacing are obvious – shorter waits and buses that are evenly filled, instead of being packed up at the front and half-empty behind. Traffic congestion is also eased.

Evidently, either the company or its drivers must prefer bunching. I continue to wonder why.

E. Peter Wright

The aforementioned citations cover four service reliability problems from a broader list: misleading timetables, missed connections, lack of amenities, and bus bunching. In a general sense, a service reliability problem can be defined as degradation to system performance because of (1) uncertainties in the operating environment, (2) lack of suitable data for efficient operations planning, (3) improper service design, (4) improper service monitoring and control, and (5) failures in executing designed schedules. This chapter attempts to highlight the main ingredients of transit service reliability and provide some insight into possible remedies to alleviate problems.

18.2 MEASURES OF RELIABILITY AND SOURCES OF UNRELIABLE SERVICE

Prior to analyzing the elements of transit reliability, we ought to view acceptable measures of both reliability and perceived reliability and the sources of unreliable service. Abkowitz et al. (1978) defined service reliability as the invariability of service attributes that influence the decisions of travelers and transit providers. This section, too, focuses its examination on the variability of attributes that are of concern to passengers and agencies.

18.2.1 Attributes and measures of reliability

Transit-related attributes that vary by time or space may be distributed. Therefore, the (statistical) characteristics of the distributions form the base for constructing measures of reliability. Measures such as mean (average) value and variance, coefficient of variation (ratio of standard deviation to mean), and percentage of observations for a value greater than the mean can represent the compactness and skewness measures of an attribute's distribution. The following are three lists of attributes: from the passenger's and agency's perspectives and exogenous attributes, all of which tend to vary by time of day, day of week, week of season, and space.

Reliability attributes of concern to passengers

- Waiting time
- Boarding time
- Seat availability
- In-vehicle time
- Alighting time
- Total travel time
- Transfer time
- Missed connections
- Pretrip information time
- Pretrip time required for changes in access path

Reliability attributes of concern to the agency

- Dispatching according to schedule adherence
- On-route schedule adherence
- Headway distribution
- Individual-vehicle headway
- Load-count distribution
- Individual-vehicle load count
- On-time pullout
- Missed trips
- Breakdowns
- Late (crew) report (arrival)
- Driver proficiency
- Dispatcher and street-inspector proficiency

Reliability exogenous attributes

- Traffic delays
- Road and other accidents
- Weather

Figures 18.1 and 18.2 illustrate the main aforementioned attributes, the exogenous attributes being associated with the agency. Figure 18.1 shows the process in which the attributes of concern to passengers take place; if each attribute's probability of complying with passengers' expectation is high, then the service is perceived as reliable. Similarly, if the probability that a given run will comply with the agency's expectation is high for each attribute in

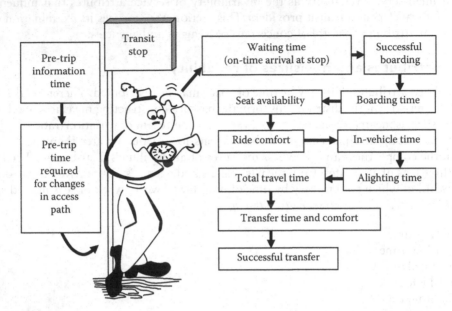

Figure 18.1 Flow diagram of reliability attributes of concern to the passenger; sources of unreliable service are characterized by high variance.

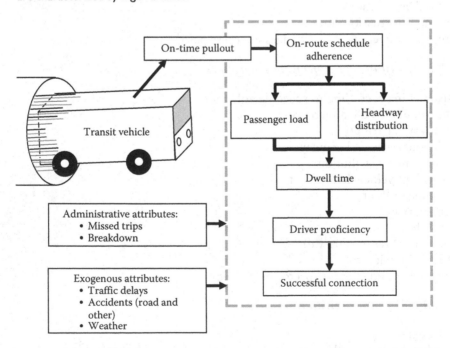

Figure 18.2 Flow diagram of reliability attributes of concern to the agency; sources of unreliable service are characterized by high variance.

Figure 18.2, then the agency may perceive the service as reliable. High probability coincides with small variance for each attribute.

Conceptually, the importance of attributes being negatively or positively perceived by individual passengers depends on their preferences. For instance, frequent users will place smaller value on reliable pretrip information time than will infrequent users. In addition, there are quality-based attributes that are difficult to quantify. These attributes especially concern passengers' satisfaction. Figure 18.1 presents ride comfort as one of these attributes; there are also expectations in regard to the existence and proper functioning of on- and off-vehicle amenities.

Concerning reliability-related standard measures, Figure 1.4 contains five such standard items, their common criteria, and corresponding remarks as follows.

Standard item	Typical criterion range[a]	Remarks
Schedule adherence[b]	Min 80% *on time* (0–5 min behind schedule) at peak period, 90% otherwise	This guideline is usually relaxed for short headways.
Timed transfer	Max of 3–8 min vehicle wait at transfer point	Used more in smaller agencies.
Missed trips[b]	Min 90%–95% of scheduled trips are OK	Missed trips may also not comply with trip-reliability criterion.
Passenger safety[b]	Max 6–10 passenger accidents per 10^6 passengers; max 4–8 accidents per $1.6 \cdot 10^5$ vehicle-km	Depends on updated safety data.
Public complaints	Limits on # of complaints per driver/pass./time period	Public comments and complaints always received.

[a] Reflects data mainly from the United States.
[b] Standards commonly in use.

These five standards are only examples, and more can be established from the lists of attributes; their criteria may vary by time, space, and agency.

Another theme affecting the reliability of system performance is maintenance reliability. Examples of maintenance-based reliability attributes are the following:

- Missed trips
- Kilometers and hours of service lost because of road calls
- Number of vehicles not available for service
- Number of late starts
- Number of vehicles overdue for inspection
- Number of operator-trouble reports
- Absenteeism

It is important to analyze these attributes historically and to disaggregate information by type of service and equipment.

18.2.2 Sources of unreliable service

Transit reliability problems can stem from a number of service factors. Abkowitz et al. (1978) divided these factors into two groups: *environmental* and *inherent*. The first group includes such factors as traffic signals, changes in traffic conditions, variation in demand, and availability of crew and vehicles on any given route and day. The second group includes setting and distribution of waiting times, headways, travel times, boarding and alighting times, and transfer times on any given route and day.

Some causes of unreliable service are chronic and known in advance; suitable planning and adjustments can address these causes. Other causes are unpredictable in nature and require real-time responses, preferably taken from a library of options. The sources of unreliable service may be found in the following lists of indicators of developing reliability problems.

Planning indicators of developing reliability problems

The items in this list, unlike those in the other lists, do not always/necessarily create reliability problems, but can serve as diagnostic indicators:

- Long urban route
- Network of routes requiring many individual origin–destination (O-D) transfers
- Lack of feeder services
- Short spacing between stops
- Problematic stop location
- Single daily average running (travel) time
- Sticking, in principle, to even headways
- Different forms of fare payment
- Poor in- and off-vehicle amenities
- Poor security

Operational indicators of developing reliability problems

- Missed trips
- Late pullouts
- Poor on-time dispatching
- Significantly late/early trips
- Bunching
- Uneven loads
- Overloaded vehicles
- Unpredictable passenger demand
- Missed transfers
- High variance of scheduled headway and/or running (travel) time
- Insufficient/extended layover time
- Absenteeism
- Road calls
- Breakdowns
- Passenger complaints
- Dispatcher complaints
- Driver complaints
- Bad press
- Road and other accidents

Maintenance indicators of developing reliability problems

- Lack of vehicles available for service
- Long hours/kilometers between preventive maintenance activities
- Large number of vehicles overdue for inspection
- Large percentage of old vehicles
- Inadequate replacement policies and contingency plans
- Poor-quality level of spare parts
- Intensive vehicle use
- Poor vehicle design
- Inadequacy of maintenance facilities

- Lack of data on vehicle histories and maintenance effectiveness
- Improper maintenance-engineering staff

Troubleshooting in all the earlier mentioned three themes in terms of suitable recurrent analysis or even in the form of a checklist can eliminate the need for diagnoses and improve service reliability.

Finally, we will describe one of the most irritating phenomena in urban bus operations, called *bunching* or *pairing* (see passenger complaints in Section 18.1). Undoubtedly, bus bunching is a major cause of unreliable service. Figure 18.3 describes three possible sources of the bunching phenomenon: (1) delayed vehicle, (2) early-dispatched and/or speeding

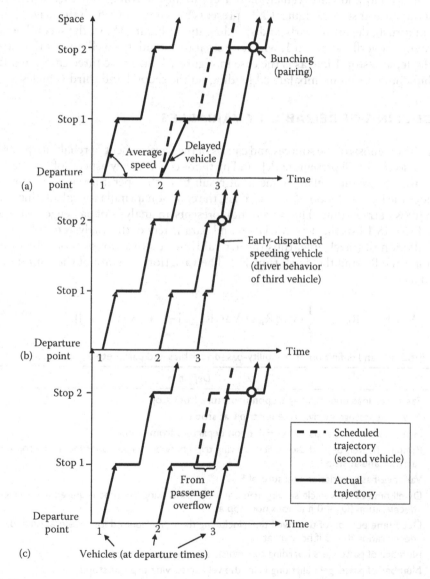

Figure 18.3 Three typical processes creating a pair of vehicles (at stop 2), notwithstanding scheduled even headways and random passenger arrivals: (a) with second vehicle delayed because of traffic, (b) with third vehicle dispatched early and speeding up because of driver behavior, and (c) with extra demand at stop 1 for the second vehicle.

vehicle, and (3) unexpected passenger overflow. Each of these sources or some combination of them can lead to the creation of two or more buses arriving nose to tail.

The base of Figure 18.3 consists of an even-headway timetable and random passenger arrivals, from which three vehicle trajectories are shown on a time–space diagram; therefore, the slope of each trajectory between two adjacent stops constitutes the average vehicle's speed. In all three cases described, bunching occurs between the second and third vehicles at stop 2. Figure 18.3a illustrates the case in which traffic conditions cause the second vehicle to slow down before stop 1. This slowing down becomes more pronounced at each successive stop, as more and more passengers accumulate after the departure of the first vehicle. Meanwhile, the third vehicle finds fewer and fewer passengers waiting (shortened headways between the second and third vehicles), and eventually, it comes together with the second vehicle, in our case at stop 2. Figure 18.3b presents a scenario in which the driver of the third vehicle is in a rush, departing early and speeding up. In Figure 18.3c, the second vehicle confronts passenger overflow at stop 1, which forces it to extend its dwell time at that stop. Thus, this vehicle departs stop 1 late as is the case in Figure 18.3a. These three cases, in addition to creating bunching, result in imbalanced loading on the second and third vehicles.

18.3 MODELING OF RELIABILITY VARIABLES

Following the diagnosis of the sources and causes of the development of reliability problems, this and the next sections will present models and methods covering key aspects of service reliability. These key aspects are time-related issues that result from the operating environment. This section provides the basic notation of the variables, the common formula for calculating dwell time, and an analysis of travel time. The next section presents an analysis of passenger waiting time at the stops. Table 18.1 lists the main variables and parameters of the analysis to follow.

The prediction of travel time, thus the arrival time at stops, consists of this information: the running time R_{ik} and the dwell time D_{ik}. The variation measure of the running time can be defined as

$$r_{ik} = \frac{1}{2} \text{Var}[R_{ik} - R_{(i-1)k}] = \frac{1}{2} \{ \text{Var}[R_{ik}] + \text{Var}[R_{(i-1)k}] - \text{Cov}[R_{ik}, R_{(i-1)k}] \} \tag{18.1}$$

Table 18.1 Notation and definition of reliability-based variables and parameters

Notation	Definition
P_{ik}	Passenger load onboard trip i, upon departure from stop k
λ_{ik}	Average passenger-arrival rate for trip i, at stop k
H_{ik}	Headway between trips i and i − 1, upon departure from stop k
R_{ik}	Running (travel) time of the vehicle serving trip i between departure time from stop k − 1 and arrival time at stop k
r_{ik}	Variance-based variation measure of R_{ik}
D_{ik}	Dwell time of the vehicle serving trip i, at stop k, including the time required for acceleration and deceleration ($D_{ik} = 0$ if it does not stop at k)
b	Dead time portion of the dwell time, including the time required for acceleration and deceleration (b = 0 if no stop at k)
B_{ik}	Number of passengers boarding the vehicle serving trip i, at stop k
A_{ik}	Number of passengers alighting from the vehicle serving trip i, at stop k
Δ_B	Marginal dwell time per boarding passenger
Δ_A	Marginal dwell time per alighting passenger

where Var[...] and Cov[...] designate the variance and covariance of the random variable in brackets. If R_{ik} and $R_{(i-1)k}$ are uncorrelated, r_{ik} is just the variance of the running time; r_{ik} is also the covariance between $(R_{(i+1)k} - R_{ik})$ and $(R_{ik} - R_{(i-1)k})$.

The correlation expressed in Equation 18.1, which exists in real life, between the running times of two consecutive vehicles can be considered in analytical or simulation models.

18.3.1 Dwell time

The dwell time depends on the number of boarding and alighting passengers. Usually, it can be expressed by the following linear model:

$$
D_{ik} = \begin{cases}
b + \Delta_B \cdot B_{ik} + \Delta_A \cdot A_{ik}, & \text{single-door vehicle,} & \text{if } B_{ik} > 0 \text{ or } A_{ik} > 0 \\
0, & \text{single-door vehicle,} & \text{if } B_{ik} = A_{ik} = 0 \\
b + \max(\Delta_B \cdot B_{ik}, \Delta_A \cdot A_{ik}), & \text{double-door vehicle,} & \text{if } B_{ik} > 0 \text{ or } A_{ik} > 0 \\
0, & \text{double-door vehicle,} & \text{if } B_{ik} = A_{ik} = 0
\end{cases}
\tag{18.2}
$$

where the dead time b is made up of the time to open and close the door(s), the time taken between sequential alighting and boarding events on single-door vehicles, and the time taken by the driver to check the traffic: it includes, by convention, the penalty for stopping because of the consequent deceleration and acceleration.

Equation 18.2 assumes that boarding and alighting passenger flows are distinct for double-door vehicles. For a given number of boarding and alighting passengers, we can assume that there is no intrinsic randomness. If there is (e.g., different groups of passengers carrying different size of luggage), its variance can be assumed to be fixed and thus added to the running-time variation in Equation 18.1.

The following factors influence dwell time:

- Form of payment
- Entrance characteristics
- Conflicts between boarding and alighting passengers
- Available space near drivers
- Traffic-flow characteristics
- Passenger characteristics

The following are practical (rounded) values of the marginal dwell time per boarding and alighting passenger; these and the dead-time component of Equation 18.2 were extracted from a number of studies, especially in the United Kingdom (Balcombe et al., 2004).

Boarding time per passenger

Δ_B = 1.5 s—pay conductor
Δ_B = 2.5 s—flat fare, with change
Δ_B = 3.0 s—flat fare, with change
Δ_B = 5.0 s—with change
Δ_B = 6.5 s—automated fare box

Alighting times per passenger

Δ_A = 1.5 s—without hand baggage and parcel
Δ_A = 3.0 s—moderate hand baggage
Δ_A = 5.0 s—considerable baggage

Dead time per stop

b = 2.0 s—pay conductor
b = 5.5 s—otherwise, but with extra space near driver
b = 7.0 s—without extra space

18.3.2 Analysis of travel time

From a reliability standpoint, one of the more crucial input elements in the planning process is vehicle travel (running) time. Often the term *travel time* is broken down into the following string: walk from origin, wait for vehicle, ride in vehicle, and walk to destination. Here, we will apply this term only to the riding or running time and exclude layover (recovery) time. The majority of the agencies do not build slack time into the scheduled travel time to compensate for the variability of vehicle travel time; average travel time is then utilized for all scheduling tasks. For convenience, let R_i be the travel time of the vehicle serving trip i across the entire transit route. This travel time depends on the trip time (hour, day, week, season), number of passengers, as well as on the habits of each individual driver.

Jeong and Rilett (2005) reviewed prediction models for travel time, including historical data-based models, regression models, time series models, and neural network models. They selected artificial neural network models to predict R_i, given real-time information on traffic congestion and transit-vehicle dwell times at stops. The researchers found that, given the availability of real-time data, their model outperformed both the historical data-based model and the regression model in terms of prediction accuracy. The increasing use of advanced public transit/transportation systems (APTS) (see Section 18.5) by transit agencies allows for better predictions of both R_i and R_{ik} in real time; this undoubtedly can be used for reducing passenger uncertainties at stops and on the vehicles and eventually for improving service reliability.

While Sections 18.5 and 18.6 focus on new technologies and techniques to improve real-time operation, the present section provides a practical, statistical-based analysis for deploying R_i at the planning level. It is based on Ceder (1981), which utilized four major procedures: (1) exclusion of outliers, (2) division of the day into intervals, (3) union of intervals, and (4) union of days for weekly and seasonal cross sections.

18.3.2.1 Description of general method

The main objective of data processing is to create a database on vehicle travel times for different cross sections of the day, week, and season (d-w-s). From a statistical viewpoint, the existence of a large data bank that can be systematically updated means that an *analysis of variance* (concerned with differences between means of groups) can be carried out (based on the normality distribution assumption) in order to estimate the effects of d-w-s (the independent variables) on the dependent variable, R_i. Afterward, the independent variables that were found to affect the dependent variable (at a desired level of significance) can be analyzed with the use of *contrasts* (comparison of the differences between pairs or combinations of means). Finally, it is possible to apply a multiple variable regression analysis.

In order to effectively carry out an analysis of variance on d-w-s, a data bank must be accumulated over a yearly period, on the assumption that there will be no physical changes along the transit route or in the location or in quantity of stops. Because it is well known that transit routes are liable to physical changes in a dynamic fashion, it is questionable whether the data bank for R_i values can rely on a yearly base. Moreover, it is desirable that the process that determines the R_i relationship allows for the intermediate involvement of those responsible for data collection. This involvement can be expressed by the identification of outliers, whose cause is known, and by the practical decisions relating to changes in statistical criteria, which are combined in the process, for different transit routes.

As a result, a number of criteria can be utilized for characterizing outliers; dividing the day into time used; dividing the week into homogeneous days; and dividing the year into seasons. If no physical changes occur for a given transit route over a yearly period and a data bank for R_i values exists, there is a possibility of carrying out an analysis of variance and contrasts for d-w-s. Once again, it should be noted that the objective is not to build a statistical model for simulation or control, but to construct a value system for R_i that can be of greater aid in realistic planning than can a single mean value over a specified period of time (e.g., year as some transit agencies do).

A possible method of analyzing R_i consists of four main components: (1) exclusion of outliers (OUTLIER procedure), (2) division of day into intervals (INTERVAL procedure), (3) union of intervals (UNIT procedure), and (4) union of days for weekly and seasonal cross sections (UWEKDAY-USEAS procedure).

The first three components relate to the data on a daily basis, and the fourth component serves as a tool for determining the division of weeks and seasons. The procedure starts with assembling data into two sets: (1) general data, which include the number of vehicles in the set, route number, origin, destination, the day (or days) of the week, and the season, and (2) specific data, which include vehicle departure time, travel time, and type of vehicle.

The outliers are deleted from these data, and then a number of methods are tested for dividing the day into intervals (division for each hour, each 2 h, each 3 h, and one daily average). Then statistical tests are conducted to determine the possibility of unifying the intervals. The procedure continues for the series of days, and then the possibility of unifying the days is examined. The next step checks the possibility of unifying a number of days from different seasons; this is possible if the division of days in each season (or at least those chosen for unification) is similar. More details are shown in Ceder (1981).

18.3.2.2 Example of bus data set

For a given bus route (in Haifa, Israel) departing at a frequency of either every 15 or 30 min, data were collected for 2 days (on the same day of the week for 2 weeks) at a total of 126 data points. The data points are shown in Figure 18.4, as well as the results obtained after the INTERVAL and UNIT components. In the OULTLIER procedure, five outliers were found and are marked on the upper portion of Figure 18.4; they also appear at the start of Table 18.2. The latter exhibits a computer record.

Following a run of the INTERVAL procedure, the method designated C in Table 18.2 was chosen, that is, a division of the day into 3 h intervals, when the first hour is considered separately as shown in the upper portion of Figure 18.4 and in Table 18.2. The record in Table 18.2 is accompanied by F values, where

$$F = \frac{\text{Found variation of the group averages}}{\text{Expected variation of the group averages}}.$$

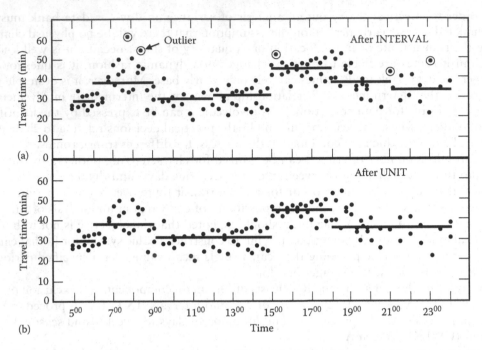

Figure 18.4 An example of travel time data points for one bus line with mean values determined by the INTERVAL procedure in (a) and by the UNIT procedure in (b).

Table 18.2 Example of data analysis using outlier, interval, and unit for a given bus route; summary of computer record

Stage	Description
Input	126 data points (see Figure 17.4)
Outlier	1. Departure 7:45, travel time = 62 min
	2. Departure 8:30, travel time = 55 min
	3. Departure 15:15, travel time = 53 min
	4. Departure 21:00, travel time = 44 min
	5. Departure 23:00, travel time = 49 min
Interval	Method A: division by 1 h
	Method B: division by 2 h periods
	Method C: division by 3 h periods
	Method D: single average for the whole day
	Steps of comparisons: (A vs. D) → (A vs. C) → (A vs. B), until significant F difference found at the 0.05 level
	1. A vs. D: F-calculated = 1.92, F-table = 1.35; conclusion: continue
	2. A vs. D: F-calculated = 1.15, F-table = 1.35; conclusion: accept
Unit	Same-means hypothesis is based on t-test at the 0.05 level

	Integer interval	Actual interval	Mean value (min)	Standard deviation (min)
	$5:00 < R_i \leq 6:00$	5:00–6:00	29.4	3.5
	$6:00 < R_i \leq 9:00$	6:15–9:00	38.2	8.1
	$9:00 < R_i \leq 15:00$	9:30–15:00	30.1	6.1
	$15:00 < R_i \leq 18:00$	15:15–18:00	44.8	4.5
	$18:00 < R_i \leq 24:00$	5:00–6:00	36.9	7.0

The data then proceed to the UNIT procedure. Of the seven intervals in method C, two are united and only five intervals remain as shown in the lower portion of Figure 18.4 and in Table 18.2. This example does not include the UWEKDAY-USEAS procedure.

The last five intervals (each with a mean and standard deviation) are transmitted to the bus company's scheduling department. This information can be transmitted either automatically or manually. Determination of the R_i value for planning purposes will depend on the degree of certainty that the bus will arrive at its destination on time. Table 18.2 contains both the mean value of R_i and its standard deviation for two possible objectives: (1) to utilize the standard deviation for introducing slack time to account for the variability of R_i and (2) to allow the scheduler more flexibility in creating/adjusting blocks, which in certain cases are already performed by a specific driver.

18.4 PASSENGER WAITING TIME AT A STOP

This section attempts to clarify some of the probabilistic issues of passenger behavior at transit stops. The section consists of two parts. First, the distribution of waiting times is analyzed for the case of random passenger arrivals. This distribution is described by the two families of headway distributions, the deterministic and the exponential, discussed in Section 11.2.2, which can approach the two extremes. Second, the mean waiting-time formulation is interpreted using observed data from two studies; in addition, two explicit expressions for the mean waiting time are derived for the case of a suburban rail station. The majority of this section follows Ceder and Marguier (1985).

18.4.1 Waiting-time distributions

Equation 11.1 presented the basic relationships between the expected (mean) waiting time $E(w)$ and the mean $E(H)$ and variance $VarH$ of the time headway. This known equation for $E(w)$ is based on two assumptions: (1) passengers can always board the first vehicle to depart (no overloading situations) and (2) the passenger random-arrival rate at the stop is independent of the vehicle-departure process and constant over the period. The following is the derivation of Equation 11.1 on a per-route basis, in which $f(H)$ is the probability density function (PDF) for headway H (from a passenger's perspective, compared to $f_H(t)$, which is from the system's perspective); $\overline{w}(\overline{H})$ is the mean waiting time per passenger for $\overline{H}$, which is H/2; and λ is the mean arrival rate of passengers, from which λH is the number of passengers arriving in the course of H minutes:

$$(\text{Total number of passengers arriving at the transit stop}) = \int_{H=0}^{\infty} \lambda \cdot H \cdot f(H) dH$$

$$\begin{pmatrix} \text{Mean waiting time} \\ \text{for passengers} \\ \text{arriving randomly} \end{pmatrix} = E(w) = \frac{\text{Total waiting time}}{\text{Total number of passengers}}$$

$$= \frac{\int_{H=0}^{\infty} \lambda \cdot H \cdot (H/2) \cdot f(H) dH}{\int_{H=0}^{\infty} \lambda \cdot H \cdot f(H) dH} = \frac{\lambda/2 \, E(H^2)}{\lambda \cdot E(H)}$$

The use of $Var(H) = E(H^2) - E^2(H)$ results in the known mean waiting-time formula shown in Equation 11.1:

$$E(w) = \frac{E(H)}{2}\left[1 + \frac{Var(H)}{E^2(H)}\right]$$

Thus, the mean passenger waiting time is minimized at half the headway when the variance of the headway is zero.

In practice, assumptions (1) and (2) may not always hold, which would then argue in favor of even-load headways, as was analyzed and proposed in Chapter 4, rather than even headways. For instance, there are surges in the passenger-arrival rate at the start and end of a factory shift and a school day.

Under the assumption of random passenger arrivals at the stop, the following quality relates the waiting-time distribution to the headway distribution: let $f_w(t)$ be the waiting time PDF:

$$f_w(t) = \frac{\int_t^\infty f_H(u)du}{E(H)} = \frac{\bar{F}_H(t)}{E(H)} = F \cdot \bar{F}_H(t) \tag{18.3}$$

where

    F is the frequency of service (vehicle-arrival rate) $\bar{F}_H(t) = 1 - F_H(t)$
    $F_H(t)$ is the cumulative distribution function of the headway H (system's perspective as in Chapter 11)

The derivation of this relationship, which is shown in Larson and Odoni (1981), is related to the phenomenon commonly referred to as *random incidence*.

Particular examples of the application of Equation 18.1 are provided by the deterministic-headway and exponential-headway cases. The following is obtained for the deterministic headway in which H = 1/F:

$$\bar{F}_H(t) = \begin{cases} 1, & t \le \dfrac{1}{F} \\[2ex] 0, & t > \dfrac{1}{F} \end{cases}$$

hence,

$$f_w(t) = \begin{cases} \dfrac{1}{F}, & t \le \dfrac{1}{F} \\[2ex] 0, & t > \dfrac{1}{F} \end{cases} \quad \text{and} \quad E(w) = \frac{1}{2F} = \frac{E(H)}{2}.$$

In this case, the waiting time is uniformly distributed between 0 and the fixed headway 1/F.

For the exponential-headway case, the following is obtained:

$$f_H(t) = F \cdot e^{-F \cdot t}, \quad \bar{F}_H(t) = e^{-F \cdot t}$$

hence

$$f_w(t) = F \cdot e^{-F \cdot t}$$

and

$$E(w) = \frac{1}{F} = E(H).$$

In this case, the waiting time is exponentially distributed with the same parameter F on the headway.

The application of Equation 11.5 to the *power* family of headway distributions can be obtained by integrating $\bar{F}_H(t)$ in Equation 11.5 into Equation 18.1

$$f_w(t) = \begin{cases} F\left(1 - \dfrac{1-C^2}{1+C^2} \cdot F \cdot t\right)^{2C^2/1-C^2}, & 0 \le t \le \dfrac{1+C^2}{1-C^2} \cdot \dfrac{1}{F} \\ \\ 0, & t \ge \dfrac{1+C^2}{1-C^2} \cdot \dfrac{1}{F} \end{cases} \tag{18.4}$$

For the second family (gamma distributed) of headway distributions, Equation 11.6 can be integrated in a close form for values of C^2, such that $1/C^2$ is an integer. In these cases, the gamma distribution is an Erlang distribution.

Nonetheless, the results for these cases also give a good sense of what the results will be for other intermediate values of C^2. Integrating $f_H(t)$ into Equation 11.6 and applying Equation 18.3 yields

$$f_w(t) = F \cdot \frac{(F/C^2)^{1/C^2}}{\Gamma(1/C^2)} \sum_{k=0}^{(1-C^2)/C^2} \frac{(F \cdot t)/C^2}{k!} \cdot e^{(-F \cdot t)/C^2} \cdot \frac{(1-C^2)/C^2!}{(F/C^2)^{1/C^2}}$$

$$= F \cdot \sum_{k=0}^{(1-C^2)/C^2} \frac{((F \cdot t)/C^2)^k}{k!} \cdot e^{(-F \cdot t)/C^2}, \quad t \ge 0 \tag{18.5}$$

The two families of waiting-time distributions obtained are illustrated in Figure 18.5 in two parts. Figure 18.5a shows the time distributions derived from the headway *power* distributions for a mean headway of 10 min and for service reliability measures C^2 of 0, 1/4, 1/3, 1/2, and 1. Figure 18.5b shows the waiting-time distributions derived from the headway gamma distributions for the same values of $E(H)$ and C^2 as in Figure 18.5a. For the extreme situations, $C^2 = 0$ and $C^2 = 1$, both families give identical curves because these values must correspond, respectively, to the deterministic-headway and exponential-headway cases.

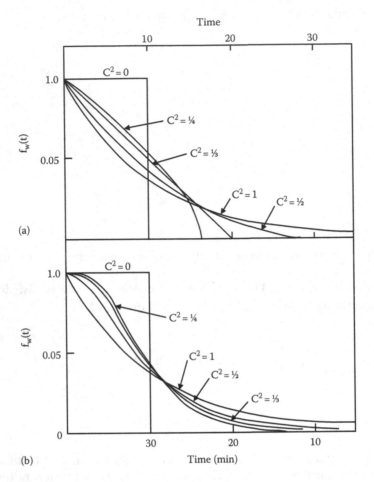

Figure 18.5 Various waiting-time distributions showing (a) the *power*-distributed headways and (b) the gamma-distributed headways for a mean headway of 10 min and different values of C^2.

For $C^2 = 0$, the uniformity of the waiting-time distribution (between 0 and 10) can be noticed. For $C^2 = 1$, the curve is exponential. For both families, the waiting-time probability densities are always decreasing. This is a general property, implied by Equation 18.3.

18.4.2 Average waiting time

Consider a transit route in which the vehicle departure times follow the timetable and the traffic flow is not congested. In such situations, which are characterized by relatively low (between days) variability, some passengers will attempt to follow the arrival pattern of vehicles in order to reduce their waiting time.

Jolliffe and Hutchinson (1975) observed buses in suburban London with published time-tables and examined between-day variations. They found no correlations between the between-day variability of bus-departure times and headways. When they considered the (between-day) expected (mean, average) waiting time for passengers arriving at the bus stop at time t, they found that the waiting-time function had a minimum wait, w_{min}, before the average departure time of each bus, the corresponding t (to w_{min}) being the optimal arrival time. The authors showed how, based on departure-time variability, it is possible to compute w_{min} theoretically. The observed average waiting time $\bar{w}$ was found to be higher than

w_{min} and lower than the expected value E(w) given by Equation 11.1, indicating that some passengers do time their arrival. It was also concluded that some passengers never wait—if they see a bus coming, they rush to board it. Jolliffe and Hutchinson (1975) proposed to represent the passenger-arrival pattern by considering three proportions: (1) passengers who arrive coincidentally with a bus (see and rush); (2) a proportion p for passengers who, based on the timetable and experience, arrive at the optimal time and wait only for w_{min}; and (3) passengers who arrive at random and wait on average for E(w).

Jolliffe and Hutchinson found the proportion of coincidental arrivals to be relatively low (16%), similar to what other studies have observed (Ceder and Marguier, 1985). The mean waiting time for non-coincidental arrivals could then be expressed as follows:

$$\bar{w} = p \cdot w_{min} + (1-p) \cdot E(w) \qquad (18.6)$$

Jolliffe and Hutchinson used w_{min} to determine the proportion p and proposed this relationship:

$$p = 1 - e^{-\alpha g} \qquad (18.7)$$

where
$g = E(w) - w_{min}$
α is a constant (found by them to be 0.131 for the peak and 0.015 for the off-peak period data)

Let us consider a suburban rail station. Assume that the service is very reliable, say $C^2 = 0$, with posted timetables, and that passengers cannot see coming trains in advance. Using the earlier approach, which should be appropriate in this case, and inserting Equation 18.7 into Equation 18.6, one obtains

$$\bar{w} = E(w) - g(1 - e^{-\alpha g}) \qquad (18.8)$$

For this deterministic headway, the expected waiting time (of randomly arriving passengers) is E(w) = (E(H))/2; and Equation 18.8 becomes

$$\bar{w} = \frac{E(H)}{2} - g(1 - e^{-\alpha g}) \qquad (18.9)$$

Huddart (1973) in London observed, as expected, that for transit services with small head-ways, passengers arrive at random. However, if the service is sufficiently infrequent, passengers may take advantage of this information to arrive at the boarding point just before the vehicle is due to arrive. Huddart considers this situation to be rare for urban bus services but fairly common at a suburban railway station.

If we consider the passenger-arrival rate $\lambda(t)$ as monotonically and continuously varying with time, as shown by the typical passenger-arrival pattern given by Huddart (1973), the mean waiting time can be expressed as

$$\bar{w} = \frac{\int_{-E(H)}^{0} -t \cdot \lambda(t)\, dt}{\int_{-E(H)}^{0} \lambda(t)\, dt} \qquad (18.10)$$

For example, the data in Huddart (1973) can be fitted into a regression analysis for an exponential function of the type $\lambda(t) = a \cdot e^{bt}$, where a, b are fitted parameters. For different passenger-arrival rates between 3 and 15 min before a train departure, the following model was calibrated: $\lambda(t) = 52.5e^{0.3}$ with $r^2 = 0.95$ (measure of regression analysis, indicating a good fit). Thus,

$$\overline{w} = \frac{1}{b} - \frac{E(H) \cdot e^{-bE(H)}}{1 - b \cdot e^{-bE(H)}} = \frac{E(H)}{2} - \frac{b \cdot E(H)^2}{12} + \frac{b^3 \cdot E(H)^4}{720} + \cdots \tag{18.11}$$

The results of the two approaches are exhibited by Equations 18.10 and 18.11. These two expressions do not have quite the same form, but both verify $\overline{w} = E(H)/2$ when passengers do not attempt to time their arrivals to reduce their waiting time, that is, for $g = b = 0$.

18.5 ADVANCED RELIABILITY-BASED DATA AND CONTROL

One of the known potential remedies for passenger complaints concerning reliability problems is the use of APTS. APTS offer new technologies to improve the mobility, convenience, and safety of transit passengers and to increase passenger demand. While it is still a challenge to use APTS efficiently in real time, an analysis of their data reveals weaknesses in regard to reliability. These weaknesses can be repaired by the transit personnel (management, planners, schedulers, dispatchers, mobile supervisors, and inspectors). This section describes the main APTS features and their required data.

The U.S. Department of Transportation (2003) issued a report on APTS based on a deployment tracking survey that was conducted over the Internet for 2002. This survey covered 593 transit agencies in the United States, representing about half of the existing agencies; they were asked about existing and planned APTS. In addition, the U.S. Federal Transit Administration issued a report by Hwang et al. (2006) on the state of the art of APTS, which emphasized the need for integrating APTS with emerging intelligent transportation systems (ITS) and ITS trends. Hwang et al. (2006), from lessons learned, stressed the need for the provision of better data, standards, and voice communication in order to (1) improve transit planning, maintenance, operations, and incident management and (2) facilitate coordination, integration, and interoperability with transportation providers and public safety organizations.

The main features contained in the U.S. survey and reports, some of which appear in Khattak and Hickman (1998), are assembled according system category in Table 17.3. Common APTS terms used in this table are AVL (automatic vehicle location), APC (automatic passenger counter), CAD (computer-aided dispatch), AFP (automatic fare payment), and ATIS (advanced traveler information system). Other known terms appearing in the literature, but not used here, are AVM (automatic vehicle monitoring), AVLC (automatic vehicle location and communication), and a different CAD (computer-assisted design).

Current practices in transit agencies show that sufficient data seldom exist for both service operations planning and improving reliability problems. Manual data collection is costly and, consequently, must be used sparingly. As a result, detailed information on passenger demand and service characteristics is generally not available at the route level. Without this information, the efficient deployment of transit service commensurate with demand is impossible. Thus, a major reason for transit agencies to be interested in the use of APTS data is the hope of gaining badly needed information at a greatly reduced unit cost. The resulting data are expected to improve utilization of vehicles and crew, as well as to resolve some reliability problems.

Table 18.3 Advanced public transit systems applications

APTS applications	System name	Features
Monitoring systems	AVL	Automatic position determination by means of dead reckoning (using the vehicle's odometer and compass), GPS, signposts (transmitted signals picked up by vehicle), ground-based radio, and real-time reporting
	APC	Automatic counts of boarding and alighting passengers (e.g., use of treadle mats or infrared beams placed by the door)
	Advanced communication	Digital radio (binary information) and/or trunked radio (computer selection of frequency)
	CAD	Dispatch software as a support tool
	Automated operations of software	Software that displays vehicle positions, vehicle and agency data, and communications information
	Silent alarm	Emergency signal triggered by the driver; possible hidden microphone for dispatcher/other listeners, onboard surveillance camera
	Vehicle component monitoring	Automatic remote measurement of engine oil pressure, engine temperature, electrical system, tire pressure, etc.
Fare payment systems	AFP	Payment by smart card, magnetic stripe card, credit/debit card, etc.
	Multicarrier fare integration	Fare scheme that covers more than one transit service provider
Traveler information systems	ATIS	Pretrip, in-terminal, and in-vehicle real-time passenger information
	Multimodal traveler information	Available information covering multiple modes (i.e., transit and traffic or different transit modes)
Multipurpose information systems	Vehicle probe	Automatic data from transit vehicles for estimating traffic travel times and speeds and flow conditions
	Mobile data terminal	Wireless device that can send and receive information
	Information sharing	Sharing of information on traffic and incidents among agencies
	Mobility manager	Coordination of travel requests and vehicle dispatching (multiple agencies)
Traffic signal control	Manual priority	Extended green—activated by driver
	Automatic priority	Automatic extension of green phase

Useful, automatically collected data are extracted from AVL, APC, CAD, and AFP systems. Their features appear in Table 18.3. These data sources have the potential to enhance schedule efficiency and to improve service reliability.

A general program for developing operations planning applications is depicted in Figure 18.6. The program as a whole intends to serve transit routes with or without AVL/APC/CAD/AFP data. Its major objectives are (1) to improve management and operations by developing, improving, validating, and testing models and procedures for transit-operation planning; (2) to improve levels of service through increased reliability resulting from better control and response; (3) to improve productivity and efficiency by better matching supply and demand; (4) to reduce data gathering, processing, and reporting costs; and (5) to develop vital components for a management information system pertaining to operations and passenger behavior.

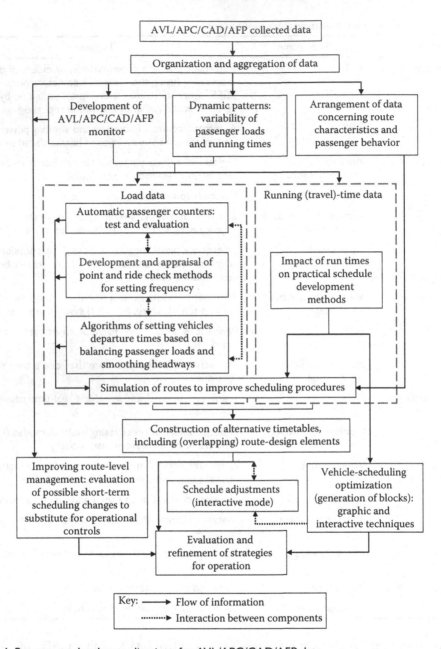

Figure 18.6 Program to develop applications for AVL/APC/CAD/AFP data.

The applications for AVL/APC/CAD/AFP data in Figure 18.6 are rooted in two streams of data flows: passenger-load data and vehicle running-time data. Some of the methods of handling the running-time data are discussed in Section 18.3. The use of load data and its appraisal appeared in Chapters 3 and 4. Moreover, AVL/APC/CAD/AFP data can improve route design, schedule adjustments, and vehicle scheduling. The outcome of best exploitation of the data is better operational strategies, which are required to reduce reliability problems.

In an implementation study utilizing AVL and APC systems, Kimpel et al. (2006) utilized operations data to improve schedules and reduce service reliability problems. The data used were recovered by the TriMet transit agency (Portland, Oregon, United States), using an automated CAD system. At TriMet, 100% of the bus fleet is equipped with AVL technology, while approximately 72% of the vehicles are equipped with APC. TriMet obtains bus-location information at regular time intervals and transmits it to two dispatch centers in real time. The data collection component of the AVL and APC systems includes arrival and departure times, dwell times, door openings, lift operations, and maximum speed since the previous stop, as well as the number of boarding and alighting passengers on APC-equipped vehicles.

The findings of Kimpel et al. (2006) show that the on-time performance measure (percentage arriving on time at stops) for 15 frequent-service bus routes rose from 73%–74% to 75.5%, which is slightly above the agency standard of 75%. In addition, they claim that TriMet's scheduling and operations management has benefited from the analysis of the AVL and APC data, especially in reports generating capabilities. The study concluded by recommending that efforts to improve schedule efficiency should center on reducing excess runs and layover times, as well as employing supervisory actions to reduce operator variability.

18.6 TECHNIQUES TO RESOLVE RELIABILITY PROBLEMS

Essentially, there are three basic mechanisms for improving transit reliability: (1) improve planning and scheduling, (2) improve priority techniques for transit vehicles, and (3) improve operations control. Abkowitz et al. (1978) classified reliability-improvement strategies as preventive and corrective/restorative. Preventive strategies are fundamentally associated with mechanisms (1) and (2) and corrective/restorative strategies, with mechanism (3). TranSystems et al. (2006), in a study on attracting and retaining transit passengers, suggested that these mechanisms should address one or more of the following parameters: travel time, convenience, comfort, perceived personal security/safety, and perceived *image* of the transit system. What follows is a description of tactics, strategies, and possible actions in conjunction with each of the three mechanisms and an extended example of a holding strategy.

18.6.1 Improved planning and scheduling

Basically, all the methods presented in this book are intended to improve transit planning and scheduling; inherently better reliability measures are incorporated into the resultant improvement. Examples of specific actions, some of which can be found in TranSystems et al. (2006), are presented in the following lists, by category.

Area coverage actions

- Increased route coverage; new routes
- Old set of routes replaced by a new set
- Service expansion; new local circulators/shuttles
- New/improved feeder services
- New/improved timed transfers; improved route coordination
- New/improved transit centers

Route restructuring actions

- Interlinings introduced
- Route shortening, extension, realignment, and removal
- Revised operating strategies
- Route splitting

- Joining routes to form a single route
- New/improved zonal, express, and local services
- Reduced number of stops

Scheduling amendment actions

- Increased span of service; longer late-night and weekend service hours
- Increased service frequency
- Average even-load headways introduced
- Adjusted departure times to suit new connections
- Modified running (travel) times by time of day
- Increased layover time
- Deadheading (DH) trips introduced for necessary reinforcement
- Shifts in departure times introduced as a corrective strategy

General system's organization actions

- New/improved geographic, O-D, route, stop, and transfer-point database
- New/improved frequency, timetable, and transfer-time database
- New/improved stop, station, transit center, and park-and-ride amenities
- New/improved passenger amenities
- New/improved real-time service information
- New/improved vehicles
- Increased security and safety

These types of actions are typically implemented in an effort to seek a better, more reliable transit service.

TranSystems et al. (2006) described examples of successful actions taken in the United States. Here are two typical examples: (1) Washington Metropolitan Area Transit Authority extended late-night weekend hours on Metrorail from 02:00 to 03:00 (a.m.) on Saturday and Sunday mornings. The trial program, which ran for 18 months, was projected to attract an additional 3000 passengers; actually performance exceeded these expectations by 20%, and it was decided to make the extension permanent. (2) Bangor (Maine) Area Comprehensive Transportation System introduced a new fleet of low-floor buses for quicker, easier boarding and alighting; this action resulted in increased passenger demand, which rose by 8% between 2003 and 2004. It is customary in the United States to have an approximately 200 m distance between bus stops, compared with about 320 m in Europe. In Europe, moreover, one can usually find the name of the stop and updated timetable information at each bus stop.

18.6.2 Improved priority techniques for transit vehicles

Transit-vehicle priority techniques are extensive, and many bus-priority strategies have been demonstrated worldwide. Traditionally, priority is granted for transit-vehicle operation at stops, at intersections, and by preferential/exclusive lanes. Usually there is a trade-off between granting priority to buses and improving traffic flow for the other vehicles. Local authorities (especially those in elected positions) are often reluctant to provide this priority. In other words, with a drop of a humor, in order to convert a street lane into strictly a bus lane (thereby increasing bus reliability and reducing traffic congestion, by helping to switch from automobile to buses), one must first prove that there is no traffic congestion on that street; this is analogous to the common experience that in order to secure a loan, one must first show that he/she doesn't need it.

Priority at stops: Transit-vehicle stop locations at intersections should normally be designed to prefer the far-side location (after crossing the intersection). In this way, the bus will not block traffic intending to turn; it will have easier pull-in and pull-out maneuvers and will experience fewer conflicts with pedestrians. However, a near-side stop is more appropriate at locations where transit vehicles make a turn and where the crossing street is one way. Other useful strategies are to prohibit parking near stops, to grant pull-out priority (reducing the merging time with the traffic), and to extend the sidewalk at the stop location (eliminating a pull-out maneuver and easing the boarding process).

Priority at intersections: At intersections, transit priority can be divided into passive and active schemes. Traffic engineers usually employ passive priority at intersections in order to utilize four measures: (1) exempt transit vehicles from turning prohibitions so as to facilitate transit routes, (2) extend the green interval at signalized intersections for nonstopping transit vehicles, (3) divide the green interval into two parts within the same cycle, and (4) provide preference to streets carrying transit routes through YIELD and STOP signs. Active priority permits transit vehicles to preempt traffic signals, using in general one or a combination of the three following procedures: (1) immediate priority upon the arrival of the transit vehicle, (2) priority dependent on the crossing-street traffic queue, and (3) priority granted only to transit vehicles with late arrivals.

Priority through preferential/exclusive lanes: Preferential treatment to transit vehicles on street lanes can be categorized according to these three lists:

Type of preferential lane

- Exclusive curb lane
- Semi-exclusive curb lane (shared only with cars about to turn)
- Exclusive median lane (with stop island)
- Exclusive lane in the center of a street
- Transit-vehicle malls (known as bus malls; limited to pedestrians and buses)
- Exclusive freeway/highway lanes
- Ramp bypass (for entering a freeway/highway during traffic congestion)
- Congestion bypass (exclusive lanes to bypass traffic bottlenecks)

Integration with traffic flow

- With-flow lane (by pavement markings and sings; enforcement problems)
- Contraflow lane (easy enforcement)
- Exclusive lanes

Period of operation

- Single-peak operation
- Two-peak operation
- Permanent operation

Some exclusive lanes for transit vehicles are shared with high-occupancy vehicles (taxis, certain minimum number of people in a car, for encouraging carpools).

Another known type of priority involves bus rapid transit (BRT): a flexible, rubber-tire form of rapid transit that combines stations, vehicles, services, priority lanes, and

Table 18.4 Examples of transit priority results in six European cities

City (population)	Benefits gained
Athens (4 million)	• Reduction of travel time and its variance on bus lanes • 10% increase in patronage on some bus routes
Vienna (1.6 million)	• Reduction in travel time on bus lanes • Possible (not clear) increase in patronage
Munich (1.2 million)	• Reduction of 19% in travel time for tram priority at one traffic signal
Dublin (1 million)	• No reduction in travel time, but a reduction in its variance on bus lanes • Reduction in boarding times • Increased patronage with new fare scale • Increase in revenue • Reduction in the variance of travel time and headways using AVL system
Turin (1 million)	• Reduction in tram travel time and more reliable tram reliability using AVL (called SIS) system
Zurich (0.36 million)	• Increase of 19 km/h in average vehicle speed because of priority at traffic signals, AVL, and passenger-information systems

intelligent-system elements into an integrated system with a unique identity. TranSystems (2006) described BRT as a system of buses with such features as signal priority, dedicated right of way, automated and off-vehicle fare collection, automated information systems, level boarding, modern vehicles, bus shelters with enhanced amenities, and unique graphics identity (painted on the bus). Two known BRT systems in South America are found in Brazil and Ecuador:

(a) *Curitiba, Brazil*: Curitiba runs one of the early BRT systems; the system features multi-application smart cards (used for transit as well as other applications).
(b) *Quito, Ecuador*: This BRT system uses controlled-access stations adjacent to the exclusive lanes.

An example of a successful BRT implementation is that run by the Los Angeles County Metropolitan Transportation Authority, which in 2000 launched the service in several of Los Angeles' heavy corridors. TranSystem reported that bus ridership in the Wilshire-Whittier and Ventura Blvd. corridors has increased by 20% and 50%, respectively, since the implementation of BRT; furthermore, up to one-third of BRT passengers had previously not been transit users.

In Europe, numerous transit-priority projects have been executed; for example, in Athens, Dublin, Munich, Turin, Vienna, and Zurich. Based on Ceder (2004), lessons can be learned from these six case studies: Table 18.4 summarizes the benefits gained from the implementation of transit-priority schemes in these cities.

18.6.3 Improved operations control

Fundamentally, there are two distinctive real-time transit-performance disruptions: (1) deviations from the schedule (timetable), but not necessarily creating an imbalance between supply and demand and (2) creation of an imbalance between supply and demand (overloaded and almost empty vehicles), but not necessarily deviating from the schedule. Given that these disruptions are known in real time (e.g., by an APTS), corrective and restorative control strategies can take place.

The main real-time control strategies are shown in the following list:

- Holding the vehicle (at terminal or at mid-route point)
- Skip-stop operation
- Adding a reserve vehicle
- Changes in speed (not above the lawful speed limit)
- Interlining operation
- DH operation
- Short-turn operation
- Shortcut operation
- Leapfrogging operation with the vehicle ahead

The first strategy on the list can be used for improving on-time performance (scheduled based), eliminating bunching (headway based), and responding to unexpected demand; however, it has an adverse effect on onboard passengers. This holding strategy is further discussed and analyzed in the next section. The skip-stop operation is meant to pass stops at which no passenger wishes to alight; however, it has an adverse effect on waiting passengers at those stops. Adding a reserve vehicle as a reinforcement action is suitable in locations where unexpected demand may appear. Changes in speeds (slowing down or speeding) can improve on-time performance and eliminate bunching; in the slowing-down cases, it may be better valued by the onboard passengers than is the holding strategy. Interlining (if commonly not in the schedule), DH, and short-turning operations can be used as corrective actions for possible no-show/lateness situations. A real-time shortcut decision to convert a local service into an express (or semi-express) service between two points is feasible if it fits the destinations of all onboard passengers. Finally, the leapfrogging operation will serve to correct a load-imbalance scenario between two following vehicles, as well as ease the bunching phenomenon.

The main drawback of possible real-time control actions is the lack of prudent modeling and software that can activate these actions by automatic/semiautomatic/manual mode. Figure 18.7 illustrates schematically a computer-aided, real-time control system; such a system can be employed in a transit-control center to allow for best exploitation of the real-time information.

Another means of improving transit-operations control can be approached from the passenger side. That is, the match between supply and demand can be improved by utilizing intelligent real-time passenger-information systems. Pretrip information has the potential to change demand by time of day and to direct the passengers to the best service. En route information can reduce passenger uncertainty, thus improving service reliability.

Current communication technology offers the opportunity to convey traveler information via a variety of media: cell phones; in-terminal systems; variable message signs at bus stops, train stations, and platforms; and in-vehicle systems, by signs or voice. Lessons may also be learned from the following four major European cities:

(a) *London*: London Transport offers several types of traveler information service, for example, the Journey Planner, bus stop-specific timetables and maps, and the Countdown bus-arrival information system and Oyster, a regional smart card system.
(b) *Paris*: Régie Autonome des Transports Parisiens (RATP) has deployed Modeus, a regional smart card payment system, with multiple-application agreements for retail, telephone calls, and more; ALTAIR, a real-time information system that provides information on board and at bus stops; and SIEL, which provides information on regional rail service.

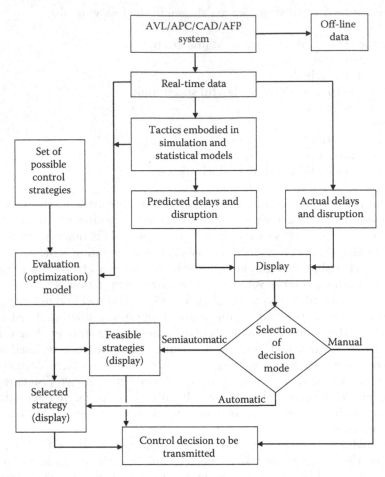

Figure 18.7 Schematic diagram of a computer-aided real-time control system.

(c) *Turin, Italy*: A public–private partnership provides transit information, including itinerary planning, variable message signs, and at-stop displays.

(d) *Helsinki, Finland*: Advanced traveler information services include the Journey Planner, the transit agency's real-time system Helsinki Electronic Library Multimedia Information (HELMI), personal mobile traveler and traffic-information service, use of smart cards, and real-time information via the Internet.

18.6.4 Holding strategy: Approximate method

Vehicle holding strategy is considered a possible remedy for reliability problems for three main reasons: (1) prevents bunches from forming, (2) ensures that scheduled connections (transfers) are made, and (3) minimizes total passenger waiting time. Holding strategies, therefore, take place at dispatching points, connection points, and major stops (e.g., max load point).

Traditionally, there are two classes of holding strategies: scheduled based and headway based. The first strategy aims at holding the vehicle until its scheduled departure time; it is especially useful for schedules with large headways. The second strategy, which usually fits schedules with short headways, has the objective of minimizing the total passenger delay at stops. This optimization framework is the result of reducing headway variability (saving delay) and creating extra delay (because of the holding strategy).

Undoubtedly, the use of a holding strategy is complicated in practice because of the uncertainty involved (in reaching its objective), its adverse effect on the onboard passengers, and its impact on real-time operations at the network level. This situation may justify the use of an approximation (isolated) method to describe the development of basic holding rules. In general, Wirasinghe (2003) describes acceptable approximate methods in transit services, including dispatching policies, headway setting, and scheduling travel times.

An early foundation for approximately analyzing dispatching and holding strategies in transit operation was set down by Newell (1971, 1977, 1982). Section 18.7 reviews chronologically the essence of control strategies to improve transit reliability. Newell (1971) and Wirasinghe (2003), in which Newell's work is discussed, present the best transit-vehicle dispatching policy for minimizing total passenger waiting time, subject to a fixed number of departures (dispatches). Newell found that the best dispatching rate was proportional to (1) the square root of the random passenger-arrival rate for large (not constrained) vehicle capacities and to (2) the random passenger-arrival rate, otherwise. Osuna and Newell (1972) presented control strategies for dispatching immediately or holding a vehicle for a hypothetical bus-route loop with only one service and control point. They used dynamic programming and queuing techniques for distributed travel times and uniform passenger-arrival rates. The following is an approximate analysis for a holding strategy inspired by Newell's work.

Figure 18.8 illustrates the process of passengers arriving at a max load stop and then departing. Figure 18.8a refers to an even-load timetable and Figure 18.8b to an even-headway timetable. In each part, the upper curve describes the accumulated number of passengers according to the scheduled departure times; the lower curve exhibits a real-life situation in which one early and one late arrival are introduced. The notations used in Figure 18.8, some of which appear in Chapters 3, 4, and 6, are these: P_m is passenger load at the max load point; d_0 is the desired occupancy on a single vehicle; t_i and t'_i are the ith scheduled and real departure times at the max load point, respectively; λ_{im} is the average uniform random passenger-arrival rate between the $(i-1)$th and the ith departures at the max load stop; H, H_i, and H'_i are the scheduled even headway, scheduled even-load headway, and real headway, respectively; and Δ_1, Δ_2 are the possible holding times for an early arrival and the extra-late arrival time, respectively.

As expected, the accumulated number of passengers in the upper curve in Figure 18.8a reaches d_0 when a departure takes place. This is not case in Figure 18.8b, assuming a different λ_{im} for the even-headway case. Improving reliability usually coincides with reducing the total waiting time at transit stops. In the lower curves in Figure 18.8a and b, the shaded area represents additional passenger-minutes delay compared to the scheduled (expected) situation. Although it is impossible to reduce the additional delay incurred by the late vehicle arrival (by Δ_2 minutes), it is conceivable that the total passenger delay can be reduced by holding the early arrival vehicle (by Δ_1 minutes). It should be noted that in real time, the situation of late arrivals may also be ameliorated or avoided by the use of skip-stop or shortcut strategies.

In an average and approximated sense, the dashed area of Δ_1 in Figure 18.8a is $H'_1 \cdot \lambda_{1m} \cdot \Delta_1$, being the additional passenger minutes incurred if the t'_1 departure is held until t_1. Otherwise, the additional approximated delay for passengers expected to board the t_2 departure is $\lambda_{1m} \cdot \Delta_1 \cdot H_2$. Consequently, a simple strategy is this: hold the vehicle only if $H'_1 \cdot \lambda_{1m} \cdot \Delta_1 < \lambda_{1m} \cdot \Delta_1 \cdot H_2$.

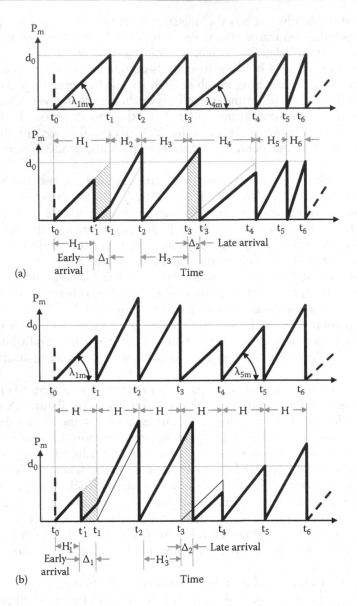

Figure 18.8 Cumulative number of passengers at the max load stop for an even-load timetable in (a) and an even-headway timetable in (b), in which the anticipated, scheduled situation in each part is in the upper curve and the real-life scenario is in the lower curve.

However, because of the uncertainty about H_2, its expected value is considered, that is, $E(H_2)$. In general, the strategy for even-load timetables is to hold the vehicle if

$$H_1' < E(H_{i+1}) \tag{18.12}$$

and to dispatch immediately, otherwise. The expected (mean) value $E(H_{i+1})$ requires data for constructing the PDF (distribution) of H_{i+1} for all departures i, i ≠ 1.

The simple rule of Equation 18.12 also applies to Figure 18.8b, the case of even headways. That is, the vehicle is held from t'_1 departure until t_1 if

$$H'_1 < E(H) \qquad (18.13)$$

and dispatched immediately, otherwise.

18.7 LITERATURE REVIEW AND FURTHER READING

The reliability of a public transit service is considered one of the main factors influencing the level of passengers' satisfaction. There is a very abundant literature on various aspects of transit reliability. This review describes, in chronological order, selected papers that focus only on the following subjects:

- Development of quantitative measures of service reliability
- Models of passengers' waiting time at stops as a function of service reliability
- Suggested control strategies for improving reliability

18.7.1 Measures of service reliability

Service reliability is an abstract term. In order to include reliability considerations in a detailed engineering design and to evaluate the differences between existing and suggested service alternatives, it is necessary to describe reliability in mathematical terms. This section includes a review of quantitative measures of service reliability.

One of the earliest discussions of transit service reliability measures was made by Polus (1978), who focused on bus services. Reliability was defined as the amount of consistency from day to day associated with any operational performance measure, the proposed reliability indicator being the inverse of the standard deviation of travel times.

Silcock (1981) mentions some traditional reliability measures:

- Number of buses taking x minutes longer than scheduled
- Percentage of buses that depart from 1 min early to 4 min late
- Average waiting time of passengers
- Excess waiting time (EWT) of passengers
- The difference between the actual average waiting time and the calculated waiting time

Turner and White (1990) formulated reliability measures in order to compare services provided by different bus sizes. They suggest the following formula for EWT as a measure of reliability:

$$\mathrm{EWT} = \frac{T}{2}\left(\frac{1}{N} - \frac{1}{S}\right) + \frac{N \cdot \mathrm{VarH}}{2T}$$

where
   T is the time duration of the reliability survey
   N is the number of observed buses during the reliability survey
   S is the number of scheduled buses during the reliability survey
   VarH is the variance of observed headways

Henderson et al. (1991) analyzed the advantages of using odd ratios of on-time performance as a measure of service reliability. An odd ratio is the percentage of trains or buses that arrive on time, divided by the percentage of those that were not on time. It should be emphasized that an odd ratio is not the ratio of on-time arrivals to total arrivals, since the denominator is not that of total arrivals. Because of this formulation of the ratio, the manner in which its value changes as the number of deviations from the schedule increases is not linear; the authors claim that this formulation gives a better representation of passengers' perception of reliability than do most other measures. For example, the reliability of a transit service with a 70% probability of arriving on time (70/30 = 2.33) is eight times worse than the reliability of a service with 95% on-time arrivals (95/5 = 19). Hence, odd ratios as reliability measures show passengers' high sensitivity to the consistency of service punctuality.

Rudnicki (1997) defined several unreliable performance measures. A suggested unreliability measure is the chance that a passenger will have to wait for a later departure because of to a too-early departure of the expected vehicle. An inconvenience measure is the average EWT. A degree-of-punctuality factor is the average value of a predetermined function, which receives a value between 0 (worse) and 1 (best) according to the deviation from the scheduled time. Another unreliability factor is the difference between the 95th percentile of actual waiting time and the planned waiting time. The paper explains which of the measures represent the interests of the passengers and which are more suitable to represent the operator's point of view.

Hallowell and Harker (1998) present a method of predicting the schedule reliability (SR) of a partially double rail line on which delays are caused by the need for one train to wait for the passing train. Using a simulation model for the design of a reliable timetable, the researchers describe measures of departure-time inaccuracy: at origin, arrival at destination, delay time from origin to destination, and slack time from origin to destination (slack time is defined as planned ride time minus free running time).

Carey (1999) developed measures of estimating service reliability in advance. The first proposed measure depends on the probability of delay for a specific train, resulting either from schedule failure of this specific train or from a delay caused by previous trains. Another measure depends not only on the chance of such delay but also on the magnitude of the delay. Both of these measures require a PDF of route delay as the input. Additionally, the author suggests two measures whose calculation does not require a given PDF. One is based on the standard deviation and the mean of headways. The other derives from the idea that a small initial delay grows larger if the delayed train, on the way to its intended platform, has to cross paths used by another train; the fewer paths crossed, the more reliable the service. The measures developed by Cary are normalized, such that they all receive values between 0 and 1, with 1 representing the highest reliability level.

Yin et al. (2002) develop a simulation-based approach to assess transit reliability, taking into account the interaction between network performance and passengers' route choice behavior. Measures are developed for three types of reliability: travel-time reliability (TTR), SR, and waiting-time reliability (WTR). TTR is the probability that the average actual travel time is less than a given threshold. SR is the probability that an actual line is frequency larger than a given threshold. WTR is the probability that passengers' average waiting time is less than a given threshold, taking into account the possibility of passengers' being unable to board the first vehicle that arrives owing to insufficient vehicle capacity. A Monte Carlo simulation approach is used, incorporating a stochastic user-equilibrium transit-assignment model with a capacity constraint and elastic frequencies. Analyzing simulation results using

Table 18.5 Characteristics of works dealing with passenger waiting time as a function of reliability measures

Source	Proposed measures	Source of input data
Polus (1978)	Inverse of standard deviation of travel times	Travel-time measurements for different days
Silcock (1981)	Number of late buses, percentage of buses that depart on time, average waiting time, excess waiting time, difference between actual and calculated wait	Vehicle performance survey/simulation
Turner and White (1990)	Excess waiting time	Vehicle performance survey
Henderson et al. (1991)	Odd ratio = percentage of on-time arrivals divided by percentage of not-on-time arrivals	Vehicle performance survey/simulation
Rudnicki (1997)	Factor of unreliability on account of unpunctuality, factor of inconvenience owing to unpunctuality, degree of punctuality as a factor of operation regularity, factor of unreliability on account of irregularity	Vehicle performance survey
Hallowell and Harker (1998)	Departure-time error at origin, arrival-time error at destination, delay time from origin to destination, slack time from origin to destination	Simulation
Carey (1999)	A measure depending on the chance of delay and a measure depending on the magnitude of delay. A measure based on standard deviation and mean of headways. A measure of the chance of delay owing to the need to wait for a crossing train	PDF of route delay. Vehicle performance survey/simulation. Information about number of rail paths crossed
Yin et al. (2004)	Probability that the average actual travel time is less than a given threshold, probability that an actual line frequency is larger than a given threshold, probability that the passengers' average waiting time is less than a given threshold	Simulation

the measures mentioned is very effective in evaluating the trade-off among alternatives for service improvement.

Characteristics of the reliability measures reviewed are summarized in Table 18.5.

18.7.2 Passenger waiting time as a function of reliability

It is often claimed that passengers are more sensitive to changes in their waiting time than to the other periods comprising the trip. Naturally, waiting time depends on the motion of the anticipated vehicle, and service is considered more reliable from this point of view if that motion is always the same. Prediction of vehicle-time variability is a complex task in itself, since it is influenced by numerous factors, such as traffic conditions and passenger flow while boarding and alighting. Waiting time also depends on other factors, some of which have to do with the passengers' behavior. Several models for estimating waiting time will now be reviewed.

The traditional formula for average passenger time, depending on headway variability, is shown in Equation 11.1 and discussed in Section 18.4.1. This formula was used by Osuna and Newell (1972), as well as by other authors. For the sake of clarity, it is presented again:

$$E(w) = \frac{\overline{H}}{2}\left(1 + C_H^2\right)$$

where

$E(w)$ is the average waiting time
$\overline{H}$ is the mean headway
C_H is the coefficient of variation in headways (standard deviation/mean).

This formula is based on the assumption that passengers arrive at a stop randomly. The average waiting time is greater than half the headway because more passengers arrive during long headway intervals (before the arrival of a delayed vehicle) than during short intervals (after the arrival of a delayed vehicle). The average wait, therefore, is longer when headways are less uniform.

Turnquist (1978) developed a model of bus and passenger arrivals at a bus stop. He investigated the impact of reliability on the average waiting time under the assumptions that the bus arrival time behaves according to a lognormal distribution and that its mean and variance are known. The bus-arrival-variance variable plays the role of the reliability indicator. Calculation of waiting time takes into account both passengers who arrive at the station randomly and passengers who plan their arrival time. If the proportion of random arrivals is known, then this model makes it easy to calculate waiting time for planning purposes.

Bowman and Turnquist (1981) developed another model for evaluating average waiting time as a function of service reliability and frequency. This model is based on more complex assumptions regarding passenger behavior: it takes into account the decision-making process that passengers go through when they choose a time to arrive at the bus stop. The model predicts the distribution of passenger arrivals when the mean and average of bus headways are given. It enables operators to make a realistic prediction of how a change in service reliability will alter the overall waiting time. This prediction is sensitive to the differences in passenger behavior among lines with different frequencies. The authors conclude that waiting time is much more sensitive to SR than to service frequency.

Guenthner and Sinha (1983) described another model for estimating waiting times. Their model is unique in that it assumes that SR is influenced by the level of vehicle maintenance: when there are not enough vehicles available to meet service demand, reliability is severely affected. The tool presented consists of three models: a maintenance model, a reliability model, and a performance model. The input for the maintenance model is the number of spare buses and the number of mechanics: its output is a dependability factor that expresses the point at which maintenance problems will lead to service failure. This factor is calculated by using simulation. The dependability value is an input for the reliability model, which determines the average waiting times both of passengers who arrive randomly and of those who come at a preplanned time. Waiting-time values are transferred to the performance model, which yields predictions of ridership and an evaluation of general system performance. Implementing the proposed set of models requires some computer tools as well as calibrated values of demand elasticities, which may not always be available.

Xuan et al. (2011) acknowledge that dynamic holding strategies based on headways alone cannot help buses adhere to a schedule. As such, the authors investigate a family of dynamic holding strategies that use bus arrival deviations from a virtual schedule at the control points. The virtual schedule is introduced whether the system is run with a published schedule or not. It is shown that with this approach, buses can both closely adhere to a published schedule and maintain regular headways without too much slack. However, the modeling is based on a long list of assumptions such as unlimited capacity, unlimited number of departures and stops, holding strategy being applied after boarding, and more. The authors show that the only data needed in real time are the arrival times of the current bus and of the preceding bus at the control point relative to the virtual schedule. The simple method was found to require about 40% less slack than the conventional schedule-based method.

Li et al. (2010) review and critique the modeling frameworks and empirical measurement paradigms used to obtain willingness to pay (WTP) for improved TTR, suggesting new directions for ongoing research. The authors review TTR studies in the context of passenger transport (car, rail, and bus). Three types of empirical models to capture TTR (i.e., the mean-variance model, the scheduling model, and the mean lateness model) were presented, including associated theoretical issues, experiment design, and practical applications. The authors estimated models to derive values of reliability, scheduling costs, and reliability ratios in the context of Australian toll roads. They used the new evidence to highlight the important influence of the way that trip time variability is included in stated preference studies in deriving WTP estimates of reliability in absolute terms and relative to the value of travel time savings. The study found that a choice experiment with multiple travel time scenarios to represent early, late, and on-time trip activity produced significantly higher reliability ratios, compared to a choice experiment with a single time component (i.e., variability).

Table 18.6 summarizes the models reviewed.

Table 18.6 Characteristics of works on passenger waiting time as a function of reliability

Source	Required input	Assumptions regarding passenger arrival at station	Assumptions regarding vehicle arrival at station
Osuna and Newell (1972)	Standard deviation and mean of headway	Random.	None
Turnquist (1978)	Standard deviation and mean of headway	Some passengers arrive randomly; others plan their arrival time.	Lognormal distribution
Bowman and Turnquist (1981)	Standard deviation and mean of headway, probability distribution function of vehicle arrival	A utility function is ascribed to each possible arrival time.	None
Guenthner and Sinha (1983)	Number of spare buses and mechanics	Some passengers arrive randomly; others plan their arrival time.	Lognormal distribution
Xuan et al. (2011)	Scheduled and actual arrival times and headways	Stationary passenger arrivals, but station dependent.	Trip times are random
Li et al. (2010)	Mean travel time, travel cost, and five equiprobable arrival scenarios	Different studies are compared with.	Different studies are compared with

18.7.3 Control strategies for improving reliability

Transit services are considered unreliable if they suffer from significant travel-time variability. Improvement in service reliability is, therefore, likely to occur if actions are taken to reduce time variability. Such actions are usually aimed either at having control over vehicle adherence to its schedule or at introducing route patterns with low time variability. This section presents several strategies for improving reliability. It should be noted that short-turn strategies, although mentioned in some papers in this context, are referred to in Chapter 16.

Osuna and Newell (1972) presented the conflict between the desire for minimal route time and the need to slow down system performance in order to increase its regularity. They formulated a control problem that aims at increasing punctuality, thereby achieving minimal waiting time. Dispatching and holding strategies are introduced: dispatching strategies help in deciding what the actual headways would be between successive departures from the starting terminal, while holding strategies mean that departure time is also controlled at other time points along the route. The performance of controlled and uncontrolled systems is analyzed, based on a PDF of travel time, which is required as input. Other basic assumptions, declaratively simplistic, refer to an idealized network. For example, only single-stop routes with one or two buses are discussed. Still, this chapter represented a significant discussion of schedule control, since it led to many other research works in this area.

Lesley (1975) suggested a procedure for designing a realistic schedule on a route along which bus progress is monitored. A reliability index is calculated for each stop on the basis of the level of headway variance. Stops at which the index is at least twice the average are chosen as time points for holding control. The required slack time at each point is calculated.

Newell (1977) developed a method for determining the slack time allocated to each time point when holding control is applied. With the aim of reducing vehicle time while keeping schedule regularity, the lower boundary of slack time is calculated. Some of the basic assumptions are simplistic, such as ignoring alighting times (which is still a realistic assumption) and supposing no correlation between adjacent buses. Passengers are assumed to arrive at stops according to a Poisson distribution, and bus delay is assumed to behave according to Fokker–Planck. Slack time is calculated for each stop, not just selected ones.

Koffman (1978) compared several real-time strategies for improving headway reliability on a single-direction bus route with a single control point. Employing a simulation model, the author examined the following strategies: holding, skipping stops, introducing bus-priority signalization, and reducing dispatch uncertainty. All control strategies are analyzed under different headway thresholds. The resulting conclusion is that the last two strategies seem the most promising.

Jordan and Turnquist (1979) discussed reliability improvement through the introduction of a special pattern for urban routes. According to this pattern, the whole path is divided into several zones, and each bus makes stops only in a specific zone. A dynamic programming problem is formulated in order to determine the optimal number of zones, points where zones switch, and the number of buses serving each zone. The objective is to find a zone structure that minimizes trip-time average and variance. The model developed includes submodels for predicting dwell time at stops, distribution of headways, mean and variance of waiting time, and mean and variance of travel time; intensive calibration is therefore required. The dynamic programming model is applied in a case study, which yields the conclusion that even a very simple zone structure can lead to a substantial improvement in reliability and a decrease in bus-fleet size.

Turnquist and Bowman (1980), using a set of simulation experiments, investigated the effect of network structure on service reliability. Transit networks on which route paths form a radial pattern are compared to grid-shaped networks. The researchers found that even though passengers were obliged to transfer more in grid networks, the uncertainty in regard to the length of the transfer delay was smaller than in radial networks. The combined travel-time uncertainty is also smaller in grid networks. The authors conclude that a concentration of transfers at the center node of radial networks has a more disruptive effect on reliability than does the dispersed distribution of transfers that occurs in grid networks.

Furth and Nash (1985) suggested a method for improving adherence to a preplanned timetable, for which all routes started at a specific terminal. According to this method, the pool of vehicles that belongs to the terminal serves all the round trips leaving that terminal in a *first in–first out* sequence, instead of being assigned to specific trips in advance. Features of this strategy are presented, and the reliability of the resulting service is estimated. A PDF of bus travel times is needed as input.

Abkowitz et al. (1986) determined which single point was optimal for holding control. They found that the location of the control point was sensitive to the passenger-boarding profile. The optimal control point is usually located just prior to a group of stops having a large number of boarding passengers, that is, toward the peak segment of the route (max load point). Many other papers published in the 1970s and 1980s describe methods of designing optimal holding strategies. Most of these methods use simulation techniques to optimize the value of a waiting-time threshold, above which a stop is declared a time point for controlling schedule adherence.

Seneviratne (1990) calibrated an expression for headway standard deviation as a function of the number of time points. This function is a second-degree polynomial, meaning that there is an optimal number of points and that increasing the points beyond that number will lead only to diminished reliability.

Li et al. (1993) presented various real-time dispatching strategies to be used at the terminal of a single route: instructing bus drivers to skip stops, directing them to cut out parts of their route, or simply adjusting layover times. The design of these strategies is expressed as a nonlinear program and as a stochastic binary integer program. Simulation is used for testing this approach, but it can also be used to evaluate preestablished schedules or decisions regarding stop-skipping strategies. Special attention is given to assessing the impact of bus-location information on the dispatching decision. The researchers find no need for an automatic vehicle-location system that covers the entire network for the purpose of real-time dispatching. Location information from a limited number of points seems to be sufficient.

Wirasinghe and Liu (1995) formed a dynamic programming model for the design of holding strategies for an isolated bus with fixed demand. The number and location of time points are determined, as is the amount of slack time at each point.

Carey (1998) formulated a problem of optimizing slack time between successive activities in a vehicle schedule. Increasing slack time reduces the risk that a delay in one activity will affect the next activity; on the other hand, allocating more spare time may cause the activity to take longer because of behavioral reasons. The author claims that some percentage of scheduled rides will always depart a few minutes late, independently of the length of the slack time before the scheduled departure. A PDF of activity time is a required input. Slack times that give minimum total operating costs are calculated. It should be noted that many other papers by this author deal with delay control and various other aspects of transit reliability.

Hallowell and Harker (1998) proposed a simulation method mentioned earlier in this section. Their method can be used for determining slack times allocated to stations along a railway line in order to increase SR. The authors compare features of real-time scheduling and master scheduling. They also discuss the difficulty in using simulation methods for improving performance; it is due to the complexity of the computation tasks involved.

Adamski and Turnau (1998) developed a tool for real-time schedule-control decision making. The input for the model is real-time information about deviations from scheduled times or from regular headways along a transit route; the output is a set of recommended decisions regarding vehicle dispatching. The tool developed includes a submodel that calculates the time spent at stops. Many other papers by Adamski deal with dispatching control.

Eberlein et al. (1998) discuss the real-time strategy of DH, that is, running empty from the terminal to the first station at which passengers are allowed to board while skipping several stations en route. A programming model is formulated in order to determine which vehicles to deadhead and how many stations to skip.

O'Dell and Wilson (1999) developed a real-time decision-support system for rail-transit operations. The system includes a deterministic model of a rail system and mixed-integer programming formulations for the choice of an optimal holding strategy. Several holding strategies are discussed: hold at all stations, hold only at the first station after the disruption occurs, and hold at a fixed chosen station. These are investigated with and without train-capacity constraints. The paper also includes a dwell-time submodel. Attention is given to assumptions on the order in which trains from different branches enter junctions; this is claimed to have a major effect on system performance. The results show that holding strategies at one station (whether the first after the disruption or any station) are almost as effective as holding at all stations. They recommend holding control at the first station after the disruption, since it is the easiest to implement.

Eberlein et al. (1999) considered several control strategies: two station-skipping strategies, holding at a given station, and combinations of the two. They formulated mathematical models for all control problems under the assumption that real-time vehicle-location information was available and investigated the advantages and disadvantages of each control type. The models are deterministic but are also examined under various stochastic conditions. The results show that a combination of station-skipping and holding strategies can have very effective results on system performance. It is also shown that combined strategies are more efficient than any single strategy although a holding strategy is slightly preferable. The authors find that the optimal holding strategy depends mainly on the route-headway pattern at the holding station and is independent of passenger demand along the route.

Shen and Wilson (2001) developed a disruption control model for rail-transit systems. The model, a deterministic, mixed-integer program, includes a dwell-time submodel. Several control strategies are compared. The main conclusion is that best system performance is achieved when holding and short-turn strategies are combined.

Liu and Wirasinghe (2001) presented another simulation model for the design of a holding control. The simulation model consists of several submodels that predict performance during various journey stages: linked travel time, passenger arrival, passenger-alighting demand, dwell time, and dispatching time at the starting terminal. The output of the overall model is the choice of time points and the slack times allocated to each of them.

Fu and Yang (2002) developed two simulation models for choosing time points for cases in which holding control should be implemented. In both models, time points are located at every stop where headways exceed some threshold. In the first method, only the headway between a bus and the preceding bus is checked; the second method checks, in addition, the headway between each bus and the following bus. The researchers find that control is most useful when the control stops chosen are those with the highest demand and close to the middle of the route. The number of control points is also investigated—scenarios with one or two or all stops with control are examined. Control at all stops reduces waiting time, but it also results in a significant increase in vehicle time. Control at two stops, one of which is the starting terminal, is found to be the optimal solution, since it enables the operator to achieve a reduction in waiting time with no increase in in-vehicle time and only a slight increase in bus-travel time.

Dessouky et al. (2003) examined simulated systems in which holding and dispatching strategies are used; they also investigated the dependence of system performance on available technologies. Systems in which communication, vehicle locating, and counting technologies are available or unavailable are compared. The analysis addresses seven different holding-strategy combinations, each of which differs in its definition of holding conditions and in its approach to the use of holding for connecting buses. The results show that advanced technology is most advantageous when there are many connecting buses; the schedule slack is then close to zero and the headway is large.

Chen et al. (2009) present an in-depth analysis of service reliability based on bus operational characteristics in Beijing. Three performance parameters are proposed in the study for the evaluation of bus service reliability: punctuality index based on routes (PIR), deviation index based on stops (DIS), and evenness index based on stops (EIS). The relationship among the three parameters is discussed using a numerical example. Subsequently, through a sampling survey of bus lines in Beijing, service reliability at the stop, route, and network levels is estimated. The authors claim that the three performance parameters allow for reliability estimation from macro- to microscale. In general, this study found that PIR > DIS > EIS. In terms of Beijing, the results imply that its transit system is operated at a relatively low reliability level. In addition the study analyzed the effects of route length, headway, the distance from the stop to the origin terminal, and the use of exclusive bus lanes.

Yu et al. (2012) present a DH strategy for transit operations to improve transit service of the peak directions of transit routes. The DH strategy takes place at different stops by letting the vehicle to turn around and continues as a DH trip; it is called partway DH. This strategy consists of two phases: reliability assessment of further transit service and optimization of partway DH operation. The reliability assessment of further transit service, which is based on the current and recent service reliability, is used to justify whether or not to implement a partway DH operation. The objective of the second phase is to determine the beginning stop for a new service for the DH vehicle by maximizing the benefit of transit system. A heuristic algorithm is also defined and implemented to estimate reliability of further transit service and to optimize partway DH operation. Then, the partway DH strategy proposed is tested with the data from a transit route in Dalian City of China. The authors analyzed the length of rolling horizon and the threshold value of service reliability and validated the results. All in all, the authors illustrated that their partway DH strategy yields the best performance measures compared with other partway DH strategies and the no-control scenario.

The major characteristics of the models reviewed are summarized in Table 18.7. For practical use, the main difference between analytic methods and methods based on simulation should be mentioned. Analytical methods require less detailed information about all network components; however, they usually rely on many assumptions. The problem is then to simplify the representation of the system performance. Methods that use simulation provide much more realistic results, but the input that they require is much more detailed.

Table 18.7 Characteristics of works on control strategies for improving reliability

Source	Control type	Method	Place of control	Suggested solution
Osuna and Newell (1972)	Dispatching, holding	Analytical	Single control point.	
Lesley (1975)	Holding	Simulation	Number and location of points are optimized.	
Newell (1977)	Holding	Analytical	All stops are controlled.	
Koffman (1978)	Real-time holding, stop-skipping, bus-priority signalization, dispatching	Simulation	Single control point (terminal).	Bus-priority signalization and reducing dispatching uncertainty.
Jordan and Turnquist (1979)	Zone scheduling	Dynamic programming		
Turnquist and Bowman (1980)	Network structure	Simulation		Grid network.
Furth and Nash (1985)	Real-time dispatching	Analytical	Terminal.	
Abkowitz et al. (1986)	Holding	Combination of simulation and analytical	Location of control point is optimized.	Optimal control point is toward the peak segment of the route.
Seneviratne (1990)	Holding	Simulation	Number of control points is optimized.	
Li et al. (1993)	Real-time dispatching/ stop-skipping	Nonlinear program/binary integer program (simulation is used for testing the method)	Terminal of a single route.	
Wirasinghe and Liu (1995)	Holding	Dynamic programming	Number and location of points are optimized.	
Carey (1998)	Dispatching	Analytical	Terminal.	
Hallowell and Harker (1998)	Holding (real time/ master)	Simulation	All stops are controlled.	
Adamski and Turnau (1998)	Real-time dispatching	Simulation	All stops are controlled	
Eberlein et al. (1998)	Real-time deadheading (stop-skipping)	Programming model	Terminal.	
O'Dell and Wilson (1999)	Real-time holding	Analytical/ mixed-integer programming	All stations/first station after disruption/fixed strategies are investigated.	Holding at first station after disruption.

(Continued)

Table 18.7 (Continued) Characteristics of works on control strategies for improving reliability

Source	Control type	Method	Place of control	Suggested solution
Eberlein et al. (1999)	Real-time stop-skipping, holding, combinations	Analytical		Combination of strategies.
Shen and Wilson (2001)	Holding, short turn	Mixed integer program		Combination of strategies.
Liu and Wirasinghe (2001)	Holding	Simulation	Number and location of points are optimized.	
Fu and Yang (2002)	Holding	Simulation	1, 2, and all strategies are investigated.	Holding at two points.
Dessouky et al. (2003)	Holding, dispatching	Simulation	Seven alternatives are compared.	
Chen et al. (2009)	Ordinary operations	Statistical analyses		Three performance indices are proposed.
Yu et al. (2012)	Deadheading (DH) strategy on route segments	Optimization and heuristics	At transit stops.	New DH strategy is proposed.

EXERCISES

18.1 Given a bus route on which passengers' mean waiting time, $E(w)$, can be estimated by $E(w) = E(H^2)/2E(H)$, where H is the headway variable,

(a) Explain and show (mathematically) how $E(w)$ can be reduced, using the waiting-time formula. Consider headway distributions ranging from regular (deterministic) to exponential (random).

(b) Outline feasible (practical) ways to reduce $E(w)$, given fixed passenger demand.

18.2 With high volumes of bus services on major arterial roads, it becomes very expensive and often ineffective for bus agencies to use curbside inspectors and mobile supervisors. The inspector's or supervisor's knowledge of situations is limited to his or her immediate vicinity, with no indication of conditions elsewhere. Accordingly, bus drivers may be instructed in some cases to act in a fashion that is counterproductive from the system point of view. This represents only one group of problems that can be resolved by introducing an AVL system. In general, the real-time control provided by an AVL system enables the operator to increase both the utilization of each bus and the service level as seen by each passenger. Furthermore, the substantial amount of off-line AVL data can be used to improve the entire bus-planning process.

(a) List all possible strategies that supervisors can accommodate in the control room while using an advanced AVL system. Note that these strategies are carried out by instructing a driver or a group of drivers to follow a certain action.

(b) Outline the major potential benefits of (adequately used) AVL off-line data.

REFERENCES

Abkowitz, M., Eiger, A., and Engelstein, I. (1986). Optimal control of headway variation on transit routes. *Journal of Advanced Transportation*, 20, 73–88.

Abkowitz, M., Slavin, H., Waksman, R., Engelstein, I., and Wilson, N. (1978). Transit service reliability. Final Report UMTA MA-06-0049-78-1. U.S. Department of Transportation, Washington, DC.

Adamski, A. and Turnau, A. (1998). Simulation support tool for real-time dispatching control in public transport. *Transportation Research*, 32A, 73–87.

Balcombe, R., Mackett, R., Paulley, N., Preston, J., Shires, J., Titheridge, H., Wardman, M., and White, P. (2004). The demand for public transport: A practical guide. TRL Report, TRL593. TRL Limited, Crowthorne, U.K.

Bowman, L.A. and Turnquist, M.A. (1981). Service frequency, schedule reliability, and passenger wait times at transit stops. *Transportation Research*, 15A, 465–471.

Carey, M. (1998). Optimizing scheduled times, allowing for behavioral response. *Transportation Research*, 32B, 329–342.

Carey, M. (1999). Ex-ante heuristic measures of schedule reliability. *Transportation Research*, 33B, 473–494.

Ceder, A. (1981). Practical methodology for determining dynamic changes in bus travel time. *Transportation Research Record*, 798, 18–22.

Ceder, A. (2004). New urban public transportation systems: Initiatives, effectiveness, and challenges. *ASCE Journal of Urban Planning and Development*, 130, 56–65.

Ceder, A. and Marguier, P.H.J. (1985). Passenger waiting at transit stops. *Traffic Engineering and Control*, 26, 327–329.

Chen, X., Yu, L., Zhang, Y., and Guo, J. (2009). Analyzing urban bus service reliability at the stop, route, and network levels. *Transportation Research*, 43A, 722–734.

Dessouky, M., Hall, R., Zhang, L., and Singh, A. (2003). Real-time control of buses for schedule coordination at a terminal. *Transportation Research*, 37A, 145–164.

Eberlein, X., Wilson, N.H.M., and Bernstein, D. (1999). Modeling real-time control strategies in public transport operations. In *Computer-Aided Transit Scheduling*. Lecture Notes in Economics and Mathematical Systems, vol. 471 (N.H.M. Wilson, ed.), pp. 325–346, Springer-Verlag, Berlin, Germany.

Eberlein, X.J., Wilson, N.H.M., and Barnhart, C. (1998). The real-time deadheading problem in transit operations control. *Transportation Research*, 32B, 77–100.

Fu, L. and Yang, X. (2002). Design and implementation of bus-holding control strategies with real-time information. *Transportation Research Record*, 1791, 6–12.

Furth, P.G. and Nash, A.B. (1985). Vehicle pooling in transit operations. *Journal of Transportation Engineering*, 111(3), 268–279.

Guenthner, R.P. and Sinha, K.C. (1983). Maintenance, schedule reliability, and transit system performance. *Transportation Research*, 17A, 355–362.

Hallowell, S.F. and Harker, P.T. (1998). Predicting on-time performance in scheduled railroad operations: Methodology and application to train scheduling. *Transportation Research*, 32A, 279–295.

Henderson, G., Adkins, H., and Kwong, P. (1991). Subway reliability and the odds of getting there on time. *Transportation Research Record*, 1297, 10–13.

Huddart, K.W. (1973). Bus priority in greater London: Bus bunching and regularity and the odds of getting there on time. *Traffic Engineering and Control*, 14, 592–594.

Hwang, M., Kemp, J., Lerner-Lam, E., Neuerburg, N., and Okunieff, P. (2006). Advanced public transportation: State of the art update 2006. Report FTA-NJ-26-7062-06.1. Federal Transit Administration, U.S. Department of Transportation, Washington, DC.

Jeong, R. and Rilett, L.R. (2005). Prediction model of bus arrival time for real-time application. *Transportation Research Record*, 1927, 195–204.

Jolliffe, J.K. and Hutchinson, T.P. (1975). A behavioral explanation of the association between bus and passenger arrivals at a bus stop. *Transportation Science*, 9, 248–282.

Jordan, W.C. and Turnquist, M.A. (1979). Zone scheduling of bus routes to improve service reliability. *Transportation Science*, 13, 242–268.

Khattak, J.A. and Hickman, M. (1998). Automatic vehicle location and computer aided dispatch systems: Commercial availability and development in transit agencies. *Journal of Public Transportation*, 2(1), 1–26.

Kimpel, T.J., Strathman, J.G., and Callas, S. (2008). Improving scheduling through performance monitoring using AVL and APC data. In *Computer-Aided Scheduling of Public Transport*. Lecture Notes in Economics and Mathematical Systems (M. Hickman, S. Voss, and P. Mirchandani, eds.), 600, 253–280. Springer-Verlag, Berlin, Germany.

Koffman, D. (1978). A simulation study of alternative real-time bus headway control strategies. *Transportation Research Record*, 663, 41–46.

Larson, R.C. and Odoni, A.R. (1981). *Urban Operations Research*, Prentice-Hall, Englewood Cliffs, NJ.

Lesley, L.J.S. (1975). The role of the timetable in maintaining bus service reliability. In *Proceedings of Operating Public Transport Symposium*. University of Newcastle-upon-Tyne, Newcastle upon Tyne, U.K.

Li, Y., Rousseau, J.M., and Gendreau, M. (1993). Real-time dispatching public transit operations with and without bus location information. *Computer-Aided Transit Scheduling*. Lecture Notes in Economics and Mathematical Systems, (J.R. daduna, B. Isabel, J.M.P Paixao, eds.), vol. 430, pp. 296–308, Springer-Verlag, Berlin, Germany.

Li, Z., Hensher, D.A., and Rose, J.M. (2010). Willingness to pay for travel time reliability in passenger transport: A review and some new empirical evidence. *Transportation Research*, 46E, 384–403.

Liu, G. and Wirasinghe, S.C. (2001). A simulation model of reliable schedule design for a fixed transit route. *Journal of Advanced Transportation*, 35, 145–174.

Newell, G.F. (1971). Dispatching policies for a transportation route. *Transportation Science*, 5, 91–105.

Newell, G.F. (1977). Unstable Brownian motion of a bus trip. In *Statistical Mechanics and Statistical Methods in Theory and Applications* (U. Landman, ed.), Plenum Press, New York, NY.

Newell, G.F. (1982). *Applications of Queueing Theory*, 2nd ed., Chapman & Hall, London, U.K.

O'Dell, S.W. and Wilson, N.H.M. (1999). Optimal real-time control strategies for rail transit operations during disruptions. In *Computer-Aided Transit Scheduling*. Lecture Notes in Economics and Mathematical Systems, vol. 471 (N.H.M. Wilson, ed.), Springer-Verlag, New York.

Osuna, E.E. and Newell, G.F. (1972). Control strategies for an idealized public transportation system. *Transportation Science*, 6, 57–72.

Polus, A. (1978). Modeling and measurement of bus service reliability. *Transportation Research*, 12, 253–256.

Rudnicki, A. (1997). Measures of regularity and punctuality in public transport operation. In *Preprints of the Eighth IFAC/IFIP/IFORS Symposium*, Chania, Greece, vol. 2, pp. 678–683.

Seneviratne, P.N. (1990). Analysis of on-time performance of bus services using simulation. *Journal of Transportation Engineering*, 116, 517–531.

Shen, S. and Wilson, N.H.M. (2001). An optimal integrated real-time disruption control model for rail transit systems. In *Computer-Aided Scheduling of Public Transport*. Lecture Notes in Economics and Mathematical Systems, vol. 505 (S. Voss, and J.R. Daduna, eds.), Springer-Verlag, Berlin, Germany.

Silcock, D.T. (1981). Measures of operational performance for urban bus services. *Traffic Engineering and Control*, 22, 645–648.

TranSystems Corp., Planner Coll., Inc., and Crikelair, T. Assoc. (2006). Elements needed to create high ridership transit systems: Interim guidebook. TCRP Report 32. Transportation Research Board, Washington, DC.

Turner, R.P. and White, P.R. (1990). Operational aspects of minibus services. *Transport and Road Research Laboratory Contractor Report*, 185, 1–5.

Turnquist, M.A. (1978). A model for investigating the effects of service frequency and reliability on bus passenger waiting times. *Transportation Research Record*, 663, 70–73.

Turnquist, M.A. and Bowman, L.A. (1980). The effects of network structure on reliability of transit service. *Transportation Research*, 14B, 79–86.

U.S. Department of Transportation. (2003). *Advanced Public Transportation Systems Deployment in the United States: Year 2002 Update*, U.S. Department of Transportation, Federal Transit Administration, Washington, DC.

Wirasinghe, S.C. (2003). Initial planning for urban transit systems. In *Advanced Modeling for Transit Operations and Service Planning* (H.K. Lam and M.G.H. Bell, eds.), Pergamon-Elsevier Science, Oxford, U.K.

Wirasinghe, S.C. and Liu, G. (1995). Determination of the number and locations of time points in transit schedule design. In *Passenger Transportation* (M. Gendreau and G. Laporte, eds.), Baltzer Science, Amsterdam, the Netherlands.

Xuan, Y., Argote, J., and Daganzo, C.F. (2011). Dynamic bus holding strategies for schedule reliability: Optimal linear control and performance analysis. *Transportation Research*, **45B**, 1831–1845.

Yin, Y., Lam, W.H.K., and Miller, A.M. (2004). A simulation-based reliability assessment approach for congested transit network. *Journal of Advanced Transportation*, 38(1), 27–44.

Yu, B., Yang, Z., and Li, S. (2012). Real-time partway deadheading strategy based on transit service reliability assessment. *Transportation Research*, **46A**, 1265–1279.

Chapter 19

Operational strategies and tactics

Skip my stop and drop me at a well-designed connection point

A.C.

CHAPTER OUTLINE

PRACTITIONER'S CORNER

The following riddle may help spread some knowledge on the substance of transit real-time control actions, especially from a sensitivity perspective, and what we need and what we don't need to pay attention to.

Assume that a tight rope surrounds the earth along the Equator. The rope is very tight. The question is: "If we add one meter to the length of the rope, can a cat then go underneath this rope, or not?" (The answer appears at the end of this practitioner's corner.)

This chapter consists of five main parts. Section 19.2 focuses on a multiagent transit system composed of the following agents: transit vehicles, passengers, road segments, and local authority (or government). This section defines each agent and their interrelationship as possible tools for real-time transit control actions. Section 19.3 analyzes the probability of the simultaneous arrival of two or more transit vehicles at a given road segment; it calculates the road-segment encounter, with any point along the segment constituting a possible encounter point, and discusses possible online tactical deployment measures. Section 19.4 develops a dynamic programming model for minimizing the total travel time, resulting in a set of preferred tactics to be deployed. An optimization model based on dynamic programming and a transit simulation that validates the benefits of such a model are described. Section 19.5 investigates how to use selected operational tactics in transit networks for increasing the actual occurrence of scheduled transfers. The model determines the impact of instructing vehicles to either hold at or skip certain stops or segments (one or more consecutive stops), on the total passenger travel time (TPTT) and the number of simultaneous transfers. This section explains how to create simulation and optimization frameworks for optimal use of the different scenarios and compare between the scenarios using a case study. In Section 19.6, an intervehicle communication (IVC)-based scheme is proposed to optimize the synchronization of planned transfers in transit networks. By using this scheme, transit drivers can share their real-time information, such as vehicle location, vehicle speed, and the number of passengers, with each other, through a central communication coordinator.

The chapter ends with a literature review.

Fundamentally, practitioners are encouraged to visit Sections 19.1 and 19.4 through 19.7, except Sections 19.5.1 and 19.6.3, and also to read the introduction and conclusions of Sections 19.2 and 19.3. The chapter offers challenges for devising alternative transit solutions that have yet (if at all) to be introduced or that may need to negotiate the barrier of implementation. Until the future actually reveals itself, practitioners can, and should, improve their readiness for future transit operations under new conditions.

We can now return to the riddle. The answer is: "Yes, indeed." The reason is that the length of the rope is not related to R, the radius of earth. The length of the original rope is known to be around 40,075,000 m = $2\pi R$. Adding 1 m will create a new radius, say $R^* = R + \Delta$; thus, $2\pi R + 1 = 2\pi(R + \Delta)$, and $\Delta = 100$ cm/$2\pi = 15.9$ cm, which is enough room for a cat to go underneath.

19.1 INTRODUCTION

Following Chapter 18, transit-service reliability is receiving increased attention as agencies are faced with immediate problems of proving credible service while attempting to reduce operating cost. Unreliable service has also been cited as the major deterrent to existing and

Table 19.1 Perception of transit options, by party

Option	Party		
	Passenger	Agency	Authority
More routes	Better service	More expensive	Better coverage
More transfers	Reduced comfort and reliability	Less expensive	More user complaints
Smart transfers	Reduced waiting and travel time	Less expensive and increased ridership	Increased use of public transit

potential passengers. Due to the fact that most of the transit attributes are stochastic, travel time, dwell time, demand, etc., the passenger is likely to experience unplanned waiting times and ride times. One of the main components of service reliability is the use of transfers as is clearly indicated in Chapters 13 and 14. Transfers have the advantages of reducing operational costs and introducing more flexible and efficient route planning. However, its main drawback is the inconvenience of traveling multilegged trips. Some works introduce synchronized (timed) timetables to diminish the waiting time caused by transfers. Their use, however, suffers from uncertainty about the simultaneous arrival of two (or more) vehicles at an existing stop. In order to alleviate the uncertainty of simultaneous arrivals, operational tactics such as hold, skip-stop, and short turn can be deployed considering the positive and negative effects, of each tactic, on the total travel time (TTT).

The arguments of Chapters 5, 12, and 15 demonstrate that the use of transfers in public transit has the advantages of reducing operation costs and introducing more flexible, efficient route planning. In contrast, the main drawback, from the passengers' point of view, is the inconvenience of traveling multilegged trips. In the attempt to diminish waiting time caused by transfers, the following sections will elaborate the new concepts.

As we noticed in Chapter 15, designing a public transit system produces conflicts among the parties involved. The agency will usually prefer a small number of routes with transfer centers in order to reduce fleet size. The passenger will prefer more routes and fewer transfers because transfers naturally decrease comfort and reliability. This conflict may intensify in areas with variable and sparse demand. The solution is to combine the advantages of a transfer-based system (lower costs) with models that increase the comfort of transfers and reduce waiting and travel time. Table 19.1 summarizes these options.

Last, using a metaphor, the only way to eat and enjoy a baguette sandwich (to arrive on time) is lengthwise (make control actions consecutively), and not across (planning all the control actions concurrently), even if you are hungry (in a rush).

19.2 MULTIAGENT TRANSIT SYSTEM

In this section, we refer to what is known as multiagent systems (MAS) and start with a brief background. Van Dyke Parunak (1997) defined MAS as collections of autonomous agents within an environment that interact with each other for achieving internal and/or global objectives. Minsky (1986) argued that an intelligent system could emerge from nonintelligent parts. His definition of the "Society of Mind" makes use of small, simple processes, called agents, each of which performs some simple action, and the combination of all these agents forms an intelligent society (system). Bradshaw (1997) classified agents by three attributes: autonomy, cooperation, and learning. He defined four agent types from these attributes: collaborative, collaborative learning, interface, and smart. The most interesting type so far as the public transit system is concerned is the collaborative agent. Such an agent is simple and can perform tasks independently, but can collaborate with other agents if necessary in order to achieve a better solution. Zhao et al. (2003) developed MAS for a bus-holding algorithm. The authors treat each bus stop as an agent; the agents negotiate with each other, based on marginal cost calculations, to devise minimal passenger waiting-time costs.

The multiagent transportation system (MATS) proposed here is based on Hadas and Ceder (2008b). The MATS will be composed of the following agents: public transit vehicles, passengers, road segments, transit agencies, and transit authorities. In order to construct the system as a whole, it is necessary to define each agent and the interrelationship of agents. Since the system is complex and cumbersome, each agent will be explored separately; this will also make it easier to define the system's elements. Figure 19.1 presents the main activities and interrelationships of the proposed system.

Passenger agent: The passenger agent plans the trip according to passenger preferences, and based on the real-time information available from road-segment agents (travel time) and vehicle agents (routes and dwell times). Passengers will input to their cell phones or personal digital assistant (PDA) every trip desired from point A to point B. The agent, referring to each passenger preference, will search for the best trip based on the public transit data

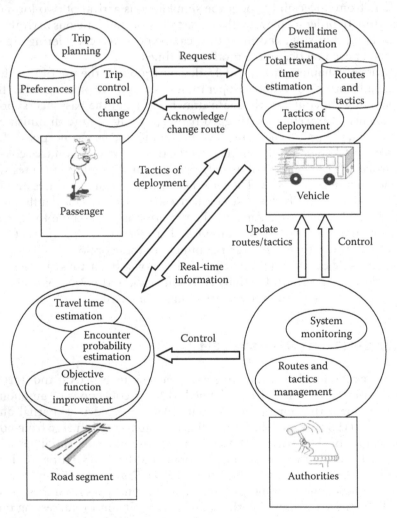

Figure 19.1 Agents' activities and relationships.

available at that time. The passenger will choose among the possible options and book a trip in the system. Then the passenger will be notified by SMS or a different means of communication about path changes owing to, for example, traffic congestion or a system-wide optimization deploying tactics that change the planned schedule of route legs. In response to the route changes, the passenger agent will try to find a better route, if possible changing the existing planned path. The agent itself can be a small software program running on a cell phone or PDA.

Road-segment agent: The road-segment agent can reside physically, as part of the road infrastructure (e.g., at a traffic light control), or virtually, as part of a multiagent software system. The agent is responsible for the following activities: estimating travel time, estimating encounter probability, improving the system's objective function, and instructing vehicles on tactical deployment. Each road-segment agent continuously collects local traffic-flow information and estimates the travel time. The agent evaluates the encounter probability (for vehicles transferring passengers), based on the adjacent road-segment travel-time estimations and vehicle locations. This probability is described in the next section. Using dynamic programming (DP), each agent or group of agents calculates the optimal tactical deployment that will optimize the system's objective function, which is the total expected travel time.

Vehicle agent: The vehicle agent can be part of the onboard automatic vehicle location (AVL) system or, virtually, part of a multiagent software system. The agent estimates the vehicle's dwell time according to booked trips and demand forecasts, using travel-time estimations (from the road-segment agents) and estimated deviations from the planned timetable schedule. The vehicle agent receives instructions for the deployment of tactics from the road-segment agent in order to improve the system's objective function.

Agency agent: This agent is responsible for designing and managing the transit network, for updating timetables, and for configuring the possible operational tactics available on each road segment for each route.

Authority agent: The authority (local authority or federal government) is responsible for monitoring system performance according to the determined/decided indicators.

The MATS offers the following benefits, which are inherent in the multiagent approach:

Extensibility: Easily allows the system's growth and adding of new resources. Each vehicle has computation power to contribute to the entire network. Adding a new vehicle is similar to plugging in a new computer to a local area network.

Fault tolerance: The proposed system will handle failures. Critical operations that are heavily dependent on computation and that are built on a standard central computing architecture must have a redundancy system in order to maintain a certain level of service. Redundant systems are expensive and cumbersome; however, MAS agents are distributed. Consequently, if some agents are down, the others will continue to perform because of their autonomous capability. Table 19.2 presents the mode of operation and outcome for different communication scenarios in case of communications disturbances.

Scalability: Distributed systems can theoretically grow without limit (e.g., the Internet).

Adaptability: Changing rules and transmitting data to the agent is quick and simple, similar to the spreading of a virus.

Efficiency: Negotiations between agents can reach an optimal or a near-optimal solution efficiently (e.g., see Raiffa, 1982).

Table 19.2 Communication scenarios

Scenario	Mode of operation	Outcome
Full communication	Online collaboration	Synchronized transfers Total travel-time reduction
Partial communication	⇕	⇕
No communication	Autonomous According to timetables	Ordinary transit system with reliability problems

Distributed problem solving: Cooperating agents can distribute subtasks to other agents that are idle and can contribute their computer time to solve these subtasks (e.g., see Smith [1980] and Davis and Smith [1983]).

Stability: The use of a closed set of operations tactics for each road segment in order to eliminate solution sets that are not stable.

The following is a simple example of the benefits of MATS. On board a shuttle heading for the train station are 30 passengers, of whom 20 plan to take the 12:20 train. The train headway is 30 min, and trains are assumed to depart on time. Along the shuttle route, four more passengers are waiting (say, for the system's callback; see Chapter 17) to take this shuttle service. According to the estimated arrival time, there is a probability of 0.60 of reaching the train station at 12:15 and a probability of 0.40 of reaching the station at 12:25. In this case, the expected total waiting time for the train will be $(0.6 \cdot 5 + 0.4 \cdot 25) \cdot 24 = 312$ min. If the shuttle deploys a shortcut tactic, the estimated arrival time probability will be increased to 0.95 for 12:15, in which case the four waiting passengers will miss the 12:20 train and will have to board the next shuttle (the average headway being 20 min). The total new waiting time for the 20 passengers will be $(0.95 \cdot 5 + 0.05 \cdot 25) \cdot 20 = 120$ min. For the existing four waiting passengers, and assuming that the next shuttle for the 12:50 train does not execute any operational tactic, the total waiting time will be $(0.6 \cdot 5 + 0.4 \cdot 25 + 20) \cdot 4 = 132$ min. In terms of system optimization, the shortcut tactic results in $120 + 132 = 252$ min total waiting time, compared with 312 min otherwise; thus, the tactic will be preferred.

19.3 VEHICLE ENCOUNTERS ON ROAD SEGMENTS

MATS encapsulates activities and processes that the agents continuously perform in parallel with one another. Some of these activities have been described in this book and can easily be integrated into MATS: estimating travel time (see Section 18.3.2), estimating dwell time (see Section 18.3.1), and finding the best route (see Chapters 15 and 17). However, this section presents a step-further method, in which MATS incorporates new modeling concepts rooted in the determination of encounter probabilities. The section is based on Hadas and Ceder (2008c).

Figure 19.2 presents the main activities of a road-segment agent. This agent is responsible, among other things, to the public transit–system optimization. Hadas and Ceder (2008b) show that modeling encounter probability, constructing a set of operational tactics, and modeling a new operational objective function of the total public transit–system travel time (travel times, transfer times, and waiting times) are essential for developing a DP model that can optimize the system.

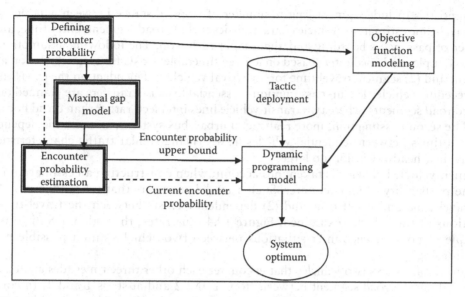

Figure 19.2 Road-segment agent's main activities.

19.3.1 Defining an encounter probability

Transit-vehicle behavior can be illustrated by a time–space (T-S) trajectory, an example of which is shown in Figure 19.3. The notations S_0 and S_3 represent the entrance and exit of the road segment, respectively; t_1 is the entry time of the vehicle to the road segment; t_2 and t_4 are the arrival times at stops S_1 and S_2, respectively; and t_3 and t_5 are the departure times from these stops. The slopes from t_1 to t_2 and from t_4 to t_5 show the average travel time (based on average speed) between two consecutive stops. The average travel speed depends on traffic-flow characteristics, speed limits, geometric design, and driver behavior.

The dwell time depends on the number of passengers boarding and alighting, vehicle configuration (number of doors, entry and exit doors, and payment methods), and driver behavior.

The probability of an encounter between two transit vehicles, especially buses, is a function of road-segment characteristics and vehicle travel characteristics. Road-segment

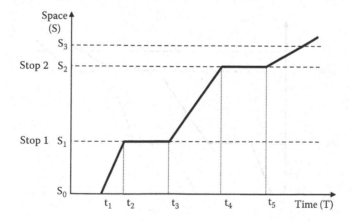

Figure 19.3 Time–space diagram illustrating transit-vehicle trajectory behavior on a road segment.

characteristics include segment length, number of stops, distance between adjacent stops, and travel-time estimation. Vehicle characteristics include road segment, entry time, and the number of passengers boarding and alighting at each stop. The following assumptions are made: (1) a planned encounter (based on a given timetable) exists for two vehicles on a road segment and (2) segment travel time for an arrival vehicle is dependent on the travel time of the preceding vehicle. The first assumption is essential because transfers are planned on pre-defined road segments (where two transit-vehicle lines intersect) rather than on ad hoc transfers. The second assumption, more realistic in urban bus systems, assumes the dependency of travel times between sequential vehicles (experiencing similar traffic characteristics) as long as their headway is not too large.

Transit-vehicle behavior is taken into account when constructing an estimation model for the probability of an encounter. Such a model considers that (1) randomness exists for travel time and dwell time and (2) dependency exists between the travel-time distributions of the vehicles examined. Figure 19.4 illustrates, through a T-S diagram, an example of possible encounter situations between two vehicles and a possible miss of encountering.

Figure 19.4a shows two vehicles that encounter each other three times (designated e_1, e_2, and e_3) along the road segment between stops 1 and 2 and at stops 1 and 2. In contrast, Figure 19.4b shows the case in which, because of a large headway, the two vehicles missed their planned encounter; as a result, the passengers who wanted to transfer missed their connection and will experience a longer wait.

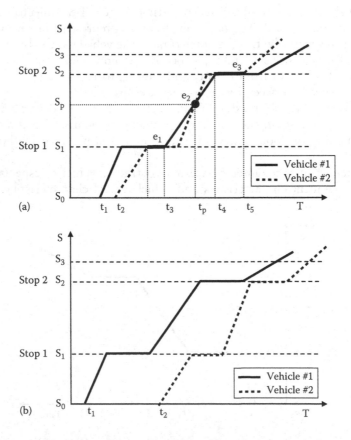

Figure 19.4 (a) Public transit vehicles making an encounter; (b) public transit vehicles missing an encounter.

T-S trajectories are constructed for each simulation run. With the use of line-intersection techniques, the number of encounters (point and trajectory intersections as in Figure 19.4a) is accumulated as is the number of missed encounters. Using this information, the following formula may be established:

$$EP = \frac{SR - ME}{SR}$$ (19.1)

where
  EP is the encounter probability estimator
  SR is the number of simulation runs
  ME is the number of missed encounters

This simulation model can be integrated into the road-segment agent. The calculation of EP will take place continuously for each group of transit vehicles planned for transferring passengers.

19.3.2 Estimating encounter probability

A JAVA-based simulation was developed for estimating encounter probabilities. The simulation emulates the behavior of two vehicles along a road segment, given road-entry time distributions, stop locations on the road segment, the number of passengers (boarding and alighting at each stop), and travel-time distribution parameters between stops. Each vehicle acts as an independent entity, using the threading capabilities of JAVA (the ability to emulate the parallel processing of the simulation entities).

19.3.2.1 Postsimulation analysis

The postsimulation analysis includes the following: (1) finding the T-S overlap of each simulation run, (2) computing the encounter's probability estimator, and (3) computing the probability of Vehicle #1 arriving before Vehicle #2 given that the vehicles did not meet along the road segment.

Let T_v^r, S_v^r be the time and space vectors for vehicle v at simulation run r, respectively, representing the T-S trajectories constructed for each vehicle and simulation run, et_v^r be the entry time of vehicle v to the road segment at simulation run r, and E_r be the encounter counter for simulation run r, ME_r be the missed encounter indicator for simulation run r, and Pv_r be an indicator whether the vehicles did not meet and vehicle v_1 arrived before vehicle v_2. Using Equations 19.2 and 19.3 as described by Shaffer and Feustel (1991), it is possible to enumerate all the encounter points and intervals (illustrated in Figure 19.4 as e_1, e_2, and e_3) for each set of lines based on vectors T_v^r, S_v^r:

$$x = \frac{\begin{vmatrix} \begin{vmatrix} x_1 & y_1 \\ x_2 & y_2 \end{vmatrix} & x_1 - x_2 \\ \begin{vmatrix} x_3 & y_3 \\ x_4 & y_4 \end{vmatrix} & x_3 - x_4 \end{vmatrix}}{\begin{vmatrix} x_1 - x_2 & y_1 - y_2 \\ x_3 - x_4 & y_3 - y_4 \end{vmatrix}}$$ (19.2)

$$
y = \frac{\begin{vmatrix} x_1 & y_1 \\ x_2 & y_2 \end{vmatrix} \quad y_1 - y_2}{\begin{vmatrix} x_3 & y_3 \\ x_4 & y_4 \end{vmatrix} \quad y_3 - y_4}{\begin{vmatrix} x_1 - x_2 & y_1 - y_2 \\ x_3 - x_4 & y_3 - y_4 \end{vmatrix}}
\tag{19.3}
$$

Moreover, it is possible to calculate the following:

$$
ME_r = 1 - \text{sign}(E_r)
\tag{19.4}
$$

$$
pv1_r = \begin{cases} 1 & \text{if } et^r_{v1} \leq et^r_{v2} \text{ and } E_r = 0 \\ 0 & \text{otherwise} \end{cases}
\tag{19.5}
$$

as well as construct the following encounter probability estimator (where SR is the number of simulation runs):

$$
EP = \frac{SR - \sum ME_r}{SR}
\tag{19.6}
$$

and these two conditional probabilities:

$$
Pr_{v1,v2} = Pr(v_1 \text{ arrives before } v_2 | \text{missed encounter}) = \frac{\sum pv1_r}{\sum ME_r}
\tag{19.7}
$$

$$
Pr_{v2,v1} = Pr(v_2 \text{ arrives before } v_1 | \text{missed encounter}) = 1 - \frac{\sum pv1_r}{\sum ME_r}
\tag{19.8}
$$

Equations 19.6 through 19.8 are the main constructors of the public transit–system objective function of minimum TTT (Hadas and Ceder, 2008b).

19.3.3 Deployment of tactics

The stochastic nature of transportation networks enforces a reduction in encounter probabilities. Hence, in order to improve encounter probabilities, a change of tactics must be imposed on the vehicles. A list and description of possible online tactics in transit operations appear in Section 18.6.3. The main real-time control tactics for the encountering-coordination problem are as follows:

- Holding the vehicle (at terminal, midroute point)
- Skip-stop operation
- Changes in speed (but not above the lawful speed limit)
- Short-turn operation
- Shortcut operation

A set of tactics will be available for deployment for each road segment and vehicle. The particular tactic or tactics deployed will be decided by the transit agency. For example, the agency may decide to employ a shortcut tactic if, according to demand forecasts, passengers seldom wait to embark along the section(s) skipped.

19.3.4 Improving encounter probabilities

Chapter 18 indicates that an important aspect of operating a public transit system is the use of online information to change vehicles routes and/or timetables dynamically. The changes are made in order to increase the system's objective functions (TTT and encounter probability for transferring passengers). Estimating an encounter probability is crucial to deciding whether or not an encounter is plausible. Moreover, information of a maximum encounter probability is important for deciding whether it is worthwhile to change tactics (which might cause increased travel time for waiting passengers).

In order to achieve maximum overlap of T-S trajectories, a time shift must be placed on a vehicle's road-segment entry time. Figure 19.5 illustrates an overlapping situation between two vehicles that depart at the same time from stop 1. Note that their dwell times at stops 1 and 2 are not the same. This S-shaped overlap has a time length of $(T_b - T_a)$ and a space length of $(S_2 - S_1)$. Two areas of Figure 19.5 are defined by A_1 and A_2 in units of space multiplied by time; A_1 is between S_1 and S_0, and A_2 is between S_3 and S_2. The simulation program calculates these areas for an average speed between each two adjacent stops for each vehicle. These areas are used to calculate an average weighted headway, H_w, in time units between the two vehicles along the road segment as follows:

$$H_w = \frac{A_1 + A_2}{S_3 - S_0} \tag{19.9}$$

The rationale for Equation 19.9 is that the smaller the H_w, the larger is the overlapping area. For two parallel trajectories, H_w will be the same as the constant headway between the two trajectories. An algorithm was constructed, first, to calculate H_w for each S-shaped overlap that exists between two adjacent stops and, second, to determine the overlapping case in which H_w is the minimum.

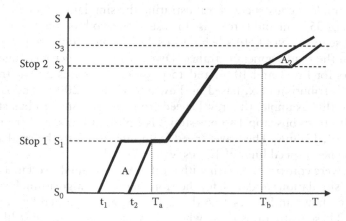

Figure 19.5 Time–space diagram illustrating an overlapping situation.

19.3.5 Finding the time gap for maximum time–space diagram overlap

The model described (Hadas and Ceder, 2008c) finds the tactics that will minimize the objective function of TTT based on a set of encounters that can be improved after calculating the current encounter probability estimator and the maximum encounter probability. The remaining-gap vector (for all encounters) is the state of the variable vector because each combination of tactics, at each step, will reduce the remaining gap. Because the effect on the objective function is calculated for each tactic, the optimal decision is known for each state.

Given road-segment characteristics, estimated time of road entry for two vehicles, and the amount of passengers to board or alight, the following algorithm will produce the time gap for maximum T-S diagram overlap (G). The algorithm calculates for each S-shaped overlap that exists within the road segment (for each adjacent bus stop) the weighted average time gap (H_w) and finds the overlap that produces $Min(H_w)$.

Let b denote the bus stop index, with b = 0 representing the segment entry point, and if n bus stops are spaced along the road segment, b = n + 1 represents the exit point. Let d_i^b denotes the dwell time of bus i at bus stop b, and let $d_i^0 = d_i^{n+1} = 0$; v_i^b is the average speed of vehicle i between bus stops b and b + 1; L_b is the distance between bus stop b and bus stop b + 1; L denotes the total segment length; t_i is the entry time of vehicle i; and A_b denotes the areas boxed by the two vehicle T-S diagrams between bus stop b and bus stop b + 1. The algorithm is as follows:

Initialization: b = 0, $T_i = t_i$, G = 0

$$\text{Loop}: G_b = |T_1 - T_2|, \quad Hw_b = \frac{\sum A_j}{L}, \quad b = b + 1, \quad T_i = T_i + L_b \cdot V_i^b + d_i^b$$

If b > n finish, else return to Loop

$$\text{Finish}: G = \{G_b \mid b : \min(Hw_b)\}$$

G will be the maximum gap that must be overcome in order to achieve maximum encounter probability.

19.3.5.1 Example

The following example is constructed for illustrating the simulation model. A road segment has a total length of 3500 m and three bus stops serving two bus routes. Bus 2 is estimated to enter the road segment N(90,50) s behind Bus 1. The bus stops are spaced, 500, 1500, and 3000 m from the road segment entrance. Dwell time is 5 s per passenger with 12, 25, and 20 passengers for Bus 1 and 10, 6, and 15 passengers for Bus 2. The travel speed has lognormal (LN) distribution of x, based on Law and Kelton (2000), and is referred to as $x \sim LN(\mu_x, \sigma_x)$. In this example, the travel speed from the entrance to bus stop 1 is in m/s and has LN(8, 10), from bus stop 1 to bus stop 2 is LN(6, 20) m/s, and from bus stop 2 to bus stop 3 is LN(10, 13) m/s. The travel speed from bus stop 3 to the road segment exit is LN(8, 5) m/s. A general speed limit of 17 m/s was enforced.

Four scenarios were constructed, each with a different S-shaped overlap and time gap, in order to provide a simulation validation for the aforementioned algorithm. For each scenario, the T-S diagram is shown in Figures 19.6 through 19.9, accompanied by 2D histograms of a T-S encounter. These histograms show where the encounters will most likely take place.

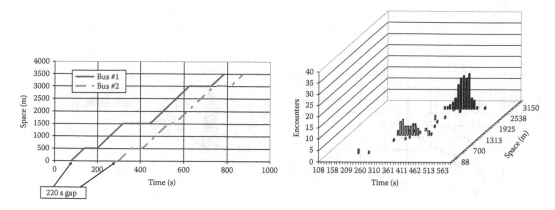

Figure 19.6 Time gap of 220 s.

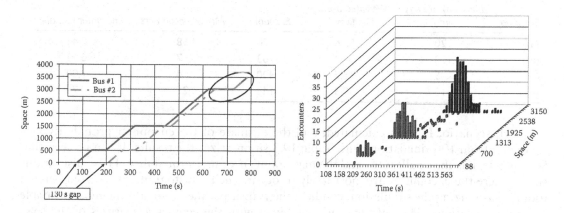

Figure 19.7 Time gap of 130 s.

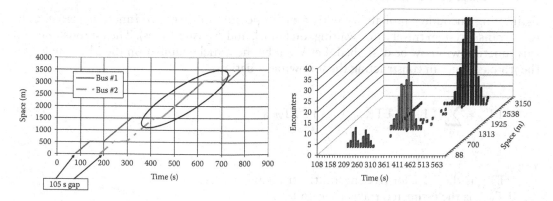

Figure 19.8 Time gap of 105 s.

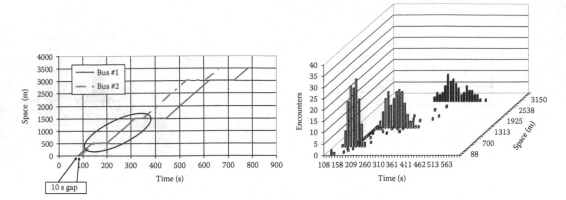

Figure 19.9 Time gap of 10 s.

Table 19.3 Simulation results

Scenario	Headway at entry point (s)	Weighted average headway (s)	Encounters	Missed encounters	Encounter probability
(a)	220	153.6	90	58	0.42
(b)	130	51.4	263	7	0.93
(c)	105	45.7	400	0	1.00
(d)	10	59.3	144	25	0.75

For each scenario, H_w was calculated, and the estimate of the encounter probability was extracted from 100 simulation runs. Table 19.3 summarizes the results. It can be seen that Scenario (c) has the smallest H_w and the maximum encounter probability. These histograms show where the encounters are most likely to occur. Such a histogram, aside from being a tool for visualizing the simulation, can help the transit planner compare the most probable encounter locations. This information, combined with the geometric features of the road segment and the existing location of the transit stops, may help in relocating the stops so as to expedite the transfer process.

19.3.6 Objective function improvement

Estimating encounter probability adds a new operational objective function, namely, the total transit-system travel time (waiting, on board, and transfer times). The establishment of this function will now be described. Let V(i, j) be the vehicle number on the ith route leg for the jth passenger in constructing the following equation:

$$\text{PTT}_{i,n} = \sum_{j=1}^{n-1} \left(\text{TT}_{V(i,j)} + \text{ETR}_{V(i,j),V(i,j+1)} \right) + \text{TT}_{V(i,n)} \tag{19.10}$$

where
  $\text{PTT}_{i,n}$ is the TTT for passenger i, the trip having n legs
  $\text{TT}_{V(i,j)}$ is the estimated travel time on leg j
  $\text{ETR}_{V(i,j),V(i,j+1)}$ is the expected transfer time between leg j and leg j + 1

The expected transfer time is

$$ETR_{V(i,j),V(i,j+1)} = EP_{V(i,j),V(i,j+1)} \cdot TR_{V(i,j),V(i,j+1)} + (1 - EP_{V(i,j),V(i,j+1)}) \cdot [Pr_{V(i,j),V(i,j+1)} \cdot W_{V(i,j),V(i,j+1)}$$
$$+ (1 - Pr_{V(i,j),V(i,j+1)}) \cdot W_{V(i,j)+1,V(i,j+1)}] \tag{19.11}$$

where

$EP_{V(i,j),V(i,j+1)}$ is the encounter probability

$TR_{V(i,j),V(i,j+1)}$ is the direct transfer time

$Pr_{V(i,j),V(i,j+1)}$ is the conditional probability that vehicle V_1 will arrive ahead of vehicle V_2, given that they will not meet along the road segment

$W_{V(i,j),V(i,j+1)}$ is the headway between two vehicles

The headway is an input to the model and is based on real-time information and estimated time of arrival of the vehicles. The TTT of the system can be expressed by

$$TTT = \min \sum PTT_i \tag{19.12}$$

Minimizing both travel time and waiting time prevents the possibilities of reducing waiting time at the cost of much longer travel time, because of the operational tactics deployment.

The objective function in Equation 19.11 can be optimized by a DP method (see Section 5.5 and Appendix 9.A for an example of DP formulation). Following the calculation of both current and maximum encounter probability estimators, operational tactics (discrete decision variables) are examined for minimizing the new objective function. Combining the distributed computing capabilities of MAS with the DP approach can produce an efficient algorithm for online use in optimizing real-world public transit systems.

19.4 SIMULATION MODEL FOR SYNCHRONIZED TRANSFERS

A brief review is introduced for describing synchronized transfers. The availability of real-time information on bus locations, estimated arrival times, number of passengers, and their destinations enabled Dessouky et al. (1999) to develop an algorithm of the bus dispatching process at timed-transfer points. Such an algorithm can intelligently decide whether or not a bus should be held in order to achieve a transfer with a late bus. In another study, based on AVL technology, Dessouky et al. (2003) present a method to forecast accurately the buses' estimated arrival times and to use bus-holding strategies to coordinate transfers. The use of advanced public transit systems on a fixed route and with paratransit operations was found to be important for improving departure times and transfers by Levine et al. (2000). In order to overcome the complexity of trip planning, Horn (2004) suggests an algorithm for the planning of a multilegged trip. Its objective is to construct a journey while minimizing the travel time, subject to time window constraints.

This section is based on Hadas and Ceder (2008a). Figure 19.10 illustrates the process of improving the TTT, which is the sum of the trip legs and waiting time of all passengers of a transit system. Use of operational tactics can help in performing transfers at point X by changing the arrival time of a vehicle (or vehicles) to the transfer point; this will decrease TTT (because of a shorter waiting time). On the other hand,

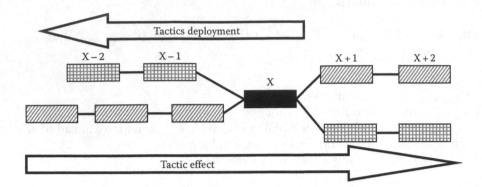

Figure 19.10 Routes of two vehicles scheduled to transfer passengers at road segment X.

the deployment of tactics may increase the travel time of passengers on board the vehicles (i.e., using hold tactic) or increase the waiting time of passengers at the bus stops (i.e., hold and skip-stop tactics).

The illustration of Figure 19.10 shows that the time change can be reached by deploying tactics along the upstream road segments of X (X − 1, X − 2, etc.) with the effect of the change on all road segments downstream to the furthest road segment in which tactics are deployed (X − 2, X − 1, X, X + 1, X + 2, etc.). The optimization process needs to balance travel-time changes in order to minimize TTT. Clearly, this process fits into the description of a DP problem that is formulated as an allocation problem with G (the estimated time gap of arrival to the transfer point between the two vehicles) as a resource, TTT as the objective function, and N road segments as stages.

Because of incorporating n stages in the model (each state is a road segment, as described is Figure 19.10), m tactics (each tactic is an optional tactic that can be deployed at a road segment), and s transfer points (state variables representing the scheduled transfers based on the synchronized timetables), its complexity will rise if the number of simultaneous transfers is high. In order to reduce the level of complexity, a distributed DP model was developed by Hadas and Ceder (2008a) breaking down the global problem into s subproblems, each with a single state variable. Such a model can be executed in a reasonable time because the complexity of each subproblem is $O(n \cdot m)$; see Section 5.5 for comprehension of the measure of complexity. This optimization model integrates the public transit network (timetables, synchronized transfer points, operational tactics) with online data (such as public transit-vehicle locations, travel-time estimation, and demand forecasts) for the reduction of TTT and increase of travel comfort. For example, if the estimated arrival time of a bus to the transfer point is delayed (because of traffic congestion or increased dwell time) and as a result the estimated encounter probability decreased significantly, the system optimization model will recommend the deployment of tactics so as to increase the encounter probability and at the same time to reduce the TTT.

19.4.1 Network simulation

The simulation model is constructed for (1) validating, evaluating, and analyzing the benefits of the optimization and (2) comparing the performance of a highly complex global-optimization problem to a low-complexity suboptimal local optimization. This simulation model is composed of two components: (1) simulation model of a public transit network and (2) DP optimization model.

19.4.1.1 Simulation principles

The simulation model is discrete. In each step, buses are moved to the next road segment in which the arrival time for each bus to each road segment is known, but all activities within the road segment are performed in a single simulation step. The simulation model can be executed in three modes:

- Not optimized, in which the operation of the buses is not altered during the simulation run; this is the basis for evaluating the optimization.
- Global optimization, in which the optimization is carried out for the whole network.
- Local optimization, in which the optimization is performed locally for each road segment.

For the three optimization modes, two parameters are in use: time horizon and space horizon. The time horizon parameter determines the forecast range of optimization. Each time optimization is carried out, all future arrivals of buses to the road segment within the time horizon are treated. This parameter is relevant for global and local optimization processes. The space horizon parameter is relevant for local optimization only. This parameter affects the collaboration of neighbor road segments. If two or more road segments are within the range of the space horizon parameter, then the optimization process is performed jointly; otherwise, the optimization is carried locally.

Figure 19.11 illustrates the behavior of the space horizon parameter on a grid-shaped transit network. The straight lines are the transit networks, the black dots are two road segments with possible transfers between transit routes, and the circles are the space horizons. In Figure 19.11a, calculations are done jointly as in global optimization because of the intersection of the space horizons; in Figure 19.11b, there are two separate calculations, without interactions, because both space horizons do not intersect.

The TTT is composed of riding time and waiting time according to the total travel-time objective function. Riding time is estimated from a riding time estimation matrix, and waiting time is estimated as follows: (1) calculate the estimated arrival time of both buses to the transfer road segment, (2) compute the direct transfer probability, and (3) compute the waiting time.

An example, shown in Figure 19.12, represents a public transit network with the main line connected to the railway and two feeding lines. It is difficult to solve such a network in a real-life situation because of the three transfer locations associated with the main line sharing all transfer areas. The example network of Figure 19.12 illustrates 14 road segments, 3 bus routes, and a train line; this network is used for evaluation and validation of the

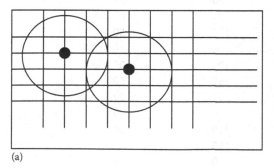
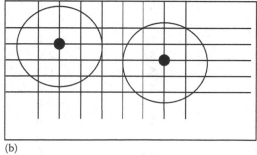

(a)   (b)

Figure 19.11 Space horizon parameters in a grid-shaped transit network: (a) joint calculation and (b) two separate calculations.

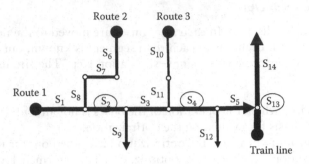

Figure 19.12 Network example.

model. The bus routes and the train line characteristics are described in Table 19.4. Planned transfers are along S2 (for routes 1 and 2) and along S4 (for routes 1 and 3). Transferring passengers to the train line is from route 1 at a fictitious road segment (S13). The transfer road segments are marked with circles in Figure 19.12. The demand between each pair of road segments appears in Table 19.5. Passenger arrival at each road segment is shown in Table 19.6. The arrival time distribution reflects the case in which most of the passengers are aware of the published timetables and arrive before the planned departure.

Riding time along each segment is distributed LN with the parameters LN(5, 2) in minutes, except for the train road segment, which is a constant of 30 min. Because the simulation model is only at the road segment level, not dealing with intersegment characteristics

Table 19.4 Route characteristics

Route no.	First departure	Headway (min)	Route layout
1	6:00	20	S_1–S_2–S_3–S_4–S_5–S_{13}
2	6:25	20	S_6–S_7–S_8–S_2–S_9
3	6:40	15	S_{10}–S_{11}–S_4–S_{12}
Train line	7:00	20	S_{13}–S_{14}

Table 19.5 Origin–destination matrix in passengers per hour

From	To		
	S_9	S_{12}	S_{14}
S_1	300	300	100
S_2	100	100	50
S_3	0	100	100
S_4	0	100	50
S_5	0	0	10
S_6	200	100	100
S_7	150	150	50
S_8	150	150	100
S_9	0	0	0
S_{10}	0	150	50
S_{11}	0	150	50
S_{12}	0	0	0
S_{13}	0	0	0
S_{14}	0	0	0

Table 19.6 Arrival time distribution

Arrival time	Percentage
Planned departure − headway	10
Planned departure − headway/2	20
Planned departure	55
Planned departure + headway/2	10
Planned departure + headway	5

(such as bus stops), a probability table for a direct transfer was constructed based on the time gap between two buses entering a road segment; that is, for a time gap of 1 min or less, the encounter probability is 0.95, whereas for 2, 3, 4, 5, and above the probabilities are 0.80, 0.60, 0.30, and 0.00, respectively.

For each pair of buses (or bus and train) and according to the estimated time gap, an encounter probability will be selected directly from that table. A number between 0 and 1 will be drawn from a uniform distribution, and if the result is less or equal to the encounter probability, then a direct encounter occurred. In the case of a direct encounter, transferred passengers will not experience any delay; otherwise, the transferred passengers will be dropped at the road segment and wait for the next bus or train. For comparison of different optimization types, the seed for the pseudorandom number generator can be selected, hence generating the same set of data. The following two discrete tactics are available: (1) hold for 1, 2, 3, 4, or 5 min and (2) skip a segment that results with travel-time reduction of 4 min.

Output: For each passenger group departing the same origin, heading toward the same destination, and arriving at the bus stop at the same time, the information of group number, origin, destination, number of passengers, and start and end of trip time will be accumulated and saved. For each trip leg, leg number, route number, bus number, transfer point, transfer type (1 = direct, 2 = arrive before second bus, 3 = arrive after second bus), and boarding and alighting time will be collected.

Scenarios: Eleven scenarios were constructed for the simulation on the basis of the parameters summarized in Table 19.7. These scenarios present the combined effect of major attributes of public transit networks on the model: (1) headway, routes with long versus short headways; (2) riding time variance, low variance (fewer disruptions to the traffic flow)

Table 19.7 List of scenarios

Scenario	Synchronized timetables	Time horizon (min)	Headway	Variance
1	Yes	10	Short	Base
2	Yes	15	Short	Base
3	No	10	Short	Base
4	Yes + high load to train	10	Short	Base
5	Yes	10	Long	Base
6	Yes	20	Long	Base
7	Yes	30	Long	Base
8	No	10	Long	Base
9	Yes	10	Short	High
10	Yes	15	Short	High
11	Yes	10	Short	Low

Table 19.8 Headways for nonsynchronized scenarios

Route	First departure	Scenario 3 Headway (min)	Scenario 8 Headway (min)
1	6:00	21	32
2	6:25	18	42
3	6:40	12	27
Train line	7:00	20	20

versus high variance; and (3) synchronized transfers, whether or not the transit network incorporates timed transfers (planned synchronization). The time horizon attribute is relevant for the performance of the model and was included for comparison purposes.

Scenarios 1 and 2 present synchronized timetables. Scenario 3, for comparison, does not have synchronized timetables; the purpose of the simulation, in this case, is to provide insights on the effect of the model in synchronized and nonsynchronized networks. Scenario 4 provides a higher load to the train (which has fixed timetables, thus cannot deploy operational tactics). Scenarios 5 through 8 are characterized by long headways, and Scenarios 9 through 11 are characterized by high and low travel-time variance. These scenarios represent typical examples of transit networks. One of the main assumptions for the model is that the transit network of routes is synchronized (at the planning stage); the objective of the optimization process is to approach better synchronization considering the changes of travel times that may be the result of traffic congestion or change in passenger demand. Hence, the simulation model was executed with synchronized timetables (Table 19.4) and nonsynchronized timetables (Table 19.8).

Few options of increased time horizon were checked. The larger the time horizon, the higher the complexity; however, it increases the ability to react earlier to more events. The space horizon for the example was set to 1 in order to force local optimization.

Results: Each scenario was executed for a time period of 6 h, once per optimization type. The full list of results appears in Hadas and Ceder (2008a). The main conclusions reached in this study were as follows: (1) by using tactics, it is possible to attain a reduction in the TTT between 3% and 17%; (2) global optimization tends to yield better results than local optimization; (3) because local optimization has only one state variable and the execution of each optimization problem is carried out in parallel, the methodology is suitable for online optimization problems; (4) aside from the reduction in travel time, the number of direct (without wait) transfers increased significantly, and this result has a significant effect on the comfort of the ride and ease of transfer; and (5) the simulation results, which depend on the scenario's characteristics as well as the optimization parameters, call for the development of a full-scale simulation system for the analysis of real-world networks and the calibration of these parameters.

19.5 HOLDING AND SKIP-STOP/SEGMENT TACTICS FOR TRANSFER SYNCHRONIZATION

The list of desirable features in designing efficient transit network appears in Chapters 12 through 15. We can then realize that in any public transit network, it is impractical to have routes between every conceivable trip origin and every conceivable trip destination. There are too many possible routes, so the service cannot be economically provided. Transfers in the network enable fewer routes to provide the same coverage, and therefore, higher-frequency and higher-quality services and any origin–destination (OD) demand can be accounted for (Mees, 2000). Having fewer routes is also easier to remember, and this, along with increased frequencies, is crucial for increasing the network attractiveness for passengers wishing to travel without having to consult timetables or travel websites.

According to a study by Barton and Tsourou (2000) and Jakob et al. (2006), transfers generally allow more flexibility in route planning and effective use of services, thus resulting in more efficient public transit and the associated environmental, economic, health, and social benefits. Knoppers and Muller (1995) argued that transfers are cited as a key reason for public transit being less attractive than cars. The stochastic nature of travel times, dwell times, and passenger demands in transit networks leads to an encounter probability of two vehicles, termed a direct transfer. A direct transfer is defined as a no-wait transfer between two vehicles meeting simultaneously at the same stop. When such a direct transfer fails (missed transfer), it can cause frustration and longer waiting times for passengers who miss connections, and a less efficient system as a whole. Transfer synchronization aims to increase the number and probability of bus encounters and thus subsequent connections. Domschke (1989), Ceder et al. (2001), and more recently Shafahi and Khani (2010) used mathematical relationships to generate timetables with the maximum opportunity for direct transfers. Dessouky et al. (1999) show that it is also possible to control the time of dispatch for a better synchronization, taking into account real-time information.

Another way of improving synchronization of transfers is by using specific operational tactics in real time (Eberlein et al., 2001; Hickman, 2001; Ceder, 2007). The specific tactics evaluated by Hadas and Ceder (2008a) in Section 19.4 were holding buses at stops in anticipation of connecting buses and instructing skip-stops and shortcuts of routes to meet subsequent connections. A recent study by Ceder et al. (2013) investigates how to use selected operational tactics in transit networks for increasing the actual occurrence of scheduled transfers. The model presented determines the impact of instructing vehicles to either hold at or skip certain stops, on the TPTT and the number of simultaneous transfers.

This section is based on Nesheli and Ceder (2014), and it is an extension of Ceder et al. (2013). The section introduces the possibility of skip segment in addition to only skip an individual stop, and for real-time operational control; the refinement takes place in the optimization formulation. That is, this section refers to three tactic scenarios: no-tactic case, holding and skip individual stops, and holding and skip segments (H-SSs). The objectives are to create simulation and optimization frameworks for optimal use of the three scenarios and compare between the scenarios using a case study.

19.5.1 Formulation and modeling framework

The methodology of work commences by the use of TransModeler simulation tool (Caliper, 2013) to represent a real-life example and to generate random input data for the proposed optimization model. Then a standard optimization software, ILOG (IBM, 2012), was used to solve optimally a range of different scenarios determined by the simulation runs. Finally, more simulation runs are made, containing the tactics determined by the optimization program, so as to validate the results attained by the model.

19.5.1.1 Model description

The model developed (Nesheli and Ceder, 2014) considers transit networks consisting of main and feeder routes. The transfers occur at separate transfer points for each route. The formulation contains all the implemented tactics using a deterministic modeling.

Analytically, the model seeks to attain minimum TPTT and to increase, in this way, the total number of direct transfers. The model formulates the tactics of holding vehicles, skipping individual stops, and skipping segments, as well as indication of missing or making a direct transfer. Thus, the components of the model are (1) the effect on TPTT of holding a vehicle, (2) the effect on TPTT of skipping a stop/segment, and (3) the effect on TPTT of vehicle's being late, or not, to a transfer.

19.5.1.1.1 State variables

N	Set of all bus stops, in which $n \in N$
R	Set of all bus routes in which $\{r, r'\} \in R$
TF	Set of all transfer points, in which $tr \in TF$
Q_r^{max}	Passenger capacity of bus of route r
l_r^n	Passengers' load of route r at stop n
b_r^n	The number of boarding passengers of route r at stop n
a_r^n	The number of alighting passengers of route r at stop n
$p_{rr'}^n$	The number of transferring passengers of route r to route r' at stop n
d_r^n	Bus dwell time of route r at stop n (*in seconds*)
h_r	Bus headway of route r
c_r^n	Bus running time of route r at stop n from the previous stop
A_r^n	Bus arrival time of route r at stop n
D_r^n	Bus departure time of route r at stop n
$\Omega(t)_r^n$	Time penalty function of route r at stop n
Γ_r^n	Time to reach a desired stop skipped of route r at stop n
T_r	Bus schedule deviation of route r
TP_r^{tr}	Transfer stop of route r at transfer point of tr
E_r	Bus elapsed time of route r from the previous stop to the current position
m_r	Maximum total number of stops of route r
k_r	Positional stop of route r for a snapshot
ω	Ratio between the average speed of a bus and the average walking speed of pedestrian (same ratio for all routes and stops)

19.5.1.1.2 Parameters

θ_r^n	The number of passengers of route r for a bus departing stop n
β_r^n	The number of passengers waiting at stops further along the routes with respect to route r and stop n (future passengers)
γ_r^n	The number of passengers who wish to have transfers at transfer points with respect to route r and stop n
λ_r^n	The waiting time per passenger at previous stops due to applied tactics

19.5.1.1.3 Decision variables

HO_r^n	Bus-holding time of route r at stop n
S_r^n	Bus skipping stop of route r at stop n; if stop, skipped = 1, otherwise = 0
$Y_{rr'}^n$	Possible transferring from route r to route r' at transfer stop n, pretactics; if a possible transfer occurs = 0, otherwise = 1
$Z_{rr'}^n$	Possible transferring from route r to route r' at transfer stop n, posttactics; if a possible transfer occurs = 0, otherwise = 1

19.5.1.1.4 Assumptions

The model is designed deterministically. Therefore, the following assumptions are made:

1. There is foreknowledge of the route information, including average travel times, average passenger demand, average number of transferring passengers, and average dwell times.
2. Passenger demand is independent of bus arrival time.
3. Vehicles are operated in FIFO manner with an evenly scheduled headway.
4. Passengers will wait at their stop until a bus arrives (none leaves the system without taking the first arrived bus).
5. The bus arriving subsequently to a bus that skipped stop cannot use any of the two tactics considered.
6. Passengers on board a bus that will skip segment will be informed on this action at the time of the decision so that they can alight before or after the skipped segment; it is to note that the formulation of optimization minimizes these type of passengers and in most cases tested it is nil.
7. Stops where passengers want to transfer cannot be skipped.

19.5.1.2 Formulation and properties of holding tactic

Holding a vehicle is a tactic considered operationally for regulating undesired scheduled deviations, reducing bunching and approaching a direct transfer at transfer points. However, using the holding tactic has some drawbacks on the TTT of the passengers. That is, the holding tactic would affect three groups of passengers: (1) those on board the bus defined as θ_r^n, (2) those waiting for the bus further along the route defined as β_r^n, and (3) those who wish to have transfers defined as γ_r^n. The following formulation can now take place. It is to note that the term "route" describes a transit service that serves a series of stops. A route is made up of a collection of "trips"; each trip represents a single run, based on a certain departure time, along the series of stops of the route. For instance, a_r^n describes the number of alighting passengers at stop n of a bus serving route r:

$$\theta_r^n = l_r^n + b_r^n - a_r^n \tag{19.13}$$

$$\beta_r^n = \sum_{i=n+1}^{m_r} \left[b_r^i + \sum_{r' \in R} \left(1 - Z_{r'r}^i\right) p_{r'r}^i \right] \tag{19.14}$$

$$\gamma_r^n = \sum_{r' \in R} \left(\left(1 - Y_{r'r}^n\right) p_{r'r}^n - p_{rr'}^n \right) \tag{19.15}$$

The effect of holding tactic on the change in total passenger travel time ($\Delta TPTT_i$) with respect to route r and stop n is

$$\Delta TPTT(\text{Holding})_r^n = HO_r^n \left[\theta_r^n + \beta_r^n + \gamma_r^n \right] \tag{19.16}$$

19.5.1.3 Formulation and properties of skip-stop and skip-segment tactics

Skip-stop: When a major disruption occurs, holding a vehicle, as the only tactic available, cannot guarantee to obviate the headway variation from the schedule. This is true even with holding the following vehicles because these actions may lead to greater schedule deviations (Sun and Hickman, 2005). Skip-stop is another tactic that can be implemented to decrease the irregularity of service and increase the number of direct transfers. The advantage of skip-stop is for passengers who already are on board the bus and those to board down-stream. On the other hand, it has an adverse effect on passengers who want to alight or board at the skipped stop.

Skip segment: The skip of individual stop tactic has the limit that no more than one stop can be skipped in a row. However, if, for instance, some stops are close to each other and only a very few passengers will be impacted from skipping those stops, another tactic can take place as is illustrated by Figure 19.13. That is, consideration of a skip-segment tactic, where a segment is defined as a group (one or more) of adjacent stops.

One of the assumptions made is that passengers on board a bus that will skip segment will be informed immediately on this action at the time of the decision so that they can alight before or after the skipped segment. This assumption is considered in the optimization formulation, and in most of the cases simulated, the amount of this type of passengers was approaching zero; it is because of minimizing the TTT. In other words, if the amount of passengers of this type is large, this tactic of skip segment would not take place. The formulation of this type of passengers is explicated as follows.

Let us consider the end and start-again service stops of the skipped segment. That is, the end stop is the last stop served before the skipped segment, and the start-again stop is the first stop served after the skipped stop. Passengers who want to alight in the skipped segment and alight at the end-service stop will have extra time to reach their destination (within the skipped segment) to be termed *forward time*, Γ (*forward*). However, to determine the actual additional travel time, the bus running time to the desired skipped stop has to be subtracted, and thus, the following formulation is used:

$$\Gamma^n_{r(forward)} = (\omega - 1) \sum_{i=1}^{n} c^i_r \left(\prod_{i=q}^{n} S^i_r \right) \quad \forall (n, q \in N)\{1 \le q < n\} \tag{19.17}$$

At the same time, passengers alighting at the start-again service stop will need to go back to their destination stop, and their time is termed *backward time*, Γ(*backward*). In this case,

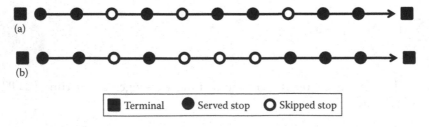

Figure 19.13 (a) Skip-stop and (b) skip-segment tactics.

additional bus running is added, and the passengers to use this way, to reach their destination, will save the dwell times of the skipped stops. These considerations yield

$$\Gamma^n_{r(backward)} = (\omega + 1)\sum_{i=n+1}^{m_r} c^i_r \left(\prod_{i=n}^{q} S^i_r\right) - \sum_{i=1}^{n} S^i_r d^i_r \quad \forall(n, q \in N)\{1 \leq n < q\} \tag{19.18}$$

It is possible that these passengers will decide to walk; thus, a *walking time* penalty function is described as

$$\Omega(t)^n_{r(walking)} = \min(\Gamma^n_{r(forward)}, \Gamma^n_{r(backward)}) \tag{19.19}$$

The alternative of walking is the waiting for the next bus to bring these passengers to their destination. The waiting time associated with upstream stops is designated γ^n_r to be

$$\lambda^n_r = \sum_{i=k_r}^{n-1} \left(S^i_r d^i_r - HO^i_r\right) \tag{19.20}$$

The *waiting time* penalty function is then determined by the following equation with consideration of schedule deviation:

$$T_r : \Omega(t)^n_{r(waiting)} = \left(h_r - T_r + \lambda^n_r\right) \tag{19.21}$$

Two penalty functions were established, walking time and waiting-time penalties, with now a new definition of *total time* penalty being the minimum of the two as follows:

$$\Omega(t)^n_{r(total)} = \min\left\{\Omega(t)^n_{r(walking)}, \Omega(t)^n_{r(waiting)}\right\} \tag{19.22}$$

Consequently, the effect of the skip-segment tactic on the change of the TTT for route r and stop n is

$$\Delta TPTT(skip - segment)^n_r = S^n_r \left[a^n_r \Omega\left((t)^n_{r(total)}\right) + b^n_r \left(h_r - T_r + \lambda^n_r\right) - d^n_r \left(l^n_r + \beta^n_r\right)\right] \tag{19.23}$$

19.5.1.4 Formulation and properties of transfers

Synchronized transfers mean that two or more buses, or other transit vehicles, meet at the same time at a transfer point. As told, this can be improved by the use of real-time operational tactics. For missed transfers, passengers have to tolerate a waiting time of $(h_r - T_r)$ where h_r is the headway at route r. Therefore, a direct transfer occurs if the following holds:

$$Y^n_{rr'} + Y^n_{r'r} = 0 \tag{19.24}$$

$$Z^n_{rr'} + Z^n_{r'r} = 0 \tag{19.25}$$

In this model, a possible transfer $Y^n_{rr'}$ (before implementing tactics) occurs if bus departure time on route of r' is after the bus arrival time on route r and vice versa for $Y^n_{r'r}$.

Same arguments apply for $Z_{rr'}^n$ and $Z_{r'r}^n$ after utilizing tactics. If the following conditions hold, direct transfers will be possible and none of the buses will be late at a transfer point:

$$A_r^n = \sum_{i=k_r}^n c_r^i - E_r \tag{19.26}$$

$$D_{r'}^n = \sum_{i=k_r'}^n c_{r'}^i - E_{r'} + d_{r'}^n \tag{19.27}$$

$$\text{if } D_{r'}^n \leq A_r^n, \text{ then } Y_{rr'}^n = 1 \quad \forall \left(p_{rr'}^n + p_{r'r}^n \geq 1 \right) \tag{19.28}$$

Subsequently, the effect of transfers on the change of the TTT for route r and stop n is

$$\Delta TPTT(transfer)_r^n = \sum_{r' \in R} \left[p_{rr'}^n \left(Z_{rr'}^n \left(h_{r'} - T_r + \lambda_r^n \right) - Y_{rr'}^n (h_{r'} - T_r) \right) \right] \tag{19.29}$$

19.5.1.5 Objective function

According to the formulations derived of the TPTT, an objective function for the proposed model can be written as

$$\min \sum_{r \in R} \sum_{n \in N} \Delta TPTT \left\{ (holding)_r^n + (skip - segment)_r^n + (transfer)_r^n \right\} \tag{19.30}$$

This objective function, of minimum TTT, is subject to the following constraints.

19.5.1.6 Constraints

Transfer points that cannot be skipped are stated as

$$S_r^n \left[\sum_{r' \in R} \left(p_{rr'}^n + p_{r'r}^n \right) \right] = 0 \tag{19.31}$$

Direct transferring would occur only if the following exists:

$$A_r^n - \lambda_r^n - D_{r'}^n - HO_{r'}^n + \lambda_{r'}^n \leq M^* Z_{rr'}^n \quad \forall \left(p_{rr'}^n + p_{r'r}^n \geq 1 \right) \tag{19.32}$$

where M is a large number.

The following is to state that tactics cannot be applied at not served stops and stops already passed:

$$HO_r^n = 0 \quad \text{when} (n < k_r) \tag{19.33}$$

$$S_r^n = 0 \quad \text{when} (n < k_r) \tag{19.34}$$

It is not allowed to skip the first stop:

$$S_r^1 = 0,$$ (19.35)

Moreover, no skipping and holding at the same stop are allowed:

$$\{S_r^n * HO_r^n = 0\}$$ (19.36)

This constraint can be simplified and reformulated. Let M denote a large number; thus, constraint (19.35) is exchanged with the following constraints:

$$\text{if } S_r^n = 1, \quad \text{then } HO_r^n \le M * \left(1 - S_r^n\right)$$ (19.37)

$$\text{if } HO_r^n > 0, \quad \text{then } M * S_r^n \le HO_r^n$$ (19.38)

The maximum holding time is not more than half the headway of the route:

$$HO_r^n \le \frac{1}{2} h_r$$ (19.39)

If a transfer occurs at pretactics situation, the same apply to with-tactics situation:

$$Z_{rr'}^n \le Y_{rr'}^n$$ (19.40)

If direct transferring is possible without the use of any tactic, there is no need to interfere; this constraint is expressed as follows where M is a large number:

$$\sum_{i=TP_r^{tr-1}}^{TP_r^{tr}} S_r^i \le \sum_{tr \in TF} M * Y_{rr'}^{tr} \quad \forall \left(TP_r^{tr} > TP_r^{tr-1} > k_r\right)$$ (19.41)

$$\sum_{i=k_r}^{TP_r^{tr}} S_r^i \le \sum_{tr \in TF} M * Y_{rr'}^{tr} \quad \forall \left(TP_r^{tr} > k_r > TP_r^{tr-1}\right)$$ (19.42)

$$\sum_{i=k_r}^{TP_r^1} S_r^i \le M * Y_{rr'}^1 \quad \forall \left(TP_r^1 > k_r\right)$$ (19.43)

If the use of tactics does not result in a transfer, no tactics are applied:

$$\sum_{i=TP_r^{tr-1}}^{TP_r^{tr}} S_r^i * \prod_{tr \in TF} Z_{rr'}^{tr} < 1 \quad \forall \left(TP_r^{tr} > TP_r^{tr-1} > k_r\right)$$ (19.44)

$$\sum_{i=k_r}^{TP_r^{tr}} S_r^{i*} \prod_{tr \in TF} Z_{rr'}^{tr} < 1 \quad \forall \left(TP_r^{tr} > k_r > TP_r^{tr-1} \right) \tag{19.45}$$

$$\sum_{i=k_r}^{TP_r^1} S_r^{i*} Z_{rr'}^1 < 1 \quad \forall \left(TP_r^1 > k_r \right) \tag{19.46}$$

Constraints (19.44 through 19.46) can be simplified. Let us denote Im as

$$Im = \sum_{tr \in TF} Z_{rr'}^{tr} \tag{19.47}$$

The maximum value of Im implies that the use of tactic does not result in a transfer and $Z = 1$ for all routes at all transfer points. Thus, the following constraint ensures the use of tactics when a transfer occurs:

$$\text{if } Im = \text{size of } Z_{rr'}^{tr}, \quad \text{then} \sum_{i=TP_r^{tr-1}}^{TP_r^{tr}} S_r^i = 0 \quad \forall \left(TP_r^{tr} > TP_r^{tr-1} > k_r \right) \tag{19.48}$$

The same applies for Constraints (19.44) and (19.45). For using skip-stop (not segment) tactic, the following constraint is for skipping only one stop in a row:

$$S_r^n + S_r^{n+1} \leq 1 \tag{19.49}$$

To restrict the number of passengers on board when the bus departs the stop, the following constraint applies:

$$\theta_r^n \leq Q_r^{max} \tag{19.50}$$

19.5.1.7 Model optimization

The formulation of the problem with the three scenarios (no tactics, holding and skip-stop [H-S], H-SS) can be solved using constrained programming (CP) technique. The CP is an efficient approach to solving and optimizing problems that are too irregular for mathematical optimization. The reasons for these irregularities as per IBM (2012) are as follows: (1) the constraints are nonlinear in nature, (2) a nonconvex solution space exists with many local-optimal solutions, and (3) multiple disjunctions exist resulting in poor returned information by a linear relaxation of the problem. A CP engine makes decisions on variables and values and, after each decision, performs a set of logical inferences to reduce the available options for the remaining variables' domains. This is in comparison with a mathematical programming engine that uses a combination of relaxations (strengthened by cutting planes) and "branch and bound."

19.5.2 Case study of real-time tactics implementation

Examination of the model developed (Nesheli and Ceder, 2014) took place with a real-life case study. The data of the case study were collected in Auckland, New Zealand. It is to

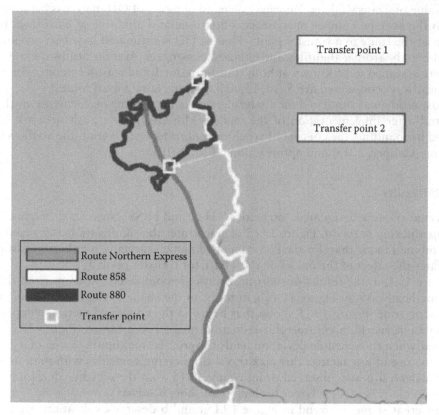

Figure 19.14 Bus system and the study routes, Auckland, New Zealand.

note that all buses of the Auckland transport network are equipped with AVL systems. The primary benefits of the AVL system are in the communication and processing of data for service monitoring, fleet management, and traveler information. In addition, AVL data are used to analyze bus travel times and schedule adherence (to implement holding tactics). This means that real-time information on bus locations can be used to predict bus arrivals at the transfer points and to determine bus schedule deviations.

The transit network of the case study consists of three bus routes and two transfer points. The first route is known as Northern Express with a dedicated lane that runs from the suburb across the Auckland CBD area and has quite a high number of passengers during peak hours. The second route, route 858, runs north–south (east of the first route), and the third route, route 880, is a loop that serves as a feeder route. Figure 19.14 illustrates the three routes and the two transfer points used.

19.5.2.1 Data

The data were recorded and available are route characteristics based, that is, stop ID, longitude, latitude, stop sequence, stop flag, stop code, stop name, route ID, user ID, point ID, and route number. A bus capacity of 60 passengers is assumed with 40 seated passengers and 20 standing. The simulation package used, TransModeler 3.0 (Caliper, 2013), considers passenger crowding for the calculation of the dwell time. That is, it takes longer to board or alight a bus if there are more standing passengers. In addition, bus dwell times were taken from Dueker et al. (2004) for simulation use. It affects whether or not the stop be skipped. The dead time (time spent at the stop without boarding or alighting) is set to 4 s as the default value of TransModeler. Passenger demand in terms of OD matrix is being estimated based on the average number of passengers boarding and alighting at each stop, from the first stop of the route to a transfer point. That is, OD is estimated based on proportions of the number of boarding, alighting, and onboard passengers. Average headways and dispatch times were assumed to be known at both transfer point 1 and transfer point 2. In addition, vehicle headways considered are 5, 10, 15, and 20 min (same for all routes).

The case study was simulated for analyzing and validating the performance of the model before and after implementation of the tactics. Moreover, the transit network has been simulated for handling the concept of synchronized transfers. Overall, the analysis includes simulation (Caliper, 2013) and optimization (IBM, 2012).

19.5.2.2 Results

Three scenarios were designated "no tactics," H-S, and H-SS. Note the difference between *skip-stop* and *skip segment*. Figure 19.15 demonstrates the significant better results for the H-SS combined tactic than for the H-S combined tactic especially for short headway cases. It illustrates the effect of the model on the TPTT for transfer point 1.

In Figure 19.15a, the results of short headways are completely different from the results of the long headways, in Figure 19.15b, in terms of the shape of the trend before and after the no-tactic zone. Figure 19.15 shows that by using the combined tactics, compared with the no-tactic scenario, a considerable reduction of TTT is attained. Figure 19.15a and b shows that when the schedule deviation tends to zero, the maximum saving of TTT occurs without the use of any tactics; this max travel-time saving coincides with max numbers of direct transfers. It is also observed, from Figure 19.15, that the schedule deviation interval, in which no tactic is used, is larger for long than short headways.

The different shapes of trend in Figure 19.15a and b deserve explication. Figure 19.15a illustrates that a larger reduction of TPTT is possible when the bus is behind schedule. This suggests that passengers waiting for a late bus prefer, in the short headway cases, to continue to wait than to find an alternative solution (walking or using another travel mode). In this case, it is worth applying the combined tactics. Figure 19.15b demonstrates an entirely different pattern; it shows that, in the long headway case, travel time is not saved much for large schedule deviations. That is, unreliable service for long headway cases cannot be compensated by the use of tactics and passengers will tend to find, for such a service, alternative solutions. However, if the deviations are reasonable, the use of tactics can save travel time and increase the number of direct transfers. It is to note that the deviations (in seconds) of Figure 19.15 are based on a few assumptions, thus cannot be translated to exact figures of being behind or ahead of the planed schedule. Instead, their trends provide a new insight of when to use the combined tactics proposed.

More on this concept of evaluation and decision of best operational tactics in real-time scenarios appears in Ceder et al. (2013) and Nesheli and Ceder (2014).

19.6 TRANSFER SYNCHRONIZATION USING IVC

This section considers another aspect related to transfer synchronization, that is, the consideration of an intervehicle communication (IVC)-based scheme to optimize the synchronization of planned transfers in transit networks. By using this scheme, transit drivers can share their real-time information, such as vehicle location, vehicle speed, and the number of passengers, with each other through a central communication coordinator. This scheme is especially useful for transit drivers from different routes and driving toward the same transfer point because

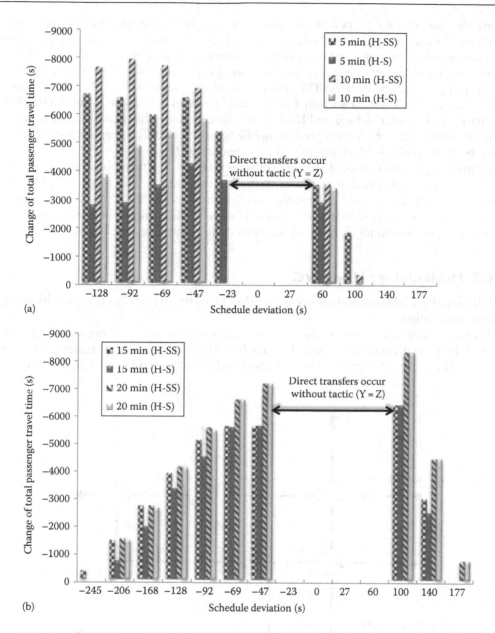

Figure 19.15 Total passenger travel time with the two combined tactics of H-S/segment at transfer point 1. (a) ΔTPTT for short headway (5 and 10 min) and (b) ΔTPTT for long headway (15 and 20 min).

they can communicate and collaborate with each other, and thus dynamically adjust their running speed on transit-route segments or holding time at transit stops. Through this cooperative driving approach, the frequency of simultaneous arrivals of transit vehicles at a transfer point can be improved. The section follows the work of Liu et al. (2014).

19.6.1 Overview of IVC

IVC enables drivers to communicate with other drivers that locate outside the range of line of sight and share real-time traffic information with each other. IVC has the ability to extend the horizon of drivers and onboard devices (OBDs) and thus has the potential to significantly improve road traffic safety, efficiency, and comfort (Luo and Hubaux, 2006). It is recognized as an important component of intelligent transportation systems (ITS) and has attracted significant research attention around the world. Since the 1980s, a number of research projects focusing on IVC have been set up, such as CarTalk 2000, Fleetnet, PATH, Car-to-Car Communication Consortium, Network on Wheels, and COMeSafety (Reichardt et al., 2002; Sichitiu and Kihl, 2008). Recent research focuses mainly on vehicular ad hoc networks, which are a kind of mobile ad hoc networks offering direct communication between vehicles (Hartenstein and Laberteaux, 2008).

The main application areas of IVC are collision warning, traffic monitoring, traffic coordination, traveler information support, traffic light scheduling, targeted vehicular communications, and car-to-land communication (Sichitiu and Kihl, 2008). These appealing applications have attracted many researches. However, the majority of current research is focused on cars. Not so much research has been done about transit vehicles.

19.6.2 Methodology Using IVC

We will describe in this section a semidecentralized control strategy for transit-vehicle transfer synchronization.

Consider a directed transit network $G = (N, R)$ with a set N of stops and a set R of routes. The transit routes are divided into a few route segments by transfer points, as is shown in the upper of Figure 19.16a. A communication control center is assigned to an

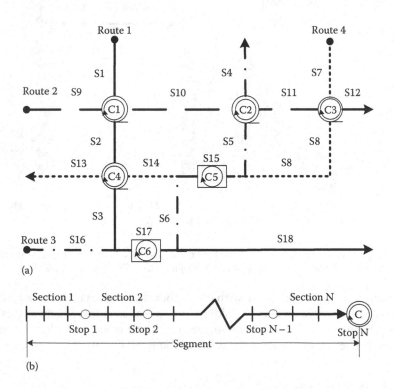

(a)

(b)

Figure 19.16 Public-transit network example illustrating network-level semidecentralized control strategy: (a) a basic transit network with communication centers and (b) an illustration of the concept of route sections.

intersection point or a shared route segment where passengers make a transfer. The communication control center is responsible for the communication coordination of vehicles on the route segments leading to it. A route segment is divided into a few route sections determined by transit stops along it, as shown in the lower of Figure 19.16b. The route section is further divided into a few smaller intervals.

An interval is a basic unit controlled by the communication control center that delivers advisory speed information to a transit vehicle when it arrives to each interval. New information is sent to the driver when moving to the next interval. A simple example is shown in Figure 19.16a. In this example network, there are 4 transit routes R_i ($i = 1, 2, ..., 4$), and the whole network is divided into 18 route segments S_j ($j = 1, 2, ..., 18$). The segments S_{15} and S_{17} are shared road-segment encounter transfers, while the other segments direct to a fixed single-point encounter transfer. Six communication control centers C_k ($k = 1, 2, ..., 6$) are assigned to the corresponding transfers and are responsible for their own route segments. For example, C_1 is in charge of the coordination of vehicles on route segments S_1 and S_9, while C_5 is in charge of S_6, S_8, and S_{15}.

Transit vehicles controlled by the same communication control center are in the same group and can communicate with their peers to share traffic information. The communication coordinator in the control center delivers advice on the real-time optimal running speed. Drivers then can adjust their running speed in a cooperative manner in order to achieve a simultaneous arrival. Once a transit vehicle passes the communication control center, it automatically joins another group of vehicles. The communication group is autonomous and self-organized. It is to note that, from the entire transit-network perspective, different communication groups can be performed at the same time. That is, the process is a decentralized and parallel process. However, from a single-route segment perspective, the communication activity is severely governed by the communication control center, and thus, it is a centralized process. Thus, this control strategy is termed semidecentralized. By doing so, drivers in the same communication group drive in a cooperative manner, and the opportunity of simultaneous arrivals improves.

19.6.2.1 Conceptual illustration

A simple example is used as an explanatory device to show the main components of the proposed IVC systems and to outline its potential benefits. The example is shown schematically in Figure 19.17, in which two transit vehicles are moving toward the same fixed single-point transfer. Both vehicles belong to the same communication group and are controlled by the same communication control center. The center comprises mainly a central server and a communication coordinator.

An OBD is installed on the transit vehicle to receive signals from global positioning system (GPS) satellites. The OBD can record information about vehicle ID, vehicle location, vehicle speed, time, route direction, and route ID. A SIM card is embedded in the OBD. The recorded data then can be transmitted to the database in the communication control center through GSM/GPRS networks. These data are visualized in geographic information system (GIS) maps. The communication coordinator has knowledge about the relatively accurate location and speed information of the vehicles of the same group. Based on the knowledge, advisory optimal speed information is disseminated to the drivers in the group so as to meet simultaneously or within a given time window at the same transfer point. The advisory information can be displayed online to the driver on the onboard variable message sign installed in the vehicle. This will allow for a peer-cooperative communication. We note that the basic assumption of the IVC system is that drivers will comply with the recommended (feasible) speed and holding time so as to materialize the direct transfers of

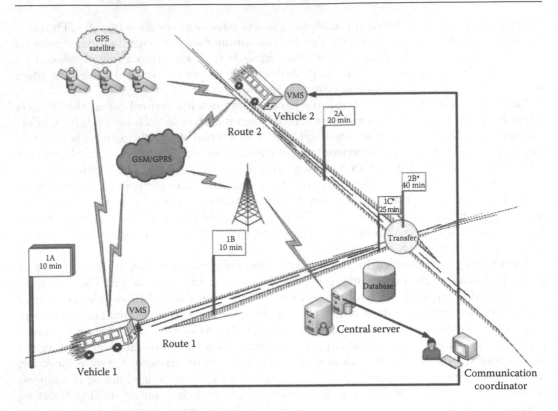

Figure 19.17 Graphical illustration of transfer synchronization of two public transit routes based on the intervehicle communication systems.

passengers without waiting time. The control center will have a record of this compliance to help in minimizing issues associated with driver behavior.

19.6.2.2 Main features

The main features of the CBVC-B are related to the whole transit-network communications comprised of different decentralized and parallel groups. The communication-based control process can be performed at the same time between different communication groups. However, technically, bus drivers of the same communication group are not exactly communicated in a direct peer-to-peer manner; it is more of a client–server way. Therefore, the whole control process is termed semidecentralized group communication. In a communication group, bus drivers leading to the same transfer point serve as clients, and the communication control center serves as the central server. Bus drivers through the central server share vehicle and passenger information with their peers. The central communication center is responsible for the communication coordination of vehicles on the route segments and delivers advice on the real-time vehicle control tactics to bus drivers. Once a bus passes the communication control center, it automatically joins another group of vehicles. The communication group is self-organized.

A small transit network is used in Figure 19.19a to illustrate the basic concepts. This network is from Ceder et al. (2001). The network is divided into four communication groups by four transfer points shown in Figure 19.18b. Each group has two routes leading to the transfer point, and a central server is assigned to be responsible for the communication between bus drivers.

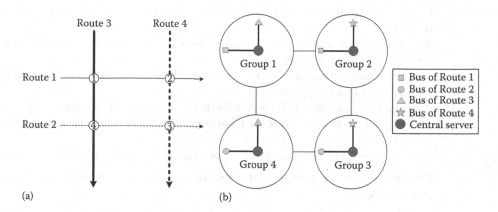

Figure 19.18 Cooperative group communication between bus drivers in a transit network: (a) an example of bus transit network and (b) semidecentralized group communication between bus drivers of the example.

19.6.3 Model

The model developed by Liu et al. (2014) assumes given GPS information of transit-vehicle speed and location. It is a distance-based dynamic speed-adjustment model based on IVC systems for vehicle running speed update under both fixed single-point encounter and flexible road-segment encounter scenarios. In this model, drivers of different routes can share their real-time travel speed and location information through the IVC communication control center. Drivers can adjust their travel speed or holding time at transfer points dynamically and cooperatively in order to achieve a preplanned direct transfer. The main objective of the model is to increase the number of simultaneous arrivals at all transfer points. This approach analyzes the impact of the changes on the TPTT, in-vehicle travel time, out-of-vehicle waiting time at stops, and transfer waiting time. These impacts are studied on a network-wide scale using the optimization model presented in Sections 19.6.3.1 and 19.6.3.2.

19.6.3.1 Notations

R	Set of all PT routes $r \in R$.
S	Set of all segments $s \in S$ of all routes $r \in R$.
N^{sr}	Set of all stops on segment s of route r, $n \in N^{sr}$.
$|N^{sr}|$	Total number of stops on segment s of route r.
PI_n^{sr}	Positional index (k) of a stop n on segment s of route r.
q_k^{sr}	Vector listing all stops on segment s of route r in order defined by positional index (k).
M^{sr}	Set of intervals on segment s of route r, $m \in M^{sr}$.
$|M^{sr}|$	Total number of intervals on segment s of route r.
p_{in}^{sr}	Number of passengers assumed to be arriving in vehicle i at stop n on segment s of route r; this is derived from a previous OD matrix; the number of the first stop on route r is the initial number of passengers on the vehicle.
e_{in}^{sr}	Number of passengers entering vehicle i at stop n on segment s of route r.
l_{in}^{sr}	Number of passengers leaving vehicle i at stop n on segment s of route r.
$d_i^{s_i r_i}$	Length of segment s of route r on which vehicle i is moving.
$d_{i,m}^{s_i r_i}$	Distance of vehicle i on segment s of route r in interval m to the nearest transfer point.

$d_s^{s_i r_i s_j r_j}$	Length of shared road segment s between segment s_i of route r_i and segment s_j of route r_j.
t_{in}^{sr}	Average running time of vehicle i on section between stop $n-1$ and stop n of segment s of route r, posttactics.
$\bar{t}_{in}^{sr}$	Average running time of vehicle i on section between stop $n-1$ and stop n of segment s of route r, pretactics.
Δt_{in}^{sr}	Average running-time difference of vehicle i on section between stop $n-1$ and stop n of segment s of route r between posttactics and pretactics, that is, $\Delta t_{in}^{sr} = t_{in}^{sr} - \bar{t}_{in}^{sr}$.
$A(t)_{iq_k^{s_i r_i}}^{s_i r_i}$	Arrival time of vehicle i of segment s_i of route r_i at transfer stop $q_k^{s_i r_i}$, posttactics.
$D(t)_{iq_k^{s_i r_i}}^{s_i r_i}$	Departure time of vehicle i of segment s_i of route r_i at transfer stop $q_k^{s_i r_i}$, posttactics.
$\Delta D_{q_k^{s_i r_i}}^m$	Dwell time of vehicle i of segment s_i of route r_i within the interval m, that is, $\Delta D_{q_k^{s_i r_i}}^m = D(t)_{iq_k^{s_i r_i}}^{s_i r_i} - A(t)_{iq_k^{s_i r_i}}^{s_i r_i}$.
$h_i^{q_{|N^{s_i r_i}|}^{s_i r_i}}$	Holding time of vehicle i of segment s_i of route r_i at transfer stop $q_{|N^{s_i r_i}|}^{s_i r_i}$.
$A(t)_{iq_k^{s_i r_i}}^{s_i r_i}$	Arrival time of vehicle i of segment s_i of route r_i at transfer stop $q_k^{s_i r_i}$, pretactics.
$\bar{D(t)}_{iq_k^{s_i r_i}}^{s_i r_i}$	Departure time vehicle i of segment s_i of route r_i at transfer stop $q_k^{s_i r_i}$, pretactics.
v_{im}^{sr}	Average speed of vehicle i in interval m on segment s of route r.
$Vmin_{i,m}^{s_i r_i}$	Minimum average speed of vehicle i in interval m on segment s of route r.
$Vmax_{i,m}^{s_i r_i}$	Maximal average speed of vehicle i in interval m on segment s of route r.
$P_{iq_{|N^{s_i r_i}|}^{s_i r_i} jq_{|N^{s_j r_j}|}^{s_j r_j}}^{s_i r_i s_j r_j}$	Number of passengers wishing to transfer from vehicle i of segment s_i of route r_i at transfer stop $q_{|N^{s_i r_i}|}^{s_i r_i}$ to vehicle j of segment s_j of route r_j at transfer stop $q_{|N^{s_j r_j}|}^{s_j r_j}$.
$twt_{iq_{|N^{s_i r_i}|}^{s_i r_i} jq_{|N^{s_j r_j}|}^{s_j r_j}}^{s_i r_i s_j r_j}$	Transfer walking time required from vehicle i of segment s_i of route r_i at transfer stop $q_{|N^{s_i r_i}|}^{s_i r_i}$ to vehicle j of segment s_j of route r_j at transfer stop $q_{|N^{s_j r_j}|}^{s_j r_j}$. It follows that if transfer stops $q_{|N^{s_i r_i}|}^{s_i r_i}$ and $q_{|N^{s_j r_j}|}^{s_j r_j}$ are a shared transfer stop, then $twt_{iq_{|N^{s_i r_i}|}^{s_i r_i} jq_{|N^{s_j r_j}|}^{s_j r_j}}^{s_i r_i s_j r_j} = 0$.

19.6.3.2 Distance-based dynamic speed-adjustment model

Consider a transfer point of two transit routes. The time–distance diagram of vehicles for fixed single-point encounter scenario and flexible road-segment encounter scenario is shown in Figure 19.19a and b, respectively. The time horizon of each route segment is divided into $|M^{sr}|$ intervals of equal size. The length of each interval is τ. Drivers will receive real-time advisory speed-adjustment information by a communication coordinator for each interval and holding time information for the last interval based on their distance to the transfer point.

Fixed single-point encounter scenario: The update rules for average running speed adjustment, for the fixed single-point encounter scenario, are given as follows:

$$\text{if } d_{i,m}^{s_i r_i} > d_{j,m}^{s_j r_j}, \; v_{i,m+1}^{s_i r_i} = v_{i,m}^{s_i r_i} + a; \; v_{j,m+1}^{s_j r_j} = v_{j,m}^{s_j r_j} - b.$$

$$\text{else if } d_{i,m}^{s_i r_i} < d_{j,m}^{s_j r_j}, \; v_{i,m+1}^{s_i r_i} = v_{i,m}^{s_i r_i} - b'; \; v_{j,m+1}^{s_j r_j} = v_{j,m}^{s_j r_j} + a'.$$

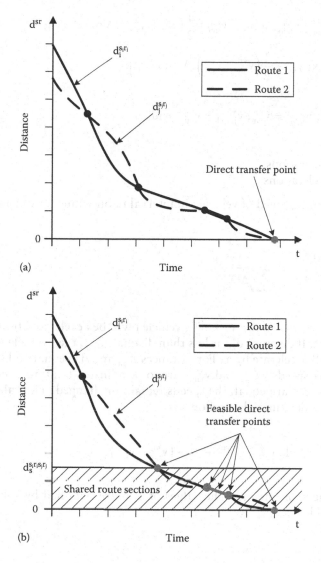

Figure 19.19 Time–distance diagrams of two vehicles moving to the same transfer point: (a) fixed single-point encounter transfer and (b) flexible road-segment encounter transfer.

else $d_{i,m}^{s_i r_i} = d_{j,m}^{s_j r_j}$.

if $v_{i,m}^{s_i r_i} > v_{j,m}^{s_j r_j}$, $v_{i,m+1}^{s_i r_i} = \max\left(v_{i,m}^{s_i r_i} - \frac{1}{2}\left(v_{i,m}^{s_i r_i} - v_{j,m}^{s_j r_j} \right), \text{Vmin}_{i,m+1}^{s_i r_i} \right)$,

$\qquad v_{i,m+1}^{s_i r_i} = \max\left(v_{i,m}^{s_i r_i} - \frac{1}{2}\left(v_{i,m}^{s_i r_i} - v_{j,m}^{s_j r_j} \right), \text{Vmin}_{i,m+1}^{s_i r_i} \right)$;

$\qquad v_{j,m+1}^{s_j r_j} = \min\left(v_{j,m}^{s_j r_j} + \frac{1}{2}\left(v_{i,m}^{s_i r_i} - v_{j,m}^{s_j r_j} \right), \text{Vmax}_{j,m+1}^{s_j r_j} \right)$.

$$\text{else if } v_{i,m}^{s_i r_i} < v_{j,m}^{s_j r_j}, \; v_{i,m+1}^{s_i r_i} = \min\left(v_{i,m}^{s_i r_i} + \frac{1}{2}\left(v_{j,m}^{s_j r_j} - v_{i,m}^{s_i r_i} \right), \text{Vmax}_{i,m+1}^{s_i r_i} \right);$$

$$v_{j,m+1}^{s_j r_j} = \max\left(v_{j,m}^{s_j r_j} - \frac{1}{2}\left(v_{j,m}^{s_j r_j} - v_{i,m}^{s_i r_i} \right), \text{Vmin}_{j,m+1}^{s_j r_j} \right).$$

$$\text{else } v_{i,m}^{s_i r_i} = v_{j,m}^{s_j r_j}, \; v_{i,m+1}^{s_i r_i} = v_{i,m}^{s_i r_i}; \; v_{j,m+1}^{s_j r_j} = v_{j,m}^{s_j r_j} . \text{ end.}$$

where
  a and a′ are accelerations
  b and b′ are decelerations

The average running speed of vehicle i in interval m on segment s of route r, v_{im}^{sr} is defined as follows:

$$v_{im}^{sr} = \frac{d_{i,m+1}^{s_i r_i} - d_{i,m}^{s_i r_i}}{A(t)_i^{m+1} - A(t)_i^{m} - \sum \Delta D_{q_k^{s_i r_i}}^{m}} \tag{19.50}$$

For distance $d_{i,m}^{s_i r_i}$ larger than distance $d_{j,m}^{s_j r_j}$, vehicle i will be accelerated by a and vehicle j will be decelerated by b. If distance $d_{i,m}^{s_i r_i}$ is less than distance $d_{j,m}^{s_j r_j}$, then vehicle i will decelerate by b′ and vehicle j will accelerate by a′. For distances $d_{i,m}^{s_i r_i}$ and $d_{j,m}^{s_j r_j}$ equal and speeds $v_{i,m}^{s_i r_i}$ and $v_{j,m}^{s_j r_j}$ not equal, the next speeds $v_{i,m+1}^{s_i r_i}$ and $v_{j,m+1}^{s_j r_j}$ are to be adjusted to attain equal speeds. If both the distance and speed are equal, the speeds remain unchanged. The values of a, b, a′, and b′ for each case are determined as follows:

$$\text{Case I: } d_{i,m}^{s_i r_i} > d_{j,m}^{s_j r_j} : \text{Let } \lambda = \left| \frac{d_{i,m}^{s_i r_i} - d_{j,m}^{s_j r_j}}{\tau} - \left(v_{i,m}^{s_i r_i} - v_{j,m}^{s_j r_j} \right) \right|,$$

where τ is the length of interval. The values of a, b are obtained by solving the following minimization problem:

$$\min \; |a + b - \lambda| \tag{19.51}$$

subject to

$$v_{i,m}^{s_i r_i} + a \leq \text{Vmax}_{i,m+1}^{s_i r_i} \tag{19.52}$$

$$v_{j,m}^{s_j r_j} - b \geq \text{Vmin}_{j,m+1}^{s_j r_j} \tag{19.53}$$

$$a \geq 0 \tag{19.54}$$

$$b \geq 0 \tag{19.55}$$

The objective function of the model is to minimize the distance gap between $d_{i,m+1}^{s_i r_i}$ and $d_{j,m+1}^{s_j r_j}$ in interval m + 1, where $d_{i,m+1}^{s_i r_i} = d_{i,m}^{s_i r_i} - v_{i,m}^{s_i r_i}\tau$ and $d_{j,m+1}^{s_j r_j} = d_{j,m}^{s_j r_j} - v_{j,m}^{s_j r_j}\tau$, in order to achieve a direct transfer. The constraints are of minimum and maximum speed limits.

Let $\rho = \left| \left(\text{Vmax}_{i,m+1}^{s_i r_i} - \text{Vmin}_{j,m+1}^{s_j r_j} \right) - \left(v_{i,m}^{s_i r_i} - v_{j,m}^{s_j r_j} \right) \right|$. Certainly, the optimal solution for this problem is $a + b = \lambda$. The rules for determining the values of a and b, to minimize TPTT, are as follows:

if $\rho \leq \lambda$, $a = \text{Vmax}_{i,m+1}^{s_i r_i} - v_{i,m}^{s_i r_i}$; $b = v_{j,m}^{s_j r_j} - \text{Vmin}_{j,m+1}^{s_j r_j}$.

else $\rho > \lambda$

if $\text{Vmax}_{i,m+1}^{s_i r_i} - v_{i,m}^{s_i r_i} \geq \lambda$, $a = \lambda; b = 0$.

else $\text{Vmax}_{i,m+1}^{s_i r_i} - v_{i,m}^{s_i r_i} < \lambda$, $a = \text{Vmax}_{i,m+1}^{s_i r_i} - v_{i,m}^{s_i r_i}$; $b = \lambda - \left(\text{Vmax}_{i,m+1}^{s_i r_i} - v_{i,m}^{s_i r_i} \right)$. end.

Case II: $d_{i,m}^{s_i r_i} < d_{j,m}^{s_j r_j}$: Similarly, let $\lambda' = \left| \left(v_{i,m}^{s_i r_i} - v_{j,m}^{s_j r_j} \right) - \dfrac{d_{i,m}^{s_i r_i} - d_{j,m}^{s_j r_j}}{\tau} \right|$, and a', b' are obtained by solving the following minimization programming problem:

$$\min \left| a' + b' - \lambda' \right| \tag{19.56}$$

subject to

$$v_{j,m}^{s_j r_j} + a' \leq \text{Vmax}_{j,m+1}^{s_j r_j} \tag{19.57}$$

$$v_{i,m}^{s_i r_i} - b' \geq \text{Vmin}_{i,m+1}^{s_i r_i} \tag{19.58}$$

$$a' \geq 0 \tag{19.59}$$

$$b' \geq 0 \tag{19.60}$$

The rules for determining the values of a' and b', to minimize TPTT, are same as those of Case I.

Flexible road-segment encounter scenario: The update rules for average running speed adjustment, for the flexible road-segment encounter scenario, when the distances $d_{i,m}^{s_i r_i}$ and $d_{j,m}^{s_j r_j}$ are not equal, are same as for the fixed single-point encounter scenario. However, when the distances $d_{i,m}^{s_i r_i}$ and $d_{j,m}^{s_j r_j}$ are equal, the update rules for travel speed adjustment are a bit different. New rules for average running speed update are given as follows:

if $d_{i,m}^{s_i r_i} = d_{j,m}^{s_j r_j} > d_s$

if $v_{i,m}^{s_i r_i} > v_{j,m}^{s_j r_j}$, $v_{i,m+1}^{s_i r_i} = \max \left(v_{i,m}^{s_i r_i} - \dfrac{1}{2} \left(v_{i,m}^{s_i r_i} - v_{j,m}^{s_j r_j} \right), \text{Vmin}_{i,m+1}^{s_i r_i} \right)$,

$v_{i,m+1}^{s_i r_i} = \max \left(v_{i,m}^{s_i r_i} - \dfrac{1}{2} \left(v_{i,m}^{s_i r_i} - v_{j,m}^{s_j r_j} \right), \text{Vmin}_{i,m+1}^{s_i r_i} \right)$;

$v_{j,m+1}^{s_j r_j} = \min \left(v_{j,m}^{s_j r_j} + \dfrac{1}{2} \left(v_{i,m}^{s_i r_i} - v_{j,m}^{s_j r_j} \right), \text{Vmax}_{j,m+1}^{s_j r_j} \right)$.

$$\text{else if } v_{i,m}^{s_i r_i} < v_{j,m}^{s_j r_j}, \; v_{i,m+1}^{s_i r_i} = \min\left(v_{i,m}^{s_i r_i} + \frac{1}{2}\left(v_{j,m}^{s_j r_j} - v_{i,m}^{s_i r_i} \right),\; \text{Vmax}_{i,m+1}^{s_i r_i} \right),$$

$$v_{i,m+1}^{s_i r_i} = \min\left(v_{i,m}^{s_i r_i} + \frac{1}{2}\left(v_{j,m}^{s_j r_j} - v_{i,m}^{s_i r_i} \right),\; \text{Vmax}_{i,m+1}^{s_i r_i} \right);$$

$$v_{j,m+1}^{s_j r_j} = \max\left(v_{j,m}^{s_j r_j} - \frac{1}{2}\left(v_{j,m}^{s_j r_j} - v_{i,m}^{s_i r_i} \right),\; \text{Vmin}_{j,m+1}^{s_j r_j} \right).$$

$$\text{else } v_{i,m}^{s_i r_i} = v_{j,m}^{s_j r_j}, \; v_{i,m+1}^{s_i r_i} = v_{i,m}^{s_i r_i}; \; v_{j,m+1}^{s_j r_j} = v_{j,m}^{s_j r_j} \cdot \text{end},$$

$$\text{else } d_{i,m}^{s_i r_i} = d_{j,m}^{s_j r_j} \le d_s, \; v_{i,m+1}^{s_i r_i} = \text{Vmax}_{i,m+1}^{s_i r_i}; \; v_{j,m+1}^{s_j r_j} = \text{Vmax}_{j,m+1}^{s_j r_j} \cdot \text{end}.$$

The change of scenario occurs when the distances of the two vehicles to the end of their shared road segment are equal and less than the length of the shared road segment. In next intervals, the vehicles will run by their maximum speeds in order to minimize the TPTT.

Holding vehicles at transfer point: Holding vehicles at transfer points can improve the opportunity of direct transfers. However, it will increase the dwell time. It has both positive and negative impacts on the TPTT. There is a trade-off between the number of transferring and onboard passengers.

Evaluation of performance: Evaluation of the performance of the proposed distance-based dynamic speed-adjustment model, IVC scheme based, takes place using two main measures: (1) total number of simultaneous arrivals and (2) ΔTPTT_i. The total number of simultaneous arrivals of an observation time horizon can be obtained directly by comparing the arrival times of vehicles at a transfer point.

For the *i*th vehicle, the ΔTPTT_i includes four elements: the change of in-vehicle passenger travel time (ΔPTT_i), the change of passenger waiting time (ΔPWT_i), the change of passenger transfer waiting time (ΔPTWT_i), and the holding time at transfer points (H_i). The ΔTPTT_i of all vehicles is defined as

$$\Delta \text{TPTT} = \sum_i \left(\Delta \text{PTT}_i + \Delta \text{PWT}_i + \Delta \text{PTWT}_i + H_i \right) \tag{19.61}$$

The ΔPTT_i is calculated by

$$\Delta \text{PTT}_i = \text{PTT}_i^* - \text{PTT}_i = \sum_r \sum_s \sum_n \left(p_{in}^{sr} + e_{in}^{sr} - l_{in}^{sr} \right) \Delta t_{in}^{sr} \tag{19.62}$$

where PTT_i^* and PTT_i denote the total in-vehicle PTT of posttactics and pretactics, respectively. The change of downstream passenger waiting time ΔPWT_i is calculated by

$$\Delta \text{PWT}_i = \sum_r \sum_s \sum_n \left(\sum_{k=\text{PI}_n^{sr}}^{\left| N^{sr} \right|} e_{ik}^{sr} \right) \Delta t_{in}^{sr} \tag{19.63}$$

The $\Delta PTWT_i$ is defined by the following equation:

$$\Delta PTWT_i = PTWT_i^* - PTWT_i \tag{19.64}$$

where PTT_i^* and $PTWT_i$ denote the passenger transfer waiting time (PTWT) at the transfer point of posttactics and pretactics, respectively. The PTT_i^* and $PTWT_i$ are obtained by Equations 19.65 and 19.66, respectively:

$$PTWT_i^* = \begin{cases} p_{iq_{|N^{s_i r_i}|}^{s_i r_i} jq_{|N^{s_j r_j}|}^{s_j r_j}}^{s_i r_i s_j r_j} \left(D(t)_{jq_{|N^{s_j r_j}|}^{s_j r_j}}^{s_j r_j} - A(t)_{iq_{|N^{s_i r_i}|}^{s_i r_i}}^{s_i r_i} - twt_{iq_{|N^{s_i r_i}|}^{s_i r_i} jq_{|N^{s_j r_j}|}^{s_j r_j}}^{s_i r_i s_j r_j} \right), & \text{if } A(t)_{iq_{|N^{s_i r_i}|}^{s_i r_i}}^{s_i r_i} + twt_{iq_{|N^{s_i r_i}|}^{s_i r_i} jq_{|N^{s_j r_j}|}^{s_j r_j}}^{s_i r_i s_j r_j} \leq D(t)_{jq_{|N^{s_j r_j}|}^{s_j r_j}}^{s_j r_j} \\[2em] p_{iq_{|N^{s_i r_i}|}^{s_i r_i} jq_{|N^{s_j r_j}|}^{s_j r_j}}^{s_i r_i s_j r_j} \left(D(t)_{jq_{|N^{s_j r_j}|}^{s_j r_j}}^{s_j r_j} + \kappa h_{r_j} - A(t)_{iq_{|N^{s_i r_i}|}^{s_i r_i}}^{s_i r_i} - twt_{iq_{|N^{s_i r_i}|}^{s_i r_i} jq_{|N^{s_j r_j}|}^{s_j r_j}}^{s_i r_i s_j r_j} \right), & \text{otherwise;} \end{cases} \tag{19.65}$$

$$PTWT_i = \begin{cases} p_{iq_{|N^{s_i r_i}|}^{s_i r_i} jq_{|N^{s_j r_j}|}^{s_j r_j}}^{s_i r_i s_j r_j} \left(D(\bar{t})_{jq_{|N^{s_j r_j}|}^{s_j r_j}}^{s_j r_j} - A(\bar{t})_{iq_{|N^{s_i r_i}|}^{s_i r_i}}^{s_i r_i} - twt_{iq_{|N^{s_i r_i}|}^{s_i r_i} jq_{|N^{s_j r_j}|}^{s_j r_j}}^{s_i r_i s_j r_j} \right), & \text{if } A(\bar{t})_{iq_{|N^{s_i r_i}|}^{s_i r_i}}^{s_i r_i} + twt_{iq_{|N^{s_i r_i}|}^{s_i r_i} jq_{|N^{s_j r_j}|}^{s_j r_j}}^{s_i r_i s_j r_j} \leq D(\bar{t})_{jq_{|N^{s_j r_j}|}^{s_j r_j}}^{s_j r_j} \\[2em] p_{iq_{|N^{s_i r_i}|}^{s_i r_i} jq_{|N^{s_j r_j}|}^{s_j r_j}}^{s_i r_i s_j r_j} \left(D(\bar{t})_{jq_{|N^{s_j r_j}|}^{s_j r_j}}^{s_j r_j} + \kappa h_{r_j} - A(\bar{t})_{iq_{|N^{s_i r_i}|}^{s_i r_i}}^{s_i r_i} - twt_{iq_{|N^{s_i r_i}|}^{s_i r_i} jq_{|N^{s_j r_j}|}^{s_j r_j}}^{s_i r_i s_j r_j} \right), & \text{otherwise;} \end{cases} \tag{19.66}$$

where

h_{r_j} is the headway of route j
κ is an integer variable that is used to minimize the PTWT

The holding time at transfer points (H_i) is defined by

$$H_i = \left(p_{iq_{|N^{s_i r_i}|}^{s_i r_i}}^{s_i r_i} + e_{iq_{|N^{s_i r_i}|}^{s_i r_i}}^{s_i r_i} - l_{iq_{|N^{s_i r_i}|}^{s_i r_i}}^{s_i r_i} \right) h_i^{q_{|N^{s_i r_i}|}^{s_i r_i}} \tag{19.67}$$

By substituting Equations 19.65 and 19.66 into Equation 19.64, and combining Equations 19.62, 19.63, and 19.67, the $\Delta TPTT_i$ is attained.

19.6.4 Case study

The effectiveness of the optimization model and its potential for implementation is assessed by a case study of a real-life transit network of Beijing, China. The transit network is illustrated in Figure 19.20; it comprises three bus routes with five transfer points. Routes 658 and 694 run from north to south and route 728 runs from west to east. Transfer points 1, 2, 3, and 4 are fixed single-point encounter transfers and transfer point 5 is a road-segment encounter transfer.

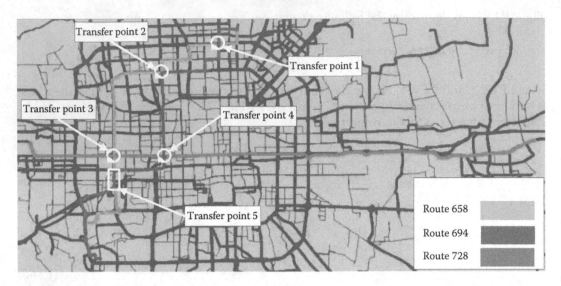

Figure 19.20 A bus network of Beijing with three bus routes and five transfer points. (Note: *Dark-coloured lines* represent the road network.)

19.6.4.1 Data collection

Data of the three routes were collected from the Beijing Public Transport Holdings, Ltd. The data include information of routes, vehicles, and passengers.

Route information: Comprises route ID, route direction, number of stops, stop ID, stop latitude and longitude, distance between stops, and preplanned timetable.

Vehicle information: All vehicles are equipped with GPS device and can share their real-time speed and location information with the control center. The information was updated using intervals of 30 s. The collected vehicle information comprises vehicle ID, driver ID, departure and arrival times at each stop, planned headway, vehicle location, and average running speed.

Passenger information: Collected from integrated circuit (IC) cards and comprises the number of passengers boarding and alighting at each stop and the number of passengers transferring at each transfer stop.

19.6.5 Results

The optimization model described (Liu et al., 2014) was applied to the case study with a comparison between the operation without using tactics and the operation with tactics. The optimization procedure for updating average running speeds, for all the five transfer points, was implemented using MATLAB®. The original time–distance diagram, of the case study, without using any operational tactics is depicted on the left-hand side of Figure 19.21 at all the five transfer points. The adjusted time–distance diagram, using the operational tactics (model outcome), is shown on the right-hand side of Figure 19.21. The interval of speed adjustment, of the case study, was 5 min. The average transfer walking time was assumed to be 1 min. The arrival time difference t is 1 min; thus, when this difference is equal or less 1 min, holding tactics are employed to increase the opportunity of direct transfers. The results of holding times of vehicles on each route at each

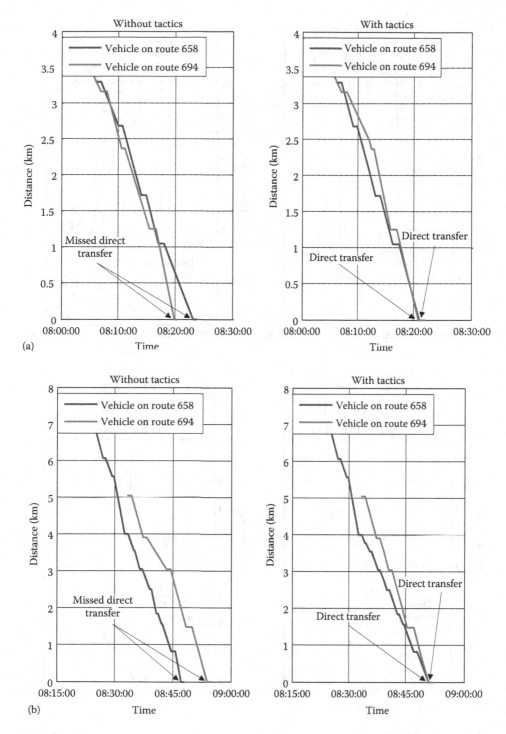

Figure 19.21   Time–distance diagrams of the five transfer points with and without tactics: (a) transfer point 1, (b) transfer point 2.   *(Continued)*

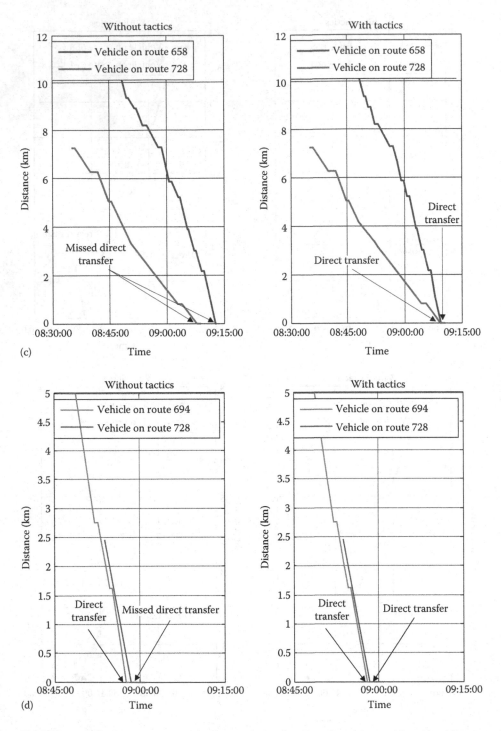

Figure 19.21 (Continued)   Time–distance diagrams of the five transfer points with and without tactics: (c) transfer point 3, (d) transfer point 4.   *(Continued)*

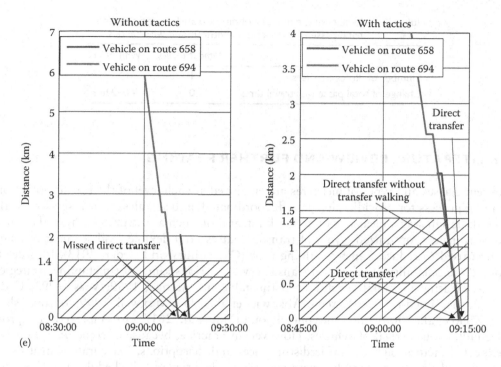

Figure 19.21 (Continued) Time–distance diagrams of the five transfer points with and without tactics: (e) transfer point 5.

transfer points are shown in Table 19.9 and it ranges between zero (simultaneous arrivals of vehicles) and 47 s.

The results of the two main measures of comparison are shown in Table 19.10. These results indicate that without using any operational tactics, preplanned direct transfers do not always occur. There is only one direct transfer that occurred at transfer point 4 for passengers transferring from route 694 to route 728. By implementing the operational tactics (model outcome), the number of successful direct transfers becomes 12, an increase of 1100%. We note that at transfer point 5 (see Figure 19.21), which is a road-segment encounter transfer, simultaneous arrivals of vehicles occur at a shared road segment and contribute largely to this significantly improved result. The TPTT with tactics is 547,340 s and reduced by 83,236 s, which is a reduction of 13.2% compared with the figure 630,576 s related to without using tactics. No doubt that both results will make the service more attractive and reliable.

These encouraging results demonstrate the potential benefits existing for public transit systems with the implementation of IVC scheme and online operational tactics.

Table 19.9 Holding times of vehicles on each route at each transfer point

Transfer point	Route 658	Route 694	Route 728
1	7 s	38 s	—
2	3 s	4 s	—
3	34 s	—	10 s
4	—	47 s	0 s
5	0 s	0 s	—

Table 19.10 Results of the number of direct transfers and change in
total passenger travel time without and with tactics

	Without tactics	With tactics
Number of direct transfers	1	12
Change of total passenger travel time	0	−83236 s

19.7 LITERATURE REVIEW AND FURTHER READING

Different classes and types of measures are employed at each level of the transit operations planning process for providing integrated, coordinated, and seamless transit service. At the network design stage, measures of network integration such as integrated physical connection of transfers, optimized layout of transfer centers, and flexible road segment transfers are first used to reduce transfer walking time (Chowdhury and Ceder, 2013). In order to improve the reliability of scheduled transit service, various kinds of timetabling strategies, such as creating maximal synchronized timetable (Domschke, 1989; Voss, 1992; Ceder et al., 2001), timed-transfer systems (Abkowitz et al., 1987; Maxwell, 1999), optimal slack time (Zhao et al., 2006), and time control point (Yan et al., 2012), are employed to improve the simultaneous arrival of vehicles. However, in practice, because of some uncertain and unexpected factors, such as traffic disturbances and disruptions, inaccurate transit driver behavior and actions, and random passenger demands, preplanned scheduled transit service is not always materialized, especially the synchronized transfer service. Missed connection of transfer service will not only frustrate current passengers but also suffer a loss of potential new users. Therefore, it is important to control the randomness existing in the movement of transit vehicles.

Osuna and Newell (1972) discussed the holding strategy at a single service point. They considered an idealized transit system of one or two vehicles and formulated the optimal control process as a DP problem. The objective is to minimize the average waiting time of each passenger. Barnett (1974) also considered holding vehicles at a chosen control stop, but with two objectives: to minimize both the average passenger waiting time at the control stop and average in-vehicle passenger delay; the author provided a simple algorithm for constructing an approximate optimal dispatch strategy. Wirasinghe and Liu (1995) investigated the determination of the number and location of time control points using DP. Eberlein et al. (1999, 2001) studied the deadheading, expressing, and holding strategies both singly and in combination under the assumption that real-time information is available. They formulated the holding problem as a quadratic mathematical programming model using heuristic algorithms to solve it. Dessouky et al. (1999) showed the potential benefits of real-time control of timed transfers using ITS technologies. Dessouky et al. (2003) examined simulated systems in which holding and dispatching strategies are used; they also investigated the dependency between system performance and available technologies. The analysis of Dessouky et al. (2003) addresses seven different holding strategy combinations for allowing transfers, each of which differs in its definition of holding and in its approach to be used for connecting buses. The results show that advanced technology is most advantageous when there are many connecting buses; the schedule slack is then close to zero, and the headway is large.

Hickman (2001) developed an analytic stochastic model and formulated the holding problem at a control stop as a convex quadratic program. Sun and Hickman (2005) formulated the real-time skip-stop strategy as a nonlinear integer programming problem using an exhaustive search method to solve it. Sun and Hickman (2008) further investigated holding vehicles at multiple holding stations. They also formulated the problem as a convex quadratic program with convex objective function and linear constraints. Fu et al. (2003) proposed a pairing dynamic-operation strategy in which the following vehicle, of a pair of vehicles, was allowed to skip some stations. Zolfaghari et al. (2004) studied the holding strategy with real-time bus-location information proposing the use of a simulated annealing-based heuristic algorithm as a solution method. Yu and Yang (2009) investigated the holding strategy considering the prediction of next stop departure time; in their study, a support-vector machine is used to predict the departure time using genetic algorithm to optimize the holding time.

Cats et al. (2011) proposed a dynamic transit simulation model, BusMezzo, to investigate the holding strategy considering the interaction between passenger activity, transit operations, and traffic dynamics. Delgado et al. (2012) investigated the holding and limited boarding strategies; the authors used a mathematical model without binary variables for real-time transit operations. Another way to examine the impact of control strategies on transit operations is through a dynamic-control approach; it is widely used for situations with uncertain dynamic characteristics. Daganzo (2009) and Daganzo and Pilachowski (2011) used adaptive-control approach to alleviate the irregular meetings of buses (bus bunching) at nontransfer stops. Xuan et al. (2011) further extended the dynamic-control approach using a family of dynamic holding strategies to improve bus schedule adherence and headway regularity.

Additional control-based approach to deal with the uncertainty of transit-vehicle movements, road-traffic disturbances, and fluctuations of passenger demand is predictive control (PC). De Schutter et al. (2002) proposed a model PC (MPC) framework to coordinate transfers of railway systems. The MPC was transferred to a nonlinear, nonconvex optimization problem. Negenborn et al. (2008) investigated the control of large-scale transportation networks using multiagent MPC scheme; two schemes, parallel and serial schemes, were studied and compared. The authors concluded that serial scheme performed better than parallel scheme with regard to the convergence speed and the quality of the solution. Sáez et al. (2012) developed a hybrid PC (HPC) framework to investigate the real-time control of transit systems using dynamic-control strategies and online information of system behavior; this is to reduce the total in-vehicle travel time and out-of-vehicle waiting time. The authors proposed genetic algorithm to solve the optimization problem. Muñoz et al. (2013) further compared the performance of the HPC and deterministic-optimization approaches under different scenarios. The authors concluded that the HPC approach performs better than the deterministic-optimization approach when vehicle capacity could not be reached frequently along the route.

The rapid development of information and communication technology allows for new real-time based modeling to attain direct/synchronized transfers in transit operations. These modeling techniques can make the use of operational tactics (Ceder, 2007) following the methodologies and examples used in this chapter. The term "tactics" in this chapter refers to real-time control actions of transit vehicles, usually taken place in a short-time manner. This is in comparison with the term "strategies," which is usually referring to a wider perspective at the transit-network level. Recent studies by Hadas and Ceder (2008a,b,c, 2010), Ceder et al. (2013), Liu et al. (2014), and Nesheli and Ceder (2014) showed that by applying selected operational tactics (holding, skip-stop/segment, speed change, short turn, flexible route encounter), the frequency of simultaneous-meeting occurrence between two or more transit vehicles can be significantly improved along with a reduction of the TPTT.

The major characteristics of the models/methods reviewed are summarized in Table 19.11.

Table 19.11 Summary of the characteristics of the methods reviewed

Source	Approach	Strategies/tactics	Comments
Osuna and Newell (1972)	Dynamic programming	Holding	An idealized transit system
Barnett (1974)	Mathematical programming	Holding	A simple model
Wirasinghe and Liu (1995)	Dynamic programming	Holding	The determination of the number and location of time control points
Eberlein et al. (1999, 2001)	Quadratic programming	Deadheading, expressing, and holding	Real-time information available
Hickman (2001)	Convex quadratic programming	Holding	An analytical and stochastic model
Dessouky et al. (1999, 2003)	Simulation	Holding	Real-time control using ITS
De Schutter et al. (2002)	Model predictive control	Holding	Transfers at railway systems
Fu et al. (2003)	Nonlinear integer programming	Skip-stop	A pairing dynamic-operation strategy
Zolfaghari et al. (2004)	Mathematical modeling	Holding	Simulated annealing-based heuristic algorithm
Sun and Hickman (2005)	Nonlinear integer programming	Skip-stop	Exhaustive search method
Yu and Yang (2009)	Mathematical modeling	Holding	Consider the prediction of next stop departure time
Sun and Hickman (2008)	Convex quadratic programming	Holding	Multiple holding stations
Hadas and Ceder (2008a,b,c)	Simulation	Flexible route segment transfer	
Hadas and Ceder (2010)	Dynamic programming	Various operational tactics	
Daganzo (2009), Daganzo and Pilachowski (2011)	Adaptive control	Speed change	
Xuan et al. (2011)	Adaptive control	Holding	
Cats et al. (2011)	Dynamic simulation	Holding	BusMezzo
Delgado et al. (2012)	Mathematical programming	Holding and limited boarding	
Sáez et al. (2012)	Hybrid predictive control	Holding	
Muñoz et al. (2013)	Hybrid predictive control and deterministic optimization	Holding	
Ceder et al. (2013)	Modeling and simulation	Holding, skip-stop	
Nesheli and Ceder (2014)	Modeling optimization and simulation	Holding, skip-stop/segment	Real-time control
Liu et al. (2014)	Modeling and optimization	Vehicle-to-vehicle communication, speed change, holding	Real-time control

EXERCISE

Given is a small IVC system depicted graphically in Figure 19.17 with two transit routes, routes 1 and 2. Suppose that the information about the number of passengers boarding, alighting, and transferring at each transit stop, together with the running time in each route section is known by using IC cards and GPS data; this information is shown in the following table of input data. The initial number of onboard passengers of route 1 is 40. The number of passengers boarding at stop 1B (see Figure 19.17) is 10 and alighting is 5. Thus, the number of onboard passengers is 45 at stop 1B. The number of passengers boarding at stop 1C* is 10 and alighting is 26 with 20 passengers wishing to transfer from route 1 to route 2.

Input data and calculated total passenger travel time of the example of the exercise

Route 1		1B		1C*		PTT (min)	PTWT (min)	TPTT (min)
		10↑ 5↓		10↑	26↓			
			5↖ 5←		6↘ 20→			
N1(pass)	40	45		29				
T1(min)	0	10		15				
Route 2		2A		2B*				
		5↑ 0↓	25↑		15↓			
			5↖ 20←		10↘ 5→			
N2(pass)	20	25	35					
T2(min)	0	20	20					

Note: ↑ = boarding, ↓ = alighting, ↖ = pure boarding, ← = transferring boarding, ↘ = pure alighting, → = transferring alighting, * = transfer point indicator.

(a) Draw the time-based load profile of the example for the condition without using any tactics. Calculate the passenger travel time (PTT) of routes 1 and 2, PTWT, and TPTT.
(b) Using the IVC system, the average running time of the first section of route 1 is changed from 10 to 15 min; the speed of the second section is unchanged. On route 2, the average running time of the two sections changes from 20 to 15 min, say because of changes of road traffic condition. Draw the new time-based load profile of the example using speed change tactic. Calculate the posttactic PTT of routes 1 and 2, PTWT, and TPTT.
(c) Calculate the ΔPTWT and ΔTPTT.

REFERENCES

Abkowitz, M., Josef, R., Tozzi, J., and Driscoll, M.K. (1987). Operational feasibility of timed transfer in transit systems. *Journal of Transportation Engineering*, 113(2), 168–177.

Barnett, A. (1974). On controlling randomness in transit operations. *Transportation Science*, 8(2), 102–116.

Barton, H. and Tsourou, C. (2000). *Healthy Urban Planning*, World Health Organization, New York.

Bradshaw, J.M. (1997). *Software Agents*, MIT Press, Boston, USA.

Caliper. (2013). *TransModeler*, Caliper, Newton, MA.

Cats, O., Larijani, A.N., Koutsopoulos, H.N., and Burghout, W. (2011). Impacts of holding control strategies on transit performance. *Transportation Research Record*, **2216**, 51–58.

Ceder, A. (2007). *Public Transport Planning and Operation-Theory, Modelling and Practice*, Elsevier, Butterworth-Heinemann, Oxford, U.K.

Ceder, A., Golany, B., and Tal, O. (2001). Creating bus timetables with maximal synchronization. *Transportation Research*, **35A**, 913–928.

Ceder, A., Hadas, Y., McIvor, M., and Ang, A. (2013). Transfer synchronization of public-transport networks. *Transportation Research Record*, **2350**, 9–16.

Chowdhury, S. and Ceder, A. (2013). Definition of planned and unplanned transfer of public transport service and user decisions to use routes with transfers. *Journal of Public Transportation*, **16**(2), 1–20.

Daganzo, C.F. (2009). A headway-based approach to eliminate bus bunching: Systematic analysis and comparisons. *Transportation Research*, **43B**, 913–921.

Daganzo, C.F. and Pilachowski, J. (2011). Reducing bunching with bus-to-bus cooperation. *Transportation Research*, **45B**, 267–277.

Davis, R. and Smith. R.G. (1983). Negotiation as a metaphor for distributed problem solving. *Artificial Intelligence*, **20**(1), 63–109.

Delgado, F., Munoz, J.C., and Giesen, R. (2012). How much can holding and/or limiting boarding improve transit performance? *Transportation Research*, **46B**, 1202–1217.

Dessouky, M., Hall, R., Nowroozi, A., and Mourikas, K. (1999). Bus dispatching at timed transfer transit stations using bus tracking technology. *Transportation Research*, **7C**, 187–208.

Dessouky, M., Hall, R., Zhang, L., and Singh, A. (2003). Real-time control of buses for schedule coordination at a terminal. *Transportation Research*, **37A**, 145–164.

De Schutter, B., Van den Boom, T., and Hegyi, A. (2002). Model predictive control approach for recovery from delays in railway systems. *Transportation Research Record*, **1793**, 15–20.

Domschke, W. (1989). Schedule synchronization for public transit networks. *OR Spektrum*, **11**(1), 17–24.

Dueker, K.J., Kimpel, T.J., Strathman, J.G., and Callas, S. (2004). Determinants of bus dwell time. *Journal of Public Transportation*, **7**(1), 21–40.

Eberlein, X.J., Wilson, N.H., and Bernstein, D. (1999). Modeling real-time control strategies in public transit operations. In *Computer-Aided Scheduling of Public Transport*. Lecture Notes in Economics and Mathematical Systems, vol. 471 (N.H.M. Wilson, ed.), pp. 325–346, Springer-Verlag, Berlin, Germany.

Eberlein, X.J., Wilson, N.H., and Bernstein, D. (2001). The holding problem with real-time information available. *Transportation Science*, **35**(1), 1–18.

Fu, L., Liu, Q., and Calamai, P. (2003). Real-time optimization model for dynamic scheduling of transit operations. *Transportation Research Record*, **1857**(1), 48–55.

Hadas, Y. and Ceder, A. (2008a). Public transit simulation model for optimal synchronized transfers. *Transportation Research Record*, **2063**, 52–59.

Hadas, Y. and Ceder, A. (2008b). Multiagent approach for public transit system based on flexible routes. *Transportation Research Record*, **2063**, 89–96.

Hadas, Y. and Ceder, A. (2008c). Improving bus passenger transfers on road segments through online operational tactics. *Transportation Research Record*, **2072**, 101–109.

Hadas, Y. and Ceder, A. (2010). Optimal coordination of public-transit vehicles using operational tactics examined by simulation. *Transportation Research*, **18C**, 879–895.

Hartenstein, H. and Laberteaux, K.P. (2008). A tutorial survey on vehicular ad hoc networks. *IEEE Communications Magazine*, **46**(6), 164–171.

Hickman, M.D. (2001). An analytic stochastic model for the transit vehicle holding problem. *Transportation Science*, **35**(3), 215–237.

Horn, M.E.T. (2004). Procedures for planning multi-leg journeys with fixed-route and demand-responsive passenger transport services. *Transportation Research*, **12C**, 33–55.

IBM. (2012). *ILOG CPLEX Optimization Studio*, IBM, Somers, NY.

Jakob, A., Craig, J.L., and Fisher, G. (2006). Transport cost analysis: a case study of the total costs of private and public transport in Auckland. *Environmental Science and Policy*, **9**(1), 55–66.

Knoppers, P. and Muller, T. (1995). Optimized transfer opportunities in public transport. *Transportation Science*, **29**(1), 101–105.

Law, A.M. and Kelton, W.D. (2000). *Simulation Modeling and Analysis*, 3rd edn., McGraw-Hill, Boston, MA.

Levine, J., Hong, Q., Hug, G.E., and Rodrigez, D. (2000). Impacts of an advanced public transportation system demonstration project. *Transportation Research Record*, **1735**, 169–177.

Liu, T., Ceder, A., Ma, J.H., and Guan, W. (2014). Synchronizing public-transport transfers using inter-vehicle communication scheme: Case study. *Transportation Research Record*, 2417, 78–91.

Luo, J. and Hubaux, J.P. (2006). A survey of research in inter-vehicle communications. In *Embedded Security in Cars* (K. Lemke, C. Paar, and M. Wolf, eds.), pp. 111–122, Springer-Verlag, Berlin, Germany.

Maxwell, R.R. (1999). Intercity rail fixed-interval, timed-transfer, multihub system: Applicability of the Integraler Taktfahrplan Strategy to North America. *Transportation Research Record*, 1691, 1–11.

Mees, P. (2000). *A Very Public Solution: Transport in the Dispersed City*, Melbourne University Press, Carlton South, Victoria, Australia.

Minsky, M.L. (1986). *The Society of Mind*, Simon and Schuster, New York, NY.

Muñoz, J.C., Cortés, C.E., Giesen, R., Sáez, D., Delgado, F., Valencia, F., and Cipriano, A. (2013). Comparison of dynamic control strategies for transit operations. *Transportation Research*, 28C, 101–113.

Negenborn, R.R., De Schutter, B., and Hellendoorn, J. (2008). Multi-agent model predictive control for transportation networks: Serial versus parallel schemes. *Engineering Applications of Artificial Intelligence*, 21(3), 353–366.

Nesheli, M.M. and Ceder, A. (2014). Optimal combinations of selected tactics for public-transport transfer synchronization. *Transportation Research*, 48C, 491–504.

Osuna, E.E. and Newell, G.F. (1972). Control strategies for an idealized public transportation system. *Transportation Science*, 6(1), 52–72.

Raiffa, H. (1982). *The Art and Science of Negotiation*, Belknap Press of Harvard University Press, Cambridge, MA.

Reichardt, D., Miglietta, M., Moretti, L., Morsink, P., and Schulz, W. (2002). CarTALK 2000: Safe and comfortable driving based upon inter-vehicle-communication. In *IEEE Intelligent Vehicle Symposium*, vol. 2, pp. 545–550, IEEE, New York.

Sáez, D., Cortés, C.E., Milla, F., Núñez, A., Tirachini, A., and Riquelme, M. (2012). Hybrid predictive control strategy for a public transport system with uncertain demand. *Transportmetrica*, 8(1), 61–86.

Shafahi, Y. and Khani, A. (2010). A practical model for transfer optimization in a transit network: Model formulations and solutions. *Transportation Research*, 44A, 377–389.

Shaffer, C.A. and Feustel, C.D. (1991). Exact computations of 2-D intersections. In *Graphics Gems III* (D. Kirk, ed.), pp. 188–192, Academic Press, Boston, MA.

Sichitiu, M.L. and Kihl, M. (2008). Inter-vehicle communication systems: A survey. *IEEE Communications Surveys and Tutorials*, 10(2), 88–105.

Smith, R.G. (1980). The contract net protocol: High-level communication and control in a distributed problem solver. *IEEE Transactions on Computers*, 29(12), 1104–1113.

Sun, A. and Hickman, M. (2005). The real-time stop-skipping problem. *Journal of Intelligent Transportation Systems*, 9(2), 91–109.

Sun, A. and Hickman, M. (2008). The holding problem at multiple holding stations. In *Computer-Aided Systems in Public Transport*. Lecture Notes in Economics and Mathematical Systems, vol. 600 (M. Hickman, P. Mirchandani, and S. Voss, eds.), pp. 339–359, Springer-Verlag, Berlin, Germany.

Van Dyke Parunak, H. (1997). Go to the ant: Engineering principles from natural multi-agent systems. *Annals of Operations Research*, 75, 69–101.

Voss, S. (1992). Network design formulations in schedule synchronization. In *Computer-Aided Transit Scheduling*. Lecture Notes in Economics and Mathematical Systems, vol. 386 (M. Desrochers and J.M. Rousseau, eds.), pp. 137–152, Springer-Verlag, Berlin, Germany.

Wirasinghe, S.C. and Liu, G. (1995). Determination of the number and locations of time points in transit schedule design—Case of a single run. *Annals of Operations Research*, 60(1), 161–191.

Xuan, Y., Argote, J., and Daganzo, C.F. (2011). Dynamic bus holding strategies for schedule reliability: Optimal linear control and performance analysis. *Transportation Research*, 45B, 1831–1845.

Yan, Y., Meng, Q., Wang, S., and Guo, X. (2012). Robust optimization model of schedule design for a fixed bus route. *Transportation Research*, 25C, 113–121.

Yu, B. and Yang, Z. (2009). A dynamic holding strategy in public transit systems with real-time information. *Applied Intelligence*, **31**(1), 69–80.

Zhao, J.M., Bukkapatnam, S., and Dessouky, M.M. (2003). Distributed architecture for real-time coordination of bus holding in transit networks. *IEEE Transactions on Intelligent Transportation Systems*, **4**(1), 43–51.

Zhao, J.M., Dessouky, M., and Bukkapatnam, S. (2006). Optimal slack time for schedule-based transit operations. *Transportation Science*, **40**(4), 529–539.

Zolfaghari, S., Azizi, N., and Jaber, M.Y. (2004). A model for holding strategy in public transit systems with real-time information. *International Journal of Transport Management*, **2**(2), 99–110.

Chapter 20

Future developments in transit operations

Frequency at which the future arrives = 60 min/h

Some people see things as they are and say: Why? I dream of things that never were and say: Why not? (George Bernard Shaw)

A.C.

CHAPTER OUTLINE

PRACTITIONER'S CORNER

Albert Einstein once said: "I never think of the future. It comes soon enough." Indeed, the rate at which the future seems to be arriving provides a good reason to plan and design ahead; this should be an integral part of any routine transit operations planning undertaking. In this last chapter, we will attempt to touch the near future while being fully aware that choice, and not chance, determines destiny.

Our times, especially the past decade, have known extreme revolutions, especially in communications, and these have brought about a whole new culture, including changes in language and behavior. The Internet and the cellular phone have altered the way we live. These two revolutions have far-reaching consequences, especially for the rapid development of technologies supporting them: fast switching, self-healing distributed wireless and mobile networks, proximity and location technologies, nanotechnologies, motion control, mobile and miniscule power supplies, new user interfaces, and user experience. All these developments are gradually trickling into the field of transit service.

This chapter starts with an introductory section describing a new (future) concept of automatic-passenger transfers along road segments rather than at transit stops and then discusses technological issues in transit-system automation. In fact the novel transfer concept described in the introduction fits some of the methodologies presented in Chapter 19. Section 20.2 moves to the technology perspective, describing existing automated transit systems and proposed future concepts, such as the dual-mode personal rapid transit system. Section 20.3 describes an innovative two-way elevated transit system. Section 20.4 provides a selected literature review, and Section 20.5 ends the book with a concluding remark.

Essentially, practitioners are encouraged to visit all sections of this chapter about possible future development. The chapter presents challenges for devising alternative future transit solutions.

Finally, one piece of advice for practitioners insofar as future developments are as follows: (1) future solutions to existing transit problems are, in fact, existing solutions (some of which may be found in this book) to future problems; (2) whatever you do, try to act like an optimistic scientist, who will never cry over spilled milk, because 88% of it is just water in any case.

20.1 INTRODUCTION AND NEW DIRECT-TRANSFER CONCEPT

Niels Bohr facetiously cautioned: "Prediction is very difficult, especially of the future." Nonetheless, future developments in transit operations are already reflected in a vast amount of articles. Although we cannot change the direction of the wind (evolution of lifestyles, land-use patterns), we can adjust the sails (create attractive transit systems that will naturally shift people from the automobile to public transit vehicles). This sail-adjustment realization relies heavily on the ongoing development of new technologies.

Innovative technologies in transit operations have three major objectives:

- To increase the productivity of transit operations, particularly through the introduction of automation
- To improve safety, performance, and service capabilities and to achieve this in a cost-effective manner
- To support transit-related priorities at the national level, such as energy conservation, safety, central city revitalization, and environmental protection

Essentially, new technologies should pursue the goal of a prudent, well-connected transit system as defined in Section 13.2: *An advanced, attractive transit system that operates reliably and relatively rapidly, with smooth (ease of) synchronized transfers, part of the door-to-door passenger chain.* Interpretation of each component of this definition appears in Section 13.2. What follows are introductory remarks of a new synchronized transfer concept, a section on transit automation, a section on a two-way elevated transit system, a section on selected literature review, and a closing remark.

20.1.1 New concept to reduce the uncertainty of synchronized transfers

Advanced transit systems, such as dial-a-ride (e.g., Dial, 1995) or other demand-responsive systems, are not widely used and have never replaced existing conventional transit systems. The flexible routing and scheduling approach can be combined into a new concept consisting of elements suitable for a mass transit system. This new concept has the following main characteristics: (a) the use of a multiagent system (MAS); (b) real-time, multilegged trip planning, based on each passenger's attributes; (c) transit stops independent of synchronized transfers; and (d) deployment of operational tactics (hold, skip stop, etc.) to overcome the stochastic nature of the transit system.

Models that treat the stochastic nature of the transit system apparently were not developed specifically for public transit, but for vehicle-routing models (e.g., Gendreau et al., 1996; Ghiani et al., 2003). A public transit model that uses variable demand was developed by Chien et al. (2001), but it does not contain the combination of variable travel time, passenger demand, and transfer probability that will be presented in this chapter.

Following the description of synchronized transfers in Chapter 19 and the reliability problems in Chapter 18, the reader will recognize the uncertainty of a planned simultaneous arrival of two vehicles at an existing stop. In order to alleviate this uncertainty of simultaneous arrivals, Section 19.3 introduces a new concept: the extension of the commonly used single-point encounter (at a single transit stop) to a road-segment encounter (any point along the road constitutes a possible encounter point). Such a change in concept will reduce the uncertainty of meeting at a certain point and will enable more flexibility in deploying the operational tactics of Chapter 19.

The new concept, which relies on online information, allows two vehicles to meet in time and space so as to eliminate the need for waiting time at transfer points. Moreover, if we do not permit conventional transit systems to obstruct our imagination, the possibility exists of physically arranging for two vehicles that arrive at the meeting point to align together either longitudinally or side-by-side in order to allow passengers an easy transfer. This is like the meeting of satellites in space. Such a perfect transfer arrangement is illustrated schematically in Figure 20.1.

Realization of the new concept can be based on MAS as described in Section 19.2. In this section, it is mentioned that MAS is defined as collections of autonomous agents within an environment that interact with each other for achieving internal and/or global objectives (Van Dyke Parunak, 1997). Moreover, Minsky (1986) argued that an intelligent system could emerge from non-intelligent parts. His definition of the *Society of Mind* makes use of small, simple processes, called agents, each of which performs some simple action, and the combination of all these agents forms an intelligent society (system). Zhao et al. (2003) developed MAS for a bus-holding algorithm. The authors treat each bus stop as an agent; the agents negotiate with each other, based on marginal cost calculations, to devise minimal passenger waiting-time costs.

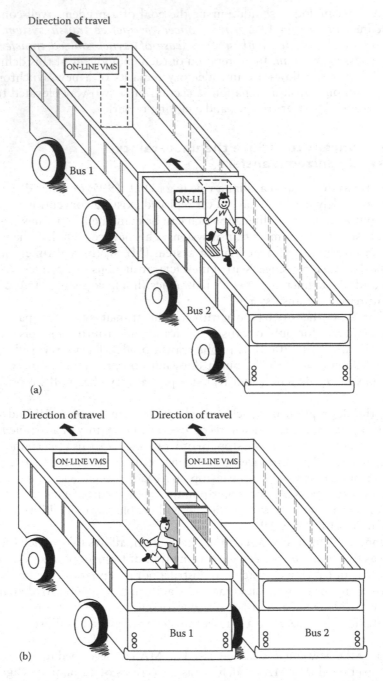

Figure 20.1 Schematic illustration of two buses aligning together longitudinally in part (a) and side-by-side in part (b), each allowing an easy transfer for passengers.

Last, as we noticed in Chapter 15, designing a public transit system produces conflicts among the parties involved. The agency will usually prefer a small number of routes with transfer centers in order to reduce fleet size. The passenger will prefer more routes and fewer transfers because transfers naturally decrease comfort and reliability. This conflict may intensify in areas with variable and sparse demand. The solution is to combine the

Table 20.1 Perception of transit options, by party

Option	Party		
	Passenger	Agency	Authority
More routes	Better service	More expensive	Better coverage
More transfers	Reduced comfort and reliability	Less expensive	More user complaints
Smart transfers	Reduced waiting and travel time	Less expensive and increased ridership	Increased use of public transit

advantages of a transfer-based system (lower costs) with models that increase the comfort of transfers and reduce waiting and travel time. Table 20.1 summarizes these options.

20.1.2 Transit-system automation

Emerging technologies, both those being implemented now and those to be implemented in the near future, are surrounding us at an increased pace. The Internet and mobile phone revolutions are creating a technological breakaway from tradition that is affecting every aspect of life; it is a revolution that cannot be stopped and so affects transit operations, as well. One of the basic claims that is frequently made to delay or reject the introduction of a new technology is this: "show us where these new systems are running in the world." The answer to this question will be given shortly, once these systems are introduced. The revolution *is* happening, and it is essential in all fields of conveyance: baggage and cargo handling, elevator and walkway systems, industrial conveyors, and of course mass transit. The quest for what new technologies are searching is echoed in the advertising of many large logistics companies, "We take Anything, Anywhere, Anytime." Even the European Union Voyager project (EU Report, 2005), the main research project on the future of transportation, concluded that what is needed and what the public demands is a high-quality, door-to-door *mobility service*.

The automation revolution has been realized in the past few years, especially in the fields of computers, telephones, and mobile networking. The automation revolution has brought technology to the point where creating an automated transit system is a workable option. Rosenbloom and Fielding (1998) in Transit Cooperative Research Program Report 28 state that single-occupant drivership can only be reduced with reverse-transit solutions, services to employers or universities, pooling incentives, and restructuring routing and feeder services. These changes are actually being implemented with the aid of the new technologies.

Another important aspect of transit-system automation is the development of special information technologies. Two technologies deserve attention: radio frequency identification (RFID) and proximity information. RFID utilizes a radio frequency transmitter and receiver to identify items and their location. Usually, these are extremely small, low-cost devices that use very little or no power and draw their needed power from the air; such devices can be used to identify transit-vehicle locations and driving paths, prevent the theft of goods, and enable cashier-less supermarkets.

Proximity information, termed the *location revolution*, is a technology that started around 2001 for finding locations, for instance, giving correct route and map directions, creating personalized advertisements according to the types of vehicles or mobile phones passing by, and even locating lost children. Together, these two technologies are paving the way to produce automated transit vehicles and driverless systems, as well as online tactics to improve transit reliability.

20.2 DEVELOPMENTS IN TRANSIT AUTOMATION

Several automatic transit systems have already been implemented, using such technologies as magnetic location, distributed networking, and linear induction motors. The most common, well-known automated transit systems are automated train operations (ATO), automated people movers (APM), elevated rail systems for automatic guided vehicle (AGV), group rapid transit (GRT), and personal rapid transit (PRT).

Examples of known ATO systems are the Paris Metro, Bay Area Rapid Transit (BART, California), Metrorail (Washington, DC), and Mass Transit Railway (MTR, Hong Kong). Known APM systems are implemented especially in airports, such as John Fitzgerald Kennedy (JFK), Newark, Düsseldorf, Frankfurt, and Schiphol in Amsterdam. APM systems are also available in light-rail metro systems in Toronto, Vancouver, Detroit, and New York.

20.2.1 Elevated rail systems

Elevated monorails have been in use for more than 120 years, or since 1890, with not one known fatality. Outside the city, elevating systems benefit nature by not creating an ecological barrier. An elevated system cannot crash with cars; it consumes less land, and it is not susceptible to traffic congestion. In some noted cases, elevated rail was the only system able to work after heavy earthquakes that brought down bridges and freeways (e.g., Red Line California 1994, Osaka 1999, Seattle 2001).

In urban areas, the problem with elevated rail systems usually concerns the right-of-way for residents who live on floors at the level of travel. Certainly, this is not a problem at airports or business districts. Heavy monorails as a solution are still extremely expensive and once installed have further difficulties.

The advanced systems—APM, AGV, and PRT—constantly failed to be realized over the past 30 years. However, some progress has recently been made in that direction. In London, Heathrow Airport has ordered a limited system that employs the Urban Light Transport (ULTRA) concept, a PRT-prototype system developed in Wales for the European Union, to take passengers from the long-term parking lot to the main airport terminal. In South Korea, a system that evolved from Raytheon's PRT is being installed. At Uppsala University in Stockholm, Sweden, the European Union is finally building a large test track for an automatic system, code-named *Vectra*. And in Dubai City, UAE, there is now a tender for PRT to work a large-scale city and hotel loop. Furthermore, a research project is in progress in Germany for an automatic urban railcar system (to operate on the NeoVAL system). Figure 20.2 illustrates several PRT systems and one AGV system related to these projects.

20.2.2 Elevator–PRT and dual-mode concepts

Often, a PRT system is seen as a limited solution because it does not take into consideration heavy cargo or large groups of people, such as big families or school classes. In some city transportation committee meetings that are debating light-rail solutions, PRT is usually brought up to show how *futuristic* a solution it is and then discarded (PRT, 2004). One possible, and faster, way to introduce PRT systems may be via the elevator industry, which has been searching for a true *door-to-door* elevator, in other words, an elevator that can go between buildings and reach parking lots. The joint elevator–PRT concept has been exhibited in several futuristic movies. Figure 20.3 illustrates such an example as visualized for the movie *Minority Report*.

Figure 20.2 PRT (upper illustrations) and AGV systems.

Finally, there is a dual-mode concept suggested by Reynolds (2006). This consists of environmentally clean transit vehicles and automobiles that use guideways at different speeds (location based) as one (main) mode and ordinary driving as a second mode. The *driving* on the guideways is done by an automatic computer-controlled system. Reynolds justified his concept with the following argument:

A hundred years ago most rural people got water into their houses by carrying it in buckets from a well in the yard. Those old buckets delivered very little water compared to modern plumbing, since a pipe can provide continuous flow. The same observation applies to transportation: Buses and trains are like one-at-a-time buckets. Cars are supposed to run or flow continuously on the highways, and they will flow continuously on the guideways. The big spaces we leave between cars on the highways are necessary because of the unsynchronized traffic, the limitations of human drivers, and the limitations of automobile braking and tire traction.

Figure 20.3 Joint elevator–PRT concept envisioned for the movie, *Minority Report*.

20.3 EXAMPLE OF A TWO-WAY MONOBEAM ELEVATED TRANSIT

This section describes, as an example, an elevated small-group automated rapid transit (ESGART) labeled as SkyCabs, designed in New Zealand (Chapman et al., 2011). SkyCabs comprises lightweight, eight-seater, automated cabs that travel up to 80 km/h on a two-way, elevated moonbeam (Chapman et al., 2011). The system is designed to run on an on-demand basis with high frequencies resulting in fast, pollution-free, unimpeded travel. Artistic impressions and schematics of SkyCabs are illustrated in Figures 20.4 and 20.5.

Chapman et al. (2011) provide in Table 20.2 a comparison between SkyCabs and other modes concerning different characteristics. Most characteristics and values for other modes

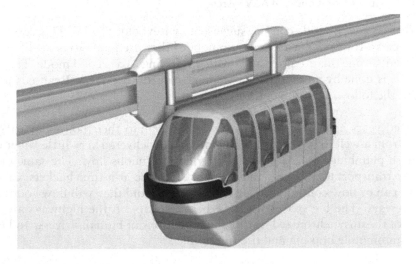

Figure 20.4 Artist's impression of SkyCabs.

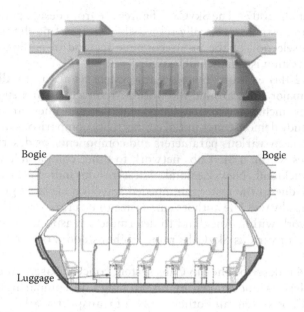

Figure 20.5 Elevation view of SkyCabs.

Table 20.2 SkyCabs comparison to other passenger travel modes

	Part busway and part bus lane	Light rail	Standard monorail	PRT	SkyCabs (ESGART)
Cost per km of two-way track	US $15–$25 million	US $31–$43.5 million	Varies to US $62 million	US $18.6–$31 million	Below US $20 million
Cost of stations and vehicles	Not included	Included	Included	Included	Included
Distance between stations	200–2000 m off-lane	200–500 m In-line	1000–1500 m In-line	500–1000 m Off-line	750 m Off-line
Speed of construction	Two opposing *on ground* lanes by motorway	Two *on ground* tracks in road	Two large elevated tracks required	Two small elevated tracks required	Single small two-way elevated beam over footpath
Average speed of travel (km/h)	30	24	42	42	60
Capacity seated	2400 people/h/ direction at 1 min intervals	3000 people/h/ direction	3000 people/h/ direction	1200–4800 people/h/ direction	4800 people/h/ direction
Capacity seated and standing	Up to 7200/h crammed at 1 min intervals	6000/h, up to 9000/h crammed	6000/h, up to 9000/h crammed	1200–4800 people/h/ direction	9000 people/h/ direction
Peak time service	2–10 min (6–30 veh/h)	2–10 min (6–30 veh/h)	4–6 min wait (10–15 veh/h)	Time to arrange passengers	30 s to max 4 min (15–120 veh/h)
Off-peak service	15–30 min wait (2–4 veh/h)	12 min wait (5 veh/h)	12 min wait (5 veh/h)	No waiting	30 s to max 4 min (15–120 veh/h)

are from Carnegie et al. (2007). The SkyCabs figures are from design parameters and consultant and manufacturer estimates. Fiscally responsible choice of public transit modes should be governed by the selection of the most economic method of adding capacity to arterial roads to ensure congestion is significantly eased.

Chapman et al. (2011) examine the SkyCabs system as a way to alleviate some of the traffic problems of major cities around the world. Architectural and engineering aspects of eight-seater SkyCabs, including cab frequency, stations, and lines, are described in detail. Furthermore, this study demonstrates initial connectivity comparisons (using the transit connectivity measurement by various parameters and components, as described in Chapter 13) of two example lines within a SkyCabs network to other comparable paths on the North Shore busway in Auckland, New Zealand. These results indicate favorable performance of SkyCabs, mainly due to the faster, unimpeded travel on elevated guideway and shorter waiting times and headways. The authors emphasize that future work should be carried out for a SkyCabs network within Auckland to determine the aspects of internal connectivity and the connectivity of various SkyCabs routes to hubs/nodes of existing public passenger transit systems.

Ceder et al. (2014a) describe the SkyCabs system concept based on three criteria. First, a comparative analysis is undertaken by researching and comparing the system characteristics of the SkyCabs system and other modes of transport available. Then, a computer simulation has been carried out to investigate the operational feasibility of the system. Finally, an economic analysis has been presented by calculating and comparing the benefit–cost ratios for each transport mode considered. The results obtained through modeling and simulation exercises clearly illustrate that SkyCabs concept has significantly lower dwell and wait times compared to the buses. Furthermore, it is revealed that the expected cab occupancy levels are lower than their respective capacity levels for every demand scenario tested. The results of the comparative analysis indicate that, overall, SkyCabs concept possesses favorable system characteristics when compared to all other modes of transport analyzed. Finally, the economic assessment indicates that SkyCabs is the most economically viable option in comparison to PRT, light rail, and GRT systems. The authors conclude that implementation of a SkyCabs system into any urban environment with similar characteristics to that of Auckland's North Shore is feasible. As such, SkyCabs concept offers a viable alternative to the traditional and other modern modes of public transit within urban areas with similar economic, social, and geographical characteristics to Auckland's North Shore.

Ceder et al. (2014b) continue to investigate the feasibility of implementing the SkyCabs system to and from Auckland central business district and Auckland international airport (in New Zealand) by examining four variables: different routes, different number of stops/stations, different passenger-demand levels, and different number of cabs in the system. The analysis made utilizes geographical information systems (GIS) and simulation tools for the various scenarios considered. The comparative analysis involved obtaining information on the following system characteristics:

- Maximum theoretical capacity
- Average travel speed
- Energy consumption
- Operation and maintenance (O&M) costs
- Average capital costs

Information on the same system characteristics was then determined by Ceder et al. (2014b) by carrying out research for each of the following traditional and modern modes of transport:

- Bus
- Light rail
- Private vehicles
- PRT
- GRT

As a basis for comparison, all the researched monetary values were converted to 2012 dollars by applying an inflation factor. They were then converted to New Zealand dollars (NZD) using the following exchange rates valid as of July 2012:

1NZD = 0.79 United States dollars
1NZD = 0.50 British pounds
1NZD = 0.78 Australian dollars
1NZD = 0.63 Euro

The results of the comparative analysis appear in Table 20.3. Generally speaking, it shows that it is possible to assess the cost–benefit of alternative routes in terms of level of service and rate of return on investment. Implementation of such a system is feasible because it has competitive pricing, increases the public transit use, and has an attractive level of service. Finally, the authors assert that due to the competitive fares and the sheer attractiveness of this new system, it is likely to experience an increased public transit patronage.

Table 20.3 Results of the comparative analysis

	SkyCabs	Bus/bus lane	Car/road lane	Light rail	PRT	GRT
Capacity seated (pass/h/dir)	4,800	3,000	2,160	3,600	2,400	7,200
Capacity including standees (pass/h/dir)	9,600	4,800	2,160	6,000	2,400	9,000
Average travel speed (km/h)	60	19.3	38.7	43	39	30
Energy consumption (MJ/pass-km)	0.7	3.15	2.38	2.08	0.55	2.16
Capital cost (NZ$/km)	$13,000,000	$53,500,000	$47,000,000	$40,000,000	$14,800,000	$44,600,000
O&M costs (NZ$/ pass-km)	$0.33	$0.63	$0.29	$0.48	$0.35	$0.39

20.4 LITERATURE REVIEW AND FURTHER READING

Recent years have seen intensive discussion in the literature about various issues relating to transit systems of the future. Naturally, much of this discussion concerns technological development; however, some conceptual issues, mode-specific issues, and others have been analyzed, as well. This section reviews only a few concepts of these developments including automated transit networks (ATN), PRT, straddling bus, and some future trends and advanced systems.

20.4.1 Automated transit networks

Larsen (2012) has presented a technical memorandum, to the Transportation and Environment Committee, on the feasibility of an ATN system to provide a connection to the airport from the city of San José (in California, United States). This report outlines specific technical and regulatory aspects of the proposed ATN system. Potential connection points between the airport and other locations within the city have been discussed. Given that the high-speed rail system is not expected to arrive in San Jose until 2027, it has been concluded that the ATN technology has merit and is worth continued pursuit.

Recently, Furman et al. (2014) discuss and present the concept of ATN in the context of urban planning/design and public policy. Also, follow-on projects on this concept have been suggested for decision makers, in order to move forward to sustainable urban transportation. The comprehensive report of Furman et al. (2014) outlines and discusses unique features of ATN, that is, (1) its ability to provide direct origin to destination services without the need for transfers or stops at intermediate stations, (2) small vehicle size that can cater exclusively to an individual or a small group, (3) demand-driven service functionality (as opposed to fixed service schedules), (4) fully automated nature of service, and (5) other aspects relating to vehicle guideways and stations within a fully connected network. The authors assert that ATN implementations require close cooperation from many kinds of local officials and most probably will be implemented through a carefully negotiated public–private partnership (PPP). Furthermore, the existing niche applications of ATN, such as within the Heathrow Airport, have been presented and the extent to which the capabilities of these existing technologies can be expanded has been discussed. Future work has been recommended to identify the ultimate capabilities of high-speed/high-capacity ATN.

20.4.2 PRT system overview and operations

The history and a comprehensive overview of PRT systems have been presented by McDonald (2013). This study traces the beginning PRT explorations all the way to the first implementation for public use of the complete paradigm change for transit. The complex interrelationship between innovative thinking in technology, communication systems, hardware, architecture, transportation, and urban design/planning spanning the globe has been discussed with specific examples from various cities.

Raney et al. (2007) explore a number of system-design perspectives for new 47-station PRT *shuttle* system for the Pleasanton's Hacienda business park (HBP) major activity center and surrounding area. This system is deemed to significantly increase the attractiveness of rapid transit by allowing carpools, vanpools, bicycles, and bus commutes. The PRT service provides elevated, electric, demand-responsive, nonstop, medium capacity, no-wait, 30 mph service for the commute's last 2 miles and services mid-day shopping and recreation trips. The authors have investigated a number of design ideas for new transit-circulator system, including horizontal mixed use; line-haul constraints; station placement; allocation of stations to passengers; ideal real-estate characteristics; high-demand, multimodal hubs; style choices; and new accessories. The subsequent insight gained from this research suggests the need for detailed *new technology product research* on horizontal mixed use for suburban resident customers. Furthermore, interview research followed by information acceleration–style surveys is suggested as future work.

Schweizer et al. (2011) provide detailed comparisons and summaries of capacity estimations and size characteristics of serial-type and *sawtooth*-type PRT stations. Station characteristics have been modeled as a complex function by incorporating geometry, vehicle dynamics, boarding strategies, and user behavior. Comparisons have been made between the capacities and space efficiencies with various load assumptions, using analytical and micro simulation techniques. The subsequent results of this study will be useful for selecting the most suitable PRT station for a given space constraint and demand scenario.

Gustafsson (2009) describes the operational characteristics of Vectus, which is at the forefront of PRT systems, currently operating in Korea and Sweden. A test track, which has been built in Sweden for validation of the novel Vectus system and for obtaining safety approval, has been considered in this study. This proposed system is able to transport a large number of people by occupying a large number of vehicles running close to each other—with headways of around 3 s between vehicles. With such a short headway, line capacities comparable to tram lines can be achieved by counteracting the limited vehicle capacity by high-frequency services. The results obtained from the test track operations have indicated that the Vectus concept for PRT fulfills all the required operating characteristics such as safety, reliability, and performance of the vehicles. This system is capable of operating without any active components in the track to detect vehicle position or speed therefore making the system more favorable in harsh conditions, in particular, with snow and ice conditions.

Villagra et al. (2012) present a novel approach to planning smooth paths and speed profiles for automated public transit vehicles in unstructured environments. The proposed approach deals explicitly with efficiency and comfort of automated transit systems by developing a planning algorithm. This algorithm comprises three layers: (1) an optimal local continuous curvature planner for obstacle-free situations; (2) a global planner that finds intermediate points to connect the configuration space to the desired degree, taking obstacles into account; and (3) a speed planner that uses the set of curves of the previous layer to compute analytically a comfort-constrained profile of velocities and accelerations. This system will therefore provide the public transit system with a priori knowledge on the shortest path within a selected area that guarantees lateral accelerations and steering wheel speeds below given preset thresholds. Furthermore, the speed profiler can utilize semantic information from the path planner to set a continuous velocity reference that considers both bounds on lateral and longitudinal accelerations consistent with comfort and also a bound on longitudinal jerk. Real-life applications of the features mentioned earlier have been undertaken and the results of the algorithms were found to be satisfactory in an automated public transit vehicle when compared with real driving maneuvers performed by human drivers. Future work has been proposed by the authors in order to include the path planner in an overall control scheme in which an adapted robust control algorithm can be used to track as closely as possible both the planned path and the planned speed.

Muller et al. (2012) present a ride-sharing methodology that can be applied to any PRT system with more than some 10 stations. The authors acknowledge that the typical PRT systems with small driverless vehicles operating at larger than initially anticipated headways, of around 3 or more seconds, have reduced capacity implications. Therefore, a ride-sharing scheme associated with such a system, even if it requires some or most passengers to no longer receive nonstop service, can be effective in increasing the system capacity. A methodology that is simple to implement with sufficient flexibility to accommodate changing demand in each station (e.g., morning peak, off-peak, evening peak, late night) has been presented in this study. Additional information and analysis have been undertaken to demonstrate the overall ride-sharing methodology and its functionality. The results of the study indicate that the overall-system capacity (in terms of passengers served) can be significantly improved over a reasonably wide range of demand levels both between and at stations.

20.4.3 Straddling bus concept

Lee (2010) and Wassener and Deng (2010) describe the novel concept of *straddling bus* proposed in China. The straddling bus has been identified to have all of the advantages of bus-rapid transit (BRT) while offering a *greener* substitution for BRT and subway systems in the future. The key innovation of this concept is that the proposed straddling buses run above the other vehicles on the road and under the overpasses. Figure 20.6 illustrates this idea. The straddling bus concept enables saving road space, one of key issues in urban transportation. Furthermore, this concept provides an efficient service, because traffic delays, caused by other vehicles on the road, can be avoided by the straddling bus. Preliminary findings have revealed that this concept can reduce up to 25%–30% traffic jams on main routes. Operating at an average speed of 40 km/h, each bus can provide high capacity of up to 1200 people at a time.

In comparison with other four main types of public transit modes in China, subway, light-rail train, BRT, and regular buses, the proposed straddling bus concept offers major advantages in a number of aspects: reductions in road space requirements, pollution levels, noise levels, traffic congestion, and delays. Another key strength offered by the straddling bus concept is its relatively short construction life cycle. It is estimated that within 1 year, up to 40 km of the system can be constructed, compared to a 3-year construction period for a 40 km long subway system. Furthermore, these buses will not require allocated bus stop spaces such as those required by regular buses. They can be parked at the stops, without affecting the passage of other vehicles on the road.

Construction of straddling buses involves two key steps: First, the roads will require remodeling by means of either laying rails on either side or creating an auto pilot technology for the buses that could simply follow a line marking along the road pavement. The second step involves provision of station platforms for passengers. This can either be

Figure 20.6 The *straddling bus* concept. (From Lee, A., Straddling" bus–a cheaper, greener and faster alternative to commute, Retrieved from http://www.chinahush.com/2010/07/31/straddling-bus-a-cheaper-greener-and-faster-alternative-to-commute/, 2010, Accessed December 10, 2014.)

through the sides of the bus or through a built-in ladder through the ceiling door. Given the typical dimensions of 6 m width by 4–4.5 m height of these buses, careful emergency exit planning strategies must be considered. The proposed emergency evacuation of passengers will be similar to those found in airplanes—where an inflated ladder will be used to slide down and away from the vehicle in case of an emergency. The proposed bus system will be powered by both municipal electricity and solar energy, thus allowing significant saving of the amounts of fuel per year and subsequently a reduction of the carbon emissions by a substantial amount.

As of 2013, Beijing's Mentougou district is testing this technology and has planned to build a 186 km straddling bus system. At these initial stages, the construction cost of a single bus including 40 km of route facilities is estimated at approximately US $7.4 million. This cost represents approximately 10% of the cost of building a subway system of the same length.

20.4.4 Future trends: Transit related

Cuddy et al. (2014) describe smart/connected cities and discuss their potential for interfacing with the emerging connected transportation environment. The authors aim to provide a well-defined foundation for the US Department of Transportation's (USDOT) Connected Vehicle Program, in order to identify and exploit opportunities to help ensure that connected vehicles and connected transportation fulfill their potential to improve safety, mobility, and environmental outcomes in a complexly interdependent and multimodal environment. Furthermore, in order to understand the potential benefits of such a concept, this report explains in detail the definition of a smart/connected city. The *intelligent infrastructure* utilized by smart cities, such as the devices and equipment that can sense the environment and/or their own status, send data, and often, receive commands, along with their potential transportation implications, has been reviewed and outlined.

Jarašūniene (2007) describes the main intelligent transport systems (ITS) enabling technologies with an emphasis on involving human factor experts at an early stage of design of ITS equipment and facilities. The author asserts that ITS technologies, such as loop detectors, are well known to transportation professionals; however, there are a number of less familiar technologies and system concepts that are keys to ITS functions. These ITS services have been conceptualized as an information chain that includes data acquisition (from the transportation system), communications, data processing, information distribution, and information utilization (for decision and control support for the ITS users).

Velaga et al. (2009) have developed a real-time, weight-based topological map-matching algorithm by improving the limitations of existing algorithms. The main novel features of this enhanced topological map-matching algorithm include (1) the selection of candidate links in the initial map-matching process and the map matching at junctions; (2) the introduction of two additional weight parameters, connectivity and turn restriction; (3) use of an optimization process to derive the relative importance of weights using data collected in different operational environments; and (4) the implementation of two consistency checks to reduce mismatches. The proposed improved map-matching algorithm has been tested using real-world field data collected in different operational environments. The results indicate that all of the novel features incorporated within the proposed algorithm have contributed to the improved performance, in that this algorithm has identified 96.36% of the road segments correctly in an urban area and 97.01% of the road segments with a horizontal accuracy of 8.9 m (95%) in a suburban area. Future work has been proposed to investigate the optimization of weights using more positioning data from each operational environment to ensure that these optimal values are transferable.

Biem et al. (2010) describe the use of IBM InfoSphere Streams, a component-based distributed stream processing platform, for tackling the challenges of scalability, extensibility, and user interaction in the domain of intelligent transportation services. A prototype system that generates dynamic, multifaceted views of transportation information for the city of Stockholm, using real vehicle global positioning system (GPS) and road network data, has been described in detail. This system is capable of continuously deriving current traffic statistics and provides useful value-added information such as shortest-time routes from real-time observed and inferred traffic conditions. The performance experiments undertaken indicate that this system is scalable; thus, it can process over 120,000 incoming GPS points per second, combine with a map containing over 600,000 links, continuously generate different kinds of traffic statistics, and answer user queries.

20.4.5 Advanced systems categorized by technology

A considerable number of papers focus on new technologies that are presently used in various transit systems. Several more recent examples follow.

20.4.5.1 World Wide Web

Lu and Shi (2007) have applied the theory of complex networks to classify public transit networks into route networks, transfer networks, and bus-station networks. The practical significance of each of the network parameters has subsequently been analyzed. Real-life application of this concept has been undertaken for the transit networks in Langfang, Jining, and Dalian in China. The results obtained have indicated that transit networks possess the features of complex networks. Furthermore, it was found that all the urban transportation network parameters significantly affect the accessibility, convenience, and terrorist security capability of the urban public transportation network.

Mallat et al. (2008) present the results obtained from a study of mobile ticketing service adoption in transit. The theoretical underpinnings of this study involve technology adoption and trust theories, which are complemented with concepts of mobile use context and mobility. The empirical findings from survey data analysis indicate that compatibility of the mobile ticketing service with consumer behavior is a major determinant of adoption. Furthermore, the adoption decision was strongly dependent on mobility and contextual factors such as budget constraints, availability of other alternatives, and time pressure in the service use situation. The authors conclude that contextual and mobile service-specific features are important determinants of mobile service adoption and should therefore be incorporated into the traditional adoption models.

20.4.5.2 Global positioning system

Liao et al. (2007) have described the foundations and experimental validation of a hierarchical model that can learn and infer a user's daily movements and use of different modes of transportation. The model uses multiple levels of abstraction in order to bridge the gap between raw GPS sensor measurements and high-level information such as a user's destination and mode of transportation. In order to achieve high levels of efficiency in the inference process, the authors have applied Rao–Blackwellized particle filters at multiple levels of the model hierarchy. The results indicate that the proposed approach can provide predictions of movements to distant goals. Furthermore, a simple and effective strategy for detecting novel events that may indicate user errors (e.g., taking a wrong bus) has been proposed by explicitly modeling activities in the context of the user's historical data. Finally, the authors discuss an application named *Opportunity Knocks* that employs the proposed techniques to assist cognitively impaired people use public transit safely.

Mazloumi et al. (2009) gave a good example of the use of GPS for transit operations analysis. The authors recognize that a major obstacle to researching the adverse impacts of day-to-day transit service reliability is the lack of comprehensive data sets on bus travel times. The authors address this issue, by utilizing a GPS data set for a bus route in Melbourne, Australia. The nature and shape of travel time distributions for different departure time windows at different times of the day along with the factors causing travel time variability of public transit vehicles has been explored using a linear regression analysis. The results indicate that in narrower departure time windows, travel time distributions are best characterized by normal distributions. In contrast, for wider departure time windows, peak-hour travel times follow normal distributions, while off-peak travel times follow lognormal distributions. The contributing factors toward the variability of travel times were identified as land use, route length, number of traffic signals, number of bus stops, and departure delay relative to the scheduled departure time. Furthermore, the travel time variability for the data set analyzed was found to be higher in the AM peak and lower in the off-peak periods.

20.4.5.3 Geographical information systems

Yigitcanlar et al. (2007) describe a comprehensive accessibility indexing model—the land use and public transport accessibility indexing model (LUPTAI). The authors demonstrate that it is possible to produce a viable accessibility model, apply the model to a major urban area, and produce a mappable accessibility index. A real-life application of this concept has been demonstrated within the city of Gold Coast in Australia. The results indicate that this proposed model is accurate, straightforward, and transparent enough to be understood by the general public and is able to be customized and replicated.

Currie (2010) identifies spatial gaps in public transit provision for people who are socially disadvantaged. This study outlines the research context for measurement of public transit supply and needs and then describes the methodology developed for an application within Melbourne, Australia. The techniques described are relatively easy to develop and apply and can be powerful in identifying spatial gaps in transit service provision in an open and defendable quantitative manner. The results of the application indicate significant gaps between services supplied and social needs for transit services. Finally, the author has outlined relevant implications of this study in transport-policy development.

20.4.5.4 Web-based geographical information systems

Hoar (2008) describes a web-based geographic information system that disseminates the same schedule information through intuitive GIS techniques. A map-based interface has been created, using data from Calgary, Canada, to show the users the routes, stops, and moving buses simultaneously. Zoom and pan controls as well as satellite imagery functions have been embedded to allow users to apply their personal knowledge about the local geography to achieve faster and more pertinent transit results. The users can utilize this application to add or remove the buses and respective stops interactively without the need to wait for responses to web-based requests; this is achieved through asynchronous requests to web services.

Pun-Cheng (2012) describes a web-map public transit inquiry system. This system is capable of providing bilingual (English and Chinese) information on all transit runs in Hong Kong in the form of interactive maps and texts. Furthermore, the users can obtain real-time derivations of optimal traveling routes in terms of multiple criteria, such as preferred mode, least changes, shortest traveling time, or lowest fare.

20.4.5.5 Automatic vehicle location

Milkovits (2008) develops and implements preprocessing techniques that estimate a dwell time model and analyzes the impact of the secondary factors. The data used for analysis have been obtained from automatic-passenger counting, automatic-fare counting, and automatic-vehicle location systems installed on Chicago Transit Authority buses. The estimates derived from the proposed model demonstrate that the advantage of smart media farecards over magnetic stripe fare tickets is about 1.5–2 s in time savings per passenger. This advantage, however, is not observed when there are more passengers onboard than the seating capacity of the bus. Therefore, it is concluded that the advantages of the new fare media technology in improving bus operations occur mostly when service levels do not experience crowding. Furthermore, the large data set obtained enable a comparison of the residuals across time of day, stop, operator, and route to look for other significant relationships. Future work has been proposed to extend the ordinary least-squares regression model to include distributions to account for the impacts of operator, time of day, stop, and other factors.

Pangilinan et al. (2008) demonstrate how to use the real-time automatic vehicle location (AVL) information to benefit service reliability on high-frequency bus routes. The decision-making process on deployment of operations control strategies can be enhanced by the knowledge on real-time headways and bus locations. The Chicago (Illinois) Transit Authority's real-time AVL pilot project for Route 20 Madison has been used as a case study for the evaluation of the effectiveness of real-time AVL to improve reliability. A simulation model of the route has been developed using archived AVL data to predict the effects on service reliability when real-time AVL information is used in bus supervision. The experiment was undertaken over a period of 1 week, in order to verify the model and to address the feasibility and scalability of the system. From the results obtained, the authors conclude that real-time AVL does indeed have great potential to improve service reliability. Service restoration strategies previously impossible to execute are now feasible because of this new information stream. Future work has been proposed by the authors in the area of network-wide implementation of the system, including the supervision communications structure and manpower deployment questions.

20.4.6 Advanced systems categorized by purpose

Some works place the main emphasis not on the type of technology used but on its purpose. The following is a review of important publications of this sort.

20.4.6.1 Passenger information

Dziekan and Kottenhoff (2007) provide a comprehensive framework of the possible effect that dynamic at-stop real-time information displays can have on public transit customers. The seven main effects identified in this study are (1) reduced wait time; (2) positive psychological factors, such as reduced uncertainty, increased ease of use, and a greater feeling of security; (3) increased willingness to pay; (4) adjusted travel behavior such as better use of wait time or more efficient traveling; (5) mode choice effects; (6) higher customer satisfaction; and finally, (7) better image. Two case studies have been presented in this work. The first study proves that perceived wait times can be reduced by 20% by employing a before–after implementation evaluation study with questionnaires on a tramline. The second study illustrates the effects of real-time displays on passenger behavior in the form of adjusted walking speeds, by using a behavior observation method in a subway station. The authors propose future work in order to account for interrelations between the effects identified.

Caulfield and O'Mahony (2007) describe provision of public transit information to the customers in Dublin, Ireland. Examination of both existing and potential methods of accessing information, with particular focus on the implementation of various ITS applications, has been undertaken. The authors have categorized the decision-making process of a passenger, when undertaking a transit trip, into four key stages: (1) *pretrip to destination*, (2) *at-stop*, (3) *onboard*, and (4) *pretrip to origin* (the return journey). Data on passenger preferences have been collected using a web-based survey. The analysis results indicate that passengers in Dublin would like to see improvements in their public transit information system and have indicated preferences to different forms of information provision.

20.4.6.2 Data collection

Pelletier et al. (2011) provide a comprehensive review of key aspects of smart card data use in the public transit context. First, the authors present and address the available advanced technologies such as the hardware and information systems required in operating these tools and privacy concerns and legal issues related to the dissemination of smart card data, data storage, and encryption. Second, the various uses of these data have been described at three distinct levels: strategic (long-term planning), tactical (service adjustments and network development), and operational (ridership statistics and performance indicators). Finally, in light of the comprehensive review, the authors have identified the following potential challenges for smart card transit operators and researchers: (1) technological improvements, (2) data validation, (3) economic feasibility, (4) journey validation, (5) new modeling approaches, and (6) new methods of analysis.

20.4.6.3 Bus priority

Liu et al. (2008) present an analytical approach for the design and evaluation of a transit signal priority (TSP) system with two fundamental TSP strategies: early-green and extended-green operations. Through graphical illustration, it has been demonstrated that the impact of early green and extended green on buses and general traffic could be quantified using D/D/1 queuing models if the issue of concern is the induced delay. A numerical application has also been presented and the results indicate that the proposed method is promising in that it requires very limited information and provides fairly reasonable results. Therefore, this analytical approach can be used independently or as a supplementary tool to simulation models for complicated situations.

Ma et al. (2010) test a different bus-priority method referred to as coordinated and conditional bus priority (CCBP). The proposed approach considers the bus arrival time at the downstream intersection and strategies are derived for early buses. Coordinated, signalized intersection groups are adopted as control objects. Buses are detected one or more cycles before their arrival at the first intersection of the control object. The proposed CCBP includes two kinds of priority strategies: increasing and decreasing bus delay. A model has been developed to generate the optimal combination of priority strategies for intersection groups in order to match the real delay of buses closely with the permitted delay defined by the bus operation system. A field application has been presented to illustrate the CCBP approach and to compare it with other two options: no priority and unconditional priority. The authors conclude that by using the CCBP approach, significant reductions on bus delay and headway deviations can be achieved. Furthermore, the results indicate that using the CCBP approach causes only minor increases in total average delay to other traffic. The authors concluded that this approach could be used to decrease bus delay deviation and enhance the reliability of bus service without significantly affecting the delay of other motor vehicles.

20.5 CONCLUDING REMARK

This is the last section of the book, and thus it deserves a final word, which will be given by way of symbolism. There is a Swedish proverb: "Worry often gives a small thing a big shadow," about which Marie Curie said: "Nothing in life is to be feared. It is only to be understood." Understanding the nature of transit operations problems is already halfway toward reaching adequate solutions. Lastly, an African proverb cautions: "Smooth seas do not make skillful sailors." Indeed, experiencing problems in transit operations provides an important lesson that, together with the solutions and thoughts proposed in this book, can hopefully end with a significantly improved transit service.

REFERENCES

Biem, A., Bouillet, E., Feng, H., Ranganathan, A., Riabov, A., Verscheure, O., and Moran, C. (2010). IBM infosphere streams for scalable, real-time, intelligent transportation services. In *Proceedings of the 2010 ACM SIGMOD International Conference on Management of Data*, pp. 1093–1104. ACM, Indianapolis, USA.

Carnegie, J.A., Voorhees, A.M., and Hoffman, P. (2007). Viability of personal rapid transit in New Jersey. Report to Department of Transportation, Bureau of Research and NJ Transit, Ewing, NJ.

Caulfield, B. and O'Mahony, M. (2007). An examination of the public transport information requirements of users. *IEEE Transactions on Intelligent Transportation Systems*, 8(1), 21–30.

Ceder, A., Hadas, Y., Wan, N.K.L., and Sundarapperuma, D. (2015). The Planning and Analysis of a New Group Transportation System: The Case of the SkyCabs Monobeam System in Auckland, New Zealand. *Transportation Planning and Technology*, 38(3), 320–334.

Ceder, A., Hadas, Y., Wan, N.K.L., and Sundarapperuma, S. (2014a). Case study analysis of SkyCabs monobeam system. *Transportation Planning and Technology*, forthcoming.

Ceder, A., Roberts, M., and Schermbrucker, R. (2014b). Investigation of SkyCabs monorail system in urban regions. *Journal of Transportation Technologies (JTTs)*, 4(1), 31–43.

Chapman, H., Chapman, M., and Ceder, A. (2011). A new architectural design of elevated small group automated rapid transit. *Journal of Public Transportation*, 14(4), 63–87.

Chien, S.I., Spasovic, L.N., Elefsiniotis, S.S., and Chhonkar, R.S. (2001). Evaluation of feeder bus systems with probabilistic time-varying demands and non additive time costs. *Transportation Research Record*, 1760, 47–55.

Cuddy, M., Epstein, A., Maloney, C., Westrom, R., Hassol, J., Kim, A., Damm-Luhr, D., and Bettisworth, C. (2014). The smart/connected city and its implications for connected transportation, U.S. Department of Transportation, FHWA-JPO-14–148, John A Volpe National Transportation Systems Center, Cambridge, MA.

Currie, G. (2010). Quantifying spatial gaps in public transport supply based on social needs. *Journal of Transport Geography*, 18(1), 31–41.

Dial, R.B. (1995). Autonomous dial-a-ride transit-introductory overview. *Transportation Research*, 3C, 261–275.

Dziekan, K. and Kottenhoff, K. (2007). Dynamic at-stop real-time information displays for public transport: Effects on customers. *Transportation Research*, 41A, 489–501.

EU Report (2005). *VOYAGER: Transportation in 2020. A Project Approach*. European Union Research, Brussels.

Furman, B., Fabian, L., Ellis, S., Muller, P., and Swenson, R. (2014). Automated Transit Networks (ATN): A review of the state of the industry and prospects for the future, Mineta Transportation Institute, MTI Report 12-31, No. CA-MTI-14–1227. College of Business, San José State University, San José, CA.

Gendreau, M., Laporte, G., and Seguin, R. (1996). Stochastic vehicle routing. *European Journal of Operational Research*, 88, 3–12.

Ghiani, G., Guerriero, F., Laporte, G., and Musmanno, R. (2003). Real-time vehicle routing: Solution concepts, algorithms and parallel computing strategies. *European Journal of Operational Research*, **151**, 1–11.

Gustafsson, J. (2009). Vectus—Intelligent transport. *Proceedings of the IEEE*, **97**(11), 1856–1863.

Hoar, R. (2008). Visualizing transit through a web based geographic information system. *Proceedings of World Academy of Science, Engineering and Technology*, **36**, 180–185.

Jarašūniene, A. (2007). Research into intelligent transport systems (ITS) technologies and efficiency. *Transport*, **22**(2), 61–67.

Larsen, H. (2012). Automated transit network feasibility study. Memo to the City of San José Transportation Environment Committee. Retrieved from: http://www. sanjoseca.gov/ DocumentCenter/View/14332. Accessed December 18, 2014.

Lee, A. (2010). "Straddling" bus—a cheaper, greener and faster alternative to commute. Retrieved from http://www.chinahush.com/2010/07/31/straddling-bus-a-cheaper-greener-and-faster-alternative-to-commute/. Accessed December 10, 2014.

Liao, L., Patterson, D.J., Fox, D., and Kautz, H. (2007). Learning and inferring transportation routines. *Artificial Intelligence*, **171**(5), 311–331.

Liu, H., Zhang, J., and Cheng, D. (2008). Analytical approach to evaluating transit signal priority. *Journal of Transportation Systems Engineering and Information Technology*, **8**(2), 48–57.

Lu, H. and Shi, Y. (2007). Complexity of public transport networks. *Tsinghua Science and Technology*, **12**(2), 204–213.

Ma, W., Yang, X., and Liu, Y. (2010). Development and evaluation of a coordinated and conditional bus priority approach. *Transportation Research Record*, **2145**, 49–58.

Mallat, N., Rossi, M., Tuunainen, V.K., and Öörni, A. (2008). An empirical investigation of mobile ticketing service adoption in public transportation. *Personal and Ubiquitous Computing*, **12**(1), 57–65.

Mazloumi, E., Currie, G., and Rose, G. (2009). Using GPS data to gain insight into public transport travel time variability. *Journal of Transportation Engineering*, **136**(7), 623–631.

McDonald, S.S. (2013). Personal rapid transit and its development. In *Transportation Technologies for Sustainability* (M. Ehsani, F.-Y. Wang, G.L. Brosch, eds.), pp. 831–850, Springer, New York.

Milkovits, M.N. (2008). Modeling the factors affecting bus stop dwell time: Use of automatic passenger counting, automatic fare counting, and automatic vehicle location data. *Transportation Research Record*, **2072**, 125–130.

Minsky, M.L. (1986). *The Society of Mind*. Simon and Schuster, New York.

Muller, P.J., Cornell, S.B., and Kubesa, M.P. (2012). Ridesharing methodology for increasing personal rapid transit capacity. In CD-ROM of the *Transportation Research Board 91st Annual Meeting* (No. 12-0882), Washington, DC.

Pangilinan, C., Wilson, N., and Moore, A. (2008). Bus supervision deployment strategies and use of real-time automatic vehicle location for improved bus service reliability. *Transportation Research Record*, **2063**, 28–33.

Pelletier, M.P., Trépanier, M., and Morency, C. (2011). Smart card data use in public transit: A literature review. *Transportation Research*, **19C**, 557–568.

PRT Cyberspace Dream Keeps Colliding with Reality. (2004). Light rail now publication. Retrieved from: http://www.lightrailnow.org/facts/fa_prt001.htm. Accessed December 22, 2014.

Pun-Cheng, L.S. (2012). An interactive web-based public transport enquiry system with real-time optimal route computation. *IEEE Transactions on Intelligent Transportation Systems*, **13**(2), 983–988.

Raney, S., Paxson, J., and Maymudes, D. (2007). Design of personal rapid transit circulator for major activity center: Hacienda business park, Pleasanton, California. *Transportation Research Record*, **2006**, 104–113.

Reynolds, F.D. (2006). The revolutionary dualmode transportation system. E-book http://faculty.washington.edu/jbs/rev/revcontents.htm. Accessed December 8, 2014.

Rosenbloom, S. and Fielding, G.J. (1998). Transit markets of the future—The challenge of change. TCRP Report 28, Transportation Research Board, Washington, DC.

Schweizer, J., Mantecchini, L., and Greenwood, J. (2011). Analytical capacity limits of Personal Rapid Transit stations. In *Automated People Movers and Transit Systems* 2011: *From People Movers to Fully Automated Urban Mass Transit*, pp. 326–338, American Society of Civil Engineers (ASCE), Reston, VA. ISBN: 9780784411933, doi: 10.1061/41193(424)30.

Van Dyke Parunak, H. (1997). Go to the ant: Engineering principles from natural multi-agent systems. *Annals of Operations Research*, **75**, 69–101.

Velaga, N.R., Quddus, M.A., and Bristow, A.L. (2009). Developing an enhanced weight-based topological map-matching algorithm for intelligent transport systems. *Transportation Research*, **17C**, 672–683.

Villagra, J., Milanés, V., Pérez, J., and Godoy, J. (2012). Smooth path and speed planning for an automated public transport vehicle. *Robotics and Autonomous Systems*, **60**(2), 252–265.

Wassener, B. and Deng, A. (2010). Straddling bus' offered as a traffic fix in China. *The New York Times*. Retrieved from: http://www.nytimes.com/2010/08/18/business/global/18bus.html?_r=0. Accessed December 3, 2014.

Yigitcanlar, T., Sipe, N., Evans, R., and Pitot, M. (2007). A GIS-based land use and public transport accessibility indexing model. *Australian Planner*, **44**(3), 30–37.

Zhao, J.M., Bukkapatnam, S., and Dessouky, M.M. (2003). Distributed architecture for real-time coordination of bus holding in transit networks. *IEEE Transactions on Intelligent Transportation Systems*, **4**(1), 43–51.

Answers to Exercises

Chapter 3

3.1 (a)

Parameter	Time period				
	6:00–7:00	*7:00–8:00*	*8:00–9:00*	*9:00–10:00*	*10:00–11:00*
A (pass/km)	444.2	1796	1186	1349	952.4
ρ	0.82	0.64	0.51	0.71	0.72
Method proposed	1[a]	1	1	1	1
χ^2 test between Methods 1 and 2 data	Calculated $\chi^2 = 2.66$ $\chi^2_{\alpha=0.05} = 9.49$				

[a] For max load methods, use Method 1.

(b)

Method	Time period									
	6:00–7:00		*7:00–8:00*		*8:00–9:00*		*9:00–10:00*		*10:00–11:00*	
	F[a]	*H*[a]	*F*	*H*	*F*	*H*	*F*	*H*	*F*	*H*
Method 1	2	30	7.80	8	6.08	10	6	10	4.44	14
Method 2	2	30	7.80	8	6.47	9	6.36	9	4.44	14
Method 3	2	30	5.85	10	4.85	12	4.50	13	3.17	19
Method 4 (20%)	2	30	7.80	8	6.08	10	6	10	4.44	14

[a] F in (veh/h), H in min.

Solutions for (c) and (d)

Time period	Method	Excess load (Passenger, km)	(c) Empty space (km)	(d) Range of x = possible reduced single tariff (¢/km)
6:00–7:00	3	0 (minimum frequency provided)	0	—
	4 (20%)	0	0	—
7:00–8:00	3	191.4	510.6	x < 8.00
	4 (20%)	0	0	—
8:00–9:00	3	161.8	420.2	x < 7.79
	4 (20%)	13.8	124.2	x < 27.00
9:00–10:00	3	179.4	378.6	x < 6.33
	4 (20%)	14.4	93.6	x < 19.50
10:00–11:00	3	119.6	258.4	x < 6.48
	4 (20%)	0	0	—

3.2 (a) 6:00–7:00, max load = 148 passengers at stop 3; 7:00–8:00, max load = 99 passengers at stop 2
   (b) F_2 (6:00–7:00) = 3.7, F_2 (7:00–8:00) = 3.0
   (c) $F_{4(40\%)}$ (6:00–7:00) = 3.2, $F_{4(40\%)}$ (7:00–8:00) = 3.0.

Chapter 4

4.1 (a) Timetable at point A using Method 3: 07:00, 07:26, 07:59
         Using Method 4 (30%): 07:00, 07:20, 07:45, 08:21.
   (b) With c = 60, only Method 3 results are changed: 07:00, 07:25, 07:55
         Method 4 results same as in (a).
4.2 (a) Even-headway timetable based on Method 2: 6:00, 6:30, 7:00, 7:08, 7:15, 7:23, 7:31, 7:38, 7:46, 7:54, 8:02, 8:11, 8:21, 8:30, 8:39, 8:48, 8:58, 9:07, 9:16, 9:26, 9:35, 9:45, 9:54, 10:05, 10:19, 10:32, 10:46, 11:00.
         Clock-headway timetable based on Method 4 (20% criterion): 6:00, 6:30, 7:00, 7:08, 7:15, 7:23, 7:31, 7:38, 7:46, 7:54, 8:02, 8:12, 8:22, 8:32, 8:42, 8:51, 9:01, 9:11, 9:21, 9:31, 9:41, 9:51, 10:01, 10:15, 10:28, 10:42, 10:56.
   (b) Clock-headway timetable based on Method 2: 6:00, 6:30, 7:00, 7:07, 7:15, 7:22, 7:30, 7:37, 7:45, 7:52, 8:00, 8:07, 8:15, 8:22, 8:30, 8:37, 8:45, 8:52, 9:00, 9:07, 9:15, 9:22, 9:30, 9:37, 9:45, 9:52, 10:00, 10:12, 10:24, 10:36, 10:48, 11:00.
         Clock-headway timetable based on Method 4 (20% criterion): 6:00, 6:30, 7:00, 7:07, 7:15, 7:22, 7:30, 7:37, 7:45, 7:52, 8:00, 8:07, 8:15, 8:22, 8:30, 8:37, 8:45, 8:52, 9:00, 9:10, 9:20, 9:30, 9:40, 9:50, 10:00, 10:12, 10:24, 10:36, 10:48, 11:00.
   (c) Total number of departures for the four timetables earlier: 28, 27, 32, and 30, respectively. Single-route fleet size is 6 and is the same for all timetables.
4.3 (a) With Method 2, the use of cumulative loads at stop 3 (6:00–7:00) and stop 2 (7:00–8:00) leads to the following timetable at stop 1: 6:17, 6:36, 6:53, 7:14, 7:34, and by extrapolation 7:54.
   (b) (i) Certainly not all the loads at each stop on each vehicle in a given hour are the same.
         (ii) On departure 6:17 at stop 2, the average load will be 52 + [(6:17–6:15)/96:45–6:150]·40 = 55, and on departure 6:53 at stop 3, the average load will be 45 (above 40); and on departure 7:14 at stop 3, the average load will be 60 (above 50).

4.4 (a) Passenger arrival rates (passengers/minute) are 240/60 = 4 and 180/60 = 3; and the even headways are 64/4 = 16 and 24/3 = 8 min, respectively, for the 2 h. Hence, the fourth departure, if at 8:00, will have 48 passengers; in order to comply with 44 passengers, it will depart at 7:59. The first departure after 8:00 will, therefore, be at 8:13. The following is the complete timetable: 7:16, 7:32, 7:48, 7:59, 8:13, 8:21, 8:29, 8:37, 8:45, 8:53, and 9:01.

(b) Using Method 2, we check for another possible departure, with 64 passengers as the desired load. Since this time will move beyond 8:00, we change the desired load to 24 and so end with a departure before 8:00. This contradiction suggests the use of the procedure at (a).

Chapter 5

5.1 The adjusted load between 6:00 and 7:00 for stop A is $25 + 65 + (25/35) \cdot 67 = 138$ and, for stops B and C, 148 and 85 passengers, respectively. Between 7:00 and 8:00, the adjusted loads are 178, 97, and 114 for stops A, B, and C, respectively. Hence, Method 2 is used for frequency determinations:

$$F_2 \ (6:00\text{--}7:00) = \max \ (148/40, 2) = 3.70 \text{ veh/h}, \quad H_2 \ (6:00\text{--}7:00) \approx 16 \text{ min}$$

$$F_2 \ (7:00\text{--}8:00) = \max \ (178/60, 2) = 2.97 \text{ veh/h}, \quad H_2 \ (7:00\text{--}8:00) \approx 20 \text{ min}$$

The results, which appear in the table that follows, are based on applying the cumulative-frequency-curve approach for the first procedure and the cumulative-load-curve approach for the two other procedures.

Observed data	Departure	6:00	6:35	7:10	7:45	8:00		
	Max load point	A	B	B	A	A		
	Number of passengers	25	72	82	84	75		
Evenly spaced headways	Departure	6:00	6:16	6:32	6:48	7:06	7:26	7:46	8:06
	Max load point	A	A	B	B	B	A	A	A
	Number of passengers	25	55	33	36	43	46	51	97
Even average load at hourly max load point	Departure	6:11	6:30	6:51	7:20	7:45	8:00		
	Max load point	A	B	B	A	A	A		
	Number of passengers	45	39	47	60	60	75		
Even average load at individual max load point	Departure	6:08	6:27	6:45	7:12	7:39	7:54	8:05	
	Max load point	A	B	B	B	A	A	A	
	Number of passengers	40	40	40	40	60	60	50	

5.2 The upper bound $Z^* = 8$. Both Synchro-1 and Synchro-2 give the optimal results of 4 meetings, with these departure times: route I at 0, 7; route II at 2, 9; and route III at 4.

5.3 The upper bound $Z^* = 15$. Synchro-1 results in 6 meetings. Synchro-2 results in the optimal solution, with 8 meetings. The optimal solution is based on these departure times: route I at 0, 8; 16; route II at 7, 13; 21; route III at 6, 14, 22; and route IV at 8, 20, 30.

5.4 *Hint*: The objective function given by Equation 5.1 is changed to

$$\text{Max} \sum_{k=1}^{M-1} \sum_{i=1}^{F_k} \sum_{q=k+1}^{M} \sum_{j=1}^{F_q} w_{ikjqn} Z_{ikjqn}$$

where w_{ikjqn} represents the number of passengers at node n who need to be transferred between the vehicle of the *i*th departure on route k and the vehicle of the *j*th departure on route q (all passenger transfers). A heuristic procedure can combine Synchro-1 with this change.

Chapter 6

6.1 The solution can use Salzborn's idea (1972) for single routes by combining the findings: construct three curves representing the frequencies for each route (*y*-axis) by time of day (*x*-axis) and the sum of the two routes. For route 1, the largest number of buses that departs in any time interval of 3 h is 26 (4–7 p.m.). For route 2, there are 28 independent departures (buses) (6–9 a.m.). However, the two time intervals do not overlap; hence, the minimum fleet size required *is not* the sum of the two. This minimum is found from the curve representing the sum of frequencies, which from 6 a.m. to 9 a.m. is 18 + 18 + 16 = 52 buses.

6.2 The augmenting-path algorithm starts by constructing a network flow similar to that in Figure 6.5. Four s–t paths can be found using the last four departures. The minimum s–t cut separates the first four departures and t from the rest of the network. Hence, Min N = n – Max Z = 8 – 4 = 4, based on Theorem 6.2. The four blocks are (by trip #) [1–5–8], [2–6–7], [3], and [4].

6.3 (i) The DF at terminal a starts with a maximum of D(a) = 3 between 7:30 and 8:00 and, at b, D(b) = 0 between 5:00 and 6:00. The URDHC procedure is then applied with the latest DH insertion possible. Based on the NT rule, two DH trips, DH_1 and DH_2, are inserted from b to arrive at a, s_{1a} at the start of the maximum interval (7:30); one DH trip, DH3, a compensatory act, is inserted from a to arrive at b at 8:30. These three DH trips result in D(a) = 1, D(b) = 0.

(ii) The sum function (of a, b, and depot), g(t, S), reveals that G(S) = 4. In order to find the minimum fleet size required of S, d(depot, t) is constructed, and it results in D(depot) = 3 between 6:40 and 9:00. Hence, G(S) indeed is the minimum fleet size required (i.e., the sum of all determined D(k), k = a, b, depot).

(iii) The four chains are [1-DH1-6], [2-5], [3-DH2-7], [4-DH3-8].

Chapter 7

7.1 Minimum fleet size is 22 vehicles for the case with only shifting, having the following FIFO blocks: [1–10–20–34], [2–11], [3–27], [4–23–43], [5–15–35], [6–17–44], [7–18–40], [8–38], [9–31], [12–32–45], [13–33], [14–29], [16–41], [19–36], [21–46], [22–47], [24], [25], [26–37], [28–42], [30], [39].

Minimum fleet size is also 22 vehicles for the case with only DH insertion trips, having the following FIFO blocks: [1–10–20–34], [2–17–44], [3–DH_3–39], [4–DH_4–30], [5–15–35], [6–23–42], [7–18–40], [8–38], [9–31], [11], [12–32], [13–33], [14–29], [16–DH_1–46], [19–36], [21–45], [22–DH_2–43], [24–47], [25], [26–37], [27], [28–41].

Minimum fleet size is 20 vehicles for the case with both shifting and DH trip insertion, having the following FIFO blocks: [1–10–20–33], [2–14–29], [3–27], [4–DH_1–30], [5–15–38], [6–17–44], [7–18–40], [8–32–45], [9–23–43], [11–34], [12–31], [13–DH_2–39], [16–41], [19–36], [21–46], [22–35], [24–47], [25], [26–37], [28–42].

7.2 (1) Minimum fleet size is six vehicles (indicated at 7:20 on the single DF).

(2) There are four shiftings required to arrive at a minimum fleet size of four vehicles: trip 3 backward by 3 min; trip 9 forward by 5 min; trip 8 forward by 5 min; but because of the shifts of trips 4 and 8, another shift is required, that of trip 10 forward by 5 min.

(3) The original shuttle-bus schedule is coordinated with the train schedule; hence, shifting the shuttle schedule indeed will have an adverse effect on this coordination.

(4) In a single terminal (terminal k) operation, there is no need for the improved lower bound, because $g(t, S) = d(t, k)$.

(5) The four FIFO blocks are [1–8–6], [2–9], [7–5], [3–4–10].

7.3 $G(S) = 3$, $G'(S') = 4$, and $G''(S'') = 5$.

7.4 (1) $G'(\bar{S}'_{sf}) = 5$ and $G''(\bar{S}''_{sf}) = 5$.

(2) $G(S_{sf}) = 3$, $G'(S'_{sf}) = 5 = 3$, and $G''(S''_{sf}) = 4$.

7.5 Follow Equations 7.6 through 7.8, established for the $\Delta^{j\tau(-)}$ criterion, and derive similarly the criterion for $\Delta^{j\tau(+)}$.

Chapter 8

8.1 (a) Three main steps are performed. First, $N_1 = 2 + 1 + 1 = 4$, $C_1 = 4 \cdot 10 = 40$, and $G''(S)$ is calculated to be 4. Second, $N_2 = 2 + 1 + 1 + 2 = 6$, $C_2 = 2 \cdot 10 + 1 \cdot 8.5 + 1 \cdot 4 + 2 \cdot 5 = 42.5$. The third and the best solution is $N = 2 + 1 + 1 + 1 = 5 \neq G''(S)$, $C = 2 \cdot 10 + 1 \cdot 8.5 + 1 \cdot 4 + 1 \cdot 5 = 37.5$. The following are the five blocks, using the FIFO rule, by trip number and type of vehicle: type I, [5], [1–8–DH4–12]; type II, [2–6]; type III, [4–DH3–7–DH1–10]; and type IV, [3–DH2–9–11].

(b) The first main step in (a) is the same. Second, $N_2 = 3 + 3 = 6$, $C_2 = 3 \cdot 10 + 3 \cdot 4 = 42$. The third and best is $N = 3 + 1 = 4 = G''(S)$, $C = 3 \cdot 10 + 1 \cdot 4 = 34$. The following are the four (FIFO) blocks: type I, [1–8–12], [2–6–DH_3–9], [3–DH_1–5–DH_4–10]; type II, [4–DH_2–7–11].

8.2 (a) The new unit costs do not satisfy condition (b) of *step* 6 in algorithm VTSP; therefore, DH_1 and DH_2 cannot be inserted. The new unit costs concerning condition (b) result in $10 - (5 + 3) = 2 > 0$. With DH_3, but without DH_4, we obtain $n_1 = 2$ (type I, $c_1 = 10$), $n_2 = 4$ (type II, $c_2 = 5$), $n_3 = 1$ (type III, $c_3 = 3$), and $C = 2 \cdot 10 + 4 \cdot 5 + 1 \cdot 3 = 43$. The following are the seven (FIFO) blocks: type I, [1–9], [3–12]; type II, [2–DH_3–7–11], [8–13], [4–10], [6]; type III, [5–14].

(b) The change starts at *step* 6 of algorithm VTSP; four new DH trips are DH_1, DH_2, DH_3, DH_4, from which we obtain $n_1 = 2$, $n_2 = 4$, $n_3 = 0$, and $C_3 = 2 \cdot 10 + 4 \cdot 11 + 0 \cdot 6 = 64$. The following are the six (FIFO) blocks: type I, [1–9], [5–DH_2–12]; type II, [2–DH_3–7–11–DH_4–14], [3–DH_1–8], [4–10], [6–13].

(c) Similar situation to (b) to obtain $n_1 = 3$, $n_2 = 2$, $n_3 = 1$, and $C_3 = 3 \cdot 12 + 2 \cdot 8 + 1 \cdot 9 = 61$. The following are the blocks: type I, [1–9], [3–DH_1–8], [5–DH_2–12]; type II, [4–10], [6–13]; type III, [2–DH_3–7–11–DH_4–14].

8.3 The three main steps are performed. First, $N_1 = 3 + 3 + 0 = 6$, $C_1 = 6 \cdot 5 = 30$, and $G''(S)$ is calculated to be 6. Second, $N_2 = 3 + 3 + 2 = 8$, $C_2 = 3 \cdot 5 + 3 \cdot 4 + 2 \cdot 3 = 33$. The third and the best solution is $N = 3 + 2 + 1 = 6 = G''(S)$, $C = 3 \cdot 5 + 2 \cdot 4 + 1 \cdot 3 = 26$. The following are the six (FIFO) blocks: type I, [1–11], [6–DH_1–9], [5–DH_3–12]; type II, [2–8–14], [4–7–DH_2–13]; type III, [3–10].

8.4 (a) Let AB travel speed = x·MB travel speed in which 0 < x < 1. For P > 30, we obtain x = 2/3. For 10% less MB travel speed, x = 11/15. By calculating the $ saving to be greater than the additional MB operational cost, we arrive at P > 60 passengers.

(b) Using the earlier definition of x with P as a variable, we obtain the inequality P > 10(9x−5)/(1−x), where x = 5/9 for P = 0. Therefore, the lower bound of AB travel speed is (5/9)·MB travel speed; and the upper bound is simply the MB travel speed.

Chapter 9

9.1 $D(k) = 9$.

(i) Algorithm $T_m F$ results in 13 joinings, from which only two (see later in bold) comply with the T_{max} constraint. The arrival–departure joinings are by trip numbers: [10–22], [11–14], [12–25], [13–16]; **[17–28]**, [18–23], [19–24], [20–27], [21–29], [25–30], [26–31], **[32–39]**, [33–40].

(ii) The use of the FIFO rule results in the following 13 joinings by trip numbers, none of which complies with the T_{max} constraint: [10–14], [11–15], [12–16], [13–22], [17–23], [18–24], [19–27], [20–28], [21–29], [25–30], [26–31], [32–39], [33–40].

9.2 The algorithm Dijkstra is applied for the two blocks. Block 1 is divided into two sets of pieces, [a-b-c-d] and [d-c-b], with a set's cost being 9 and 4, respectively (total cost of block is 13). Block 2 is divided into three sets of pieces, [a-b], [b-c], and [c-a-d], with a set's cost being 3, 3, and 12, respectively (total cost of block is 18).

9.3 The Roster procedure ends with d_4^3 and d_3^6 left untreated; thus, a new roster is manually created: R_4 covering Wed and Sat. The minimum number of drivers derived is five—three with R_1, one with R_3, and one with R_4—that is, $\left[d_1^1 - d_1^2 - d_1^3 - d_4^4 - d_3^5 \right]$, $\left[d_2^1 - d_2^2 - d_2^3 - d_2^4 - d_1^5 \right]$, $\left[d_4^1 - d_3^2 - d_3^3 - d_3^4 - d_2^5 \right]$, $\left[d_1^6 - d_4^7 \right]$, $\left[d_3^6 - d_4^3 \right]$.

Chapter 10

10.1 Revenue = $20q - 0.04q^2$ with a maximum of $2500 per h at q = 250 tickets/h and p = $10. The function p – q is inelastic for p between 0 and 10 and elastic for p between 10 and 20; for p = 0, it is perfectly inelastic; for p = 20, it is perfectly elastic; and for p = 10, we obtain a unit elastic point.

10.2 (a) Price elasticity of demand for transit trips is −0.3 (1% reduction in fare would lead to a 0.3% increase in transit patronage). A reduction of 33.33% (from $1.20 to $0.80) results in 22,000 hourly passengers; the company will then lose 24,000–17,600 = $6,400.

(b) Automobile price cross-elasticity of demand is 0.2; a $0.60 rise is 15% of $4, leading to an increase in patronage, from 20,000 to 20,600 (+3%).

10.3 (a) Bus, light-rail, and automobile utilities are −4.3, −5.35, and −5.35, respectively. Utilizing Equation 10.10 results in a modal split of 58.8%, 20.6%, and 20.6% for the bus, light-rail, and automobile, respectively.

(b) For an imposed gasoline fee of $1.20, the utility of automobiles becomes −10.15; the modal split is then 73.92%, 25.87%, and 0.21% for the bus, light-rail, and automobile, respectively.

Chapter 11

(a) There are five possible *single* travel options: [route 1(A-B)]; [route 2(A-X) and route 3(X-B)]; [route 2(A-Y) and route 3(Y-B)]; [route 2(A-X), route 3(X-Y), and route 4(Y-B)]; [route 2(A-Y) and route 4(Y-B)]. In addition, these options may be combined; for example, the passenger may decide to take the first route that arrives at A (either route 1 or 2). In the combined option, we assume that the passenger will take the route according to its frequency proportion (inverse of the headway).

The following table summarizes the possible sets of combinations for each node.

Node	Attractive routes (Route-exit node)	Waiting time (min)	Route probabilities			
			1	2	3	4
A	1-B	6.0	1	—	—	—
A	2-X	6.0	—	1	—	—
A	2-Y	6.0	—	1	—	—
A	1-B, 2-X	3.0	0.5	0.5	—	—
A	1-B, 2-Y	3.0	0.5	0.5	—	—
X	2-Y	6.0	—	1	—	—
X	3-Y	15.0	—	—	1	—
X	3-B	15.0	—	—	1	—
X	2-Y, 3-Y	4.3	—	0.71	0.29	—
X	2-Y, 3-B	4.3	—	0.71	0.29	—
Y	3-B	15.0	—	—	1	—
Y	4-B	3.0	—	—	—	1
Y	3-B, 4-B	2.5	—	—	0.17	0.83

(b) The expected travel times for each option can be calculated using this table, assuming that travel times are composed of waiting and in-vehicle times only. For example, the first row in the table, corresponding to the option of taking route 1, gives an expected travel time of 31 min (6 + 25). The expected travel time for each combination is found by averaging the route probabilities. For example, the option indicated in the fourth row (either route 1 or 2), combined with boarding route 3 at X, gives an expected travel time of $(3 + 0.5 \cdot 25) + (3 + 0.5 \cdot (7 + 15 + 8)) =$ 33.5 min.

(c) The minimum expected travel time is obtained by the following combination: Take either route 1 to B or 2 to Y; at Y, take either route 3 or 4. The expected travel time is $(3 + 0.5 \cdot 25) + (3 + 0.5 \cdot (13 + 2.5 + 0.17 \cdot 4 + 0.83 \cdot 4)) = 15.5 + 12.75 = 27.5$ min.

Chapter 12

12.1 In the beginning, $D(a) = 2$, $D(b) = 2$, $D(c) = 1$, and $P(a) = 2$, $P(b) = 2$, $P(c) = 3$. Shifting trips 5 and 2 reduces $D(c)$ by one. Shifting trip 1 to the right reduces $P(b)$ by one, but $P(a)$ increases, hence requiring the shifting of trip 4. *Solution*: (a) $D(a) = 1$, $D(b) = 2$, $D(c) = 1$; (b) $P(a) = 1$, $P(b) = 1$, $P(c) = 3$.

12.2 In the beginning, $D(a) = 2$, $D(b) = 1$, $D(c) = 0$, $D(d) = 2$, and $P(a) = 2$, $P(b) = 1$, $P(c) = 2$, $P(d) = 2$. It is impossible to reduce $D(k)$, k = a, b, c, d. Shifting trips 6 and 7 reduces $P(a)$ by one, but $P(c)$ and $P(b)$ increase, hence requiring the shifting of trip 3 to the left and trip 9 to the right. *Solution*: (a) same as in the beginning; (b) $P(a) = 1$, $P(b) = 1$, $P(c) = 2$, $P(d) = 2$.

12.3 In the beginning, $D(a) = 2$, $D(b) = 2$, $D(c) = 0$, and $P(a) = 2$, $P(b) = 2$, $P(c) = 2$. It is impossible to reduce $D(k)$, $k = a, b, c$. Shifting trips 6 and 7 reduces $P(a)$ by one, but $P(c)$ increases, hence requiring the shifting of trip 3 to the left and also inserting a DH trip from c (at 8:30) to reduce $P(c)$. *Solution*: (a) same as in the beginning; (b) $P(a) = 1$, $P(b) = 2$, $P(c) = 1$.

12.4 (a) Minimum of two stops is required; they can be located at nodes 5 (a stop for nodes 1, 3, 4, 6) and 2 (a stop for node 2), as well as around these stops in a variety of possibilities.

  (b) The minimax distance for four stops is $\ell = 250$ m, which is determined by the stop located at the midway point on arc (4, 6); the locations of the three other stops are as follows: on arc (1, 3), between 50 m from node 1 and 50 m before node 3; on arc (3, 4), between 150 m from node 3 and 150 m before node 5; and around node 2, by 250 m from node 2 on arcs (1, 2), (2, 3), and (2, 4).

  (c) The single stop will be located on arc (2, 3) at a distance of 1050 m from node 2 and 50 m away from node 3; thus, the minimax is $\ell = 1050$ m, which will be fully used by the demand at nodes 2, 4, and 6.

Chapter 13

13.1 (a) Max 1–8 flow is 6, where for the min cut $(X, \bar{X})$, $X = \{1, 2, 3, 4, 5\}$ and $\bar{X} = \{6, 7, 8\}$.

  (b) Max 4–5 flow is 5, where for the min cut $(X, \bar{X})$, $X = \{4, 6, 7, 8\}$ and $\bar{X} = \{1, 2, 3, 5\}$.

13.2 (a) Find the arcs associated with all of the min cuts $(X, \bar{X})$; there are 3 min cuts $(X, \bar{X})$ as follows: (i) $X = \{1, 2, 3\}$ and $\bar{X} = \{4, 5, 6\}$ with crossing arcs (2, 4), (3, 5); (ii) $X = \{1, 2, 3, 4\}$ and $\bar{X} = \{5, 6\}$ with crossing arcs (3, 5), (4, 5), (4, 6); and (iii) $X = \{1, 2, 3, 4, 5\}$ and $\bar{X} = \{6\}$ with crossing arcs (4, 6), (5, 6).

  (b) Find the minimum number of arcs to increase all of the three cuts in (a) and check all of their possible combinations to be [(2, 4) and (4, 6)], [(3, 5) and (4, 6)], and [(3, 5) and (5, 6)].

Chapter 14

14.1 (1) *Attitude*: Attitude is defined as the individual's positive or negative evaluation of performing the intended action. Attitude is formed from personal norms. Personal norms are internalized social norms that have assisted in shaping the attitude toward the action.

  Social norm: Social norm is composed of two types of norms: injunctive and descriptive norms. Injunctive norms refer to rules about what is morally approved or disapproved of, what ought to be done. Descriptive norms describe what is typical or normal, what most individuals do in a given situation.

  Perceived behavioral control: Perceived behavioral control (PBC) is defined as the perceived volitional control an individual believes he or she has to successfully complete the intended behavior. PBC is formed from *control beliefs*. Control beliefs are based on barriers that are perceived by the individual to undertake an action. When individuals perceive a higher level of control, the control factor is no longer important and the emphasis is placed on whether the individual intends to perform the action, thus strengthening the intention–behavior relationship. When control beliefs are weak, the individual is unlikely to have the intent to perform the action, thus weakening the intention–behavior relationship.

(2) The theory of planned behavior assumes that intention is the main direct determinant of behavior, for majority of the situations. The rationale for a direct link between perceived behavioral control and behavior is that given a sufficient degree of actual control over the behavior, individuals are expected to carry out their intentions when requisite opportunities and resources (e.g., time, money, skills) are available.

(3) (i) Improve *personal safety* at out-of-vehicle times; (ii) improve *reliability* of transfer; and (iii) minimize *transfer duration*.

14.2 (1) Cumulative prospect values are 0.14 for Scenario 1 and -0.71 for Scenario 2. Public transit users are most likely to pick Scenario 1.

(2) *Step 1 Fuzzification*: The data will need to be converted from their *crisp* (aggregated) value form into *fuzzy linguistic* variables. Membership functions of the fuzzy input and output variables will need to be developed. In this example, the fuzzy input is the weighted waiting times and the output is the proportion of users selecting each scenario.

Step 2 Fuzzy inference: The fuzzy rules in the form of *if–then* logic statements need to be developed. The number of fuzzy rules is dependent on the combination of input variables. A fuzzy inference method, such as the max–min composition method, needs to be selected.

Step 3 Defuzzification: The final step is to convert the fuzzy inference outputs into a *crisp value*. A common approach is the center of gravity method.

14.3 Five attributes need to achieved for this intermodal transfer (between train and bus) to be considered a *planned* transfer. These are network integration, integrated timed-transfer, integrated physical connection of transfers, information integration, and fare and ticketing integration. For network integration, the train and bus routes need to intersect. At this route intersection point, the intermodal transfer can occur. Physical integration between the two route services need to be provided for the transfer waiting and walking times. This means at least sheltered seating areas and walkways. Full information provisions must be made available for users to plan the complete trip in detail (start, transfer, end). Operators need to have scheduled their services to minimize transfer waiting times and provide integrated ticketing systems to reduce any additional cost for transfers.

Chapter 15

15.1 (a) Two alternative settings: routes A and B, with 14 and 11 departures, respectively, and routes A and B, with 15 and 10 departures, respectively. The second alternative is selected with headways of 8 (route A) and 12 (route B) minutes because the headways are integers; for route B, they are clock headways.

(b) Minimum fleet sizes required for routes A and B are 13 and 7 vehicles, respectively.

(c-1) There are four combinations of single routes: 1–2–4–3; 1–2–3–4; 1–4–2–3; and 1–4–3–2.

(c-2) Adequate criterion: minimum passenger-hours.

(c-3) Load profiles for routes 1–2–4–3, 1–2–3–4, 1–4–2–3, and 1–4–3–2 are [1500, 960, 840], [1500, 960, 600], [1500, 1780, 840], and [1500, 1780, 1380], respectively.

(c-4) Minimum fleet size, 30, 30, 25, and 30 vehicles, for routes1–2–4–3; 1–2–3–4; 1–4–2–3; and 1–4–3–2, respectively.

(c-5) Route 1–4–2–3 is selected, with 64,600 passenger-min compared with 67,800 for the two existing routes.

15.2 (a) The calculated measures are given in this table:

Route	PHr (pass-h)	Her (pass-h)	DPHr (pass-h)	wr (pass-h)
1–2–4–6	670.8	236.7	75	53.6
4–5–3	383	72	1.7	27.2
1–8–7	117	21.7	7	14.4

    (b) Total DPH, including transfers, is 338.7 pass-h.

15.3 (a) The calculated frequency (from load profile) in veh/h and headway in minutes for fast train and fast ferry is, respectively, [F = 3, H = 20] and [F = 4, H = 15]. Thus, the average waiting time is 10 min for the train and 7.5 for the ferry; the hourly cost, therefore, is $3120 for the train and $2340 for the ferry.

    (b) Lost cost for empty seat-hours (utilizing the load profile) is $14,700 for the train and $12,600 for the ferry; the lost cost for travel time is $3,660 and $6,630 for the train and ferry, respectively.

    (c) Hourly income is $67,700 and $53,200 for the train and ferry, respectively; the profit (using the loss in (a) and (b) previously), therefore, is $46,220 and $31,630 for the train and ferry, respectively.

    (d) Thus from (c), the preferred mode will be the fast train.

    (e) Some of the neglected cost elements in the actual analysis are as follows: vehicle cost, operational cost (equipment, labor, maintenance, fuel, etc.), real-estate and land costs, terminal/berth construction cost, and road/sea accident cost.

Chapter 16

16.1 (a) Constructing the deficit functions $d(a, t)$ and $d(c, t)$ results in $D(a) = 6$ and $D(c) = 5$ vehicles, with a single DH trip from c (at 7:00) to a, $D(a) = 5$; thus, $N_{min} = 10$ vehicles. The minimax H procedure is then applied, and this results in minimax H = 14 min ($a \rightarrow c$, 6:00–7:00), = 18 ($a \rightarrow c$, 7:00–8:00), = 15 ($c \rightarrow a$, 6:00–7:00 and 7:00–8:00). Hence, in the $a \rightarrow c$ direction, four trips will end at b (c is the max load point); and in the $c \rightarrow a$ direction, three trips will start at b (max load point) while creating the timetable with maximum short-turn trips. The arrivals at b for the $a \rightarrow c$ direction are as follows, with a single asterisk for arrivals ending at b: [6:20; 6:27*; 6:32; 6:40; 6:47; 6:55*; 7:01; 7:08*; 7:15; 7:25*; 7:33; 7:40; 7:50; 7:55; 8:03]. The departures at b for the $c \rightarrow a$ direction are as follows, with a single asterisk for departures starting at b: [6:25; 6:35; 6:43*; 6:50; 6:57*; 7:05*; 7:15; 7:20; 7:35; 7:43; 7:50; 7:57; 8:05; 8:13*; 8:18]. Creating $d(a, t)$, $d(b, t)$, and $d(c, t)$ for the schedule, with maximum short-turn trips, results in $D(a) = 6$, $D(b) = 0$, and $D(c) = 5$, and with a single DH trip from b (at 7:08) to c, $D(c) = 4$; thus, $N'_{min} = 10$ vehicles $= N_{min}$. In this particular exercise, therefore, it doesn't make sense to introduce short-turn trips.

    (b) Listing the trips in increasing order of departure times, with departures at a coming before c for the same departure times, results in a list of 31 trips. Note that trip number 18 on the list is a DH trip between 7:00 and 7:28 from c to a. Applying the FIFO procedure results in the following 10 blocks, by trip number determined in the list of trips: [1, 11, 22], [2, 12, 21, 31], [3, 13, 24], [4, 14, 23], [5, 16, 26], [6, 15, 25], [7, 18, 27], [8, 17, 28], [9, 20, 29], and [10, 19, 30].

16.2 (a) Using component FIRST from Chapter 6 at the single synchronization point b results in max (5, 8, 6, 7) ≤ d ≤ minimum (10, 13, 15, 20); thus, the minimum (even) headway is d = 8 min. Setting the first departure from terminal c at 6:00 will determine the remaining departures with 8 min headways. That is, the first departure from a will be at 6:05, from e at 6:15, and from k at 6:05, and all will meet at b at 6:25. This is for maximizing the number of meetings at b. The 12 departures from terminal a are [6:05; 6:13; 6:21; 6:29; 6:37; 6:45; 6:53; 7:01; 7:09; 7:17; 7:25; 7:33] and from c are [6:00; 6:08; 6:16; 6:24; 6:32; 6:40; 6:48; 6:56; 7:04; 7:12; 7:20; 7:28].

(b) Constructing $d(a, t)$ and $d(c, t)$ results in $D(a) = 5$ and $D(c) = 7$ vehicles, with no possibility of inserting DH trips; thus, $N_{min} = 12$ vehicles. The minimax H procedure is then applied, and this results for all hours and directions of travel in minimax H = 16 min. Hence, four trips in the $a \to c$ direction will start at b (max load point), and four trips in the $c \to a$ direction will end at b (c is the max load point) while creating the timetable with maximum short-turn trips. Because b is a meeting point, the departure and arrival times at b are the same for both directions; these times appear in the following list, with a single asterisk for departures starting at b and a double-asterisk for arrivals ending at b: [6:25; 6:33; 6:41*; 6:49**; 6:57*; 7:05**; 7:13; 7:21(*)**; 7:29; 7:37; 7:45(*)**; 7:53]; note that 7:21 and 7:45 are both starting and ending times for short-turn trips. Creating $d(a, t)$, $d(b, t)$, and $d(c, t)$ for the schedule with maximum short-turn trips results in $D(a) = 3$, $D(b) = 1$, and $D(c) = 7$; thus, $N'_{min} = 11$ vehicles $< N_{min}$. The last step is an attempt to extend, that is, minimize, the short-turn trips, which results in extending the 6:41 departure at b to start at 6:21 from a and extending the 7:05 arrival at b to arrive to c at 7:20.

Chapter 17

(a) The following are the 22 impedance ratios, z_{ij}, with their corresponding links in parenthesis: 2(1, 2); 2(2, 1); 1.33(2, 3); 2(3, 2); 7(2, 8); 1.5(8, 2); 0.66(3, 4); 2(4, 3); 1.66(4, 5); 2(5, 4); 2.5(4, 9); 2(9, 4); 2.5(5, 6); 1.33(6, 5); 1.25(6, 7); 1(7, 6); 1(6, 9); 1(9, 6); 1.66(7, 8); 2(8, 7); 1.33(8, 9); 1.5(9, 8).

(b) The 22 links in the network are arranged in parenthesis in decreasing order according to the following pd_{ij} units:

6 (3, 4); 5 (7, 6); 4 [(8, 2), (6,7), (9, 6)]; 3 [(2, 3), (7, 8), (8, 9), (6, 9), (4, 5), (6, 5)];

2 [(9, 8), (3, 2), (4, 3), (5, 4), (4, 9), (9, 4), (8, 7), (5, 6)]; 1 [(2, 1), (1, 2), (2, 8)]

The highest pd_{ij} group selected, with five links, contains units with pd_{ij} = 6, 5, 4.

Given the initial group of five links and p = 2, there are 20 arranged (the order of links considered) combinations (subgroups) of links that may be considered for inclusion in the circular route: [(3, 4), (7, 6)]; [(3, 4), (8, 2)]; [(3, 4), (6, 7)]; [(3, 4), (9, 6)]; [(8, 2), (6, 7)]; [(8, 2), (9, 6)]; [(9, 6), (6, 7)]; [(8, 2), (7, 6)]; [(6, 7), (7, 6)]; [(9, 6), (7, 6)]; [(7, 6), (3, 4)]; [(8, 2), (3 ,4)]; [(6, 7), (3, 4)]; [(6, 7), (8, 2)]; [(9, 6), (8, 2)]; [(6, 7), (9, 6)]; [(7, 6), (8, 2)]; [(7, 6), (6, 7)]; [(7, 6), (9, 6)]; [(9, 6), (3, 4)].

For each pair, a circular route is constructed using the shortest-path criterion of z_{ij} between the links and the train. From the efficient routes found, each complying with the constraint of T = 30 min, the best route is determined (by node number): 1–2–3–4–9–8–2–1. The one selected has a 26 min circular travel time and a pd_{ij} sum of 19.

Chapter 18

18.1 Answers to (a) and (b) can be found within the chapter itself.
18.2 Same as 18.1; answers to (a) and (b) can be found within the chapter itself.

Chapter 19

(a)

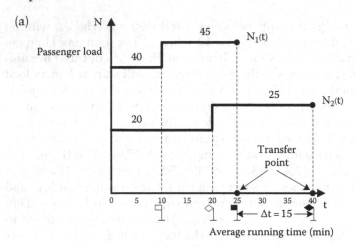

PTT of route 1 = 1075 min; PTT of route 2 = 900 min; PTWT = 300 min; TPTT = 2275 min

(b)

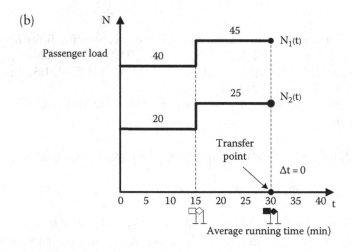

PTT of route 1 = 1275 min; PTT of route 2 = 675 min; PTWT = 0 min; TPTT = 1950 min

(c) ΔPTWT = –300 min; ΔTPTT = –325 min

Index

Printed in the United States
by Baker & Taylor Publisher Services